changing the way the world learns

To get extra value from this book for no additional cost, go to:

http://www.thomson.com/wadsworth.html

thomson.com is the World Wide Web site for Wadsworth/ITP and is your direct source to dozens of on-line resources. *thomson.com* helps you find out about supplements, experiment with demonstration software, search for a job, and send e-mail to many of our authors. You can even preview new publications and exciting new technologies.

thomson.com: *It's where you'll find us in the future.*

Universe
ORIGINS AND EVOLUTION

Theodore P. Snow
Kenneth R. Brownsberger
THE UNIVERSITY OF COLORADO

Wadsworth Publishing Company
I⬦T⬦P® An International Thomson Publishing Company

Belmont, CA • Albany, NY • Bonn • Boston • Cincinnati • Detroit • Johannesburg
London • Madrid • Melbourne • Mexico City • New York • Paris
San Francisco • Singapore • Tokyo • Toronto • Washington

Editor: Denise Simon
Development Editors: Theresa O'Dell, Janine Wilson
Production and Design: Ann Rudrud
Cover Design: Doug Abbott
Illustrations: David Aguilar, Scott Kahler, & Teshin Associates
Promotion Managers: Ann Hillstrom and Ellen Stanton
Copy Editor: Pat Lewis
Proofreader: Heather Jones, Thrushcross Editing
Cover Photographs: NASA/STScI
Compositor: Parkwood Composition Service
Printer: West Publishing Company

COPYRIGHT © 1997 by Wadsworth Publishing Company
A Division of International Thomson Publishing Inc.

I(T)P®   The ITP logo is a registered trademark under license.

Printed in the United States of America
1 2 3 4 5 6 7 8 9 10

For more information, contact Wadsworth Publishing Company, 10 Davis Drive, Belmont, CA 94002, or electronically at http://www.thomson.com/wadsworth.html

International Thomson Publishing Europe
Berkshire House 168-173
High Holborn
London, WC1V 7AA, England

International Thomson Editores
Campos Eliseos 385, Piso 7
Col. Polanco
11560 México D.F. México

Thomas Nelson Australia
102 Dodds Street
South Melbourne 3205
Victoria, Australia

International Thomson Publishing Asia
221 Henderson Road
#05-10 Henderson Building
Singapore 0315

Nelson Canada
1120 Birchmount Road
Scarborough, Ontario
Canada M1K 5G4

International Thomson Publishing Japan
Hirakawacho Kyowa Building, 3F
2-2-1 Hirakawacho
Chiyoda-ku, Tokyo 102, Japan

International Thomson Publishing GmbH
Königswinterer Strasse 418
53227 Bonn, Germany

International Thomson Publishing Southern Africa
Building 18, Constantia Park
240 Old Pretoria Road
Halfway House, 1685 South Africa

Library of Congress Cataloging-in-Publication Data

Snow, Theodore P. (Theodore Peck)
    Universe : origins and evolution / Theodore P. Snow, Kenneth R.
  Brownsberger.
      p.   cm.
    Includes bibliographical references and index.
    ISBN 0-314-09838-0 (soft : alk. paper)
      1. Astronomy.   I. Brownsberger, Kenneth R.   II. Title.
  QB43.2.S667 1997
  520--dc20
                                                                      96-14610
                                                                          CIP

Contents

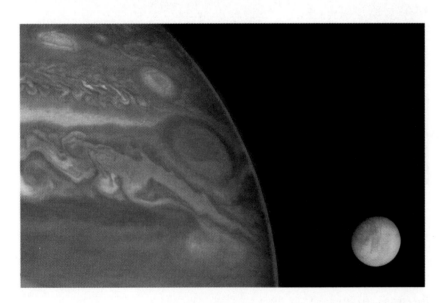

Preface

In writing *Universe: Origins and Evolution,* we have drawn much from experience and wisdom gained from four editions of *The Dynamic Universe* and four editions of *Essentials of the Dynamic Universe.* In this new text we have distilled the best from its predecessors while stirring in several new elements: a refreshed and reinvigorated writing style; usage of modern learning tools such as the Internet; and, of course, coverage of the many recent developments in astronomy.

Universe is built around the twin themes of origins and the connection between science and everyday life. Origins has become a focal point in today's research in astronomy, as incredible images of young suns and preplanetary disks and the detection of extrasolar planets, along with new views of the young universe, have stimulated research as well as the public imagination. Both the National Science Foundation and the National Aeronautics and Space Administration have established major programs in origins, uniting research efforts in cosmology, galaxy formation, the births of stars, and origins of planetary systems. This textbook mirrors these grand themes by emphasizing origins and evolution throughout its progression from the Earth and the solar system to the stars, the galaxy, and the universe.

Emphasizing the link between science and everyday life is a central purpose of teaching science to non-science majors in college, and in *Universe* we explicitly serve this purpose through the inclusion of commentaries, called "Science and Society" boxes, which detail many different ways in which the student and science interact outside of the classroom. These inserts, which range in topic from the technology of solar energy (Chapter 13) to orbital properties of communications satellites (Chapter 5) to the cost and value of basic research (Chapter 6) to astronomically-based terminology in the English language (Chapter 19) to the contrast between theory and faith (Chapter 20), will help students place their science class into the perspective intended. Upon completion of a course based on this text, every student should understand not only something of the modern science of astronomy, but also the role of science in society.

The book is streamlined and compact, while providing complete coverage of the field. Apart from the "Science and Society" boxes and very brief marginal commentaries on the use of the Internet and the World Wide Web, the text is uninterrupted. Mathematical treatments are included within the body of the chapters, rather than being set aside in boxes where students can either ignore them, be traumatized by them, or both. Simple algebraic treatments are interwoven with the text, allowing students to see the relationships and to become familiar with quantitative reasoning. Chapter-end problems support this by providing examples for the students to complete on their own, thus reinforcing the methodology used in the text.

The new information technology of the Internet is brought to the students through a number of mechanisms. First, in each chapter there is a "Web Activity", with link information provided, leading the students to guided activities on the World Wide Web that are related to the subject matter in that chapter. Second, throughout the book we provide link information to additional resources on the WWW which will be useful or informative for the students. The guided Web activities are intended to help the student go beyond merely "surfing" the Web for amusement, instead providing direction so that the truly useful aspects of the Web can be brought out. Examples include eclipse hunting on the Web (Chapter 2), finding links to telescopes and observatories (Chapter 6), a study of image resolution (Chapter 9), tracking data on incoming comets (Chapter 11), monitoring daily images showing solar activity (Chapter 13), finding models and animations of stellar evolution and remnants (Chapters 15 and 16), carrying out a multi-spectral study of the Milky Way (Chapter 17), and classifying galaxies based on images found on the Web (Chapter 18) Apart from these guided Web activities, the Web resources provided throughout the book (primarily in figure captions) lead to on-line reproductions of images from the text, additional, related images, or other Web resources such as sites containing additional information or links.

In order to avoid the common problem of disappearing or moving Web sites, we have taken the innovative approach of channeling all Web references through our own WWW site, so that we can maintain and update the links. Thus every Universal Record Locator (URL) given in the book will always remain valid, leading to a page on our site where we provide further information and links. This guarantees that the information provided in the book on this ephemeral, evolving resource will remain valid even as the Web changes from day to day.

The WWW home page that we have created to support and enhance the learning experience for users of this book can be found at:

http://universe.colorado.edu/.

This site can be accessed by anyone, but is especially tailored to complement our book. The site is organized on a chapter-by-chapter basis, with additional illustrations and images as noted above, and also including the review questions and problems from the chapter ends (with an e-mail option so that students can send their answers to their instructor). The guided Web Activities are also located under each chapter, and can be reached either directly through the URL given in the book, or by following links from the home page to the chapter pages. In addition to the chapter-by-chapter material, the WWW site also includes an Astronomy Update feature, where we will add information on late-breaking developments in astronomy (cross-referenced to the book), including summaries, images, and links, so that instructors and students using the book can keep current on new developments. The electronic mail addresses of the authors will be posted, so that comments or questions may be directed specifically to us.

The organization of *Universe* follows the traditional Earth-to-the-universe format which has been so successful with our previous books. There are some new elements, however. Following the opening section on the nighttime sky (Chapters 1 and 2), history (Chapters 3 and 4), and tools of astronomy (Chapters 5 and 6), the opening chapter on the planets (Chapter 7) provides an overview of the solar system, in which the basic principles of planetary science are introduced, setting the stage for the detailed discussions of the planets and interplanetary bodies that follow (Chapters 8, 9, 10, and 11). The exploration of the solar system is followed by another overview chapter (Chapter 12) which sets the stage for studies of stars and stellar systems in much the same way that Chapter 7 does so for the planetary discussions. Chapter 12 exemplifies the theme of origins and evolution, introducing the stars and bringing out the intimate connection between star formation and planetary systems, and including a substantial discussion of techniques and recent results in the search for planets orbiting other suns. The stellar section begins with a detailed look at the Sun (Chapter 13), followed by discussions of stellar properties (Chapter 14), stellar structure and evolution (Chapter 15), and stellar remnants (Chapter 16). Next come the chapters on the universe at large, beginning with a thorough look at our own galaxy (Chapter 17) before proceeding to a general treatment of galaxies and the distribution of matter in the universe (Chapter 18), active galaxies and quasars (Chapter 19), and the uni-verse as a whole (Chapter 20). Again the theme of origins and evolution is stressed, reflecting modern perspectives gained through the fantastic images of distant galaxies, active galactic nuclei, and galaxy formation as seen directly with the deep images obtained by the largest telescopes on the ground and in space, and new insights into the origin of the cosmos itself that have come from a combination of observations and theory. The final chapter (Chapter 21) offers an overview of current knowledge of life in the universe, a subject that has gained new urgency in view of the many extrasolar planets that have been found, as well as the exciting possibility that evidence for life on Mars has been found in fossilized form. The student who follows a course through this sequence of topics will be well versed in the current state of our understanding of the universe we live in, and at the same time will have gained a deeper appreciation of the nature of science and inquiry.

As mentioned previously, *Universe* is intended to be compact. With only 21 chapters, it is well-suited to a one-term course in general astronomy, while at the same time providing the framework for a full year course for the instructor who desires to take a slower and more thorough pace. The text within each chapter is virtually uninterrupted, without the numerous sidebars and boxes that can be an obstacle to continuity. The ends of the chapters include summaries, review questions, problems, an "Astronomical Activity", and an updated list of references for further reading. The summaries are intended to provide the student with a quick overview of the main points, as an aid in reviewing and studying. The review questions, for the most part, are meant to stimulate the student to recall material from the text, but some are thought-provoking questions for which the answer requires some synthesis of information and ideas. The problems are quantitative exercises which help clarify methods used in the text, and which help students gain some confidence in their ability to take an analytical approach to problem-solving.

The illustrations and images in *Universe* have been upgraded from our earlier books. We have not only brought in many new images from recent observations by the *Hubble Space Telescope,* large telescopes on the ground, and new space-based observatories, but we have also added a new feature: full-page "posters", colorful montages designed to convey the visual excitement of astronomy and at the same time help reinforce the unifying concepts of the text.

At the back of *Universe* is the most complete set of Appendices in any book on the introductory astronomy market. These Appendices have been expanded from the already-comprehensive list in the *Dynamic Universe,* with the addition of a description of the Internet and the World Wide Web (Appendix

1), a summary of mathematical techniques and unit conversion factors (Appendix 2), a listing of space-based telescopes and planetary probes (in addition to the existing summary of major ground-based telescopes; Appendix 7), and a description of the principal astronomical coordinate systems (Appendix 11). The remaining Appendices, updated and expanded from earlier versions, include data on physical constants (Appendix 4), temperature scales (Appendix 5), radiation laws (Appendix 6), planetary and satellite data (Appendix 8), solar composition (Appendix 9), stellar data (Appendix 10), the constellations (Appendix 12), a mathematical treatment of stellar magnitudes (Appendix 13), a summary of the nuclear reactions that occur in stars (Appendix 14), a list of all detected interstellar molecules (Appendix 15), and data on galaxies (Appendix 16) and the Messier catalog (Appendix 17). This large set of data not only provides information of interest for student and instructor alike, it also gives the instructor a basis for making quantitative assignments in which students are required to look up information and tabulate or analyze it. As an example, the data on nearby stars (Appendix 10) can be used in an assignment designed to demonstrate the relative numbers of stars of different spectral types in the galaxy.

Ancillary materials available with *Universe* include a CD-ROM, a set of slides, a collection of transparencies, and an Instructors Manual (written by Victoria Alten and Kamran Sahami of the University of Colorado). There is also a student Study Guide, authored by noted astronomical educator C. Gregory Seab of the University of New Orleans, which contains many new approaches, such as a "concept map" for each chapter.

The CD-ROM is a departure from the rapidly-developing tradition of merely repeating the content of the book in electronic form, with a few cross-links inserted. On the contrary, the *Universe* CD-ROM does not even include the text; instead, it provides a number of value-added features that the student cannot get from the book. The CD-ROM, which is packaged with every copy of the book at no additional cost, is organized on a chapter-by-chapter basis, and includes the chapter summaries, with links to explanatory material or new illustrations (many of them animated); a self-quiz, in multiple-choice format with feedback and references to the appropriate section of the text if answered incorrectly; and a wide range of extra figures, illustrations, and animations, as well as a glossary with links from other sections of the CD-ROM. We regard the CD-ROM as an extra teaching tool, rather than a duplication or expansion of the book itself.

In the course of developing this book, we have called upon many scientists for assistance, in finding resources such as photos and data, and in reviewing

the manuscript at several stages. It is a pleasure to acknowledge the help of these individuals, with apologies to anyone we may have inadvertently omitted: John Bally, David Devine, and Jon Morse of the University of Colorado, for their help in providing images of star formation and young stellar objects; Debra Meloy Elmegreen (Vassar College), Bruce Elmegreen (IBM), and Philip Seiden (IBM) for discussions and images of galactic spiral arm development; Rogier Windhorst and Sebastian Pascarelle (Arizona State University) for assistance with images of primeval galaxies; S. Alan Stern (Southwest Research) for data and images of Pluto; Debra Meloy Elmegreen and Barbara Welther (Smithsonian Astrophysical Observatory) for information on women in astronomy; John Stocke (University of Colorado) for consultations on active galactic nuclei and quasars; Richard Wainscoat (University of Hawaii) for his marvelous aerial photo of the summit of Mauna Kea; Ralph Doty (North Little Rock, Arkansas) for his images of Comet Hyakutake; Jun-ichi Watanabe (National Astronomical Observatory of Japan) for providing images of Comet Hale-Bopp; Marcus Price and Helen Sim (CSIRO) for photos of the Parkes radio telescope; George Dimitoglou and Joe Gurman (NASA) for help in obtaining high-resolution *SOHO* images; Gary Linford (Lockheed) for providing access to high-resolution *Yohkoh* images; J. McKim Malville (University of Colorado) for his Indian Sun temple photo; Bruce Jakosky (University of Colorado) for assistance in locating a good *Viking* image of Martian water channels; Gisbert Winnewisser (Max Planck Institute, Germany) and Veronica Bierbaum (University of Colorado) for help with the list of interstellar molecules in Appendix 15; John Kormendy (University of Hawaii) for data on black hole candidates; Harry Ferguson (Space Telescope Science Institute) and Renee Kraan-Korteweg (Observatoire de Paris-Meudon) for help in obtaining the image of Dwingeloo 1; Hans Betlem (Dutch Meteor Society) for his image of a bright meteor. And especially we want to acknowledge and thank those who helped us with two of the most significant collections of images that we tapped for this book: the marvelous photographs from the Anglo-Australian Observatory, provided through the courtesy of David Malin and Coral Cooksley of the AAO; and the high-resolution digital versions of the *Hubble Space Telescope* images, obtained through the help of Carol Christian, Zoltan Levay, and Cheryl Gundy of the Space Telescope Science Institute. We are also indebted to David Aguilar (VISION-1) for his creative work in assisting with art for the book and for his leadership in creating the CD-ROM.

In the course of preparing a new book such as this, the publisher relies on reviews by several instructors from the astronomical community, and these reviews provided us with valuable insights into the direction we should take, in addition to helping check us for accuracy. Those who reviewed the manuscript at various stages were:

Rayford L. Ball
Tarrant County Community College

John H. Beiging
University of Arizona

William J. Boardman
Birmingham-Southern College

Bernard W. Bopp
University of Toledo

Eugene R. Capriotti
Michigan State University

James N. Douglas
University of Texas, Austin

David L. DuPuy
Virginia Military Institute

Alex Fillipenko
University of California, Berkeley

Vasken Hagopian
Florida State University

Steven D. Kawaler
Iowa State University

Shiv S. Kumar
University of Virginia

Robert J. Leacock
University of Florida

Marie E. Machecek
Northeastern University

Bruce Margon
University of Washington

David Murdock
Tennessee Technological University

Tina Riedinger
University of Tennessee

James A. Roberts
University of North Texas

Robert D. Schmidt
University of Nebraska at Omaha

J. Scott Shaw
University of Georgia

Caroline E. Simpson
Florida International University

Michael L. Sitko
University of Cincinnati

John A. Soules
Cleveland State University

Charles R. Tolbert
University of Virginia

David Wingert
Georgia State University

Donald Witt
Nassau Community College

George W. Wolf
Southwest Missouri State

We are grateful to all of these people for their insights and suggestions, which have vastly improved the book while helping us to be sure we will respond to the needs of the students and instructors for whom it is designed.

We wish also to acknowledge with gratitude the contributions made by the editorial staff at West Publishing, who are ultimately responsible for this book. We especially wish to thank Denise Simon of West (now Wadsworth), who has been editor and mentor to the senior author since his first textbook project was started some 17 years ago. Denise has guided, cajoled, and encouraged us from the start, and the result has been a far better book than we might otherwise have produced. Similarly, we thank production editor Ann Rudrud of West (now Wadsworth as well) for her creativity, cheerfulness, and ever-helpful support of this project. The design, style, and artistic content of the book are due largely to her efforts. Theresa O'Dell and the late Janine Wilson of West have helped immeasurably with the ancillaries, the reviews, and the production of this book. David Aguilar, with assistance from Scott Kahler, created the dynamic chapter-opening art and in-text visual summaries. Copyeditor Patricia Lewis exercised excellent judgement; our text is clearer and more succinct thanks to her. Heather Jones of Thrushcross Editing greatly improved our accuracy by reading the galley proofs with the eye of a copy editor trained in physics.

Finally, we wish to thank our students, past, present, and future, for their inspiration and curiosity, which we hope will be stimulated and expanded by this book. We hope that they will evolve as does our knowledge of the cosmos, along with the universe itself.

Theodore P. Snow
Kenneth P. Brownsberger
November, 1996

Universe

ORIGINS AND EVOLUTION

Chapter 1
The Essence of Astronomy

Check out the Web site for this chapter:
http://universe.colorado.edu/ch1

For I dipped into the Future,
    far as human eye could see;
Saw the vision of the world,
    and all the wonder that would be.

Alfred Lord Tennyson, 1842

3

The principal point of this opening chapter is to give you a sense of what astronomy is and how we explore the nature of the universe. In the process, you will learn something about science: how we learn about things by seeking the best explanation for what we observe, and how we test our explanations by making predictions and comparing them with our observations. You will also learn something about the nighttime sky, as some of the most prominent objects are introduced. Along the way you will gain some feeling for the enormous range of sizes and distances in the universe and a better understanding of just how small our familiar planet is amid the vastness of the cosmos.

The oldest of all sciences is perhaps also the most beautiful. No artificial light show can rival the splendor of the heavens on a clear night, and few intellectual concepts can compare with the beauty of our modern understanding of the cosmos.

Today we study astronomy for a variety of reasons, some technological, some practical; but no one loses sight of the underlying majesty, of the human instinct for intellectual satisfaction. To study astronomy is to ask the grandest questions possible, and to find hints at their answers is to satisfy one of humankind's most deeply ingrained yearnings.

In this text we explore astronomy in the modern context, which is highly technical and sophisticated, but we will endeavor to retain the sense of wonder and beauty that has motivated the science since the beginning. Although some may argue that astronomy has little practical use, we will see that its origins are rooted in very practical requirements for methods for keeping time and maintaining calendars. In this chapter we begin our study by defining astronomy and introducing some simple terminology that will assist us in later chapters.

What Is Astronomy?

Astronomy is the science that has the entire universe as its subject. It is the science in which we derive the properties of celestial objects and from these properties deduce the laws by which the universe operates. It is the science of everything.

Technically, we might say that astronomy is the science of everything except the Earth or, more specifically, the study of everything beyond the Earth's atmosphere, since the Earth and its atmosphere fall into the purview of other disciplines such as geophysics and atmospheric science. We will find, however, that the study of astronomy necessarily includes an examination of the properties and the evolution of the Earth and its atmosphere.

In the modern sense, astronomy is probably more aptly called **astrophysic**s. Ever since the time of Isaac Newton (the late seventeenth century), the universe has been explored through the application of the laws of physics—most of them derived from earthly experiments and observations—to celestial phenomena. Other scientific disciplines enter into our discussions as well: to study the planets, for example, we must know something of geology and geophysics; to analyze molecules in space, we must understand the principles of chemistry; and to discuss the possibility that life might exist elsewhere, we must know something about biology.

Figure 1.1 An ancient astronomical site. Stonehenge, a monument in England, was built around 2100 B.C. Alignment of some of the stones with the midsummer sunrise and midwinter sunset suggests that Stonehenge was used for seasonal festivals. (Left: Michael Howell, right: L'aura Colan, Photonica) http://universe.colorado.edu/fig/1-1.html

Humans were making astronomical observations and keeping records at least as early as the time of the most ancient recorded history **(Fig. 1.1)**, and we believe that they studied the skies and pondered their cyclic motions long before that. In the earliest times, astronomers had a practical motivation: knowledge of motions in the heavens made it possible to predict and plan for certain significant events, such as the changing of the seasons.

Along with the practical came the whimsical and the spiritual. In ancient times, events in the heavens were thought to exert some influence over the lives of people on Earth, and many early astronomers practiced what today we call *astrology*. Many a monarch retained the services of an astronomer, not only to foretell the seasons, but also to provide advice on strategies for war, love, politics, and business. In many cultures, religion and astronomy were intimately linked, so astronomers attained the importance that major religious leaders have today.

The rich and diverse Greek civilization that flourished for many centuries before and around the time of Christ developed several sciences—astronomy included—beyond the mystical and the spiritual. The Greeks had many preconceived notions about the nature of the universe, but they also made many rational advances in understanding the heavens, setting the stage for the development of modern science many centuries later. Following the era known as the Dark Ages, astronomy (like many other disciplines) experienced the pangs of rebirth during the Renaissance, becoming a rational and methodical science. Although battles would still be fought between religion and science, the course was set during the Renaisance. The work of pioneers such as Copernicus, Galileo, Kepler, and especially Newton placed astronomy on a firm physical basis.

Modern astronomy still has a practical aspect, but few who pursue the science do so for that reason. Today we study astronomy primarily for the sake of expanding our knowledge; it is research of the purest sort. Even so, many practical and concrete benefits derive from astronomy; witness, for example, the many technological spin-offs from the space program and the multitude of new physical and chemical processes, applicable here on Earth, that have been discovered through astronomical observation.

No matter how analytical we may be in modern astronomy, we never lose sight of the same basic human feelings that inspired our ancestors. The modern astronomer, who may use a large telescope and a variety of complex electronic instruments **(Fig. 1.2)**, still treasures the moments spent outside the telescope building, simply watching the skies with the same tools the ancients used.

The Philosophy of Science and the Nature of Theory

Astronomy has evolved from a mixture of folklore, objective observation, and speculation to become a true *science*, meaning that it operates under certain ground rules aimed at ensuring objectivity and that its practitioners are willing to change ideas as demanded by observation. During the Renaissance, particularly through the work of Galileo, astronomy came to be at the leading edge of the development of scientific methods and procedures. The basic scientific method developed then is still used today.

Science progresses by hypothesis and test, by trial and error. Rather than adopting a hypothesis and then attempting to ignore or rationalize data that contradict this preconceived notion, a scientist is willing to revise a theory in the light of new information. If a scientist adopts anything as an article of faith, it is that no theory is above the possibility of modification—that our knowledge always has room to grow.

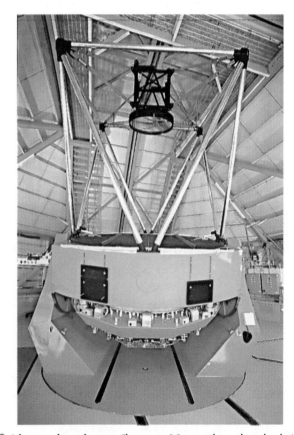

Figure 1.2 **A large modern telescope.** This is a new 3.5-meter telescope located at the Kitt Peak National Observatory in Arizona. It is called the WIYN Telescoope, in recognition of the four institutions (Unviersity of Wisconsin, University of Indiana, Yale, and NOAO) that built and support it. (National Optical Astronomy Observatories) http://universe.colorado.edu/fig/1-2.html

This willingness to change does not mean that scientific theories are wrong or invalid. A scientific theory should be regarded simply as the best explanation available that is consistent with the known facts. A theory is modified or replaced when more facts become known or when a better explanation is found.

A theory usually starts as a **hypothesis,** an initial suggestion made to explain some observation that raises a question. As new information is gathered, the hypothesis develops into a **theory,** a framework allowing for prediction and test. Many theories quickly fall by the wayside because new information contradicts them, while others persist and become widely accepted because they are consistent with new data that become available. One feature of any theory is that it leads to specific predictions about what will happen in new situations. Part of the job of testing a new theory is to make predictions from it and then make observations or perform experiments to test those predictions.

A theory must meet three general criteria. First, it must be consistent with the known facts, as revealed through observation or experimentation. Second, it must be capable of making predictions that can be tested. An untestable theory is not a theory at all, but a belief system. Third, a theory should satisfy **Occam's razor,** the postulate that nature is simple, and that an explanation that requires many complex, unprovable assumptions is less likely to be correct than one that requires few or no such assumptions.

One of the most important goals of this text is to help you understand how scientific progress is made —how theories are developed, tested, and revised. Our descriptions of astronomical discoveries and how they were made will help illustrate how scientists develop, test, and modify theories in the light of new information. As you will see, even the best scientists make errors, but they all share one characteristic— their willingness to discard incorrect hypotheses when conflicting evidence is found or a better explanation appears.

You will also see that scientists do not always know what to look for, but that our knowledge of the universe grows because scientists seek the best explanation for what they observe, rather than trying to modify their observations to conform to existing theories. Our hope is that these descriptions will help you develop a sense of how science works and a better perspective on new theories and controversies that you may encounter outside the classroom.

What Is an Astronomer?

Throughout this text we will be referring to astronomers, scientists, astrophysicists, and physicists. In a book that endeavors to summarize all that we know about the universe and its contents, we can hardly omit a description of the people who devote their time to developing this knowledge.

In the United States alone, thousands of people study astronomy either as a profession or as a hobby **(Fig. 1.3)**. Representative of the latter are the amateur astronomers who engage in a variety of astronomical activities, often on their own, but in many cases through local and even nationwide organizations. These activities include telescope making, astrophotography, long-term monitoring of variable stars, public programs, and just plain stargazing. If you wish to join such a group, get advice on buying a telescope, or learn techniques such as photography of celestial objects, your best bet is to get in touch with an amateur astronomy group. Local clubs can be found in most major cities, and regional associations are everywhere. These groups may not be listed in the telephone book, but a telescope shop or planetarium is bound to know whom to contact. You may also find contact information on the World Wide Web, where many astronomy clubs and publications have sites.

The amateur astronomers in the United States outnumber the professionals. The American Astronomical Society, the principal professional astronomy organization in the United States, has some 6,500 members. As a rule, these people fall into a limited number of categories: some do research and teach at colleges and universities; others do research and work at government-sponsored institutions, such as the national observatories and federal agencies like the

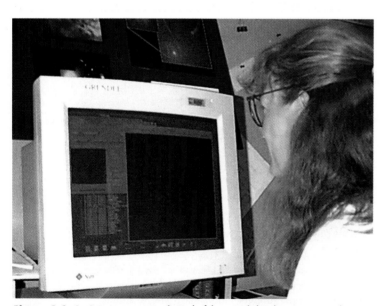

Figure 1.3 An Astronomer at work. Much of the research done by astronomers today including telescope operations, as well as data analysis, is done on computers.. (Scott Kahler)

Can the Stars Influence Your Life?

Throughout this text you will find commentaries such as this one on the relationship between science and society. Some of these will discuss technological advances ("spin-offs") that have come about as a result of scientific progress, while others will be aimed at helping you to think critically about things you may see, hear, or read. This first commentary falls into the latter category.

There is an unfortunate tendency in modern society to confuse astrology and astronomy or, worse yet, to consider one a legitimate alternative to the other. Behind this tendency lies an even more unfortunate misunderstanding of what science is and of the distinction between scientific theory and beliefs adopted on faith.

Astrology, the belief system based on the premise that events and configurations in the sky influence human affairs and activities on Earth, arose at a primitive stage of human development, at a time when the Earth was still thought to be a flat disk under the dome of the heavens. Although ancient Greek astronomers did much to raise the sutdy

of the heavens to a scientific level, the astrological reationale for study of the heavens persisted through the subsequent Dark Ages.

Belief in the predictions of astrology began to waver during the Renaissance, when the true nature of the heavens and the motions of astronomical objects were untangled, but even so, substantial interest in astrology was maintained. In fact, some of the most eminent and forward-looking astronomers of the time practiced astrology. Unfortunately, even today, some people continue to profess faith in astrology.

One of the basic lessons we have learned about the universe is that it is easy to make mistakes unless we are careful to be objective and to accept only conclusions that can be verified by repeated observations or experiments, or by making predictions that can be tested. Astrology fails to meet these criteria. Researchers have made serious attempts to test astrological lore by statistical analysis of people born under different signs, but no trace of a correlation has ever been found. One particularly interesting test, which was conducted with the help of a national organization of astrologers and astrological researchers, showed that astrologers were unable to match actual and predicted personality traits in a large sam-

pling of people. The success rate for predictions based on astrological data was no better than random chance (see the magazine *Nature*, December 5, 1985).

Astrology fails to meet any of the criteria for a scientific theory, in that it fails to fit observations. Some would argue that this is unfair, because astrological predictions refer only to people's tendencies and cannot be applied to specific individuals. This argument, however, amounts to saying that astrology cannot make predictions that can be tested, which disqualifies it as a scientific theory.

Also, according to Occam's razor, when presented with alternative possible interpretations, nature usually favors the one requiring the fewest arbitrary assumptions or unprovable postulates. Certainly, it is far simpler to admit that astrology does not work than to accept all the complex and arbitrary rules (many of which do not command universal agreement even within the astrological community!) that are required in order to believe in astrology.

National Aeronautics and Space Administration (NASA); and some perform research and related engineering functions in private industry, most often with companies involved in aerospace activities. A growing number of astronomers dedicate their careers to teaching, often at small colleges where research programs are not supported.

Most of the funding for research in astronomy, even for those not working directly for federal labs and observatories, comes from the government. A significant function of an astronomer on a university faculty is to write proposals, usually to the National

Science Foundation or NASA, asking for support for projected research programs.

The United States has national observatories that provide astronomers with access to large telescopes for visible-wavelength observations (through the National Optical Astronomical Observatories) and radio observations (through the National Radio Astronomy Observatory). For other wavelengths that cannot be observed from the ground, the U.S. government supports observations through NASA, which funds and operates various space-based observatories.

The situation is similar worldwide. Most nations that can support basic research have professional astronomical organizations, some of them with very long and distinguished histories (for example, the Royal Astronomical Society in England and its parent organization, simply called the Royal Society). The activities of astronomers in these nations typically are similar to those of U.S. astronomers, and funding mechanisms are comparable, although the details vary with the form of government and the degree of commitment to the field. Many countries have national observatories, supported individually or as part of an international consortium (for example, several European nations have formed the European Southern Observatory, which has telescopes in Chile, and a different combination of European countries constitutes the European Space Agency, which supports space-based astronomy).

An international organization of astronomers, called the International Astronomical Union (IAU), coordinates activities and agreements on terminology (such as a recent, highly publicized discussion of whether or not Pluto should be considered a true planet). The world's astronomers engage in a very high degree of communication and cooperation, including day-to-day research collaborations, hirings across international boundaries, and conferences and symposia, which are often organized by the IAU and attended by people from many nations.

The terms **astronomer** and **astrophysicist** have come to mean much the same thing, although historically they were different. An astronomer studied the skies, gathering data but doing relatively little interpretation; an astrophysicist was primarily interested in understanding the physical nature of the universe and therefore carried out comprehensive analyses of astronomical data or did theoretical work, in both cases applying the laws of physics to phenomena in the heavens. Today, however, nearly all astronomers do astrophysics to some degree. Consequently, the two terms now tend to be used interchangeably. Many modern astronomers call on the fields of engineering (for instrument development), chemistry (for studying planetary and stellar atmospheres and the interstellar medium), geophysics (for probing interior conditions in planets and other solid bodies), and sometimes even biology, but always with an underlying foundation in physics.

The unifying characteristic of all astronomers, regardless of vocation, is a deep curiosity about the skies, coupled with an esthetic appreciation of the beauty of the celestial phenomena and the relationships among them. Every astronomer or astrophysicist enjoys the appeal of a clear night under a starry sky.

A Typical Night Outdoors

Let us imagine that we are sitting outdoors on a fine, clear night, far from city lights and other distractions. This is easy to do and highly recommended; by simply getting out into the countryside, we can see many of the beautiful objects that we will be studying in more detail throughout this text.

The most obvious objects in the sky, assuming that the Moon is not in a bright phase, are the stars. They appear in profusion, scattered across the heavens, displaying a wide range of brightness and subtle variations in color. They twinkle, giving the appearance of vitality. Here and there we may see concentrations of stars in a cluster **(Fig. 1.4)** or possibly a dimly glowing gas cloud **(Fig. 1.5)**.

If it is a moonless night, we see a broad, diffuse band of light across the sky (in the Northern Hemisphere, this is most easily seen during summer). This is the Milky Way, our own galaxy of stars, seen edgewise from an interior position **(Fig. 1.6)**. The Milky Way consists of billions of stars intermixed with patchy clouds of interstellar gas and dust. We may also see a few bright, steady objects that do not appear to twinkle. These are the planets, and as many as five of them may be visible on a given night, distributed along a great arc through the sky **(Fig. 1.7)**. Careful observation over several weeks will reveal that the planets are all moving gradually along this arc, changing their positions relative to the fixed stars.

The most prominent object in the nighttime sky is usually the Moon **(Fig. 1.8)**, which shines so brightly when near its full phase that it drowns out the light of all but the brightest stars and planets. The Moon, about one-fourth the diameter of the Earth but some 387,000 km (or about 60 times the radius of the Earth) away, presents various appearances to us, depending on how much of its sunlit portion we see. The Moon is always found somewhere along the same east-west strip of the sky, called the **zodiac**, where the planets travel.

Along with these permanent objects, temporary visitors to the nighttime sky may be visible as well. Occasionally, we may see brief flashes or trails of light called **meteors.** These "shooting stars" can be spectacular events, particularly when they arrive with great frequency, as they do during a meteor shower.

A TYPICAL NIGHT OUTDOORS

Clockwise ➤ **Figure 1.4 (upperr left) A cluster of stars.** This group of stars is held together by mutual gravitational attraction. This cluster, called the Pleiades, is easily visible to the unaided eye in the autumn and winter. (© 1985 Anglo-Australian Telescope Board. Photo by D. Malin) ➤ **Figure 1.5 (upper right) A gaseous nebula.** Clouds of gas and tiny dust particles such as this are the birthplaces of stars. This is an area known as the Eagle nebula, containing many dense cloud knots where stars could form, as imaged by the Hubble Space Telescope. (NASA/STSLI) ➤ **Figure 1.6 (middle right) The Milky Way** The composite photograph shows the hazy band of light across the sky known as the Milky Way. It is a cross-sectional view of the disk of our galaxy, which contains roughly 100 billion stars. (Mt. Wilson and Las Campana Observatories. Carnegie Institution of Washington) ➤ **Figure 1.7 (lower right) A Planet.** This is an image of the ringed planet Saturn, photographed by one of the Voyager probes. (NASA/JPL) ➤ **Figure 1.8 (lower left) The Moon.** Astronauts on one of the Apollo missions made this photograph from space. The portion at the left in this image is not visible from Earth. (NASA) ➤ **Figure 1.9 (middle left) A comet.** This is a view of Comet Kohoutek during its 1973 passage through the inner solar system .(NASA)

Comets are occasional visitors to our sky **(Fig. 1.9)**. Perhaps once or twice a year astronomers find a comet that is bright enough to be seen without binoculars or a telescope. These largely gaseous bodies orbit the Sun, as do the Earth and other planets, but in very elongated paths that bring them close enough to the Sun to heat up and glow visibly only for brief periods of days or weeks. In ancient times some of the brighter, more spectacular comets were interpreted as harbingers of catastrophe.

The sky displays a complex pattern of motions, some of them evident to an alert watcher in an hour or so, others requiring careful observations over hours, days, or weeks. The most obvious motion is the steady rotation of the entire sky; we see objects rise in the east and set in the west as the Earth rotates on its axis. (The terms **rotation** and **revolution** are sometimes interchanged in everyday conversation, but here rotation means the same thing as spin, whereas revolution means orbital motion, such as the motion of the Earth around the Sun or of the Moon around the Earth.) Another motion that can be discerned readily is that of the Moon with respect to the stars. As it orbits the Earth, the Moon moves a distance in the sky equal to its own apparent diameter in just one hour, so it is possible to see its position change with respect to the background stars in a short time.

Observing other cyclical motions, such as those of the planets as they gradually travel along their orbits about the Sun, requires more patience and care. It is noteworthy, however, that ancient astronomers noticed many of the regular patterns, some of them quite subtle, in the motions of the heavenly bodies. That they did so is a testimony to the care and diligence they applied to their studies of the skies.

The View from Earth

When we look at the sky, we do not see it in three dimensions because there are no obvious clues to tell us the distances to the objects we see. Long ago this fact led to the concept of the celestial sphere **(Fig. 1.10)**, in which the stars and other objects in the sky are envisioned as lying on the surface of a sphere that is centered on the Earth. Although we no longer think of this as literal truth, the celestial sphere may still serve as a convenient device for discussing and visualizing the heavens.

Directions and separations of objects on the celestial sphere are measured in angular units because, lacking knowledge of distances to objects, we have no easy way to determine their actual separations in true distance units such as meters or kilometers. Thus we may specify where a star is by saying how many degrees, minutes, and seconds of angle it is from another star or from a reference direction.

We have noted that the Sun and stars rise and set with the daily rotation of the Earth. This gives us a natural basis for timekeeping, and our standard units of time are based on the Earth's rotation. The length of the day is equivalent to the rotational period of the Earth (a more specific definition is given in the next chapter).

Anyone who has traveled from one hemisphere to the other may have noticed that the visible stars in the Northern Hemisphere are not the same as those in the Southern **(Fig. 1.11)**. The portion of the sky that we can see depends on our latitude (our distance, in degrees, north or south of the equator). For those of us living in the Northern Hemisphere, a large region of the southern sky is hidden from view. We see different constellations as we travel north or south, a fact that was well known to early astronomers, who deduced from this and other evidence that the Earth is round. One consequence is that astronomers must have telescopes in both hemispheres in order to study the entire sky.

We usually cannot see celestial objects during daylight, so we must observe at night. This limits our view; we can observe only in the direction away from the Sun. But because of the Earth's motion around the Sun, the part of the sky that we see during the night gradually changes **(Fig. 1.12)**. Therefore, if we are patient, we can observe any part of the sky at some time during the year.

From the Earth to the Universe: The Scale of Things

We have been describing the appearance of the sky to the unaided eye, which has necessarily limited us to nearby objects such as the Sun, the Moon, the planets, and the stars in our part of the local galaxy. It is interesting now to expand our horizons and to try to comprehend the scale of the universe beyond this local neighborhood. Some of the distance scales we discuss are listed in **Table 1.1 (page 14)** .

Even the nearest star is much farther from the Earth than any solar system object. If we take the average Sun-Earth distance as a unit of measure (we call this the **astronomical unit** or **AU**), the nearest star is nearly 300,000 of these units away from us. The most distant known planet, Pluto, is only about 40 AU from the Sun, so clearly the stars are much more widely dispersed in space than the objects within our solar system.

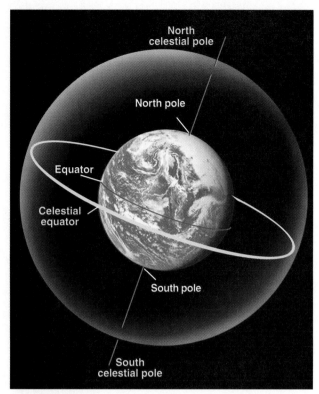

Figure 1.10 **The celestial sphere.** Because we see the sky in only two dimensions, it is useful and convenient to visualize it as a sphere centered on the Earth, with the stars and other bodies set on the surface of the sphere. We measure positions of objects on the celestial sphere in angular units because the actual distances are not directly known. (Throughout this text, we will learn about distance measurements in astronomy.)

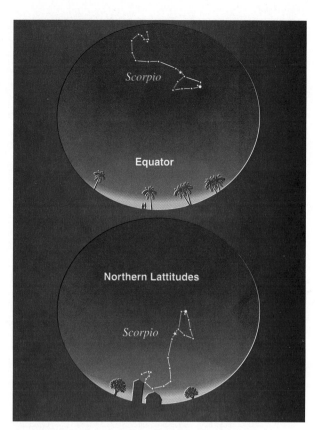

Figure 1.11 **Latitude and view of the sky.** The portion of the sky that we see depends on where we are with respect to the Earth's equator. These drawings show how the apparent position of the constellation Scorpio varies, as seen from a northern latitude versus the equator.

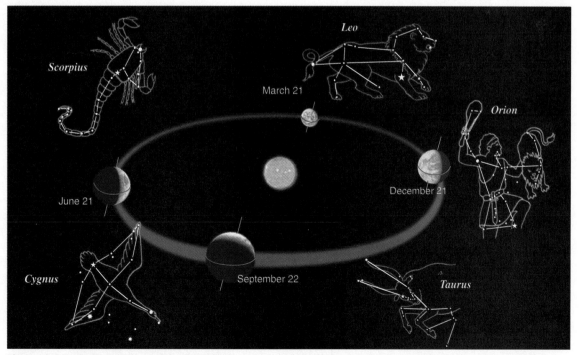

Figure 1.12 **The changing view of the sky with the seasons.** As the Earth orbits the Sun, the portion of the sky that we can see at night changes. The constellations visible at any time of the year are those that lie in the direction opposite the Sun, as seen from the Earth. For example, Orion is a prominent wintertime constellation.

Proxima Centauri
4.2 light-years away

Pluto & Charon
8 light-hours from Earth

The farther away we look in
space, the farther back in time
we travel.

Distant Galaxies
30–100 million light-years away

Quasars
11 billion light-years away

Andromeda Galaxy
2.2 million light-years away

THE SCALE of THINGS

Table 1.1
Size Scales in the Universe

Object or Phenomeonon	Size[a]
Atomic nucleus	10^{-15} m
Atom	10^{-10}
Virus	5×10^{-8}
Interstellar dust grain	5×10^{-7}
Bacterium	10^{-6}
Human body cell	5×10^{-5}
Human	1.8
Planet Earth	1.3×10^{7}
Sun	1.4×10^{9}
Sun-Earth distance	1.5×10^{11}
Distance to nearest star	4.1×10^{16}
Milky Way galaxy	9×10^{20}
Distance to Andromeda galaxy	2×10^{22}
Local Group of galaxies	5×10^{22}
Rich cluster of galaxies	10^{23}
Supercluster	10^{24}
The universe	$>10^{26}$

[a]The sizes listed are meant to be typical, to illustrate the relative scales. For round objects, the diameter is used; for irregular objects, an approximate average dimension is given.

Figure 1.13 A galaxy similar to the Milky Way. We can not obtain an exterior view of our own galaxy, but this one is believed to resemble ours. Most of the stars lie in a disk, seen face-on in this photo. (Photo courtesy NOAO)

http://universe.colorado.edu/fig/ 1-13.html

Reducing the scale of the solar system might help us to visualize the relative distances. For example, if we let the Earth be the size of a basketball (with a diameter of 0.3 m, we can convert the rest of the solar system to the same scale. The Sun would be 32.7 m in diameter, and the distance from the Sun to the Earth would be 3.53 km. Pluto would be an object the size of a tennis ball roughly 140 km away. The distance from the Sun to Alpha Centauri, the nearest star, would be over 1 million km, or more than twice the actual Earth-Moon distance!

Now consider our galaxy, the vast collection of stars to which our Sun belongs **(Fig. 1.13)**. The Milky Way galaxy contains roughly 100 billion (10^{11}) stars, arranged in a huge disklike structure having a diameter of about 100,000 (10^5) light-years. (A light-year, the distance light travels in a year at its speed of 300,000 km per second, is equal to about 9,500 bil-lion km). The distance to Alpha Centauri is about 4 light-years, and the most distant stars easily seen with the unaided eye are several hundred light-years away. (The majority of the brightest stars in the sky are actually quite nearby by galactic standards; see Appendix 10.) When light from these stars reaches our eyes, it has been traveling for hundreds of years, and light from the far side of the galaxy takes about 100,000 years to reach us.

The nearest galaxies beyond the limits of the Milky Way are the Magellanic Clouds, two irregularly shaped, fuzzy patches of light visible only from the Southern Hemisphere **(Fig. 1.14)**. The Magellanic Clouds lie between 150,000 and 210,000 light-years from the Sun, so they are not very far outside the galaxy. They are considered satellites of the Milky Way, orbiting it in a time of several hundred million years. Light from the Magellanic Clouds takes more

Figure 1.14   The Magellanic Clouds. (left) These two small, irregularly shaped galaxies lie just outside the Milky Way galaxy and orbit it, taking hundreds of millions of years for each complete orbit. (Fr. R. E. Royer) http://universe.colorado.edu/ fig1-14.html

Figure 1.15   The Andromeda galaxy. (below) At a distance of over 2 million light-years, this galaxy is the most distant object visible to the unaided eye. Without a telescope and a time-exposure photograph, the eye sees only an extended, fuzzy patch of light, rather than the detailed view shown here. The full extent of the galaxy, as revealed by a time exposure such as this, covers about 2.5° on the sky, or five times the angular diameter of the Moon. http://universe.colorado.edu/fig1-15.html

Figure 1.16   A cluster of galaxies. (below, left) The faint, fuzzy objects in this photograph are galaxies belonging to a cluster that lies over a billion light-years from the Milky Way. Like our own, each galaxy contains billions of individual stars. (Palomar Observatory, California Institute of Technology).

than 150,000 years to reach us. The most distant object visible to the unaided eye is the Andromeda galaxy **(Fig. 1.15)**, which lies about 2.3 million light-years from Earth. When we look at the Andromeda galaxy, we are receiving light that has been traveling for more than 2 million years!

Even the distance to the Andromeda galaxy is insignificant compared with the scale of the universe itself. The Milky Way, the Magellanic Clouds, the Andromeda galaxy, and a number of other galaxies all belong to a concentrated grouping, or cluster, of galaxies. Most of the other galaxies in the universe also belong to clusters **(Fig. 1.16)**, whose diameters can

be as large as tens of millions of light-years. Between clusters of galaxies, space is relatively empty. (Actually, this point is controversial, as we shall see; all we can say for certain is that there are relatively few visible galaxies between clusters.)

Clusters of galaxies are themselves grouped into larger conglomerates called **superclusters,** whose size scales are significant, even considering the scale of the universe itself. A supercluster typically may have a diameter measured in the hundreds of millions of light-years. (*Diameter* is probably not a good word; superclusters seem to be sheetlike or filamentary structures, not rounded like clusters of galaxies.) It is

Becoming Familiar with the Sky

ASTRONOMICAL

ACTIVITY

Many students taking astronomy for the first time may not be familiar with the nighttime sky. As a starting exercise, then, it is a good idea to go outdoors on a clear night and make some simple observations. Doing this will help acquaint you with the sky and will impart to you some of the sense of awe and wonder that inspires astronomers to pursue careers in this field.

To begin, you must have a clear night, preferably with no Moon (that is, you should choose a time when the Moon is between its third quarter and new phases, so that it will not rise until very late at night, leaving the sky dark during the early evening). You will be able to see a lot more if you can get away from city lights—bright lights "night-blind" your eyes, and scattered light from streetlights, glowing neon signs, and the like obscures the sky. Take a sky chart with you (the ones inside the front and back covers of this book will do, or you can find monthly charts in magazines like *Sky and Telescope*). If you bring a flashlight to see the chart, tape red cellophane or plastic over it; red light does not ruin your night vision the way white light does.

You can gain some concept of angular measure quite easily. Extend your arm and note how wide your fist appears against the sky. The angular width of your fist at arm's length is about 10°. The width of one of your knuckles is about 2°. If the Moon is up, you should find that its angular diameter is about a quarter of a knuckle width, or about 0.5°. You can experiment with these measures; for example, the distance from the horizon up to the zenith (the direction straight overhead) should be about 9 "fists."

Continue your familiarization with the sky by just looking up and allowing your eyes to become comfortable with the grand view of thousands of stars. In time you will begin to notice patterns of relatively bright stars; you can then start to match these with the chart you have brought. These patterns are the same as those seen by the earliest humans, and various cultures have attached a great deal of significance to them.

Learning the constellations, as the patterns are called, and some of the mythology associated with them is fun. See which constellations you can identify. The particular constellations you will see depend on the time of the year when you look. During the year, you should go out every month or two and identify the new constellations that have shifted into view during evening hours.

Once your eyes have adapted to the darkness, on a spring or summer night you should be able to make out a hazy, bright band stretching across the entire sky. This is the Milky Way, our home galaxy. You are seeing a cross-sectional view of a great disk; the solar system is located within the disk, about two-thirds of the way out from the center. Later you will learn that the Milky Way contains

---

difficult to imagine an organized object or collection of objects so large that light takes hundreds of millions of years to travel across it.

Beyond the size scale of the superclusters, we approach the scale of the universe itself. It is apparent from a variety of lines of evidence that the observable universe has an overall size scale measured in tens of billions of light-years. Light reaching us now from the farthest reaches of the universe has been traveling for many billions of years. This has the fascinating implication that when we observe very distant objects, we are looking back in time, viewing the universe as it was long ago. This fact is an important benefit for astronomers who study the origins of the universe.

Considering the sizes and distances of objects in the universe gives us a sobering perspective on ourselves and our tiny planet. It can be quite a revelation to see how much we presume to explain about the universe and how much we think we have learned about the origins, present state, and future evolution of the cosmos and all it contains. It is wise, however, to keep in mind that there is very much that we do not know.

Summary

1. Astronomy is the science in which the entire universe is studied. The study of astronomy requires knowledge of several sciences, including physics, chemistry, geology, and biology.
2. The essence of science is to find the best explanation of observed phenomena. A theory is an explanation that may fit, but is always subject to

more than 100 billion stars, and that this great pinwheel of stars is just one of hundreds of billions in the universe. At this point in your nighttime viewing session, you may begin to gain some appreciation for how small the Earth is on the grand scale of things.

At almost any time of the year, you will see at least one or two planets. These will be obvious because they do not "twinkle" the way stars do, and they will be among the brightest objects you see. Venus and Jupiter are usually brighter than any star, as is Mars at its most favorable position; Saturn also ranks with the brightest stars. Mercury, the other planet that can be seen without binoculars or a telescope, is also quite bright, but is always very close to the Sun.

You may notice that the planets fall along a line that stretches across the sky from east to west; this is the ecliptic, and it represents an edge-on view of the disk of the solar system (the specific definition is given in Chapter 2). If the Moon is near third quarter when you are observing, it will rise around midnight. As it does, you will see how its scattered light makes the entire sky glow faintly. It soon becomes impossible to see the faintest stars, and the Milky Way is drowned out as well. The Moon itself is a fascinating body to look at; even without binoculars or a telescope, you can make out surface markings. You can also see the Moon's motion relative to the fixed stars; in a time of only one hour, the Moon moves a distance approximately equal to its own diameter. You should be able to observe this by noting the Moon's position relative to stars near it on the sky (the planets also move with respect to the fixed stars, but much more slowly, so that days or weeks are required to detect the motion).

During your evening outdoors, you may be lucky enough to see other celestial phenomena, such as meteors. It is not unusual to see several over a few hours. Meteors are caused by small particles of rock (often no bigger than grains of sand) that enter the Earth's atmosphere from space and are burned up by the heat of friction as they collide with air molecules.

During your observing session, you will see the same view of the sky as your ancient ancestors, but with a couple of differences. No matter how far you go from city lights, your view is affected, at least a little, by human influences. Even if scattered light is minimized, the air overheard is less clear than it was before the Industrial Revolution due to global air pollution (this effect is actually very small in remote areas, but can be very significant near cities). In addition, you very likely will see artificial satellites moving across the sky; these are quite obvious because they do not twinkle and move rapidly (most move in the west-to-east direction, but you may see some in polar orbits, which move from north to south). And almost certainly you will see the lights of aircraft as they carry people across the sky.

In modern times, as in the past, viewing the sky on a clear night is a refreshing experience and one that you may enjoy long after your astronomy course is over. Perhaps a lasting appreciation for the pleasure of doing so will be one of the most valuable lessons you learn from your course in astronomy.

---

revision if observations reveal it to be incorrect. An essential element of any theory is its ability to make predictions that can be tested.

3. Professional astronomers today are more properly called astrophysicists, because they apply laws of physics to celestial phenomena in order to better understand them. Most professional astronomers work at colleges or universities, at government-sponsored research centers or observatories, or in private industry.

4. On a clear, dark night, the unaided eye can see up to five planets, the Moon, several thousand stars, and the occasional meteor or comet.

5. The Earth's rotation causes all objects in the sky to undergo daily motion, rising in the east and setting in the west.

6. Because we cannot directly determine how far away objects are, we must measure their positions or separations in angular units. For convenience, we visualize a celestial sphere, centered on the Earth, on which the astronomical bodies lie.

7. The part of the sky that can be seen depends on the latitude of the observer, and the portion visible at night depends on the time of year.

8. The most distant planet in our solar system is about 40 times farther from the Sun than the Earth is, whereas the nearest star is some 300,000 times the Sun-Earth distance, or 4 light-years, away. Our galaxy is about 100,000 light-years in diameter, nearby galaxies lie between 150,000 and 2.3 million light-years away, and clusters of galaxies are typically separated by tens of millions of light-years. The size of the observable universe is measured in billions of light-years.

9. The light we receive from a distant object has been traveling toward us for as long as billions of

years (in the case of the most distant galaxies), so we can observe the universe as it was long ago.

Review Questions

1. Express in your own words the nature of scientific theory, and explain the differences between theory, speculation, and faith.

2. Suppose there were two competing theories of how the solar system formed. In one theory, the planets formed as the result of a collision between two stars, which created debris (streamers of gas) that then condensed to form planets. In the other theory, a disk of gas and dust surrounding a young star was created as a natural part of the Sun's formation process, and this disk then condensed to form planets. Which of these two theories is favored by the principle of Occam's razor? Explain.

3. Discuss how you might test the validity of astrology; that is, describe an experiment or observation that would show whether predictions made by astrologers are correct or incorrect.

4. Why do you think it is best to get away from the city to view the nighttime sky?

5. We have noted that the planets appear to travel along the same strip across the sky. What does this tell us about the orbits of the planets around the Sun?

6. Explain why positions of objects in the sky are measured in angles rather than in units of linear distance such as meters or kilometers.

7. Suppose two objects on the sky have the same angular diameter (the Sun and the Moon are an example of such a pair). What does this tell us about the *actual* diameters of the two objects? Explain.

8. Can you think of any observational evidence or proof that the Earth moves around the Sun? Discuss this in the context of early beliefs that the Sun moves around the Earth.

9. Explain why it is necessary to have observatories in both the Southern and Northern Hemispheres. At what latitude is the largest possible fraction of the sky visible?

Problems

1. An angle of one degree contains 60 minutes of arc, and a minute of arc contains 60 seconds of arc. How many arcseconds are in a full circle (360°)?

2. The Sun's angular diameter is 30 arcminutes. What is the diameter in degrees and in arcseconds?

3. The Moon takes 29.5 days to orbit the Earth, as seen by an observer on the Earth. How far must the Moon travel each day, in units of degrees per day? How far must it travel each hour? How does this hourly distance compare with the angular diameter of the Moon, which is 0.5°? Would this motion be easily observed?

4. Suppose the scale of the solar system were changed so that the Earth was the size of a marble (with a diameter of 1 cm). On this scale, what would be the diameter of the Sun? What would be the Earth-Moon distance? What would be the Sun-Earth distance? How far from the Sun would Pluto be? How far away would the nearest star be? (Note: To answer these questions, you need to know the true sizes and distances involved. Most are given in the text of this chapter, but many can also be found easily in Appendix 7.)

5. Using information in Appendix 7, calculate the light-travel time from the Sun to Earth, to Jupiter, and to Pluto. (You will find it easiest to use the speed of light in metric units, which is 300,000 km per second.)

Additional Reading

We can further appreciate the essence of astronomy, as well as its beauty, by reading a wide range of books and periodicals. Many bookstores contain volumes of astronomical photographs, in addition to numerous books on astronomy written for the layperson.

Periodicals that are particularly well suited for students using this text include *Sky & Telescope, Mercury,* and *Astronomy. Sky & Telescope* is strongly recommended for students wishing to carry out projects such as telescope building or astrophotography, because it includes monthly sky charts showing the positions of the stars and planets as they change throughout the year, and because it carried articles on practical (observational) astronomy for the amateur.

Bahcall, J. M. 1991. U.S. Astronomy's Next Decade. *Sky & Telescope* 81(6):584.

Carney, B., *et al.* 1996. The Future of Astronomy (a series of articles on all aspects of the field). *Mercury* 25(1):8–36.

Culver, R. 1984. *Astrology, True or False: A Scientific Evaluation.* New York: Prometheus Books

DeRobertis, M. and P. A. Delaney. 1993. The Roots of Astrology. *Mercury* 23(5):21.

Dobson, A. K. and K. Bracher 1992. Urania's Heritage: A Historical Introduction to Women in Astronomy. *Mercury* 21(1):4.

Fraknoi, A. 1990. Scientific Responses to Pseudoscience Related to Astronomy. *Mercury* 19(4):144.

Sagan, C. 1993. Why We Need to Understand Science. *Mercury* 22(2):52.

Trimble, V. and R. Elson 1991. Astronomy as a National Asset. *Sky & Telescope* 82(5):485.

Chapter 2
The Nighttime Sky

*The stars hang bright above,
silent, as if they watch over
the sleeping Earth.*

Samuel Taylor Coleridge

o many people, the essence of astronomy is the naked-eye view of the sky. Forget about abstract mathematical analyses, never mind high-tech gadgetry; just go out on a clear night and *look*. Not many people do that today, but in pretechnological times, everyone was aware of the nighttime sky. Not only did ancient peoples know the stars, but they also developed a keen awareness of changes in the sky, of movements and patterns that repeat over time. The shifting positions of the stars, the Sun and Moon, and the planets may seem chaotic at first, but with patience and care, regularity and repetition become obvious.

Many of the observed motions are not actual motions of the observed bodies but reflections of our own travels through space, because we view the heavens from a moving platform. The Earth spins while it travels around the Sun in a nearly circular path **(Fig. 2.1),** and these motions create both daily and yearly cycles of celestial events as seen from the Earth's surface. The other eight planets all behave in similar fashion, and most, including the Earth, have one or more satellites orbiting them. Because we are viewing the skies from a moving vantage point and because all celestial objects have motions of their own, our impression is that the heavens are very complex.

In this chapter, we will learn about the nighttime sky as it appears to the unaided eye. We will develop a modern understanding of the objects that can be seen without a telescope, and we will see how their simple motions create complex paths through the sky as seen from the Earth. In the process, we will develop an appreciation for the task of the ancient philosophers who strove to comprehend the workings of the universe, for the view of the heavens described here is the sum of all the evidence available to pretechnological cultures.

The Constellations

One of the most persistent concepts of astronomy, dating back to the earliest recorded origins of the field, holds that star patterns in the sky have significance. These patterns, called **constellations,** arose from early Greek mythology. Contrary to common misperception, the Greeks did not think of the constellations as literal representations of their heroes and gods. Instead the Greeks named certain recognizable patterns of stars in honor of various figures from their legends. Other cultures (such as the ancient Chinese and Native Americans) developed their own legends of the sky and their own constellations. It is the Greek constellations and related terms that have carried through into our own culture, however.

The original constellation names were translated from Greek to Latin when the Roman Empire rose to dominance, and the designations we use today come from the Latin. On the other hand, many of the modern names for prominent individual stars come not from Latin but from Arabic. These names were assigned by Arab astronomers following the downfall of the Greek civilization and often consist of Arabic translations of the original Greek designations. For example, Betelgeuse, the bright red star in the shoulder of Orion the Hunter, translates literally to "armpit of the giant."

Ancient and nonscientific though the constellations are, they have significant impact on the terminology of modern astronomy by providing a reference frame or map of the sky. The modern constellations are officially designated and charted by the International Astronomical Union (IAU), and they define very specific regions of the sky. Every bit of the sky is covered by the 88 "official" constellations (Appendix 12), each including the pattern of stars associated with it in the ancient legends. Thus the modern constellation of Orion contains within it the hunter of ancient lore, along with an extended region surrounding the figure. A map of the constellations looks a bit like a map of the western United States, where the boundaries are generally straight lines but the shapes of the states are irregular **(Fig. 2.2).**

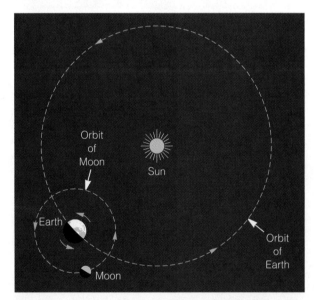

Figure 2.1 Motions of the Earth and Moon. The Earth and Moon spin and move along their orbits at the same time. The combination of motions affects the apparent paths of the Sun and the Moon as seen by an observer on the spinning Earth. (Not to scale.) http://universe.colorado.edu/fig/2-1.html

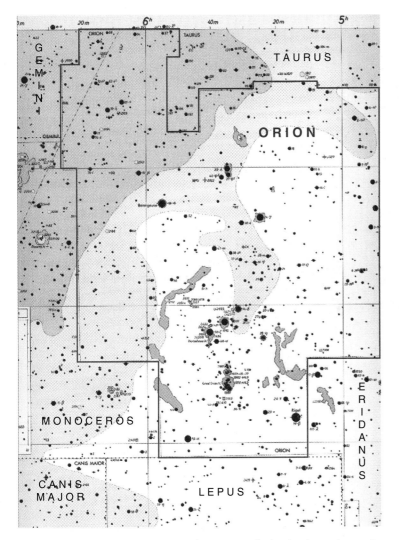

Figure 2.2 Map of the Constellations. This is a portion of a sky atlas, showing the constellation Orion with its modern day boundaries. Every spot on the sky is included in one of the 88 "official" constellations. [Wil Tirion 1981, Sky Atlas 2000.0 (Cambridge, Mass: Sky Publishing Corporation)]

Table 2.1
Periods of Significant Motions[a]

Motion	Period
Sidereal day	$23^h56^m4.098^s$
Mean solar day	$24^h00^m00^s$
Tropical year (equinox to equinox)	$365^d5^h48^m45^s$
Sidereal year (fixed stars)	$365^d6^h9^m10^s$
Synodic month	$29^d12^h44^m3^s$
Sidereal month	$27^d7^h43^m11^s$
Mercury: Sidereal period	87.969^d
Synodic period	115.88^d
Venus: Sidereal period	224.701^d
Synodic period	583.92^d
Mars: Sidereal period	1.88089^y
Synodic period	2.1354^y
Jupiter: Sidereal period	11.86223^y
Synodic period	1.0921^y
Saturn: Sidereal period	29.4577^y
Synodic period	1.0352^y
Uranus: Sidereal period	84.0139^y
Synodic period	1.0121^y
Neptune: Sidereal period	164.793^y
Synodic period	1.00615^y
Pluto: Sidereal period	248.5^y
Synodic period	1.0041^y

[a]Units for lunar and planetary motions are mean solar days or tropical years.

Many students of astronomy, professional and amateur alike, learn the constellations by heart as an aid to identifying specific stars and also as a link to the more human side of astronomy. If you are interested in doing this, several useful sky guides can be found in any bookstore, and, of course, a local astronomy club or planetarium can give you excellent guidance. For our purposes in this text, we will refer to constellations from time to time as we use the standard terminology of astronomy, but we will not devote significant time to the study of the constellations for their own sake.

Rhythms of the Cosmos

The observed motions of celestial bodies result from a combination of rotational and orbital motions, including the spin of the Earth and its annual movement around the Sun, as well as the individual motions of the Moon and the planets. Many of the motions in the heavens are cyclical; that is, they repeat regularly in a well-defined period. The periods of many solar system motions are summarized in **Table 2.1.**

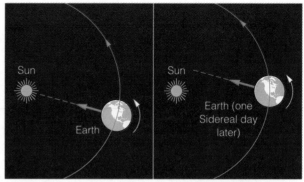

Figure 2.4 **The contrast between solar and sidereal days.** The blue arrow indicates the overhead direction from a fixed point on the Earth. From noon one day (left), it takes one sidereal day for the arrow to point again in the same direction, as seen by a distant observer. Because the Earth has moved, however, it will be about 4 minutes later when the arrow points directly at the Sun again; hence the solar day is nearly 4 minutes longer than the sidereal day. (Not to scale.)

Figure 2.3 **Star trails.** The curved streaks shown in this ten-hour exposure are the trails left by stars on the photographic film as the Earth rotated. This is a view of Mt. Rainer in Washington state, as seen from the south. (© Rick Morley Photography)

In science in general, and especially in astronomy, we must always be careful to note that what we observe depends on the frame of reference in which our observations are made. Our view of the sky from a location on the Earth's surface is strongly affected by our reference frame and its own motions. Physicists have learned that there is no absolute reference frame; in any situation, we must define the frame in which we observe. In astronomy we can use the reference frame of the Earth, as we are doing now in discussing our view of the nighttime sky, but we use more general frames in other contexts, such as the framework established by the (approximately) fixed stars or perhaps the reference frame of distant galaxies.

Daily Motions

The most obvious motion that we can observe in the sky is the daily east-to-west cycle of all celestial objects caused by the rotation of the Earth. The Earth spins on its axis, so we earthbound observers see a continuously changing view of the heavens. We see the Sun rise and set, along with the Moon, the planets, and most of the stars **(Fig. 2.3).** Even though we understand that these daily, or **diurnal motions** are the result of the Earth's spin, we still refer to them as though the objects in the sky themselves were moving. The rotation of the Earth means that a person standing on the equator covers the entire circumfer-

ence (about 40,000 km) in 24 hours; yet our senses give us no feeling of motion.

In the reference frame of the Earth, the Earth's rotation forms the basis for our timekeeping system, since the length of the day is a natural unit of time on which to base our lives—one to which, indeed, nearly all earthly species have adapted. The day is divided into 24 hours, each containing 60 minutes, each of which in turn consists of 60 seconds. These divisions are based on the numbering system developed several thousand years ago by the Babylonians.

Careful observation shows that from our earthbound point of view the Sun and the Moon take longer than the stars to complete their daily cycle. If we time precisely how long the Sun, the Moon, and the stars take to make one full cycle, returning to the same spot overhead, the Sun and the Moon take longer than the stars. This is because both the Sun and the Moon move in the same direction as the Earth's rotation, opposite to the daily rising and setting of the stars. The daily motions are always toward the west as seen from the Earth's surface, but the Sun and the Moon move toward the east against the background of the stars. Therefore, from our point of view, the Sun and the Moon fall behind the stars, taking a little longer to return to the same position each day **(Fig. 2.4).** The Sun rises about 4 minutes later each day by comparison with the stars, and the Moon rises almost an hour later each day. The planets also move toward the east most of the time, but so slowly that their position shifts relative to the stars can take weeks or months to become noticeable without careful measurement.

The commonly used reference for measuring the length of the day is the **meridian.** This is the north-south line extending from pole to pole and passing directly overhead at the observer's location. The meridian is a well-defined local reference frame for time measurements. To measure the time when a celestial object crosses the meridian, an astronomer uses a special telescope called a **transit circle (Fig. 2.5),** which is fixed so that it can point only at objects on the meridian. Technically, noon occurs when the Sun crosses the meridian, although the adoption of standardized time zones, along with the semiannual shift into and out of daylight saving time, causes astronomical noon to differ from noon on the clock.

The length of time from one noon to the next is called the **solar day.** As we have just seen, this is different from the length of time the stars take to go through a complete cycle, which is called the **sidereal day.** The solar day is about 4 minutes longer than the sidereal day, which is the rotation period of the Earth with respect to the fixed stars.

Our timekeeping system is based on the solar day. If we used the sidereal day instead, clock time and Sun time would soon be out of synchronization, and eventually the Sun would be setting when the clock says it is morning and rising during the clock's night. Because the Earth's orbital speed is not precisely constant, the length of the solar day varies a little throughout the year. It would be inconvenient to allow the hour, minute, and second to vary along with the solar day, so the average length of the solar day, called the **mean solar day,** has been adopted as our timekeeping standard. The mean solar day is 3 minutes and 56 seconds longer than the sidereal day.

The point about which the stars circle, directly over the Earth's North Pole, is the **north celestial pole,** the point where the Earth's rotational pole is projected onto the celestial sphere. Similarly, the projection of the South Pole onto the celestial sphere is the **south celestial pole.** A bright star called Polaris lies very close to the position of the north celestial pole and is therefore almost stationary (in the reference frame of the Earth) throughout the night. The projection of the Earth's equator onto the sky is called, naturally enough, the celestial equator. Positions of celestial objects are mapped using a coordinate system that is based on the celestial poles and equator (see Appendix 11).

We have mentioned that the stars rise and set as the Earth rotates, but some stars never set. These are the stars that lie so close to one of the celestial poles that they can be seen throughout the night. From any latitude on the Earth's surface (except right at the equator), it is possible to see an area of the sky around one of the poles that is never obscured by the Earth. From a latitude of 30° N, for example, we can see the sky to an angular distance of 30° beyond the

Figure 2.5 **A transit circle.** Such a device normally points straight up. It is used to record the times when certain reference stars pass over the meridian (the north-south line that passes directly overhead) and is therefore helpful in measuring the sidereal day. (U.S. Naval Observatory)

north celestial pole. Therefore, we can always see the part of the sky that lies within 30° of the north celestial pole (but we can never see the sky within 30° of the south celestial pole), and stars in this part of the sky circle the pole but do not set (see Fig. 2.3). The same thing happens around the south celestial pole for observers in the Southern Hemisphere.

Basing sky maps on the celestial poles and equator presents one complication, however. The Earth's rotational axis slowly wobbles in a motion called **precession (Fig. 2.6),** and this causes the celestial poles and equator to move slowly through the sky. The Earth takes some 26,000 years to complete one cycle of this motion, which is exactly like the wobbling of a play top or gyroscope. Our coordinate system therefore undergoes a very gradual motion that causes star positions to shift extremely slowly. Despite the small magnitude of the effect, observers noticed it more than 2,000 years ago, and modern astronomers must allow for it when planning observations. Catalogs or

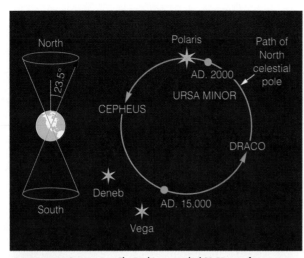

Figure 2.6 Precession. The Earth's axis is tilted 23.5° away from perpendicular to the orbital plane, and it wobbles on its axis, so that an extension of the axis describes a conical pattern *(left)* in a time of about 26,000 years. The north celestial pole therefore follows a circular path on the sky *(right)*.

charts of star positions always specify the date for which they are valid so that astronomers know how much precession to allow for.

Annual Motions

Now let us think about motions in the sky that are caused by the Earth's motion as it orbits the Sun. We have already mentioned the apparent daily eastward motion of the Sun with respect to the stars, which is the reason for the difference between the solar day and the sidereal day. We can also consider the apparent position of the Sun with respect to the stars. It is difficult to directly measure the Sun's position relative to the stars because we cannot see the stars when the Sun is up, but the Sun's position can easily be inferred by noting which constellations lie just ahead of it or behind it near sunrise and sunset.

The Sun's Path through the Sky

The changing solar position relative to the stars is a reflection of the motion of the Earth as it orbits the Sun. An outside observer would see that it is the Earth, not the Sun, that is moving, but to us the Sun appears to be moving gradually eastward with respect to the stars. The Earth travels in a fixed plane, so the Sun appears to follow the same line through the constellations each year (**Fig. 2.7,** see next page). The Sun's path is called the **ecliptic.** The sequence of stellar con-

stellations through which the Sun passes is called the **zodiac.** The zodiac includes 12 principal constellations, which were identified in antiquity and once thought by astronomers to have significance for our daily lives (actually there are 13 major constellations along the ecliptic, but astrologers have tended to ignore one of them, Ophiuchus).

Because we can see stars only in the nighttime sky, the constellations most easily visible to us at any given time are the ones at least a few degrees to the east or west of the Sun. This means that most stars, except those near the poles, cannot be observed year-round. Astronomers planning observing programs must take into account the best time of year for their target objects to be observed.

The orbits of the other planets and of the Moon lie in planes that are closely aligned with the plane of the Earth's orbit. Thus the solar system has an overall disklike structure. This means that from our point of view, the planets and the Moon are always seen near the ecliptic and therefore pass through the constellations of the zodiac. Hence it is not surprising that ancient astronomers attached great significance to this band of constellations.

The Seasons

Besides causing the apparent annual motions of the Sun and planets, the Earth's orbital motion has a second, and far more significant, effect on us: it creates our seasons. The Earth's spin axis is tilted relative to the plane of its orbit, so during the year the exposure to sunlight at any location of the Earth will vary. As the Earth orbits the Sun, the orientation of the Earth's spin axis is fixed relative to the stars, but our orbital motion causes this orientation to change relative to the Sun **(Fig. 2.8).** Summer in the Northern Hemisphere occurs when the North Pole is tilted toward the Sun; winter occurs during the opposite part of the Earth's orbit, when the North Pole is tilted away from the Sun (see Fig. 2.8). The large seasonal variations in climate at intermediate latitudes are caused by a combination of two effects: (1) the length of time the Sun is up varies, so in summer the Sun has more time to heat the Earth's surface; and (2) the Sun's rays strike the ground at a more nearly perpendicular angle in the summer **(Fig. 2.9),** so the Sun's intensity is much greater, heating the surface more efficiently. In the winter the days are shorter, and the Sun's radiation is less intense because the sunlight strikes the ground at a more oblique angle.

The Earth's axis is tilted approximately 23.5° from the perpendicular to the orbital plane. Therefore, during the year, the Sun, as seen from the

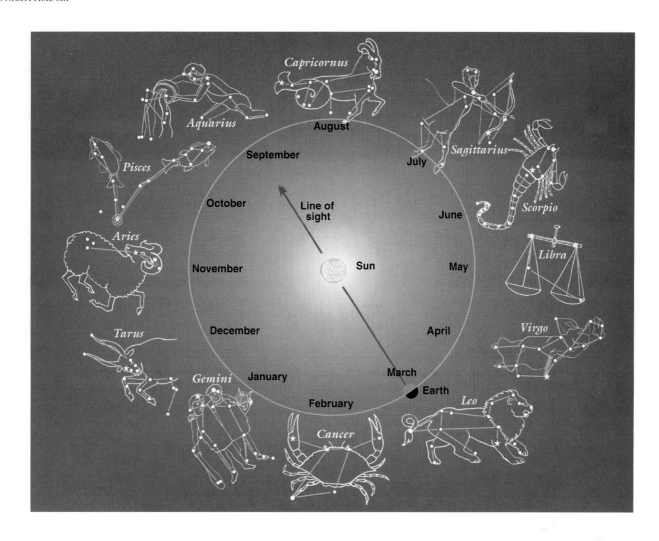

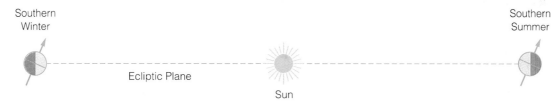

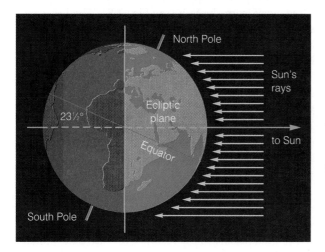

Figure 2.7 **(above) The path of the Sun through the constellations of the zodiac.** The dates refer to the position of the Earth each month. To see which constellation the Sun is in during a given month, imagine a line drawn from the Earth's position through the Sun; that line will extend to the Sun's constellation. For example, in March the Sun is in Aquarius. http://universe.colorado.edu/fig/2-7.html

Figure 2.8 **(middle) Seasons.** The Earth's tilted axis retains its orientation as the Earth orbits the Sun. Thus, at opposite points in the orbit, each hemisphere has winter or summer, depending on whether that hemisphere is tipped toward the Sun or away from it. http://universe.colorado.edu/fig/2-8.html

Figure 2.9 **(left) Effect of the Earth's tilted axis.** The Sun's rays strike the ground at varying angles, depending on the latitude and the time of the year. Therefore, the solar heating, which depends on how directly the rays reach the ground, varies with the season. Here it is summer in the Northern Hemisphere; the Sun's rays are nearly perpendicular to the Earth's surface at low to moderate northern latitudes, but they strike the surface very obliquely at southern latitudes.

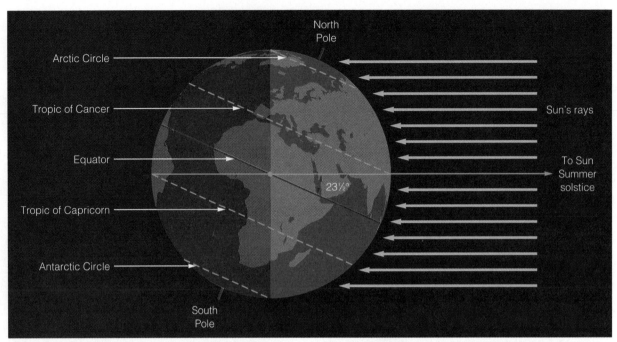

Figure 2.10 **The definition of latitude zones on the Earth.** At summer solstice, the Sun is overhead at 23.5°N latitude, its northernmost point. This defines the Tropic of Cancer, the northern limit of the tropical zone. At the same time the entire area within 23.5° of the North Pole is in daylight throughout the Earth's rotation; the boundary of this region is the Arctic Circle. Similarly, the Antarctic Circle receives no sunlight at all during a complete rotation of the Earth. Six months later, the Sun is overhead at the Tropic of Capricorn.

Earth's surface, can appear directly overhead (that is, at the **zenith**) as far north and south of the equator as 23.5°, defining a region called the **tropical zone (Fig. 2.10).** For people who live outside the tropics, the Sun can never be directly overhead. When the Northern Hemisphere is tilted most nearly in the direction toward the Sun, an occasion occurring around June 21 and called the **summer solstice,** the Sun passes directly overhead at 23.5° N latitude at local noon, but it passes to the south of the zenith for anyone at more northerly latitudes. On this occasion, sunlight covers the entire north polar region to a latitude as far as 23.5° south of the pole. This defines the **Arctic Circle** (see Fig. 2.10), and at the time of the solstice, the entire circle has daylight for all 24 hours of the Earth's rotation. At the pole itself there is constant daylight for 6 months.

During winter in the Northern Hemisphere, the Sun does not rise as high above the southern horizon at local noon as it does during the summer because the South Pole is now tilted toward the Sun. At the **winter solstice** (around December 21), when the Southern Hemisphere is pointed most nearly in the direction of the Sun, the Sun's midday height above the horizon, as viewed from the Northern Hemisphere, is the lowest of the year.

If we follow the Sun's motion north and south of the equator throughout the year, we find that it follows a graceful curve as it traverses its range from 23.5° N latitude to 23.5° S latitude **(Fig. 2.11).** The Sun crosses the equator twice in its yearly excursion, at the times when the Earth's North Pole is pointed in a direction 90° from the Earth-Sun line. At these times, the lengths of day and night in both hemispheres are equal. These occasions, which are referred to as the **vernal** (spring) and **autumnal** (fall) **equinoxes,** occur on about March 21 and September 23, respectively.

Calendars

We have referred to the association of astronomy and timekeeping, particularly in ancient times, when astronomers were responsible for predicting major natural events such as changes of season. Today everyone knows when these events will occur and generally keeps track of dates for all purposes with a modern calendar **(Fig. 2.12),** which can be purchased well in advance for any given year. It is interesting to realize, however, that even today our calendars—and hence the annual pattern of our daily lives—are determined by the motions of bodies in the heavens.

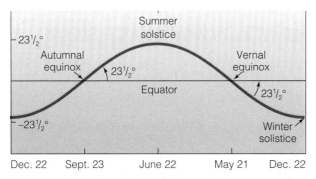

Figure 2.11 **The path of the Sun through the sky.** Because of the Earth's orbital motion and the tilt of its axis, the Sun's annual path through the sky has the shape illustrated here.

SUNDAY	MONDAY	TUESDAY	WEDNESDAY	THURSDAY	FRIDAY	SATURDAY
1	2 ○ Full Moon	3	4	5	6	7
8	9	10 ◗ Last Quarter	11	12	13	14
15	16	17 ● New Moon	18	19	20	21
22	23	24 ◖ First Quarter	25	26	27	28
29	30	31 ○ Full Moon				

Figure 2.12 **Modern calendar showing the Moon's phases.**

Different ancient cultures established various calendars, all of which were based to some degree on the length of the year. In several cases people deemed it unlikely or unacceptable that the year should not be evenly divisible into a round number of lunar months or even days, so some of the old calendars adopted lengths of the year that differed from the actual period of the Earth's orbit of the Sun. All such calendars had to be adjusted every so often or fall badly out of agreement with the seasons.

The length of the year that is important in calendar making is not the sidereal period of the Earth's orbit, as we might expect. A slight adjustment (about 21 minutes) is required because of precession, which creates a slight difference between the length of time required to repeat the pattern of seasons (the **tropical year**) and the Earth's actual orbital period around the Sun. The year based on the seasons and used in calendars is called the tropical year. Precise values for the lengths of the sidereal and tropical years are included in Table 2.1.

The year does not contain a whole number of days. Thus a truly accurate calendar would have 365.242 days, and we would start the new year at a different time of day each year. This would be inconvenient, so most calendars are based on the assumption that the year includes a whole number of days. The errors that build up are allowed to accumulate until an extra day is inserted to bring the calendar back into agreement with the seasons. The Babylonians had a calendar with 360 days and thus had to deal with an extra 5.242 days that accumulated each year (they handled this by adding a whole month every few years). The Roman Julius Caesar adopted a calendar (now called the Julian calendar) in which the length of the year was taken to be 365¼ days, which was only a few minutes in error. Every fourth year (a **leap year**) an extra day (February 29) was inserted, a practice that we still follow. The year was divided into 12 months that did not correspond exactly to lunar months, so that at least a year would have a whole number of months. Caesar was also responsible for starting the year on January 1; it had begun in March before his reform, which took place in 46 B.C.

In 1582, to correct for the few minutes' error that had crept into the Julian calendar because the tropical year is not precisely 365¼ days long, Pope Gregory XIII ordered additional reform. As a result, in the current calendar (called the Gregorian calendar), leap year is sometimes *not* observed. Leap year is ignored (the extra day is not inserted after February 28) in century years not divisible by 400. Thus there will be a leap year in the year 2000, which is divisible by 400, but not in the years 2100, 2200, and 2300. With this refinement, the Gregorian calendar builds up an error of only 1 day in 3,300 years, a sufficient degree of accuracy for most people's appointment books!

Motions of the Moon

Except for the Sun, the Moon is the brightest object in the sky. Like the Sun, the Moon also has its own complex motions as seen from the Earth. The Moon's motion with respect to the stars is much more rapid than that of the Sun and is therefore more easily

SCIENCE AND SOCIETY

Keeping in touch with the Stars

The passage of time affects us all. We measure our progress through a day, a week, or a lifetime using clocks and calendars. All the while, we rarely stop to think about the connection between timekeeping and astronomy.

The basis for timekeeping is the rotation of the Earth, as measured through the observations of stars. To measure the Earth's rotation period with respect to the stars (that is, the sidereal day), astronomers refer to the meridian, the imaginary north-south line passing overhead at an observer's position. As the Earth rotates, each star crosses the meridian once every sidereal day. Special instruments called **transit circles** (see Fig. 2.5) are used to measure the precise moment when a star crosses the meridian, thus measuring sidereal time. Allowance for the Earth's variable orbital speed must be made in determining the length of the mean solar day from these measurements.

Today relative time is measured precisely with atomic clocks, but the starting point for the measurement of time is still kept synchronized with the Earth's rotation. Atomic clocks measure intervals of time very accurately because they are based on the vibration frequencies of certain kinds of atoms, which are very constant. These clocks have several advantages over measuring sidereal time based on observations of stars. For example, atomic clocks keep a continuous record of the passage of time, so they can be referred to whenever

needed. Also, identical clocks can be placed in many locations and kept coordinated, a practice that allows many laboratories to measure the time precisely.

Scientists around the world have agreed on a standard time, called **coordinated universal time,** which, as the name implies, is coordinated among all nations so that we can all agree on the time. While atomic clocks keep track of **physical time,** the precise duration of time intervals, coordinated universal time must be adjusted occasionally to maintain synchronization with the Earth's rotation. Such adjustments are needed because occasionally the Earth's rotation period changes very slightly.

One persistent change that always occurs in the same direction is a very gradual slowing of the Earth's rotation rate, probably because of internal frictional forces exerted by the tidal force of the Moon. The combined effect of all the changes in the past century has increased the length of the day by about 0.0014 seconds. Over time, this slowing of the Earth's rotation rate reduces the number of days in a year, since each day is longer than it used to be. Fossil evidence indicates that the year once contained about 400 days, instead of the current 365¼.

In addition to this very gradual overall slowing, the Earth's rotation rate undergoes short-term random changes that can either increase or decrease the length of the day. Geophysicists think that the flow of molten material deep in the Earth's core is responsible for some of these changes, but the causes of most of them are not known. Some of them are probably related to a slow shrinking

of the Earth as its interior gradually cools; these minor effects act to speed up the rotation temporarily.

Over long periods of time, however, the overall trend is toward a decrease in the rotation rate and a lengthening of the day. To keep coordinated universal time in synchronization with the Earth's spin, therefore, it is occasionally necessary to add some time to the clocks. Hence, every few years, a "leap second" is added to all official atomic clocks to bring their timekeeping into accord with the Earth's rotation, that is, with sidereal time.

The occasional insertion of an extra second is distinct from the leap year, when a full day is added every fourth year. This added day is inserted because the standard calendar has exactly 365 days in a year, whereas the year's actual length is closer to 365¼ days. Thus a full day must be added every 4 years to make up for the quarter days that have been dropped. Interestingly, if the length of the day keeps increasing as expected, in a few million years the year will contain exactly 365 days, and for a while no leap years will be needed.

You may occasionally notice small announcements in newspapers about the addition of "leap seconds" to official clocks. These small changes in official time make little practical difference in our daily lives, but now you understand that they are part of a large-scale astronomical process, with important long-term effects.

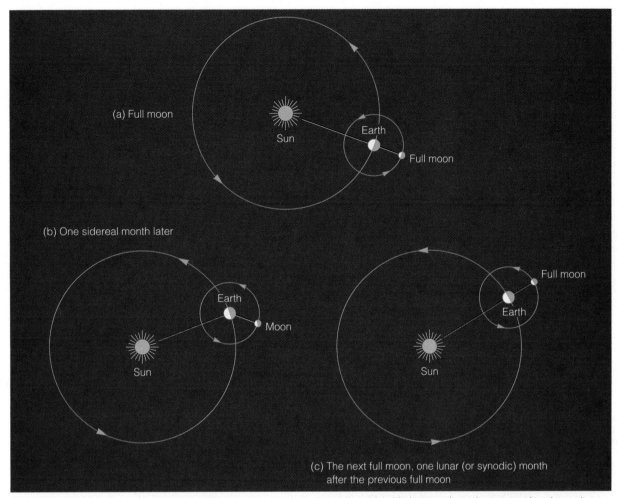

Figure 2.13 Sidereal and synodic periods of the Moon. Because the Earth moves in its orbit while the Moon orbits it, the Moon goes through more than one full circle (as seen by an outside observer) to go from one full moon to the next. Hence, the lunar (or synodic) month is about two days longer than the Moon's sidereal period.

noticed. It is not difficult, in fact, to observe this lunar motion during the course of an evening. The Moon moves about 13° across the sky every day, or about 1° every 2 hours. Its angular diameter is about 0.5°, which means that the Moon moves a distance in the sky about equal to its own diameter every hour.

The Month

The plane of the Moon's orbit is closely aligned with that of the Earth's orbit, so the Moon stays near the ecliptic, as seen from the Earth. In reference frame of the fixed stars, it takes the Moon a little over 27 days ($27^d7^h43^m11.5^s$) to make one trip around the Earth; this is the Moon's **sidereal period,** sometimes called the **sidereal month.** The Earth moves at the same time, however, so in the Earth's reference frame the Moon appears to take longer than 27 days to make one complete circuit. From our point of view on the

moving Earth, the Moon takes $29^d12^h44^m2.8^s$ to complete its full cycle of phases **(Fig. 2.13).** This defines the **synodic period** of the Moon, more commonly called the **lunar month.** The difference is related to the distinction between the solar and sidereal days discussed earlier in the chapter. In both cases, it is the Earth's motion about the Sun that lengthens the time required to complete a full cycle as we see it.

Synchronous Rotation

The Moon always keeps the same side facing the Earth because its rotation period is equal to its orbital (sidereal) period **(Fig. 2.14).** The idea that the Moon does not rotate is a common misconception. It should be clear that the Moon is rotating in the reference frame of the stars, because if it did not, we would see all sides of the Moon as it circled the Earth. The fact that the orbital and spin periods are

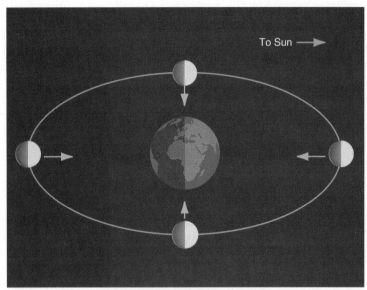

Figure 2.14 Synchronous rotation of the Moon. The arrow, fixed to a specific point on the Moon, illustrates that the Moon spins once during each orbit of the Earth. Thus the Moon keeps the same side facing the Earth at all times. (Not to scale.)

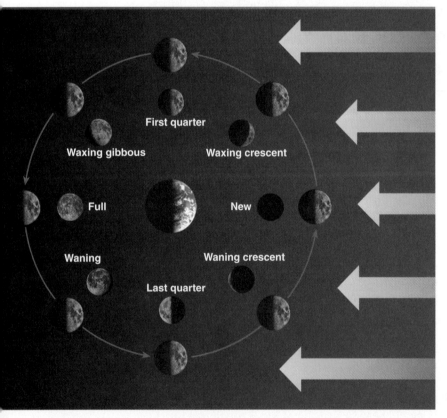

Figure 2.15 Lunar phases and configurations. As the Moon orbits the Earth, we see varying portions of its sunlit side. The phases sketched here (inside the circle representing the Moon's orbit) show the Moon as it appears to an observer in the Northern Hemisphere. http://universe.colorado.edu/fig/2-15.html

equal is not a coincidence, but is a result of the tidal forces exerted on the Moon and the Earth by their mutual gravitational pull (this is discussed in Chapters 5 and 8). The phenomenon of matching orbital and spin periods is called **synchronous rotation,** and it is common in the universe, both within the solar system and in double stars.

Phases of the Moon

The Moon's changing appearance is one of the most noticeable of all the celestial cycles and is obvious even to casual observers. The **phases of the Moon** are caused by a lighting effect. As we see varying portions of the Moon's sunlit side during the month, the Moon's apparent shape changes drastically **(Fig. 2.15).** The full cycle of phases is completed during one synodic period, or lunar month, of about 29½ days.

The extremes of the cycle are represented by the full moon (occurring when the Moon is directly opposite the Sun, so we see its entire sunlit hemisphere) and the new moon (when the Moon is between the Earth and the Sun, with its dark side facing us). The new moon cannot be observed because it is very dim and is only up during daytime. Hence, our nighttime sky is moonless at the time of the new moon. There are several intermediate phases between new and full moon (see Fig. 2.15), the most significant of which are **first quarter** and **third quarter.** At the times of these phases, we see exactly half of the Moon's sunlit side, so from our point of view the Moon is a half circle.

Just as we speak of phases of the Moon, which really refer to its apparent shape as seen from the Earth, we also can speak of the Moon's **configurations,** which describe its position with respect to the Earth-Sun direction. For example, a full moon occurs when the Moon is at **opposition** (it is in the direction opposite to that of the Sun), and a new moon occurs at **conjunction** (when the Moon lies in the same direction as the Sun). The quarter phases of the Moon occur when it is at **quadrature,** meaning that its direction is perpendicular to the Earth-Sun line. Figure 2.15 shows how the Moon's phases and configurations are related to each other.

Eclipses of the Sun and Moon

An eclipse of the Sun, or **solar eclipse,** occurs when the Moon passes directly in front of the Sun, as seen from the Earth. A **lunar eclipse,** by contrast, occurs when the Moon passes through the Earth's shadow, so that for a brief period the Moon is not directly illuminated **(see following pages).** In both cases, the Moon's direction relative to the Sun (that is, the Moon's configuration) is responsible for creating the eclipse. A solar eclipse occurs when the Moon is at conjunction (aligned with the Sun), and a lunar eclipse takes place when the Moon is at opposition (in the direction opposite the Sun).

It may seem that the Moon should pass directly in front of the Sun on each trip around the Earth and through the Earth's shadow at each opposition, producing alternating solar and lunar eclipses at 2-week intervals. This is not the case, however, because the Moon's orbital plane does not lie exactly in the ecliptic. Instead it is tilted by about 5°. Therefore the Moon usually passes just above or below the Sun as it goes through conjunction and similarly misses the Earth's shadow at opposition. On each trip around the Earth, the Moon passes through the ecliptic only at the two points where the plane of its orbit intersects the Earth's orbital plane. Because the Moon's orbital plane wobbles slowly in a precessional motion similar to that of the Earth's spin axis, the line of intersection with the Earth's orbital plane slowly moves around. The combination of this motion, the Moon's orbital motion, and the movement of the Earth around the Sun creates a cycle of eclipses, with the same pattern recurring every 18 years (see **Table 2.2** for a list of future solar eclipses). This cycle of eclipses, called the **saros,** was recognized in antiquity.

It is purely coincidental that the Moon and the Sun have nearly equal angular diameters, so that the Moon neatly blocks out the disk of the Sun during a solar eclipse **(Figs. 2.16** and **2.17, next page).** The **angular diameter** of an object is inversely proportional to its distance, meaning that the farther away the object is, the smaller it looks. The Sun is much larger than the Moon, but it is also much more distant. The two objects have almost exactly the same angular diameter because the ratio of the Sun's diameter to that of the Moon just happens to be almost the same as the ratio of the Sun's distance to that of the Moon **(Fig. 2.18, next page).**

If a total solar eclipse occurs at the time when the Moon is farthest from the Earth in its slightly noncircular orbit, the Moon does not quite block all of the Sun's disk and instead leaves an outer ring of the Sun visible. This is called an **annular eclipse (Fig. 2.19, next page).** Because a total (or annular) solar eclipse requires precise alignment of the Sun and the Moon, an eclipse will appear total (or annular) only along a well-defined, narrow path of the Earth's sur-

Table 2.2
Total Solar Eclipses of the Future

Date of Eclipse	Duration (min)	Location
March 9, 1997	2.8	Siberia, Arctic regions
February 26, 1998	4.4	Central America
August 11, 1999	2.6	Central Europe, central Asia
June 21, 2001	4.9	South Atlantic, South Africa
December 4, 2002	2.1	South Africa, Indian Ocean
November 23, 2003	1.9	Antarctica
March 29, 2006	4.2	South Atlantic, Africa, Middle East
August 1, 2008	2.5	Greenland, North Atlantic, Russia
July 22, 2009	6.7	Indonesia, South Pacific
July 11, 2010	5.3	South Pacific
November 13, 2012		South Pacific
March 20, 2015		North Atlantic
August 21, 2017		United States, Atlantic Ocean

face **(Fig. 2.20, next page).** In a wider zone outside the path, the Moon appears to block only a portion of the Sun's disk; people in this zone see a partial solar eclipse.

During a lunar eclipse, when the Moon passes through the Earth's shadow, observers everywhere on the nighttime side of the Earth see the same portion of the Moon eclipsed. If the entire Moon passes through the **umbra** (the dark inner portion of the Earth's shadow), the eclipse is total, as no part of the Moon's surface is exposed to direct sunlight. If the Moon is only partly immersed in the umbra, it undergoes a partial eclipse. A **penumbral eclipse** occurs when the Moon passes through the **penumbra,** the outer portion of the Earth's shadow. Penumbral eclipses are almost unnoticeable.

WEB ACTIVITY

Solar and Lunar Eclipses
In this web based activity we will learn more about solar and lunar eclipses. We will see a variety of images and movies from previously observed eclipses, we will learn about the dates and locations of future eclipses, and we will travel to other WWW sites around the world that have information about eclipses. Our starting point is the following URL: http://universe.colorado.edu/ch2/web.html

ECLIPSES OF THE SUN

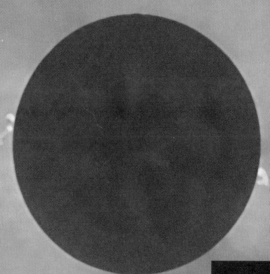

(background) Figure 2.16 A total solar eclipse. This occurs when the Moon entirely blocks our view of the Sun. At this time we see the Sun's corona. This photo is a sum of seven individual exposures made from Baja California during the eclipse of July 11, 1991. Special processing was used to bring out details of coronal structure. (E. E. Barnard Observatory; photo by Steven Albers, Dennis DiCicco, Gary Emerson, and David Sime)

▼ **Figure 2.19** This photograph shows an eclipse that was seen from southern California in 1992. Because the Moon was relatively distant from the Earth at the time of the eclipse, its angular diameter was a little smaller than that of the Sun. (National Optical Astronomy Observatories)

Total Eclipse

The moon's shadow

Annular Eclipse

▲ **Figure 2.20** This photograph, taken from space, shows the shadow of the Moon on the Earth during a solar eclipse. The eclipse appeared total only to observers on the Earth who were located directly in the center of the shadow's path (that is, in the umbra). (NASA)

Figure 2.17 The Moon moving across the Sun during the 1991 solar eclipse at La Paz, Mexico (A. Fujii, H. Tomioka, and Y. Shiono)

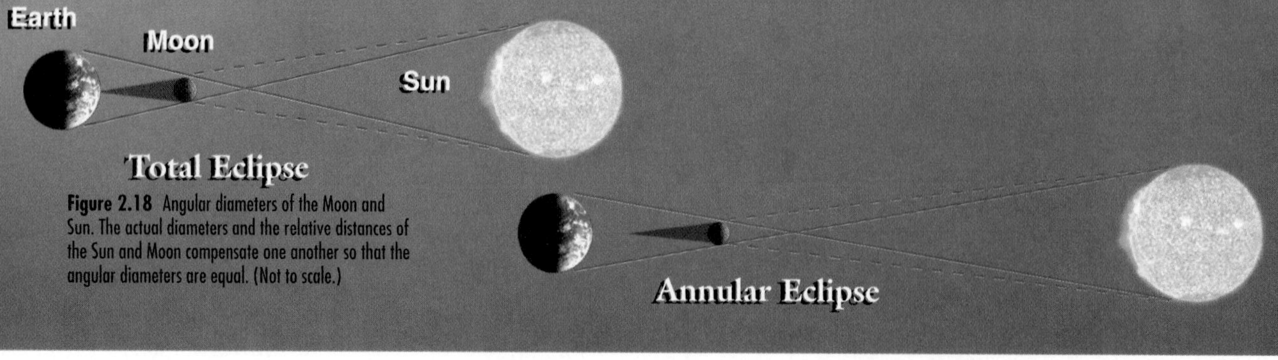

Earth
Moon
Sun

Total Eclipse

Figure 2.18 Angular diameters of the Moon and Sun. The actual diameters and the relative distances of the Sun and Moon compensate one another so that the angular diameters are equal. (Not to scale.)

Annular Eclipse

ECLIPSE OF THE MOON

Moon

Earth

Sun

Total lunar eclipse

This composite image superimposes a photo of a
lunar eclipse (top) and a solar eclipse (bottom),
with a lunar landscape in the foreground. These
phenomna would not appear together from any real
viewpoint.

Planetary Motions

Careful observation reveals that, like the Sun and the Moon, the planets move with respect to the background stars. Indeed, this fact gave the planets their generic name, since **planet** is the Greek word for "wanderer." The planets all orbit the Sun in the same direction as the Earth does and in nearly the same plane, so they appear to move through the constellations of the zodiac.

Planetary Configurations

The two planets orbiting closer to the Sun than the Earth are called **inferior planets** and can never appear far from the Sun in our sky **(Fig. 2.21)**. For Mercury, the greatest angular distance from the Sun, called the **greatest elongation,** is about 28°, whereas Venus can be seen as far as 47° from the Sun. Like the Moon, the planets have specific configurations, referring to their positions with respect to the Sun-Earth line. An inferior planet is said to be at **inferior conjunction** when it lies between the Earth and the Sun (or in **transit** if the planet crosses directly in front of the Sun's disk) and at **superior conjunction** when it is aligned with the Sun but on the far side.

The outer planets, or **superior planets,** can be seen in any direction with respect to the Sun, including opposition, when they are in the opposite direction from the Sun (Fig. 2.21). Conjunction for a superior planet can occur only when the planet is aligned with the Sun but on its far side, a configuration analogous to superior conjunction for an inferior planet. Quadrature occurs when a superior planet is seen 90° from the direction of the Sun.

Sidereal and Synodic Periods

Each planet has a sidereal period and a synodic period. The sidereal period is the time required for one full orbit as seen in the fixed framework of the stars; the synodic period is the length of time the planet takes to pass through a complete cycle as seen from the Earth **(Fig. 2.22)**. The situation is much like that of two runners on a track. The time required for the faster runner to overtake the slower one is analogous to the synodic period, whereas the time required simply to circle the track corresponds to the sidereal period.

It is possible to calculate the synodic period from the sidereal period, or vice versa. Consider a pair of planets, an inner one whose sidereal period in days is P_i and an outer one whose sidereal period is P_o. In a day each moves along its orbit by an angular distance equal to $360°/P$ **(Fig. 2.23)**. Since the outer planet

moves more slowly, it moves a smaller angular distance in a day than the inner planet. During the course of one day, this difference equals $360°/S$, where S is the synodic period (in days) of either planet as seen from the other. The difference equals $360°/S$ because during the course of one synodic period, the difference grows to a full 360° as the inner planet catches up with the outer planet. Setting the difference between the daily angular motions of the inner and outer planets equal to the angular synodic motion yields:

$$\frac{360°}{S} = \frac{360°}{P_i} - \frac{360°}{P_o},$$

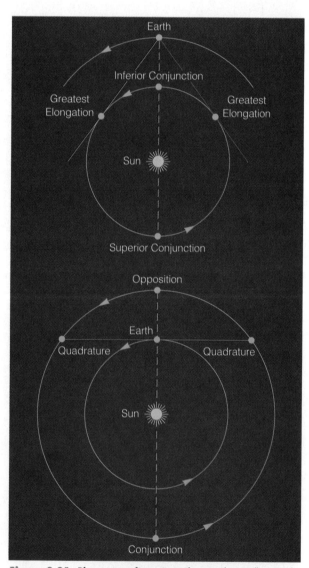

Figure 2.21 Planetary configurations. The upper drawing illustrates the configurations (planetary positions relative to the Sun-Earth line) for an inferior planet; the lower drawing shows the configurations for a superior planet.

which leads to

$$\frac{1}{S_i} = \frac{1}{P_i} - \frac{1}{P_o}.$$

Up to now we have assumed that the periods are measured in days, but the equation still holds if we use any other unit of time. It is convenient to use Earth years, because in that case if the Earth is the inner planet, then $P_i = 1$ and we find

$$\frac{1}{S} = 1 - \frac{1}{P_o}.$$

This formula can be used to calculate the synodic period *(S)* of any superior planet, given its sidereal period *(P_o)*, or it can be used to calculate the sidereal period if the synodic period is known, which is more likely.

Similarly, if we now assume that the Earth is the outer of the two planets, we substitute $P_o = 1$ and find

$$\frac{1}{S} = \frac{1}{P_i} - 1.$$

We can use this formula to find the synodic period of an inferior planet if S is its sidereal period or to find its sidereal period if its synodic period is known.

We can check our formula by calculating the synodic period of one of the planets and comparing our answer with Table 2.1. If we choose Jupiter, we use the first version of the formula and substitute $P_o =$

11.86223. This yields

$$\frac{1}{S} = 1 - \frac{1}{11.86223} = 0.91570,$$

or

$$S = 1.09206,$$

which agrees with the value in the table.

As noted, in normal practice these formulas are more commonly used to calculate the sidereal period of a planet from its synodic period. We can easily measure the synodic period from the Earth, but the sidereal period cannot easily be determined from observations.

Retrograde Motion

The motion of the Earth has one very important effect on the observed planetary motions. As we go outward from the Sun, each successive planet has a slower speed in its orbit (see the discussion of Kepler's laws of planetary motion in Chapters 4 and 5). This means that the Earth is moving faster than the superior planets and therefore passes each of them at regular intervals (the synodic period). As the Earth overtakes one of the superior planets, for awhile the line of sight to that planet sweeps backward with respect to the background stars, making the planet appear to

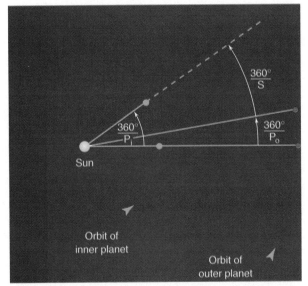

Figure 2.23 Angular motion in one day. This shows the angular motion of an inner and outer planet in one day, starting (for convenience) when they are aligned in the same direction relative to the Sun. In one day, the inner planet moves through an angle of $360°/P_i$, where P_i is its sidereal period (in days), while the outer planet moves through an angle of $360°/P_o$. The difference is the angle $360°/S$, where S is the synodic period (also in days). During the synodic period, the inner planet will "catch up" with the outer one, so that the difference equals a full $360°$.

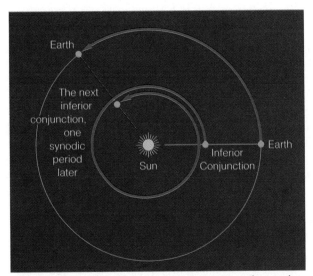

Figure 2.22 The synodic period of an inferior planet. The inner planets travel faster than the Earth in their orbits and therefore "lap" the Earth, much as a fast runner laps a slower runner on a track. This illustration shows approximately the situation for Mercury, which has a synodic period of about 116 days, or roughly one-third of a year.

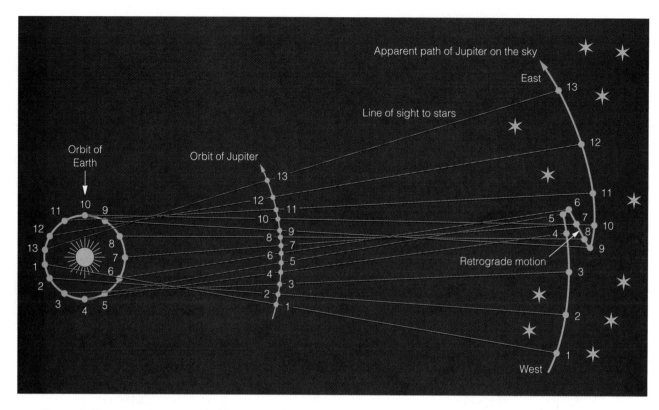

Figure 2.24 Retrograde motion. As the faster-moving Earth overtakes a superior planet in its orbit, the planet temporarily appears to move backward with respect to the fixed stars. This sketch illustrates the modern explanation of a phenomenon that ancient astronomers did not understand correctly for a long time. http://universe.colorado.edu/fig/2-24.html

be moving backward. This is like passing a slower car on a highway; for a brief moment the other vehicle appears to move backward with respect to the fixed background. Ancient astronomers thought this apparent backward movement, called **retrograde motion,** was a real motion of the superior planets rather than merely a reflection of the Earth's motion **(Fig. 2.24).** This mistaken belief greatly complicated many of the early models of the universe and gave rise to a whole class of theories in which the planets were thought to move in small circles called epicycles, which in turn orbited the Earth. These ancient views of the universe are discussed briefly in the next chapter.

Summary

1. The concept of the celestial sphere, along with the celestial equator and the celestial poles, provides a convenient mechanism for visualizing the sky.

2. The constellations, which are based on ancient Greek mythology, have been modernized and are still used as a convenient reference system for the sky.

3. The Earth's rotation causes diurnal motions: the daily rising and setting of the Sun, the Moon, the stars, and the planets.

4. Timekeeping is based on the length of the solar (synodic) day rather than the sidereal day.

5. The orbital motion of the Earth about the Sun causes annual motions, such as the apparent motion of the Sun through the constellations of the ecliptic, and is responsible, along with the tilt of the Earth's axis, for our seasons.

6. Precession of the Earth's rotation axis causes a gradual shifting in star positions relative to the reference system established by the celestial poles and equator.

7. Modern calendars are based on the tropical year, the length of time required for a full cycle of seasons, and must be adjusted for the fact that the length of the year is not a whole number of days.

8. The Moon orbits the Earth, while the Earth orbits the Sun. The Moon's synodic period, or the lunar month, is the length of time required for one full cycle of lunar phases; the sidereal month is the time required for the Moon to orbit the Earth in the reference frame of the stars.

9. The Moon is in synchronous rotation, always keeping the same side facing the Earth.

10. The phases of the Moon are created by the varying portion of the sunlit side of the Moon that can be seen from the Earth as the Moon orbits.

11. Solar and lunar eclipses occur when the Moon passes directly in front of the Sun and through the Earth's shadow, respectively. The occurrence of these alignments is affected by the tilt and precession of the Moon's orbit.

12. The planetary orbits lie nearly in the same plane, so they are always seen near the ecliptic in various configurations with respect to the Sun-Earth line.

13. Each planet has a synodic period, the time required to return to a given position relative to the Sun as seen from the Earth, and a sidereal period, the time required for the planet to orbit the Sun as seen by an outside observer (that is, with respect to the stars).

14. The planets go through temporary retrograde motion because of the relative speed with which they pass or are passed by the Earth.

Review Questions

1. Explain the difference between solar time, sidereal time, and the time kept by atomic clocks.

2. A constellation is a grouping of stars that appear near each other on the sky. Are the stars in a constellation all at the same distance from us? Explain.

3. For each of the following types of motion, state whether the motion is due entirely to movement of the observed object, due entirely to the Earth's (observer's) motion, or due to a combination of the object's motion and the Earth's motion: annual motion of the Sun; daily rising and setting of the stars; retrograde motion of a superior planet; and daily rising and setting of the Sun. Briefly explain each answer.

4. Why do we have a north pole star, but not a south pole star? Will we always have a north pole star? Explain.

5. Consider the meridian, the imaginary north-south line that passes overhead at your location. A star crosses the meridian at a certain time one night and crosses it again the next night. Does the star cross the meridian at exactly the same time each night? Explain. Exactly how long is the period between crossings?

6. How would our seasons be different if the Earth's axis were not tilted (that is, if the axis were perpendicular to the Earth's orbital plane)? What would the seasons be like if the Earth's axis were tipped 90° instead of 23.5°?

7. Use Figure 2.7 to compare the constellation where the Sun was when you were born with your sign of the zodiac according to astrology. Explain why they are not the same.

8. Explain why the Moon is never visible on certain nights.

9. Explain why a solar eclipse is often followed by a lunar eclipse about two weeks later.

10. Table 2.1 lists the synodic periods for the planets. Note that the synodic periods for the outer planets are close to 1 year. Can you think of an explanation for this?

Problems

1. How fast, in kilometers per hour, does a point on the Earth's equator travel due to the Earth's rotation?

2. The Moon moves eastward on the sky an angular distance of about 12° per day. How much later does it set each evening, compared with the evening before?

3. The difference between a solar day and a sidereal day is $3^m 56^s$. Over the course of one year, how many days does this difference amount to? Discuss the significance of your answer.

4. Due to precession, the celestial poles shift position relative to the stars, undergoing a full cycle in 26,000 years. Each pole travels over a circular path on the sky during this time. How far, in angular units, does each pole shift in one year?

5. Similarly, precession causes the Sun's position at any given date to move gradually along the zodiac. If each major constellation occupies about ¹⁄₁₂ of the zodiac (that is, the constellations are centered about 30° apart), how long does it take for the Sun's position to shift from one zodiacal sign (constellation) to the next?

6. Suppose you adopted a calendar that contained exactly 365 days instead of the more accurate

Comparing Solar and Sidereal Days

ASTRONOMICAL ACTIVITY

The text says that the sidereal day is 3^m56^s shorter than the solar day. You can check this for yourself with a little care. All you need is a watch that is accurate to a few seconds per day (most modern digital watches will do) and a means of measuring the time when a particular star passes a certain position relative to you each night.

Your watch measures solar time. As explained in the text, timekeeping is synchronized with the Sun, so your watch will tell you when a star passes your reference point in mean solar time for your time zone. The goal in this exercise is to time how long it is between passings of a given star on consecutive nights. Therefore the actual time of night when the star passes over is not important; you want to measure the *difference* between the time of passage from one night to the next. You should find that the star arrives at the same position about 3^m56^s earlier each night.

To see this, you will need a method for pinpointing the exact moment (to a few seconds' accuracy) when your chosen star passes some reference point. Astronomers use devices called transit circles for this, but you can do something much simpler since you only care about one particular star. Probably the easiest method is to find a building with a north-south wall along its east side and a clear view of the sky along that wall as you look toward the south. Stand at the northeast corner of the building and sight along the wall toward the south. You will see that as the Earth turns, stars moving from east (on your left) to west (on your right) are eclipsed by the wall, as seen from your point of view. It is possible to time the moment when a star passes out of sight behind the building (you may need a flashlight to read your watch).

Find a prominent star that you can identify again the next night, and note the exact time at which this star disap-

pears from sight as it moves behind the building from your point of view. Come out the next night and time the disappearance of the same star. You should find that it disappears about 4 minutes earlier, according to your watch.

If the sky is cloudy the second night, don't worry—come out on the third night, or the fourth, or any night, and time the disappearance of your reference star. It should go behind the building about 4 minutes earlier for each night that has passed since your first observation. In fact, your measurement of the daily time advance will become more accurate as you make the measurements over a longer period of time, because the error in your individual time measurements will become smaller relative to the total time interval.

It may not be necessary to find a building with a suitable north-south wall. Any fixed reference pointer will do; all you need is something that allows you to determine with some accuracy the time at which a particular star passes a fixed point or direction. For example, overhead wires might work, if you can observe a star's passage across a particular wire from precisely the same point on the ground each night. Or you can stand at the bottom of a vertical pole or wall and look straight up, timing the moment when a selected star passes directly overhead. All that is necessary is that you observe along precisely the same direction each night.

This exercise should help illustrate the simple concept of solar and sidereal days. It will also serve to familiarize you further with the sky because you will have to pick out a particular star that you can easily identify on separate nights. You will undoubtedly accomplish this by memorizing a pattern of stars in the sky, either one of the traditional constellations or a new one that you invent.

As a further exploration of timings and motions, make the same measurement of the daily time of passage of the Moon or of a planet. What do you find?

365.242 days of the tropical year. How long would it take for an error of one full day to accumulate? How is this related to the leap year that we observe every 4 years?

7. How often would you need to have leap year if the length of the year were 365.200 days?

8. Suppose you live at a latitude of 40° N. You are building a house, and you want to make the south-facing roof (with solar panels) face directly toward the Sun at the time of the winter solstice, when the Sun is 23.5° south of the celestial equator. At what angle, relative to the ground, should you set your roof? (Hint: You may find it useful to make a sketch of the situation.)

9. At its closest approach to the Earth (that is, at opposition), Mars is about 0.5 AU away, and when it is farthest from the Earth (at superior conjunction), it is roughly 2.5 AU away. If Mars's angular diameter at opposition is 18^s, what is its angular diameter when it is at superior conjunction?

10. Suppose a new superior planet is discovered and found to have a synodic period of 6.85 years. What is its sidereal period?

Additional Reading

A number of magazines, including *Mercury, Sky & Telescope,* and *Astronomy,* contain practical information for the sky watcher, such as planetary positions and the seasonal appearance of the constellations. There are also annually published handbooks with similar data. One of the most widely-used of these is the *Observer's Handbook,* by Roy L. Bishop (Toronto: Royal Astronomical Society of Canada). In addition, occasional articles related to the nighttime sky may appear; a few are listed below.

Gurshtein, A. 1995. When the Zodiac Climbed into the Sky. *Sky & Telescope* 90(4):28.

Itano, W. M. and N. F. Ramsey 1993. The Keeping of Precise Time. *Scientific American* 269(1):56.

Schaefer, B. E. 1994. Solar Eclipses That Changed the World. *Sky & Telescope* 87(5):36.

Web Connections

The Review Questions and Problems also appear at the following URLs:
http://universe.colorado.edu/ch2/questions.html
http://universe.colorado.edu/ch2/problems.html

Chapter 3
The Early History
of Astronomy

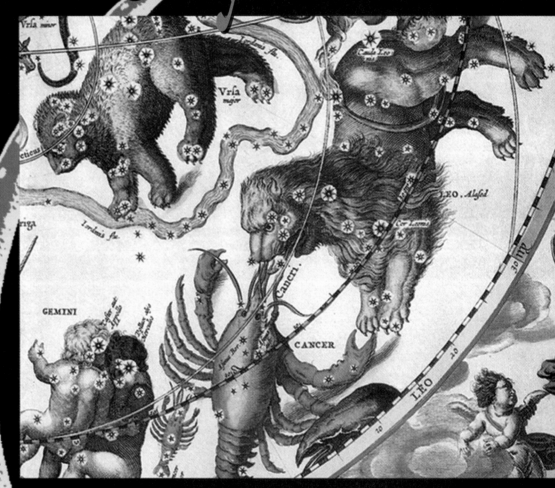

Chapter web site: http://universe.colorado.edu/ch3

The oldest picture book in our posession is the nightime sky.

E. Walter Maunder

ow that we know something about the motions and cycles in the nighttime sky that can be seen without instruments except for our eyes, we can examine what the ancient peoples knew and thought about these phenomena. In all likelihood, people were preoccupied with the heavens from the time they first became aware of their environ-ment. We certainly know this is so for cultures that left records of their knowledge, but we must speculate about the astronomical sophistication of early societies that did not have written languages. The following discussion emphasizes developments that occurred around the shores of the Mediterranean **(Fig. 3.1),** especially the areas of modern Iraq, Greece, Turkey, and Italy, because these achievements laid the foundation for the modern understanding of the universe. Indeed, the roots of modern science can be traced to this region.

Nevertheless, it is important to recognize that parallel developments occurred in many parts of the world, including Asia, India, and the Americas. We should not overlook the astronomical insights gained by cultures in other parts of the world. Many important and, in some cases, rather sophisticated discoveries were made in regions and continents far removed from the Mediterranean, so we will also discuss some of the more significant of these developments in this chapter.

Babylonian Astronomy

Along the valleys of the Tigris and Euphrates Rivers, in what is now Iraq, arose one of the earliest civilizations known to have left written records. The first people in this region were the Sumerians, who occupied the southern portion by 3000 B.C. and invented a system of writing called cuneiform script that was engraved on clay tablets **(Fig. 3.2).** Cuneiform was used for writing about all aspects of Sumerian life, including astronomical knowledge. These tablets have survived rather well and provide modern scholars with a detailed and rich account of life and culture in ancient Sumeria.

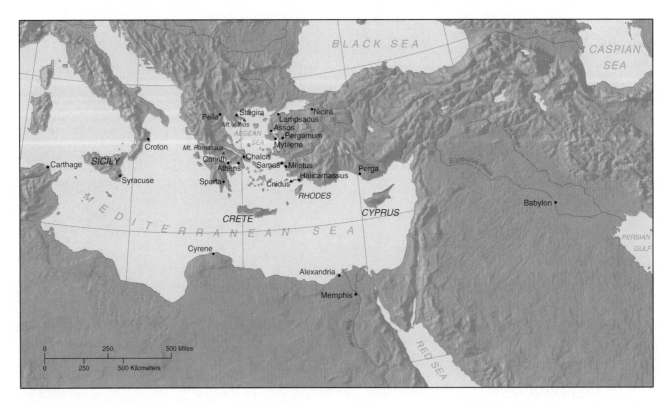

Figure 3.1 The ancient Mediterranean. This map shows the locations of many of the sites mentioned in the text. To the right is the region of Babylonia, now Iraq. On the northern shores of the Mediterranean Sea lay most of the Greek empire, in what is now Italy, Greece, and Turkey. (C. Ronan, *The Astronomers* [New York: Hill and Wang, 1964])

The entire region was united around 1700 B.C. under the leadership of King Hammurabi, who established his capital at the city of Babylon (not far from the present Baghdad), from which the kingdom took the name Babylonia. The clay tablets from that time show that the Babylonians attributed significance to the planets. Many different celestial configurations and events were thought to be omens of things to come on Earth; apparently, Babylonian life was governed by astrology. The intense interest in what the heavens had to say about life on Earth led the Babylonians to develop a very detailed observational record, which was maintained continuously over several centuries, an achievement matched by few other cultures.

Thanks to this devotion to observational data, the Babylonians developed very accurate knowledge of many astronomical phenomena. The length of the year was known to an accuracy of about 4 minutes, for example. Nevertheless, the Babylonian calendar divided the year into 12 equal months of 30 days each, thus containing 360 days. The Babylonians developed a numbering system based on 60 (much as our own system is based on 10) and thought that the even division of 12 into 60, as well as the even division of 60 into 360, was very significant. They developed a system of angular measure in which the entire sky (that is, a full circle) was divided into 360 equal parts (degrees), corresponding to the 360 days in their calendar, and they divided each degree into 60 minutes and each minute into 60 seconds. Similarly, they divided each hour into 60 minutes of time and each minute into 60 seconds. They had a 24-hour day because they believed in symmetry between day and night and thought that each should have 12 equal parts. The number 12 came into the picture because there were 12 lunar cycles (months) during their year. Of course, the Babylonians recognized that they had made several inaccurate approximations in developing their calendar; they simply inserted an extra month every few years to bring the calendar back into synchronization with the seasons. Clearly, for better or worse, we owe our modern systems for measuring time and angles to the Babylonians.

Over the centuries several empires came and went in the land occupied by the Babylonians, which was known to the Greeks as Mesopotamia. The Assyrians, who dominated the region from around 800 B.C. until 600 B.C., preserved both the astronomical traditions of the Babylonians and their clay tablets. Around 200 B.C., after the center of power had moved west to Greece, a massive amount of data was compiled by the then-occupants of Babylonia, the Chaldeans. Their tables of planetary motions were sufficiently accurate that the Chaldeans could predict solar, lunar, and planetary positions, as well as the occurrence of eclipses.

The primary motivation throughout these times was astrology. The Babylonians and their successors in Mesopotamia seem to have speculated very little about the nature of the universe, apparently envisioning the heavens as a dome supported by the mountains surrounding their region. No evidence has been uncovered for any theories of the motions of heavenly bodies, that is, *why* the planets move as they do.

Some astronomical development also occurred during the same period in Egypt, but there the main motivation for studying astronomy was essentially practical: the Egyptians were very concerned with predicting the seasons, which were central to their agricultural civilization.

As Babylonian science developed and the region underwent successive conquests, a separate civilization began to rise along the shores of the Mediterranean in Greece. Greek civilization eventually reached the greatest heights of scientific accomplish-

WEB Activity

Astronomy on the WWW
In this exercise, we'll learn to use various tools that exist to locate astronomical information on the Internet. We'll then use these tools to find information on several topics discussed in this chapter. Our starting point is the following URL: http://universe.colorado.edu/ch3/web.html

Figure 3.2 A cuneiform tablet. This stone tablet, from around 870 B.C., depicts the sun god Samas. The Babylonians developed very accurate knowledge of many astronomical phenomena. (The Granger Collection)

ment in ancient times, but the Greeks' early astronomical awareness was greatly influenced by the achievements of the Babylonians.

Greek Developments

The ancient Greek civilization not only gave rise to very sophisticated knowledge and theories of the heavens, but also made unprecedented advances in other sciences, in the arts, and even in government. The Greek philosophers, who were the equivalent of the scientists of today in some respects, laid the groundwork for modern thought and methods of investigating nature. In astronomy, as in other arenas, we trace the origins of our modern understanding to the Greeks. **Table 3.1** summarizes several of the most important individuals and their contributions.

The Early Greeks

The foundations of the Greek civilization were established some 5,000 years ago in the eastern Mediterranean. There arose a seafaring culture, whose earliest home was the island of Crete, where legends were born that gave us the names of most of our constella-

Table 3.1
Notable Greek Achievements

Date	Name	Discovery or Achievement
c. 900–800 B.C.	Homer	*Iliad* and *Odyssey;* summaries of legends
c. 624–547 B.C.	Thales	Rational inquiry leads to knowledge of universe
c. 611–546 B.C.	Anaximander	Universal medium; primitive cosmology
c. 570–500 B.C.	Pythagoras	Mathematical representation of universe; round Earth
c. 500–400 B.C.	Philolaus	Earth orbits central fire
c 500–428 B.C.	Anaxagoras	Moon reflects sunlight; correct explanation of eclipses
c. 428–347 B.C.	Plato	Material world imperfect; deduce properties of universe by reason
c. 408–356 B.C.	Eudoxus	First mathematical cosmology; nested spheres
c. 384–322 B.C.	Aristotle	Concept of physical laws; proof that Earth is round
c. 310–230 B.C.	Aristarchus	Relative sizes, distances of Sun and Moon; first heliocentric theory
c. 273–? B.C.	Eratosthenes	Accurate size of Earth
c. 265–190 B.C.	Apollonius	Introduction of the epicycle
c. 200–100 B.C.	Hipparchus	Many astronomical developments; full mathematical epicyclic cosmology
c. A.D. 100–200	Ptolemy	*Almagest;* elaborate epicyclic model

tions. This civilization prospered around 1600 B.C. during the reign of King Minos and is known as the Minoan culture for that reason. Much of what we know about this era is gleaned from Homer's epic poems the *Iliad* and the *Odyssey.* These were probably written around 900–800 B.C., however, long after the Minoan culture had passed its prime.

The cosmology of this time viewed the heavens as a dome over a disklike Earth floating on water. This was similar to the Babylonian view, which was developed at about the same time. But the Greeks went further, hypothesizing, for what were essentially theoretical reasons, that there was an underworld comparable in scope and complexity to the heavens. Apparently, these people had a sense of aesthetics and unity in the universe, which shows that their thinking went beyond mere description of the visible skies.

The first formal scientific thought is associated with the philosopher Thales (624–547 B.C.) and his followers, who lived in Miletus, on the eastern shore of the Aegean Sea. The principal contribution of Thales was the idea that rational inquiry can go beyond merely describing the universe and lead ultimately to understanding it. This general concept formed the central creed of the Ionian school of thought, named for the part of Greece (now Turkey) surrounding Miletus. One of Thales' most significant claims to fame as an astronomer was his supposed prediction of a solar eclipse that took place in 585 B.C., a feat that some historians believe did not actually occur. Much of the Greek knowledge of the motions of the heavens at that time came from the Babylonians, and it is not clear that they were able to predict eclipses.

The Pythagoreans

Pythagoras (c. 570–500 B.C.) was born on the island of Samos, off the Aegean coast not far from Miletus. His early life is obscure, but at around the age of 40, following several years of travel and study, he established a school at Croton, on the southern tip of Italy, in which scientific and religious teachings were intermixed. The most important contribution of this school is the idea that all natural phenomena can be described by numbers. This laid the foundation for geometry and trigonometry and established a fundamental philosophy that is followed to this day in science. Pythagoras himself is thought to have been the first to claim that the Earth is round and that all heavenly bodies move in perfect circles, ideas that became fundamental beliefs of the influential Greek philosophers who followed.

Most of what we know about Pythagoras, who himself left no written records, comes from the works of his follower Philolaus (c. 500–400 B.C.), who added some new ideas of his own. Both Philolaus and Pythagoras, for largely philosophical or religious reasons, believed the planetary distances correspond to the lengths of vibrating strings that produced harmonious musical notes. This led to the concept of the celestial harmony, or "music of the spheres," that later had considerable influence on philosophical thought.

The Pythagorean school had numerous followers who in various ways refined this view of the universe. One of particular note was Anaxagoras (c. 500–428 B.C.), who is credited with the realization that the Moon shines by reflected sunlight, rather than by its own power. This awareness, based on observation and deduction rather than personal philosophy, allowed him to correctly attribute solar and lunar eclipses to the passage of the Moon in front of the Sun or through the Earth's shadow, respectively.

Plato and His Followers

On the Greek mainland, a number of democratically ruled city-states arose, of which Athens had become the most prominent by about the fifth century B.C. Even after 404 B.C. when Athens suffered a military defeat at the hands of its rival city-state Sparta, Athenian culture continued to flourish. It was in this environment that Plato (428–347 B.C.; **Fig. 3.3**)

reached maturity, having studied philosophy under the guidance of Socrates. In about 387 B.C., Plato established an academy outside Athens, where he taught his ideas of natural philosophy. His basic concept was that what we see of the material world is an imperfect representation of ideal creation.

One consequence of this idea, in Plato's view, was that we can learn more about the universe by reason than by observation, since observation can give us only an incomplete picture. Hence Plato's ideas of the universe, described in his *Republic,* were based on certain idealized assumptions that he found reasonable. One of the most important, derived from the teachings of Pythagoras, was that all motions in the universe are perfectly circular and that all astronomical bodies are perfect spheres. Thus he adopted the Pythagorean view that the Sun, the Moon, and planets move in various combinations of circular motion about the Earth. Apparently, Plato thought of these objects as affixed to clear, ethereal spheres that rotated. Plato's teachings, particularly the premise that the mysteries of the cosmos should be solved through reason rather than observation, dominated much of Western thought until the Renaissance, some 2,000 years later.

One of Plato's students was Eudoxus (c. 408–356 B.C.), who made the first attempt to put Plato's view of the universe on a firm mathematical footing. Eudoxus postulated that the Sun, Moon, and planets moved on nested rotating spheres **(Fig. 3.4).**

Figure 3.3 Plato and Aristotle. Plato's (right) beliefs that the universe and all celestial objects were perfect and immutable had great influence on later philosophers. Aristotle (left) refined this view to include the concept of physical laws that were universal. (Detail from The *School of Athens,* by Raphael, 1509. Art Resource, N.Y.)

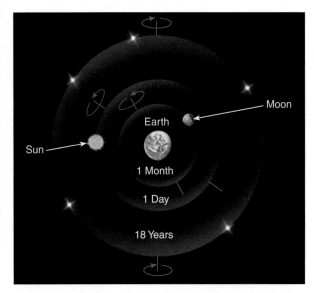

Figure 3.4 The spheres of Eudoxus. This shows the relative orientations and directions of rotation for spheres thought to be responsible for the monthly motions of the Moon, as well as its 18-year eclipse cycle.

Plato's most renowned student, however, was Aristotle (c. 384–322 B.C., see Fig. 3.3) who, after Plato's death in 347 B.C., established his own small academy across the Aegean Sea at Assos. He then moved to Pella, where he tutored Alexander, son of King Phillip of Macedonia, leader of the Greek empire. Aristotle later founded a school in Athens, where he taught his own view of the universe, which, like Plato's, was based on reason rather than observation.

Aristotle was the first to adopt physical laws and then use those laws to explain how the universe works. He taught that circular motions are the only natural ones and that the center of the Earth was the center of the universe. He also believed, without any experimental or observational basis, that the world is composed of four elements: earth, air, fire, and water. He used his laws to demonstrate that both the universe and the Earth are spherical. He had three proofs of the latter claim:

1. *Only at the surface of a sphere do all falling objects seek the center by falling straight down.* (Another of his premises was that falling objects are following their natural inclination to reach the center of the universe.)

2. *The view of the constellations changes as one travels north or south.*

3. *During lunar eclipses, it can be seen that the shadow of the Earth is curved* **(Fig. 3.5).**

By relating his theories to observation in this manner, Aristotle broke with Plato's tradition to some

extent, although he approached the problem in the same manner, letting reason rather than observation guide the way.

Aristotle also concluded that the universe is finite in size (this led directly to his idea that the heavenly bodies can follow only circular motions, because otherwise they might run into the edge of the universe). He believed that the heavenly bodies were made of a fifth fundamental substance, called the aether.

The Geometric Genius of Aristarchus and Eratosthenes

After Aristotle's time, the center of Greek scientific thought moved across the Mediterranean Sea to Alexandria, the capital city established in Egypt in 332 B.C. by Aristotle's former pupil Alexander the Great. The most prominent astronomer of this era, however, was born on the Mediterranean island of Samos, and historians are not sure where he lived and worked, although it may have been in Alexandria. This astronomer was Aristarchus (c. 310–230 B.C.), the first scientist to adopt the idea that the Sun, not the Earth, is at the center of the universe. Aristarchus reached this conclusion some 1,700 years before the time of Copernicus, who is usually given credit for this concept.

The **heliocentric** (Sun-centered) hypothesis of Aristarchus failed to attract many followers at the time, however. Probably the main reason for this was that there were no obvious problems with the prevailing picture (the **geocentric** model) in which the Earth was the center. There was no evidence that the Earth was moving, and observations of solar, lunar, and planetary motions seemed to fit well enough with the general idea that all of these bodies moved around the Earth. As is true of human nature today, people then were not inclined to change their thinking unless the old ways became obviously incorrect.

It is interesting to consider how Aristarchus reached his revolutionary conclusions. He used geometric measurements and arguments to deduce the relative sizes and distances of the Moon and the Sun. Even though his results were very inaccurate, he did conclude correctly that the Sun is much larger than either the Moon or the Earth. Accepting this conclusion, he found it difficult to imagine that the Sun orbited the smaller Earth, so he adopted the opposite idea—that the Earth must orbit the larger Sun.

A younger contemporary of Aristarchus was Eratosthenes (c. 273–? B.C.), who worked in Alexandria during his long career of research in astronomy. His greatest achievement was the determination of the size of the Earth by a method of great simplicity **(Fig. 3.6).** It was known that at the time of the summer solstice, sunlight penetrated all the way to the bot-

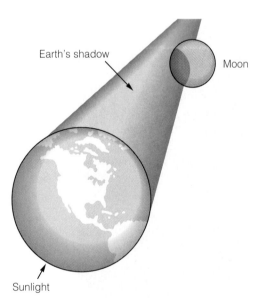

Earth's shadow

Moon

Sunlight

Figure 3.5 Curvature of the Earth's shadow on the Moon. Only a spherical body can cast a circular shadow for all alignments of the Sun, Moon, and Earth.

tom of a deep well located at Syene, near present-day Khartoum. On the same day in Alexandria, the Sun was observed at midday to lie 7° south of the zenith. This indicated that the distance between Syene and Alexandria corresponded to 7° of the 360° circle of the Earth's circumference (note that this is based on the assumption that the Sun's rays are parallel when they strike the Earth at different points). In mathematical form, this corresponds to

$$\frac{7°}{360°} = \frac{d}{C}$$

where d is the distance between Alexandria and Syene and C is the circumference of the Earth. Solving for C, we get

$$C = d\left(\frac{360°}{7°}\right) = 51.4d \,.$$

Hence the circumference of the Earth was equal to $360/7 = 51.4$ times the distance between Syene and Alexandria. This distance was about 4,900 **stadia** (singular: **stadium,** a unit of measure equal to about one-tenth of a mile). Therefore Eratosthenes calculated the circumference of the Earth at about 252,000 stadia. Modern scholars do not agree on the exact size of the stadium, but under one widely accepted value, the circumference found by Eratosthenes was within 2 percent of the correct value. If so, this was partly due to luck because Eratosthenes could not have measured the Sun's angular separation from the zenith with sufficient precision to justify a claim of 2 percent accuracy, nor could he measure the distance between Syene and Alexandria to

this degree of accuracy. Nevertheless, the method he used was basically correct, and it is a testimony to both the creativity and the diligence of Eratosthenes that he found a very accurate value for the size of the Earth.

Geocentric Cosmologies: Hipparchus and Ptolemy

By the third century B.C., the need for more precise mathematical models of the universe had become apparent, as better observing techniques were developed. A mathematical concept that provided the needed precision while still preserving the ideology of Aristotle was the **epicycle (Fig. 3.7).** The astronomer Apollonius (c. 265–190 B.C.) put the epicycle on a firm geometric footing, thereby providing the first mathematically reasonable alternative to the idea of spheres that had been developed by Eudoxus. An epicycle is a small circle on which a planet moves; the center of the epicycle in turn orbits the Earth following a larger circle called the **deferent.** Adjusting the sizes and rates of rotation of the epicycle and the deferent made it possible to closely match the observed motions of the Sun, Moon, and planets. The epicycle also provided a clear advantage for explaining retrograde motions, something that had confounded Eudoxus's theory.

The epicyclic motions of the celestial bodies were refined further by Hipparchus **(Fig. 3.8, next page),** who lived during the middle of the second century B.C. (very little is known about his life, not even the dates of his birth and death). Hipparchus, who did most of

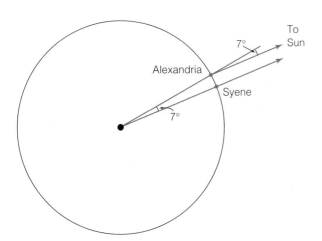

Figure 3.6 The method of Eratosthenes for measuring the size of the Earth. Eratosthenes deduced the portion of a circle that corresponded to the distance between Alexandria and Syene. From this he was able to calculate the circumference of the Earth.

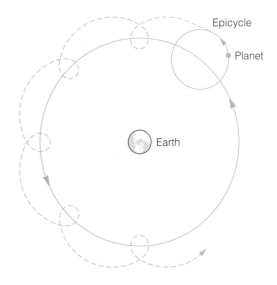

Figure 3.7 The epicycle. Ancient astronomers realized that planetary motions could be represented by a combination of motions involving an epicycle, which carries a planet as it spins while orbiting the Earth.

his work at his observatory on the island of Rhodes, was one of the greatest astronomers of antiquity. He was responsible for many advances in science and mathematics, including the first application of trigonometry to astronomy, the refinement of instruments for measuring star positions, improvements in the geometric methods of Aristarchus, and some major innovations in the study of stars. These were incorporated into a major catalog containing positions and brightness classifications of some 850 stars. Hipparchus expressed the brightness in terms of **magnitudes,** a ranking system still in use today (but now expressed in precise mathematical terms: see Chapter 14). Perhaps his most impressive discovery was the precession of star positions, which he was able to detect by comparing his observations with some that were made 160 years earlier (for a discussion of precession, see Chapter 2).

Although Hipparchus did not add much that was new to the then-current models of planetary motions, he did compile an extensive series of observations of solar motion. These led him to a refinement of the epicyclic theory, namely, that the Earth was off-center in the large circle on which the Sun moved. This idea of an eccentric, or off-center, orbit accounted rather accurately for the observed variations in the speed of the Sun in its annual motion.

Despite the earlier suggestions that the Sun was the central body in the universe, Hipparchus clung to the Earth-centered view. Apparently his main reasons for doing so were that the geocentric picture was quite capable of explaining the observations and that there was no evidence that the Earth moved. On the contrary, in fact, there was an observational argument that the Earth did not move. Hipparchus knew that if the Earth moved, nearby stars should appear to move back and forth on the sky as a reflection of the Earth's orbital motion **(Fig. 3.9),** but no such stellar parallax could be detected. (The shifts do occur, but they are far too small to have been measured with the crude instruments of the Greek astronomers.)

After the great work of Hipparchus, almost three hundred years passed before any significant new astronomical developments occurred. Claudius Ptolemaeus, or simply Ptolemy **(Fig. 3.10),** lived in the middle of the second century A.D. and undertook to summarize all the world's knowledge of astronomy. He did this in a treatise called the *Almagest,* which was based in part on the work of Hipparchus but also contained some new material devised by Ptolemy himself. The 13 books of the *Almagest* include a summary of the observed motions of the planets, a detailed study of the motions of the Sun and Moon, descriptions of the workings of all the astronomical instruments of the day, a reproduction of the star catalog of Hipparchus, and, most importantly, detailed models of the planetary motions **(Fig. 3.11).** Here Ptolemy made his greatest personal contribution, for these models predicted the planetary positions so accurately that they were used for the next thousand years.

Figure 3.8 A fanciful rendition of Hipparchus at his observatory. Hipparchus made important discoveries based on his observations and on his advanced methods for analyzing data. (The Bettmann Archive)

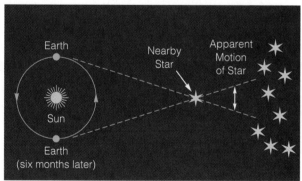

Figure 3.9 Stellar parallax. As the Earth orbits the Sun, our line of sight toward a nearby star varies, causing the star's position (with respect to more distant stars) to change. The change of position is greatly exaggerated in this sketch; in reality, even the largest stellar parallax displacements are too small to have been measured by ancient astronomers.

To provide this accuracy, Ptolemy had to adopt and extend Hipparchus's idea of placing the Earth a bit off-center (see Fig. 3.11). To satisfactorily account for all of the irregularities in the motion of a planet, Ptolemy had to assume not only that the center of the deferent was off-center from the Earth, but also that the rate of motion of the epicycle as it orbited the Earth was constant about another off-center point, called the **equant,** on the opposite side of the Earth from the geometrical center of the deferent. This combination of offsets, with properly chosen dimensions and rotational speeds, allowed for good representation of the observed motions while preserving the hypothesis that all heavenly motions are circular. It is interesting to note that Ptolemy did not strictly follow other doctrines of Aristotle, such as the precise centering of all motions on the Earth.

Some of Ptolemy's writings suggest that he may not have thought of his model as a literal representation of how the universe actually works. Instead he may have viewed it as simply a mathematical device for predicting planetary positions. As such, it would have been an **empirical model,** which is something that a modern scientist might use as a descriptive tool when a full theory for something has not yet been developed. In this sense, we might consider Ptolemy to have been like a modern scientist. Whether or not Ptolemy believed that his model represented physical reality, it seems that everyone else did—for centuries to come. The concept of a geocentric universe with only circular motions dominated Western thought for some 14 centuries after Ptolemy's time, and even then the idea was not easily uprooted.

Developments in Other Cultures

New developments in Greek astronomy came to an end with the work of Ptolemy. His era was followed by a period known as the Dark Ages, when Western civilization was largely dormant. During this time, Arab astronomers preserved many of the Greek traditions, so that the ancient teachings were still firmly entrenched centuries later, when the first stirrings of a new spirit of inquiry began to be felt in Europe. In the meantime, many other areas of the world gave rise to sophisticated observations and understandings of the heavens.

Early Astronomy in Asia

One of the most ancient recorded astronomies outside the Middle East arose in China. Chinese legends refer to astronomical activities as far back as 2000 B.C., although the oldest writings are much more recent. Ancient China had a tumultuous history, with periods of peace, prosperity, and cultural advance interspersed with eras of disarray resulting from bar-

Figure 3.10 Ptolemy. Ptolemy summarized all of the Greek knowledge of astronomy, including significant contributions of his own, in the *Almagest.* (The Granger Collection)

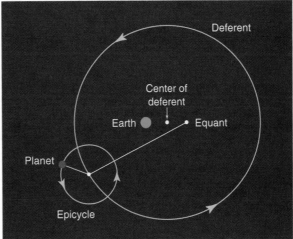

Figure 3.11 The cosmology of Ptolemy. To account for nonuniformities in the planetary motions, Ptolemy devised a complex epicyclic scheme. The deferent, the large circle on which the epicycle moves, is not centered on the Earth and has a rotation rate that is constant as seen from another displaced point called the equant. This meant that from the Earth, the planet appeared to move faster through the sky when on one side of its deferent than when on the other. Each planet had its own deferent and epicycle, with motion and offsets adjusted to reproduce the observations.

barian invasions and civil unrest. Historical records were sometimes lost during the transitions, so historians are not sure when the first significant astronomical activity began. Legend refers to a pair of astronomers named Hi and Ho who were put to death after they failed to predict a solar eclipse that occurred in 2137 B.C.

We know that timekeeping was well established in China by the fourth century B.C. because records from that era indicate that the length of the year and the lunar month were known to considerable accuracy. The principal roles of astronomers were to keep the time and to officiate at state functions, where they ensured that all was done in accordance with the rules of the gods. Astrology played a major role, and the constellations were identified with the emperor and with government organization **(Fig. 3.12)**. The north polar region of the sky, for example, was thought to represent the seat of the emperor, and within a particular constellation, stars were given royal rank according to their brightness. The astronomer Shi-Shen compiled a catalog of some 809 stars in the fourth century B.C., about 200 years earlier than the catalog of Hipparchus.

The Chinese concept of the universe, however, was not as sophisticated as that of the Greeks. The motions of the planets were given little significance, and the universe was thought to consist simply of the sphere of the sky, which rotated about the Earth, carrying the stars, Sun, and planets with it. The poles were considered to be the exalted regions of the sky, whereas the celestial equator and the ecliptic seem to have been given no great importance. Heaven and Earth were thought to be closely related, with events in one affecting those in the other. Thus irregularities in the heavens were regarded as omens of calamities on Earth, a parallel with the astrological beliefs of the West.

In later centuries observations became more accurate, as simple devices for measuring angles were employed. By the first century A.D., accurate tables of lunar motions had been constructed, and eclipses could be predicted. Apparently Chinese astronomers of this era were familiar with precession as well.

The Chinese came into contact with the Western world during the first few centuries A.D. as a result of Chinese conquests toward the West and the eastward diffusion of Greek culture, which reached China by way of India. This intermingling of cultures makes it difficult for modern historians to distinguish Chinese developments from those of the West, but records indicate that the Chinese made numerous important advances, particularly in the improvement of measuring devices.

Records of celestial events were kept continuously until medieval times, providing later generations with important data that can be found nowhere else. Among the events mentioned in the Chinese records is the appearance of a "guest star" in A.D. 1054, identified in modern times as the supernova explosion that created the Crab nebula (this nebula and the explosive process that created it are discussed in Chapters 15 and 16). The explosion occurred during the European Dark Ages when no one there was recording what was seen in the sky.

Thus Chinese achievements in astronomy were significant. Nevertheless, because of China's distance from other cultures, Chinese astronomical developments had little direct influence elsewhere.

Hindu Astronomy

In India, as in China, some astronomical lore had evidently developed by very early times, but our knowledge of the oldest accomplishments is based on ancient legends and on the astronomical orientations of some buildings **(Fig. 3.13)**. Written records of lunar motions date back to about 1500 B.C., and by 1100 B.C. a calendar based on the solar year had been established. Apparently, all the prominent celestial bodies were named for gods or goddesses and were thought to represent their characters.

Figure 3.12 Ancient Chinese depiction of the celestial sphere. Places near the North Pole were assigned to nobility, with lesser beings relegated to lower latitudes. This image dates from the T'ang Dynasty (600–800 A.D.) (The Granger Collection)

Indian astronomy flourished during and immediately after the time of the great prince Rama, who was born in 961 B.C. During this period, tables of planetary motions were constructed for use in astrology, since the Hindus thought that heavenly occurrences represented earthly affairs. A very complex calendar was adopted that was based on an observed 247-year cycle of solar and lunar motions. The construction of this calendar and the discovery of precession show that Hindu astronomers made use of positional records that must have been maintained over hundreds of years.

Sometime in the first millennium B.C., India came into contact with the Middle Eastern civilizations of the Babylonians and the Greeks. After that time, it is difficult to determine the direction in which information flowed. Some people claim that the entire basis of Babylonian and Greek astronomy was borrowed from India in about the eighth century B.C., while others argue that the developments occurred independently in the two regions. In any case, by A.D. 500, Western influence had come to dominate Hindu astronomy, and few new developments occurred in India after that time.

Ancient Astronomy in the Americas

A modern appreciation of ancient astronomical developments in the Americas has been slow in coming. The principal reasons for this were the nearly complete destruction of the writings of the Mayan civilization during the Spanish conquest and the fact that the native cultures of North America had no written language.

Studies of the ruins of ancient civilizations in Central and South America have shown that very sophisticated cultures flourished there before the time of Columbus. The ancient cities of the Mayan and Incan civilizations, now in ruins, show highly complex and organized structures requiring sound knowledge of astronomical phenomena. In some cases, major buildings and avenues were aligned with significant astronomical directions, such as the rising Sun on the day of its passage through the zenith. Much of what we know about astronomy in these ancient cultures is gleaned from the record of their architecture, rather than from their writings.

The center of the Mayan civilization was located in Mesoamerica, a region that included what is now Mexico and several Central American countries to the south. In addition to the Maya, the Olmec, Zapotec, and Aztec cultures also flourished in this region at various times between 500 B.C. and A.D. 1500.

Figure 3.13 Ancient building in India. This is Konarak, a temple (c. 1240 A.D.) located on the Bay of Bengal, south of Calcutta. It is dedicated to Surya, the Hindu sun god, and is constructed with astronomical alignments. (J. M. Malville).

Figure 3.14 The Venus Tables in the Dresden Codex (Maya). These tables show that the Mayans attached particular significance to the planet Venus. (Historical Pictures Service)

As in other ancient cultures, the principal motivations behind Mesoamerican astronomy were timekeeping and astrology. The Maya developed a very accurate calendar and measured the motions of the celestial bodies with great precision. Particular significance was attached to Venus, which inspired sacrificial rites and other ceremonies. The motions of Venus were extensively tabulated in a surviving document known as the *Dresden Codex* **(Fig. 3.14)**.

In some locations, special astronomical observatories were built, usually with many significant alignments in the directions of important celestial events or phenomena. The most famous of these observatories is the Caracol tower at Chichén Itzá, in the Yucatán peninsula. The Caracol tower **(Fig. 3.15)** is a building whose strange architecture defied explanation until its numerous astronomical alignments were discovered in the twentieth century.

Mayan cosmology apparently consisted of a layered universe, each level of which contained one type of celestial body. Above the Earth was the domain of the Moon, then the clouds, the stars, the Sun, Venus, comets, and so on. Similarly, the Maya believed in an underworld consisting of nine levels, which provided symmetry with the heavens, an idea parallel to the early Greek cosmologies described by Homer. It is fascinating that such similar philosophies developed independently of each other in cultures that had no mutual contact.

In South America, in the Andean territory of modern-day Peru, the Incan civilization flourished in the century or so before the Spanish conquest. The Incas, too, paid great attention to astrological omens and devoted much of their astronomical effort to constructing accurate timekeeping systems. The Incas devised a rather complex calendar that had 12 months, but was otherwise very unlike our modern calendar. The weeks had different numbers of days, for example. Like their Mesoamerican neighbors to the north, the Incas often laid out their temples and cities in accordance with astronomical reference points. In at least one case, they used markers set out over an entire valley as a framework for computing dates.

The astronomy practiced by North American natives in ancient times is much more difficult to assess because they had no written language and few records of any kind survive, except for a few scattered rock drawings and artifacts such as mounds, religious symbols, and artwork **(Fig. 3.16)**. It seems clear, however, that the peoples of North America were intimately familiar with the skies and significant events such as the solstices that foretold the changing of the seasons. Among the few records that do exist are illustrations carved or painted on rocks **(Fig. 3.17)** in the Southwest, some of which may depict the appearance of the supernova of A.D. 1054, an event also recorded by Chinese astronomers. The Anasazi culture, centered in the Four Corners region of the American Southwest (particularly in Chaco Canyon, New Mexico), built many structures that have significant astronomical alignments and left behind several notable astronomical pertoglyphs. Among the best-known of these is the Sun Dagger at Chaco Canyon **(Fig. 3.18),** which consists of a spiral marking high on the side of Fajada Butte that is bisected by a dagger-like ray of bright sunlight just before noon on the summer solstice.

Some buildings that survive appear to have astronomical significance, including the Sun temple at Mesa Verde and other ruins in the Southwest that may have been influenced by the Mesoamerican culture to the south. In addition, many earthen mounds of various sizes and shapes, located in many parts of the midwestern United States, seem to follow astronomically significant alignments.

In the American West are several circular monuments, called medicine wheels, marked out in rock. These were probably left behind by the Plains Indians and show astronomical influences to varying degrees. Among the best studied is the Bighorn medicine wheel, located in the Bighorn Mountains of Wyoming **(Fig. 3.19)**. It is a circular structure some 30 m in diameter with rock cairns around its perimeter. Lines through these piles of rock coincide with a

Figure 3.15 **The Caracol Tower at Chichén Itzá.** This is one of the most significant of the many astronomically oriented structures in Mesoamerica. (J. Sidaner, Photo Researchers)

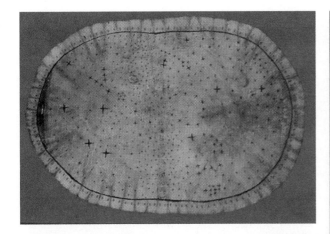

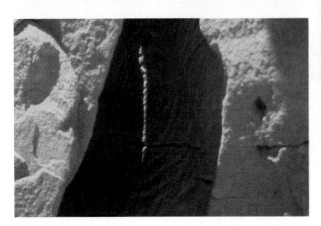

number of astronomically significant directions, including a primary line pointing in the direction of the rising Sun on the morning of the summer solstice.

Unfortunately, it is impossible to be certain when these monuments were built, so we do not know just how long ago our predecessors on this continent developed their understanding of the heavens.

Figure 3.16 Pawnee Indian sky map. (Top left) This chart, embossed on hide, appears to depict constellations of the Northern Hemisphere skies. (From Von Del Chamberlain 1982, *When Stars Came Down to Earth: Cosmology of the Skidi Pawnee Indians of North America* [Ballena Press: Los Altos, Calif.] Skidi Pawnee chart of the heavens, Field Museum of Natural History, photograph by Von Del Chamberlain)

Figure 3.17 North American Native Petroglyphs. (Above) This Anasazi drawing on a ledge in Chaco Canyon, New Mexico, is thought to depict the great supernova of 1054 A.D. (J. A. Eddy)

Figure 3.18 Sun Dagger. (Lower Left) On Fajada Butte in Chaco Canyon is this spiral pattern, which is bisected by a sliver of light at the time of the summer solstice. (National Park Service)

Figure 3.19 The Bighorn Medicine Wheel. (Middle left) This prominent and well-studied North American Indian medicine wheel is located in the Bighorn Mountains of Wyoming. (U.S. Forest Service, provided by J. A. Eddy)

SCIENCE AND SOCIETY

Astronomical Alignments in Modern Buildings

In this chapter we have learned that ancient structures in many areas of the world were designed to have astronomical orientations. These structures range from Hindu temples to Stonehenge in England to the many ancient temples and observatories of Mesoamerica. Yet you may not be aware that many buildings in today's society are still constructed and oriented along astronomical lines.

Probably the most common examples of astronomically aligned structures today are businesses and homes that are designed to be energy-efficient and therefore are built to take maximum advantage of sunlight. A house might be oriented with many windows facing south (if the house is in the Northern Hemisphere), so that during winter sunlight will enter through these windows and heat the interior. A well-designed solar house would have a stone or brick wall inside the south-facing windows, so that the Sun's heat would be stored up during the day and then be reradiated by the wall during the night. In a region where the summer Sun is very hot, an overhang or awning would be carefully placed above the windows, so that when the Sun reaches the high altitudes of summer, the overhang blocks sunlight from entering the same south-facing windows that welcome it during winter. The use of solar energy in this way is called *passive solar heating.*

Another approach to using sunlight for energy in the home or business is to build an *active solar heating system,* which generally incorporates pipes for the circulation of water that has been heated by sunlight. Again, the orientation of the building, or at least of the solar panels on its roof, depends on astronomical alignments. A house that is heated actively usually has solar panels set so that the incidence angle of the Sun's rays is most direct during winter. For example, at a latitude of 40° N, where the Sun's altitude above the southern horizon at the winter solstice is 26.5°, the panels would be set so that they faced a direction of about 26.5° above the southern horizon. Thus the panels would receive maximum intensity of sunlight at the time of the year when the days are shortest and the Sun's light hits the ground at its most oblique angle.

Archaeologists of the future may discover and ponder the astronomical orientations of buildings of the late twentieth century, just as we unearth and speculate about the alignments of ancient temples. Perhaps those investigators from times yet to come will realize why our buildings incorporate these alignments, or perhaps they will assume that the ancient societies (meaning ours) had religious or mystical motivations such as those we ascribe to the Hindu, Chinese, and Mesoamerican cultures of our past.

Astronomy in the Dark Ages

Let us now return to the main historical thread, with a brief mention of events after the time of Ptolemy. As Christian influence came to be felt throughout what had been the Greek (and subsequently the Roman) empire, people began to turn away from the belief that astronomical events determined their fates. As a result, they had less incentive to improve their understanding of the heavens. The prophet Muhammad, the founder of Islam, was born in the mid-seventh century A.D., and within a century his Arab followers had conquered most of the eastern and southern shores of the Mediterranean and taken parts of southern Europe, including Spain. In Egypt they had burned what was left of the library at Alexandria.

The need for accurate calendars led to a revival of interest in astronomy and encouraged the Arabs to resurrect a great deal of the Greek knowledge of the sky. During this period, our modern numbering system, brought from India by the Arabs, was established, and new mathematical techniques including algebra were developed. Although the Arabs made some new astronomical observations, their principal contribution was to keep alive the Greek knowledge of astronomy and to pass it on to the western European civilization in the twelfth century through Spain.

In the thirteenth century, the Spanish king Alfonso X ordered the construction of new tables of planetary motions, as the old tables of Ptolemy had finally become too inaccurate. The new tables, compiled by a group of astronomers using the epicyclic model of Ptolemy, became known as the *Alfonsine Tables* and were used for the next 300 years.

Meanwhile, the philosophical teachings of Aristotle, which had been adopted by the Islamic astronomers, were accepted by the European Christians as well. Thus, as the sixteenth century dawned, European astronomy was based almost entirely on the teachings and observations of the ancient Greeks.

Summary

1. The earliest recorded astronomical records come from the Sumerian culture, which occupied the southern part of the area now known as Iraq, beginning around 3000 B.C. The region was unified by King Hammurabi, who founded the kingdom of Babylonia about 1700 B.C.

2. The Greeks, who occupied the shores of the Mediterranean as early as 5000 B.C. and whose written records date back to the time of Homer, made the first methodical inquiry into the nature of the cosmos and developed the first cosmological models.

3. Starting around 570 B.C., Pythagoras and his followers developed the idea that the universe can be represented mathematically and introduced the beliefs that the Earth is spherical and that the motions of celestial bodies obey simple geometric laws.

4. Plato and his pupil Aristotle adopted idealized principles that could not be tested by observation and constructed an Earth-centered cosmology based on those principles, which included the postulate that the center of the Earth is the center of the universe, and that all celestial bodies are spherical and follow perfectly circular paths.

5. In the third century B.C., Aristarchus developed geometrical arguments that led him to believe that the Earth must orbit the Sun, but this idea was not widely accepted for two reasons: there was no widely recognized problem with the prevailing Earth-centered view, and there was no evidence that the Earth was moving.

6. During the same period, Eratosthenes used simple geometric arguments to make an accurate calculation of the size of the Earth.

7. In the second century B.C., Hipparchus made several important advances (improved instruments, use of trigonometry, stellar brightness estimates, an extensive star catalog, recognition of precession) and also developed a sophisticated Earth-centered cosmology based on epicycles.

8. In the second century A.D., Ptolemy collected all of the astronomical knowledge of the Greeks in a series of volumes called the *Almagest* and further refined the epicyclic, geocentric model of Hipparchus. Ptolemy's tables of planetary positions were to be used for the next thousand years.

9. Astronomical developments in China and India paralleled many of the discoveries and beliefs of the Greeks, and it is not clear to what extent or when mutual interaction may have occurred between the Western and Eastern civilizations. Both the Chinese and the Hindu astronomers of India were aware of long-term cyclical patterns in the sky, and both developed sophisticated calendars and timekeeping systems.

10. In the Americas, similar developments took place in Mesoamerica (southwestern United States, Mexico, and Central America), in the Incan civilization of Peru, and among the Native American cultures of North America. All of these societies recognized and attached great significance to astronomical events such as solstices and constructed buildings with astronomical alignments.

11. In North Africa and Spain, Arab astronomers preserved and carried forward the astronomical traditions of the Greeks, setting the stage for new discoveries and developments following the Dark Ages. Further advances in astronomy, based in Europe, were to arise from these traditions.

Review Questions

1. Choose one of the beliefs of the ancient Greeks, and discuss how well it fits the definition of a scientific theory, as described in Chapter 1.

2. Compare and contrast the motivations of the ancient cultures for studying the sky with those of modern astronomers. What are the similarities, and what are the contrasts?

3. From the viewpoint of the origins of modern science, why is it significant that the Greeks developed the concept of an underworld that was comparable to the heavens?

4. How was Plato like a modern scientist, and how did he differ from one?

5. Do you find any contradiction between Aristotle's adherence to the teachings of Plato on the one hand and his proof that the Earth is a sphere on the other? Explain.

Understanding Parallax

Stellar parallax is a special case of a more general phenomenon. There are many examples in everyday life where the apparent position of something varies with the observer's direction. For example, different people reading the same wall clock from widely separated positions in a room might get different times because of the small separation between the hands of the clock and the clock face, which causes the hands to appear at different positions depending on the angle from which the observer looks at the clock.

A simple way to demonstrate the principle of parallax to yourself is to hold a pencil in a vertical position at arm's length. Close one eye and look at the pencil, taking note of where it is positioned relative to the background. Now close the eye that was opened and open the other one. Again note the position of the pencil relative to the background. Do this repeatedly. Because of the separation between your eyes, you should see the pencil appear to shift back and forth as you switch eyes. This shifting of position is analogous to the shifting of a nearby star's position as seen from the Earth during the course of a year.

Make a scale drawing of the triangle formed by the distance between your eyes (the baseline) and the pencil. This triangle will have a base of about 5 cm and a height (long dimension) of about 60 cm, so it will be a long, skinny triangle. Using a protractor, measure the angle of the apex (point) of this triangle. How does this angle correspond with what the ancient astronomers could measure? How would this angle change if the length of your arm were increased?

You can use simple scaling arguments to find out what would happen if you extended the triangle you drew to simulate stellar parallax. First, suppose the distance between your eyes is increased to match the diameter of the Earth's orbit about the Sun, which is 2 AU. Instead of 5 cm, you would now have a baseline of 3×10^{11} m, an increase by a scale factor of 6×10^{12}. How far away is the pencil now, if its distance is increased by the same scale factor?

Your answer for the rescaled distance of the pencil should be about 4×10^{12} m, or around 25 AU. This is similar to the distances of Uranus and Neptune from the Sun. The distance to the nearest star, Alpha Centauri, is approximately 275,000 AU, however. What would happen to the apex angle of your triangle if you drew it to scale with a baseline of 2 AU and a height of 275,000 AU? Can you estimate the angle by scaling from the value you measured for your eyes and the pencil? Would the ancient astronomers have been able to measure this angle? Would they have been able to detect stellar parallax if the nearest star had been only 25 AU away? (The measurements of the ancient Greeks were accurate to about 2 arcminutes at best.)

6. Explain why few people accepted the arguments of Aristarchus that the Earth must orbit the Sun.

7. Compare the epicycle model of Hipparchus and Ptolemy with a modern scientific theory. Did the geocentric model conform to observations? Did it make predictions that could be tested? Did it represent the simplest explanation of the observed phenomena?

8. In what ways did the complex geocentric model of Hipparchus and Ptolemy violate the teachings of Plato and Aristotle?

9. Explain parallax in your own words, and describe some everyday examples of this phenomenon.

10. Why do we say that the astronomical developments of the Greeks set the stage for modern science, when several other cultures (e.g., the Chinese, the Hindu, and several societies in the Americas) achieved similar levels of knowledge and beliefs about the sky? In other words, why do we point to Greek science and philosophy as the foundation for modern science and astronomy?

Problems

1. The Babylonian calendar contained 12 months of 30 days each, so that the length of the year was exactly 360 days. How often did the Babylonians have to add a "leap month," if they did so every time the accumulated error reached 30 days?

2. Aristarchus measured the angular diameters of the Sun and Moon, finding each to be about 2°, whereas the correct values are about 0.5°. What would be the length of 1 AU (in kilometers) if the Sun actually were close enough to have an angular diameter of 2°?

3. Aristarchus also estimated that the Earth is three times larger in diameter than the Moon and that the Sun is 20 times more distant than the Moon. What would be the true diameter of the Sun in this case, assuming that the Earth's diameter is 12,736 km (the modern value)? Recall that the Sun and Moon have equal angular diameters.

4. Suppose that when Eratosthenes determined the size of the Earth, the distance between Syene and Alexandria was 3,300 km and the zenith angle of the Sun at Alexandria was 30° on the day when the Sun's light reached the bottom of the well at Syene. What would you find for the circumference and diameter of the Earth in this case? How do your answers compare with modern values?

5. The angle (in degrees) of the yearly shift in a star's apparent position due to stellar parallax is approximately equal to 57.3 times the ratio of 1 AU to the star's distance in AU (that is, the angle is given by $\theta = 57.3/d$, where d is in AU). If Hipparchus could measure angular positions of stars to an accuracy of 20 arcminutes, how close would a star have to be for Hipparchus to have been able to detect its parallax motion? How does this compare to actual stellar distances? How does this compare to the distance from the Earth to the outer solar system? (Hint: See Appendix 2.c and use the conversion factor 1 parsec = 206,265 AU.)

Additional Reading

Ahmad, I. A. 1995. The Science of Knowing God: Astronomy in the Golden Era of Islam. *Mercury* 24(2):28.

Aveni, A. F. 1977. *Native American Astronomy*. University of Texas Press.

Aveni, A. F. 1986. Archeoastronomy: Past, Present, and Future. *Sky & Telescope* 72(5):456.

Aveni, A. F. 1984. Native American Astronomy. *Physics Today* 37(6):24.

Bobrovnikoff, N. T. 1990. *Astronomy Before the Telescope*. Tucson: Pachart.

Heath, T. L. 1991. *Greek Astronomy*. New York: Dover.

Krupp, E. C. 1983. *Echoes of the Ancient Skies*. New York: Harper and Row.

North, J. D. 1994. *Norton's History of Astronomy*. New York: Norton.

Shukla, K S. (ed.) 1987. *History of Oriental Astronomy*. Cambridge University Press.

Web Connections

The Review Questions and Problems also appear at the following URLs:

http://universe.colorado.edu/ch3/questions.html
http://universe.colorado.edu/ch3/problems.html

Chapter 4
Renaissance Astronomy

The stars are not so strange
as the mind that studies them.

H. E. Fosdick

The fifteenth century saw the beginning of a reawakening of the intellectual spirit in Europe. Some scientific studies were undertaken at the major universities; increased maritime explorations brought demands for better means of celestial navigation; and the art of printing was developed, opening the way for widespread dissemination of information.

Major advances in all the sciences accompanied the new developments in other fields of human endeavor. In this chapter we discuss the principal achievements in Renaissance science that led to the development of modern astronomy. In doing so, we discuss the accomplishments of four major figures: Copernicus, Brahe, Kepler, and Galileo.

Copernicus: The Heliocentric View Revisited

Nineteen years before the epic voyage of Columbus, Niklas Koppernigk **(Fig. 4.1)** was born in Torun, in northern Poland. As a young man, he attended the University of Cracow, where his fondness for Latin,

Figure 4.1 Nicolaus Copernicus. Copernicus's view that the Sun, rather than the Earth, occupied the central place in the cosmos had great influence on later scientists. (The Granger Collection) http://universe.colorado.edu/fig/4-1.html

the universal language of scholars, led him to change his name to Nicolaus Copernicus. At Cracow he developed an avid interest in astronomy, becoming fully acquainted with the Aristotelian view as well as the Ptolemaic model of planetary motions.

Early Belief in the Heliocentric Theory

Copernicus continued his academic pursuits in Italy, studying ecclesiastical law at Bologna and Ferarra and medicine at Padua before returning to Poland to assist his uncle, the bishop (and therefore chief government official) of the region of Ermland. During all this time, Copernicus persisted in his studies of astronomy, and by 1514 he had developed some doubts about the validity of the accepted system.

The reasons for his doubts have been the subject of some uncertainty and misconception. It was long assumed that Copernicus was encouraged to adopt the Sun-centered view of the universe because he recognized shortcomings in Ptolemy's geocentric model. There is, however, no evidence of widespread dissatisfaction with the Ptolemaic system, nor are there any records indicating that Copernicus himself found serious inaccuracies in it. His reasons for adopting the heliocentric viewpoint were more subtle.

The basis for his conversion was primarily philosophical. Copernicus believed that his new system presented a pleasing and unifying model of the universe and its motions. While he was no doubt encouraged by the climate of change and cultural revolution that was sweeping Europe with the advent of the Renaissance, he did not adopt the heliocentric view just to be different or to improve the accuracy of the accepted view or reduce its complexity. The Copernican model was, in fact, no more accurate in predicting planetary motions than the Ptolemaic system, and it was just as complex. Copernicus adhered to the notion of perfect circular motions and was obliged to include small epicycles to make his model conform to the observed planetary positions.

The Advantages of the Heliocentric Model

To Copernicus, one of the most pleasing aspects of the Sun-centered system was the fact that the relative distances of the planets could be deduced **(Fig. 4.2)**. In the geocentric model based on epicycles, it was possible to find many different combinations of values for the sizes and rotation speeds of the various deferents and epicycles and still match the observed motions of the planets. Copernicus found this disturbing, for it meant that the system was not really unique; the riddle was not solved unambiguously. He did not think the cosmos was created along arbitrary or random lines. On the contrary, in the Sun-centered theory that he devised, the relative distances of the planets from the Sun were fixed uniquely. There was nothing arbitrary about the relative distances, which were easy to deduce from simple observations (see Fig. 4.2). The spacings had a pleasing regularity, with smoothly increasing separations between planetary orbits with increasing distance from the Sun. To Copernicus, this model seemed more like nature's grand plan.

Copernicus was also able to determine the relative speeds of the planets in their orbits, finding that each planet moves faster than the next one farther out from the Sun. Like Aristarchus long before him, Copernicus also recognized that the Sun is the largest body in the solar system, which he considered a strong argument for placing it at the center. Finally, Copernicus was bothered by the need for epicycles and other complexities of the ancient geocentric cosmology and hoped initially that his Sun-centered theory would allow a return to perfect circular motions about a common center, without need for complexities such as equants and epicycles. But as we have seen, in this goal he was frustrated because his model still required some of these complications.

Copernicus was able to provide natural explanations for some of the phenomena that had caused the complexities of the epicyclic model. For example, retrograde motion, the main reason for having epicycles in the geocentric model, was now explained as due to the motion of the Earth, particularly to the relative speeds of the Earth and other planets in their paths around the Sun (see Fig. 2.24). The seasons were also readily explained by invoking a tilt of the Earth's axis. Copernicus realized that the stars should undergo parallax motion if the Earth orbits the Sun, but explained the lack of observed parallax with the daring assumption that the stars were too far away for the parallax to be detected (in this he was correct; when parallax was eventually detected, astronomers found that the largest stellar shifts are at least a factor of 100 smaller than could have been measured in Renaissance times).

A Reluctant Author

Copernicus first circulated his ideas informally sometime before 1514 in a manuscript called *Commentoriolis,* which drew increasing attention over the next several years. He was reluctant to publish his findings in a more formal way for fear of arousing controversy, and he continually rechecked his calculations and otherwise put off going public with his views. Eventually, a young Protestant scholar known as Rheticus encouraged him to publish his work. Persuaded that Copernicus was right, Rheticus had published his own summary of the heliocentric theory. Now Rheticus delivered the manuscript of Copernicus to a Lutheran priest named Osiander, who assumed responsibility for publishing it.

Fearful of strong religious opposition, Osiander added an unsigned preface in which he stated that the ideas in the treatise did not represent physical reality, but should be used only as a mathematical tool for predicting planetary motions. There is no doubt that Copernicus himself believed in the literal truth of his heliocentric model, however. The book was titled *De Revolutionibus Orbium Coelestium (On the Revolution of the Celestial Sphere).* Copernicus died just about the time his work was published, so he never knew of its impact upon the world.

The impact was profound, although not immediately. The many important improvements in understanding the solar system that were embodied in the new theory eventually made a strong enough impression on other scientists to win them to this point of view. Among the most persuasive aspects of the theory were the natural ordering of the planets (the relationship between orbital period and distance from the Sun) and the simple explanation for the differing sizes of the retrograde loops that the planets go through. The extent of the backward motion observed for each planet, if it is the result of the Earth's relative motion as Copernicus believed, depends on how far away the planet is. The ability of the Copernican model to explain the seasons and the lack of stellar parallax, which we mentioned earlier, were additional attractive features of the theory.

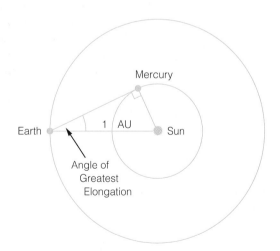

Figure 4.2 Copernicus's method for finding relative planetary distances from the Sun. For an inferior planet such as Mercury or Venus, Copernicus knew the angle of greatest elongation and was therefore able to reconstruct the triangle shown, which provided the Sun-Mercury (or Sun-Venus) distance relative to the Sun-Earth distance (that is, the astronomical unit). For superior planets, similar but slightly more complicated considerations provided the same information.

Tycho Brahe: Advanced Observations

Although Copernicus himself had contributed little observationally and had not insisted upon a close match between theory and observation as long as general agreement was found, there was, in fact, a strong need for improved precision in astronomical measurements. The next influential character in the historical sequence met this need. If he had not, further progress would have been seriously delayed, as would the eventual acceptance of the Copernican doctrine.

Tycho's Astronomical Development

Tycho Brahe **(Fig. 4.3)** was born in 1546, some three years after the death of Copernicus, in what is now the extreme southern portion of Sweden (the region was part of Denmark at the time). Of noble descent, Tycho spent his youth in comfortable surroundings and was well educated, first at Copenhagen University, then at Leipzig, where he insisted on studying mathematics and astronomy despite his family's wish that he pursue a law career. Wars and epidemics of plague caused Tycho to change universities several times, with most of his formal training eventually taking place at the university in Wittenberg. He developed a strong reputation as an astronomer (and astrologer—the distinction was still scarcely recognized) and attracted the attention of the Danish king, Frederick II. In 1575 the king ceded to Tycho the island of Hveen, about 14 miles north of Copenhagen, along with enough servant support and financial assistance to allow him to build and maintain his own observatory **(Fig. 4.4).**

The event that established Tycho's reputation as an astronomer was the appearance of a **nova,** or new star, in 1572. His own observations, along with information gained from others and from his travels around Europe, showed that the position of the nova remained fixed among the stars regardless of the location of the observer on the Earth. Tycho realized that if the nova were a nearby object, there should have been a parallax effect. Because there was none, he concluded that the nova belonged to the realm of the stars. The fame he garnered from this analysis brought him to the attention of King Frederick and led to Tycho's appointment as court astronomer and the establishment of his observatory.

Advances in Observational Techniques

Even before this time, Tycho had shown an acute interest in astronomical instruments, and with the grant to build his observatory, this interest bore fruit. He devised a variety of instruments that did not encompass any new principles, but were capable of more accurate readings than any before his time. The largest was a quadrant built into a wall of his observatory **(Fig. 4.5),** with a radius of approximately 10 m. A quadrant is a quarter circle with graduated markings indicating degrees and minutes of arc around its circumference. A sight located at the center allows the observer to read the angular heights of celestial objects above the horizon. Tycho's instruments improved on those of his contemporaries in that the angular scales were more precisely marked, and that a new sighting design allowed the scales to be read more accurately. Tycho was able to make positional measurements with an accuracy of about 1 arcminute with his instruments.

Challenges to the Established System

In 1577, soon after the observatory was established, Tycho observed a comet and once again combined his observations with those of others to demonstrate a lack of parallax. Thus both the nova seen five years earlier and now the comet lay beyond the Earth's atmosphere (in fact, Tycho's analysis showed that they were beyond the orbit of the Moon). The nova had represented a mild challenge to the ancient doctrine because it indicated change among the immutable stars, but the comet was a more serious blow, for Aristotle had taught that comets were atmospheric phenomena. Worse, the comet moved and therefore passed right through the crystalline spheres of the planets and the stars! This proved that the spheres could not be the solid, material structures envisioned by the ancient Greeks.

Despite these apparent flaws in the established doctrine, Tycho was unable to accept the heliocentric view, primarily because he could find no evidence that the Earth was moving. He tried and failed to detect stellar parallax, which he thought he should see with his accurate observations if the Earth really moved. Furthermore, as a strict Protestant, he found it philosophically difficult to accept a moving Earth when the Scriptures stated that the Earth is fixed at the center of the universe.

On the other hand, he realized that the Copernican system had advantages of mathematical simplicity over the Ptolemaic model, and in the end he was ingenious enough to devise a model that satisfied all of his criteria. His model featured a fixed Earth orbited by the Sun and the Moon, while all the other planets orbited the Sun **(Fig. 4.6).** Mathematically, this model is equivalent to the Copernican system in terms of accounting for the motions of the planets as seen from the Earth. The idea never won much

Figure 4.3 Tycho Brahe. Tycho's principal contribution, a massive collection of accurate observations of planetary positions, was crucial to subsequent advances in understanding planetary motions. (The Granger Collection)
http://universe.colorado.edu/fig/4-3.html

Figure 4.4 Detail of Tycho's Observatory at Hveen. (Photo Researchers, Inc.)

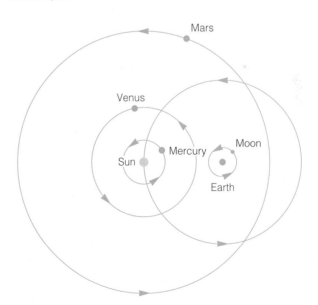

Figure 4.5 The Great Mural Quadrant at Hveen (left). This instrument was used to measure the angular positions of stars and planets with respect to the horizon. Its large size made the angle markings easy to read accurately. (Photo Researchers, Inc.)

Figure 4.6 Tycho's model of the universe (above). Tycho held that the Earth was fixed and that the Sun orbited the Earth. The planets in turn orbited the Sun. This system was never worked out in mathematical detail, but it successfully preserved both the advantages of the Copernican system and the spirit of the ancient teachings. It also accounted for the lack of observed stellar parallax, since the Earth was fixed.

acceptance, however, and Tycho is remembered primarily for his fine observations.

Data of Unprecedented Quality

Tycho's observational contributions were significant for both the unprecedented accuracy of his data and the completeness of his records. Until his time, astronomers generally had recorded the positions of the planets only at notable points in their travels, such as when a superior planet comes to a halt just before beginning retrograde motion. Tycho made much more systematic observations, recording planetary positions at every opportunity, thereby building up a continuous record of planetary motion over a span of time exceeding 20 years.

Tycho did something else that was unprecedented and also very important. He made multiple observations of planetary positions, rather than taking only a single reading. When observing Mars on a given night, for example, he might record five or six measurements in a row and enter all of them into his log. He understood that small, unavoidable errors creep into all measurements and that by making several readings, these errors would tend to cancel each other out. You can see for yourself that this is so: use

a ruler to make several measurements of the width of a table or desk, and you will find that the measurements differ, particularly if you try to measure to the smallest unit available on the ruler (say, 1 mm).

Tycho himself did not attempt any detailed analysis of the collection of data that he amassed; for example, he did not take the step of averaging his multiple measurements together to improve their accuracy, nor did he attempt to fit his planetary positions into any model of the cosmos. The task of unraveling the secrets of planetary motion contained within Tycho's data was left to those who followed him.

In the 1590s, after the death of King Frederick, Tycho moved to Prague to assume the position of court astronomer to Emperor Rudolf of Bohemia. There he continued his observations of the planets. In 1600 he took on a young assistant named Johannes Kepler. A year later, in 1601, Tycho died of a urinary infection resulting from overindulgence at a royal party.

Johannes Kepler and the Laws of Planetary Motion

Until now it has been possible to follow developments in a straight sequence, but at this point significant astronomical advances began to occur at nearly the same time in different locations. To complete the thread begun with our discussion of Tycho Brahe, this section will examine the contributions of Johannes Kepler, who worked briefly with Tycho himself and then spent many years analyzing the great wealth of observational data that Tycho had accumulated. We must keep in mind, however, that during the same period, Galileo Galilei was at work in Italy, and he and Kepler were both aware of the other's accomplishments.

The Young Kepler

Kepler **(Fig. 4.7)** was born in Weil, Germany, in 1571, just one year before Tycho's nova was observed. A sickly youngster, he seemed headed for a career in theology, but he had difficulty conforming to the fundamentalist religious philosophy of the university he attended in Tübingen. There he encountered a professor of astronomy who inspired in him a strong interest in the Copernican system.

After completing his education, Kepler accepted a teaching position in the Austrian city of Graz in 1594. There he published his first scientific work, a book called the *Mysterium Cosmographicum (Cosmic Mystery)*, in which he outlined several arguments favoring the Copernican system and stated his own

Figure 4.7 Johannes Kepler. Kepler's fascination with numerical relationships led him to discover the true nature of planetary motions. (The Bettmann Archive) http://universe.colorado.edu/fig/4-7.html

view that the distances of the planets from the Sun were determined by a series of regular geometric solids separating the crystalline spheres of the planets **(Fig. 4.8)**. Kepler's reasons for adopting this notion and the Copernican system as well were based on his love of mathematical simplicity and his intuitive feeling that the universe was constructed according to a harmonious, unified plan. He devoted his life to seeking out the underlying principles of this universal harmony, a task that, as we shall see, led him to a number of profound discoveries. The hypothesis put forth in the *Mysterium Cosmographicum* was, of course, incorrect, but the book had a major impact anyway, especially through its eloquent arguments in favor of the Sun-centered universe.

In 1598 a decree banning all Protestants forced Kepler to leave Graz, and in 1600 he accepted a post in Prague as chief assistant to Tycho. Tycho died the next year, and Kepler succeeded him as court astronomer to Emperor Rudolf. After some dispute with Tycho's heirs about the ownership of Tycho's data on planetary positions, Kepler was granted the right to use the data. Kepler's main mission, due both to Tycho's wishes and to his own interests, was to develop a refined understanding of the planetary motions and to upgrade the tables used to predict their positions.

Kepler's Mathematical Approach

Realizing that the Sun must play a central role in making the planets move, Kepler was inspired to look for a physical link between the Sun and the planets. In this sense, he was like a modern scientist and quite unlike other astronomers of his time and earlier. Whereas they sought only a mathematical device for predicting motions, Kepler believed in a true physical machinery of the solar system and sought its underlying principles. He believed that the Sun must exert a force on the planets and even deduced that this force must diminish in strength with increasing distance from the Sun. But he did not know what this force was. The only kind of force known to operate at a distance was magnetism, so Kepler suggested that magnetic forces emanating from the Sun were the driving mechanism for the planets.

Kepler experimented extensively with different geometrical interpretations of the planetary motions. He tried many arrangements using circular motions, including offsets to account for the observed variability of planetary orbital speeds, and some of his solutions came quite close to matching the observed motions. But he demanded a perfect match (to within the accuracy of the observations) and would not accept any model that did not provide it. This approach, which arose from Kepler's belief that the

universe is a physical system that can be described in true physical terms, was very different from that of his predecessors, including Hipparchus, Ptolemy, and Copernicus, for whom an approximate mathematical representation was good enough.

Kepler chose first to work with the extensive data on the motion of Mars. This decision was fortunate, because Mars has one of the more noncircular orbits among the planets, affording Kepler a better chance of determining its true shape. He might have been frustrated for a long time had he chosen Venus or Jupiter for his initial attack on the problem, for both of these planets have more nearly circular orbits.

Kepler's first task was to separate the motion of Mars from that of the Earth. By a complex process, he was able to derive the path of Mars in the reference frame of the stars. He was then able to analyze the properties of the orbit of Mars without confusion due to the motion of the Earth.

By 1604 Kepler had determined that the orbit of Mars was some kind of oval, and further experimentation revealed that it was fitted precisely by a simple

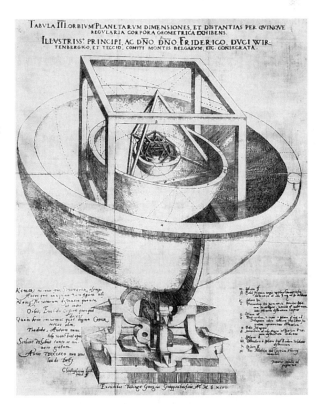

Figure 4.8 Kepler's nested geometric solids. Kepler expended considerable effort in exploring the possibility that the planetary distances from the Sun were determined by the spacings of nested regular geometric solids (that is, figures whose faces were regular polyhedrons; the number of gaps between possible solids of this sort was equal to the number of known planets). Kepler eventually abandoned this particular scheme, but never relented in his search for regularity in the solar system (O. Gingerich).

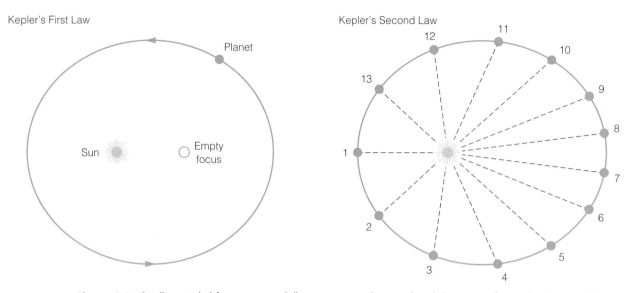

Kepler's First Law

Kepler's Second Law

Figure 4.9 **The ellipse.** At the left is an exaggerated ellipse representing a planetary orbit with the Sun at one focus; at the right is a similarly exaggerated ellipse with lines drawn to illustrate Kepler's second law. If the numbers represent the planet's position at equal time intervals, the areas of the triangular segments are equal.

geometric figure called an **ellipse (Fig. 4.9).** He was led to this shape in part because of his realization that Mars moves more rapidly when close to the Sun than when it is farther away (in this he anticipated his second law of planetary motion, which we will get to shortly).

The Nature of the Ellipse

An ellipse is a closed curve defined by a fixed total distance from two points called **foci** (singular: **focus**), and Kepler found that the Sun was at the location of one focus of the ellipse (the other focus is empty). One way to make an ellipse, and in the process gain an understanding of some of its properties, is to loop a piece of string around two pegs or tacks on a board **(Fig. 4.10).** Then use a pencil to draw a figure around the pins, keeping the string taut. The fixed length of the string (which is equal to the major axis of the ellipse) ensures that the distance from one focus to any point on the ellipse, and then from there to the other focus, is always the same. The long axis of the ellipse is the **major axis;** one-half of that distance is the **semimajor axis.** The semimajor axis may be thought of as the average distance of a planet from the Sun.

Ellipses can range in shape from circular to very elongated. The key parameter is the **eccentricity,** which is related to the separation between the foci as compared with the semimajor axis. The eccentricity is defined mathematically as

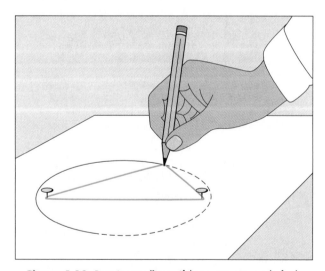

Figure 4.10 **Drawing an ellipse.** If the two pins represent the focal points of an ellipse, then the fixed length of the string ensures that the total distance from F_1 to any point on the ellipse and from there to F_2 is always the same. The length of the loop of string is equal to 2a, the major axis of the ellipse (to see that this is so, consider the case where the string connects to a point at either end of the major axis of the ellipse).

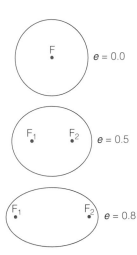

Figure 4.11 Ecentricity of an ellipse. The eccentricity (usually denoted e) is equal to the ratio of the distance between the foci ($\overline{F_1F_2}$) to the length of the major axis (2 a, where a is the semimajor axis). The eccentricity is a measure of how elongated an ellipse is. For a circle (which is ot elongated at all), the distance between foci is 0 and the eccentricity is e = 0. For a very elongated orbit, such as the orbits of comets (see Chapter 11), the eccentricity approaches a value of e = 1 (but is always less than 1 for a close orbit).

$$e = \frac{\overline{F_1F_2}}{2a},$$

where $\overline{F_1F_2}$ is the distance between the foci and a is the semimajor axis **(Fig. 4.11)**. If the two foci are so close together that they merge into a single point, then $\overline{F_1F_2} = 0$, so the eccentricity is zero, and the figure is a circle. In that case, the semimajor axis is the radius of the circle. On the other hand, if the foci are widely separated relative to the semimajor axis, then the ellipse is very elongated (that is, very eccentric). The orbits of the planets have small eccentricities, meaning that they are nearly circular. But as mentioned, Mars has a relatively noncircular orbit (its eccentricity is 0.0934), which helped Kepler to determine the orbit's shape.

Another mathematical expression related to eccentricity is very useful. The point of closest approach between a planet and the Sun, called **perihelion,** is related to eccentricity by

$$P = a(1 - e),$$

where e is the eccentricity, P is the perihelion distance, and a is the semimajor axis as before. Similarly, the point of greatest separation between a planet and the Sun, the **aphelion,** is related to the eccentricity by

$$A = a(1 + e),$$

where A is the aphelion distance. Using these expressions, we can visualize the shape of an orbit, given its eccentricity. For example, given that the eccentricity of Mars's orbit is $e = 0.0934$ and its semimajor axis is 1.524 AU, we see from these expressions that the distance of Mars from the Sun varies between 1.38 AU (at perihelion) and 1.67 AU (at aphelion). Thus the orbit of Mars is quite noticeably noncircular.

The First Two Laws of Planetary Motion

Kepler's analysis of the motion of Mars revealed a second characteristic that we have already alluded to: The planet moves fastest in its orbit when it is nearest the Sun and slowest when it is on the opposite side of its orbit, farthest from the Sun. Mathematically, Kepler's second discovery was that a line connecting Mars to the Sun sweeps out equal areas in space in equal intervals of time (see Fig. 4.9). As we shall see, Isaac Newton later realized that this is a statement of the conservation of **angular momentum,** a property combining the circular motion, the mass, and the radius of the orbit (this is discussed in the next chapter).

These two properties of the orbit of Mars were later generalized as Kepler's first two laws of planetary motion, applicable to all of the planets. Kepler's results on the orbit of Mars were published in 1609 in a book entitled *Astronomia Nova (The New Astronomy: Commentaries on the Motions of Mars).* The book attracted a great deal of attention.

Kepler's first two laws of planetary motion, in general terms, can be stated as follows:

First law: The orbits of the planets are ellipses with the Sun at one focus.

Second law: For each planet, a line connecting the planet to the Sun sweeps out equal areas in space in equal intervals of time.

Kepler's Third Law

In 1612 Emperor Rudolf, who had been supporting Kepler's research, died, and Kepler moved to Linz, in northern Austria. There his work on the planetary motions continued, as he still sought the true nature of the relationship among all of the planets. In 1619 he published a book entitled *Harmonices Mundi (The Harmony of the World),* in which he reported his discovery of a simple relationship between the orbital periods of the planets and their average distances from the Sun. Now known as Kepler's third law or simply the harmonic law, it states:

Third law: The square of the sidereal period of a planet is proportional to the cube of the semimajor axis.

Put in other terms, this says that

$$P^2 \propto a^3,$$

where P is the sidereal period of a planet and a is the semimajor axis of its orbit. The symbol $\propto$ means "is proportional to," meaning that the two quantities vary together. If the period is expressed in units of years and the semimajor axis in astronomical units, then for the Earth $P^2 = 1$ and $a^3 = 1$ and the relationship becomes an equality; that is,

$$P^2 = a^3.$$

Table 4.1 lists values for P^2 and a^3 for the planets, confirming that these two quantities are equal.

Undoubtedly, the harmonic law was a very pleasing discovery for Kepler because it represented the mathematical glue to hold the solar system together that he had been trying to find. Now he had a relationship that all the planets obeyed, confirming that indeed the system does have an underlying unity and supporting his notion that the Sun must somehow control the motions of the planets.

Table 4.1
Testing Kepler's Third Law[a]

Planet	a (AU)	P (years)	a^3	P^2
Mercury	0.387	0.241	0.058	0.058
Venus	0.723	0.615	0.378	0.378
Earth	1.000	1.000	1.000	1.000
Mars	1.523	1.881	3.533	3.538
Jupiter	5.203	11.86	140.85	140.66
Saturn	9.539	29.46	867.98	867.89
Uranus	19.18	84.01	7,055.79	7,057.68
Neptune	30.06	164.8	27,162.32	27,159.04
Pluto	39.44	248.4	61,349.46	61,762.56

[a]The minor disagreement between P^2 and a^3 seen here for the outermost planets do not indicate failures of Kepler's third law; instead they reflect inaccuracies in the measured values of P and a.

Kepler's Final Works

In another major work, the *Epitome of the Copernican Astronomy*—published in parts in 1618, 1620, and 1621—Kepler presented a summary of the state of astronomy at that time, including Galileo's discoveries. In this book Kepler generalized his laws, explicitly stating that all of the planets behaved similarly to Mars, as he had clearly believed all along.

The Roman Catholic Church had not taken a strong position on the heliocentric theory at the time of Copernicus, but by the time the *Epitome* was published, it had joined the Lutheran Church in taking a very intolerant stance toward what were considered to be heretical teachings. Kepler's treatise soon found itself on the *Index of Prohibited Books*, along with *De Revolutionibus*.

In 1627 Kepler published his last significant astronomical work, a table of planetary positions based on his laws of motion, which could be used to predict planetary motions accurately. These tables, which he called the *Rudolfine Tables* in honor of his former benefactor, represented an improvement in accuracy over any previous tables by a factor of nearly 100, a resounding and remarkable confirmation of the validity of Kepler's laws. In a very real sense, the *Rudolfine Tables* represented Kepler's life's work, since with their publication he completed the task set before him when he first went to work for Tycho. Kepler died in 1630 at the age of 59.

Galileo, Experimental Physics, and the Telescope

Very strong contrasts can be drawn between Kepler and his great contemporary, Galileo Galilei **(Fig. 4.12)**. Where Kepler was fascinated with universal harmony and therefore with the underlying principles on which the universe operates, Galileo was primarily concerned with the nature of physical phenomena and was less interested in finding fundamental causes. Galileo wanted to know how the laws of nature operated, whereas Kepler sought the reason for their existence.

Galileo's approach was levelheaded and rational in the extreme. He used simple experiment and deduction in advancing his perception of the universe and has frequently been cited as the first truly modern scientist, although others of his time probably deserve a share of that recognition.

Galileo was born in 1564 in Pisa, in north-central Italy. Although his family had planned a career for him in a trade, his intellectual brilliance soon became obvious, and in 1581 he began formal university training in Pisa. There his reluctance to accept dogmatic ideas without question made him infamous, and his tendency toward debate earned him nickname "The Wrangler." Always poor, he earned extra income during his university years by tutoring other students. He stayed on in Pisa for several years as an instructor, but sought appointment to a university with more prestige. Through carefully cultivated political connections, he achieved this goal in 1592, when he was appointed to the University of Padua. There he stayed until 1610.

Galileo's Mechanics

During his early academic years, Galileo carried out numerous experiments in physics, particularly in the study of the motions of bodies. Both his method and his findings were revolutionary. Whereas a follower of Plato and Aristotle would proceed by rational deduction from standard unproven assumptions, Galileo thought it much more sensible to begin with experiment or observation and work toward a recognition of the underlying principles. When Galileo argued a point, he based his position on what could be demonstrated, although he certainly mixed in a large share of logical deduction. To him, the supreme test of a hypothesis was whether it fit the observed phenomena, not whether the observed phenomena fit a previously adopted hypothesis. In developing this approach, Galileo established an entirely new basis for scientific inquiry, an achievement in many ways more profound than his contributions to astronomy, which were considerable.

Galileo's discoveries in physics, having to do with the motions of objects, were published in his later years and are considered by many to be his most important scientific work. Considering the impact that Galileo was to have on astronomy, this is a very strong statement. Galileo's early interest in the science of motion, now called **mechanics,** was what lured him away from medicine, the subject of his first studies.

Among his most important contributions in mechanics was his understanding of a principle called **inertia,** the tendency of a moving object to continue moving or of a stationary object to remain at rest. Unlike his Aristotelian predecessors, Galileo realized that on a level surface an object would require a force to stop it from moving. He deduced the presence of **friction,** a force that always acts against motion when two bodies are in contact with each other. A key experiment in demonstrating the truth of inertia was to drop a ball from the top of the mast of a moving ship and show that the ball would fall straight down to the foot of the mast, rather than being left behind

due to the ship's forward motion while the ball was falling. Inertia would require that the ball continue the forward motion imparted by the ship. Though this experiment was widely recognized as an important test of Galileo's hypothesis, it was very difficult to perform, and for years the result was debated (dropping a ball from a mast may not sound difficult, but the problem lay in finding smooth enough water so that the pitching of the ship did not deflect the fall of the ball. Winds also created difficulties).

Galileo also understood the concept of **acceleration,** or the rate of change of velocity, at least in the limited situation of a falling object. Here Galileo first got some basic points wrong, but with continued experimentation he eventually reached the correct understanding. He found that falling objects accelerated for a while, then seemed to continue at a steady speed. Initially, he thought that the rate of fall was related to the composition of the falling object, and that each type of material had its own specific final speed. Further study showed that all objects accelerate at the same rate, independent of mass or composition, but that shape played a role. Galileo was

Figure 4.12 Galileo Galilei. From his numerous experiments and observations, Galileo deduced the nature of the cosmos. His methodical attacks on traditional beliefs led to personal troubles for him but greatly influenced contemporary thinking. (The Granger Collection) http://universe.colorado. edu/fig/4-12.html

exploring the properties of air resistance and ultimately understood that in the absence of air resistance all bodies would accelerate at the same rate as they fell. Of course, it was impossible for him to devise an experiment that avoided air resistance; for this reason some historians argue that the famous experiment of dropping balls of unequal weight from the Leaning Tower of Pisa probably never took place. If Galileo had gathered a crowd and performed this demonstration as legend has it, he might not have convinced anyone of his claims, because the balls almost certainly would have hit the ground separately (not only would they have experienced differing amounts of air resistance, but it is also very difficult for a person to drop two objects of unequal weight at exactly the same instant).

Among Galileo's other contributions to mechanics was his discovery of the principle of the pendulum. He noticed that a hanging weight would swing back and forth at a steady rate (frequency), regardless of how far it swung from side to side. If a hanging lamp began to sway in a breeze, it would start by moving fairly far from side to side, but as its range diminished, the time of each oscillation would stay the same. Galileo realized that this could be used as the mechanism in a clock, and in his old age he tried unsuccessfully to build one (he ultimately had to abandon this project because of failing eyesight).

Galileo's Astronomical Discoveries

Galileo's astronomical discoveries were quite sufficient to earn him a major place in history, and his flair for debate and his habit of ridiculing those

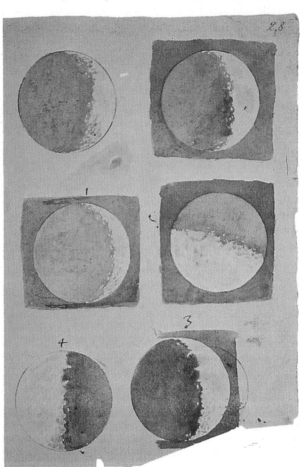

Figure 4.13 Galileo's sketches of the Moon. Galileo's observation of mountains, craters, and "seas" led him to believe that the Moon was not the perfect heavenly body envisioned in the ancient Greek teachings. (The Granger Collection)

whose arguments he disproved made him famous in his own time, though not universally loved. This mixed reaction to Galileo was later to prove fateful for him, as power struggles took place between his supporters and his detractors.

Galileo's first astronomical contribution was the observation of a nova in 1604, one also observed and analyzed by Kepler. (Modern astronomers have deduced that the novae of 1572 and 1604 were actually **supernovae,** extremely powerful and rare stellar explosions. It is quite remarkable that two supernovae occurred within 32 years of each other, for there has not been another as bright or as close to the Earth in the nearly 400 years since.) Both men showed that the lack of parallax placed the nova in the realm of the stars, as Tycho had done for the nova of 1572.

In 1609 Galileo learned of the invention of the telescope and devised one of his own, which he soon put to use in systematic observations of the heavens. The popular belief that Galileo invented the telescope is a misconception. He did not; the use of magnifying glasses had been known for a long time, and the trick of placing two lenses in a series to make distant objects appear closer was invented by a Dutchman named Hans Lippershey. News of this discovery led Galileo to assemble his own telescopes. What set him apart from others was his realization that this new instrument could be used to improve one's view of the universe and gain new information about the cosmos. Thus his was the first *astronomical* telescope.

Despite the poor quality of the instrument, Galileo made a number of important discoveries almost at once from his observation tower in Padua and reported them in 1610 in a publication called *Siderius Nuncius (The Starry Messenger).* Here Galileo showed that the Moon was not a perfect sphere but was covered with craters and mountains

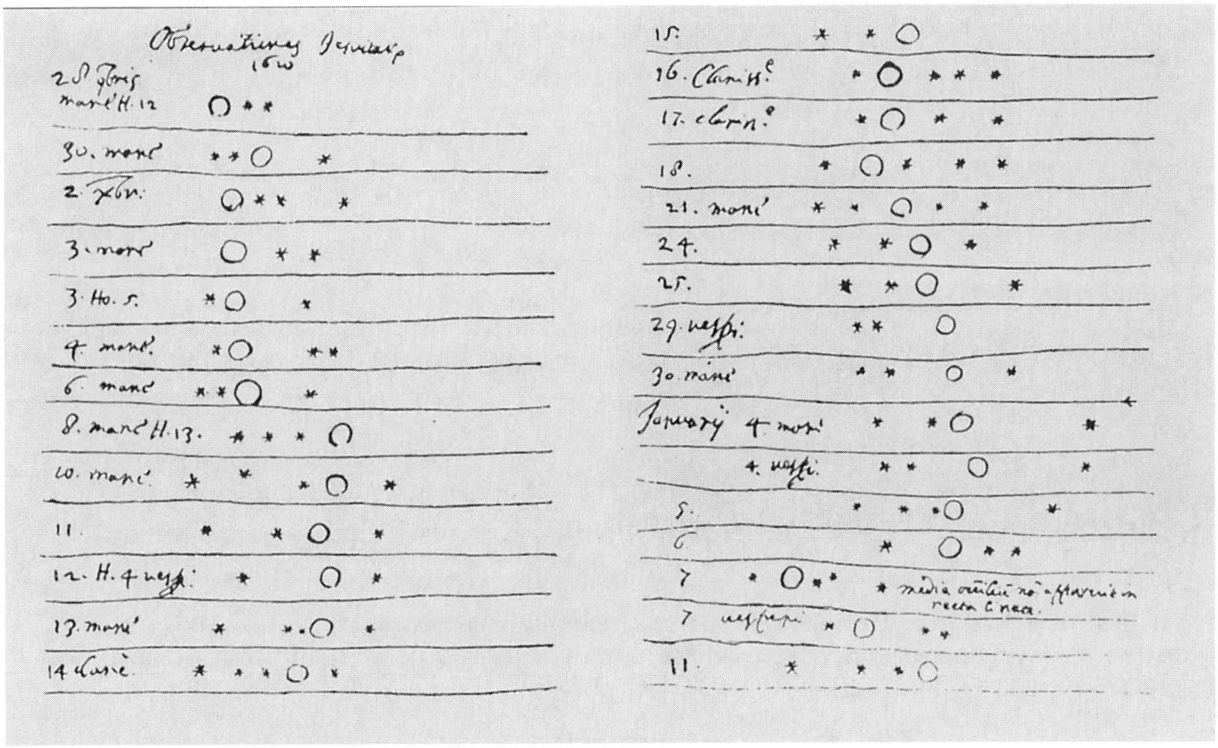

Figure 4.14 Galileo's sketches of the moons of Jupiter. This series of drawings of Galileo's Jovian observations is often attributed to *The Starry Messenger,* but in fact was made some years later. Galileo's discovery of moons orbiting Jupiter showed that there are heavenly bodies that do not orbit the Earth. (Yerkes Observatory)

(Fig. 4.13). He also reported that the broad band of the Milky Way, previously thought to be a cloud, consisted of countless stars, which presented difficulties for the conventional view that the number of fixed stars was finite. Most significant of all, Galileo found that Jupiter was attended by four satellites, whose motions he observed long enough to establish that they orbited the parent planet **(Fig. 4.14).** All of these discoveries, but especially the latter, violated the ancient conception of an idealized universe centered on the Earth. The satellites of Jupiter helped convince Galileo that there were centers of motion other than the Earth.

Once Galileo began his astronomical observations, he became fascinated with the structure of the universe, and for a time he put aside his studies of mechanics. He spent most of the next several years in pursuit of evidence for the Copernican model **(Table 4.2).**

Mostly as a result of the reputation he earned with the publication of *The Starry Messenger,* Galileo was able to negotiate successfully for the position of court mathematician to the grand duke of Tuscany, and in 1610 he moved to Florence, where he spent the remainder of his long career. Once established in Florence, Galileo continued his observations and

Table 4.2	
Some of Galileo's Arguments for the Heliocentric Theory	
Discovery	**Argument**
Many faint stars	Difficult to reconcile with idea of stars as points attached to crystalline sphere
Craters on the Moon	Moon is not perfect, immutable heavenly body
Moons of Jupiter	A body other than Earth as center of motion
Phases of Venus	Explained only if Venus orbits the Sun and shines by reflected sunlight
Sunspots	Spots on the solar surface, showing that the Sun is not perfect
Variable planetary sizes	Angular size variations explained by motion of planets around Sun

soon added new discoveries to his list. Finding that Venus changes its appearance much as the Moon does, he explained these phases as the result of the motion of Venus around the Sun, which causes varying portions of the sunlit side to be visible from

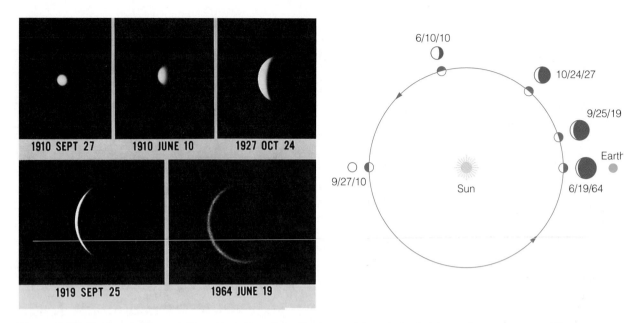

Figure 4.15 **The phases of Venus.** As Venus orbits the Sun and its position changes relative to the Sun-Earth line, its phase varies as we see differing portions of its sunlit side. In addition, its apparent size varies because of its varying distance from Earth. (Photos from NASA) http://universe.colorado.edu /fig/4-15.html

Figure 4.16 **Galileo's analysis of sunspots.** The dashed lines indicate the apparent (angular) motion of a sunspot, as seen from Earth, during two equal time intervals. The variation in obvserved speed of a sunspot led Galileo to argue that sunspots actually lie on the Sun's surface and were not shadows of planets passing between the Earth and the Sun.

Earth **(Fig. 4.15).** This discovery had two important implications: it showed that Venus shines by reflected sunlight rather than by its own power; and it demonstrated that Venus orbits the Sun rather than the Earth, directly contradicting the geocentric model. Further evidence came from the fact that the apparent (angular) size of Venus changes as the planet goes through its phases; it is largest when appearing as a crescent (when it is on the near side of the Sun) and smallest when appearing nearly full (on the far side of the Sun).

Galileo also noticed that the angular diameter of Mars varies as this planet moves in its orbit. In this case the variation is quite large. Galileo argued that such large changes in angular diameter, which must be caused by variations in the distance from the Earth to Mars, would be impossible to explain if Mars orbited the Earth in the manner envisioned by the epicyclic model. As in the case of Venus, Galileo used the variations in the angular size of Mars to argue for the heliocentric model.

At about the same time, Galileo entered into an ongoing debate over the properties of the dark spots on the Sun and soon was able to show that they truly lie on the solar surface, rather than being small planets orbiting in the foreground. He showed this by noting that the spots move rather slowly when near the edge of the Sun's disk, which he attributed (correctly) to foreshortening as the Sun's rotation carried them toward the Earth on the approaching edge of the Sun's disk and away from the Earth on the receding edge. More rapid sideways motion as seen from

SCIENCE AND SOCIETY
Can Religion and Science Coexist?

The case of Galileo clearly brought out the contrast between faith and science and led to confrontation between the two. Galileo was not the only person to incur the wrath of the church for his scientific views during that time, as the case of Giordano Bruno illustrates. Bruno was a former Catholic priest who repudiated the authoritarian teachings of Aristotle; he traveled through Europe teaching that the Sun was one among many stars, all of which were surrounded by inhabited planets just as the Earth and the other planets orbit the Sun. In 1591 he was lured to Venice on a pretext and handed over to the Roman Inquisition. He was eventually put on trial, convicted of heresy, and burned at the stake in 1600.

Today we do not see such violent conflict between faith and science in Western society, but nevertheless the potential for confrontation between science and religion remains. In the United States, the most prominent modern battleground is probably the dispute over creationism, the belief that the universe, the Earth, and all life-forms were created spontaneously a few thousand years ago, as described in the Old Testament. There is no evidence for this view, it makes no predictions that can be tested, and in general it fails to meet the criteria for a scientific theory. Yet in many communities, proponents of creationism insist that it should be taught in the schools as science. For creationists the distinction between faith and science is blurred, and they see conflict where others, including most of the established churches, do not.

Ironically, it is the Roman Catholic Church that has recently offered new insight into the relationship between religion and science and how they can coexist without conflict. And Galileo's case provided the occasion for the church to do so.

The Roman Catholic Church recently revisited the confrontation with Galileo and his treatment nearly 400 years ago.

Galileo's books (along with those of Copernicus and Kepler) were removed from the *Index* in the nineteenth century, but Galileo's condemnation by the Inquisition still stood. In the interest of clearing his record and, perhaps more importantly, clarifying the relative roles of science and faith in the modern church, new hearings were held. In 1981 Pope John Paul II appointed a special commission to investigate Galileo's case. The commission spent a decade researching and analyzing all available documents and reported its findings to the pope in 1992. The pope not only issued a statement officially rehabilitating Galileo, but also used the occasion to clarify the distinction between faith and science, as Galileo had tried to do. Pope John Paul went on to say that there should be no conflict between science and faith, for the two operate in entirely different realms.

Thus, perhaps in an indirect way, Galileo has made yet another important contribution to modern science. Perhaps this lesson can be applied to other areas where faith and science appear to be in conflict.

(Note: In late 1996, Pope John Paul issued a similar statement about evolution thoery, again arguing that no conflict exists between science and religion.)

---

the Earth occurred as the spots moved across the central portion of the solar disk **(Fig. 4.16).** Not only did this analysis run counter to established doctrine, but it also offended leading scientists of the day because of the manner in which Galileo usurped their authority over the interpretation of sunspots. Galileo made enemies who were to play a role later in his censure by the Roman Catholic Church.

Confronting the Establishment

Galileo came under increasingly strong pressure from the Roman Catholic Church, which represented authority and held the official view that the Aristotelian geocentric cosmology was not to be questioned. Galileo, a devout Catholic, did not want to offend the church and tried to persuade officials that

it was harmless to accept his findings as only a convenience for mathematical representation of motions in the heavens, totally avoiding the question of whether these views represented physical reality. He tried to argue that there is a distinction between scientific construction and religious faith and that a person could separate the two. The church did not agree, and in 1616 Galileo was instructed to refrain from any statements favoring the Copernican hypothesis. Galileo agreed, but subsequently it turned out that he had misunderstood just what the restrictions were on his public statements.

The conflict came to a head in 1632, with the publication of Galileo's greatest book on astronomy. Following a conventional format of the day, Galileo cast his book in the form of a dialogue among three characters. One of the three, named Salviati, espoused Galileo's own views. Another, Simplicio,

Inquisition and put on trial. His books were rounded up and heavily censored and then were listed on the *Index of Prohibited Books,* along with the works of Copernicus and Kepler.

The Inquisition stopped short of excommunicating Galileo, but did take harsh steps. Galileo was ordered to be isolated from all outside contact with others, including his family (later some visits were allowed), and he was placed under house arrest at his estate near Florence. There he spent the remainder of his life.

Galileo was permitted to work while in exile, however, and in 1638 he published his major work on mechanics (with a friend's help, he smuggled the manuscript out of Italy to get it published). This book later was very influential and was well known to Isaac Newton, who developed mechanics to a level still taught in today's university physics classes (see the next chapter). Galileo became blind during his last years and was frequently ill. His health continued to fail, and he died in 1642.

Despite Galileo's suppression by the church, his work and that of Copernicus, Brahe, and Kepler had raised enough questions that serious consideration of the cosmos thereafter centered on the heliocentric system. The stage was now set for the discovery of the missing link, the physical cause of the motions in the heavens.

Figure 4.17 **The first edition of Galileo's *Dialogue.*** (Erich Lessing, Art Resource, NY)

took the stance of official doctrine, while the third, Sagredo, was ostensibly neutral (but was always quick to see the wisdom of Salviati's arguments). In the book, entitled *Dialogue on the Two Chief World Systems* **(Fig. 4.17),** Galileo systematically laid waste to Aristotelian cosmology, using all of the arguments gathered from his years of astronomical observation to show how much more sensible it was to suppose that the Earth orbits the Sun rather than the other way around. As though to rub salt into the wounds he was creating in church doctrine, Galileo published the book in Italian rather than the scholarly Latin, thereby making its message more readily accessible to the public.

Galileo believed that he could avoid violating his agreement with the church by including a lengthy preface to the book, where he claimed that he personally did not believe in the Copernican hypothesis and that it could be considered merely a device to help envision the motions of heavenly bodies. Galileo badly misjudged the church's mood, however, and this disclaimer in the preface did not spare him from drastic retribution. Conventional scientists whom Galileo had offended over the years helped persuade the church officials that serious punishment should be forthcoming. Galileo was called before the Roman

Summary

1. Copernicus developed the heliocentric theory primarily because to him it provided a more aesthetically and philosophically pleasing picture of the universe than the geocentric model, not because it was more accurate or simpler than the Earth-centered theory.

2. Copernicus's new system enabled him to calculate the relative distances of the planets from the Sun, showing the concentric pattern of the solar system, which he found to be very satisfying.

3. Copernicus showed that his system provided natural and simple explanations for retrograde motion, the seasons, and precession, and he asserted that the lack of stellar parallax was due to the great distances of the stars.

4. Tycho Brahe made vast improvements in the quantity and quality of astronomical observations of stars and of the planets, accomplishing this by making more precise and complete measurements than his predecessors.

5. Tycho observed planets at every opportunity instead of recording their positions only at signifi-

Drawing Ellipses

ASTRONOMICAL

ACTIVITY

The properties of an ellipse can be described in many ways, including mathematical formulas that are not used in this text. But one definition of an ellipse that is used here lends itself to an easy method for drawing the figure. Recall that the total distance from any point on an ellipse to the two foci is always the same. To use this fact to draw an ellipse, you will need a board (a cork board will do nicely), two pins or nails, and some string and a pencil. Stick the two pins into the board to represent the two foci of the ellipse. Tie the ends of the string together to make a loop, and then use the pencil to trace out an ellipse by hanging the loop of string over the pins and keeping the string taut as you move the pencil around the loop. (See Figure 4.9; note that this is easier to do if you make a loop of string by tying the ends together and hanging it over the two pins. Then you can trace the full ellipse without getting the string tangled. The extra length of string connecting the two foci is fixed; thus the total distance from the foci to the ellipse also is constant.) The fixed length of the string ensures that the total distance from any point on the ellipse to the two foci remains constant. The length of the lopp is equal to $2a$; i.e. the major axis of the ellipse.

Now you can experiment by changing the length of the string relative to the distance between the foci (you can do this by changing the length of the string, by moving the pins closer together or farther apart, or by a combination of both). If the string is only a little longer than the distance between the foci, you get a very elongated, thin ellipse; if the string is very much longer than the distance between the foci, the ellipse is nearly a circle (think about what would happen if the two foci were moved so close

together that they occupied the same spot; in that case, you would be drawing a precise circle).

The distance between the foci relative to the length of the major axis defines the eccentricity of an ellipse. In an intuitive sense, we say that a very elongated ellipse (with a major axis only a little longer than the separation between the foci) has a high eccentricity; on the other hand, an ellipse whose major axis is much greater than the separation between its foci (that is, an ellipse that is nearly circular) has a small eccentricity. Thus the nearly circular orbits of the planets have small eccentricities (see Appendix 8), while the elongated orbits of comets (discussed in Chapter 11) have large eccentricities.

Now that you have a little experience drawing ellipses, you can make properly scaled sketches of the orbits of the planets. Given the eccentricity of an orbit (Appendix 8), you can decide how far apart to make your foci relative to the length of the loop of string (which is equal to $2a$). One definition of eccentricity, given in the text, is $e = \frac{\overline{F_1 F_2}}{2a}$, where $\overline{F_1 F_2}$ is the separation between the foci. In your drawings, this translates to

$$e = \frac{\text{distance between pins}}{\text{length of loop}}$$

Thus to draw an ellipse with a specified eccentricity, separate the pins representing the foci by the correct distance so that the ratio of the separation between the pins to the length of the loop of string equals the desired value of e.

Using this technique, make scale drawings of the orbits of Mercury, the Earth, Mars, and Pluto. Discuss the variations in orbital shape that you find. Which orbits look almost circular? Which orbits are clearly not a circle?

---

cant points in their motions, and he made multiple measurements each time instead of single sightings. He accumulated a large set of data on planetary positions over a period exceeding 20 years.

6. Kepler sought the underlying harmony among the planets by analyzing Tycho's data, and in the process, he discovered his three laws: each planet orbits the Sun in an ellipse; a line connecting a planet with the Sun sweeps out equal areas in space in equal intervals of time; and the square of the period of a planetary orbit is proportional to the cube of its semimajor axis.

7. Kepler's tables of planetary motion, based on his laws, were far more accurate than any previous tables, which gave great impetus to the heliocentric theory and to the idea that the Sun somehow drives the motions of the planets.

8. Galileo used telescopic observations and deductive reasoning to argue for the heliocentric concept of the universe. Among his discoveries were the phases of Venus, the satellites of Jupiter, the varying angular size of Venus and Mars, and the fact that sunspots are dark regions on the solar surface.

9. Despite church opposition, Galileo brought his ideas before the public in the form of a fictional

dialogue between characters of opposing points of view. This book, combined with Kepler's work, essentially ended the reign of Aristotle's ideas among Western scientists.

10. Though under house arrest for the last several years of his life, Galileo published a book on his discoveries in mechanics, the science of motion; this book was to have great influence on later scientists such as Isaac Newton.

Review Questions

1. Summarize the reasons why Copernicus adopted the heliocentric view.

2. Describe how Copernicus's beliefs conformed to the teachings of Aristotle and how they violated those teachings.

3. A principle adopted by most scientists holds that the simplest explanation for something, the one requiring the fewest unprovable assumptions, is the best. Copernicus and Hipparchus had different explanations of retrograde motion. Which explanation better satisfies the principle of simplicity? Explain.

4. Summarize the ways in which Tycho Brahe's observations were superior to those made by earlier astronomers.

5. If Kepler had adopted Tycho's cosmology in which the Sun orbits the Earth while the other planets orbit the Sun, would he still have derived the same laws of planetary motion? Explain. What observational tests might we conduct to distinguish between Tycho's model and the Sun-centered model?

6. Discuss Kepler's philosophical outlook. In what ways was he like the followers of Plato and Aris-

totle, and in what ways was he like a modern scientist?

7. Why was it fortunate that Kepler chose to work first with the data on Mars when he attempted to deduce the nature of planetary motions?

8. Suppose you push a cart, let it roll, and observe that it soon comes to a stop. Aristotle and Galileo would have offered different explanations for the fact that it stops. Discuss.

9. In what ways did the telescopes used by Galileo help him to argue against the Earth-centered cosmology?

10. Do you think that modern scientists are ever subjected to censure and suppression of their views, as Galileo was? Explain.

Problems

1. Suppose astronomers find a new inferior planet with a greatest elongation of 36°. Using the method of Copernicus, determine the distance of this planet from the Sun. (Hint: Make a sketch similar to Figure 4.2. If you are not familiar with trigonometry, use a protractor and a ruler to make the drawing, and measure the distance of the planet from the Sun relative to the Sun-Earth distance.)

2. Find the speed of the Earth in its orbit by dividing the circumference of the orbit (in kilometers) by the length of the sidereal period (in seconds). Then do the same for Venus, Mars, Jupiter, and Saturn. (Hint: Assume the orbits are circular, so the circumference is given by $2\pi a$, where a is the radius.) Discuss the relative speeds of these planets as they move in their orbits, and comment on the role this calculation

played in the adoption of the Sun-centered universe by Copernicus.

3. Tycho's great quadrant consisted of a quarter circle with a radius of 9.6 m, covering an angle of 90°. Marks on the quadrant corresponded to degrees and minutes of arc. How far apart, in centimeters, were the degree marks? The minute marks?

4. Suppose a certain star's true altitude above the horizon was 27°14'. Suppose further that Tycho's measurements of the star's altitude were 27°12', 27°13', 27°15', 27°13', and 27°16'. How close is the average of these values to the true value? Discuss the relationship between the average of several measurements and the accuracy of individual measurements.

5. The Earth's distance from the Sun ranges from 147 million to 152 million km. Calculate the eccentricity of the Earth's orbit?

6. The perihelion distance for Pluto is 29.58 AU and the aphelion distance is 49.30 AU. Find the eccentricity of Pluto's orbit. Compare your answer with the value given in Appendix 8, and discuss the contrast between the values for Pluto and the Earth.

7. If asteroid Grumpy has an orbital eccentricity of 0.47 and a perihelion distance of 3.54 AU, how far is it from the Sun at aphelion?

8. What is the period of a planet whose semimajor axis is 3 AU? What is the period of a planet whose semimajor axis is 25 AU?

9. What is the semimajor axis of a planet whose period is 11.18 years? What would the semimajor axis be if the period were 267 years?

10. Find the eccentricity and the orbital period for a comet whose closest approach to the Sun is 0.14 AU and whose greatest distance from the Sun is 456 AU.

Additional Readings

The references listed here are primarily sources of biographical information and historical data on the people discussed in this chapter. Many of these works contain additional lists of references. Readings on the principles of physics described in this chapter can most easily be found in elementary physics texts, which are available at all levels ranging from completely nonmathematical to any degree of mathematical sophistication desired.

Beer, A. and P. Beer (eds.) 1975. *Copernicus. Vistas in Astronomy* Vol 17. New York: Pergamon Press.

Beer, A. and K. A. Strand (eds.) 1975. *Kepler. Vistas in Astronomy* Vol 18. New York: Pergamon Press.

Drake, S. 1980. Newton's Apple and Galileo's Dialogue. *Scientific American* 243(2):150.

Gingerich, O. 1973. Copernicus and Tycho. *Scientific American* 229(6):86.

Gingerich, O. 1983. The Galileo Affair. *Scientific American* 247(2):132.

Gingerich, O. 1993. How Galileo Changed the Rules of Science. *Sky & Telescope* 85(3):32.

Web Connections
The Review Questions and Problems also appear at the following URLs:
http://universe.colorado.edu/ch4/questions.html
http://universe.colorado.edu/ch4/problems.html

Chapter 5
The Laws That Govern Motions

Chapter web site: http://universe.colorado.edu/ch5

Newton's idea of gravity has done nothing but
astonish our imagination.

Frederick the Great, 1780

uring the half-century after the publication of Galileo's *Dialogue on the Two Chief World Systems* in 1632, few major discoveries were made, but substantial progress occurred nevertheless. The use of telescopes to observe the heavens continued. One of the biggest issues of the period was the nature of Saturn, which had a curiously elongated shape as seen by Galileo, and whose appearance changed with time. Finally, in 1655 the Dutch astronomer Christiaan Huygens deduced correctly that Saturn was girdled by a ring system and that the planet's tilt caused the rings to be seen edge-on periodically, at which times they were invisible because of their thinness. Huygens also discovered Titan, Saturn's giant moon, thereby adding Saturn to the list of objects other than the Sun that could be centers of motion.

Other astronomers used telescopes to map the surface of the Moon in detail. Studies of the Earth gained momentum as well, and accurate new measurements demonstrated that it is not a perfect sphere, but instead bulges at the equator. Meanwhile the New World, as well as Asia and even the South Pacific, were being explored and colonized. It was a time of growing intellectual activity and excitement.

Figure 5.1 Isaac Newton. Newton's experiments and brilliant mathematical intuition led him to profound new understandings of physics and mathematics. (The Granger Collection) http://universe.colorado.edu/fig/5-1.html

The Life of Isaac Newton

In the year 1643, a few months after the death of Galileo, Isaac Newton **(Fig. 5.1)** was born in Woolsthorpe, England. Newton was born into a prosperous family of sheep ranchers and farmers, but had a lonely childhood. His father had died before he was born, and when his mother remarried three years later, young Isaac was placed in the care of his maternal grandparents. He spent most of his boyhood years at his grandparents' country home or at boarding school and never had many friends. This was due in part to his own intellectual aloofness, for it became clear early on that Newton far surpassed his contemporaries in mental capabilities.

Newton was almost the first in his family line to be able to read and write. On his father's side, he was the first Newton to be able to sign his own name, and if his mother's family had not had a stronger educational tradition, he might never have attended school. But he did attend, and he excelled so impressively and enjoyed intellectual pursuits so thoroughly that his mother's attempt to have him return to country life upon finishing preparatory school failed. His teachers urged him to continue his studies, and Newton himself was eager to go on, so in June 1661, at age 18, he set off to attend Cambridge University. There his incredible intellect, as well as his tendency to be a loner, reinforced his isolation from his peers.

At Cambridge Newton quickly left behind the Aristotelian philosophy that was still being taught there and explored the worlds of science, philosophy, and mathematics on his own. He was much impressed by the geometrical works of the great French mathematician René Descartes, and he also absorbed the earlier geometry of the Greek mathematician Euclid. Newton desired a permanent position at Cambridge upon graduation, but first he had to pass a series of examinations in the traditional subjects that he had ignored. He did pass in 1664, although it is not clear whether he did so because he had mastered the topics, or because he had the behind-the-scenes support of a couple of senior professors. Newton became an instructor at Cambridge but spent much of the period from 1665 to 1667 at his home in Woolsthorpe, because the plague, or Black Death, had made city living rather dangerous. During this time he made a remarkable series of discoveries in the fields of physics, astronomy, optics, and mathematics in what surely must have been one of the most intense and productive periods of individual intellectual effort in human history. On his return to Cambridge in 1667, Newton was appointed a Fellow at Trinity College, and three years later, at the age of 27, he was named Lucasian Professor of Mathematics. He was to occupy this position for many years, and he maintained his association with Cambridge for most of his life.

Newton had a tendency to exhaust a subject, become bored with it, and go on to new fields without publishing his results. As a consequence, some of his discoveries were repeated (and reported) independently by others. Naturally, this led to some acrimonious debates over credit for several of his breakthroughs, and Newton was not above engaging in unseemly arguments. An early example was his longstanding battle with the British scientist Robert Hooke, who himself worried constantly that his colleagues did not give him proper credit for his work. Hooke (and some others) hit upon the idea that gravity might diminish with the square of the distance from a body but did not know how to show this concept mathematically. As it turned out, Newton had done so some years earlier but had not bothered to publish his result. When Newton eventually did publish, Hooke insisted that he should receive credit for the idea. Similarly, Newton derived the basics of calculus at an early time, but he made little mention of it and did not publish anything about his work until much later. The great German mathematician Gottfried Wilhelm Leibniz later developed the calculus independently and published his results, putting him in position to claim that Newton had plagiarized the idea when it later appeared in Newton's writings. The battle over priority in the discovery of calculus consumed both men in their later years and ended only when Leibniz died in 1716. Today both are recognized for having found the principles of calculus independently, but the form developed by Leibniz is the one most commonly used.

Newton first became prominent as a result of his work in optics. He discovered that sunlight can not only be separated into colors by a prism, but that the colors can then be recombined, re-creating white light. This discovery demonstrated for the first time that all the colors are contained in white light. Newton studied and developed laws of refraction and reflection and postulated that light consists of particles (**refraction** is the bending of light when it passes through a surface between different substances, as when it passes from air into glass or water). The opposing idea, that light was a wave phenomenon, was promoted by the Dutch astronomer Christiaan Huygens (it was demonstrated eventually that light has both wave and particle properties; see Chapter 6). During the course of his investigations in optics, Newton invented a telescope that brought light to a focus using a concave mirror rather than a lens.

When he presented one of his reflecting telescopes to the scientists and philosophers of the Royal Society of London, they immediately elected him a member, beginning an affiliation that was to last for the rest of Newton's life.

In the mid-1680s, several of the noted scientists in London agreed that elliptical orbits, such as Kepler had shown the planets to follow, were the result of a force that diminished as the square of the distance from the central body. The noted astronomer and architect Christopher Wren offered a prize to anyone who could demonstrate this mathematically, but no one could. In 1684 a young astronomer named Edmond Halley visited Newton in Cambridge to ask about this problem. He found that Newton had solved it years before. The interaction with Halley stimulated Newton to revisit his earlier work on orbits and, with constant prodding (and financial support for publication costs) from Halley, to publish his masterwork, a three-volume book on motion and gravitation. This massive work, called *Philosophiae Naturalis Principia Mathematica* (**Fig. 5.2**), now usually referred to simply as the *Principia*, appeared in 1687. In it, Newton established the science of

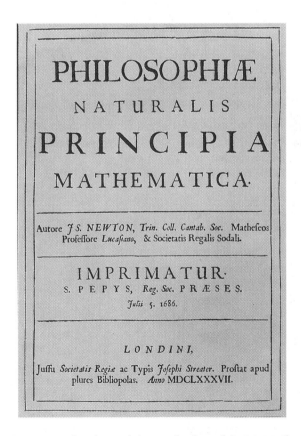

Figure 5.2 The title page from an early edition of the *Principia*. This massive volume is still considered one of the greatest and most influential books ever written. (The Granger Collection)

mechanics (which he viewed as merely background material and relegated to an introductory section) and applied it to the motions of the Moon and the planets, developing the law of gravitation as well. His work in optics was published separately in 1704, although it was probably written much earlier.

The *Principia* attracted great notice, particularly in England. As a result, Newton's later life was a public one, with various government positions. In 1695 he accepted a post at the Royal Mint becoming Master of the Mint in 1699; in 1703 he was elected president of the Royal Society a post he held until his death in 1727; and in 1705 he was knighted by Queen Anne, making him the first scientist to be so recognized. As a result of his public obligations, Newton had less and less time for his own scientific investigations, although he resumed some of his earlier studies in alchemy (yet another area of profound interest to him!). With the help of younger associates, he did revise the *Principia* on two occasions, in 1713 and 1726, making some improvements each time. Newton died in 1727 at the age of 84. His influence lives on in our modern understanding of physics and mathematics. Newton's conclusions on the nature of motions and gravity are still viewed as correct, although scientists now realize that in certain circumstances more comprehensive theories (such as Einstein's theory of relativity) must be used.

The Laws of Motion

Newton postulated three laws of motion, principles he considered so self-evident that he included them in the introductory section of the *Principia,* along with the background material. In this section we discuss the three laws and show how they are applied.

Inertia

The first of Newton's laws states the principle of **inertia,** a concept first recognized by Galileo, who realized that an object in motion tends to stay in motion unless something acts to slow or stop it. This was completely contrary to the teachings of Aristotle, who held that the natural tendency of any moving object was to stop, and that it would continue moving only if a force were applied. Aristotle was misled by his failure to recognize friction, a force that tends to stop motion in most everyday situations.

Newton expanded the concept of inertia, recognizing it as just one in a series of physical principles that govern the motions of objects, and adding the all-important notion of **mass.** The mass of an object reflects the amount of matter it contains, which in turn determines other properties such as weight and momentum. The other properties are easily observed, but the mass is not, so mass can be a difficult concept. Mass and weight are particularly easy to confuse because at the Earth's surface, a given amount of mass will always have the same weight (because the gravitational force is nearly the same everywhere on the Earth). The distinction between weight and mass is easier to see if we consider locales other than the Earth's surface. For example, astronauts in space may be weightless, but they contain just as much mass as they do on the ground.

Mass is usually measured in units of grams or kilograms. One gram is the mass of a cubic centimeter of water, and a kilogram, which weighs about 2.2 pounds at sea level, is the mass of 1,000 cubic centimeters, or one liter of water.

Inertia is very closely related to mass because the mass of a body determines how much resistance to a change in motion a body has; that is, the mass of a body determines its inertia. The more massive an object is, the more inertia it has, and the more difficult it is to start the object moving or to stop or alter its motion once it is moving **(Fig. 5.3).** For example, imagine two crates of equal size, one containing books and the other pillows. Each crate has wheels on the bottom, minimizing friction. You may be able to move the crate of pillows rather easily, but not the crate of books. The crate of books contains more mass and therefore has more inertia, or resistance to being moved. Similarly, a massive object in motion is more difficult to turn or stop than a less massive one. Con-

Figure 5.3 Inertia. An object in motion tends to remain in motion. The more massive an object is, the more inertia, or resistance to being moved, it will have. Similarly, a massive object in motion is more difficult to turn or stop than a less massive one. (H. Roger Viollet)

sider, for example, the deftness required to bring a large ship to rest at a pier without destroying the pier.

Newton summarized the concept of inertia in what has become known as his first law of motion:

A body at rest or in a state of uniform motion tends to stay at rest or in uniform motion unless an unbalanced outside force acts upon it.

As we have seen, Galileo had stated the principle of inertia some time before Newton, but Galileo applied it only to the limited situation of objects in motion at the surface of the Earth. Newton stated the principle in much more general terms that included the Earth itself, the planets, and all other celestial bodies. Here "uniform motion" means motion in a straight line at a constant speed. "Outside force" means a force exerted on a body externally. Friction is a very common example of an outside force.

Newton made the giant leap in understanding required to comprehend that the laws of motion are universal. One very important consequence of applying the law of inertia to celestial objects is that the planets would fly off into space in straight lines if they were not affected by outside forces. We will return to this point shortly.

Force and Acceleration

Having stated that a force is required to change an object's state of rest or of uniform motion, Newton went on to determine the relationship between force and the change in motion that it produces. To understand this, we must discuss the idea of acceleration.

Acceleration is a general word for any change in the motion of an object. Acceleration occurs when a moving object is speeded up or slowed down or when its direction of motion is altered. Acceleration also occurs when an object at rest is put into motion. A planet orbiting the Sun would be undergoing acceleration even if its speed were constant; otherwise it would fly off in a straight line.

When we change the motion of an object, we change its velocity. **Velocity** in physics is defined as having both a speed and a direction, and acceleration is defined formally as the rate of change of velocity, meaning a change in speed , direction, or both. Acceleration is expressed in units of change in velocity per second. Thus, if a car accelerates from rest to 80 km per hour in 10 seconds, its acceleration is 8 km per hour per second. In astronomy and physics, velocities are usually expressed in meters per second or kilometers per second, so the units of acceleration can be either m/sec/sec or km/sec/sec, commonly written as m/sec^2 or km/sec^2.

An object falling in the Earth's gravitational field at the Earth's surface increases its downward speed by 9.80 m/sec for every second it falls (in the absence of air resistance), so we say that the acceleration due to gravity at the Earth's surface is $9.80 \ m/sec^2$. Because of air resistance, of course, a falling object reaches a constant speed that, as Galileo realized, depends on its size and shape. Only in a vacuum would the acceleration continue as long as the object fell.

Since Newton's first law specifies that an object will remain in a state of rest or constant motion (i.e., constant velocity) unless an unbalanced outside force acts on it, we can restate this to say that for an object to be accelerated, an unbalanced force must act on it. Note that this force must be an unbalanced one; that is, acceleration will occur only when a force is applied to an object with no other force to counteract it. Thus, a crate sitting on the floor is subject to the downward force of gravity, but this force is balanced by the upward force due to the floor, so there is no acceleration. Newton's second law spells out the relationship among an unbalanced force, the resultant acceleration, and the mass of the object:

The acceleration of an object is equal to the unbalanced force applied to it divided by its mass.

This may be written mathematically as

$$a = \frac{F}{m} \ ,$$

where *a* is the acceleration, *F* is the unbalanced force, and *m* is the mass. More commonly, it is written in the equivalent form

$$F = ma.$$

We can visualize simple examples to help illustrate the second law. If one object has twice the mass of another, for example, and equal forces are applied to the two objects, the more massive one will be accelerated only half as much. Conversely, if unequal forces are applied to objects of equal mass, the one to which the greater force is applied will be accelerated to a greater speed.

Action and Reaction

Newton's third law of motion is probably more subtle than the first two, although in some circumstances it is quite obvious. It states:

For every action there is an equal and opposite reaction.

In other words, when a force is applied to an object, it pushes back with an equal force **(Fig. 5.4).** This may

Figure 5.4 **Action and reaction.** Some manifestations of Newton's third law are rather subtle, while others are not. One boy exerts a force on another, and the second boy exerts an equal and opposite force on the first. This situation is static; there is no motion. When a rifle is fired, however, the bullet is accelerated one way and the rifle the other. The same force is applied to both the rifle and the bullet, but the rifle has more mass than the bullet and is therefore accelerated less, as Newton's second law states. (left: FPG International; right: UPI/Corbis-Bettmann)

Figure 5.5 **Newton's third law applied to the launch of a rocket.** Hot gases are forced out of a nozzle (or several, as in this case), and in return they exert a force on the rocket that accelerates it. (NASA)

sound confusing because the acceleration is not necessarily equal, and because in most common situations other forces such as friction complicate the picture. Furthermore, many situations are static: the forces are balanced and no acceleration occurs.

Note that the "action" force and the "reaction" force always act on different bodies. This law is sometimes stated another way, by saying that forces always occur in pairs; a body that exerts a force on another always has an equal and opposite force exerted on it.

The third law can be most easily visualized by considering situations in which friction is not important. As an example, imagine standing in a small boat and throwing overboard a heavy object, such as an anchor. The boat will move in the opposite direction to the anchor, because the anchor exerts a force on you as you throw it. The "kick" of a gun when it is fired is another example of action and reaction. Technically speaking, when a person jumps off the ground by pushing against the Earth, both she and the Earth are accelerated by the forces they exert on each other, but, of course, the immensely greater mass of the Earth prevents it from being accelerated noticeably.

The third law of motion states the principle on which a rocket works. In this case hot, expanding gas is allowed to escape through a nozzle, creating a force on the rocket. The gas is accelerated in one direction, and the rocket is accelerated in the opposite direction **(Fig. 5.5).** Anyone who has inflated a balloon and then let go of it, allowing it to zoom through the air, is familiar with the operating principle of a rocket.

Gravitation and Orbits

We return now to a point raised earlier—namely, that the planets would fly off along straight lines if no force were acting upon them. Newton's first law says that this should happen, yet obviously it does not. Newton realized that the planets must be undergoing constant acceleration toward the center of their orbits, that is, toward the Sun **(Fig. 5.6).** He set out to understand the nature of the force that creates this acceleration. In doing so, he was on the trail of the elusive driving mechanism of the solar system that had so frustrated Kepler.

If you feel confused about the direction of the force that prevents the planets from flying off into space, remember what you have just learned about acceleration and inertia: A planet needs no force to keep it moving, but it does require a force to keep its path curving as it travels around the Sun. What is needed is something to pull the planet inward, toward the Sun. This force can be compared to the tension in the string tied to a rock that you whirl about your head; if you suddenly cut the string, the rock will fly off in whatever direction it happens to be going at the time.

What, then, is the "string" that keeps the planets whirling about the Sun? Newton realized that the Sun itself must be the source of this force, and he made use of Kepler's third law, as well as observations of the Moon's orbit and falling objects at the Earth's surface (such as the famous apple), to discover its properties. He was led to formulate his law of universal gravitation:

Any two bodies in the universe are attracted to each other with a force that is proportional to the product of the masses of the two bodies and inversely proportional to the square of the distance between them.

The law of gravitation is one of the fundamental rules by which the universe operates. As we shall see in later chapters, it explains the motions of stars about each other or about the center of the galaxy, the movements of the Moon and planets, and the motions of galaxies about one another. Gravity, in fact, appears to be the dominant factor that will determine the ultimate fate of the universe. **Table 5.1** shows the relative gravitational forces on a person on Earth due to various bodies in the universe.

It is useful to consider a few examples illustrating how the law of gravitation is applied. The mass of a spherical body acts as though it were concentrated at a single point at the center, something that Newton showed, but only after laborious efforts and the development of new methods of calculus. This means that the weight of an object on the Earth is simply the gravitational force between it and the Earth when they are separated by a distance equal to the Earth's radius. If, for example, we visited a planet with the same mass as the Earth but twice the radius, our weight would decrease by a factor of $2^2 = 4$. On another planet with the same mass as the Earth but only one-third the radius, we would weigh 9 times more than we do on the Earth. If we climb to the top of a high mountain on the Earth, our weight decreases, but not very much, because even the highest mountains are small compared with the radius of the Earth. The Earth exerts a gravitational force on an astronaut in orbit, but the spacecraft and the

Table 5.1
Gravitational Forces Acting on a Human Standing on Earth

Source of Force	Relative Strength of Force
Earth	1.0
Moon	3.4×10^{-6}
Sun	8.6×10^{-4}
Venus (at closest approach)	1.9×10^{-8}
Jupiter (at closest approach)	3.3×10^{-8}
Nearest star	1.4×10^{-14}
Milky Way galaxy	2.1×10^{-11}
Virgo cluster of galaxies	10^{-15}

astronaut are accelerated together, so the astronaut feels no acceleration relative to the spacecraft. It is commonly said that an astronaut is weightless, but in fact this is true only in the local environment of the spacecraft; the Earth's gravity still attracts the astronaut, who has weight in the reference frame of the Earth. A better way of looking at this situation is to say that the astronaut and the spacecraft are in free fall, meaning that they are accelerating together so that there is no relative acceleration between them

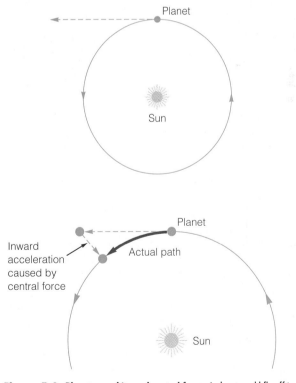

Figure 5.6 Planetary orbits and central force. A planet would fly off in a straight line if no force attracted it toward the center of its orbit. From diagrams like the lower one shown here, Newton was able to determine the amount of acceleration required by an orbiting body to keep it in its orbit.

(Fig. 5.7). One can achieve the same state of "weightlessness" at the Earth's surface by falling in a container such as an elevator (but this might not be a very safe experiment!).

The law of gravitation states that the force also depends on the masses of the two objects that are attracting one another. It is easy to imagine that a person's weight would double if his or her mass doubled; similarly, the force between the Earth and the Moon depends on the masses of these two bodies. If we triple the mass of the Moon, the force is tripled; if we triple the mass of the Moon but decrease the mass of the Earth by a factor of 3 at the same time, the force remains unchanged.

Although it is useful for the purpose of illustration to work out simple thought examples as we have just done, a more quantitative approach is needed to deal with more realistic situations. As we might expect, the law of universal gravitation can be written in the form of an equation:

$$F = \frac{Gm_1m_2}{r^2} \, ,$$

where F is the strength of the force between two objects whose masses are m_1 and m_2 and whose separation is r. The symbol G represents a constant that is required for the value of F to be expressed in normal units of force. In the system of units most commonly used by physicists (called the **Système Internationale,** or **SI** system), force is measured in **newtons,** where 1 newton (N) is the force exerted on a mass of 1 kilogram under an acceleration of 1 meter per second per second. A person who weighs 150 pounds (or has mass of 67 kg) has a weight in this system of 656.6 N, so we see that a newton is a rather small unit of force. The value of the constant G is 6.67×10^{-11} N·m^2/kg, where the masses are expressed in kilograms, the separation between them in meters, and the force in newtons.

The law of universal gravitation can be used to show something that Galileo had postulated long before Newton's time: namely, that the acceleration due to gravity is independent of the mass of an object. Suppose, in the previous equation, that m_1 represents the mass of the Earth, and m_2 the mass of an object falling at the Earth's surface. In that case, r is the Earth's radius because, as already noted, the mass of a spherical body acts as though it were concentrated at a central point.

Now recall that the acceleration on a body is equal to the force applied to it, divided by its mass; hence, the acceleration of the falling object is F/m_2. If we call this acceleration g, because it is due to gravity, then we have

$$g = \frac{F}{m_2} = \frac{Gm_1}{r^2} = \frac{GM}{R^2} \, ,$$

since the symbols M and R are generally used for the mass and radius of a planetary body. At the surface of the Earth, the value of g is 9.80 m/sec^2, as discussed earlier in this chapter. To find the surface gravity on some other planet or satellite, the same expression for g can be used, along with the appropriate values for the mass of the planet M, and its radius R (**Table 5.2**). For example, the Moon has 0.2732 times the radius of the Earth and 0.0123 times the mass. Thus, the acceleration of gravity at the Moon's surface is $0.0123/(0.2732)^2 = 0.165$ that at the surface of the Earth. Therefore, astronauts on the Moon weigh approximately one-sixth as much as they do on the Earth.

Figure 5.7 Weightlessness. Although astronauts orbiting the Earth are subject to the Earth's gravitational force and are "falling around" the Earth along with their spacecraft, they are weightless because they experience no gravitational acceleration relative to their surroundings. This photo shows two astronauts aboard *Skylab*, an orbiting scientific research station. (NASA)
http://universe.colorado.edu/fig/5-7.html

Energy, Angular Momentum, and Orbits: Kepler's Laws Revisited

Even though Newton made use of Kepler's third law in deriving the law of gravitation, the latter is in fact more fundamental. It was soon possible for Newton to show that all three of Kepler's laws follow directly from Newton's laws of motion and gravitation. Kepler's studies of planetary motions had revealed the result of the laws of motion and gravitation, while Newton found the *cause* of the motions. To appreciate how he accomplished this, we must further discuss some basic physical ideas.

An important concept in understanding not only orbital motions but also many other aspects of the universe is **energy.** In an intuitive sense, energy may be defined as the ability to do work. Energy can take many possible forms, such as electrical energy, chemical energy, heat, and others. All forms of energy can be classified as either **kinetic energy,** which is the energy of motion, or **potential energy,** which is stored energy that must be released (i.e., converted to kinetic energy) if it is to do work. A speeding car has kinetic energy because of its motion; a tank of gasoline has potential energy in the form of its chemical reactivity, a tendency to produce large amounts of kinetic energy if ignited. Thus a car operates when this potential energy is converted to kinetic energy in its cylinders.

The units used for measuring energy can be expressed in terms of the kinetic energy of specified masses moving at specified speeds. The kinetic energy of a moving object is $\frac{1}{2}mv^2$, where m is its mass and v its speed. In the SI system, the unit of energy is the joule, a small amount of energy that is equal to the kinetic energy of a mass that is accelerated over distance of 1 meter by a force of 1 newton. A 67 kg person walking at a brisk pace (2 m/sec) has a kinetic energy of 135 joules.

We often speak in terms of **power,** which is simply energy expended per second. The SI system unit for power is the **watt,** which is 1 joule per second. Thus we will speak of the power (or, equivalently, the **luminosity**) of a star in terms of its energy output in watts, or joules per second. In this system the Sun has a luminosity of about 4×10^{26} watts.

Using our understanding of energy, we can now discuss orbital motions in a much more general way than in our previous discussions. Two objects subject to each other's gravitational attraction have kinetic energy due to their motions and potential energy due to the fact that each experiences a gravitational force. Just as a book on a table has potential energy that can be converted to kinetic energy if it is allowed to fall, an orbiting body also has potential energy by virtue of the gravitational force acting on it.

Newton's laws and the concepts of kinetic and potential energy can be used to show that many types of orbits are possible when two bodies interact gravitationally **(Fig. 5.8).** Not all the orbits are ellipses, because if one of the objects has too much kinetic energy (exceeding the potential energy due to the gravity of the other), it will not stay in a closed orbit. Instead it will follow an arcing path known as a **hyperbola** and will escape after one brief encounter.

Table 5.2	
Surface Gravities on Solar System Bodies	
Body	**Surface Gravity** $(g)^a$
Earth	1.0
Sun	27.9
Moon	0.17
Mercury	0.38
Venus	0.90
Mars	0.38
Ceres (largest asteroid)	0.000167
Jupiter	2.64
Saturn	1.13
Uranus	0.89
Neptune	1.13
Pluto	0.08?

$^a g = 9.80$ m/sec^2.

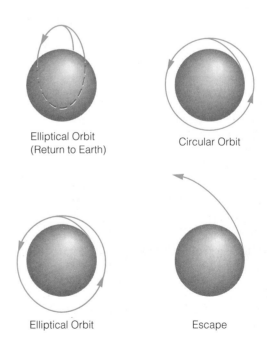

Elliptical Orbit (Return to Earth)

Circular Orbit

Elliptical Orbit

Escape

Figure 5.8 **Orbital and escape velocities.** A rocket launched with insufficient speed for a circular orbit would tend to orbit the Earth's center in an ellipse, but would intersect the Earth's surface. Given the correct speed, the rocket will follow a circular orbit. A somewhat higher speed will place it in an elliptical orbit that does not intersect the Earth. Given enough speed so that its kinetic energy is greater than its gravitational potential energy, however, the rocket will escape entirely.

Communications
Satellites

Kepler's third law can be used to develop a concept that has had profound influence on modern society: the communications satellite. Perhaps you have a satellite dish in your backyard, carefully aligned to receive television signals from one of these satellites, or perhaps you subscribe to a cable company that uses large dishes for the same purpose. Even if you do not watch television, you may make telephone calls that are relayed by communications satellites, or you may read newspapers containing text and images that were transmitted from around the world via these satellites. We live in an age of instant communication, made possible by satellites orbiting high above the Earth.

The key to the communications satellite concept is the **geosynchronous orbit,** in which the orbital period of the satellite is synchronized with the rotation of the Earth; that is, the orbital period is 24 hours. Thus the satellite hovers over a fixed location on the Earth's surface, keeping pace with the spin of the planet as it moves in its orbit. This idea was first published in the 1950s by the science fiction author Arthur C. Clarke, who later lamented that he could have gotten rich by

patenting the concept, except that at the time there was no way to demonstrate it. Normally, the orbital plane chosen for a communications satellite parallels the equator; otherwise the satellite would undergo alternate excursions to the north and south of the equator during each orbit, and this would negate the advantage of the 24-hour period. What is this advantage? It lies in the fact that the satellite remains fixed in the sky from the vantage point of someone on the Earth. This is why your backyard satellite dish only has to be aligned once and then will maintain its pointing angle toward the satellite that is targeted. Furthermore, the signal you receive has to be sent up to the satellite in the first place and then relayed to the ground. Because the satellite hovers over a specific spot, the transmitter that sends the signal up to the satellite can be fixed in location as well.

We can use Kepler's third law to calculate the altitude of a communications satellite. We know that the period has to be 24 hours, or 86,400 seconds. We also know that the term for the sum of the masses, which represents the mass of the Earth plus the mass of the satellite, is essentially equal to the mass of the Earth alone, or 5.974×10^{24} kg. Now we can solve the equation for a, the semimajor axis of the satellite's orbit:

$$a = \left(\frac{GMP^2}{4\pi^2} \right)^{1/3},$$

where M is the Earth's mass and P is the period of the satellite. Note that raising a quantity to the ⅓ power is the same as taking the cube root. Substituting the values for P and M given above, we find that

$$a = 4.224 \times 10^7 \text{ m} = 42,240 \text{ km.}$$

This is the altitude relative to the center of the Earth. Subtracting the Earth's radius of 6,367 km, we find that the communications satellite orbits nearly 35,900 km (22,200 miles) above sea level. This is more than 6.5 times the Earth's radius from the Earth's center, or more than one-tenth of the way from the Earth to the Moon.

A typical orbital altitude for the space shuttle or other near-Earth satellites is about 500 km, and the orbital period is around 90 minutes. Needless to say, a communications satellite that fails cannot be rescued by shuttle astronauts. On the other hand, at such high altitudes there is no residual atmosphere, and an orbit can remain stable indefinitely (sometimes low Earth-orbiting satellites experience drag that causes them to spiral into the upper atmosphere).

The next time you view a program on television that is "brought to you by satellite," you may want to contemplate the mechanics involved and realize how seemingly esoteric concepts such as Kepler's third law can help you understand the technology by which we live.

Some comets have so much kinetic energy that after one trip close to the Sun, they escape forever into space, following hyperbolic paths.

If the kinetic energy is less than the potential energy, as it is for all of the planets, then the orbit is an ellipse, as Kepler found. It is technically correct to say that a planet and the Sun orbit a common **center of mass (Fig. 5.9),** rather than saying that the planet orbits the Sun. The center of mass is the point in space between the two bodies where their masses are

essentially balanced; more specifically, it is the point where the products of mass times distance from this point for the two objects are equal. An Earth-based analogy is a seesaw that is balanced by two people sitting on it; the balance point is closer to the heavier (more massive) person. Since the Sun is so much more massive than any of the planets, the center of mass for any Sun-planet pair is always very close to the center of the Sun, so the Sun moves very little, and we do not easily see its orbital motion (only

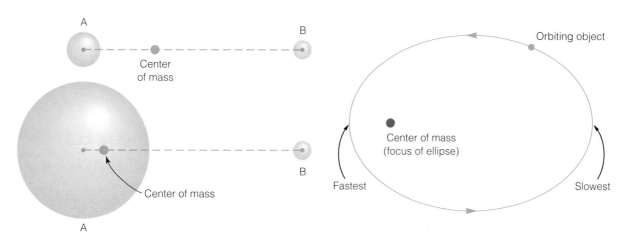

Figure 5.9 **Center of mass.** The upper sketch depicts a double star where star A has twice the mass of star B, so the center of mass, about which the two stars orbit, is one-third of the way between the centers of the two stars. In the lower sketch, star A has 10 times the mass of star B, so the center of mass is very close to star A. The Sun is so much more massive than any of the planets that the center of mass for any Sun-planet pair is near the center of the Sun, so the Sun's orbital motion is very slight. http://universe.colorado.edu/fig/5-9.html

Figure 5.10 **Conservation of angular momentum.** An object moves in an elliptical orbit with varying speed, and the product of its mass times its velocity times its distance from the center of mass (that is, its angular momentum) is constant (this statement is precisely correct only for a circular orbit; for an ellipse, the velocity in question is the component that is at right angles to the Sun-planet line). Kepler's second law of planetary motion is a statement of this fact.

Jupiter is sufficiently massive that the center of mass between it and the Sun lies barely above the Sun's surface). It is true, however, that the Sun's position wiggles a little as it orbits the centers of mass established by its interaction with the planets, especially the most massive ones. In a double star system, where the two masses are more nearly equal, it is easier to see that both stars orbit a point in space between them.

Thus Kepler's first law as he stated it requires a slight modification:

Each planet has an elliptical orbit about the center of mass between it and the Sun, with the center of mass at one focus.

The second law can also be restated in terms of Newton's mechanics. Any object that either rotates or moves around some center has **angular momentum.** This depends on the object's mass, its velocity, and its distance from the center of mass. In the simple case of an object in circular orbit, the angular momentum is the product mvr, where m is its mass, v its speed, and r its distance from the center of mass.

The total amount of angular momentum in a system is always constant. Because of this, a planet in an elliptical orbit must move faster when it is close to the center of mass than when it is farther away, so that its velocity compensates for the changes in distance **(Fig. 5.10).** Thus a planet moves faster in its orbit near perihelion (its point of closest approach to the Sun) than at aphelion (its point of greatest distance from the Sun). Kepler's second law is really a state-

ment that angular momentum is constant for a pair of orbiting objects.

Newton also revised Kepler's third law—a revision that is especially important for our studies. Newton discovered that the relationship among the period and the semimajor axis depends on the masses of the two objects. Kepler had not realized this, primarily because the Sun is so much more massive than any of the planets that the differences among the masses of the planets have only a very small effect. Kepler's form of the third law can be written $P^2 = a^3$, where P is the planet's period in years and a is the semimajor axis in astronomical units. Newton revised this to

$$(m_1 + m_2)P^2 = a^3,$$

where m_1 and m_2 represent the masses of the two bodies, for example, the Sun and one of the planets. The masses must be expressed in terms of the Sun's mass in this equation. If we use other units, such as kilograms for the masses, seconds for the period, and meters for the semimajor axis, the equation is complicated by the addition of a numerical factor and is written

$$(m_1 + m_2)P^2 = \frac{4\pi^2}{G} a^3,$$

where G is the gravitational constant.

Watching Satellites
In this exercise, we'll examine on-line information about some of the many satellites that orbit the earth. We'll learn about the different orbits that they travel in, and how the shapes of these orbits can be determined by the simple laws of physics discussed in the chapter. Our starting point for satellite watching is the following URL
http://universe.colorado.edu/ch5/web.html

The practical importance of Newton's generalized revision of Kepler's third law is that it can be used to determine the masses of distant bodies. In any situation where the period and the semimajor axis of an orbiting object can be observed directly, the equation can be solved for the sum of the masses (again using solar units):

$$m_1 + m_2 = \frac{a^3}{P^2}.$$

For example, consider how Newton could have derived the mass of Saturn. Titan, the largest of Saturn's moons, has an orbital period of $P = 15.945$ days $= 0.04365$ years and a semimajor axis of $a = 1,222,000$ km $= 0.00816$ AU. Substituting these values for P and a in the above expression for the sum of the masses yields $m_1 + m_2 = 0.000285$ solar masses; in other words, the sum of the masses of Saturn and Titan is 0.000285 times the mass of the Sun. Since Titan is very much smaller than Saturn, we can assume that its mass is negligible and conclude that we have found the mass of Saturn (compare this with the value for the mass of Saturn in Appendix 8).

The Laws of Motion Applied to Gas Particles

Now that we have an understanding of the laws of motion and gravitation, we can see how they may be applied in several instances that will be important for our later discussions. Among these are environments, such as the atmospheres of planets, where the motions of atomic or molecular particles become important.

Escape Speed

The consideration of orbital motions in terms of kinetic and potential energy leads to the concept of an **escape speed.** If an object in a gravitational field has greater kinetic than potential energy, it can escape the gravitational field entirely. To launch a rocket into space (that is, completely free of the Earth) therefore requires giving it enough upward speed at launch to make its kinetic energy greater than its potential energy due to the Earth's gravitational attraction (see Fig. 5.8). The speed required to accomplish this is the same for an object of any mass, and in equation form it is

$$v_e = \sqrt{\frac{2GM}{R}},$$

where v_e is the escape velocity, G is the gravitational constant, and M and R are the Earth's mass and radius, respectively. For the Earth, this speed is 11.2 km per second, or just over 40,000 km per hour. Escape speeds for all the planets are given in Appendix 8.

Molecular Motion, Temperature, and Atmospheric Escape

The particles in a gas such as the Earth's atmosphere move all the time, with a range of speeds. The range of speeds, and the average and most probable speeds as well, are closely related to the temperature of the gas **(Fig. 5.11).** In a hot gas, the atoms and molecules move more rapidly than in a cool gas. This is why a heated gas expands; the individual particles move more rapidly and therefore exert greater pressure on their surroundings. In a strict sense, temperature can be defined in terms of molecular motion (this leads directly to the concept of **absolute zero,** the temperature at which all molecular activity ceases). We can speak of the kinetic energy of individual particles in a gas or, more realistically, of the average kinetic energy, which in turn is related to the average particle speed. For any gas, the temperature is proportional to the average kinetic energy.

In physics and astronomy, temperatures are usually expressed in terms of the absolute scale, in which zero is absolute zero, and the degrees, equal to one one-hundredth of the difference between the freezing and boiling points of water, are called **kelvins** (K). Water freezes at 273 K and boils at 373 K.

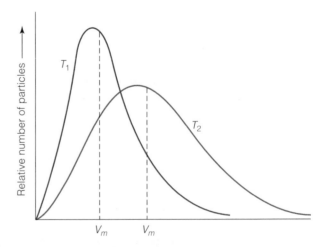

Figure 5.11 Particles speeds in a gas. These curves show the relative number of gas particles moving at different speeds. The curve farther to the right represents a hotter gas, with higher speeds. The vertical lines show the average particle speed in each case. Note that in each case there are many particles moving faster than the average speed.

Comfortable room temperature is about 295 *K*. For details, see Appendix 5.

When we discuss the planets, we will be concerned with the likelihood that specific gases will escape into space, thus leaving the atmosphere devoid of those gases. The probability that a gas will escape depends on how the average speed of the particles compares with the escape speed. The average speed of particles in a gas depends on the temperature, as we have just seen, and also on the mass of the particles. The hotter the gas, the faster the average speed; but the higher the particle mass, the lower the average speed. At a certain fixed temperature, the lighter-weight particles will be moving faster on average than the heavier ones, so the lightweight gases may escape an atmosphere while the heavier ones remain.

The graph in Figure 5.11 shows that some particles are always moving at higher speeds than the average. Thus, even if the average speed is well below the escape speed for a planet, some particles will exceed the escape speed and will leave. Particle collisions will replace the particles that leave; that is, other particles in the gas will be accelerated to high speeds, as the gas remains in balance with its temperature. Thus additional particles will escape. Detailed calculations show that a gas will eventually escape completely if the average particle speed is greater than one-sixth of the escape speed. Later we will apply this criterion when we discuss why certain gases remain while others have escaped the atmospheres of the planets.

Tidal Forces

Tides are another important astronomical phenomenon that can be understood in terms of the laws of motion and gravitation. To understand tides, however, we have to think not of how a gravitational force affects a body as a whole, but rather of how it affects different points on a body.

We have seen that the gravitational force due to a distant body decreases with the square of the distance. This means that an object subjected to the gravitational pull of such a body feels a stronger pull on the side nearest that body and a weaker pull elsewhere. For example, the side of the Earth facing the Moon feels a stronger attraction toward the Moon than do other points on the Earth, and a point on the opposite side from the Moon feels a weaker force than a point on the near side. The Earth is therefore subjected to a **differential gravitational force,** which tends to stretch it along the line toward the Moon **(Fig. 5.12)**. Of course, the Sun exerts a similar

stretching force on the Earth, but because of its distance, it does not have as strong an effect as the Moon (even though the Sun's *total* gravitational force on the Earth is much greater than that of the Moon). A **tidal force,** as differential gravitational forces are called, depends on how close one body is to the other, because the key is how rapidly the gravitational force drops off over the diameter of the body subject to the tidal force, and it drops off most rapidly at small distances. (In contrast to the total gravitational force, the tidal force decreases as the third power, not the square, of the separation between two bodies.)

The Earth is a more or less rigid body, so it does not stretch very much due to the differential gravitational force of the Moon. Nevertheless, the tidal forces exerted on the Earth by the Moon create net forces that tend to make the liquid oceans flow toward the points facing directly toward the Moon and directly away from it. As the Earth rotates, the water in the oceans tends to follow the tidal forces created by the Moon, so that in effect the oceans have two bulges of water that remain fixed in position relative to the Earth-Moon direction, while the Earth rotates. The effect that we see, as we move with the Earth's rotation, is two huge ridges of water that appear to flow around the Earth as it rotates. Since the two ridges of water are on opposite sides, and the Earth rotates in 24 hours, one of these ridges passes any given point on the Earth's surface every 12 hours. Thus we have the ocean's tides, with high tides occurring at any given location about every 12 hours, separated by low tides. The time between successive high tides is actually a bit longer than 12

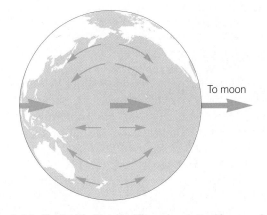

Figure 5.12 The Earth's tides. The differential gravitational force caused by the Moon tends to stretch the Earth (large arrows). Seawater at any given point on the Earth is subjected to a combination of vertical and horizontal forces, causing it to flow toward either the side of the Earth facing the Moon or the side opposite it (curved arrows).

http://universe.colorado.edu/fig/5-12.html

hours because the Moon moves along in its orbit while the Earth rotates.

It is interesting to consider what is happening to the Moon at the same time. It is subjected to a more intense differential gravitational force than the Earth because the Earth is more massive than the Moon. Even though the Moon is a solid body, its shape is deformed by this force, and it has tidal bulges. The Moon is slightly elongated along the line toward the Earth. This has had drastic effects on the Moon's rotation, causing it to keep one side facing the Earth **(Fig. 5.13).** The original spin of the Moon was probably much faster than it is today, but tidal forces have slowed the spin so that the rotation period and the orbital period have become equal. This situation, called **synchronous rotation,** is also characteristic of many other satellites in the solar system.

The tides are also gradually slowing the rotation of the Earth. Given enough time, the Earth will come into synchronous rotation with the Moon. But this will take a very long time; the rate of slowing is only about 0.001 seconds per century (this was discussed in the Science and Society box on timekeeping in Chapter 2).

There are many other examples of tidal forces, both in our solar system and outside it. The satellites of the massive outer planets are subjected to severe tidal forces, and in some double star systems, the two stars are so close together that they are stretched out of round. We will discuss these in more detail later along with star clusters and galaxies that are affected by tidal forces, sometimes even to the extent of tearing each other apart.

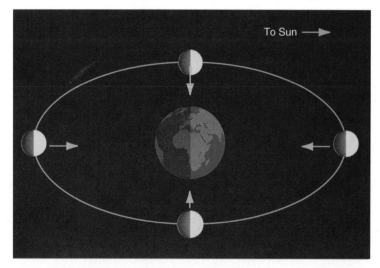

Figure 5.13 Synchronous rotation of the Moon. The arrow, fixed to a specific point on the Moon, illustrates that the Moon spins once during each orbit of the Earth. Thus the Moon keeps the same side facing the Earth at all times. (Not to scale.)

Summary

1. The British scientist Isaac Newton made important advances in optics, mechanics, mathematics, and astronomy, establishing the mathematical framework in which the motions of celestial bodies could be explained.

2. Newton's optics, particularly the discovery that white light consists of all the colors and his invention of the reflecting telescope, first brought him fame even though most of his work on gravitation and mechanics had already been completed earlier.

3. In the mid-1660s, Newton developed the laws of motion, invented the calculus, and applied both to the motions of celestial bodies. He derived the law of universal gravitation as a result of comparing the acceleration required to keep the Moon and planets in their orbits with the acceleration of falling bodies at the Earth's surface.

4. Newton's first law of motion describes the concept of inertia, the tendency of a body to remain in a state of rest or uniform motion unless an unbalanced outside force acts on it.

5. Newton's second law states that acceleration is proportional to the force exerted on a body and inversely proportional to its mass; the law is often rewritten mathematically as $F = ma$.

6. The third law states that for every action there is an equal and opposite reaction, meaning that forces occur in pairs such that when a force is applied to a body, the body exerts an equal force in the opposite direction.

7. The law of universal gravitation states that any two objects in the universe attract each other with a force that is proportional to the product of their masses and inversely proportional to the square of the distance between them.

8. Newton's laws, along with the concepts of kinetic and potential energy and angular momentum, can be used to explain orbital motions. Newton was able to derive all of Kepler's laws of planetary motion and discovered new properties of the laws that Kepler had not known.

9. Newton expanded Kepler's third law to show that the relationship between the period and the semimajor axis of an orbit is dependent on the sum of the masses of the two objects. This is an important tool for measuring the masses of distant objects.

10. Every body such as a planet or a star has an escape velocity, the speed at which a moving object has more kinetic energy than potential energy and will therefore escape into space. The speeds of atmospheric particles, which depend on the temperature and the particle mass, determine which gases may escape from a planet's atmosphere.

11. Differential gravitational forces are responsible for tides on the Earth and in the interiors of other planets and satellites. One common phenomenon that results from tidal forces is synchronous rotation.

Review Questions

1. Both Galileo and Newton became involved in somewhat public disputes concerning their scientific discoveries. Compare and contrast the controversial aspects of the work of these two men.

2. Explain, in your own words, the difference between weight and mass.

3. Why does the law of inertia require that there must be a force attracting each planet toward the Sun?

4. Explain the difference between acceleration and velocity and also between velocity and speed.

5. Explain how a book lying on a table can be stationary, even though it exerts a force on the table and the table exerts a force on it.

6. Explain why an astronaut in an orbiting spacecraft is said to be "weightless." Is there really no force of gravity acting on the astronaut?

7. A rocket is launched from the Earth, but falls back without escaping. What can you say about the relationship between its kinetic and potential energy at the time of launch? What would the relationship be if the rocket were able to escape from the Earth's gravity?

8. Why did Kepler not discover that his third law depends on the sum of the masses of the Sun and each planet?

9. Describe the roles of temperature and surface gravity in the ability of a planet to retain an atmosphere.

10. Explain why the Earth's oceans have two tidal bulges on opposite sides of the Earth.

Problems

1. Suppose you are a construction worker, laboring in Earth orbit to build the Space Station. You are free-floating (space-walking) outside the station, and you have to move a massive beam, also free-floating. The beam has a mass of 1,000 kg (10 times your mass), and you push it with a force of 200 N. What is the acceleration of the beam? What is your acceleration after you push the beam?

2. Jupiter is about five times farther from the Sun than the Earth is, and its mass is about 300 times the Earth's mass. Compare the gravitational force between Jupiter and the Sun with that between the Earth and the Sun. Perform the same calculation for Saturn, which has about 100 times the mass of the Earth and orbits about 10 times farther from the Sun.

3. Calculate the weight, in newtons and pounds, of a person whose mass is 65 kg. To find the weight in newtons, use the law of universal gravitation, with the mass of the Earth and the mass of the person as the two masses and the radius of the Earth as the distance between them. Then convert from newtons to pounds, using the conversion factor 1 N = 0.225 lb. Finally, compare your answer with the value you would have gotten by simply converting kilograms to pounds using the conversion factor 1 kg = 2.2 lb.

4. How much would you weigh on the summit of Mt. Everest, which has an altitude above sea level of 29,000 feet (8.84 km), if your sea-level weight is 150 pounds? Find the radius of the Earth in Appendix 8.

5. What would you weigh on the surface of Titan, the giant satellite of Saturn? Use data from Appendix 8.

6. Suppose you drop a rock from a high cliff. How fast is the rock falling after 3 seconds? After 10 seconds? If the mass of the rock is 2 kg, what is its kinetic energy after 3 seconds of falling? After 10 seconds?

7. Calculate the mass of Jupiter, in solar units, using Kepler's third law and data from Appendix 8 on the orbit of Callisto. Convert your answer from solar units to kilograms. How does your answer compare with the value given in Appendix 8?

Web Connections
The Review Questions and Problems also appear at the following URLs:
http://universe.colorado.edu/ch5/questions.html
http://universe.colorado.edu/ch5/problems.html

Center of Mass and Orbits

ACTIVITY

A planet does not orbit the Sun; instead it orbits a point in space called the *center of mass*. So does the Sun. The center of mass for any Sun-planet pair is close to the center of the Sun, so the Sun's orbital motion is small. In a double star system, where the masses of the two orbiting bodies are more nearly equal, the center of mass lies closer to the midpoint between the two stars.

To find the center of mass between two bodies, we use the formula

$$m_1 r_1 = m_2 r_2,$$

where m_1 and m_2 are the two masses and r_1 and r_2 are their distances from the center of mass. This can be rearranged to read

$$\frac{r_1}{r_2} = \frac{m_2}{m_1}.$$

From this you see that the center of mass is closer to the more massive of the two bodies. It may help to think of two people of very different weights, balancing on a see-saw. The heavier person will have to be closer to the pivot point in order to achieve a balance. Similarly, in a pair of orbiting bodies, the more massive will be closer to the center of mass.

If one mass is twice the other, then the ratio of distances is 1:2, and the center of mass is one-third of the way from the more massive body to the less massive one. If the ratio of masses is 3:1, then the center of mass is one-fourth of the distance between the two bodies. In general, the center of mass is $\dfrac{1}{1 + (m_2/m_1)}$ of the way from body 2 toward body 1.

For the Sun and Jupiter, the most massive planet, the ratio of masses is approximately 1:1,047, so we find that the ratio of distances from the center of mass is 1,047:1.

Thus the Sun's distance from the center of mass is approximately $1/(1 + 1,047)$ of the Sun-Jupiter distance, which is 5.2 AU. This places the center of mass 5.2 AU/1,048 = 0.005 AU, or 7.5×10^8 m, from the center of the Sun. The Sun's radius is 7.0×10^8 m, so the center of mass of the Sun-Jupiter pair lies just outside the solar surface. As Jupiter and the Sun orbit, Jupiter follows a nearly circular path with a radius of 5.2 AU and a period of 11.86 years, while the Sun follows a nearly circular path about a point lying just outside its surface, also with a period of 11.86 years. From a distance the motion of Jupiter is quite obvious while that of the Sun is not.

You can construct an artificial representation of orbital motion about a fixed center of mass, but it is easier to do for a pair of bodies whose mass ratio is not as large as the Sun-Jupiter pair. Use a lightweight wooden meter stick (one that you won't mind damaging). Attach weights at the ends, taping coins or heavy metal washers in a ratio that will represent the ratio of the masses of your two bodies. The demonstration will work best if you have a mass ratio of at least 2 or 3. Then find the balance point of the meter stick and drill a hole through it at this point, which represents the center of mass.

Now hold the meter stick parallel to the ground and put a pencil or nail (or a finger, if the hole is big enough) through the hole you made. Give the stick a spin and watch the motions of the ends where the weights are located. You should observe that (1) the center of mass remains fixed, with no unbalanced force (as long as you hold it steady against gravity) acting on it; (2) the ratio of the distances of the weights from the center of mass is inversely proportional to the weights (this will be only approximately true, as the weight of the meter stick is not zero); (3) the weights describe circular orbits about the center of mass, with orbital radii that are inversely proportional to the masses; and (4) the weights are always opposite each other in their orbits. This is a good representation of the motion of a pair of stars in a binary system, except that there is no material equivalent of the meter stick in a binary star system.

8. Calculate the escape speed for Mars, and compare it with the escape speed for the Earth.

9. Consider a planet with a mass 2.3 times the Earth's mass and a radius 1.6 times the Earth's radius. If the average speed of hydrogen molecules in the atmosphere of this planet is 2.9 km/sec, will all of the hydrogen escape into space? Explain.

Additional Readings

There are many books on conceptual understandings of physics in which more detailed explanations of the principles discussed in this chapter may be found. Many are used as textbooks in introductory physics courses, and should be easy to locate in a bookstore or library. Recent articles:

Christianson, G. E. 1987. Newton's *Principia:* A Retrospective. *Sky & Telescope* 74(1):8.

Drake, S. 1980. Newton's Apple and Galileo's Dialogue. *Scientific American* 243(2):150.

Chapter 6
Light and Telescopes

*The microscope enlarges worlds we are too big to step into . . . .
Telescopes reveal worlds too young for us to fit into.*

Dusty Joseph Partello, 1907

Chapter web site: http://universe.colorado.edu/ch6

ome of the tools for unlocking the secrets of the universe became available with the publication of Newton's *Principia* in 1687, but others had to wait two hundred years or more to be discovered. The laws of motion allowed astronomers to understand how the heavenly bodies move, and they were of fundamental importance in unraveling the clockwork mechanism of the solar system.

The only information we can obtain on the nature of a distant object is conveyed by the light that reaches us from it. Fortunately, this light contains an enormous amount of information, and astronomers have learned a lot about how to extract it. Thus, to understand the essential nature of a distant object—to learn what it is made of and what its physical state is—one must understand what light is and how it is emitted and absorbed. It is also necessary to have tools that can capture the light and analyze it.

The Nature of Light

By the late seventeenth century, several experiments on the properties of light were being performed, most notably by Newton and Huygens, who developed opposing points of view. Newton developed his "corpuscular" theory of light, which holds that light consists of a stream of tiny particles, while Huygens, noting some similarities between the behavior of light and waves in water, adopted the view that light consists of waves. More than two centuries would pass before scientists reached an understanding of these contrasting viewpoints.

Wave or Particle?

One characteristic of light is that it acts like a wave **(Fig. 6.1).** It is possible to think of light as passing through space like ripples on a pond (although, as we will discuss shortly, the picture is actually somewhat more complicated). The distance from one wavecrest to the next, called the **wavelength,** distinguishes one color from another. Red light, for example, has a longer wavelength than blue light. It is possible to spread out the colors in order of wavelength, using a prism to obtain the traditional rainbow. Newton was the first to discover that sunlight contains all the colors, and he did so by carrying out experiments with a prism **(Fig. 6.2).** Whenever light is spread out by wavelength, the result is called a **spectrum;** that is, a spectrum is an arrangement of light according to wavelength. The science of analyzing spectra is called **spectroscopy** and will be discussed later in this chapter.

The concept of **frequency** is often used as an alternative to wavelength in characterizing light waves. The frequency is the number of waves per second that pass a fixed point. It is determined by the wavelength and the speed with which the waves move. The speed of light, usually designated c, is constant, and the frequency of light with wavelength λ is $f = c/\lambda$. The standard unit for measuring frequency is the **hertz** (Hz), one hertz being equal to one wave per second. The frequency of visible light is typically about 10^{14} Hz.

In some situations, light is more like a stream of particles than a wave. As noted previously, Newton and Huygens adopted different points of view on the nature of light, Newton arguing for a particulate form (a stream of particles), while Huygens favored a wave nature. There are good arguments for both

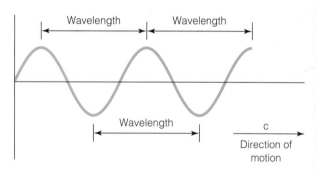

Figure 6.1 Properties of a wave. Light can be envisioned as a wave moving through space at a constant speed, usually designated c. The distance from one wavecrest to the next is the wavelength, often denoted by the Greek letter lambda, (λ). The frequency f is the number of wavecrests that pass a fixed point per second and is related to the wavelength and the speed of light by $f = c/\lambda$.

Figure 6.2 Prism dispersing light. The rainbow of colors seen here is a spectrum. (Runk/Schoenberger, Grant Heilman Photography)

points of view. For example, the manner in which light waves seem to bend as they pass obstacles (called **diffraction**) or pass through a boundary from one medium to another **(refraction),** and the way in which they **interfere** with each other, are wave characteristics. On the other hand, light can carry only discrete, fixed quantities of energy and can travel in a complete vacuum rather than requiring a medium as waves do, and these are properties of particles.

Out of a variety of seemingly contradictory evidence has developed the concept of the **photon.** A photon is thought of as a particle of light that has a wavelength associated with it. The wavelength and the amount of energy contained in the photon are intimately linked; in general terms, the longer the wavelength, the lower the energy. Thus a red photon carries less energy than a blue one. Mathematically, the energy can be expressed as $E = hc/\lambda$, where h is a constant (called the Planck constant), c is the speed of light, and λ is the wavelength. It is important to understand that a photon carries a precise amount of energy, not some arbitrary or random quantity, and that when light strikes a surface, this energy arrives in discrete bundles like bullets, rather than as a steady stream. When a photon is absorbed, this energy can be converted into other forms, such as heat.

Nonvisible Radiation and the Electromagnetic Spectrum

Let us consider for a moment what lies beyond red at one end of the spectrum or violet at the other end. By the mid-1800s, experiments had demonstrated that there is invisible radiation from the Sun at both ends of the spectrum. At long wavelengths, beyond red, is **infrared radiation,** and at short wavelengths is **ultraviolet radiation.** Later it was shown that the spectrum continues in both directions without limit. Going toward long wavelengths, after infrared light come **microwaves** and **radio waves;** going toward short wavelengths, after ultraviolet come **X rays** and then **gamma (γ) rays.** All of these different kinds of radiation are just different forms of light, distinguished only by their wavelengths, and together they form the **electromagnetic spectrum (Fig. 6.3).** Electromagnetic radiation is a general term for all forms of light, whether it is visible light, X rays, radio waves, or anything else.

The reason for the name **electromagnetic radiation** is that the wave motion associated with this radiation consists of oscillating electric and magnetic fields **(Fig. 6.4).** These fields propagate through space (or a medium) as they alternate, with the electric and magnetic fields always oscillating in planes

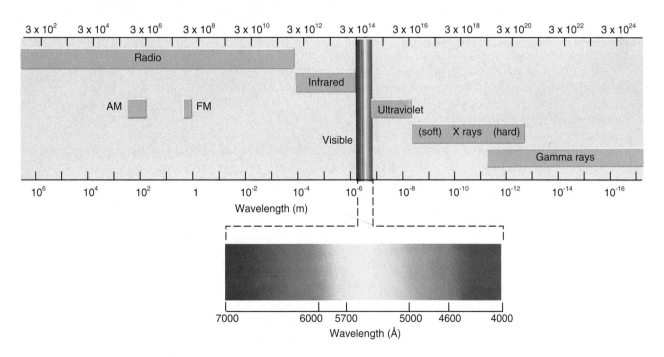

Figure 6.3 The electromagnetic spectrum. All of the indicated forms of radiation are identical except for wavelength and frequency. http://universe.colorado.edu/fig/6-3.html

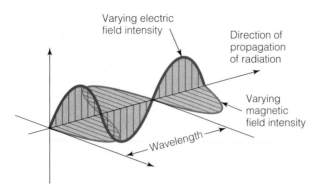

Figure 6.4 An electromagnetic wave. This illustrates how the electric and magnetic fields oscillate in an electromagnetic wave. The directions of the fields and the direction of wave travel always have the relative orientations shown here.

Table 6.1		
Wavelengths and Frequencies of Electromagnetic Radiation		
Type of Radiation	**Wavelength Range**[a]	**Frequency Range**
Gamma rays	$< 10^{-10}$	$> 3 \times 10^{18}$ Hz
X rays	$1-200 \times 10^{-10}$	$1.5 \times 10^{16}-3 \times 10^{18}$
Extreme ultraviolet	$200-900 \times 10^{-10}$	$3.3 \times 10^{15}-1.5 \times 10^{16}$
Ultraviolet	$900-4000 \times 10^{-10}$	$7.5 \times 10^{14}-3.3 \times 10^{15}$
Visible	$4000-7000 \times 10^{-10}$	$4.3 \times 10^{14}-7.5 \times 10^{14}$
Near infrared	$0.7-20 \times 10^{-6}$	$1.5 \times 10^{13}-4.3 \times 10^{14}$
Far infrared	$20-100 \times 10^{-6}$	$3.0 \times 10^{12}-1.5 \times 10^{13}$
Radio	> 0.0001	$< 3 \times 10^{12}$
(Radar)	$(.02-0.2)$	$(1.5-15 \times 10^9)$
(FM radio)	$(2.5-3.5)$	$(85-120 \times 10^6)$
(AM radio)	$(180-380)$	$(540-1600 \times 10^3)$

[a]All wavelengths are given in meters; recall that 1×10^{-10} m = 1 Å and 1×10^{-6} m = 1 micron.

that are perpendicular to each other. The speed of propagation in a vacuum is always the same (almost exactly 300,000 km per second, or about 186,000 miles per second), but it is slightly slower in a medium such as air or glass. The electric and magnetic fields are said to be **in phase** with each other, meaning that both reach their maximum and minimum values at the same time as the wave propagates, with the relative orientation shown in Figure 6.4.

Polarization

The light from a typical source, such as a lightbulb or a distant star, consists of vast numbers of photons, each containing oscillating electric and magnetic fields. Each photon has a characteristic orientation as it travels through space; that is, the planes in which its electric and magnetic fields oscillate remain fixed as the photon travels. Normally, the orientations in a collection of photons from a source are random, but in some circumstances, they are not. When light is **polarized,** there is a preferred orientation of the electric and magnetic fields; that is, most of them have the same alignment. Polarization of light from a source can be caused by reflection from a surface that reflects more efficiently in one orientation than others, by passage through a medium that tends to absorb photons of a certain orientation, or by an emission mechanism that preferentially creates photons of one orientation. As it turns out, all of these processes occur naturally, so polarization of light is an important phenomenon in astronomy.

Measures of Wavelength

Astronomers tend to identify different portions of the electromagnetic spectrum by specialized names such as infrared or ultraviolet, but we must keep in mind that all forms of radiation are the same. All

consist of alternating electric and magnetic fields and differ only in wavelength (or frequency).

The range of wavelengths from one end of the electromagnetic spectrum to the other is immense **(Table 6.1).** Visible light has wavelengths ranging from 0.0000004 (4×10^{-7}) m to 0.0000007 (7×10^{-7}) m. For visible light, and also ultraviolet and to some extent infrared radiation, a special unit called the angstrom (Å) is used, defined such that 1 m = 10,000,000,000 = 10^{10} Å, or 1Å = 0.0000000001 m = 10^{-10} m. Thus visible light lies between 4000 Å and 7000 Å in wavelength. Recently, physicists and astronomers have begun to use the nanometer (nm), which is equal to 10^{-9} m. In terms of this unit, visible light lies between 400 and 700 nm.

Infrared light has wavelengths between 7000 Å and a few million Å; that is, between 7×10^{-7} m and 2 or 3×10^{-4} m. Microwave radiation (which includes radar wavelengths) lies roughly between .001 and 0.5 m; no well-defined boundary separates this region from radio waves, which simply include all longer wavelengths, up to many meters or even kilometers. At the other end of the spectrum, ultraviolet light is usually considered to lie between 200 Å and 4000 Å, while X rays are in the range of 1–200 Å; any shorter wavelengths are considered gamma rays.

Laws of Continuous Radiation

Researchers who followed Newton discovered that the Sun's spectrum contains a number of dark lines, each corresponding to a particular wavelength. These **spectral lines** provide a great deal of information about a source of light such as a star, as does the **continuous radiation (Figs. 6.5** and **6.6),** the smooth distribution of light as a function of wavelength. Here we discuss continuous radiation, and in a later section we consider the spectral lines.

Thermal Radiation and Blackbodies

Any object with a temperature above absolute zero emits radiation over a broad range of wavelengths, simply by virtue of the fact that it has a temperature. Thus not only stars, but also such commonplace objects as the walls of a room or a human body, emit radiation. The continuous radiation produced because of an object's temperature is called **thermal radiation,** and in this chapter we will restrict ourselves to this type of continuous radiation. In other chapters we will discuss nonthermal sources of radiation. Some of the properties of thermal radiation were discovered experimentally in the nineteenth century, before the underlying physical basis for the observed phenomena was developed.

It has been found experimentally and theoretically that for a certain class of emitters, the radiation emitted depends *only* on the surface temperature. The composition does not matter. The bodies that conform to this principle glow only by emitted light; they reflect none of the radiation that may be striking them. Thus they are perfect absorbers and appear pure black to the human eye unless they are hot enough to emit at visible wavelengths. For this reason, such sources are called **blackbodies.** As we are about

to see, many simple laws govern the radiation from a blackbody. Although no pure blackbodies exist in nature, we will find that stars and planets can approximate them to a reasonable degree. Therefore it is fruitful for astronomers to apply laws of blackbody radiation to the objects they observe.

Wien's Law and the Colors of Stars

Glowing objects such as stars emit thermal radiation over a broad range of wavelengths (technically, in fact, at least some radiation is emitted at *all* wavelengths), with a peak in intensity at some particular wavelength. One of the experimental discoveries about thermal radiation was made in 1893 by Wilhelm Wien, who found a simple relationship between the wavelength of maximum emission and the temperature of an object. This relationship now known as **Wien's law:**

The wavelength of maximum emission is inversely proportional to the absolute temperature.

In mathematical terms, this can be expressed as

$$\lambda_{max} = \frac{W}{T},$$

where λ_{max} is the wavelength at which the maximum energy is emitted (corresponding to the peak in Fig. 6.6), T is the surface temperature on the absolute scale (i.e., in units of kelvins), and W is the Wien constant, whose value is 0.0029 m·deg when the wavelength λ_{max} is in units of meters (or $W = 2.9 \times$

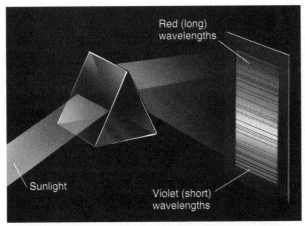

Figure 6.5 Continuous spectrum and spectral lines. When sunlight is dispersed by a prism, the light forms a smooth rainbow of continuous radiation; one color gradually merges into the next, with a maximum intensity in the yellow portion of the spectrum. Superimposed on this continuous spectrum are numerous spectral lines, wavelengths at which little or no light escapes the Sun.

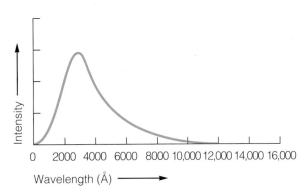

Figure 6.6 An intensity plot of a continuous spectrum. This kind of diagram shows graphically how the brightness of a glowing object varies with wavelength. The curve shown roughly represents a star of surface temperature 10,000 K, whose continuous radiation peaks near 3000 Å in the ultraviolet.

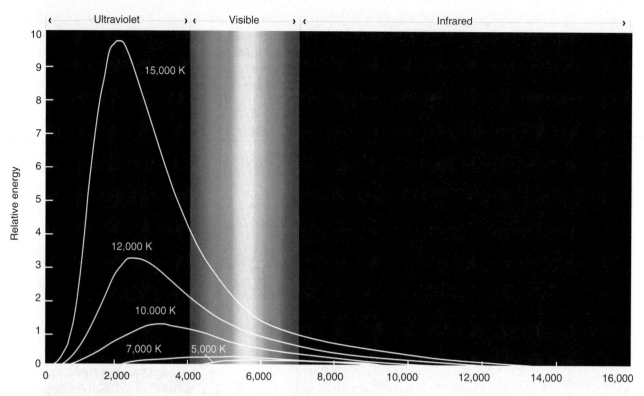

Figure 6.7 **Continuous spectra for objects of different temperatures.** This diagram illustrates Wien's law, which says that the wavelength of maximum emission is inversely proportional to the temperature (on the absolute scale). It also illustrates another property of thermal emission: The hotter of the two objects is brighter at *all* wavelengths. http://universe.colorado.edu/fig/6-7.html

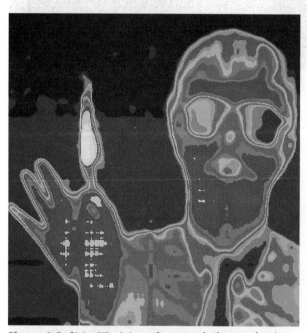

Figure 6.8 **Giving Wien's Law a human touch.** This is an infrared image of a person holding a match. Colors are used here to represent different temperatures: white is the hottest (the flame); deep red, the warmest portions of the person (the palm of the hand and the chin area); and blue the coolest regions (eyeglass and necktie). (NASA/JPL)

10^7 Å·deg for λ_{max} in angstroms). This expression can be used to calculate the wavelength of peak emission for any object that is glowing thermally; that is, whose radiation is due solely to its surface temperature. It is a good approximation for many astronomical objects such as planets and stars.

The hotter an object is, the shorter its wavelength of peak emission, and the cooler it is, the longer its wavelength of peak emission **(Figs. 6.7** and **6.8)**. Thus the variety of stellar colors is due to a range in stellar temperatures **(Fig. 6.9)**. A hot star emits most of its radiation at relatively short wavelengths and thus appears bluish, whereas a cool star emits most strongly at longer wavelengths and appears red. The Sun is intermediate in temperature and in color.

When speaking of the colors of stars, we must keep in mind that a star emits light over a broad range of wavelengths, so we do not have pure red or pure blue stars. Our eyes receive light of all colors, and stars therefore are all essentially white. Our impression of color arises from the fact that there is a wavelength (given by Wien's law) at which a star emits more strongly than at other wavelengths. But the star does not emit *only* at that wavelength.

The Stefan-Boltzmann Law

A second property of glowing objects, known as either **Stefan's law** or the **Stefan-Boltzmann law,** has to do with the total amount of energy per second that is emitted over all wavelengths, and how this total energy is related to the temperature of an object:

The total energy radiated per second per square meter of surface area is proportional to the fourth power of the temperature.

This shows that the total energy emitted per second is very sensitive to the temperature; if you change the temperature by a little, you change the energy by a lot. If, for example, you double the temperature of an object (such as the electric burner on your stove), you increase the total energy per second, or the **power,** that it radiates by a factor of $2^4 = 2 \times 2 \times 2 \times 2 = 16$. If one star is three times hotter than another, it emits $3^4 = 3 \times 3 \times 3 \times 3 = 81$ times more energy per second per square meter of surface area.

To calculate emitted energy in basic units such as joules per second (i.e., watts), a numerical constant term is needed, and the expression for the energy per second, or power, that is emitted per square meter of surface area is

$$E = \sigma T^4,$$

where σ is called the Stefan-Boltzmann constant and has a value of 5.67×10^{-8} W/m²·deg⁴.

Notice that we have been careful to express this law in terms of energy emitted per second per square meter of surface area. This indicates that the *total* energy emitted per second by an object, usually called the **luminosity** of the object, depends on how much surface area it has (note that luminosity is the same thing as **power;** see Chapter 5). For spherical objects such as stars or planets, the surface area is $4\pi R^2$, where R is the radius. Then the luminosity is equal to the surface area times the energy emitted per second per square meter of surface. In mathematical form, this translates to

$$L = 4\pi R^2 \sigma T^4,$$

where the term $4\pi R^2$ represents the surface area and the term σT^4 represents the energy emitted per second per square meter of surface area.

The meaning of this law can be illustrated by considering two stars, one of which is twice as hot but has only half the radius of the other. The hotter star emits $2^4 = 16$ times more energy per square meter of surface but has only $(½)^2 = ¼$ as much surface area; hence, it is $16 \times ¼$ or 4 times more luminous overall. If, on the other hand, this star were twice as hot and three times as large in radius as the other, it would be $2^4 \times 3^2 = 144$ times brighter.

As before, we can also do calculations in basic energy units such as watts. For example, the Sun has a radius of about 7×10^8 m and a surface temperature of approximately 5,780 K; if these values are substituted into the expression above, we find that the solar luminosity is $L = 3.9 \times 10^{26}$ W (for a more exact value, see Appendix 4).

Both Wien's law and the Stefan-Boltzmann law were first derived experimentally, in much the same manner as Kepler discovered the laws of planetary motion. In the case of planetary motions, it remained for Newton to find the underlying reasons for the laws, and he was able to derive them strictly on a theoretical basis. Analogously, Max Planck, the great German physicist who was active early in the twentieth century, found a theoretical understanding of

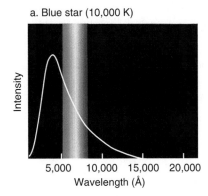
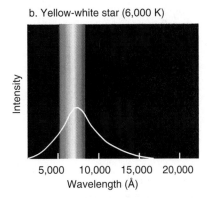
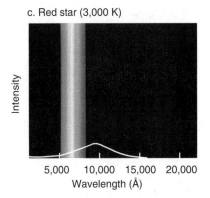

Figure 6.9 The colors of stars. The figure shows intensity plots for stars of different temperatures in comparison to the colors perceived by the human eye. The variation in peak emission illustrates why stars have a range of colors as seen by the eye.

thermal emission and was able to derive Wien's law and the Stefan-Boltzmann law from purely theoretical considerations. The basis of Planck's new understanding was the **quantum** nature of light; that is, light has a particle nature and carries only discrete, fixed amounts of energy. Both Wien's law and the Stefan-Boltzmann law can be derived mathematically from an equation called the **Planck function,** which is a general formula describing the spectrum of a blackbody. This equation is presented in Appendix 6.

The Inverse Square Law

Finally, in addition to taking into account the temperature and size of a glowing object, we must consider the effect of its distance. So far we have discussed the energy as it is emitted at the surface, but not how bright it looks from afar. What we actually observe, of course, is affected by our distance from the object. For a spherical object that emits in all directions, the brightness decreases as the square of the distance **(Fig. 6.10).** Thus, if we double our distance from a source of radiation, it will appear $(\frac{1}{2})^2 = \frac{1}{4}$ as bright. If we approach the object, reducing the distance by a factor of two, it will appear $2^2 = 4$ times brighter. As we will see in later discussions of stellar properties, we must know something about the distances to stars before we can compare other properties connected with their brightness.

The Atom and Spectral Lines

As we have just seen, the laws of continuous radiation tell us much about the overall properties of an emitting object. Surface temperature, radius, and luminosity are global properties due to the collective behavior of all the particles in a body that emit light.

The spectral lines, by contrast, have much to say about detailed processes that occur at the atomic level.

Observations of the Lines

The first detailed cataloging of the Sun's spectral lines was done in the early 1800s by Josef Fraunhofer, and the most prominent lines are known to this day as **Fraunhofer lines (Fig. 6.11).** In the late 1850s, the German scientists Robert Wilhelm Bunsen and Gustav Kirchhoff performed experiments and developed theories that made clear the importance of the Fraunhofer lines. Bunsen observed the spectra of flames created by burning various substances and found that each chemical element produces light only at specific places in the spectrum **(Fig. 6.12).** The spectrum of such a flame in this case is dark everywhere except at these specific places, as though the flame were emitting light only at certain wavelengths. The bright lines seen in this situation are called **emission lines** for that reason.

It was soon noticed that some of the dark lines in the Sun's spectrum coincide exactly in position with some of the bright lines seen by Bunsen in his laboratory experiments. Kirchhoff studied these lines in detail and was able to show that a number of common elements such as hydrogen, iron, sodium, and magnesium must be present in the Sun because of the coincidence in wavelengths of the lines. This was the first hint that the chemical composition of a distant object could be determined.

Spectral Lines and the Atom

Further studies of the spectral lines revealed various regularities in the arrangement of the lines from a given element. Although scientists suspected that these regularities must reflect some aspect of the

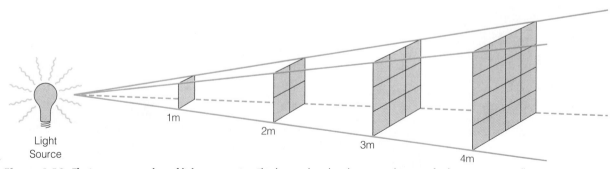

Figure 6.10 The inverse square law of light propagation. This diagram shows how the same total amount of radiation energy must illuminate an ever-increasing area with increasing distance from the light source. The area to be covered increases as the square of the distance: hence, the intensity of light per unit of area decreases as the square of the distance.

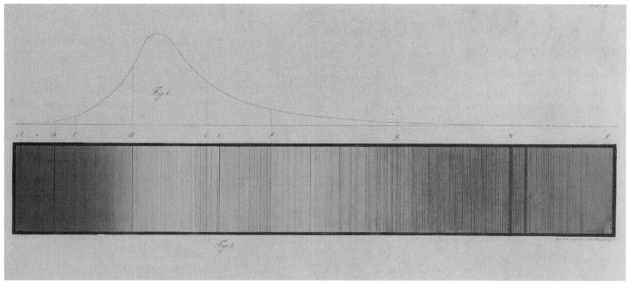

Figure 6.11 **Lines in the Sun's spectrum.** This is a drawing of the solar spectrum made by Josef Fraunhofer, the first to systematically study the dark lines in the spectrum (now called Fraunhofer lines). The colors illustrate roughly the human-eye response to the wavelengths in the spectrum. The graph at the top shows the relative intensity of sunlight as a function of wavelength. (Deutsches Museum, Munich) http://universe.colorado.edu/fig/6-11.html

structure of atoms, the true relationship was not discovered until 1913, by the Danish physicist Niels Bohr. By that time it had been established that an atom consists of a small, dense nucleus surrounded by a cloud of negatively charged particles called **electrons.** The nucleus contains positively charged particles called **protons** and neutral particles called **neutrons.** In a normal atom, the number of protons in the nucleus is equal to the number of electrons orbiting it, and the overall electrical charge is zero. The electrons are held in orbit by electromagnetic forces, with laws of the science called **quantum**

mechanics governing their motions. Bohr found that the electrons were responsible for the absorption and emission of light, and that they did so by gaining energy (absorption) or by losing it (emission) in the form of photons.

Evidence had been accumulating that the electrons could exist only in specific, fixed orbits, or energy levels. Bohr was able to develop a model for the emission and absorption of spectral lines that was based on this concept. The key to the fixed pattern of spectral lines for each element lay in the fixed pattern of energy levels the electrons could be in **(Fig. 6.13).** Each kind of atom (each element) has its own characteristic number of electrons, and in each case the electrons have a certain set of orbits. It may be helpful to visualize a ladder, with each rung representing an orbit, or energy level. The ladder for one element —hydrogen, for instance—has different spacings between the rungs than the ladder for some other element **(Table 6.2).** The energy associated with each level increases the higher up the ladder, or the farther from the nucleus, the electron goes.

An electron can absorb a photon of light if the photon carries precisely the amount of energy needed to move the electron to some higher level than the one it is in. Returning to our ladder analogy, the electron can jump up only if it will land precisely on a higher rung; that is, it can only absorb a photon whose energy will just boost it to another fixed energy level. Since the wavelength of a photon is

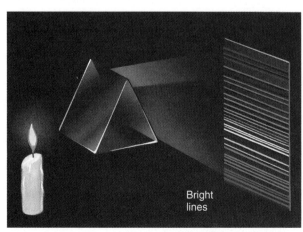

Figure 6.12 **Emission lines.** The spectrum of a flame is devoid of light except at specific wavelengths where bright emission lines appear.

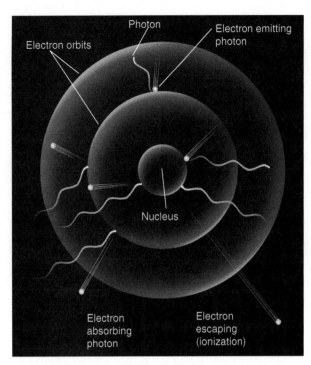

Figure 6.13 The formation of spectral lines. An electron must be in one of several possible orbits, each representing a different energy level. If an electron absorbs a photon carrying precisely the amount of energy needed to jump to a higher (more outlying) orbit, it may do so. This is how absorption lines are formed: the wavelength of the photon absorbed corresponds to the energy difference between the two electron orbits, and that difference is fixed for any given kind of atom. Conversely, when an electron drops to a lower orbit, it emits a photon whose wavelength corresponds to the energy difference between the orbits. Emission lines are formed in this way. Note that it is possible for an electron to gain sufficient energy to escape from the atom altogether. This process is called ionization and is discussed in the next section.

Table 6.2

Principal Lines in the Spectrum of Hydrogen

Series	Lower Level	Line	Wavelength (Å)
Lyman (ultraviolet)	1	Alpha	1,216
		Beta	1,025
		Gamma	972
		Delta	949
		Epsilon	937
		(Limit)	(912)
Balmer (visible)	2	Alpha	6,563
		Beta	4,861
		Gamma	4,340
		Delta	4,101
		Epsilon	3,970
		(Limit)	(3,646)
Paschen (infrared)	3	Alpha	18,751
		Beta	12,818
		Gamma	10,938
		Delta	10,049
		Epsilon	9,545
		(Limit)	(8,204)
Brackett (infrared)	4	Alpha	40,512
		Beta	26,252
		Gamma	21,656
		Delta	19,445
		Epsilon	18,175
		(Limit)	(14,585)

determined by its energy, this means that an electron can only absorb photons with certain wavelengths. Because each kind of atom has its own unique set of energy levels, each has a unique set of wavelengths at which it can absorb photons.

Thus each kind of atom has its own pattern of spectral lines. A spectral line is an absorption line when the electron receives energy from a photon and jumps to a higher level, and it is an emission line when an electron drops from a high level to a lower one, releasing a photon. For a given atom, the absorption and emission lines occur at the same wavelengths because the spacing of the energy levels, which does not change, determines the wavelengths. Whether the lines are emission or absorption lines depends simply on whether the electrons are dropping in energy level, giving off photons, or climbing in energy level, absorbing photons.

We can now state three rules of spectroscopy, first discovered experimentally by Kirchhoff **(Fig. 6.14)** and later put in the context of atomic structure:

1. *In a hot, dense gas or a hot solid, the atoms are crowded together so that their energy levels overlap and their lines are all blended together, and we see a continuous spectrum.*

2. *In a hot, low-density gas, the electrons tend to be in high energy states and create emission lines as they drop to lower levels.*

3. *In a relatively cool gas in front of a hot continuous source of light, the electrons tend to be in low energy levels and create absorption lines as they absorb radiation from the background continuous source.*

Deriving Information from Spectra

A great wealth of information is stored in the spectrum of an object. From the analysis of the spectral lines, we can learn such things as the chemical composition of a source of light, the temperature and density of a gas, and the velocity of a glowing object with respect to the Earth.

If an electron gains enough energy, it can fly free from the atom altogether **(Fig. 6.15;** see also Fig.

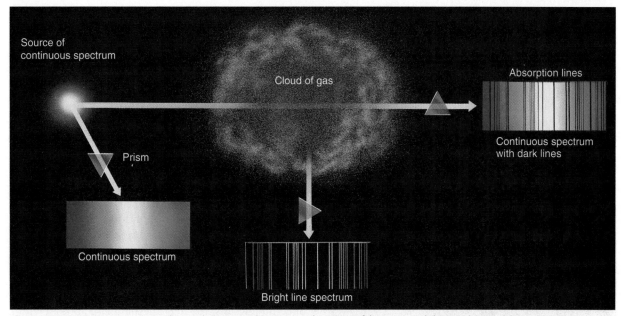

Figure 6.14 Continuous, emission-line, and absorption-line spectra. The positions of the emission and absorption lines match because the same element emits or absorbs at the same wavelengths. Whether it emits or absorbs depends on the physical conditions, as described by Kirchhoff's rules.

6.13). This event is called **ionization,** and the result is a free electron and an ion, which is a particle with a net positive electrical charge. If one electron is lost from an atom, the electron carries away a single negative charge, leaving the atom (i.e., the ion) with a single positive charge. If two electrons are lost, then the remaining ion will have a charge of +2 in units of the charge on an electron.

The degree of ionization is an important physical property of a gas. In ordinary gases that we are familiar with here on Earth, little or no ionization exists; the air we breathe is entirely in un-ionized, or neutral, form. But in other environments, often encountered in astronomy, sufficient energy may be available to ionize gas. The gain in energy that frees an electron can come either from the absorption of a photon with energy exceeding that of the highest electron energy level of the atom, or from a collision between atoms. The likelihood of collisional ionization depends on the temperature of the gas, since the speed of collision depends on temperature. The hotter the gas, the more violent the collisions between atoms, and the greater the degree of ionization.

The spectrum of an atom changes drastically when it has been ionized because the arrangement of energy levels is altered and different electrons are now available to absorb and emit photons. The spectrum of atomic helium, for example, is quite different from that of ionized helium, so the astronomer not only can see that helium is present in the spectrum of a star, but also can tell whether it is ionized or not. This provides information on the temperature in the

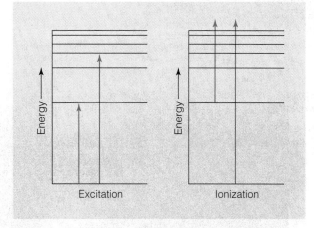

Figure 6.15 Excitation and ionization. In the energy-level diagram at the left, electrons are jumping up from the lowest level to higher ones. This process is called excitation, and it can be caused either by the absorption of photons, or by collisions between atoms or between atoms and the free electrons. On the right, electrons are gaining sufficient energy to escape altogether, a process called ionization.

outer layers of the star where the absorption lines are formed. By analyzing the degree of ionization of all the elements seen in the spectrum of a star, it is possible to determine the gas temperature quite precisely.

When an electron moves to an energy state above the lowest possible one, it is said to be in an excited state (see Fig. 6.15). This can happen as the result of the absorption of a photon of light with appropriate energy or as the result of a collision that is not energetic enough to cause ionization. The degree of excitation of a gas affects its spectrum because the observed spectral lines created in the gas depend on which energy levels the electrons are in. Therefore, analysis of the spectrum of a gas can tell us how highly excited the gas is, which in turn pro-

vides information on other properties such as density (the density affects the frequency of collisions between atoms, which in turn determines how many electrons are in excited levels).

From careful analysis of a star's spectrum, then, it is possible to determine both the temperature of the outer layers of the star (from the ionization) and the density (from a combination of excitation and ionization). To carry out either kind of analysis requires computations in which the temperature and density are adjusted until the calculated ionization and density match those observed. The mathematical basis for this was developed in the 1920s by the Indian physicist M. N. Saha, and to this day, it is a standard part of the education of every graduate student in astrophysics.

The Doppler Effect

Spectral lines can tell us something else about a distant object in addition to all the physical data that we have been discussing. We can also learn how rapidly an object such as a star or a planet is moving toward or away from us.

When a source of light such as a star is approaching, the lines in its spectrum are all shifted toward shorter wavelengths than if the light source were at rest, and if the source is moving away from the observer, the lines are all shifted toward longer wavelengths **(Fig. 6.16)**. These two cases are called blueshift (approach) and redshift (recession), respectively, because the spectral lines are shifted toward either the blue or the red end of the spectrum.

A general term for any wavelength shift due to relative motion between source and observer is **Doppler shift,** in honor of the Austrian physicist who first explored the properties of such shifts. The effect applies to other kinds of waves than light. Most of us, for example, have noticed the Doppler shift in sound waves when a source of sound passes by. The whistle on an approaching train suddenly changes to a lower pitch at the moment the train passes us, because the wavelength we receive suddenly shifts to a longer one and the frequency drops.

In the case of the Doppler shift of light, it is possible to determine the speed with which the source of light is approaching or receding from the simple formula

$$v = c \left(\frac{\Delta\lambda}{\lambda} \right),$$

where v is the relative velocity between source and observer, $\Delta\lambda$ is the shift in wavelength (the observed wavelength minus the rest, or laboratory, wavelength

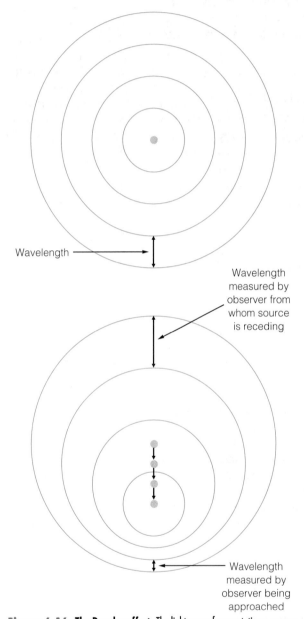

Wavelength

Wavelength measured by observer from whom source is receding

Wavelength measured by observer being approached

Figure 6.16 The Doppler effect. The light waves from a stationary source (upper) remain at constant separation (that is, constant wavelength) in all directions, whereas those from a moving source get "bunched up" in the forward direction and "stretched out" in the trailing direction (lower). This causes a blueshift or a redshift for an observer who approaches or recedes, respectively, from a source of light. Note that it does not matter whether the source or the observer is moving. **http://universe.colorado.edu/fig/6-16.html**

of the same line), λ is the laboratory wavelength of the line, and c is the speed of light. If the observed wavelength is greater than the rest wavelength, we find a positive velocity, corresponding to a **redshift;** if the observed wavelength is less than the rest wavelength, the result is a negative velocity, indicating a **blueshift.**

It is important to notice that the Doppler shift tells us only about *relative* motion between the source and the observer. It is not possible to distinguish whether it is the star or the Earth that is moving or a combination of the two (which is most likely the case). It is also important to keep in mind that the Doppler shift tells us only about motion directly toward or away from the Earth, called the **radial velocity.** There is no Doppler shift due to motion perpendicular to our line of sight. If a star is moving with respect to the Earth at some intermediate angle, as is usually the case, then we can determine the part of its velocity that is directed straight toward or away from us (i.e., the radial velocity), but we cannot determine its true direction of motion or its speed transverse to our line of sight **(Fig. 6.17).**

Telescopes: Tools for Collecting Light

Now that we know how to analyze light from distant objects, we can turn our attention to the methods for collecting light. The universe is filled with radiation of all wavelengths, so our discussion will necessarily include telescopes designed for all forms of radiation, not just visible light. Telescopes come in a variety of

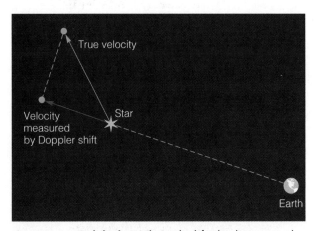

Figure 6.17 Radial velocity. The Doppler shift only indicates motion along the line of sight, called the radial velocity. Here a star is moving at an angle to the line of sight, so the Doppler shift measures only a fraction of its true velocity.

sizes and designs **(Fig. 6.18),** and in some sense they represent the material monuments to the science of astronomy. Certainly, large telescopes are fascinating, complex technological achievements that combine huge masses and dimensions with incredible precision and are usually located in remote, exotic parts of the world. If you get a chance to visit a major observatory where some of the instruments are on public display, do so—you will almost certainly be impressed.

The Need for Telescopes

Telescopes offer several basic benefits. For one thing, telescopes collect light from a large area and bring it to a focus, so that fainter objects than the eye could see unaided can be observed. The human eye has a collecting area only a tiny fraction of a meter in diameter, whereas the largest telescopes are 4 to 10 m in diameter. The **light-gathering power** of a telescope depends on the area of its collecting surface; since the area of a circle of radius r is equal to πr^2, a 2 m diameter telescope collects 4 times as much light as a 1 m telescope, for example.

Another basic advantage of telescopes over the unaided eye is that the telescope can be equipped with instruments to record and analyze light in ways that the human eye cannot. Cameras can record images, and other kinds of instruments can measure brightnesses or disperse light to create spectra.

An additional, very powerful advantage of telescopes that are equipped with such instruments is that they can record light over a long period of time by using photographic film or an electronic detector. The eye has no capability to store light, but instead captures and sends images to the brain about 60 times per second. A telescope and a camera or other instrument may collect and store light continuously for several hours and reveal objects far too faint to be seen with the eye, even by looking through the same telescope. By combining large size with the capability of making long exposures, the largest telescopes can detect objects more than 10 billion times fainter than the unaided eye can see even under the best conditions.

Telescope Resolution and Interferometry

Another major advantage of large telescopes is that they have superior **angular resolution,** the ability to discern fine detail. The resolution of a telescope is normally expressed in terms of the smallest angular separation between a pair of objects that can be discerned by the telescope. Thus small resolution is good. The limiting factor in the resolving power of a telescope is interference between rays of light that are

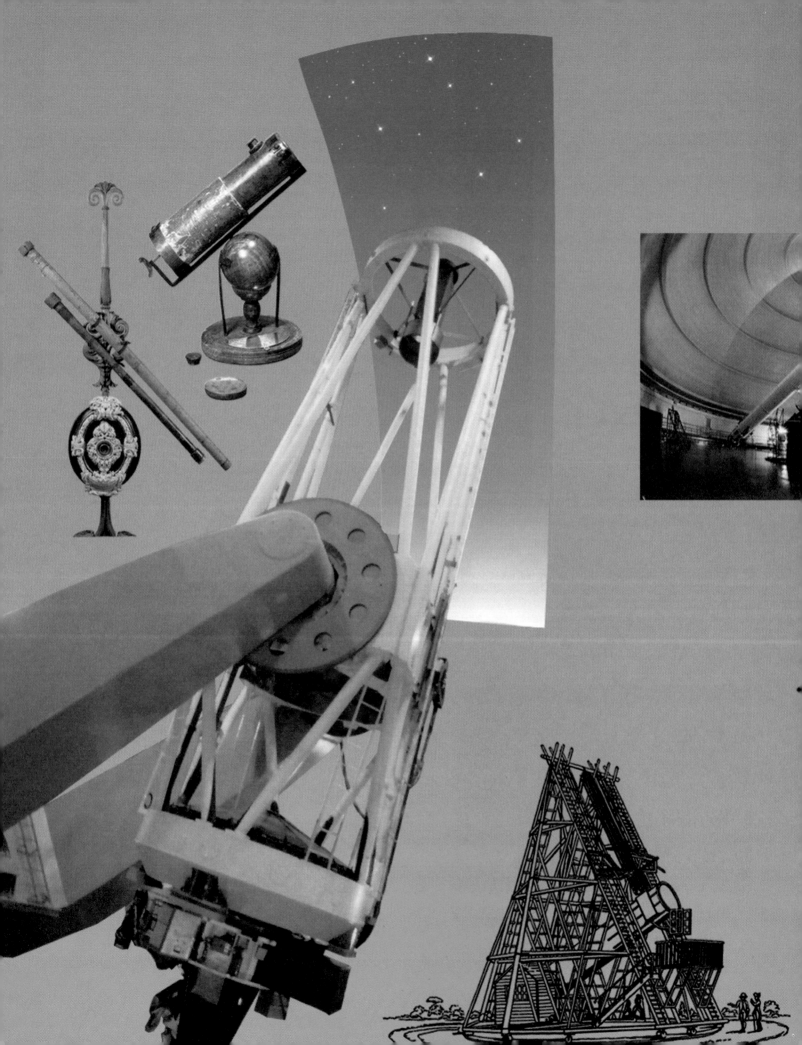

TELESCOPES

Figure 6.18 Telescopes. Telescopes come in a variety of sizes and designs, and in some sense they represent the material monuments to the science of

bent, or diffracted, as they pass the edges of the telescope opening. The theoretical limit on resolution is therefore called the **diffraction limit.** To achieve diffraction-limited resolution is the ultimate goal of the telescope maker. The resolution at the diffraction limit is proportional to the wavelength being observed and inversely proportional to the diameter of the telescope; in equation form, this is

$$\theta = 2.5 \times 10^5 \left| \frac{\lambda}{D} \right|,$$

where θ is the angular separation in seconds of arc of the closest objects that can be seen as separate objects, λ is the observed wavelength, and D is the telescope diameter expressed in the same units as the wavelength.

For visible light (assume a wavelength of 5500 Å = 5.5×10^{-7} m), a telescope 0.1 m in diameter (about 4 inches) has a resolution of about 1.4 arcseconds. (This is really quite good; an arcsecond is about the angular diameter of a dime seen at a distance of 2 miles!) The human eye, with a pupil diameter of about 0.002 m (2 millimeters), has a resolution of roughly 70 arcseconds, or about 1.1 arcminutes (this is why pretelescopic positional measurements of planets and stars were only accurate to about 1–2 arcminutes). In principle, a large instrument such as the 5 m *Hale Telescope* on Mt. Palomar has a resolution far smaller than 1 arcsecond, but in practice the Earth's atmosphere limits the resolution that is actually achieved. Normally, it is impossible to distinguish objects separated by less than about 0.5 arcsecond when looking up through the turbulent atmosphere, and often this blurring effect, called "seeing," makes resolution much worse.

Consider the resolution of a radio telescope. Here the observed wavelength might be a meter or longer, and while the instrument can be quite large, the resolution is still very poor. For a wavelength of 1 m and a diameter of 100 m (typical of the world's largest radio telescopes), the formula above yields a resolution of more than *40 minutes of arc*. Thus a radio telescope provides only a very blurry view of the universe.

Astronomers have developed a very successful technique for surmounting this problem, however. If two or more radio telescopes are used in combination, radio waves arrive at the separate antennae at slightly different times. These small timing differences can be measured by combining the waves arriving at separate telescopes and allowing them to interfere with each other. Whether a given pair of waves interferes constructively (wavecrests adding) or destructively (wavecrests canceling each other out)

depends on the timing difference, which depends in turn on the exact angle from which the waves arrived at the separate telescopes. By using this technique, called **interferometry,** astronomers can reconstruct angular information from the sky in astonishing detail. In effect, the resolution that can be achieved is what you would calculate from our formula for a single radio dish whose diameter equals the separation between dishes. Interferometry has been used successfully with radio telescopes separated by a few meters up to thousands of kilometers. The *Very Large Array (VLA),* operated in the high plains of New Mexico by the U.S. National Radio Astronomy Observatory, is the largest concentrated collection of antennae for this purpose (see Fig. 6.18), but recently an intercontinental array called the *Very Long Baseline Array (VLBA)* has been completed.

Principles of Telescope Design and Construction

To illustrate the general principles of telescope design, we begin by describing telescopes built for visible light. We will then see how those designed for other wavelengths have essentially the same features but differ in some details.

It is possible to construct a telescope, called a **refractor,** that focuses light with lenses **(Fig. 6.19),** and many of the earliest telescopes were of this type. Today, however, nearly all telescopes are **reflectors,** which use mirrors to focus light. Reflectors can be built much larger and are not hampered by certain technical difficulties associated with lenses, such as the fact that a lens refracts different wavelengths of light at different angles. In a reflector, light is focused by a concave mirror, called the **primary mirror (Figs. 6.20** and **6.21)**. Because it is usually inconvenient to look at or record an image formed inside the telescope tube, most often a telescope has a

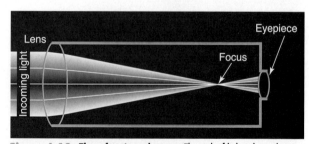

Figure 6.19 The refracting telescope. The path of light is bent when it passes through a surface such as a glass lens. A properly shaped lens can thus bring parallel rays of light to a focus, where the image of a distant object may be examined. A second lens is used to magnify the image.

secondary mirror that deflects the image outside the tube in one of various arrangements **(Fig. 6.22)**.

The variations include the **prime focus**, in which the light is received and recorded at the focus of the primary mirror; the **Newtonian focus** (Newton's original concept), in which the light is reflected to a focus outside the telescope at the side; the **cassegrain focus**, in which a convex secondary mirror reflects the light to a focus behind the primary mirror (requiring a small hole in the center of the primary); and the **coudé focus**, in which a series of mirrors is used to bring the focused light to a location remote from the telescope where a large, immobile instrument can be used to analyze it. All of these designs have either a mirror or an instrument mounted inside the telescope that blocks some of the incoming light, but the percentage of the light lost in this way is usually very small.

Reflectors offer several advantages over refractors, but probably the most important is that a mirror can be supported from behind, whereas a lens must be mounted only by its edges. Thus mirrors can be much larger and still be rigid enough to maintain image quality, while a large lens will sag and distort its shape as the telescope is pointed in different directions. Until the 1970s, the strategy for making ever-larger telescopes was simply to make larger and larger molds, pour molten glass in, hope that it hardened (annealed) without cracking, and then grind it to shape. The largest successful result is the 5 m *Hale Telescope*, the biggest in the world for some 30 years starting in 1947 (in the late 1970s, the Soviet Union produced a 6 m telescope, but its primary mirror was flawed, and the bad portions had to be screened off to avoid blurring images).

In recent years, new techniques have been developed for making larger telescopes. The first to be tried is the multiple-mirror design **(Fig. 6.23)**, in which several mirrors are mounted together and aligned so that they focus light at a single point. The advantage is that the cost and technical challenges of making

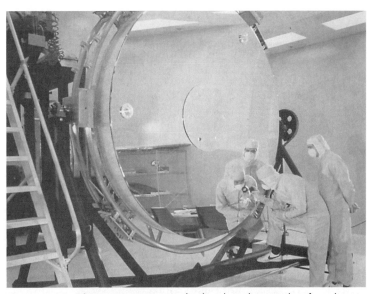

Figure 6.21 **A large primary mirror.** As this photo shows, the proper shape for a telescope mirror is not very highly concave. In this case only the distorted reflections reveal that the mirror is not flat. This is the 2.4 m primary mirror for the *Hubble Space Telescope.* (Perkin Elmer Corporation)

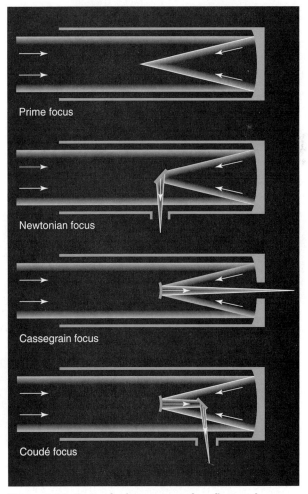

Prime focus

Newtonian focus

Cassegrain focus

Coudé focus

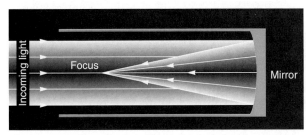

Incoming light

Focus

Mirror

Figure 6.20 **The reflecting telescope.** A properly shaped concave mirror can be used instead of a lens to bring light to a focus. Usually, an additional mirror is used to reflect the image outside the telescope tube.

Figure 6.22 **Various focal arrangements for reflecting telescopes.**

several smaller mirrors are less than the aggregate agony of making a single large one (by the old methods, anyway). The multiple-mirror design was used for several years by the *Multiple-Mirror Telescope (MMT;* see Fig. 6.23), but ironically this is now being replaced by a single 6.5 m mirror constructed by another of the new techniques.

This second method of large mirror construction utilizes a hollow honeycomb structure for the back side of the mirror, which makes the mirror both lightweight and rigid. In addition, the mirror is cast in a rotating oven whose spin rate is calculated to

make the surface of the glass concave, thereby producing a mirror that is already close to the desired final shape **(Fig. 6.24).** Spin-cast mirrors of this type with diameters as large as 8 m are under construction. The largest put into operation so far has a 3.5 m diameter and works beautifully, so there is optimism that monolithic mirrors as big as 8 m or even larger are possible using this technique.

Another new method makes use of a **meniscus mirror,** a very thin mirror that makes no pretense of being rigid. Instead the mirror is expected to sag, and to compensate for this, a system of push-rods, controlled by computer, is installed behind the mirror to adjust its shape as the telescope is pointed in different directions. Star images are monitored during observations, and the mirror is continually adjusted to keep the images sharp. So far the largest activated mirror of this type in operation is the 3.5 m *New Technology Telescope (NTT),* located in the Chilean Andes and managed by a consortium of European countries. Several huge telescopes using similar techniques are under construction. These include the four 8 m mirrors of the so-called *Very Large Telescope,* a project being built in Chile by the same consortium that built the NTT; the two 8 m *Gemini* telescopes being built in Chile and Hawaii by the United States in collaboration with a few other countries; and the 8 m Japanese national telescope, dubbed *Subaru,* under construction in Hawaii.

The largest telescope in operation today, however, uses a combination of the multiple-mirror

Figure 6.23 The *Multiple-Mirror Telescope,* Mt. Hopkins, Arizona. This observatory is operated by the University of Arizona and the Smithsonian Institution. Six 1.8 m mirrors bring light to a common focus, giving the instrument the light-collecting power of a single 4.5 m telescope. (Multiple-Mirror Telescope Observatory, University of Arizona and the Harvard-Smithsonian Center for Astrophysics)

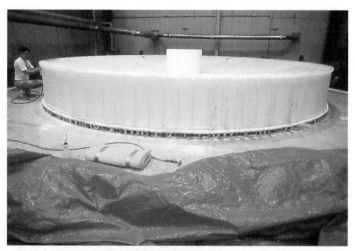

Figure 6.24 Casting a giant mirror in a spinning oven. These two photos show steps in the creation of the 6.5 m mirror that is being installed in the *Multiple-Mirror Telescope* in place of the existing six 1.8 m mirrors. At the left, chunks of glass are being placed in the mold before heating. Note the hexagonal pattern in the bottom of the mold, which creates a hollow honeycomb structure in back of the mirror. At the right is a photo of the mirror after casting, just after removal from the oven. Note that the surface already has a concave shape, which is very close to the final curvature that is needed. (University of Arizona)

and meniscus mirror techniques. This is the 10 m *Keck Telescope* **(Fig. 6.25)** atop Mauna Kea in Hawaii. The primary mirror consists of thirty-six 2 m panels of hexagonal shape, mounted together in a single unit. Thus this is a single 10 m mirror, but it is made of many panels, so it is like a multiple-mirror design at the same time. The individual panels are computer-controlled to retain the overall shape and keep the images sharp, so in that sense the telescope operates like a meniscus-mirror design. The *Keck Telescope* went into full operation in 1993 and has demonstrated that it can obtain very fine images, often achieving resolutions of better than 1 arcsecond.

The use of multiple mirrors or mirror segments, or of a flexible meniscus mirror, is related to a general technique called **active optics,** a system in which the mirror is continually flexed to correct for atmospheric blur. A number of experiments are now being carried out in the attempt to refine this technique. To make rapid corrections in response to atmospheric fluctuations virtually as they occur, astronomers need an artificial star whose intrinsic image shape is perfectly known. The artificial star is continually monitored and the telescope mirror adjusted so that the image of the artificial star remains exactly as it should be. Then the images of other objects in the field of view will also be perfectly corrected. What do astronomers use as artificial stars? Most efforts today are using powerful lasers to project intense spots of light onto the upper atmosphere, where they can be observed by the telescope. The characteristics of the laser beam are precisely controllable, so the spot it creates in reflection has a well-known intrinsic shape and the effect of the atmosphere can be accurately monitored and eliminated. It is hoped that this technique will soon allow ground-based telescopes to obtain diffraction-limited images.

Another big challenge for telescope design is to use neighboring instruments to do interferometry at infrared or even visible wavelengths. We have seen that this technique is very successful for arrays of radio telescopes, but it is much more difficult for the shorter wavelengths of infrared and visible light. One of the goals of the four-instrument *Very Large Telescope* in Chile will be to attempt interferometry, but before that happens a second *Keck Telescope* (dubbed ***Keck II***) will be in operation on Mauna Kea, right next door to the original Keck instrument (i.e., ***Keck I***). So by 1997, two 10 m telescopes will be operating side-by-side on Mauna Kea, and infrared interferometry is one of the goals.

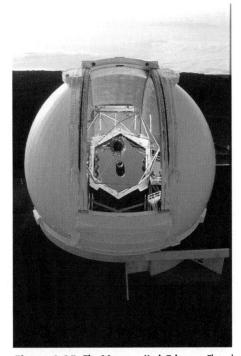

Figure 6.25 The 10-meter *Keck Telescope*. The only ways to obtain a full view of this enormous instrument are either to hover above the dome in a helicopter, as was done in the left image, or to use a time-exposure photograph while the dome is rotated, as in the right image. (© 1992 Roger H. Ressmeyer/Starlight)

http://universe.colorado.edu/fig/6-25.html

SCIENCE AND SOCIETY

The Cost and Value of Basic Research

Very rarely has research in astronomy been done in pursuit of some specific, practical goal. The driving motivation for astronomers has always been curiosity, the search for knowledge for its own sake. There have been exceptions, especially in early times when the need to predict seasonal changes was important or astrological omens were considered significant, but even then curiosity was the principal rationale. Astronomy today is the purest form of basic research, meaning that no one goes into it expecting any practical benefits other than simply knowing more about the universe. Yet many very valuable spin-offs have come from astronomical research, and we can expect more to come in the future. But we have no way of knowing when or how.

At this point, when we are discussing the most expensive elements of astronomical research (telescopes and space-based observatories), it is appropriate to examine the costs and the benefits. The U.S. National Science Foundation is spending about $90 million on the *Gemini* project to build two 8 m telescopes, and NASA's largest program to date, the *Hubble Space Telescope*, cost $1.5 billion for development and launch. Who pays, and why?

Astronomy has always had patrons, private individuals or foundations with the wealth and the inclination to support this science. For example, the twin *Keck telescopes* are being built with more than $150 million in grants to Cal Tech from the Keck Foundation. Other observatories have been built with privately donated funds. On the other hand, because there is little expectation of practical benefit, few profit-making corporations are willing to pay for astronomical research.

The majority of the expense of our science is borne by the governments of the world. In Renaissance times, most governments were monarchies, and the prominent astronomers of the day were supported by kings and emperors. Certainly, this patronage had an astrological element to it, but it was also motivated by curiosity and the desire to learn more about the cosmos. Universities provided another means of support for basic research, as they do today, but now government grants and contracts are usually the ultimate source of this support. In modern times, basic research in the United States and elsewhere is funded almost entirely by national government programs. Instead of a benevolent monarch, the general public pays for research through tax-supported programs. In a time of budgetary stress worldwide and in the United States, it is appropriate to ask whether the public is getting its money's worth.

What is the value of basic research? Both philosophical and practical arguments can be made. The philosophical value arises, as it always has, from the desire to *know,* from the human need to explore. Many people say this is sufficient reason why the public should support basic research. They point to the enormous interest in new discoveries in astronomy and other sciences and the popularity of astronomical images and posters. For many people, the feelings of awe and wonder they experience when they learn of supermassive black holes in the cores

Instruments for Recording and Analyzing Light

At the focus of any telescope (except small ones used only for simple viewing of the sky) is an instrument that records light. This could be a simple camera, for making images of the sky, or a photometer, an instrument that measures intensities of light using a photocell (a device that converts light into an electrical current). A commonly used instrument is the spectrograph, which disperses light according to wavelength so that the spectrum can be recorded and analyzed. Auxiliary instruments used in astronomy range from rather compact, lightweight devices that can easily be mounted directly on the telescope for use at the prime or cassegrain focus to large, heavy instruments (usually spectrographs) that must be used at the coudé focus.

Film has traditionally been used to record the light in cameras and spectrographs, but today many kinds of electronic detectors are being developed, and film is now rarely used. Electronic detectors offer many advantages over film, both in accuracy and in such practical matters as storing the data in electronic form so that they can be transmitted directly into computers for analysis. More specifically, an electronic detector has the following advantages over film: it is more sensitive, recording a higher fraction of all the photons that strike it; it has greater range, being able to simultaneously record the images of stars that might differ in brightness by a factor of a million or more; it measures the brightness (technically, the

of distant galaxies or read about the discovery of a planet orbiting another star are worth the expense. So long as scientists share what they learn with the public, for many this public fascination is sufficient justification of the expenditures.

Basic research also offers practical benefits, unpredictable though they may be. The basic understanding of atomic structure, the discovery of new forms of matter that have earthly value, the observation of chemical processes in environments that cannot be reproduced on Earth, the understanding of nuclear structure and reactions—these are fundamental building blocks of modern technology, and all came in whole or in part from basic astronomical research. Specific spin-offs have occurred as well. Teflon, freeze-dried foods, laptop computers, and microelectronics were all developed for the space program and have found enormous application on the ground as well. The laser arose from the maser, which was invented by astronomers and invokes processes that are observed in interstellar clouds. Modern image-processing techniques, developed by astronomers to analyze their pictures of the cosmos, have been applied to the diagnosis of breast cancer in X-ray images, making this disease detectable at much earlier stages than was previously possible. And the list goes on.

But much astronomical research provides no such practical benefits. And we have no way to predict which projects will produce practical results and which will not. Certainly, no scientist would try to make such predictions. How then do we decide what basic research to pay for, and how much to pay?

It may be instructive to answer the latter question first. In the United States, federal support for basic research in astronomy comes almost entirely from two agencies: NASA and the National Science Foundation (NSF). In 1996 NASA's total budget was about $13.7 billion per year, and the NSF's budget was $3.2 billion. Only a small fraction of this money goes to support basic astronomical research, however; NASA's Office of Space Science, which funds research grants to astronomy and related sciences and pays for scientific missions such as the *HST* and planetary probes, has an annual budget of $880 million. The NSF's astronomy section has a current annual budget of $106 million. Thus the total expenditure by the U.S. government on basic astronomical research is just under $1 billion. This constitutes about 0.07% of the total federal budget and amounts to just over $4 per year for every person in the U.S. population.

How are these budgets determined? In the United States, funding for research is allocated through the same political process as all other federal expenditures: Congress writes bills that determine the budget, first in pieces and then in total. Funds for basic astronomical research appear as individual items in the overall budgets for NASA and the NSF, which are scrutinized by various congressional subgroups, first by committees on science and technology, and later by the committees for appropriations and authorizations. Thus the real decisions on the value of basic research are made in these committees. Therefore an important function of scientists and scientific organizations is to ensure that their opinions about the importance of basic research are made known to the members of these committees. Even those pursuing basic research for the sake of curiosity must become involved in the real world of politics.

intensity, or incident energy per square meter) of light more accurately (film can have a fairly nonuniform response to light); and as already noted, it converts intensities of light directly into electronic signals that can be analyzed by computer (when film is used, the process of converting images to digital form is very laborious and often inaccurate).

The major advantage of film is that it has a much larger format than current electronic detectors, so film is still used for wide-angle images of the sky. The largest electronic detectors currently in astronomical use are about 8 cm square, whereas photographic plates larger than 40 cm square have been used. Film has the further advantages of low cost and simplicity of use, particularly for people observing on their own as a hobby.

The most common type of electronic detector in use today is the **charge-coupled device,** or **CCD.** CCDs are the same as the "chips" used to record light in a typical video camera such as you may have in your home. The major difference is that the CCDs used at astronomical observatories are generally of higher quality; in particular, they have much lower background noise levels. A CCD has an array of positions (called picture elements, or **pixels**) at which electrical charges accumulate when photons strike the silicon matrix of which the CCD is made. At frequent intervals (usually many times per second), the charge is allowed to transfer off to one side of the chip, row by row, and the pattern of charge on each row is recorded, thus preserving the entire image of the chip.

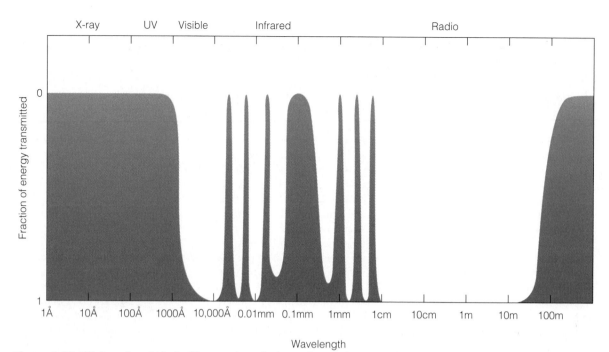

Figure 6.26 Windows through the Earth's atmosphere. This diagram illustrates the many wavelength regions where the atmosphere blocks incoming radiation from space (shaded regions). The atmospheric "windows" allow ground-based observations only in visible and radio wavelengths and some portions of the infrared spectrum. All other wavelengths must be observed from space.

CCDs (and a few other types of electronic detectors) have revolutionized modern observational astronomy by providing more accurate data than ever before and by recording data for much fainter objects than could be readily observed previously. Along with the advent of space-based astronomy and, more recently, the development of new techniques for building ultralarge telescopes on the ground, the development of electronic detectors ranks as one of the most significant advances in astronomy of our times.

Telescopes for Nonvisible Wavelengths

Telescopes for wavelengths other than visible light have the same basic components as the telescopes just described. They have an element such as a mirror for collecting and focusing light and an instrument at the focus for analyzing and recording the radiation.

Telescopes for ultraviolet and infrared wavelengths are virtually identical in design to those for visible light, with a couple of exceptions. Ordinary mirrors do not reflect well in the ultraviolet, so special surfaces and chemical coatings have to be used; and infrared telescopes glow in the wavelengths they are built to observe, so the instrument must be cooled to reduce the glow (usually by circulating liquid nitrogen or liquid helium through the part of the instrument that contains the detector). Of course, ultraviolet light and large portions of the

infrared spectrum do not penetrate the Earth's atmosphere **(Fig. 6.26),** so these telescopes have to be launched into space. Therefore, the fields of ultraviolet and infrared astronomy are relatively new, with most research occurring only in the past 30 years or so. Some existing and planned space observatories are listed in Appendix 7.

For wavelengths beyond infrared—that is, the radio portion of the spectrum—the radiation reaches the ground, and large Earth-based telescopes are used. Radio telescopes usually consist of very large dishes made of metal plating or a wire mesh, and they reflect radio waves to a focus at a point above the center of the dish **(Fig. 6.27).** A **receiver,** an electronic device that records the radio waves and turns them into electrical impulses, is suspended at the focus. Because radio telescopes have poor resolution due to the long wavelengths being observed, they tend to be very large, and much of the radio astronomy done today uses the technique of interferometry we already discussed earlier.

For the shortest wavelengths, below ultraviolet, it is no longer possible to build a simple concave reflector to focus light. Extreme ultraviolet and X-ray radiation normally do not reflect well from any kind of mirror. These wavelengths can be reflected, however, if they strike a surface at a very oblique angle; this is called a **grazing incidence reflection,** and an X-ray telescope can be constructed using an optical design based on this kind of reflection. Such a telescope has one or more ringlike primary mirrors at the front and can even have a secondary mirror to bring the radiation to a focus where a detector is located **(Fig. 6.28).** Like ultraviolet light, X rays do not penetrate the Earth's atmosphere, and observations must be made from space.

The shortest wavelength radiation, the gamma rays, are very difficult to focus. Even grazing incidence reflections do not work for these extremely high energy photons, and the few gamma-ray telescopes built so far have consisted simply of detectors, with no primary light-gathering element to bring the radiation to a focus from a large collecting area. A device called a **collimator** is placed in front of the detector, so that only gamma rays from a certain direction are recorded; without this, there would be no way of knowing where the observed gamma rays came from. Again, gamma rays must be observed from space because they cannot pass through the Earth's atmosphere.

Major Observatories for All Wavelengths

Observatories for visible light, some portions of the infrared spectrum, and radio observations can be built on the Earth's surface because all these wavelengths reach the ground (see Fig. 6.26). Even so, most major observatories are located in remote regions for a number of reasons. To minimize absorption of light by the atmosphere, it is desirable to locate an observatory on a high mountain, above as much of the atmosphere as possible (this is especially important for infrared observations, since atmospheric water vapor absorbs much of the incoming infrared radiation). Of course, a site with generally clear weather is usually imperative, for only radio telescopes can peer through clouds. It is also best to locate observatories well away from large cities because the diffuse light from a densely populated area can drown out faint stars. Another consideration is the latitude of the site, which should not be too far north or south of the equator, since that would exclude a large portion of the sky.

Another factor that affects the quality of an observing site is the cleanliness of the air, for pollution can reduce the transparency of the atmosphere. The amount of turbulence in the air above the site is also important because turbulence creates the blurring effect known to astronomers as "seeing," as we discussed earlier. (Interestingly, it has recently been found that air currents inside the telescope or the dome building are sometimes the major cause of poor seeing, so modern telescope design seeks to minimize these currents.)

The largest observatories in the world tend to be located along the western portions of North and

Figure 6.27 A radio telescope and receiver. This radio dish, part of the U.S. National Radio Astronomy Observatory at Green Bank, West Virginia, has a receiver mounted on a tripod at the focal point of the reflector. (National Radio Astronomy Observatory)

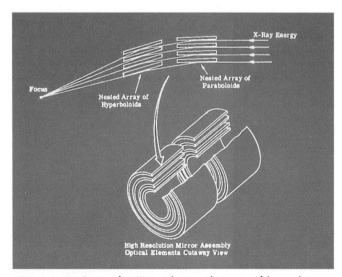

Figure 6.28 Design of an X-ray telescope. This is a view of the nested X-ray reflections that formed a primary mirror assembly for the *Einstein Observatory,* a NASA spacecraft that operated in the early 1980s. (NASA/Harvard-Smithsonian Center for Astrophysics)

Figure 6.29 **Telescopes on Mauna Kea.** This aerial view shows most of the large telescopes currently located on top of Mauna Kea. At the top of the ridge in the rear we see, from left to right, the Japanese Subaru telescope (under construction) and the two Keck 10-m telescopes (the two white domes). Other domes include the NASA Infrared Telescope Facility (IRTF; small silver dome to the right of the Kecks), the bulbous-looking Canada-France-Hawaii Telescope building (CFHT; at far right), the University of Hawaii 2.2-m telescope building (white building in the foreground, on the same ridge as the CFHT but to its left in this view), and the United Kingdom Infrared Telescope (UKIRT; the silver dome closest to the viewer, almost in front of the silver dome of the IRTF). At left are two submillimeter telescopes, the UK-Dutch James Clerk Maxwell Telescope (JCMT) in the rear, and the California Institute of Technology's CalTech Submillimeter Observatory (CSO) in front. One of the Gemini 8-m telescopes will be built atop a large cinder cone that is just out of the field of view, in the background and to the left. The true summit of Mauna Kea, where no telescopes can be built, is just out of the field of view to the right. (Richard Wainscoat, University of Hawaii) http://universe.colorado.edu/fig/6-29.html

South America, in Hawaii, and in Australia; a fine site has also been developed in the Canary Islands by the British government (see Appendix 7). Through the National Science Foundation, the U.S. government supports large observatories in North America (Kitt Peak National Observatory, about 80 km west of Tucson, Arizona) and in South America (the Cerro Tololo Inter-American Observatory on the crest of the Andes, some 300 km north of Santiago, Chile). The summit of Mauna Kea on the island of Hawaii is recognized as one of the finest sites in the world. Many of the most recently built major telescopes are there **(Fig. 6.29),** and several more are under construction, as noted earlier.

The major radio observatories of the world are located in Green Bank, West Virginia; near Socorro, New Mexico (the site of the *VLA*); and in England, Holland, and Australia. The majority of the ultraviolet, X-ray, gamma-ray, and far-infrared observatories built so far have been developed and launched into space by NASA (see Appendix 7). Recently, the European Space Agency (ESA), a consortium of several nations, has also been active in

space astronomy. In ultraviolet astronomy, the *International Ultraviolet Explorer* (a collaborative effort by NASA, ESA, and the United Kingdom) has been in successful operation for 18 years, and the *Hubble Space Telescope (HST)* has been in operation since 1990 (the *HST* also operates in visible wavelengths where the space environment allows better resolution without the blurring effects of the atmosphere and greater sensitivity without atmospheric absorption).

X-ray astronomers are studying data from the recent *ROSAT* mission, a joint U.S.-German-British project, and are awaiting the *Advanced X-Ray Astrophysics Facility (AXAF),* now under development and scheduled for launch around the turn of the century. Higher-energy observations have been carried out by the *Compton Observatory* (a gamma-ray mission). The most recent addition to the family of active space observatories is the *Extreme Ultraviolet Explorer,* a U.S. telescope that began surveying the sky in the shortest ultraviolet wavelengths in mid-1992. The *Infrared Space Observatory (ISO),* another U.S.-European collaboration, was launched in late 1995, and plans are under way for a U.S. infrared orbital observatory called the *Space Infrared Telescope Facility (SIRTF)*. Infrared astronomers will also benefit from a planned airborne observatory called *SOFIA,* which will consist of a Boeing 747 aircraft modified to carry a large infrared telescope to high altitudes above most of the atmospheric water vapor. In 1998 a new ultraviolet observatory called the *Far Ultraviolet Spectroscopic Explorer (FUSE)* will be launched and will cover ultraviolet wavelengths not accessible to the *Hubble Space Telescope.*

The Hubble Space Telescope

By now most people are at least somewhat aware of the impact the *Hubble Space Telescope* has made on astronomy and on science. The many delays before the telescope's launch in May 1990, its high cost, the problems with its optics, and its many spectacular scientific discoveries have made the *HST* the most visible and notorious astronomical experiment of all time.

The concept of the *HST* was first proposed in the 1950s by Lyman Spitzer, a visionary American astronomer who anticipated the many advantages of placing telescopes above the Earth's atmosphere. The idea seemed fanciful at the time, but as space programs in the United States and elsewhere progressed, the possibility of placing a large telescope in orbit became more and more feasible. In the meantime, smaller-scale space observatories were developed and launched, starting with sounding rocket experiments in the 1950s and moving on to small satellites in the 1960s and beyond. Many of these more recent missions were mentioned in the preceding section.

The first serious planning for the *HST* began in the late 1970s, when NASA contracted for the development of the spacecraft, the telescope and its 2.4 m mirror, and the first scientific instruments. From the beginning, the planners envisioned a multipurpose observatory in space with several different cameras and spectrographs **(Fig. 6.30)**, and once NASA had committed to the manned space shuttle program, the plans also incorporated the possibility of on-orbit servicing and repair. The immensity and complexity of the project and then the *Challenger* disaster led to many delays, and the projected launch date slipped many times. Finally, the launch took place on May 20, 1990.

Immediately, there were successes, but also problems. The telescope proved impossible to focus properly, and within weeks it was determined that there was an error in its shape. Investigation showed that a test facility used at the company where the

Figure 6.30 *The Hubble Space Telescope.* This is a schematic drawing of the 2.4 m telescope launched into Earth orbit in May 1990. The open end of the telescope is pointed away from us in this view; the near end houses control systems, cameras, and other scientific instruments. The large "wings" are solar panels that provide electrical power for the spacecraft. (NASA) http://universe.colorado.edu/fig/6-30.html

mirror was polished had been set up improperly, with the result that the mirror was ground to match incorrect specifications. The images it produced were flawed, with a substantial fraction of the light being spread out instead of being concentrated in a point. In short, the mirror failed to produce diffraction-limited images. Despite this, many important discoveries were made: the spectrographs were relatively unaffected by the poor image quality, and techniques for data processing were developed that allowed images from the cameras to be sharpened considerably. Even at its worst, the *HST* produced images far superior to anything obtainable from the ground.

In late 1993 the first service mission to the *HST* took place **(Fig. 6.31)**. Two instruments were removed, and a new camera and a special optical device designed to correct the images were installed. The results were very gratifying, as the images were now fully diffraction limited; that is, the telescope was now performing as well as had ever been hoped. The results since then have been spectacular, including observations of developing storm systems in the atmospheres of the outer planets, discoveries of new volcanoes on one of Jupiter's satellites, images of possible planetary systems in the process of forming, the verification that supermassive black holes probably exist in the cores of some galaxies, and the

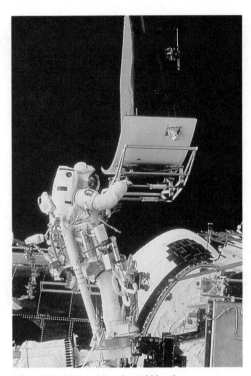

Figure 6.31 Servicing the Hubble. This is an astronaut replacing one of the cameras on the *Hubble Space Telescope* in late 1993. (NASA)

observation of galaxy formation at early times in the history of the universe.

The saga of the *HST* has provided many lessons in what to do and what not to do in space astronomy. It is unlikely that another project of its magnitude will be attempted soon, but on the other hand, the world of astronomy continues to anticipate with bated breath every new observation it makes. If all continues to go well, there will be new servicing missions and new instruments installed in early 1997, 1999, and 2002. We can expect the *HST* to continue to produce marvelous new views of the universe for some time to come.

Summary

1. Light has properties associated with both waves and particles, and these aspects are combined in the concept of the photon.
2. Visible light is just one part of the electromagnetic spectrum, which extends from gamma rays (high-energy, short wavelengths) to radio waves (low-energy, long wavelengths).
3. The continuous radiation from a star provides information on its temperature, luminosity, and radius through the use of Wien's law, the Stefan-Boltzmann law, and the more general Planck's law.
4. The observed brightness of a glowing object is inversely proportional to the square of the distance between the object and the observer.
5. Spectral lines are produced by transitions of electrons between energy levels in atoms and ions. Absorption occurs when an electron gains energy, and emission occurs when it loses energy; in both cases, the wavelength corresponds to the energy gained or lost as the electron changes levels.
6. Each chemical element has its own distinct set of spectral line wavelengths; therefore the composition of a distant object can be determined from its spectral lines.
7. The ionization and excitation of a gas can be inferred from its spectrum and analyzed to provide information on the temperature and density of the gas that produces the observed spectrum.
8. Any motion along the line of sight between an observer and a source of light produces shifts in the wavelengths of the observed spectral lines, and measurement of this Doppler effect can be used to determine the relative speed of source and observer.
9. The principal reasons for using telescopes are to collect radiation from a large area, to allow radiation from a source to be collected over longer periods of time than is possible with the unaided eye, and to provide angular resolution, that is, the ability to discern fine detail.
10. At all wavelengths, but most commonly in the radio portion of the spectrum, several telescopes can be used together to produce high-resolution images by interferometry.
11. The basic telescope design consists of a large mirror or reflecting surface to bring radiation to a focus, where an instrument containing film or an electronic detector records it. This general concept works at all wavelengths, from X ray to radio, but the details vary from one wavelength region to the next.
12. The instrument that receives the radiation at the telescope focus may be used to record images, measure the brightness of objects, or analyze the spectrum of the radiation.
13. Visible-light and radio telescopes can be located on the Earth's surface, but those for the gamma-ray, X-ray, ultraviolet, and much of the infrared spectrum must observe from above the Earth's atmosphere. On the ground, observatory sites must be chosen for clear weather, high atmospheric transmissivity, low turbulence in the air overhead, minimal pollution, remoteness from city lights, and the proper latitude for viewing the desired part of the sky.

The Spectroscope

ASTRONOMICAL

ACTIVITY

With only a modest investment in supplies, you can easily build your own spectroscope. A spectroscope is a device that spreads out light according to wavelength so that when you look through it, you see the spectra of light sources (a spectrograph is a similar device except that it records the spectra on some kind of detector). The element of a spectroscope that spreads the light out according to wavelength (i.e., the part that *disperses* the light) can be a prism, as described in the text, or it may be a surface with closely spaced grooves called a *reflection grating* (this accounts for the colors you see when looking at light reflected from a grooved surface such as a phonograph record or a laser disk). Another type of grooved surface that disperses light is called a transmission grating; in this case the grooves are embedded in a transparent material so that light can shine through. Both types of gratings disperse light due to its wave nature; as light is reflected from or passes through the grooved surface, the waves create a pattern of interference in which they alternately cancel and reinforce each other. The transmitted light is brightest where the waves reinforce each other, and the direction in which this occurs depends on the wavelength; as a result, different wavelengths of light pass through the grating in different directions. Thus the light passing through the grating is dispersed according to wavelength.

Small transmission gratings are inexpensive and usually easy to obtain; a science supply store or scientific catalog will have them, or perhaps your instructor can get some. All that you need to build a spectroscope is a small piece of transmission grating (2-inch squares are often available), a cardboard tube (an inch or two in diameter and about a foot long), some opaque tape (such as electrical tape or duct tape), some flat cardboard (such as the back of a tablet of paper), and a pair of razor blades. Tape the grating across one end of the tube, being sure to seal it around the edges so that no light can seep in except by passing through the grating (see the drawing). Tape a piece of cardboard with a tiny slit formed by the razor blades across the other end of the tube so that light can pass only through the slit. The slit must be aligned in the same direction as the grooves in the grating. To make the slit, first cut a slit about an inch long and a quarter inch wide in the cardboard, then tape the razor blades over it so that the blades form new edges that are close together (it is helpful to insert a piece of cardboard between the blades while you are taping them to ensure that the spacing is both small and uniform).

Once the spectroscope is all sealed with tape, you are ready to try it out. To use it, point the slitted end toward a light source, and look through the grating at the other end, toward the slit. Try the spectroscope on a regular (incandescent) lightbulb first; you should see a continuous spectrum, or rainbow (you will find that you have to look a little off to the side to see this). Then, at night, go outside and look at street lights, the Moon, stars, and any other light sources. What do you see? How does it relate to the discussions in this chapter about emission lines, continuous spectra, and absorption lines?

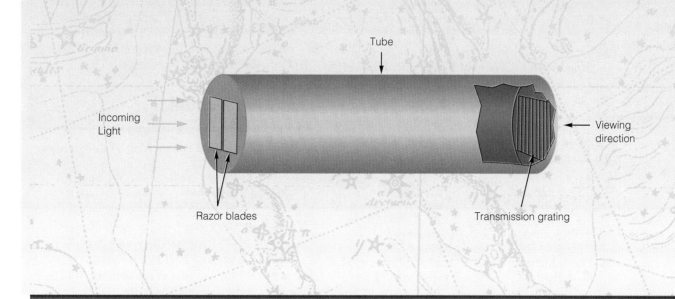

Tube

Incoming Light

Viewing direction

Razor blades

Transmission grating

14. The *Hubble Space Telescope* and other orbiting observatories have allowed astronomers to observe the universe at many wavelengths throughout the electromagnetic spectrum, providing a more complete picture than could be obtained using visible light alone.

Review Questions

1. Explain why we say that light has properties of both waves and particles.

2. Discuss possible reasons why the human eye is sensitive to the particular set of wavelengths that we know as visible light. (Hint: Consider the wavelengths that might be seen by life-forms living on a planet orbiting a much hotter or much cooler star than the Sun.)

3. Explain how the color of a star is related to its surface temperature.

4. Why are infrared-sensitive cameras useful for spotting people in the dark?

5. Explain how the spectrum of a particular type of atom is related to the atom's structure.

6. Explain how the same atom can form emission lines in one situation and absorption lines in another.

7. What is ionization and how does it occur?

8. Why does a measurement of the Doppler effect only describe the component of motion along the line of sight between an object and an observer?

9. Why do telescopes that observe in the X-ray, ultraviolet, and infrared portions of the spectrum need to be placed in space? Why is the design of the primary mirror for X-ray telescopes so different in shape from the mirror in ordinary optical telescopes?

10. Summarize the information that can be derived from the analysis of light from a distant object such as a planet or a star.

Problems

1. Calculate the frequency and energy of a photon whose wavelength is $\lambda = 5560$ Å.

2. Find out the frequency of your favorite radio station, and calculate the wavelength of its transmissions in cm, m, nm, and Å.

3. How long does it take to send a radio message from Earth to a spacecraft flying by Saturn? (Assume Saturn is at opposition.) With this time

frame in mind, would it make more sense to control the probe directly from Earth as it flies by Saturn or to preprogram the probe's maneuvers before the planetary encounter?

4. An average body temperature for an adult human is 98.6°F (~310 K). When the person is sick with a fever, however, the body temperature can rise to around 102°F (~312 K). Assuming a human radiates like a blackbody—i.e., using the Stefan-Boltzmann equation—how much more energy would a person emit when feverish than when healthy?

5. Calculate the wavelengths of maximum emission for a star of surface temperature 25,000 K; a star of surface temperature 2,500 K; the Sun's corona at a temperature of 2,000,000 K; and a human body at temperature 310 K. What kind of telescope would be best suited to observe each of these bodies?

6. Jupiter is approximately five times farther from the Sun than is the Earth. Compare the intensity of sunlight reaching these two planets.

7. Calculate the luminosity of a star whose surface temperature is 10,000 K and whose radius is 1.6×10^9 m. Compare your answer with the luminosity of the Sun and discuss the difference.

8. The rest, or laboratory, wavelength of a particular transition of hydrogen called H-α is 6563 Å. Suppose the H-α transition is observed in a star at a wavelength of 6565 Å. What can you say about the relative motion of this star along your line of sight?

9. How much greater is the light-collecting power of a 10 m telescope than that of a 4 cm telescope?

10. Suppose you purchase a reflecting telescope with a 20 cm diameter mirror. Ignoring the effects of atmospheric "seeing" and assuming you have perfect optics, what is the smallest angle you can resolve with your new telescope?

Additional Readings

Bond, B. 1994. 100 Years on Mars Hill. *Astronomy* 22(6):28.

Bond, P. 1996. Lunar Windows to the Heavens. *Astronomy* 24(9):50.

Borra, E. F. 1994. Liquid Mirrors. *Scientific American* 270(2):76.

Bowyer, S. 1994. Extreme Ultraviolet Astronomy. *Scientific American* 271(2):32.

Brunier, S. 1993. Temples in the Sky Part I—The Far South. *Sky & Telescope* 85(2):18.

Brunier, S. 1993. Temples in the Sky Part II. *Sky & Telescope* 85(6):26.

Brunier, S. 1993. Temples in the Sky Part III—The New World. *Sky & Telescope* 86(6):18.

Bunge, R. 1993. Big Scopes: Dawn of a New Era. *Astronomy* 21(8):48.

Davidson, D. 1994. Attacking the Atmosphere. *Mercury* 23(3):6.

Davis, J. 1992. The Quest for High Resolution. *Sky & Telescope* 83(1):29.

Eicher, D. J. New Visions from CCDs. *Astronomy* 19(2):70.

Englart, B.-G., M. O. Scully, and H. Walther 1994. The Duality in Matter and Light. *Scientific American* 271(6):86.

Fugate, R. Q. and W. J. Wild. Untwinkling the Stars—Part I. *Sky & Telescope* 87(5):24.

Fugate, R. Q. and W. J. Wild. Untwinkling the Stars—Part II. *Sky & Telescope* 87(6):20.

Gehrels, N., C. E. Fichtel, G. J. Fishman, J. D. Kurfess, and V. Schonfelder 1993. The Compton Gamma Ray Observatory. *Scientific American* 269(6):68.

Gillette, F. C., I. Gutlye, and D. Hollenbach 1991. Infrared Astronomy Takes Center Stage. *Sky & Telescope* 82(2):148.

Hardy, J. W. 1994. Adaptive Optics. *Scientific American* 270(6):60.

Jastrow, R. and S. Baliunas. 1993. Mount Wilson: America's Observatory. *Sky & Telescope* 85(3):18.

Kondo, Y., W. Wamsteker, and D. Stickland. 1993. *IUE:* 15 Years and Counting. *Sky & Telescope* 86(3):30.

Krisciunas, K. 1987. Two Astronomical Centers of the World: Mauna Kea and LaPalma. *Mercury* 18(2):34.

Krisciunas, K. 1994. Science with the Keck Telescope. *Sky & Telescope* 88(3):20.

McLean, I. S. 1995. Infrared Arrays: The Next Generation. *Sky & Telescope* 89(6):18.

McPeak, W. 1992. Building the Glass Giant of Palomar. *Astronomy* 20(12):30.

Mims, S. S. 1980. Chasing Rainbows: The Early Years of Astronomical Spectroscopy. *Griffith Observer* 44(8):2.

Powell, C. S. 1991. Mirroring the Cosmos. *Scientific American* 265(5):112.

Ressmeyer, R. 1995. Tradition & Technology Yerkes Observatory. *Sky & Telescope* 90(3):32.

Ridpath, I. 1990. The William Herschel Telescope. *Sky & Telescope* 80(1):136.

Robinson, L. J. 1992. Spinning a Giant Success. *Sky & Telescope* 84(1):26.

Spradley, J. L. 1988. The First True Radio Telescope. *Sky & Telescope* 76(1):28.

Svec, M. T. 1992. The Birth of Electronic Astronomy. *Sky & Telescope* 83(4):496.

Sweitzer, J. A. 1993. The Last Observatory on Earth. *Mercury* 22(5):13.

Tucker, W. 1995. *ROSAT* in Review. *Sky & Telescope* 90(2):35.

Tucker, W. and K. Tucker 1986. *The Cosmic Inquirers: Modern Telescopes and Their Makers.* Harvard University Press.

Wearner, R. 1992. The Birth of Radio Astronomy. *Astronomy* 20(6):46.

Web Connections

The Review Questions and Problems also appear at the following URLs:
http://universe.colorado.edu/ch6/questions.html
http://universe.colorado.edu/ch6/problems.html

Chapter 7
Overview of the Solar System

*Pluto is the last little outpost before
stepping out to the stars.*
Stefan Finch

 s we prepare to explore the planets and interplanetary bodies that make up the solar system, it helps to begin with an overall description of the system and the general processes that shape the planets. Today this overview is very different from what we might have said only a few decades ago, because a major era in planetary exploration has recently overturned our view of the planets and the processes that have shaped them. The *Voyager* missions to the outer planets were the culmination of a 20-year period when all the planets except Pluto were visited by probes from Earth.

We start this chapter with a general description of the solar system and follow this with a summary of the methods by which we learn about the planets and interplanetary bodies. The rest of the chapter is then devoted to an outline of the basic processes that mold the planets. We will refer repeatedly to these general processes in the chapters that follow, as we discuss the individual objects that make up the solar system.

Overview of the Solar System

The solar system is dominated by the Sun, which contains more than 99% of the total mass and provides the major source of energy that heats the surfaces of most of the planets. A distant observer would have great difficulty even detecting the small, dim objects orbiting this rather ordinary star. Recall from our discussion in Chapter 1 how small and close together the planets are, relative to interstellar distances. Yet the planets are individual worlds of great

complexity, and at least one of them serves as the only home for life that we know of. Thus we are justified in devoting some time and effort to learning what we can about the planets, and this chapter will serve as our introduction to them.

There are nine known planets **(Table 7.1)**, which orbit the Sun at average distances ranging from about 58 million km (just under 0.4 AU) to almost 6 billion km (almost 40 AU). To many people, this listing of the nine planets is a complete inventory of the solar system. But one of the revolutionary new understandings of recent years is that the solar system actually contains a continuum of bodies orbiting the Sun, with the major planets representing the high-mass end of the distribution. There are innumerable smaller objects as well, some of them only discovered within the past few years. These include the asteroids or minor planets, most of which orbit between Mars and Jupiter; the comets, which appear to fall into two general classes having somewhat different origins; the interplanetary dust disk, a tenuous medium that is confined to the plane of the planetary orbits; and assorted other bodies such as the icy asteroid-like objects occasionally found among the outer planets. In addition to this family of Sun-orbiting bodies, the solar system contains many satellites, moonlets, and rings, which orbit planets rather than the Sun.

The solar system as a whole has a disklike structure, much like a record on a turntable, except that in this case the record is not rigid but instead acts like a fluid. In contrast to the behavior of a rigid disk, each planet and interplanetary object is a free body and follows its own distinct orbit about the center. The plane of the solar system is very thin compared with its overall diameter **(Fig. 7.1)**. Only the orbits of Mercury and Pluto deviate from the plane established by the Earth's orbit (i.e., the ecliptic) by more than

Table 7.1

The Planets

Planet	Semimajor Axis (Au)	Sidereal Period (yr)	Mass[a]	Diameter[a]	Density (g/cm³)	Temperature (K)
Mercury	0.387	0.241	0.0553	0.378	5.427	100–700
Venus	0.723	0.615	0.815	0.950	5.204	730
Earth	1.000	1.000	1.000	1.000	5.518	200–300
Mars	1.524	1.881	0.107	0.532	3.933	145–300
Jupiter	5.203	11.86	317.83	10.97	1.326	165
Saturn	9.539	29.46	95.162	9.14	0.687	134
Uranus	19.182	84.01	14.536	3.981	1.318	76
Neptune	30.06	164.79	17.147	3.865	1.638	74
Pluto	39.53	247.68	0.0021	0.1785	2.050	40

[a]The masses and diameters of the planets are expressed relative to the mass and diameter of the Earth, which are 5.974×10^{24} kg and 12,756 km, respectively.

about 3°, and even these two have orbital planes lying within a few degrees of the ecliptic (the tilt of Mercury's orbit is about 7°, and that of Pluto is just over 17°).

Despite the individual idiosyncrasies of the planets, they conform to an overall pattern in several general ways. All of the planets orbit the Sun in the same direction, for example, and most of them spin in the same direction as well. We call this direction, which is counterclockwise as seen from above the North Pole, the **prograde** direction **(Fig. 7.2).** The rare exceptions in which planetary or satellite motions go against this trend are referred to as retrograde motions (this is similar to the retrograde motion discussed in Chapter 2, because it involves motion in the opposite direction from the norm, but here we are talking about actual motion as seen by an outside observer— we might think of it as *sidereal* retrograde motion).

Most of the planetary orbits are nearly circular; that is, they are ellipses with small eccentricities (see Appendix 8). The paths of Pluto and Mercury deviate significantly from circular, but even these two planets have orbits that are far closer to circular than is the path of a typical comet. The comets, especially the long-period ones, usually follow very elongated elliptical paths around the Sun, an indication that the formation of these interplanetary bodies was significantly different from the origin of the planets.

The spin axes of most of the planets and their moons are oriented approximately perpendicular to their orbital planes, although there is quite a large range of **obliquities,** as the tilts of the planetary rotation axes are called. Except for Uranus and Pluto, all of the planets have obliquities of less than 30° (the Earth's tilt is 23.5°). The unusual tilts of Pluto and Uranus (both over 90°) will require some singular explanation, and may be telling us that the formations of these two planets differed from the formation of the others.

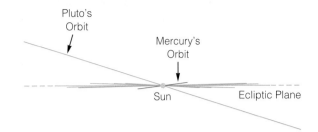

Figure 7.1 Alignment of planetary orbits. As this edge-on view of the planes of the planetary orbits illustrates, seven of the nine planets have orbital planes closely aligned (within about 3° or less). Two planets, Mercury (7°) and Pluto (17°) deviate significantly. The relative sizes of the orbits are not drawn to scale here.

Most of the interplanetary bodies, such as asteroids, the periodic comets, and interplanetary dust grains, obey the same general rules as the planets, and follow prograde orbits that lie near the plane of the ecliptic. There are some exceptions, however: the long-period comets (discussed in Chapter 11) trace out paths that are randomly oriented, and some of the interplanetary dust has apparently been disturbed enough to deviate from strict alignment with the ecliptic plane.

Classifying the Planets

Of the nine planets, eight fall cleanly into two general categories, based on their general characteristics. The innermost four are called the **terrestrial planets** because they are Earth-like in their overall nature. Like the Earth, these planets are relatively small (compared with the giant planets), have high densi-

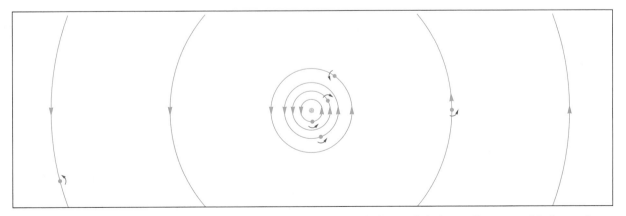

Figure 7.2 Prograde motions. In this scale drawing of the inner six planets, arrows show the direction of orbital motion (blue arrows) and the direction of spin (red arrows). All of these motions are in the same direction, except that Venus (the second planet from the Sun) spins backward (i.e., in the retrograde direction).

The Terrestrial Planets

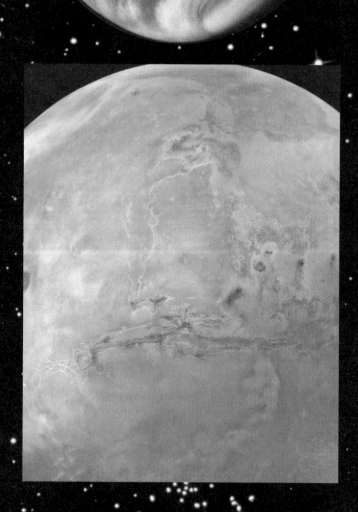

Figure 7.3   All of these images were obtained from space: Mercury (upper left) and Venus (upper right) were both photographed by *Mariner 10;* Earth (lower left) by one of the *Apollo* missions; and Mars (lower right) by one of the *Viking* orbiters. The terrestrial planets are all relatively small and dense with rocky surfaces. (NASA) http://universe.colorado.edu/fig/7-3.html

ties indicating a rocky or metallic interior, and have hard, rocky surfaces **(Fig. 7.3).** The Moon **(Fig. 7.4)** is sometimes classified as a terrestrial planet due to its large size, its similar overall characteristics, and its similar geological history.

The terrestrial planets are all **differentiated,** meaning that the denser materials of which they are made have sunk to their centers. Differentiation can occur only if a planet is fully or partially molten inside, so we conclude that each of the terrestrial planets underwent some melting at some time in its past. In some the interiors are still quite hot today, and the Earth, for one, is still partially molten.

The terrestrial planets can have atmospheres if their surface gravities are sufficiently strong to trap gas particles. (Their ability to retain atmospheres depends also on the surface temperature, as discussed later in this chapter.) Only the Moon and Mercury fail to retain any significant atmosphere.

The next four planets beyond Mars are the **gaseous giant planets (Fig. 7.5),** also called the **Jovian planets** after the largest and closest one, Jupiter. These planets are radically different from the terrestrial planets, being very much larger, having no solid surfaces and having much lower densities. Where the terrestrial planets have densities ranging from around 3 g/cm^3 to over 5 g/cm^3, the densities of the gaseous giants range from 0.7 to 1.6 g/cm^3 (for com-

parison, the density of water is 1 g/cm^3, and that of ordinary rocks is in the range of 2–3 g/cm^3).

All of the giant planets have atmospheres; in fact, as noted already, they do not have solid surfaces, but consist instead of a fluid (gaseous and then liquid) structure nearly all the way to their centers (it is thought that each of the giants has a small, rocky core). At the highest levels, these atmospheres are very cold and contain, in addition to the dominant hydrogen (in the form of H$_2$), more complex molecules consisting primarily of the simplest elements such as hydrogen, carbon, nitrogen, and oxygen (there is also a good deal of helium in atomic form). By contrast, the terrestrial planets contain very little hydrogen and helium in their atmospheres because these lightweight gases can easily escape the low gravitational fields of the small, warm planets.

The ninth planet, Pluto, does not fit readily into either category, but instead more closely resembles some of the giant moons of the outer planets **(Fig. 7.6).** These bodies consist of ice and rock in comparable proportions (giving them overall densities ranging from 1 to 2 g/cm^3). They have hard surfaces and are generally differentiated, and some of them (Pluto included, at least some of the time) have gaseous atmospheres. Current models of the formation of the planets suggest that Pluto was formed by a rather different process than either the giant planets or their satellites. Current thinking among astronomers is that perhaps Pluto should be removed from the list of *bona fide* planets and instead be designated as the largest of the nonplanetary bodies orbiting the Sun (see the Science and Society feature in this chapter).

Methods for Observing the Planets

In Chapter 6, we discussed general principles of observational astronomy along with the properties of light. In those discussions, we assumed that our only means of studying objects in the universe is to collect and analyze light that reaches the Earth. This passive approach is indeed our only choice for most of the universe, but for the planets we can apply other, more direct techniques.

Obviously, the most direct method of studying a planet is to go there and examine it at close hand. This direct sampling method, with or without human interaction, has been applied to just five bodies in the solar system, the Earth, the Moon **(Fig. 7.7),** Venus, Mars, and Jupiter (the latter three have been sampled by robot probes). Naturally, we find that these are the best-studied objects, though in the case of the

Figure 7.4 The full Moon. This is a high-quality photograph taken through a telescope on the Earth. Note that in this telescopic view, the Moon is upside down and reversed left-to-right, compared to the view with binoculars or the unaided eye. (Lick Observatory, University of California) http://universe. colorado.edu/fig/7-4.html

The Jovian Planets

Figure 7.5 The giant planets. The images of Jupiter (upper left), Saturn (upper right), Neptune (lower left), and Uranus (lower right) were all obtained at close range by the *Voyager* probes. The giant planets are very different from the terrestrial planets, being much larger and far less dense and having no solid surfaces. (NASA) http://universe.colorado.edu/fig/7-5.html

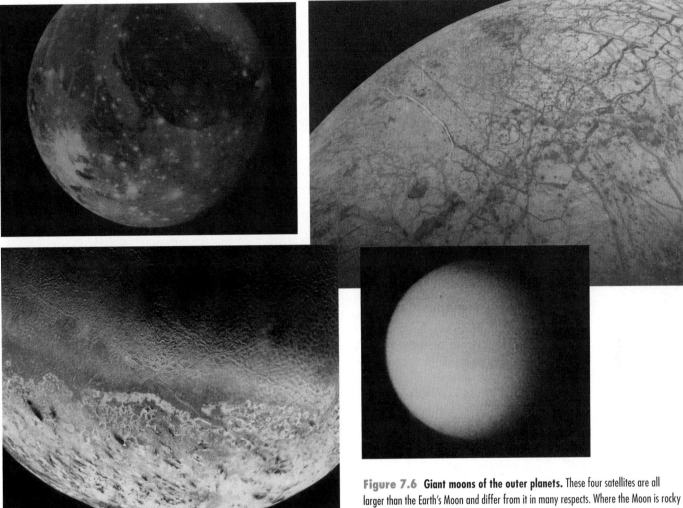

Figure 7.6 Giant moons of the outer planets. These four satellites are all larger than the Earth's Moon and differ from it in many respects. Where the Moon is rocky throughout, these giant satellites are partially composed of ice. Whereas the Moon is inactive geologically, some of the giant satellites are quite active. Here we see Ganymede of Jupiter (upper left), Europa of Jupiter (upper right), Triton of Neptune (lower left), and Titan of Saturn (lower right). Ganymede and especially Europa show fault lines due to crustal shifting; Triton has active ice geysers and a dimpled crust due to slumping of the surface; and Titan has a thick atmosphere, possibly created by outgassing that is still active. (NASA)

Figure 7.7 Human exploration of the Moon. This is astronaut Harrison Schmidt on the last *Apollo* mission to the Moon (*Apollo 17*), collecting samples of small lunar rocks. (NASA)

Moon at least, perhaps not necessarily the best understood. Direct analysis of surface samples has been achieved for Mars through the U.S. *Viking* missions in the mid-1970s **(Fig. 7.8)** and for Venus through the Soviet *Venera* landers (see Chapter 9).

Even if it is impractical for us to visit the rest of the planets personally, we can send unmanned probes to visit them and make observations from close range. This has now been accomplished for all of the planets except Pluto; the *Voyager 2* flyby of Neptune in August 1989 marked the completion of a remarkable era of planetary exploration. Both the U.S. and the Soviet space programs made stunning discoveries through their robot missions to the planets. Several

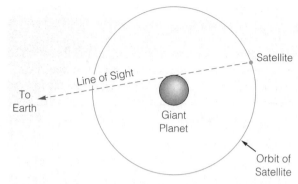

Figure 7.9 **Using a satellite to probe a planetary atmosphere.** As the satellite passes behind the disk of its parent planet, the satellite is blocked by the planet. As the satellite passes behind the edge of the planet's disk as seen from the Earth, its light passes through the planetary atmosphere on its way to us. Scientists on the Earth can analyze the effects of the atmosphere on the light from the satellite to gain information on the gases in the atmosphere of the planet.

Figure 7.8 **Robots on Mars.** This is a U.S. *Viking* lander in a simulated Martian environment. Two of these robotic vehicles landed successfully on Mars in 1976, and carried out numerous experiments and observations. A mechanical arm was able to pick up soil samples and deposit them in test chambers where experiments were conducted in search of microscopic life-forms. *Viking*'s cameras took many panoramic photographs of the surface, and kept operating long enough to show how the dust and soil deposits vary with the seasonal winds. (NASA)

advantages can be gained by making close-up observations: vastly more detail can be seen; objects can be observed from different angles (this is particularly useful in interpreting light-scattering properties); charged particles trapped in planetary magnetic fields can be captured and measured directly; the planetary gravitational field can be measured directly, and from those measurements the internal structure of the planet may be determined; and it may be possible to measure weak emissions (such as radio noise from particle belts) that cannot be detected from the Earth.

Even though the "grand tour" of the outer planets by *Voyager 2* is over, several other sophisticated planetary probes are in action or planned. The *Galileo* probe arrived at Jupiter in late 1995. It dropped a probe into the atmosphere while the main spacecraft went into orbit, allowing for long-term exploration of Jupiter and its major moons. Exciting new data and images were obtained, and will be described in Chapter 10. The United States is planning a new mission to Mars to achieve some of the goals of the lost *Mars Observer* spacecraft, and the *Cassini* mission, a joint U.S.-European project, will send an orbiter to Saturn and probes into the

atmosphere of its giant moon Titan. Around the turn of the century, the United States may launch a mission to Pluto that will reach the most distant planet about eight years later.

Despite the enormous successes of the manned and unmanned probes to the Moon and the planets, much can still be learned from Earth-based observations. Of course, for a long time, telescopic data were all we had on any of the planets, so our first "explorations" of each body were of the traditional astronomical kind. From these observations, scientists were able to learn a great deal: surface temperatures could be derived from Wien's law (Chapter 6); surface compositions could be roughly analyzed from spectroscopy of reflected light (particularly at infrared wavelengths); and atmospheric composition could be deduced, again from spectroscopy. One of the great advantages of Earth-based observations, still important even in the age of space exploration, is the ability to make observations repeatedly over long time periods. This enables us to gather detailed information on seasonal variations, for example, and also allows us to use orbiting satellites as probes of the atmospheres of the planets **(Fig. 7.9)**. Earth-based radio observations complement the visible and infrared data by providing information on charged particle belts (and indirectly on planetary magnetic fields, which create the belts).

Even space-based observations from the vicinity of the Earth are important. Orbiting spacecraft such as the *International Ultraviolet Explorer (IUE)* have obtained extensive ultraviolet spectroscopic observations of several of the planets, and sounding rocket observations have done ultraviolet work as

Pluto and the Press

As you have learned from reading this chapter, astronomers believe that there are very many more bodies in the solar system than the traditional nine planets. The inventory of solar system bodies will no doubt continue to grow as searches reveal more comets and additional members of the family of outer solar system bodies left over from the time of solar system formation. The possibility that the known membership of the solar system might grow may not seem very surprising, but recently astronomers have discussed the possibility of *reducing* the number of recognized planets. This discussion has been highly publicized, and it helps bring home a couple of interesting points about how science works and how society perceives science.

As you have seen in this chapter and will learn in more detail later (Chapter 10), Pluto defies most of the normal guidelines for planethood. It is by far the smallest of all the planets (being only about two-thirds the diameter of our own Moon), and it does not fit neatly into the category of either terrestrial planet or gaseous giant. Pluto lies in the outer reaches of the planetary domain beyond the region of gas giant planets, yet it is small and has a hard surface. Unlike a terrestrial planet, however, Pluto is composed in large part of ice, having only about half the average density of a terrestrial planet such as Mars. Where most of the planetary orbits are roughly circular, Pluto has such an eccentric orbit that it actually spends part of its time closer to the Sun than Neptune.

Even as Pluto's shortcomings as a planet have been revealed, astronomers have come to recognize a large class of objects orbiting the Sun between 30 and 50 AU. These icy bodies are thought to be remnants of the formation of the solar system, essentially icy planetesimals that have survived because of their remoteness from the Sun.

Given all of these inconsistencies between Pluto and the other planets and the discovery of the numerous subplanetary bodies in the outer solar system, it is not surprising that astronomers are beginning to suggest that Pluto is more properly regarded as a member of the family of primordial planetesimals rather than as a legitimate planet in its own right. Pluto's classification is not a topic of daily discussion or debate among astronomers, however; no one is losing sleep over this issue. In due course, the appropriate subcommittee of the International Astronomical Union (IAU) will discuss the question and make a recommendation, and we may begin to think of our solar system as having eight planets instead of nine. But whatever astronomers officially decide about Pluto, nothing will change: the solar system and Pluto will be the same regardless of the decisions we humans may make about which object falls into what category.

The general public, on the other hand, with ample encouragement from the media, have perceived this as a burning issue for astronomy. The national television networks all featured the story in February 1996, interviewing astronomers, schoolchildren, and science educators for reaction. Opinions ranged from disappointment at the prospect of "losing" one of our planets to expressions of affection for Pluto, whatever it is, to arguments for maintaining Pluto's planetary status (few seemed to favor demoting Pluto). Some of the coverage was done in a humorous vein, which was perhaps appropriate, given the low priority professional astronomers are placing on the question.

The important insight one might have gained from the notorious Pluto affair of 1996 is not so much whether Pluto is a legitimate planet, but that many people perceive science as a structure made of facts. Many people seem to think that what astronomers do is classify things—decide firmly what something is to be called—and then move on. Either Pluto is a planet or it is not; there is little room in this view for the possibility that Pluto does not quite fit into any category. To people with that view of science, the question of how to define Pluto appears to be very important indeed. But to a scientist, the question of what we call Pluto is almost irrelevant; it is far more important to understand how Pluto formed and evolved and what it has to tell us about the history and evolution of our solar system.

An additional lesson to draw from the public discussion of Pluto is the tendency of the mass media to portray science in oversimplified terms. The next time you read a newspaper report about a controversy in science or see a televised opinion poll on some scientific controversy, remember the Pluto debate of 1996. Science is not about black and white, and what to call something is almost never an issue of substance in any scientific controversy.

well. The **Hubble Space Telescope,** orbiting the Earth since May 1990, is making enormous contributions to our knowledge of the planets; it is not only providing both extended wavelength coverage (ultraviolet as well as visible) and unprecedented clarity of images, but is also revealing surface details and allowing them to be monitored over time. This last factor is especially important, because a rapid flyby encounter by a robot spacecraft does not provide any information about a planet's seasonal changes or other kinds of phenomena that change with time. The *HST,* planned to have at least a 15-year operational lifetime, has already shown its usefulness in studying planetary weather and seasons.

Even though the completion of the *Voyager* mission signaled the end of a particular phase of planetary exploration, the wide assortment of ongoing and planned studies of the planets will continue to produce a steady stream of new information. There is no doubt that future astronomy textbooks will require extensive revisions as the exploration of the solar system continues.

Origins of the Planetary System

A more complete discussion of the formation of planetary systems will be found in Chapter 12, but here it is useful to have some idea of the basic processes by which the solar system formed. Certain general processes apply to all of the planets, but there are some differences as well, particularly between the terrestrial and giant groups.

The solar system apparently formed from a disk of gas and dust that orbited the young Sun at the

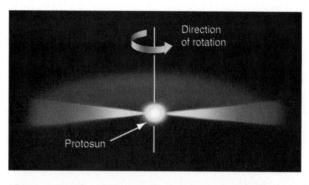

Figure 7.10 **The solar nebula.** This is an edge-on view of the disk of gas and dust from which the Sun and planets formed. Rotation of a collapsing interstellar cloud caused the flattening. Gravitational compression created a hot concentration of gas at the center that was to become the Sun. The disk of surrounding material became clumpy as gas began to condense into solid form, and the clumps then merged, eventually creating the planets.

time of its formation **(Fig. 7.10).** This disk, composed of material gathered together by gravitational forces, was initially dominated by hydrogen, the most abundant element in the universe. The composition (by mass fraction) of the original preplanetary material was approximately 73% hydrogen and 25% helium, with all the other elements composing the remaining 2% of the mass. This is the composition of the Sun and most of the stars as well as of the interstellar matter in the galaxy today, and astron-omers think it was about the same 4.5 billion years ago, when the solar system formed.

At an early time during the formation of the system, the innermost part of the disk, the central core from which the Sun was to form, became hot as the gas in the core was compressed, causing the release of gravitational potential energy. This heating resulted in the emission of intense infrared radiation, which in turn heated the gas and dust surrounding the core.

The disk of atoms and small, solid particles that orbited the young Sun had a natural tendency to become clumpy, which caused the condensation of larger solid objects than the original tiny dust grains. With time, these larger objects built up so that first snowball-sized, then basketball-sized, then even larger bodies were formed. The rate of growth of these objects increased until the disk became a collection of planetesimals, or preplanetary bodies, with diameters up to several thousand kilometers. During the buildup, or **accretion** process, some elements were driven out by heat, particularly in the inner solar system. The lightweight **volatile elements** (i.e., those most easily vaporized) were the least likely to become trapped in the solid state. Hence, the planetesimals that formed in the inner solar system never contained large quantities of lightweight elements such as hydrogen and helium.

The planetesimals themselves collided occasionally, often with small relative speeds (because they orbited the Sun in the same direction and at nearly the same speed at a given distance from the Sun). These collisions caused the planetesimals to coalesce, or stick together, eventually producing the planets of today. The entire accretion process is thought to have taken place fairly rapidly by astronomical standards; astronomers believe that the planets formed within the first few hundred million years after the Sun was born.

Note that the planets that formed in the inner solar system started out with small quantities of the volatile elements. These elements were unable to condense into solid form in the first place due to the high temperatures in the inner part of the system caused by compression of the core object and the radiation that this proto-sun consequently emitted. In the outer solar system, however, the temperature was so low that even hydrogen and helium could become trapped; thus the giant planets started out with a radically different composition from the terrestrial planets. The ability of the cold outer planets to trap these abundant gases gave the outer planets much larger initial masses than the inner planets. These larger masses helped the outer planets to gravitationally capture additional quantities of gas from their surroundings. Thus the outer planets grew into much larger, more massive bodies than the inner planets and had rather different compositions as well.

In the inner solar system, proximity to the Sun created strong tidal forces (see Chapter 5 as well as further discussion later in this chapter) that prevented the planets from capturing surrounding gas and dust, and the higher temperature of this material made it more difficult to trap anyway. The result was a set of inner planets that contained mostly heavy elements and turned out to be dense and rocky, and a set of outer planets that were cold and far more massive and contained ample quantities of lightweight gases. The trapped material surrounding the outer planets formed into extensive systems of rings and moons, while the inner planets tended not to have significant satellites (except the Earth, whose large moon is thought to have formed largely by accident; see Chapter 8).

During the latter phases of planet formation, many planetesimals and smaller leftovers remained in solar orbit. Some of these collided with the young planets, creating anomalies such as the extreme tilt of Uranus's axis, the slow backward spin of Venus, and the formation of the Earth's moon. Other bodies were ejected from the inner solar system by near collisions with planets, and a large number of small bodies were moved into orbits farther from the Sun by this process. A residual disk still survives beyond the orbit of Neptune; it contains thousands of icy cometary bodies and a few larger objects as well. Pluto is the largest of these, but others with sizes up to a few hundred kilometers have been spotted.

The inner solar system is now relatively free of leftover debris but not completely so. The asteroids, which orbit the Sun between the paths of Mars and Jupiter, are thought to be planetesimals that survived without merging into the larger planets. These bodies, along with the icy objects orbiting beyond Neptune, represent primitive material from the early days of solar system formation.

All of the planets and many of their satellites, as well as the largest of the asteroids, are round. This is because the inward force of gravity controls the

shapes of these bodies. Gravity presses inward from all sides, resulting in compression in the interior. As the material inside a developing planet is compressed, pressure builds up that counteracts gravity, eventually stopping the contraction. The result is a spherical body, denser in the center than in the outer region, in which gravity and pressure are balanced at every point. Only if there is another force to counteract gravity will a body end up in a nonspherical shape. For very small objects (i.e., the smaller asteroids or the tiniest moons), the strength of the material of which the object is composed may exceed the inward force of gravity, allowing the body to retain a nonspherical shape. Thus some of the smallest objects in the solar system have irregular shapes, while all of the larger ones are spherical.

While a planet formed, gravitational forces squeezed it together tighter and tighter, causing compression and heating in its interior. Radioactive elements in the interior of a planet release additional heat as they decay, helping to keep the interior of the planet hot, even to present times. In the terrestrial planets, this heating caused the interior to be partially molten over long periods of time, which in turn allowed the bulk of the heaviest elements (such as iron, nickel, cobalt, and the superheavy species such as lead, uranium, and so on) to sink to the center of the planet through differentiation. In some of the terrestrial planets (including the Earth and probably Mercury and Venus), the core remains at least partially molten, whereas others (probably Mars and the Moon) have cooled enough to be essentially solid throughout **(Fig. 7.11)**.

In the outer planets, composed as they are of lightweight gases, most of the interior remained

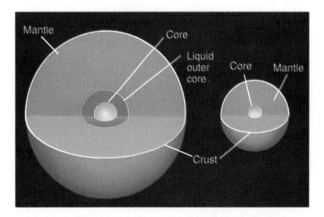

Figure 7.11 **Internal structure of the terrestrial planets.** At the left is a cross-sectional view of a planet like the Earth, which has a liquid zone in its outer core. Other terrestrials, particularly the small ones such as Mars and the Moon, have internal structures more like the sketch at the right, with no liquid zone.

The Origin of the Solar System

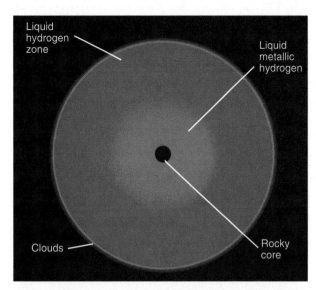

Figure 7.12 Internal structure of a gas giant. Beneath the visible cloud tops, the desnity and temperature increaseinward. As the density increases, hydrogrn is forced into unusual forms, which are discussed in Chapter 8. Because the planet is fluid throughout, differentiation occurs readily, and virtually all of the planet's heavy elements sink to the center, forming a small, rocky core.

fluid. Thus these planets have no solid outer surfaces, and differentiation was able to segregate out the heavy elements very thoroughly, forming small, dense solid cores deep within **(Fig. 7.12).**

Evolution of the Planets: Planetary Processes

Once formed, the planets have not remained static, but instead have evolved and changed with time. This is more true of some than others, as the terrestrials have been subjected to a wider range of processes that alter their nature than have the gas giants. But even the cold outer planets have gradually changed with time, as the heat of their formation has dissipated.

In the following subsections, we review the major processes that work on planets as they age. These basics of planetary science will then be applied to the individual planets in the chapters that follow.

Planetary Energy Budgets

The surface conditions on a planet, and indeed the interior conditions as well, are determined by a balance between energy gains and losses. A planet starts out with substantial internal heat due both to its formation and to radioactivity, and this heat is dis-

sipated through various processes. Additional heat energy is deposited at the surface through the absorption of sunlight, and energy is radiated away, obeying the laws of thermal radiation (discussed in Chapter 6). The relative rates of gain and loss determine the temperature, while other properties of a planet in turn depend on the temperature. Thus, in a very real sense, the planetary energy budget controls the basic properties of a planet, inside and out.

The major sources of energy for a planet are internal heat and external radiation. The internal heat has two components: stored heat from the formation of the planet, and radioactivity due to naturally unstable elements. Technically, the heat energy stored during formation is gravitational potential energy that was converted to heat when the planetesimals came together to form the planet. In practice, you may think of this as the heat of collision and compression during this merging process. This is *old* energy by now, since the solar system has an age of about 4.5 billion years, but as we will see, the loss of heat from a planetary interior is an inefficient, slow process.

It may be surprising to you that natural radioactivity is an important source of internal heat for a planet, since we normally don't feel the heat of radioactive elements as we roam over the Earth's surface. The reason we don't feel this heat is that most of the radioactive elements that occur naturally are quite heavy and exist primarily in the deep interiors of the planets, as the result of differentiation.

The internal heat energy stored inside a planet decreases with time for at least two reasons. First, as heat is transported outward to the surface, it is gradually lost due to radiation into space. Second, the radioactive elements gradually decay, reducing their heat production.

A key to understanding the current physical state of any planet is the amount of heat it has retained, which depends on the rate of heat loss. A fundamental law of physics is that heat will always flow from a region of high temperature (such as a planetary interior) to one of lower temperature (such as the space surrounding a planet), so a planet will inevitably cool with time. The question is how fast it cools.

The rate of heat loss depends on the size of the planet. Large planets have more mass and hence more capacity to store heat, so we might expect large planets to remain hot for longer than small planets. On the other hand, a large planet has more surface area from which to radiate, so its rate of heat loss is higher than for a smaller planet. The mass is proportional to the cube of the radius, however, while the surface area is proportional to the square of the radius, so the heat retention ability of a planet grows with radius. Therefore, as a general rule, a larger planet will retain its internal heat longer than a smaller one.

The details matter, though, particularly the mechanism by which heat is transported from the interior to the surface. The most efficient transport mechanism is **convection.** This is a gradual overturning of a fluid that is heated from below, as hot fluid (which is less dense than its surroundings) rises while cooler fluid descends. As we will see, convection plays a major role in the circulation of atmospheric gases. But how can it operate inside a planet, particularly a solid terrestrial body? It is easy to imagine convection inside the gas giants, but not so easy for the Earth or the other terrestrials. Inside the terrestrials are zones that are not exactly fluid in the normal sense, but are sufficiently pliable and plastic that a very slow convective overturn can occur; where it does, this is the main mechanism for heat flow from the interior outward. Heat flow is less efficient and therefore slower in the more rigid terrestrial bodies, which tend to be the smaller and cooler ones. In those cases, heat flows primarily by **conduction,** which is generally a very slow process.

The surface of a planet receives incident radiation from the Sun, which causes some heating to supplement the flow of heat from the interior. The intensity of the solar radiation depends on the distance of the planet from the Sun (according to the inverse square law) and also on the fraction of the sunlight that is absorbed. A cloud-covered planet like Venus reflects a large fraction of the light that strikes it, while a dark body such as the Moon or Mercury reflects a much smaller fraction. The **albedo** of a planet is the fraction of incident radiation that is reflected; values for the terrestrial planets range from 0.06 (for the Moon) to 0.65 (for Venus). For the gas giants, the albedos are generally around 0.5. The albedo is an indicator of the surface energy balance on a planet, with the highly reflective planets (i.e., those with high albedos) receiving less solar heating than those with low reflectivities (low albedos).

Another factor that affects the energy balance of a planet is its atmosphere, as well as the ability of the atmosphere to retain heat. Here a phenomenon called the **greenhouse effect** comes into play, and its importance varies a great deal from planet to planet. When sunlight strikes the surface of a planet and heats it, the surface responds by emitting radiation. For typical planetary surface temperatures, the wavelength of maximum emission (according to Wien's law; see Chapter 6) lies in the infrared part of the

spectrum. Many molecular gases strongly absorb infrared radiation, so the radiation from the surface of a planet may be absorbed in its atmosphere. If so, heat builds up in the atmosphere, raising the temperature. The process is called the greenhouse effect because the glass walls of a greenhouse act in the same way, trapping infrared radiation due to solar heating inside the greenhouse. The high temperature inside a closed car in the summer is due to the same mechanism.

Carbon dioxide (CO_2) is a very efficient absorber of infrared radiation, so greenhouse heating is particularly important on Venus, which has a dense atmosphere of nearly pure carbon dioxide. Mars also has a carbon dioxide atmosphere, but a much thinner one than Venus, so the effect on Mars, while strong, is not as important as on Venus. On the Earth, water vapor, methane, and carbon dioxide (including a growing portion introduced by industrial activities) are greenhouse gases, but they exist only in trace amounts and our surface temperature is only moderately enhanced by the greenhouse effect.

Many other factors, some quite individualized, also affect the rates of energy gain and loss for a planet; hence, the bodies of the solar system exhibit a variety of internal and surface conditions. In each case a balance is reached, but it can be a very different balance from one planet to the next. Thus we have a rich variety of physical conditions; furthermore, we can understand how changes can occur with time if any of the rates of energy loss or gain are altered. Energy balance has a great deal to say about the evolution of the planets, and we will keep this in mind as we discuss them individually.

Geological Processes in the Terrestrial Planets

The terrestrial planets are still alive geologically in varying degrees, depending largely on the quantity of stored heat that remains and that is trying to escape the interior. The Earth appears to be the most active, while the Moon is probably the most inert. Venus may be active internally, but apparently does not undergo the same kinds of surface activity (such as continental drift) as the Earth, and Mars appears to have had little geological activity since its early days, more than 3 billion years ago.

The general term for the dynamic interior processes and the associated surface phenomena (such as volcanic eruptions and earthquakes) that can take place in a terrestrial planet is **tectonic activity** or **plate tectonics.** In brief, the crust is a thin, brittle surface layer with a lower density than the interior, and internal stresses can literally move the crust around. On the Earth the crust has fragmented into plates that are mobile, whereas on Venus and Mars such motion is limited or nonexistent. It is thought that the driving force for Earth's tectonics is slow-moving fluid flows in the outer portion of the mantle, the zone of the interior extending from just below the crust all the way to the outer core, about halfway to the center of the Earth. As mentioned in the preceding section, these flowing motions in the mantle are thought to be driven by convection. The mantle of the Earth is usually considered to be solid and rigid, but in fact it is somewhat plastic and deformable, so that very slow flowing motions are possible.

The other terrestrial planets, as already noted, do not appear to have such active crustal motions. The reasons for this will be discussed as we treat each planet individually, but here we can make some general comments. The more massive a planet, the greater the degree of internal heat retention and the higher the residual temperature today. Thus we might expect that the two most massive terrestrial planets, Venus and the Earth, would be the ones most likely to be undergoing tectonic activity today. Indeed, this appears to be the case; we have already mentioned some of the activity occurring on the Earth, and we will find that Venus has had extensive outpourings of lava and may even be volcanically active in modern times. On the other hand, the crust of Venus has undergone little or no global motion, and we will have to try to understand that. Mercury apparently has a molten zone in its interior, while Mars probably does not have a molten core and shows little evidence of current activity.

Because the giant planets have no real surfaces and no solid interiors (except for their cores), we do not speak of geological or tectonic activity in the same manner as we do for the terrestrial planets. On the other hand, some similar processes are taking place. For example, all of the giant planets have strong magnetic fields, indicating that electrical currents are flowing in their interiors. The presence of these currents is not surprising because all of these planets are fluid throughout most of their interiors, and all of them rotate very rapidly.

Surface Processes on the Terrestrials

In addition to tectonic activity, which can modify the surface of a terrestrial planet through quakes and volcanic eruptions, there are other forces that can modify the hard surfaces of the terrestrial planets. The surfaces of these planets are largely composed of rock, which may be covered by water over large

regions (on today's Earth, for example, and perhaps on the early Mars and the young Venus as well) and decomposed into soil, dust, or sand in other regions.

On the Earth, we recognize three general types of rock, based on origin: (1) **igneous rocks,** which formed from molten material that has hardened; (2) **sedimentary rocks,** which formed from the deposition of mud and clay deposits on the floors of oceans and lakes; and (3) **metamorphic rocks,** which have been altered by heat and pressure resulting from large-scale crustal movements. Continual geological activity can gradually change rocks from one of these forms to another. For example, a rock may first form as an igneous rock as the result of an eruption, but later be broken down by wind and water into sand that is deposited underwater and forms sedimentary rock; still later this rock may be uplifted by geological forces and squeezed and compressed so that it becomes a metamorphic rock.

Apart from its origin, we may also classify a rock according to its chemical composition. Thus we define minerals based on composition. By far the most common minerals on the surface of the Earth and the other terrestrial planets are the **silicates,** which contain compounds of silicon and oxygen. The common basalts and granites on the Earth are silicates of igneous origin. Other common mineral types include the **carbonates,** which are dominated by carbon-bearing compounds, and **oxides,** consisting largely of oxygen compounds.

We have seen that tectonic activity can alter the face of a terrestrial planet significantly, as pieces of the crust move about, and at the same time that local processes may alter the nature of surface rocks. Additional mechanisms are at work as well. One such mechanism is **impact cratering.** The solar system still contains a large number of small bodies orbiting the Sun between the planets, and occasionally these objects strike the surface of a planet, as happened recently when comet Shoemaker-Levy 9 struck Jupiter in July 1994. Such an impact leaves no permanent scar on a gaseous giant planet, but it forms a crater when it occurs on a body having a solid surface. The surfaces of all the terrestrial planets, as well as the satellites of the giant planets, are scarred by the craters that result **(Fig. 7.13).** On the Earth, weathering processes such as wind and water erosion, as well as tectonic activity, have obliterated most of these craters, except those that have formed most recently.

On the Earth and Mars, erosion by water has been an important force affecting the structure of the surface. The flowing water that once carved channels and floodplains into the Martian surface **(Fig. 7.14)** has not been present for a very long time (perhaps 3

Figure 7.13 Impact craters throughout the solar system. No planet or satellite is safe from impacts due to objects from space. Here we see a recent crater on the Earth, near Winslow, Arizona (upper left); a very ancient crater on the Moon (upper right); a crater on Venus (lower left), discovered by recent *Magellan* radar images; and ancient craters on Saturn's icy moon Dione (lower right). Impact craters are characterized by high rim walls and central peaks, which can be seen in the examples shown here; volcanic craters may not have high walls and never have central peaks. (upper left: Meteor Craters Enterprises, Inc.; upper right: NASA; lower left: NASA/JPL; lower right: NASA/JPL) http://universe.colorado.edu/fig/7-13.html

billion years), but on Earth, water is still a very important factor in surface modification. In addition, the atmospheres of the Earth and Mars alter the surface through winds and chemical processes, and the atmosphere of Venus affects its surface as well.

The planets are all believed to be the same age, having formed at about the same time some 4.5

billion years ago. On the other hand, the surfaces of the planets have been altered to varying degrees. Some surfaces, such as that of the Earth, are relatively young, since the time scale for complete reorganization of the continents due to plate motions is around 200 million years, while other surfaces are very old (for example, parts of the lunar surface appear to have been unchanged for nearly 4 billion years). Thus, throughout this text, we will speak of the ages of planetary and satellite surfaces in contrast to the ages of the bodies themselves.

Evolution of Planetary Atmospheres

Most of the planets have atmospheres. An atmosphere is an envelope of gas held in place by the planetary gravitational field. Only Mercury is without

Figure 7.14 **An ancient river channel on Mars.** This is one of the most striking examples of evidence for flowing water in the Martian past. (NASA)

an atmosphere (and even here traces of gas are present). The terrestrial planets, as we have seen, have solid surfaces that mark a clear boundary between the atmosphere and the interior, whereas the giant planets are essentially fluid throughout, with no clear-cut boundary. Pluto's atmosphere is apparently seasonal, being partially frozen into solid form for most of its 249-year-long "year" and existing in gaseous form only during the time of the planet's closest approach to the Sun (which occurred most recently in 1989, so at present Pluto has an atmosphere).

A planet gains an atmosphere through a combination of two processes: **capture,** or **accretion,** of gases from its surroundings and the **escape,** or **outgassing,** of volatile elements from its interior. For the terrestrial planets, outgassing during the early, partially molten phase was probably the more important process, with the possible exception that a major portion of the water supply, and perhaps much of the carbon on the surface, may have been accreted through the impacts of cometary bodies, which consist largely of ice and carbon-rich compounds. In the case of the giant planets, accretion probably played the major role in forming the atmospheres, as these bodies captured enormous quantities of gas from their surroundings when they formed.

Once formed, an atmosphere may still undergo changes. On the Earth, for example, the early atmosphere was dominated by hydrogen-bearing compounds, and only later did the present-day oxygen-nitrogen atmosphere develop. (Interestingly, the formation of life on the Earth's surface apparently had a lot to do with this, providing some of the nitrogen and all of the oxygen.) On Mars, we find evidence of an early atmosphere that was much denser than today's thin carbon dioxide atmosphere, and as noted earlier in this chapter, evidence also indicates that Mars once had a lot of liquid water on its surface. Venus has the densest atmosphere among the terrestrials, and a very unpleasant one as well; although the main constituent is carbon dioxide, other compounds, such as sulfuric acid, make this planet a very inhospitable place.

The atmospheres of the giant planets are fairly uniform in composition except for variations created by local conditions. In all cases, hydrogen is the most common element, making up nearly three-fourths of the mass of the atmospheres, and helium is the next most abundant element. The hydrogen is mostly in its molecular form (H_2), but some hydrogen also appears in trace forms such as water vapor (H_2O), ammonia (NH_3), and methane (CH_4), while the helium is completely in the form of single atoms (helium does not easily combine chemically with other elements). Depending on the temperature in the outer layers of these atmospheres, more complex molecules are present in varying degrees.

We have already noted that the heavy elements in the giant planets have undergone differentiation and sunk to the planetary cores. In addition, some slow differentiation may still be taking place in at least one case (Saturn), as helium atoms gradually sink through the lighter-weight hydrogen.

So far, we have described how gases enter an atmosphere, but not how they can leave it. For the terrestrial planets, the escape of certain elements has been an important factor in determining their compositions today, whereas very little has escaped the strong gravitational fields of the giant planets. The likelihood that a particular kind of gas will escape depends on how massive the individual particles are, how strong the planetary gravitational field is, and how hot the atmosphere is. The dependence on the planet's gravity is probably obvious: for a given planetary mass and radius, there is a specific escape speed, which is easily calculated (see Chapter 5). The escape speed is the upward speed needed for a particle to escape the planet's gravitational field completely and fly away into space. The escape speed itself does not depend on the mass of the escaping

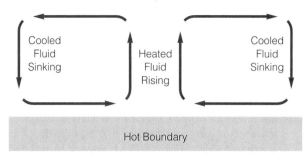

Figure 7.15 **Convective overturn.** A fluid (i.e., a gas or a liquid) subjected to heating from below (in the presence of a gravitational field) may experience convective overturn, as shown here. Heated fluid, less dense than its surroundings, rises to higher levels, where it cools, becomes denser, and sinks. The result is a steady, overturning motion. In planetary atmospheres, the source of heat from below may be absorbed sunlight or internal heat from within the planet.

particle, but, as explained in Chapter 5, the likelihood that a gas particle will reach escape speed does depend on its mass, because at a given temperature the lightest particles travel the fastest on average. We saw in Chapter 5 that if the average particle speed for a certain gas is at least one-sixth of the escape speed, then that gas will eventually escape altogether.

Thus some gases have managed to escape from each of the terrestrial planets, which have lower escape speeds and higher temperatures than the gas giants. For the Moon and Mercury, the combination of high temperature and low escape speed has ensured that all gases satisfy the one-sixth criterion, so these two bodies do not have permanent atmospheres, with the exception of a thin transient layer of gas on Mercury that is continually being replenished by outgassing from the interior of the planet. On the Earth, Venus, and Mars, however, some gases have been lost (most notably, the lightest-weight ones, hydrogen and helium) while others have been retained. For these planets, the masses of the gas particles have been instrumental in determining which ones will have average speeds exceeding one-sixth of the escape speed, and which ones will not.

As already mentioned, the giant planets have lost little or none of their original atmospheres. This is due to a combination of two factors: the high escape speeds of these massive planets and the very low temperatures, and hence low average particle speeds, of their outer atmospheres.

In addition to escape, another loss mechanism for atmospheric gases in the terrestrial planets can be surface chemical reactions. For example, carbon dioxide can be dissolved in water and deposited in carbonate rocks, so a planet with extensive oceans has a mechanism for getting rid of its atmospheric carbon

dioxide. One of the major contrasts between the atmosphere of the Earth and those of Venus and Mars is that the atmospheres of the latter are both dominated by carbon dioxide, whereas this gas is present only in trace quantities in the Earth's atmosphere. This difference is caused not only by the presence of oceans on the Earth, but also by the existence of life-forms, especially marine algae, that metabolize carbon dioxide.

Atmospheric Circulation

All planetary atmospheres undergo flowing motions, and in each case the basic causes of these motions are the same. There are two general driving forces for atmospheric circulation: heat-driven convection and planetary rotation.

Convection has already been mentioned as a heat transport mechanism in the Earth's mantle and as the probable driving force for continental drift on the Earth. In any situation in which a fluid under the influence of a gravitational field is heated from below, bubbles of the fluid may become buoyant and rise, creating an overturning motion **(Fig. 7.15).** For the atmospheres of the terrestrial planets, the source of heat is the surface, which is warmed by the Sun's radiation. For the giant planets, sunlight has some effect, but internal heat also makes an important contribution. These planets have hot interiors because of gravitational compression, and this heat slowly makes its way outward, helping to create convection in the outer atmospheres.

The rotation of a planet forces atmosphere flows to veer sideways, creating circular flow patterns. Consider a stream of gas flowing northward from the Earth's equator **(Fig. 7.16).** Because of the Earth's rotation, this air current is being carried rapidly to the east at the same time it flows northward. The rotational speed of the Earth's surface at the equator is about 1,670 km/hr, but the speed decreases with distance from the equator. Hence, as the air current makes its way north, it finds itself moving eastward relative to the surface, because it still has the eastward velocity given to it by the Earth's equatorial rotation but is moving over regions having a slower speed of rotation. A current moving southward in the Southern Hemisphere would also veer to the east as it moves away from the equator. In both hemispheres, air currents flowing toward the equator veer toward the west. The resulting motion of the flowing gas is a rotary pattern, as the horizontal flow is converted into circular flow. The "force" that causes these veers is called the **Coriolis force,** and the result is a circular motion for atmospheric flows everywhere on the Earth.

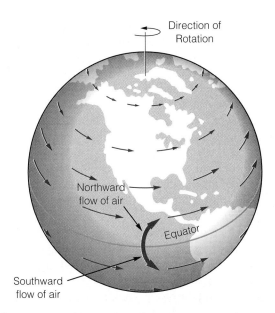

Figure 7.16 **The Coriolis force.** The lengths of the thin arrows indicate the relative speed of rotation of the Earth at different distances from the equator. (On a rigid sphere, points at the equator move fastest because they have farther to travel in a single rotation.) Wind flowing away from the equator is curved in the direction of rotation, because it comes from a region of higher rotational velocity.

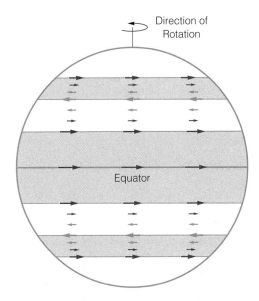

Figure 7.17 **Atmospheric flow on a giant planet.** The lengths of the arrows indicate relative wind speed. The fastest winds are typically in the equatorial zone, where the flow is in the direction of rotation. North and south of this zone are circulation patterns that stretch all the way around the planet. These are driven by a combination of convection and the Coriolis force created by the rapid planetary rotation.

The effect of the Coriolis force depends on how rapidly a planet rotates. The Earth and Mars have fairly rapid rotations, whereas Venus rotates extremely slowly. Thus the Earth and Mars have well-defined circular flows in their atmospheres, while on Venus these patterns are weaker. The giant planets all rotate extremely rapidly, and the Coriolis force has a huge effect. There the circular flow patterns are stretched out very far in the east-west direction, resulting in elongated circulation patterns that stretch all the way around these planets **(Fig. 7.17).**

Tidal Forces

One other phenomenon affects many of the planets and their satellites. As discussed in Chapter 5, the gravitational force exerted on one body by another is not equal at all locations, but is stronger on the near side than on the far side **(Fig. 7.18).** This creates a differential gravitational force, commonly called a tidal force. In Chapter 5, we saw how this force, exerted on the Earth by the Moon, creates tides in the Earth's oceans.

Tidal forces also come into play in many other situations in the solar system. Only the Earth and Pluto have satellites large enough to cause significant tidal effects on their parent planets. On the other hand, many of the satellites in the solar system are strongly affected by tidal forces due to their parent planets, and one planet, Mercury, is influenced by tidal forces due to the Sun.

Tidal forces can cause internal stress in bodies and can lead to alterations in their motions. The stretching force that is exerted on a satellite by its parent

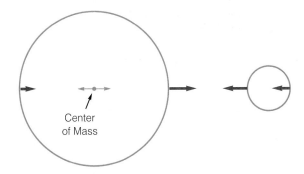

Figure 7.18 **Tidal force.** This diagram shows two spherical bodies near each other in space, such as a planet and its moon. The lengths of the red arrows indicate the relative strength of the gravitational force at different points on each body due to the other. Note that the side nearer the other body feels a stronger force than the far side; the difference between these forces acts as a stretching force. The sketch of the larger body shows how this differential gravitational force acts as though it pulls on opposite sides of the center of mass.

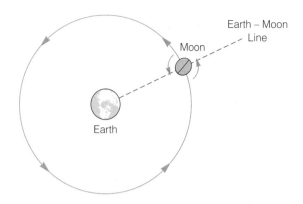

Earth – Moon
Line

Moon

Earth

Figure 7.19 **Tidal effect on the Moon's spin.** Here we see the Moon, its shape elongated by the tidal force due to the Earth, at a time in the past when the Moon's spin was faster than it is today. The spin of the Moon causes its elongated axis to point away from the Earth-Moon line, as shown. The Earth's gravity acts to keep the Moon's bulge pointed directly at the Earth; thus the Earth's gravity acts to slow the Moon's spin (the bulge actually lags behind, as shown here). Eventually, the slowing led to synchronous rotation, with the same side of the Moon always pointing toward the Earth. (Not to scale.) http://universe.colorado.edu/fig/7-19.html

planet can be enough to cause internal heating and, consequently, geological activity such as active volcanic eruptions. Io, the innermost large moon of Jupiter, is perhaps the most famous example; the *Voyager* spacecraft found that this satellite was undergoing almost continuous eruptions. There is evidence that at least one satellite of Saturn may be undergoing similar activity and that part of the interior of Europa, the second large moon of Jupiter, is warm and liquid. Tidal forces due to the Earth stretch our own Moon out of round, but do not heat its interior enough to cause current volcanic activity.

The tidal stretching of the Moon has had another effect, also mentioned in Chapter 5. The Moon used to spin much more rapidly than it does now, but its elongated shape has caused its spin to slow until the same side always faces the Earth. As long as the Moon spun more rapidly, the Earth kept tugging at its elongated sides, trying to force them to point directly along the Earth-Moon line **(Fig. 7.19)**. In time, the Earth succeeded, "locking in" the Moon's rotation rate to equal its orbital period. This situation is called *synchronous rotation.*

Most of the satellites in the solar system are in synchronous rotation because of the tidal effects of their parent planets. In one case—the Pluto-Charon system—the parent planet itself is also locked into synchronous rotation, so that the planet and its satellite perpetually keep the same sides facing each other. This means that at a certain point on the surface of Pluto, Charon is overhead permanently, since Pluto's day equals the orbital period of the

satellite (this is similar to the situation of communications satellites placed in Earth orbit with 24-hour periods).

The result of the Sun's tidal force on Mercury is a bit more complex than simple synchronous rotation. Here the rotation has been altered by tidal forces, but not to the point where the rotational and orbital periods are equal. Instead, the rotation period is equal to exactly two-thirds of the planet's orbital period, so that the same side of Mercury faces the Sun at a particular point in every other orbit. The reasons for this unusual situation are discussed in Chapter 9.

Not only can tidal forces deform bodies and alter their spins, but they can also prevent them from forming in the first place. The myriad small particles that make up the ring systems of the giant planets might have merged to form satellites, except that they are too close to the parent planet to do so. Within a certain distance called the **Roche limit,** the tidal forces keeping particles apart can exceed the gravitational forces trying to pull them together. This is discussed further in Chapter 10.

Summary

1. The principal bodies in the solar system are the Sun and nine planets, each orbiting the Sun in the same direction and nearly in the same plane. Most of the spin motions are also in the same direction, and most of the orbits are nearly circular. In addition to the planets, there are myriad minor bodies, including asteroids and comets.

2. The planets (except Pluto) fall into two general categories: (a) the terrestrial, or rocky, planets; and (b) the giant planets, which are fluid throughout (except for small rocky cores) and have much lower average densities than the terrestrial planets. Pluto is similar to certain giant planetary satellites.

3. Planetary data can be collected through direct sampling (by robot or manned probes), through closeup observations by spacecraft, and by Earth-based and Earth-orbiting telescopic observations.

4. The planets are thought to have formed by the coalescence of planetesimals, which were the first solid bodies to condense out of the disk of gas and dust that orbited the young Sun. The inner planets contain few lightweight volatile

elements because they formed under high-temperature conditions, whereas the outer planets, formed under low-temperature conditions, were able to retain even the volatile elements.

5. In all the planets, heavy elements have sunk to the cores in a process called differentiation. This can occur only in a fluid, which implies that even the terrestrial planets have undergone molten or partially molten phases.

6. The terrestrial planets have similar internal structures, in each case consisting of a thin, rocky crust overlying a mantle (which may be somewhat elastic) and a core. These planets are either now undergoing or once underwent tectonic activity to varying degrees; this activity may involve volcanic eruptions, earthquakes, and the shifting of crustal plates. Tectonic activity may be driven by convection in the mantle.

7. Rocks, which make up the surfaces of the terrestrial planets, are classified according to origin as igneous, sedimentary, or metamorphic. Rocks are also classified according to their mineral composition.

8. The surfaces of the terrestrial planets are modified by several processes in varying degrees: (a) water erosion, (b) meteoroid impacts, (c) wind erosion, and (d) geological processes such as continental drift and uplift. Because of these processes, the age of a planet and the age of its surface can be very different.

9. Planetary atmospheres formed through a combination of outgassing and accretion. The composition of an atmosphere is determined by which gases are supplied and which can escape (which depends, in turn, on temperature and molecular mass, as well as planetary escape speed). The terrestrial planets that have atmospheres retain only relatively heavy elements, whereas the giant planets have retained even lightweight species in their atmospheres.

10. The circulatory motions of all planetary atmospheres are driven by two forces: (a) convection due to heating from below and (b) planetary rotation, which creates circular motions through the action of the Coriolis force.

11. Tidal forces play many important roles in the solar system, including the creation of synchronous rotation in many satellites; the inducement of internal heating and volcanic activity in some satellites; and the prevention of satellite formation close to the outer planets, contributing to ring systems there.

Review Questions

1. At the beginning of this chapter, we outlined several general trends among the planets, relating to their motions and the locations and orientations of their orbits. As we noted, however, there are several distinct exceptions to the general rules. Using information in this chapter and in Appendix 8, list the planetary and satellite motions and orbits that do not conform to the general trends.

2. Summarize the distinctions between terrestrial and gas giant planets.

3. Explain differentiation in your own words. Can you think of everyday examples of differentiation, perhaps in your kitchen?

4. Compare the benefits of making close-up observations of a planet with a space probe with the advantages gained by making long-distance observations using Earth-based telescopes. Why is it useful to make both types of observations?

5. Explain why hydrogen is not a dominant component of the atmospheres of the terrestrial planets, even though it is the most common element in the universe and in the solar system.

6. List the processes that modify the surface of the Earth. Which ones also work on Mercury? On Venus? On Mars? On the Moon?

7. Explain the difference between the age of a body and the age of its surface. On which terrestrial planets are the two ages essentially the same?

8. Explain why a large planet retains internal heat longer than a small planet. Include a summary of the heat loss mechanisms that operate in a planet.

9. Summarize the factors that determine the composition of a planet's atmosphere. Include all of the atmospheric gain and loss mechanisms you can think of.

10. Explain in your own words how tidal forces are created and why the Earth has two tidal bulges on opposing sides of the planet.

Problems

1. Add up the masses of the planets (Appendix 8), and compare the total with the mass of the Sun (Appendix 9), as a means of checking the claim made in this chapter that the Sun contains more than 99% of the total mass of the solar system.

Drawing Scale Models of Planetary Orbits

ASTRONOMICAL

ACTIVITY

In order to better appreciate both the relative sizes of the planetary orbits and their shapes, it is helpful to construct accurate scale drawings of a few of the orbits in the solar system. If you choose Mercury, the Earth, Mars, and Saturn, you will see how the distances increase as you go outward from the Sun, and you will see how circular most orbits are while appreciating how much Mercury and Mars deviate from circularity.

First, decide on a scale, keeping in mind that the semimajor axes you are dealing with range from 0.37 to nearly 10 AU. Clearly, you will need a large, flat surface on which to do this, such as a large table. You will also need some string (enough to slightly exceed the major axis of your largest orbit, plus extra for the distance between foci) and a couple of pegs or paperweights to be used as the foci (these need to be movable). You will also need a large sheet of posterboard or heavy paper, large enough to accommodate your scaled drawing of the largest orbit you plan to include.

Use Appendix 8 to find the data you need: semimajor axes and eccentricities of the orbits you intend to draw. Start by converting the semimajor axes of the orbits to your scaled values. For example, if you choose a scale where 1 AU = 10 cm, then the semimajor axis of Mercury's orbit will be 3.7 cm, while that of Saturn will be about 1 m.

Next you may need to review information in Chapter 4 on the relationship between the eccentricity and the semimajor axis of an ellipse. The simplest form of the equation to use is

$$e = \frac{\overline{F_1 F_2}}{2a}$$

where e is the eccentricity, $\overline{F_1 F_2}$ is the distance between the two foci (one of which resides at the location of the Sun), and a is the semimajor axis. Using this formula for each planet, you can determine the separation between foci on the scale you have chosen by solving for $\overline{F_1 F_2}$. You should get the separations in centimeters or inches; that is, in the same units you used to calculate your scaled values for the semimajor axes. Now place a pin or weight at the location of the Sun, and locate the position of the other focus for each of the orbits. Note that the second foci need not be aligned on the same side of the Sun.

Your task now is to create a loop of string equal in length (circumference) to the major axis, plus the distance between foci for each orbit (you can use the same string; start with the largest orbit, draw it, then make a smaller loop for the next orbit, and so on). Loop the string over the two pins or weights, and trace out an ellipse using a pencil and keeping the string taut (but make sure the pins or weights don't move). The result will be accurate scale drawings for each of the orbits, showing not only the correct relative size but also the correct shapes.

As an interesting final step, you can calculate the diameters of the Sun and planets on the same scale that you used for the orbital drawings. You can even try to indicate the disks of the Sun and planets on your drawing, but you may find this difficult due to the small sizes of the planets on this scale.

This exercise should serve to illustrate some aspects of the regularity of the solar system, as well as impressing you with the vastness of the spaces between the planets relative to their sizes. (As an addendum to the project, you might want to try calculating the size and the positions of the foci for Pluto's orbit on the same scale. If you have enough room and enough string, try adding Pluto's orbit to your drawing. Even if you can't fit it in, doing the calculation will further illustrate how large the system is relative to the inner solar system.)

2. Using the Stefan-Boltzmann law (Chapter 6), calculate the luminosities of the planets and compare them with the luminosity of the Sun. The data needed (surface temperatures and radii) are given in Table 7.1. Discuss the implications of your answers for the problem of detecting planets orbiting other stars.

3. Suppose the core of a terrestrial planet contains half of the mass of the planet and extends halfway out from the center. How much denser is the core than the average density of the planet? (Hint: Write the formula for the density of the core and the formula for the density of the entire planet, which is the same except that the radius and the mass are both twice as great as for the core. Divide the two expressions to get the ratio of the densities. The terms for mass and radius should cancel out.)

4. What is the wavelength of maximum emission from the sunlit surface of Mercury if the temperature is 700 K? Compare this with the wavelength of maximum emission from the cloud tops of Jupiter, where the temperature is 130 K. What kinds of telescopes would be needed to directly observe this thermal emission from each of these planets?

5. A new planet is discovered in our solar system. Its orbital period is 3.6 years, its mass is 4.0×10^{25} kg, and its radius is 12,500 km. Calculate the orbital semimajor axis and the density of this body, and use your results to decide whether it is a terrestrial or gas giant planet. Explain how you reached your conclusion.

6. Calculate the escape speed from the surface of the new planet described in Problem 5. If the average speed of hydrogen molecules in this planet's atmosphere is 4.5 km/sec, would you expect to find hydrogen in the atmosphere of the planet? Explain.

7. Suppose that major meteoroid impacts occur at the average rate of one per square kilometer per 100 million years, and that this rate has been constant since the planets formed. If the Moon has 45 craters per square kilometer, how old is the Moon? Note that you had to assume the Moon has retained all the craters ever formed on its surface. Now assume that the Earth has only one crater for every million square kilometers of surface area. How old is the surface of the Earth? You should get an unrealistically small value for the age of the Earth's surface; how can you explain this?

Additional Readings

Beatty, J. K., B. O'Leary, and A. Chaikin 1990. *The New Solar System*. 2nd Ed. Cambridge University Press.

Kasting, J. F., B. Toon, and J. B. Pollack 1988. How Climate Evolved on the Terrestrial Planets. *Scientific American* 258(2):90.

Kross, J. F. 1995. What's in a Name? *Sky & Telescope* 89(5):28.

Morrison, D. 1992. Planetary Astronomy in the 1990s. *Sky & Telescope* 83(2):151.

McLaughlin, W. I. 1989. Voyager's Decade of Wonder. *Sky & Telescope* 78(1):16.

Web Connections

The Review Questions and Problems also appear at the following URLs:
http://universe.colorado.edu/ch7/questions.html
http://universe.colorado.edu/ch7/problems.html

Chapter 8
The Earth-Moon System

Chapter Web site: http://universe.colorado.edu/ch8

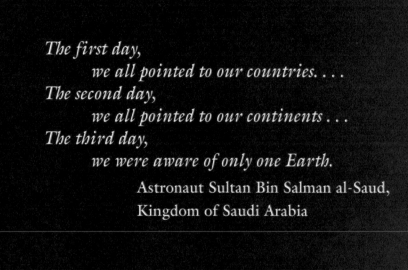

The first day,
*   we all pointed to our countries. . . .*
The second day,
*   we all pointed to our continents . . .*
The third day,
*   we were aware of only one Earth.*

Astronaut Sultan Bin Salman al-Saud,
Kingdom of Saudi Arabia

f all the bodies in the heavens, the Earth **(Table 8.1)** is, of course, the best studied, although many mysteries remain. Understanding the planet we live on has practical as well as philosophical importance. Furthermore, working at such close quarters with the object of study has numerous advantages, including the ability to observe in great detail and over long periods of time and the possibility of making direct experiments by probing and sampling the surface.

One of the premier questions is whether the Earth is unique, or whether we have reason to expect simi-lar planets to have formed in orbits around other stars. We will find that none of the other terrestrial planets in our system matches the Earth in detail, yet we will argue later that Earth-like conditions were essential for the beginning of life. Thus the question of Earth's uniqueness will bear heavily on our speculation about life elsewhere.

Detailed knowledge of the Earth is also important for understanding the other terrestrial planets, largely because we can learn so much about the evolution of the Earth, about the processes that have shaped and altered it over the ages. When we study another plan-et, we usually get only a "snapshot," a brief picture of what it is like today. But for the Earth we can analyze structures formed at different times in the past so that we can unravel the geological processes that formed them; we can interpret fossil

remains to learn about the past history of the atmosphere; and we can deduce the internal structure of our planet and extrapolate from it to a better understanding of the internal properties of the other terrestrials.

The Earth is a planet that has had a very active evolution and is still dynamic today. In contrast, the Moon's development was arrested early, and the Moon itself has changed very little for billions of years. Thus, in studying the Earth-Moon system, we have an opportunity to learn about two objects that represent the extremes of evolutionary behavior for terrestrial bodies in our solar system. This will aid us greatly when we discuss the other terrestrial planets, which may be considered as intermediate cases whose properties lie somewhere between those of the Earth and of the Moon.

Overview of the System

Planetary scientists consider the Earth and the Moon together **(Fig. 8.1)** to be a double planet because the two are more similar in mass and diameter than most planet-satellite systems in the solar family. The Moon's diameter is over one-fourth (specifically, 0.273) the diameter of the Earth, and its mass is more than 1% (0.0123) of the Earth's mass. By con-trast, Phobos, one of the satellites of Mars, has a diameter only 0.003 times that of its parent planet and a mass

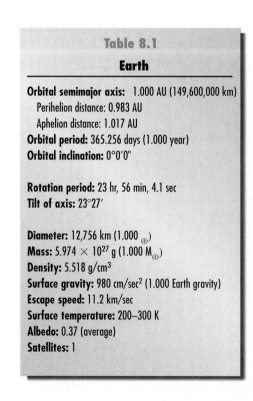

Table 8.1

Earth

Orbital semimajor axis: 1.000 AU (149,600,000 km)
  Perihelion distance: 0.983 AU
  Aphelion distance: 1.017 AU
Orbital period: 365.256 days (1.000 year)
Orbital inclination: 0°0'0"

Rotation period: 23 hr, 56 min, 4.1 sec
Tilt of axis: 23°27'

Diameter: 12,756 km (1.000 $_\oplus$)
Mass: 5.974 × 10²⁷ g (1.000 $M_\oplus$)
Density: 5.518 g/cm³
Surface gravity: 980 cm/sec² (1.000 Earth gravity)
Escape speed: 11.2 km/sec
Surface temperature: 200–300 K
Albedo: 0.37 (average)
Satellites: 1

Figure 8.1 The Earth and Moon as seen from space. (NASA) http://universe.colorado.edu/ fig/8-1.html

only 0.000000015 (i.e., 1.5×10^{-8}) as large. The satellites of the giant planets are all very small and low in mass compared to their parent planets. In the entire solar system, only the Pluto-Charon pair is more nearly equal than the Earth-Moon system.

The Earth and the Moon are unusual in other ways. No other terrestrial planet has a moon nearly as massive as the Earth's satellite, and most have no moons at all (the total census of natural satellites among the terrestrial planets is three: the Moon and the two tiny satellites of Mars—Mercury and Venus have no moons). Another oddity of the Earth-Moon system is the large separation between them relative to their diameters. The Moon's average distance from the center of the Earth is approximately 60 times the radius of the Earth; Phobos orbits Mars at a distance from the center of Mars of only about 2.8 times the radius of the planet, and Ganymede, one of the giant satellites of Jupiter, orbits at about 15.5 Jovian radii. One consequence of the large mass of our Moon and its wide separation from the Earth is that the Moon carries a large amount of angular momentum with it (for a discussion of angular momentum, see Chapter 5). This has significant implications for the formation of the Earth-Moon system, which will be discussed at the end of this chapter.

Figure 8.2 Full disk Earth. (NASA)

The Earth

The Earth itself is unusual compared to the other terrestrials. It is the largest among them (only a little larger than Venus in both mass and diameter, but much larger than Mercury and Mars), it has the highest average density, and it is the only one to have liquid water on its surface and an oxygen-bearing atmosphere. As we will see, the Earth is also the most active geologically. As we study the evolutions of all of the terrestrial planets, we will come to see how these facts are interrelated and how they came about.

The Earth as seen from space is a brilliant sight **(Fig. 8.2).** Its overall color is blue due to its oceans and to scattering of light in its atmosphere. Much of it is white, where there is cloud cover, and here and there brown landmasses can be seen. The Earth is so bright because it reflects a large fraction of the incident sunlight. The **albedo** of the Earth, the fraction of incoming light that is reflected back into space, is 0.37 on average (it varies from place to place because continents, oceans, and clouds have differing reflection efficiencies; so to speak of an overall albedo, we must refer to an average value). The Earth's albedo is much higher than is typical for bodies without atmospheres. The Moon has an albedo of only 0.07, so

the Earth appears far brighter when the two are viewed together from space (see Fig. 8.1).

Composition of the Earth's Atmosphere

The atmosphere of the Earth is a thin blanket of gases that are bound to our planet by gravity. Though it is thin (its height is only 1 to 2% of the Earth's radius), the atmosphere is extremely important for the Earth and especially for the life-forms that inhabit its surface. The atmosphere traps heat at the surface, and its weathering action erodes surface features, modifying the structure of the planet's crust. The atmosphere has not always been the same as it is today; evidence suggests that it has undergone at least one complete replacement early in the Earth's history.

The Earth's atmosphere is composed of a variety of gases **(Table 8.2),** as well as a distribution of suspended particles called **aerosols.** The gases, which are evenly mixed together up to an altitude of about 80 km, are primarily nitrogen and oxygen in the form of the molecules N_2 and O_2. Nearly 80% of the gas is nitro-gen, about 20% is oxygen, and all the other species exist only in relatively small quantities, although in many cases these minor constituents are

Partners in Space
The Earth & Moon

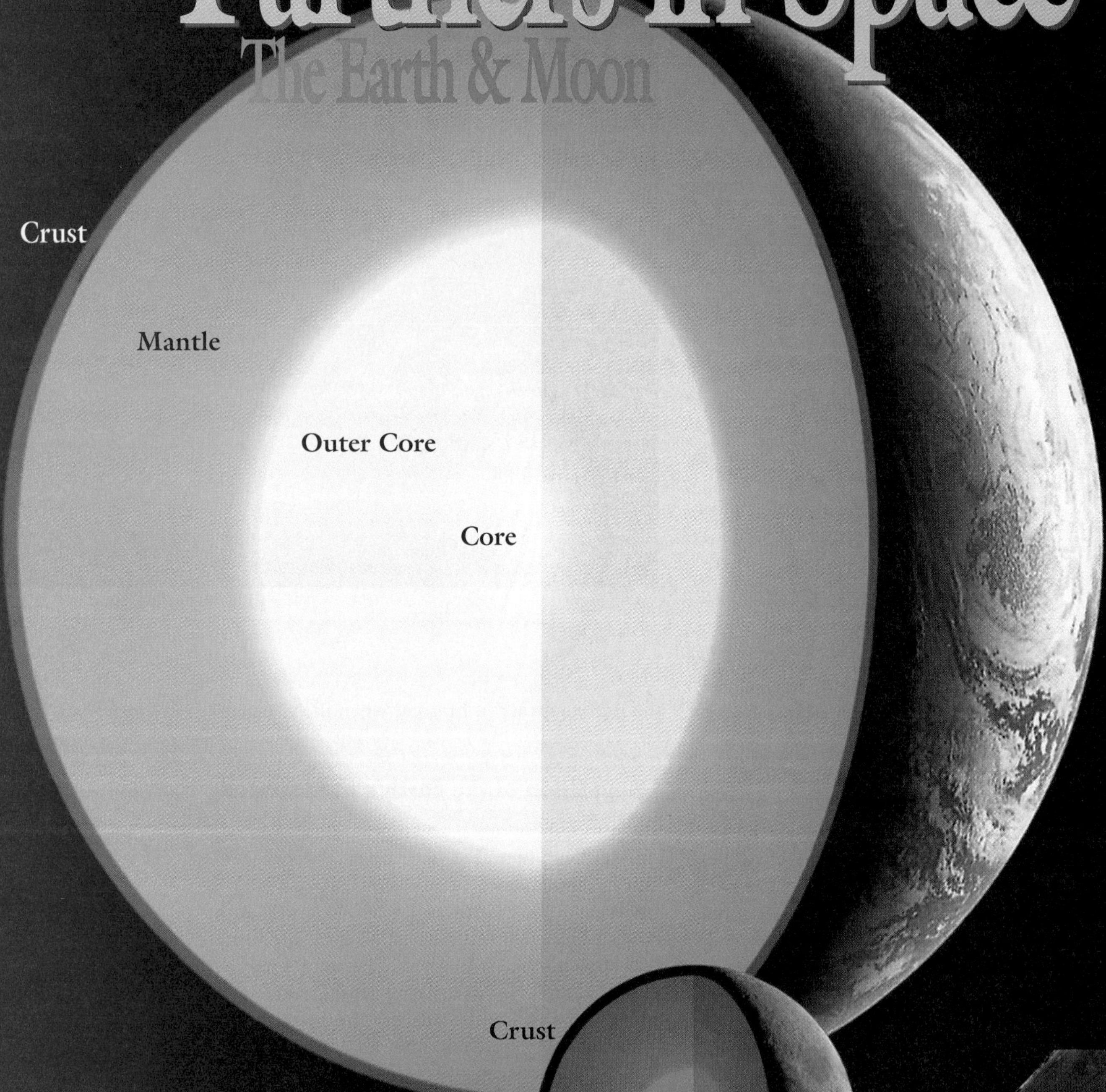

Crust

Mantle

Outer Core

Core

Crust

Core

Solid
Lithosphere

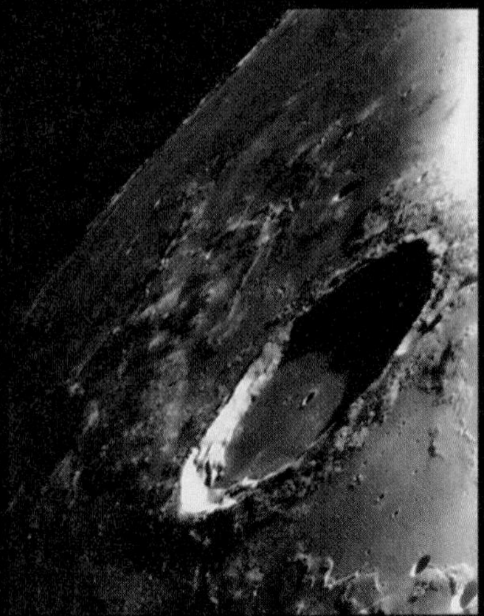

The interior structures of the Earth and the Moon, said by
some scientists to be a double-planet system, are shown
(to correct relative size) in these cut-away illustrations.

Table 8.2

Composition of Earth's Atmosphere

Gas	Symbol	Fraction[a]
Nitrogen	N_2	0.77
Oxygen	O_2	0.21
Water vapor	H_2O	0.001–0.028
Argon	Ar	0.0093
Carbon dioxide	CO_2	3.3×10^{-4}
Neon	Ne	1.8×10^{-5}
Helium	He	5.2×10^{-6}
Methane	CH_4	1.5×10^{-6}
Krypton	Kr	1.1×10^{-6}
Hydrogen	H_2	5×10^{-7}
Ozone	O_3	4×10^{-7}
Nitrous oxide	N_2O	3×10^{-7}
Carbon monoxide	CO	1.2×10^{-7}
Ammonia	NH_3	1×10^{-8}

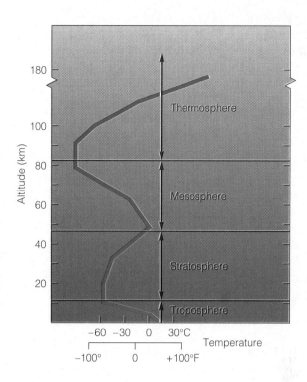

Figure 8.3 **The vertical structure of the Earth's atmosphere.** The distinct zones are defined according to the temperature variation with height, which influences atmospheric motions.

important. Among these, water vapor (H_2O) is essential because it plays a direct role in supporting life. Others, such as ozone and carbon dioxide, are also important but for less direct reasons. Ozone (designated O_3) is a form of oxygen in which three oxygen atoms, instead of two, are bound together. Unlike ordinary oxygen (O_2), ozone in the upper atmosphere acts as a shield against ultraviolet light from the Sun, which could be harmful to life-forms if it penetrated to the ground.

The importance of carbon dioxide (CO_2), which is produced by respiration in animals, by fires and decomposition of organic matter, and by many industrial processes, lies in its ability to trap the Sun's radiant energy within the atmosphere, creating over-all heating by a process called the **greenhouse effect.** This heating mechanism has been much more effective on Venus than on the Earth (and will be discussed more fully in Chapter 9).

The principal atmospheric constituents, nitrogen and oxygen, are both released into the air by life-forms on the surface, although nitrogen has nonbiological origins as well. Thus biological contributions have been important in the evolution of the Earth's atmosphere, which leads to the expectation that other planets in the solar system are unlikely to have the same atmospheric composition, although we do find nitrogen elsewhere because it can have nonbiological sources. Nitrogen is released into the Earth's atmosphere by the decay of biological material and by emission from volcanic eruptions. Oxygen comes almost exclusively from the photosynthesis process in plants, in which carbon dioxide is converted into oxygen with the assistance of radiant energy from the Sun.

Heating and Atmospheric Motions

Although there are obvious fluctuations in the temperature of the atmosphere, particularly near the surface, the temperature has a definite, stable structure as a function of height **(Fig. 8.3).** This structure includes several distinct layers in the atmosphere: the **troposphere,** from the surface to about 10 km in altitude, where the temperature decreases slowly with height and where the phenomena that we call weather occur; the **stratosphere,** between 10 and 50 km, where the temperature increases with height; the **mesosphere,** extending from 50 to 80 km, where the temperature again decreases; and the **thermosphere,** above 80 km, where the temperature gradually rises with height to a constant value above 200 km. In layers where the temperature decreases with height, there can be substantial vertical motion of the air, whereas in layers where it gets warmer with height, the air is stable, with primarily lateral motions.

The principal source of heating throughout the atmosphere is the Sun's radiation, but the energy is deposited in different ways at different levels. Much of the heat (that is, infrared radiation) is absorbed by the ground. This heating, which is concentrated near the equator, is responsible for driving the large-scale motions of the atmosphere, which produce the global wind patterns. Heat is also deposited in the stratosphere where ozone (O_3) absorbs ultraviolet light from the Sun. This absorption is responsible for the increase in temperature with height in the stratosphere.

The atmosphere exhibits many scales of motion, ranging from small gusts a few centimeters in size to continent-spanning flows thousands of kilometers in scale. Here we will discuss only the large-scale motions, since generally we are only able to study motions of similar scale in the atmospheres of other planets.

The primary influences on the global motions in the Earth's atmosphere are heating from the Sun, as already noted, and the rotation of the Earth. The heating creates regions where the air rises. The air must later cool and fall somewhere else, and a pattern of overturning motion called convection is created (see Chapter 7). Generally, air rises in the tropics and descends at more northerly latitudes, although the situation is more complicated than that. For example, the continents tend to be warmer than the oceans in the summer, so that high-pressure regions (characterized by descending air) preferentially lie over water, whereas low-pressure regions (where air rises and cools) tend to be over land. In winter this pattern is reversed.

These tendencies, combined with the rotation of the Earth, create horizontal flows at several levels in the atmosphere. Air that is rising or falling is forced into a rotary pattern by the Coriolis force, as described in Chapter 7. Where air rises in a low-pressure zone, the resultant swirling motion is called a **cyclone (Fig. 8.4).** In the Northern Hemisphere, the motion of a cyclone is counterclockwise; south of the equator, it is clockwise. Descending air flows in high-pressure regions are forced into oppositely directed circular patterns, called **anticyclones.** Cyclonic flows sometimes intensify into storms of great strength; hurricanes and typhoons are examples. The net result of the vertical motions caused by the Sun's heating and the spiral motions caused by the Earth's rotation is the creation of a complex pattern of flows **(Fig. 8.5)** with vertical and horizontal components.

Seasonal shifts in the distribution of the Sun's energy input cause changes in the flow patterns, creating our well-known seasonal weather variations. Apparently, other factors can influence the location of the high- and low-pressure regions because the climate undergoes longer-term fluctuations whose cause is not well understood. Some of these may be related to cyclic behavior in the Sun (see Chapter 13).

In later chapters, when we examine the flow patterns in the atmospheres of the other planets, we will see that in many ways they behave like the Earth, except for differences in the amount of solar energy input and the rate of rotation.

Figure 8.4 A storm system on Earth. This low-pressure region is in the Northern Hemisphere, so the circulation is counterclockwise. (NASA) http://universe.colorado. edu/ fig/8-4.html

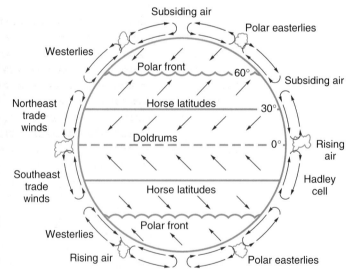

Figure 8.5 The general circulation of the Earth's atmosphere. The circulation is made even more complex than shown here by temperature contrasts between seas and land and by surface features such as mountains. (Based on Fig. 7.3, p. 150, F. K. Lutgens and E. J. Tarbuck 1979. *The Atmosphere: An Introduction to Meteorology* (Englewood Cliffs: N.J.: Prentice-Hall.)

SCIENCE AND SOCIETY

Technology and the Atmosphere

Life-forms have had a profound impact on the past evolution of the Earth's atmosphere, as we have observed in the text. Much of the nitrogen that dominates the composition of the atmosphere comes from the decay of life-forms, and virtually all of the oxygen is the by-product of photosynthesis in plants. Since life first appeared on the Earth, the interplay between biological activity and atmospheric processes has been rich and diverse. Today the interaction continues, with the added impact of our technological society.

Human activities result in many kinds of emissions into the atmosphere that would not occur naturally. These emissions include both gases and tiny particles, or aerosols. The atmosphere is a complex, dynamic entity, and scientists have difficulty predicting how it will react to these emissions. Furthermore, significant global changes take a long time to occur, so it is not easy to see right away what the effects of a change in conditions might be. Despite these difficulties, two gases in particular have been identified as potentially capable of changing the atmosphere enough to alter the Earth's climate, and many others are known or suspected to have other kinds of effects.

One of the gases having known effects is carbon dioxide (CO_2), a by-product of burning wood and fossil fuels (i.e., oil, gas, and coal). Carbon dioxide is an effective absorber of infrared radiation, which means that a high abundance of CO_2 in the atmosphere can trap heat near the Earth's surface, increasing the temperature of the lower atmosphere through the greenhouse effect. The Earth's atmosphere already has substantial greenhouse heating due to natural levels of CO_2 and other trace gases, but studies indicate that the relative abundance of CO_2 has increased by more than 50% in the past century due to human sources. The resulting temperature increase in the lower atmosphere could have far-reaching effects. For example, the abilities of various regions to support farm crops might be vastly altered. In addition, even a relatively small increase could, over time, result in substantial melting of the ice in Arctic regions, which in turn would raise sea levels worldwide enough to inundate coastal areas.

As an illustration of the complexity of atmospheric processes, it is noteworthy that not all scientists agree on what the effects of increased CO_2 would be. Some argue that it would increase the cloud cover, which would block incoming sunlight sufficiently to decrease the heating at the Earth's surface.

An even more controversial technological addition to the atmosphere is created by the so-called chlorofluorocarbons, complex molecules containing chlorine and other elements. These are commonly used as refrigerants (such as freon) and as propellants in making foam products. In this case, scientists are concerned about complex chemical reactions that may occur in the upper atmosphere involving by-products of the chlorofluorocarbons. One of these by-products is chlorine oxide (ClO), which can react with ozone (O_3), destroying the ozone. For the past decade or so, there has been a growing worry that the release of chlorofluorocarbons into the atmosphere has already decreased the abundance of ozone in the stratosphere (at an altitude of about 16 to 18 km). If so, this could have serious consequences because the ozone layer is our primary shield against ultraviolet radiation from the Sun, and an increase in the amount of ultraviolet reaching the ground could be quite harmful to life on the Earth.

Recently, scientists have discovered that there is indeed a deficiency in the quantity of ozone in the atmosphere, at a particular time and at a particular place. During the spring (October) in the Southern Hemisphere, the quantity of ozone over the south polar region diminishes. To learn more about this south polar ozone deficiency, teams of scientists spent several weeks in Antarctica in late 1986 and will continue to do so for some years.

The first of these intensive studies produced detailed measurements of the quantity of ozone and of various molecular species known to react with ozone. Balloons and other devices made measurements at various altitudes in the atmosphere. As a result, we know that the quantity of ozone over the pole has dropped by about 50% since the late 1960s, confirming that the ozone hole is either a new phenomenon or one that varies, perhaps with the Sun's cycle of activity (see Chapter 13).

Some of the scientists involved in the study have concluded that the ozone hole is caused by chlorofluorocarbons. This conclusion is based on measurements of other molecular species, inconsistencies in other hypotheses, and direct measurements of freon in the atmosphere, showing that its abundance has increased by 50% over the same time period when the south polar springtime ozone abundance has decreased. All of these arguments are circumstantial and not completely conclusive, but they are consistent with the idea that human technology has begun to erode the Earth's protective ozone shield. Many important questions remain about whether this conclusion is valid and, if

(Continued next page)

so, whether the ozone deficiency will be confined to the south polar region or will eventually spread.

Very recently, new studies of ozone over the north polar region were begun, and scientists found that a deficiency may also exist in that region. It now appears that thinning of the ozone layer

is occurring globally and at a faster rate than was expected. This has prompted the U.S. government to hasten its plans to phase out all use of chlorofluorocarbons in the hope that the ozone layer will stop eroding away and may perhaps even begin to recover.

The changes in the Earth's atmosphere caused by natural processes related to life-forms took millennia to occur, and they took place in concert

with the development and evolution of those life-forms. Now it seems we must take measures to prevent much more rapid changes due to unnatural processes from altering our atmosphere in ways that would make the Earth inhospitable to life.

The Earth's Interior

Learning about the deep insides of the Earth is not easy, and at first it may seem surprising that we know much about it at all. But a great deal has been learned, largely through indirect techniques. The primary method for probing the Earth's interior is to observe and analyze the vibrations in the Earth that are created as the result of earthquakes. These vibrations travel through the Earth in the form of **seismic**

waves. These waves can travel completely through the Earth and are detected at sites located all around the globe after any significant earthquake.

Seismic waves take three possible forms, called **P, S,** and **L waves.** Studies of wave transmission in solids and liquids have shown that P (primary) waves, which are the first to arrive at a site remote from the earthquake location, are compressional waves, which means that the oscillating motions occur parallel to the direction of motion of the wave, creating alternating regions of high and low density without any sideways motions **(Fig. 8.6).** Sound waves are examples of compressional waves. The S (secondary) waves are transverse, or shear, waves, in which the vibrations occur at right angles to the direction of motion (see Fig. 8.6). These waves will only pass through a material that has some rigidity, and unlike the P waves, they cannot be transmitted through a liquid. The L waves (or Love waves, named after the scientist who pioneered their study) travel only along the surface of the Earth and thus do not provide much information on the deep interior.

By measuring both the timing and the intensity of these seismic waves at various locations away from the site of an earthquake, scientists can determine what the Earth's interior is like. The speed of the P waves depends on the density of the material they pass through, and the distribution of the P and S waves reaching remote sites provides data on the location of liquid zones in the interior **(Fig. 8.7).**

The general picture that has emerged from these studies is of a layered Earth **(Fig. 8.8** and **Table 8.3).** At the surface is a **crust** whose thickness varies from a few kilometers beneath the oceans to perhaps 60 km under the continents. There is a sharp break between the crust and the underlying material that is called the **mantle.** The mantle transmits S waves, so it must be

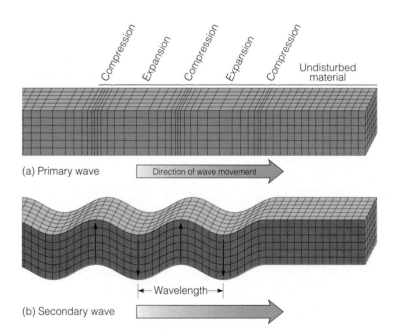

Figure 8.6 S and P waves. Compressional, or P, waves (top) have no transverse motion but consist of alternating dense and rarefied regions created by motions along the direction of propagation. Sound waves are P waves. An S, or shear, wave (bottom) consists of alternating motions transverse (perpendicular) to the direction of propagation. Waves in a tight string or on the surface of water are S waves.

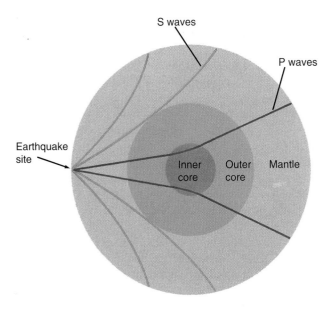

Figure 8.7 Seismic waves in the Earth. This simplified sketch shows that P waves can pass through the core regions (although their paths may be bent), whereas S waves cannot. This led to the deduction that the outer core is liquid, because S waves, which require an elastic or solid medium, cannot penetrate liquids.

solid; on the other hand, it undergoes slow, steady, flowing motions in its uppermost regions. Perhaps it is best viewed as a plastic material that has some rigidity, but can be deformed given sufficient time (this is discussed more fully in the next section).

The uppermost part of the mantle and the crust together form a rigid zone called the **lithosphere.** Just below the lithosphere is the part of the mantle where the fluid motions occur; it is called the **asthenosphere.** Below the asthenosphere is a more rigid portion of the mantle that extends nearly halfway to the center of the Earth. The lower mantle is called the **mesosphere** (not to be confused with the level in the Earth's atmosphere bearing the same name).

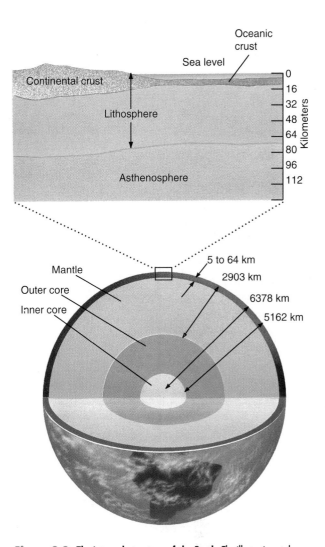

Figure 8.8 The internal structure of the Earth. The illustration at the bottom shows the layers that are defined by contrasts in density or physical state, as deduced from observations of seismic waves. The expanded view at the top illustrates the relative thickness of the crust in continental areas as compared with the seafloor.

Table 8.3				
Earth's Interior				
Layer	**Depth (km)**	**Density (g/cm³)**	**Temperature (K)**	**Composition**
Crust	0–30	2.6–2.9	300–700	Silicates and oxides
Mantle				
Lithosphere	30–70	2.9–3.3	700–1,200	Basalt, silicates, oxides
Asthenosphere	70–1,000	3.9–4.6	1,200–3,000	Basalt, silicates, oxides
Mesosphere	1,000–2,900	4.6–9.7	3,000–4,500	Basalt, silicates, oxides
Core				
Outer core	2,900–5,100	9.7–12.7	4,500–6,000	Molten iron, nickel, cobalt
Inner core	5,100–6,378	12.7–13.0	6,000–6,400	Solid iron, nickel, cobalt

Beneath the mantle is the **core,** consisting of an **outer core** (between 2,900 and 5,100 km in depth) and an **inner core.** The S waves are not transmitted through the outer core, which is therefore thought to be liquid. As a result, the inner core can be probed only with the P waves. Since these waves travel more rapidly through the inner core, this innermost region is thought to be solid and to have a density greater than that of the outer core.

The overall density of the Earth is 5.5 g/cm³. The rocks that make up the crust have typical densities of less than 3 g/cm³, and the mantle material is thought to have relatively low density also, perhaps 3.5 g/cm³. From these facts, scientists deduce that the core must have a density of roughly 13 g/m³. The most likely substance that could give the core this high density is iron, probably combined with some nickel and perhaps cobalt. These elements, as well as the much more rare superheavy species such as lead and uranium, are concentrated at the Earth's core due to differentiation, the sinking of heavy elements toward the center that occurred early in the Earth's history (discussed in the previous chapter).

The Earth's Magnetic Field

It has been known for a long time that the Earth possesses a magnetic field. It is a **dipolar field,** meaning that it has two poles connected by magnetic lines of force. A simple bar magnet also forms a di-polar field. Other structures are possible for magnetic fields, as we shall see when we discuss the outer planets.

It is convenient to visualize the Earth's magnetic field in terms of the magnetic lines of force connecting the two poles. In cross-sectional view, the Earth's

magnetic field is reminiscent of a cut apple, but one that is lopsided because of the flow of charged particles from the Sun that constantly sweep past the Earth **(Fig. 8.9).** The region enclosed by the field lines is called the **magnetosphere,** and it acts as a shield, preventing the charged particles from reaching the Earth's surface.

The axis of the Earth's magnetic field is aligned closely (within 11.5°) with the rotation axis of the planet, but this has not always been the case. The past alignment of the magnetic field can be ascertained from studies of certain rocks that contain iron-bearing minerals whose crystalline structure is aligned with the direction of the magnetic field at the time the rocks solidified from a molten state. Thus, traces of the Earth's ancient magnetic field, called **paleomagnetism,** can be detected from the analysis of magnetic alignments of rocks. Such studies reveal that the magnetic poles have moved about during the Earth's history, and that the north and south magnetic poles have completely and rather suddenly reversed from time to time. When a reversal occurs, the north magnetic pole moves to the south and vice versa. These flip-flops of the magnetic poles seem to have happened at irregular intervals, typically thousands to hundreds of thousands of years apart.

The Earth's magnetic field is generated in its interior, but the exact nature of the mechanism that is responsible is not fully understood. It does appear certain that the field must arise in a molten zone, and therefore that it forms in the outer core region of the interior. A magnetic field is produced by flowing electrical charges, as in the current flowing through a wire wound around a metal rod, which is the basis of an electromagnet. Convection and the Earth's rotation may combine to create systematic flows in the liquid outer core, giving rise to the magnetic field, if the core material carries an electrical charge. This general type of mechanism is called a **magnetic dynamo.** The cause of the reversals of the poles is

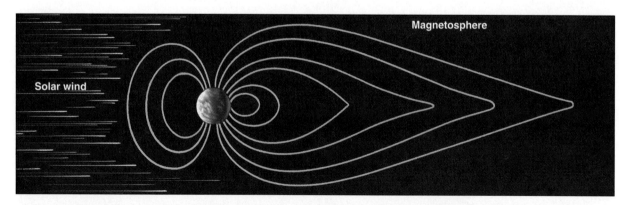

Figure 8.9 The Earth's magnetic field structure. This cross-sectional view shows the Earth's field and how its shape is affected by the stream of charged particles from the Sun known as the solar wind. http://universe.colorado.edu/ fig/8-9.html

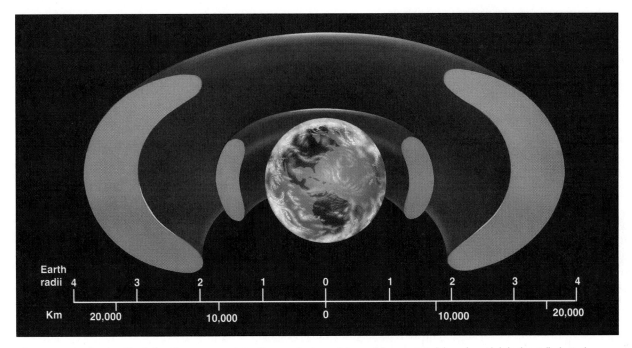

Figure 8.10 **The Van Allen belts.** This cross-sectional view illustrates the sizes and shapes of the two principal charged-particle belts that girdle the Earth.

not well understood but presumably could be related to occasional changes in the direction of the flow of material in the core.

We have referred to the magnetosphere and its ability to control the motions of charged particles. When the first U.S. satellite was launched in 1958, zones high above the Earth's surface were discovered to contain intense concentrations of charged particles, primarily protons and electrons. There are several distinct zones containing these particles, and they are now called the **Van Allen belts (Fig. 8.10)** after the physicist who first recognized their existence and deduced their properties.

The charged particles, or **ions,** in the Van Allen belts were captured from space (primarily from the solar wind) and are forced by the magnetic field to spiral around the lines of force. The same phenomenon occurs in the uppermost portion of the outer atmosphere, called the **ionosphere,** which extends upward from a height of about 60 km. When a reversal of the Earth's magnetic poles is taking place, the magnetic field is much weakened for a short period of time, and consequently the Van Allen belts and the ionosphere undergo a major disruption. The magnetosphere is greatly diminished, and charged particles from space are more likely to penetrate to the ground. These particles, particularly the very rapidly moving ones called **cosmic rays,** can cause important effects on life-forms, including genetic mutations. The sporadic reversals of the Earth's magnetic field

may have played a major role in shaping the evolution of life on the surface of our planet.

The ionosphere has important effects for us on the surface, including enhanced radio communications and the beautiful light displays **(Fig. 8.11)** known as **aurora borealis** (northern lights) and **aurora australis** (southern lights). The aurorae are caused by charged particles that enter the atmosphere and collide with atoms and molecules, exciting them so that they emit light. The aurorae occur most commonly near the poles, where the magnetic field lines allow particles to penetrate closest to the ground. The ionosphere contributes to radio communications because it reflects signals in the short-wave band, enabling them to travel around the Earth. When there are fluctuations in the solar wind, particularly following solar flares, enhanced fluxes of charged particles entering the ionosphere from space can disrupt radio communications.

A Crust in Action

Nearly three-fourths of the Earth's surface is covered by water, while the rest takes the form of several major continents. As we explained in Chapter 7, the basic substance of the Earth's crust is rock, and rocks are classified into three basic groups according to their origin: igneous, sedimentary, and metamorphic. Igneous rocks, formed from volcanic activity, consist of cooled and solidified magma, the molten material

Figure 8.11 The aurorae. The photograph at left shows the aurora borealis, or northern lights, as seen from the ground at middle latitudes in the Northern Hemisphere. The image at right shows the aurora australis, or southern lights, as photographed from space by astronauts aboard the space shuttle *Discovery*. The Earth's magnetic field prevents the ions that create the aurorae from reaching the surface except near the poles, where the magnetic field reaches low altitudes. (Left: © Ned Haines, Photo Researchers; right: NASA) http://universe.colorado.edu/ fig/8-11.html

that flows to the surface during volcanic eruptions. Sedimentary rocks are formed from deposits of gravel and soil that have hardened, usually in layers where old seabeds or coastlines lay. Metamorphic rocks have been altered in structure by heat and pressure created by movements in the Earth's crust. All three forms can be changed from one to another in a continuous recycling process. Rocks are also classified according to their chemical compositions as minerals of various types. The most common minerals are the silicates, which make up some 90% of all rock on the Earth's surface.

Because of various evolutionary processes, the surface of the Earth is continuously being renewed. While the age of the Earth is thought to be some 4.5 billion years, the ages of most surface rocks can be measured in the millions or hundreds of millions of years. As we pointed out in Chapter 7, the distinction between the age of a planet and the age of its surface is an important one, and it will reappear as we discuss the surfaces of other bodies.

The crust of the Earth is not static but rather is in constant motion. The continents themselves move about, and the world map is variable on geological time scales **(Fig. 8.12).** The idea that continental drift occurs dates back to the beginning of the twentieth century, when maps such as the one in Figure 8.12 were drawn simply on the basis of the obvious fit between continents on opposite sides of the Atlantic Ocean. Additional evidence came from the match of rock formations and even fossils on opposing sides of the ocean, but despite this compelling evidence for

continental drift, it was not widely accepted until the 1960s, when additional clues were found. Among the most important of these clues was seafloor spreading away from mid-ocean ridges, where residual magnetic fields embedded in the seafloor rocks showed clearly that the floor was youngest near the ridges and grew progressively older with distance, just what would be expected if the seafloor was forming from material erupting along the ridges and then spreading away from them.

Even more significant than the new evidence for continental drift, a possible *cause* was finally understood, a key element that had been missing. From all of this evidence has arisen the theory of **plate tectonics,** which postulates that the Earth's crust (the lithosphere) is made of a few large, thin pieces that float on top of the asthenosphere. Due to flowing motions in the asthenosphere, these plates constantly move about, grinding and bumping against each other. The rate of motion is only a few centimeters per year at the most, and the major rearrangements of the continents have taken many millions of years to occur. (The Americas and the European-African system became separated from each other between 150 and 200 million years ago.)

Slowly moving convection currents in the region of the asthenosphere are thought to be the driving force behind the plate movements **(Fig. 8.13).** The speed of the convective overturning motions in any fluid depends largely on the **viscosity** of the fluid, that is, the degree to which it resists flowing freely. The Earth's mantle, as we have already seen, is sufficiently rigid to transmit S waves and must therefore have a high viscosity. Hence, if convection is occurring in the mantle, it is reasonable to expect that the motion is very slow. The principal reason for uncertainty about whether convection is the driving force behind continental drift is that it is not clear how heat is transferred from the core to the upper mantle at a sufficient rate to cause the convection. Therefore, other mechanisms, such as gravitational dragging of regions of differing density, are under discussion as well.

Whether convection is the cause or not, the effects of tectonic activity are becoming well known. Maps of the locations of earthquakes and volcanoes on the Earth reveal a fascinating pattern. Both of these relatively violent geophysical events occur most frequently along specific lines girdling the globe **(Fig. 8.14)**. While the tendency for these occurrences to take place along such zones has been known for a long time (for example, the term "Ring of Fire" refers to the frequency of volcanic eruptions around the shores of the Pacific), the interpretation has only been developed recently. The zones of activity are

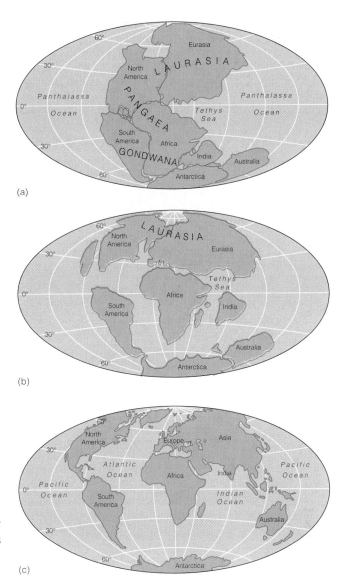

(a)

(b)

(c)

Figure 8.12 (right) Continental drift. These maps show the distribution of the continents over the last 240 million years. (top) The Earth about 240 millions years ago. (middle) The Earth about 70 millions years ago; and (bottom) the modern Earth.

Figure 8.13 (below) Schematic of the mechanisms of continental drift. This sketch shows how continental drift is responsible for uplifted mountain ranges and parallel undersea trenches, where one crustal plate sinks below another (subduction zone). It also shows how a mid-ocean ridge is built up where two plates move away from each other. The arrows indicate the convection currents in the upper mantle that are thought to be responsible for continental drift.

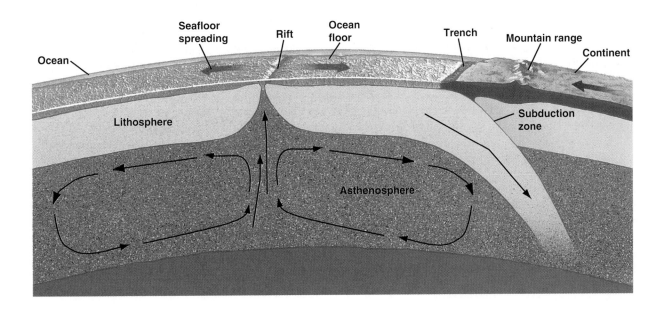

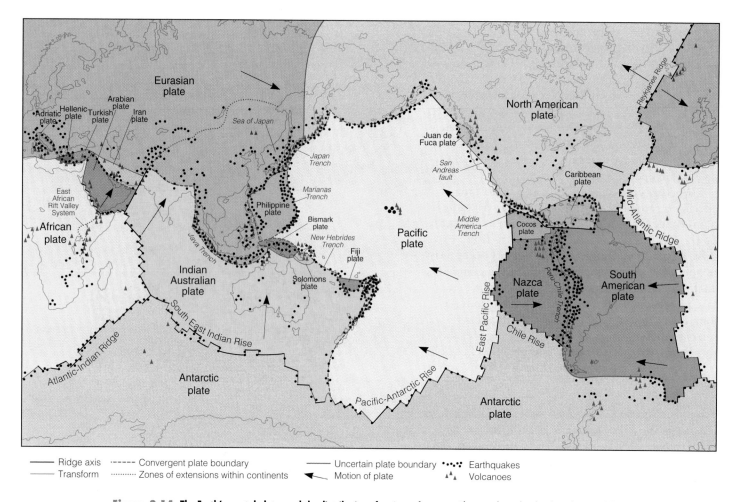

Ridge axis ----- Convergent plate boundary ———— Uncertain plate boundary •••• Earthquakes

——— Transform ·········· Zones of extensions within continents ◄—— Motion of plate ▲▲ Volcanoes

Figure 8.14 **The Earth's crustal plates and the distribution of active volcanoes.** This map shows the plate boundaries and the direction in which the plates are moving. It also shows the locations of volcanoes that have been active in recent years. Note that the volcanoes are concentrated along the boundaries of the Earth's crustal plates, particularly around the edges of the huge Pacific plate. http://universe.colorado.edu/ fig/8-14.html

now understood to be the borders of the enormous crustal plates that are floating atop the mantle (see Fig. 8.14). At the locations where the plates abut one another, many phenomena take place, including earthquakes (as the plates stick and slip as they slide past each other; **Fig. 8.15**), volcanoes (where seafloor plates slide beneath continental plates, creating molten zones near the surface; **Figs. 8.16** and **8.17**), and mountain uplift (where two continental plates collide, forcing crustal material upward; **Fig. 8.18**).

Chains of volcanoes, such as the Hawaiian Islands, are also attributed to the action of plate tectonics. In several locations around the world, volcanic hot spots lying deep below the surface are fixed in place and bring molten rock to the surface. As the crust passes over one of these hot spots, volcanoes are formed and then carried away, creating a chain of mountains as new material keeps coming to the surface over the hot spot. This process is still active, and the Hawaiian

Islands, for example, are still growing. Volcanic activity is occurring on the southeastern part of the largest (and youngest) island, Hawaii, and a new island, already named Loihi, is rising from the seafloor about 20 km south of Hawaii. Its peak has already risen 80% of the way to the surface and has only about 1 km to go, but it will take an estimated 50,000 years to break through.

Plate tectonics, then, accounts for many of the most prominent features of the Earth's surface. Of course, other processes, such as running water, wind erosion, and glaciation, are also important in modifying the face of the Earth. Above all, it is important to note that these are ongoing evolutionary processes; the Earth is still being modified today. This is true of the other terrestrial bodies as well, but to a lesser degree; the Earth is the most active and dynamic of all the inner, rocky planets.

Figure 8.15 **Crustal shifting during an earthquake.** This abrupt tear in the Earth's crust was the result of an earthquake that struck the Andes Mountains of Peru. (Carl Frank, Photo Researchers)

Figure 8.16 **A recent volcanic eruption.** This is a photograph of the explosive eruption of Mt. St. Helens in Washington State during its outburst in 1980. Events such as this provide graphic evidence that the Earth is very active geologically. (John H. Meehan/Science Source, Photo Researchers)

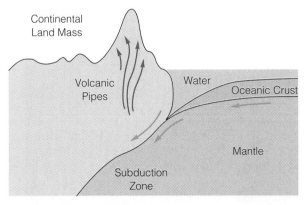

Figure 8.17 **A subduction zone.** Here oceanic crust (seafloor) is shown sinking below continental crust as two adjacent crustal plates move toward each other. Along the line of intersection of the two plates (i.e., along the subduction zone), the oceanic crust slips below the continental crust, creating a deep undersea trench.

Figure 8.18 **The result of a collision between crustal plates.** This is a portion of the Himalaya Mountains, the Earth's tallest range, including Mt. Everest (at right). These mountains are being uplifted by pressure created in a slow collision between the Indian plate and the Eurasian plate and are growing taller each year. (G. J. James/BPS)

The Moon

Seen from afar, the Moon **(Table 8.4)** is a very impressive sight, especially when full **(Fig. 8.19)**. Its 0.5° diameter is pockmarked with a variety of surface features, the most prominent of which are the light and dark areas. The latter, thought by Renaissance astronomers to be bodies of water, are called **maria** (singular: **mare**) after the Latin word for seas. Closer examination of the Moon with even primitive telescopes also revealed numerous circular features called **craters** after similar structures found on volcanoes. The lunar craters are not volcanic, however, but were formed by the impacts of bodies that crashed onto the surface from space.

The pattern of surface markings on the face of the Moon is quite distinctive, and because it never changes, observers recognized long ago that the Moon always keeps the same face toward the Earth. The far side cannot be observed from the Earth, but has now been thoroughly mapped by spacecraft **(Fig. 8.20)**.

Global properties of the Moon

The lunar surface in general can be divided into lowlands, primarily the maria, and highlands, which are vast mountainous regions not as well organized into chains or ranges as mountains on the Earth. With no atmosphere (because the escape velocity is so low that all the volatile gases were able to escape long ago), erosion due to weathering does not occur on the Moon, so large features such as craters and mountain ranges retain a jagged appearance when seen from afar. The surface everywhere consists of loosely piled rocks ejected and redistributed by impacts. Other types of features seen on the Moon include **rays,** light-colored streaks emanating from some of the large craters, and **rilles,** winding valleys that resemble earthly canyons.

As noted earlier in this chapter, the Moon's mass is only about 1.2% of the mass of the Earth, and its radius is a little more than one-fourth the Earth's radius. The Moon's density, 3.34 g/cm³, is below that of the Earth as a whole and more closely resembles the density of ordinary surface rock. This suggests that the Moon probably does not have a large, dense core, which in turn indicates that the Moon is not as highly differentiated as the Earth. As we will see, another reason for the Moon's lower overall density is a lower abundance of iron and other heavy elements than the Earth.

Even before the space program allowed close-up probes of the Moon, infrared measurements showed that the lunar surface is subject to hostile temperatures, ranging from 100 K (−279°F) during the two-week night to 400 K (261°F) during the lunar day. It

Figure 8.19 The full Moon. This high-quality photograph was taken through a telescope on the Earth. (Lick Observatory, © Regents University of California)

Figure 8.20 The lunar farside. This full-disk view, obtained by the U.S. *Clementine* spacecraft, shows that there are few maria on the side of the Moon that faces away from the Earth. (NASA)

Table 8.4
Moon

Mean distance from Earth: 384,401 km (60.4 R$_\oplus$)
   Closest approach: 363,297 km
   Greatest distance: 405,505 km
Orbital sidereal period: $27^d7^h43^m12^s$
Synodic period (lunar month): $29^d12^h44^m3^s$
Orbital inclination: 5° 8'43"

Rotation period: 27.32 days
Tilt of axis: 6°41' (with respect to orbital plane)

Diameter: 3,476 km (0.273 D$_\oplus$)
Mass: 7.35 × 10^{25} g (0.0123 M$_\oplus$)
Density: 3.34 g/cm^3
Surface gravity: 0.165 Earth gravity
Escape speed: 2.4 km/sec
Surface temperature: 400 K (day side); 100 K (dark side)
Albedo: 0.07

Figure 8.21 **Man on the Moon.** The *Apollo* missions, six of which included successful manned landings on the Moon, are humankind's only attempt so far to visit another world. (NASA) http://universe.colorado.edu/ fig/8-21.html

was also obvious that the Moon has no significant atmosphere; based on our discussions of atmospheric escape (Chapter 7), we can understand why.

Exploration of the Moon

Of course, all of the long-range studies of the Moon were almost instantly made obsolete by the space program, which has featured extensive exploration of the Moon, first by unmanned probes and then by manned landings **(Figs. 8.21** and **8.22).** The historic first manned landing occurred on July 20, 1969. This was the *Apollo 11* mission; it was followed by five more manned landings, the last being *Apollo 17,* which took place in late 1972. Each mission incorporated a number of scientific experiments, some involving observations of the Sun and other celestial bodies from the airless Moon, but most devoted to the study of the Moon itself. Little space-based exploration of the Moon has occurred since the end of the *Apollo* program, but recently there has been a renewal of interest. The possibility of establishing permanent scientific facilities on the Moon has been discussed, and new images of portions of its surface (particularly the south polar region) were recently obtained by the U.S. *Clementine* and *Galileo* missions **(Fig. 8.23),** which revealed the largest impact basin on the Moon and may have shown that ice

Figure 8.22 **The lunar rover.** The later *Apollo* missions used these vehicles to travel over the Moon's surface, allowing the astronauts to explore widely in the vicinity of the landing sites.

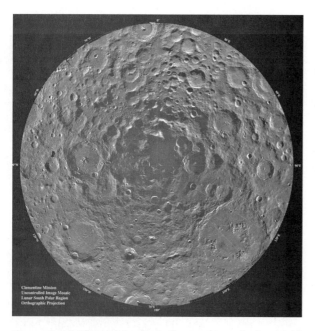

Figure 8.23 **The south lunar pole.** This view of the south polar region of the Moon, obtained by the *Clementine* spacecraft, reveals a large, previously unknown impact basin near the pole, at lower right in this view. (NASA)

Figure 8.24 **The lunar "seas."** Here is a broad vista encompassing portions of three maria. Mare Crisium (foreground); Mare Tranquilitatis (beyond Mare Crisium); and Mare Serenitatis (on the horizon at the upper right). These relatively smooth areas are younger than most of the lunar surface, having been formed by lava flows after much of the cratering had already occurred. (NASA)

persists on the lunar surface in the bottom of a crater near the pole, where direct sunlight never penetrates.

The Lunar Surface and Interior

Viewed on any scale from the largest to the smallest, the Moon's surface is irregular, marked throughout by a variety of features. We have already mentioned the maria, the large, relatively smooth, dark areas **(Fig. 8.24).** The maria appear darker than their surroundings because they have a relatively low albedo (that is, they reflect less sunlight). Despite their smooth appearance relative to the more chaotic terrain seen elsewhere on the Moon, the maria are marked here and there by craters.

Outside the maria, much of the lunar surface is covered by rough, mountainous terrain. Even though the maria dominate the near side of the Moon, there are almost none on the far side, and the highland regions actually cover most of the lunar surface.

Craters are everywhere **(Figs. 8.25** and **8.26).** They range in size from huge craters hundreds of kilometers in diameter to microscopic pits that can be seen only under intense magnification. In some regions, the craters are so densely packed together that they

overlap. The lack of erosion on the Moon allows craters to survive for billions of years, providing plenty of time for younger craters to form within the older ones. Relatively few craters are seen in the maria, an indication that the surface in these regions is younger than in the highlands, where the crater densities are higher.

All of the craters on the Moon are impact craters, formed by collisions of interplanetary rocks and debris with the lunar surface rather than by volcanic eruptions. This conclusion is based on the shapes of the craters, the central peaks in some of them, and the trails left by **ejecta,** or cast-off material created by the impacts **(Fig. 8.27).** The rays stretching away from craters are strings of smaller craters formed by the ejecta from the large, central crater.

Some of the lunar mountain ranges reach heights greater than any on Earth. They are more jagged than earthly mountains because there is no erosion, and they lack the prominent drainage features usually found in terrestrial ranges.

The rilles, which resemble dry riverbeds, are rather interesting features **(Fig. 8.28).** Apparently, they were formed by flowing lava rather than water. In some cases, there are even lava tubes that have par-

Figure 8.25   Craters on the Moon. This photograph, taken by *Apollo 16* astronauts, shows cratered terrain on the lunar far side. A portion of the gamma-ray spectcrometer carried by *Apollo 16* is visible at the right. (NASA)

Figure 8.26   Large lunar craters.   (NASA)

Figure 8.27   A crater with ejecta. This crater on the lunar far side is a good example of a case in which material ejected by the impact has created rays of light-colored ejecta. Close examination of such features often reveals secondary craters, formed by the impacts of debris blasted out of the lunar surface by the primary impact. (NASA)

Figure 8.28   Rilles. This photograph shows Hadley rille meandering through the Hadley-Appenine area. One of the *Apollo* landings was close enough to Hadley rille, to allow the astronauts to explore it. (NASA)

Figure 8.29 **A field of boulders.** Rocks in a wide variety of sizes are strewn over much of the Moon's surface. Most have been blasted out of the surface by impacts. (NASA)

Figure 8.30 **A large boulder.** Rocks on the lunar surface range in size from tiny pebbles to massive objects like this. (NASA)

Figure 8.31 **Moon rocks.** One of the thousands of lunar samples brought back to Earth by the *Apollo* astronauts, the rock at the left is an example of a breccia. At the right is a thin slice of a Moon rock, illuminated by polarized light, which causes different crystal structures to appear as different colors. (NASA)

tially collapsed, leaving trails of sinkholes. The rilles and the maria indicate that the Moon has undergone stages when large portions of its surface were molten.

In many areas, the *Apollo* astronauts found a surface strewn with loose rock, ranging in size from pebbles to boulders as big as a house **(Figs. 8.29 and 8.30).** The rocks generally have sharp edges, due to the lack of erosion, and occasional cracks and frac-

tures. In most cases, the large boulders appear to have been ejected from nearby craters and are therefore thought to represent material from beneath the surface.

The *Apollo* astronauts brought almost 800 pounds of small lunar rocks back to Earth, giving scientists an opportunity to study the lunar surface characteristics in as much detail as is possible for

Earth rocks and soils **(Fig. 8.31)**. Thus, both extensive chemical analysis and close-up observation of rocks in place on the Moon were possible. The rock samples from the Moon are currently housed in numerous scientific laboratories around the world, where analysis continues. Specimens have also found their way into museums, and visitors to the Smithsonian's Air and Space Museum in Washington, D.C., can touch a lunar rock.

The lunar soil, called the **regolith,** consists of loosely packed rock fragments and small glassy mineral deposits probably created by the heat of meteor impacts. In addition to the loose soil, a few distinct types of surface specimens have been recognized on the basis of morphology. The most common of these are the **breccias** (see Fig. 8.31), which consist of small rock fragments fused together and resembling chunks of concrete. Similar kinds of rock are found on Earth, except that the Earth breccias are formed in streambeds where water plays a role in shaping them. The lunar breccias contain jagged, sharp rock fragments and were probably fused together by heat and pressure created in meteorite impacts.

All lunar rocks are pitted on the side that is exposed to space, with tiny craters called **micrometeorite craters.** These are formed by the impact of tiny bits of interplanetary material no bigger than grains of dust.

Radioactive dating techniques show the lunar rocks to be very old by earthly standards—as old as 3.5 to 4.5 billion years. Rocks from the maria are not quite so old, but still date back to 3 billion years ago or earlier. The great ages of the lunar rocks are among the strongest arguments that the Moon's evolution came to a halt a very long time ago, and that today the Moon is not geologically active. Clearly, no process is at work restructuring and renewing the Moon's surface the way tectonic activity and weathering are doing for the surface of the Earth.

Some of the experiments done on the Moon by the *Apollo* astronauts were aimed at revealing the interior conditions by monitoring seismic waves caused by quakes on the Moon. Therefore, the astronauts carried with them devices for sensing vibrations in the lunar crust; because it was not known whether natural quakes occurred frequently, they also brought along devices for thumping the surface to make it vibrate. It turned out that natural moonquakes do occur, although not with great energy. The seismic measurements continued after the *Apollo* landings, with data radioed to Earth by instruments left in place on the lunar surface.

The measurements showed that the regolith is typically about 10 m thick and is supported by a

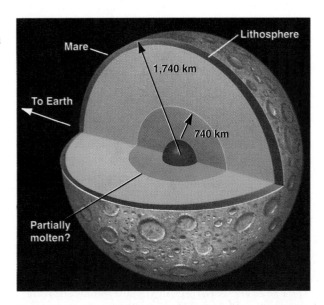

Figure 8.32 The internal structure of the Moon. This cross-sectional view shows the interior zones inferred from seismological studies. The existence of a dense core is not certain. Note that the maria lie almost exclusively on the side facing the Earth, where the lunar crust is relatively thin.

thicker layer of loose rubble. The crust is 50 to 100 km thick at the *Apollo* sites and may be somewhat thicker than this on the far side where there are no maria **(Fig. 8.32).** Beneath the crust is a mantle, consisting of a well-defined lithosphere, which is rigid, and beneath that an asthenosphere, which is semiliquid. The innermost 500 km consists of a relatively dense core, but not as dense as that of the Earth.

The seismic measurements indicate no truly molten zones in the Moon at the present time, although temperature sensors on the surface discovered a substantial heat flow from the interior. This is probably caused by radioactive minerals below the surface.

The Moon has no detectable overall magnetic field, which is further evidence against a molten core. The chief distinction among the zones in the lunar interior is density, with the densest material being closest to the center. Thus moderate differentiation has occurred in the Moon, implying that it was once at least partially molten.

There is no trace of present-day lunar tectonic activity, perhaps the greatest single departure from the geology of the Earth. The force that has had the greatest influence in shaping the face of the Earth has no role today on the Moon. Instead, the lunar surface is entirely the result of the way the Moon was formed (which may have involved some tectonic activity in the early stages) and the manner in which it has been altered by lava flows and by the incessant bombardment of debris from space.

The Evolution of the Earth-Moon System

While the formation of the Earth (and of planets in general) is reasonably well understood, there has been a great deal of uncertainty about the Moon's formation. The main reasons for the uncertainties arise from the many peculiarities about the Moon, discussed at the beginning of this chapter. We will discuss the Earth's formation first and then delve into the complexities of the Moon's origin.

Formation of the Earth and Its Atmosphere

The age of the Earth-Moon system, estimated from geological evidence, is about 4.5 billion years. The Earth is thought to have formed from the coalescence of a number of planetesimals, the solid bodies that were the first to condense in the earliest days of the solar system. Most or all of the Earth was molten at some point during the first billion years. The Earth became highly differentiated during its molten period, as the heavy elements tended to sink toward the planetary core. At the same time, volatile gases were emitted by the newly formed rocks at the surface. These gases, including hydrogen (H_2), helium (He), ammonia (NH_3), methane (CH_4), and water vapor (H_2O), formed the earliest atmosphere of the Earth. Thus the initial atmosphere of the Earth consisted largely of hydrogen and its compounds and helium, a very common element in the universe, but one that does not easily form molecular bonds and is therefore almost always found in its atomic form.

It is possible that the Earth lost its atmosphere, perhaps more than once, during its early history. Major impacts almost certainly occurred several times as the solar system formed, and the Earth would not have escaped the bombardment. Some of the impacts may have been energetic enough to blow away the atmosphere, which then would have reformed gradually as outgassing continued. Other, less severe impacts would have helped rebuild the atmosphere, by adding gases. Such impacts may have created the Earth's seas.

According to some theories of solar system formation, the solid material from which the Earth condensed did not contain much water vapor. It is possible that most of the water originated in comets instead, which were much more numerous in the inner solar system during the first billion years than they are today. Hence, some scientists believe that most of the water in the Earth's oceans was accreted through the impacts of comets early in the Earth's history, rather than being present from the very beginning.

Whatever its origin, water collected on the surface, and from its earliest days our planet had seas. Carbon dioxide (CO_2) was probably abundant for a time but was eventually removed from the atmosphere by the oceans and deposited in carbonate rocks. This absorption process, of critical importance to the further evolution of the Earth's atmosphere, requires liquid water; if the early Earth had not had oceans, the carbon dioxide might never have left the atmosphere. This would have had fateful consequences for the Earth, which will become clear in our discussion of the evolution of Venus (Chapter 9).

Nearly all of the hydrogen and helium in the atmosphere had escaped into space by the time the Earth was about a billion years old (about when the first simple life-forms apparently appeared), while the heavier gases that now dominate the atmosphere were too massive to escape. While atomic helium escaped due to its low mass, no other helium was preserved on Earth in the form of compounds with heavier elements, because helium does not form chemical bonds. Consequently, helium is so rare on Earth today that it was not discovered until spectroscopic measurements revealed its presence in the Sun. Yet it is the second most abundant element in the universe, following hydrogen.

The critical reactions that led to the development of life on Earth must have taken place before all the atmospheric hydrogen escaped, because the types of reactions that were probably responsible involve this element. The earliest fossil evidence for primitive life dates back at least 3 billion years, when some hydrogen was still left in the atmosphere. This point is significant: if much oxygen had been present, the extremely high reactivity of this element would have prevented the critical reactions necessary to form life. Ironically, the element that is crucial to the existence of our form of life today would have poisoned and prevented the development of life, had it been present in the early atmosphere.

Essentially no free oxygen was present in the atmosphere until the development of life-forms that released this element as a by-product of their metabolic activities. Most plants release oxygen into the atmosphere during photosynthesis; as a supply of this element built up, the opportunity arose for complex animal forms to evolve. Besides providing the oxygen necessary for the metabolisms of living animals, the buildup of oxygen created a reservoir of ozone (O_3) in the upper atmosphere. The ozone in turn began to screen out the harmful ultraviolet rays from the Sun, allowing life-forms to move onto the exposed land.

Once life had gained a toehold on the continental landmasses, the process of converting the atmosphere to its present state began to accelerate. Soon nitrogen from the decomposition of organic matter began to be released into the air in large quantities (some nitrogen was already present due to volcanic activity), and by the time the Earth was perhaps 2 billion years old, the atmosphere had reached approximately the composition it has today.

By this time also, the mantle had solidified and the crust had hardened. (The oldest known surface rocks are nearly 3.5 billion years old.) The interior has remained warmer than the surface because heat escapes slowly through the crust and because radioactive heating of the interior has taken place over a long period of time and is still effective today.

Origin of the Moon

The Moon's beginnings are much more obscure than the Earth's despite the close-up examination of the Moon afforded by the *Apollo* missions. The reasons for the difficulty in explaining how the Moon formed lie in the very unusual chemical composition of the Moon and the large amount of angular momentum of the Earth-Moon system, noted earlier in this chapter. Recall that angular momentum is related to the mass, orbital speed, and separation of two bodies in mutual orbit. Compared with other planet-satellite pairs in the solar system, the Moon is very large relative to its parent planet and quite distant, so that the Earth-Moon system has a high angular momentum. This has been difficult to explain in any theory postulating that the Moon formed by splitting off from the Earth, for example, or that the Earth and Moon simply formed together as a double planet. In other words, scientists have had difficulty explaining why the material that formed the Moon remained separate from the Earth instead of coalescing with it.

The chemical composition of the Moon is characterized by a low overall abundance of heavy elements, such as iron, and a higher relative abundance of the **refractory elements,** which do not vaporize easily. The readily-vaporized volatile elements are underabundant in comparison with the Earth, and water is almost completely absent. These contrasts might suggest that the Moon formed somewhere else in the solar system rather than near the Earth, yet there are some strong similarities, such as the relative proportions of different isotopes of oxygen (forms of oxygen with varying numbers of neutrons in their nuclei), that argue for formation in the same general vicinity. Seemingly, most of the chemical properties of the Moon could be explained if the Earth and

Moon formed out of very similar material originally, but then somehow the Moon lost much of its iron and was subjected to substantial heating early in its history.

Traditionally, three classes of models for lunar formation have been advanced: (1) the **fission hypothesis,** in which the Moon somehow split off from the Earth; (2) the **capture theory,** in which the Moon formed elsewhere and was then captured by the Earth's gravitational field; and (3) the **coeval formation,** or **double planet,** model, in which the two bodies formed together. Each of these theories has always had serious difficulties. The fission theory lacks a mechanism for causing the split of the young Earth, lacks an explanation for the high angular momentum of the Earth-Moon system, and cannot easily explain the low iron content of the Moon or the other chemical dissimilarities. The capture hypothesis can easily account for the high angular momentum, since the Moon would have approached the Earth with a substantial velocity that would have been transformed into orbital angular momentum upon capture, but that same velocity is the downfall of the theory, for it is difficult or impossible for an initially free Moon to be slowed enough to be caught by the Earth's gravity. This model also has difficulty explaining the chemical similarities between the Earth and the Moon, as well as the low iron content of the Moon. The coeval formation hypothesis has grave problems with the chemical contrasts between Earth and Moon, since this view holds that the two formed together from the same material, and this theory also founders on the angular momentum problem.

A recent suggestion appears to resolve all of the problems related to angular momentum and chemical composition and is therefore becoming the favorite among lunar scientists. This idea includes some elements of the capture model, in that it invokes a large body approaching the Earth, but there the similarity ends. In this model, the body that approaches is a large planetesimal, at least 10% the mass of the Earth (hence at least the size of Mars). It is proposed that this large body, already differentiated and having an iron core, collided with the young Earth at a grazing angle **(Fig. 8.33)** and was partially vaporized by the energy of the impact. The nickel-iron core of the impacting body merged with the Earth, enriching its iron content, while material from the body's iron-poor mantle formed a hot disk around the Earth, which subsequently coalesced to form the Moon. The result is a Moon having a high angular momentum, derived from the speed (about 10 km/sec) of impact; a low overall iron abundance due to the fact that the incoming planetesimal had already differentiated and only its mantle material formed the Moon; and a very low content of volatile elements, since these were vaporized by the heating that occurred on impact and were lost to space.

Invoking a singular event such as a collision with a Mars-size planetesimal may seem contrived, but studies of the process by which early solar system material condensed into solid form and then built up planetesimals shows that this hypothesis is quite reasonable. There should have been a period of time when most of the preplanetary material was concentrated in a few rather large planetesimals, all orbiting the Sun in the same direction. The final formation of the planets came about as these large planetesimals merged, so the early history of each planet would have involved collisions between large bodies. Evidence for this is found in some other anomalous planetary characteristics: Venus has a backward, slow rotation, for example, and Uranus is tipped over on its side. Modern planetary science includes impacts as a major force affecting the development and evolution of the solar system.

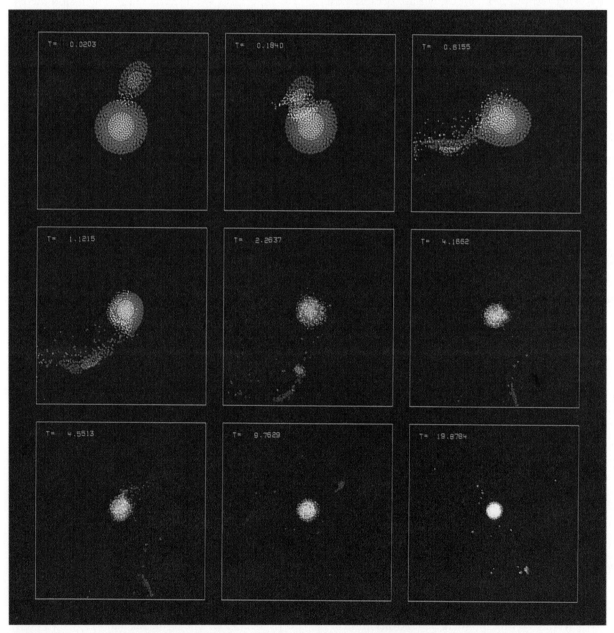

Figure 8.33 A computer simulation of the Moon's formation. The newest theory of lunar formation envisions that a planetesimal as massive as Mars or larger collided with the young Earth and was partially vaporized on impact. This sequence of computerized illustrations shows how the resulting debris could then have coalesced to form the Moon. (W. Benz and W. Slattery, Los Alamos National Laboratory)

History of the Moon

The Moon's evolution after its formation has not been as difficult to deduce. Detailed study of the surface structure of the Moon, along with analyses of surface rocks and information on the interior structure gleaned from seismic data, has revealed a comprehensive picture.

The Moon underwent a period when it was molten to a depth of about 400 km (this factor was also more easily explained by the planetesimal-collision formation model than by the others). As the crust cooled and solidified, impacts began to sprinkle it with craters. Within the first 1 or 2 billion years, extensive lava flows occurred on the surface, welling up in regions where the crust was relatively thin. Much of one side of the Moon was covered by the flows, which we know today as maria. The rate of impacts from space began to taper off at about the time the maria formed; hence, they have fewer craters than the highlands, where the high concentration of craters records the high frequency of collisions suffered by the young Moon. Impacts have continued throughout the history of the Moon, but following that early, intense period, the rate has been much slower and quite steady.

The Moon was much closer to the Earth when it formed than it is today, but tidal stresses exerted on it by the Earth have gradually slowed the Moon's spin, causing it to retreat from the Earth as angular momentum due to its spin has been converted into orbital angular momentum. The Moon's rotation slowed to the point where it keeps one face toward the Earth as it orbits; this is the face containing the maria, whose high density causes the distribution of the Moon's mass to be slightly nonsymmetric. The relatively higher density of the side with the maria caused this side to "lock in" facing the Earth; thus it is no coincidence that the near side is the face with the maria. This scenario explains why the lunar far side, having almost no maria and a lower crustal density, has a very different appearance than the side facing the Earth.

Geologically, the Moon now is quite inert, having no molten core and insufficient internal heat to drive such processes as tectonic activity. The only changes that are expected to occur over the next few billion years are the addition of new craters.

Summary

1. The Earth's atmosphere is 80% nitrogen and about 20% oxygen (by number of particles), with only traces of water vapor and carbon dioxide.
2. The atmosphere is divided vertically into four temperature zones: the troposphere, the stratosphere, the mesosphere, and the thermosphere.
3. Heat from the Sun, combined with the Earth's rotation, creates the global wind patterns.
4. The Earth's structure, explored with seismic waves, consists of the solid inner core, the liquid outer core, the mantle, and the crust.
5. The Earth has a magnetic field, probably created by currents in the molten core; the magnetic field traps charged particles in zones above the atmosphere called radiation belts.
6. The crust is broken into tectonic plates that shift around, which accounts for continental drift and for most of the major surface features of the Earth.
7. The lunar surface consists of relatively smooth areas called maria and mountainous regions; it is marked everywhere by impact craters.
8. The lunar soil is called the regolith, and rocks on the surface are all igneous, mostly silicates, with low abundances of volatile gases.
9. Seismic data show that the Moon has a crust 50 to 100 km thick, a mantle, and a core extending about 500 km from the center.
10. The Moon has no present-day tectonic activity and no magnetic field, which indicate that it probably does not have a liquid core.
11. The Earth's evolution from a largely molten planet with a hydrogen-dominated atmosphere to its present state was caused by the presence of liquid water on its surface; the loss of lightweight gases into space; and the development of life-forms on its surface, which helped convert the atmospheric composition to nitrogen and oxygen.
12. The Moon has significant chemical contrasts with the Earth, but also some similarities, arguing for a distinct origin at a similar distance from the Sun. The formation mechanism that appears to provide the best explanation of the Moon's properties is that the Moon formed as the result of a collision between the young Earth and a large planetesimal.

Observing the Moon

Even with the unaided eye, it is easy to see that the surface of the Moon has bright and dark regions. The bright regions were called *terrae*, or land, by medieval astronomers who thought the dark regions were maria, or seas. Today we know that the terrae are rugged, mountainous regions while the maria are low-lying lava plains that reflect light less efficiently. Without a telescope or binoculars, it is difficult to make out much detail, but with even a modest pair of binoculars, many of the more prominent features are easy to see.

The photographs on the following page will help you identify features on the Moon. Note that these are inverted with respect to the lunar photos shown in the chapter (Fig. 8.20), which show the view as seen through an astronomical telescope whose optics invert images (top-to-bottom and left-to-right). When binoculars (or a so-called terrestrial telescope) are used, however, the image is not inverted. Thus the images and drawings shown here are oriented the same way you will see the Moon with the unaided eye or through binoculars.

It is preferable to view the Moon at first quarter or third quarter rather than when it is full. The reason is that when the Moon is full, sunlight strikes its surface from approximately the vertical direction, and no shadows are cast. This makes it difficult to see surface relief. At first and third quarter, sunlight strikes the Moon from an oblique angle, and surface features create long shadows. Mountains and crater walls are much more prominent at these times.

Use the photos provided here to see how many of the major features on the Moon you can identify. To view the western hemisphere, you will need to observe during first quarter; the eastern hemisphere is visible at third quarter. Since the Moon does not rise until midnight when it is at third quarter, you may have to sacrifice some sleep to chart this portion of its surface.

13. The Moon's evolution consisted of a molten state, followed by hardening of the crust and subsequent large-scale lava flows that created the maria. Since that time (about 1 billion years after the Moon's formation), the Moon has been geologically quiet.

Web Connections

The Review Questions and Problems also appear at the following URLs:
http://universe.colorado.edu/ch8/questions.html
http://universe.colorado.edu/ch8/problems.html

Review Questions

1. Explain how heat from the Sun and the rotation of the Earth create the global circulation pattern of the atmosphere. How do the oceans influence the circulation?
2. How do we know anything about the Earth's interior? Do you think the same methods that are used to study the Earth can be applied to other planets?

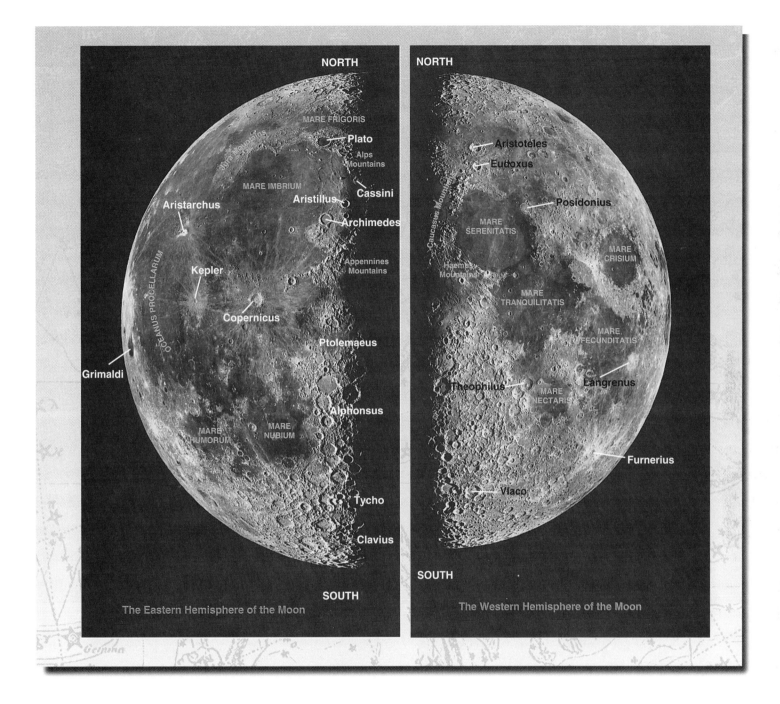

The Eastern Hemisphere of the Moon

The Western Hemisphere of the Moon

3. Explain the differences between transverse and compressional waves. Can you think of everyday examples of each kind? Explain how we know that the Earth's core has a liquid component from studying how these waves travel through the Earth's interior.

4. If the density of typical surface rocks is 3.5 g/cm³, and the Earth's average density is 5.5 g/cm³, what does this tell us about the density in the deep interior? How did this situation arise?

5. Compare and discuss the ages of the surfaces of the Earth and the Moon, as opposed to the ages of the bodies themselves.

6. Summarize the effects of life-forms on the evolution of the Earth's atmosphere.

7. What is the evidence that the craters on the Moon are formed by impacts, rather than by volcanic eruptions?

8. Summarize the similarities and contrasts between lunar rocks and those found on the Earth's surface.

9. Why does the Earth not have as many craters as the Moon?

Problems

1. Use data from Appendix 8 on the semimajor axis of the Moon's orbit and its sidereal period to calculate the mass of the Earth. Compare your answer with the value given in the appendix.

2. Find the location of the center of mass of the Earth-Moon system using information from Chapter 5 (note especially the Astronomical Activity box at the end of the chapter). Express your answer in terms of the Earth's radius; i.e., say how many Earth radii the center of mass is from the center of the Earth.

3. Table 8.3 lists values for the depth and the density of the various distinct layers in the Earth's interior. Use those figures to determine the fraction of the Earth's mass and volume that is occupied by the Earth's core (include both inner and outer core). Use the average value of the density for each layer; i.e., the middle value within the range given in the table.

4. When a solar flare occurs, a burst of charged particles enters the solar wind and eventually reaches the Earth's orbit. How long does it take for the particles to reach the Earth, if the average speed of the solar wind is 300 km/sec? What strategy does this suggest for predicting "magnetic storms" (disruptions of radio communications caused when a burst of solar wind particles strikes the Earth's magnetic field)?

5. If North America is approaching Japan at a rate of 3 cm/yr, and the present distance between North America and Japan is 5,000 km, how long will it take for the two to collide?

6. Calculate the escape velocity from the surface of the Moon. Comment on the relevance of your answer to the absence of any significant atmosphere on the Moon.

7. If meteorites have rained down on the Moon at a constant rate since it formed, and the maria are one-tenth the age of the highlands, how should the density of craters compare? If the observed crater density in the maria is actually one-fourth of the density in the highlands, what does this tell you about the rate of meteorite bombardment over the lifetime of the Moon?

Additional Readings

Allégre, C. J. and S. H. Schneider 1994. The Evolution of the Earth. *Scientific American* 271(4):66.

Beatty, J. K. 1995. New Measures of the Moon. *Sky & Telescope* 90(1):32.

Benningfield, D. 1993. The Odd Little Moons of Mars. *Astronomy* 21(12):48.

Bonatti, E. 1994. The Earth's Mantle Below the Oceans. *Scientific American* 270(3):44.

Broadhurst, L. 1992. Earth's Atmosphere: Terrestrial or Extraterrestrial? *Astronomy* 20(1):38.

Bruning, D. 1994. *Clementine* Maps the Moon. *Astronomy* 22(7):36.

Burnham, R. 1993. What Makes Venus Go? *Astronomy* 21(1):40.

Chaikin, A. 1994. The Moon Voyagers. *Astronomy* 22(7):26.

Dalziel, I. W. D. 1995. The Earth Before Pangea. *Scientific American* 272(1):58.

Goldman, S. J. 1994. *Clementine* Maps the Moon. *Sky & Telescope* 88(1):20.

Goldreich, P. 1972. Tides and the Earth-Moon System. *Scientific American* 226(4):42.

Green II, H. W. 1994. Solving the Paradox of Deep Earthquakes. *Scientific American* 271(34):64.

Hoffman, K. A. 1988. Ancient Magnetic Reversals: Clues to the Geodynamo. *Scientific American* 258(5):76.

Jeanloz, R. and T. Lay 1993. The Core-Mantle Boundary. *Scientific American* 268(5):48.

Kasting, J. F., B. Toon, and J. B. Pollack 1988. How Climate Evolved on the Terrestrial Planets. *Scientific American* 258(2):90.

Kates, R. W. 1994. Sustaining Life on the Earth. *Scientific American* 271(4):114.

Kirshner, R. P. 1994. The Earth's Elements. *Scientific American* 271(4):58.

Kross, J. F. 1995. What's in a Name? *Sky & Telescope* 89(5):28.

Murphy, J. B. and R. D. Nance 1992. Mountain Belts and the Supercontinent Theory. *Scientific American* 266(4):84.

Sampson, R. 1992. Fire in the Sky. *Astronomy* 20(3):38.

Spudis, P. D. 1996. The Giant Holes of the Moon. *Astronomy* 23(5):50.

Stern, A. 1993. Where the Lunar Winds Blow Free. *Astronomy* 21(11):36.

Taylor, G. J. 1994. The Scientific Legacy of Apollo. *Scientific American* 271(1):40.

Taylor, S. J. and S. M. McLennan 1996. The Evolution of Continental Crust. *Scientific American* 274(1):76.

York, D. 1993. The Earliest History of the Earth. *Scientific American* 268(1):90.

Chapter 9

EARTH'S SIBLINGS
The Terrestrial Planets

*The beauteous planet Venus,
which invites to love, makes
all the Orient laugh, outshining
the light of the fishes, which
followed close behind.*

Dante, 1316

Chapter Web site: http://universe.colorado.edu/ch9

The Earth is the largest and most geologically active of the terrestrial bodies, and the Moon is the smallest and least active. Now as we go on to discuss the other three terrestrial planets, we will find that in most properties they are intermediate between the Earth and the Moon. Here again we will find that the contrasts can be understood in terms of evolution: Mercury, Venus, and Mars all appear to have undergone more extensive periods of geologic activity than the Moon, but less extensive than the Earth. In this chapter we will seek an explanation for these contrasts.

Collective Properties of the Terrestrials

The general properties of Venus, Mars, and Mercury are given in **Table 9.1,** and recent portraits are presented in **Figures 9.1, 9.2,** and **9.3.** Venus is the most massive of the three and, in fact, is nearly identical to the Earth in diameter, mass, and therefore density. Mars is intermediate in diameter and mass, and Mercury is the smallest of the three. As we will see, there is probably a connection between these overall characteristics and the degree of geological activity each planet has undergone.

Venus, Mars, and Mercury have all been the subject of intensive observations from Earth since antiquity, and all three have been visited by unmanned probes from Earth in modern times (Appendix 7). All three are easily visible to the naked eye, but because Mercury and Venus are inferior planets, they never are seen far from the Sun and must be observed during the twilight of early morning or early evening. It is especially important to catch Mercury just at greatest elongation (see Astronomical Activity at the end of this chapter), and many people never see this planet because they do not know when to look.

Earth-based observations of the terrestrials have revealed surprises. Whereas telescopic studies can tell us little about the surfaces of Mercury and Venus (Mercury is too small and too close to the Sun; Venus is perpetually shrouded in clouds), some detail can be discerned on the surface of Mars. Mercury and Venus have been probed using radar techniques, in which intense microwave signals are beamed at the planet and the reflected signal provides information on surface structures and planetary rotation (via the Doppler effect; **Fig. 9.4**). Both Mercury and Venus have very long days, and the rotation of Venus is retrograde, that is, backward relative to most motions in the solar system. In addition, radio data showed that Venus is very hot, with a surface temperature of about 750 K (about 900°F), attributed to intense greenhouse heating. Earth-based observations of Mars showed that it has polar ice caps and seasonal color variations, contributing to speculation that life may exist there.

The major highlights of the unmanned exploration of the terrestrials have been the 1974 *Mariner 10* mission to Mercury; the *Pioneer-Venus* and *Magellan* orbiters of Venus, as well as the Soviet *Venera* probes; and the *Mariner 9* and *Viking* missions to Mars. *Mariner 10* encountered Mercury three times, providing detailed images of much of its surface. *Pioneer-Venus* operated in orbit about Venus from 1978 until 1994, allowing scientists to monitor changes in its atmosphere, while the *Magellan* orbiter (1991–1993) mapped the surface of Venus in great detail, using radar reflection techniques. The *Venera* probes made a series of soft landings on Venus, surviving for as long as an hour under the hellish conditions there, and succeeded in sending back pictures and data from the surface. The *Mariner 9* orbiter provided the first global maps of Mars, and the *Viking* missions, which included orbiters and landers, gave us our first images from the surface of Mars and also carried out robotic investigations of the Martian soil.

By the end of this era of exploration, the three terrestrial planets had been revealed as individual,

Table 9.1
Basic Properties of Mercury, Venus, and Mars

	Mercury	Venus	Mars
Orbital semimajor axis (AU)	0.387	0.723	1.524
Perihelion distance (AU)	0.313	0.716	1.404
Aphelion distance (AU)	0.459	0.731	1.638
Orbital period (yr)	0.241	0.615	1.881
Orbital inclination	7°0′15″	3°23′40″	1°51′0″
Rotation period (days)	58.785	-243.7[a]	1.029
Tilt of axis (obliquity)	0°	177°	23°59′
Diameter (relative to Earth)	0.383	0.950	0.532
Mass (relative to Earth)	0.0553	0.815	0.107
Density (g/cm³)	5.427	5.204	3.933
Surface gravity (relative to Earth)	0.378	0.887	0.377
Escape speed (km/sec)	4.3	10.36	5.03
Surface temperature (K)	700/100[b]	750	145–300
Albedo (average)	0.106	0.65	0.15
Satellites	None	None	2

[a]The minus sign indicates that the rotation of Venus is retrograde; i.e., in the direction opposite the spins of most of the other planets.

[b]The temperature of the side facing the Sun is 700 K; the temperature of the dark side is 100 K.

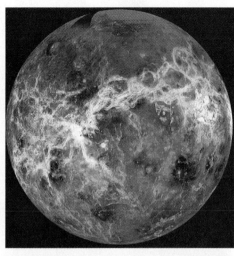

Figure 9.1 **Venus.** At the left is a visible-light image of the cloudtops of Venus, obtained by the *Pioneer-Venus* orbiter; at the right is a full-disk image of Venus based on radar data from the *Magellan* mapper. The color of the *Magellan* image is synthetic, with areas of relative brightness and darkness indicating radar reflectivities and variations in elevation. The *Magellan* mapper, orbiting Venus since 1990, has now nearly completed its coverage of the entire surface of the planet. (NASA/JPL)
http://universe.colorado.edu/fig/9-1.html

Figure 9.2 **Mars.** At the left is an image of Mars obtained from Earth orbit by the *Hubble Space Telescope;* at the right is a mosaic of *Viking* orbiter images showing the full disk of Mars in great detail. (left: NASA/Space Telescope Science Institute; right: U.S. Geological Survey; data from NASA)
http://universe.colorado.edu/fig/9-2.html

Figure 9.3 **Mercury.** On the left is an Earth-based photograph of Mercury, in which little detail can be seen. On the right is a mosaic of *Mariner 10* images covering a large portion of the sunlit portion of the planet at the time of the *Mariner* encounters. (NASA) http://universe.colorado.edu/fig/9-3.html

Reflected radar
shifted to longer
wavelength

Observer's
Location

Reflected radar
shifted to shorter
wavelength

Figure 9.4 **Using Doppler shifts to determine rotation speed.** The reflected radar waves are Doppler-shifted if they bounce off a portion of the planet that is moving away from us or toward us. By measuring the amount of the shift, the speed of approach or recession can be determined. Then the rotation period can be found if the planet's diameter is known.

The Terrestrials

Mercury

Mars

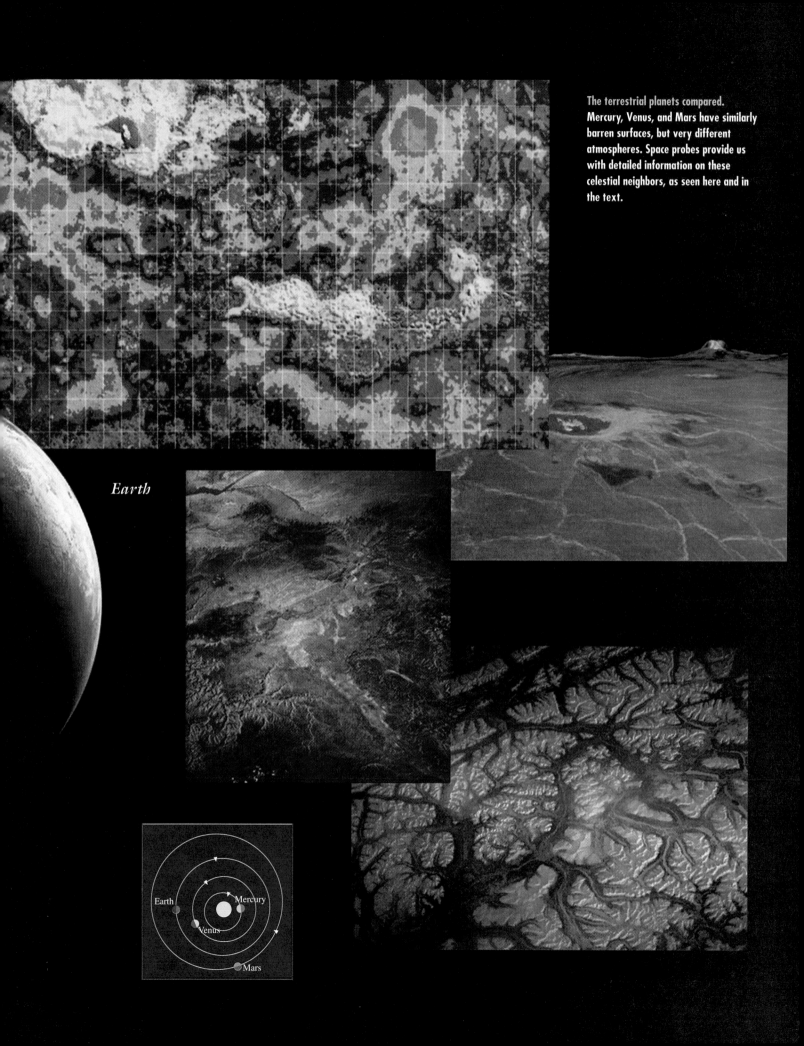

The terrestrial planets compared. Mercury, Venus, and Mars have similarly barren surfaces, but very different atmospheres. Space probes provide us with detailed information on these celestial neighbors, as seen here and in the text.

Earth

Earth

Mercury

Venus

Mars

unique worlds. Mercury, so hot that it cannot hold a permanent atmosphere, resembles the Moon, at least superficially. Venus was revealed as very much unlike the Earth in both its surface conditions and its geology, even though in bulk properties such as density the two are very similar. Mars was found to be more barren than hoped by those who expected to find life there and is geologically dormant as well.

Atmospheres of the Terrestrial Planets

In general, we might expect that the more massive a planet is, the more extensive the atmosphere it can maintain (recall our discussion of escape speeds and the loss of atmospheric particles in Chapter 7). This expectation must be modified by the temperature of the planet, however, because we know that the hotter it is, the higher the speeds of individual atmospheric particles and the greater the likelihood that they will escape into space. From these principles, we might expect that the Earth and Venus would have more extensive atmospheres than either Mars or Mercury. Indeed, Earth and Venus both have atmospheres, Mars has a far less dense atmosphere, and Mercury has virtually none at all. This makes sense, because the Earth and Venus are the most massive of the terrestrial planets, Mars is next, and Mercury is the least massive (and also very hot).

The picture we have just painted is a bit too simple, however. For one thing, the atmosphere on Venus is far more massive and dense than that of the Earth, although Venus is much hotter than the Earth and has a slightly lower surface gravity and a lower escape speed. Furthermore, the atmospheres of Venus and Mars both have a composition very different from that of the Earth. Whereas the Earth's atmosphere is dominated by nitrogen and oxygen, those of Venus and Mars are almost pure carbon dioxide (CO_2).

Table 9.2
The Atmospheres of Venus and Mars

Gas	Symbol	Fraction (by Number)	
		Venus	Mars
Carbon dioxide	CO_2	0.96	0.95
Nitrogen	N_2	0.035	0.027
Argon	Ar	0.00007	0.016
Oxygen	O_2	—	0.0013
Carbon monoxide	CO	0.0004	0.0007
Water vapor	H_2O	0.0001	0.0003
Neon	Ne	5×10^{-6}	2.5×10^{-6}
Sulfur dioxide	SO_2	0.00015	—
Krypton	Kr	—	3×10^{-7}
Ozone	O_3	—	1×10^{-7}
Xenon	Xe	—	8×10^{-8}
Hydrogen chloride	HCl	4×10^{-7}	—
Hydrogen fluoride	HF	1×10^{-8}	—

The explanation for Venus's apparent deviation from the general principles we have learned lies in the very different composition of its atmospheric gases compared with those of the Earth (**Table 9.2**), which in turn is related to the differences in the formation and evolution of the two planets. The Earth is actually the misfit among the terrestrial planets, differing from them in two major ways: (1) the Earth has had liquid water on its surface for billions of years; and (2) life has evolved on the Earth and modified its atmosphere. Mars may have had oceans or at least extensive lakes of water, but these have long since disappeared; Venus apparently has long been too hot to allow liquid water to survive on its surface, although there is evidence that Venus, too, may once have had seas. The lack of oceans on Mars and Venus today has allowed carbon dioxide to persist in the atmospheres of these planets, whereas on the Earth carbon tends to be dissolved in seawater and then deposited in carbonate rocks. Thus, on the cooler Earth, much of the carbon dioxide is in the ground or in the oceans, but on Venus it resides in the atmosphere. Furthermore, the Earth's atmosphere has been modified over time because life-forms have contributed vast quantities of oxygen (largely through photosynthesis by plants, especially the many forms of algae that live in the oceans) and much of the nitrogen as well (through the decay of organic matter).

The carbon dioxide in the atmosphere has also helped to heat the surface of Venus through the greenhouse effect. The greenhouse effect is so strong on Venus because of its thick carbon dioxide atmosphere, which absorbs infrared radiation very efficiently (recall from Chapter 7 that carbon dioxide is one of the most effective greenhouse gases, and even the minute traces of it in the Earth's atmosphere have a significant effect).

Mars also has a carbon dioxide atmosphere, but such a thin one that the greenhouse heating it creates is insignificant compared to that of Venus. Even so, Mars is a much warmer place than it would have been without its thin greenhouse blanket. Water vapor and other trace gases in the Earth's atmosphere create some greenhouse heating, and there is reason for concern that the injection of additional greenhouse gases by human industrial and agricultural activity may cause large-scale climatic changes.

The primary forces driving the circulation pattern in a planet's atmosphere are heating, which causes convection to occur, and rotation of the planet, which causes flows to curve due to the Coriolis force. These mechanisms are at work on the terrestrial planets. We have already discussed the Earth's atmos-pheric circulation in some detail in the previous chapter, where we found that heating occurs at the surface and in the upper stratosphere (where ozone absorbs ultraviolet energy from the Sun). This heating, which is greatest near the equator, causes air flows from low latitudes toward the poles, and these flows are turned into rotary patterns by the Earth's rotation. Hence, the general pattern of atmospheric motion on the Earth consists of rotary flows with a strong dependence on latitude.

In general, the pattern of atmospheric motion on Mars is similar to that of the Earth. Currents flow from warm regions to cooler ones and are forced into rotary patterns by the Coriolis force due to the planet's rotation (the Martian day is just half an hour longer than an Earth day). The major difference occurs at certain times of the Martian year when very strong temperature variations from place to place on the surface cause intense horizontal winds to develop. The temperature variations that drive these winds are caused by seasonal effects.

Venus, as we might expect, has a very different pattern of atmospheric motions than do the Earth and Mars. On the one hand, the very hot surface might be expected to create strong convection, causing an overturning motion, but on the other hand, the very dense atmosphere tends to suppress flows. The probes that have landed on the surface of Venus have discovered very little wind there; as yet not much is known about vertical motions at higher elevations. The rotation of Venus is very slow, so we might not expect to find strong rotary patterns in its atmosphere, and indeed we do not. At high elevations (above the clouds, where it is possible to observe the motions), there are rapid horizontal motions, which are driven by the planet's rotation (the entire upper atmosphere flows uniformly in the same direction as the planet's rotation; **Fig. 9.5**).

Mercury has only a trace of an atmosphere—it is so thin that it defied detection until recently. The elements hydrogen, oxygen, sodium, and potassium have been measured spectroscopically. Astronomers believe these originate from the planet's crust and are released through meteorite impacts, venting from the interior, and gradual erosion due to the scouring effect of charged particles flowing away from the Sun.

Internal Properties and Tectonic Activity

According to the principles discussed in Chapter 7, we would expect that the heat energy contained in the interiors of the terrestrial planets should be related to planetary mass and diameter. A small planet has a higher ratio of surface area to volume than a larger planet and will lose its internal heat more rapidly. Thus we might expect to find that the Earth, the largest of the terrestrials, has the hottest

interior, with Venus very similar and Mars and Mercury having less internal heat. In turn, we might predict that tectonic activity, which is driven by the outflow of heat from the interior, should be most energetic for the Earth and Venus and less so for the smaller planets.

These expectations are largely met. Earth is certainly the most active tectonically, having continual overturn of its mantle and persistent volcanic activity on its surface. The situation for Venus is not so clear and is controversial as well. There is no doubt that tectonic activity has occurred there; the surface is largely covered with lava flows, and there are regions showing clear indications of stress and strain as subsurface motions of molten material have occurred. Crater counts indicate that the age of the Venusian surface is about 500 million years, suggesting that a major tectonic episode may have occurred about that long ago. But scientists disagree as to whether activity has continued on Venus or whether the events of 500 million years ago represented an isolated time of unusual activity.

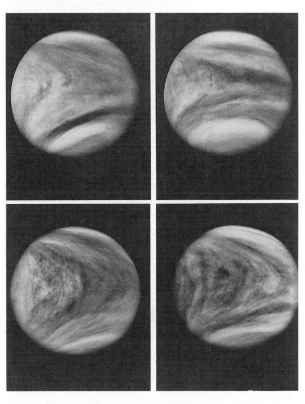

Figure 9.5 Circulation of the atmosphere of Venus. These four ultraviolet views show two rotations of the planetary cloud cover. The upper two are separated by one day; the lower two were obtained about a week after the first pair and are also separated by a day. The atmospheric motion is from right to left in these images. (NASA)

Mars shows signs of former tectonic activity, sporting several enormous volcanoes on its surface as well as a huge uplifted region (the Tharsis plateau) covering about one-fourth of its surface. But crater counts indicate that all major activity on Mars diminished billions of years ago and that the planet is geologically inert today. Mercury, the smallest of the terrestrials except for the Moon, is probably inert also, having a very old surface and little sign of geological activity.

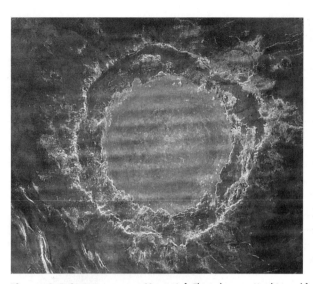 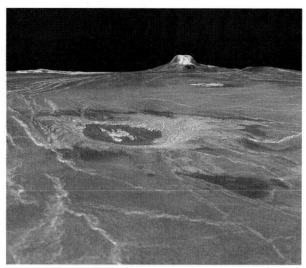

Figure 9.6 Impact craters on Venus. Left: This is the crater Mead (named for American anthropologist Margaret Mead), the largest impact crater on Venus, having a diameter of 275 km. Right: In the foreground is the impact crater Cunitz; in the background is the volcano Gula Mons. Cunitz crater is roughly 48.5 km in diameter; it was named after the astronomer and mathematician Maria Cunitz. (NASA/JPL)

The densities of the planets reveal that at least some differentiation has occurred in each case. Remember that ordinary rock typically has a density of about 3 g/cm³, so a planet with an average density higher than this must have a concentration of mass in its core. Earth, Venus, and Mercury have very similar densities, while those of Mars and the Moon are lower. This suggests that the Earth, Venus, and Mercury have undergone more differentiation than the other bodies. But Mercury, as the smallest terrestrial body other than the Moon, would not be expected to undergo much differentiation, so some other explanation of its high density may be needed. Magnetic fields are indicators of internal conditions because a mol-ten zone is needed in order to produce a field. Again we would expect the planets that have retained the most internal heat to be most likely to have magnetic fields, and Earth bears this out. But other factors besides internal heat are important here. Venus, for example, with presumably similar internal structure to the Earth, has no detectable magnetic field. This is attributed to Venus's very slow rotation, which apparently is insufficient to drive the internal fluid flows needed to produce a field. Mars has no detectable field despite having a rotation period very similar to that of the Earth, and this is thought to be an indicator that its interior lacks an extensive molten region. Once again Mercury produced a surprise when probed by *Mariner 10,* which revealed a small but definite magnetic field despite the small size and slow rotation of the planet. This discovery has prompted scientists to hypothesize that Mercury must have a very extensive molten core, sufficient to generate a field despite the slow rotation. This, along with its high density, is a strong clue that Mercury may have an unusual history of formation and heating.

Surface Characteristics

Like the Earth, the other terrestrial planets have highland and lowland regions on their surfaces. Most of Venus consists of rolling plains, but lowlands cover about a quarter of the surface, and there are three isolated highland regions (covering less than 10% of the surface). These highland regions may be likened to the continental areas of the Earth, and the lowlands in some ways resemble the seafloors of the Earth. Everywhere there are lava flows and mountains

Figure 9.7 Craters on Mars. This mosaic of *Viking* images shows a portion of a cratered highland region on Mars (foreground). See also Figure 9.2. (U.S. Geological Survey; data from NASA)

that indicate past volcanic activity, and many regions bear the signs of uplift and horizontal stresses that indicate that tectonics have played a role in shaping this surface. Venus has impact craters **(Fig. 9.6)** but only relatively large ones because small bodies coming in from space are burned up by the thick atmosphere. The density of craters, as noted earlier, indicates that the surface is about 500 million years old.

Impact craters are much more numerous on Mars because its surface is older **(Fig. 9.7).** Mars has mountainous regions as well, and a major portion of the surface consists of a huge uplifted region called the Tharsis plateau. This area is quite old (around 3 billion years) and appears to be the result of an early period when convection in the mantle of Mars created uplift. Near the Tharsis plateau are several huge volcanic mountains, again dating from the early period of intense activity (there is some uncertainty about whether volcanic eruptions have occurred more recently as well; the controversy is based on interpretations of detailed shapes of mountain slopes, and some have argued for eruptions in rather recent times). These giant volcanoes on Mars include Olympus Mons **(Fig. 9.8),** with a height above its base of some 27 km (90,000 feet, about three times that of Mt. Everest) and a base about 700 km across (about the size of the state of Colorado!). Olympus Mons and the other giant volcanoes are thought to have formed from the same kind of upward lava flow that created the Hawaiian Islands on the Earth, except

Figure 9.8 Olympus Mons. This is the largest mountain known to exist in the entire solar system. Its base is comparable in size to the state of Colorado, and its height is about three times that of Mt. Everest. (NASA)

that on Mars there was no continental drift to carry the newly formed mountains away as lava continued to be forced up from below. Thus the effect is as though all the molten material that formed the Hawaiian Islands were piled together into one gigantic mountain instead of a chain of smaller ones.

Another major feature of the Martian surface is a huge canyon called Valles Marineris **(Fig. 9.9);** this canyon was named after the *Mariner 9* probe whose close-up photos first revealed its size. This valley is five to ten times larger than the Grand Canyon in all dimensions, stretching across an expanse of some

Figure 9.9 Valles Marineris. At the left we see a full-disk photomosaic of Mars, with the great trench of Valles Marineris stretching across the lower center. At the right is a detailed view of a portion of the valley called the Candor Chasm. In the right-hand view, note the sections of canyon wall (center right) that have subsided or slumped; you may also be able to see a section of layered terrain (just left of center, above a dark region) that is thought to be an ancient lakeshore. (U.S. Geological Survey; data from NASA)

Figure 9.10   A photomosaic of Mercury. This is a composite image of a large portion of the side of Mercury that was observed by *Mariner 10*. The brownish color has been artificially introduced to simulate the actual color that would be seen by the human eye. (NASA)

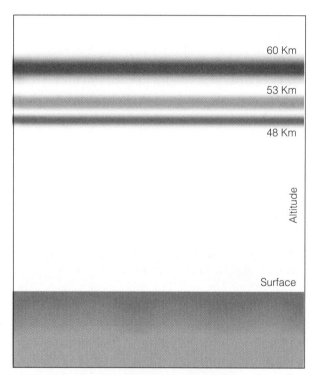

Figure 9.11   The clouds of Venus. The sulfuric acid clouds are separated into three distinct layers. These layers occur at altitudes where the combination of temperature and pressure causes sulfuric acid to condense.

4,500 km. Valles Marineris appears to have formed due to crustal fracturing that may have occurred at the time the Tharsis plateau was uplifted.

Mercury's surface is dominated by impact craters, giving the planet an uncanny resemblance to the Moon **(Fig. 9.10).** There are some subtle differences, though, which will be discussed later in the chapter. The high density of craters indicates that the surface of Mercury is quite old, much older than the surfaces of the Earth, Venus, or Mars.

Venus: A Closer Look

Now that we have described the terrestrial planets as a class and examined their similarities and differences in terms of large-scale planetary processes, it is interesting to look a bit closer to see how some of the details vary from one body to another. We begin with Venus, the closest to the Earth, both literally and in its overall properties.

As we have already seen, the atmosphere of Venus consists mostly of carbon dioxide (Table 9.2), with traces of other species, primarily nitrogen. Some of the trace species play important roles, however: sulfur dioxide (SO_2) for example, acts as an absorber of solar ultraviolet radiation at high elevations in the atmosphere and therefore plays a role very similar to that of ozone in the Earth's atmosphere; and sulfuric acid (H_2SO_4) forms droplets at certain levels, creating the clouds of Venus and sulfuric acid rainfall, thus duplicating the role of water vapor in the Earth's atmosphere. The clouds lie at three distinct levels **(Fig. 9.11)** and appear featureless except when observed at ultraviolet wavelengths, which reveal dark streaks due to sulfur dioxide absorption (see Fig. 9.5). As mentioned previously, the clouds flow around the planet in the same (retrograde) direction as the planetary rotation.

At the surface of Venus, the atmospheric pressure is about 90 times that at sea level on the Earth (i.e., it is equivalent to the pressure some 3,000 feet deep in the Earth's oceans!), and the temperature is about 750 K due to greenhouse heating. Photos obtained at the surface by the Soviet *Venera* probes show that sunlight penetrates there, and that the surface rocks appear sharp-edged **(Fig. 9.12),** indicating a lack of wind erosion. The rocks have been found to be silicates, probably volcanic basalts.

The Surface as Seen by *Magellan*

The detailed surface maps obtained by *Magellan* have given scientists the first chance to carry out true geological studies of Venus. One of the key questions

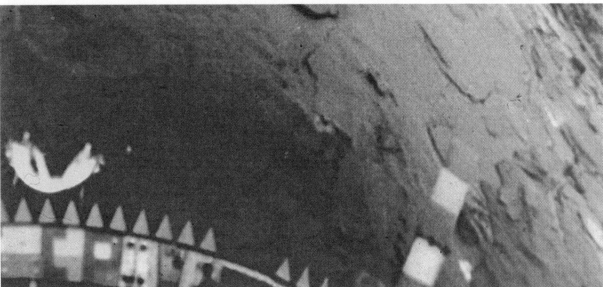

Figure 9.12 **Color photos of surface rocks.** These photographs were obtained by the Soviet *Venera 13* and *14* landers. The yellowish color seen here is an artifact of the camera response; reprocessing of these images has shown that the surface rocks are gray, like similar rocks on Earth. (TASS from SOVFOTO)

before *Magellan's* advent was the amount of volcanic activity on Venus. Some of the mountains detected by earlier probes (including a lower-resolution radar mapper aboard the *Pioneer-Venus* orbiter) were clearly volcanic in origin, but it was not known how recently the activity had taken place. Indirect evidence suggested that Venus might be active currently; the *Pioneer-Venus* ultraviolet spectrometer found varying levels of atmospheric sulfur dioxide (SO_2), which is an abundant product of volcanic eruptions. The *Magellan* maps indicate clear-cut evidence of volcanic lava flows, craters, and mountains

(Figs. 9.13 and **9.14),** but do not definitively answer the question of whether eruptions are taking place at this time. Some of the volcanic structures are apparently quite young, however, so it is reasonable to think that eruptions are taking place sporadically.

The *Magellan* maps have revealed some very unusual surface features. The **tesserae** are large expanses of rock that are fractured by regular rectangular cracks **(Fig. 9.15),** probably as a result of uplifting after lava flows had hardened. Crater counts indicate that these regions are among the oldest on the surface of the planet. Circular structures **(Fig. 9.16)** called

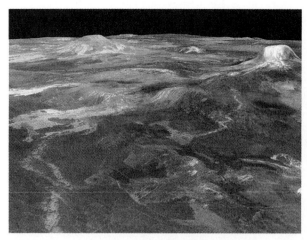

Figure 9.13 **Volcanoes on Venus.** This is a *Magellan* reconstruction of a region in Eistla Regio. At the left is the volcano Gula Mons; at the right is Sif Mons. Lava flows extend from Gula Mons toward the lower right. (NASA/JPL)
http://universe.colorado.edu/fig/9-13.html

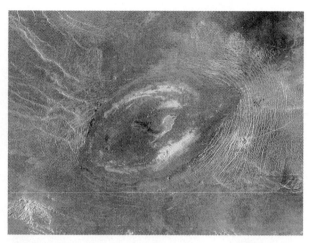

Figure 9.14 **A large volcanic caldera.** This is a depression called Saca-jawea, in the Lakshmi Plenum region of Venus. The enormous caldera is 1 to 2 km deep, 120 km wide, and 215 km long. It is thought to have formed as the result of drainage and collapse of a large underground magma chamber. (NASA/JPL)

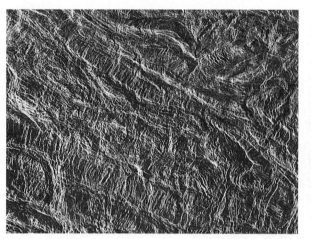

Figure 9.15 **Tesserae.** This complex pattern of intersecting ridges and cracks is thought to be the result of repeated episodes of horizontal motion. (NASA/JPL)

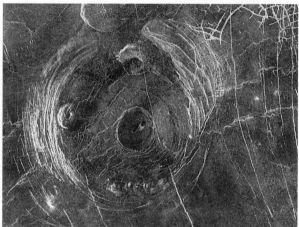

Figure 9.16 **A corona.** The large, circular feature seen here is a corona. It is a raised structure, approximately 200 km in diameter, thought to be the result of uplift due to upwelling magma from below. The smaller circular feature is a "pancake" dome, about 35 km in diameter, formed by the eruption of very viscous lava. Another pancake is seen to the left; these features are seen only on Venus. (NASA/JPL)

coronae are unique to Venus and are thought to be the result of upward flows of lava beneath the surface. The upflow causes a bulge, which subse-quently collapses, leaving behind a large circular feature. It is not clear why similar structures are not seen on the Earth, where subsurface lava flows like those thought to create the coronae on Venus also occur.

On the question of tectonic activity, again it is apparent that Venus is active: its features include fracture zones that appear to have been created by subsurface motions that have acted to stretch and compress the crust in opposing directions **(Fig. 9.17)**, and uplifted regions where it appears that magma has risen from below and pushed upward on the crust.

But instead of having rigid crustal plates that move about as they do on the Earth, the crust of Venus simply stretches or is compressed while staying in place. One possible explanation is that the high temperature of the surface has maintained a high buoyancy of the crustal material, preventing subduction, the slipping of one crustal plate under another. Without subduction, the crust of Venus is unable to break up into plates that can become mobile by sliding over one another.

There is uncertainty, even controversy, about how tectonically active Venus is today. The fact that most of the surface is about 500 million years old

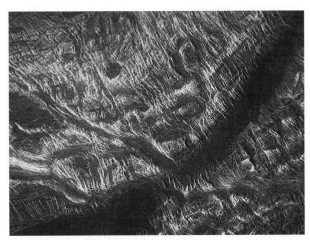

Figure 9.17 The effects of compression. This is the highland region Ovda Regio, near the equator of Venus. The complex structure seen here is thought to be the result of compression (in the upper left–lower right direction) followed by stretching (upper right to lower left), caused by flowing motions in the mantle of Venus that are similar to the motions that cause continental drift on the Earth. (NASA/JPL)

indicates that significant overturning of the surface took place about that long ago. Some scientists have suggested that Venus may undergo occasional outbreaks of tectonic activity, when the internal temperature becomes so high that the heat must escape, and major convective overturn takes place for a time. As the internal temperature drops, the rigidity and buoyancy of the crust bring the tectonics to a halt until the heat begins to build up again and another cataclysmic episode takes place. In this view, the difference between the Earth and Venus is that the Earth can get rid of its internal heat continuously through its steady-state tectonic overturn, while Venus can do so only through drastic episodes every few hundred million years. One way to determine whether this picture works is to find out for sure whether there has been recent activity such as lava flows or volcanic eruptions, so the *Magellan* maps are being scrutinized very carefully for such evidence. If it is demonstrated that volcanoes are currently active or that recent convective overturn has occurred, this evidence will counter the picture just described, suggesting that Venus's activity is continual instead of only sporadic.

The Earth and Venus: So Near and Yet So Far

In this chapter we have been stressing the similarities between Venus and the Earth, and indeed many geological similarities can be found. But what of the enormous contrasts in surface conditions and atmospheric properties? We can begin to understand these differences by taking a look at the environments in which the two planets formed.

Like the Earth, Venus is thought to have formed from rocky debris orbiting the infant Sun in a great disk. The early evolution of Venus must have been similar to that of the Earth, with Venus undergoing a molten period during which its dense elements sank to the center, leaving a lighter crust composed largely of silicates and carbonates. Before the crust cooled, volatile gases escaped from the surface, forming a primitive atmosphere of hydrogen compounds and carbon dioxide, much like the earliest atmosphere on the Earth. At this point the evolution of the Earth and Venus began to diverge in a major way. On Earth, as oxygen escaped from the rocks and combined with hydrogen to form water, much of it could persist in the liquid state, and our planet had oceans from this time onward. Like the Earth, Venus may have had large quantities of liquid water early in its history due either to outgassing or to the impact of many water-bearing comets (the role of comets in supplying the initial water for the oceans of the Earth is currently the subject of some controversy as well; some scientists suggest that comets may have been a more important source than outgassing).

At this point a significant environmental difference between Venus and the Earth came into play. Venus is closer to the Sun, and the extra solar heating that resulted was apparently enough to prevent liquid water from remaining on the surface of the planet. Venus is 0.72 AU from the Sun, so the intensity of sunlight at its surface is $(1/.72)^2 = 1/.52 = 1.92$ times greater than at the surface of the Earth, according to the inverse square law (Chapter 6). Thus Venus receives almost twice as much solar energy per square meter as the Earth.

As water on Venus evaporated due to the intense sunlight, water vapor was added to the atmosphere. Water vapor causes a mild greenhouse effect, so the temperature of the surface began to rise further. In addition, the lack of liquid oceans meant that there was no water in which atmospheric carbon dioxide could be dissolved, so the concentration of carbon dioxide continued to rise as this gas escaped from the planet's interior. Meanwhile, on the Earth the carbon dioxide was being removed from the atmosphere by reactions with ocean water and ended up deposited in carbonate rocks, where it still remains. The development of life on the Earth (at a later time) also had a major effect that was absent on Venus: on the Earth plant life gradually converted carbon dioxide to oxygen through photosynthesis. The present-day atmosphere of the Earth, dominated by nitrogen and oxygen, does not have a strong greenhouse effect, and the atmosphere has stabilized with a moderate temperature.

In contrast, the large quantity of carbon dioxide in the atmosphere of Venus caused intense greenhouse heating, which helped make the atmosphere very hot at a time when a lot of water vapor was present. The high temperature began to destroy the water vapor, separating (dissociating) the water molecules into hydrogen and oxygen atoms. The hydrogen then escaped into space (as discussed in Chapter 7), while the heavier oxygen became trapped in surface rocks through a process called oxidation. At the same time the surface of Venus probably continued to follow a geological evolution much like that of the Earth, with the exception of the differences noted in the previous section.

In contrast to this picture of a Venus that never had liquid oceans, some scientists have argued that the planet may once have had extensive, very hot seas. In this scenario, the oceans were able to persist because of the high atmospheric pressure; the boiling point of water rises with increasing pressure, so it is speculated that the high atmospheric pressure on Venus might have compensated for the high temperature and allowed oceans to remain. If so, these oceans would have eventually caused their own demise, however, because they would have gradually reduced the atmospheric pressure by gobbling up carbon dioxide. In due course the boiling point would have been reduced enough so that the oceans would have boiled away, allowing the atmosphere to evolve as described above. If this scenario is correct, then it is possible for a planet to lose oceans if the greenhouse effect becomes severe enough; therein lies a warning to us as we try to understand the effects of our own alterations of the Earth's atmosphere.

Terrestrial Planets and Resolution of Images
There are several WWW sites that contain vast collections of data from NASA space missions and other sources. We'll begin by locating these sites, and then follow specific links to find information about Mercury, Venus and Mars. By examining electronic pictures of these planets, we'll investigate the concept of the 'resolution' of an image. Our starting point is the following URL:
<http://universe.colorado.edu/ch9/web.html>

Mars: Another Watery World?

The question of water comes up again as we turn our attention to Mars, the next planet beyond the Earth as we go outward from the Sun. In its overall characteristics, Mars is quite distinct from the Earth, but at the same time its surface conditions are closer to those of the Earth than any other member of the solar system. Unlike the Earth, Mars has a carbon dioxide atmosphere (Table 9.2), but its daily cycles and seasonal variations closely resemble those of the Earth—at the height of the Martian summer, the temperature can reach comfortable values. Other Earth-like features include the polar ice caps, which exhibit seasonal variations, and the thin, wispy clouds that can be seen in the atmosphere.

The surface of Mars is distinctly reddish in color (this color is one reason for the planet's name; Mars was the god of war in ancient mythologies). The color is apparent in naked-eye and telescopic observations from the Earth (**Fig. 9.18**) and in close-up photos taken by the *Viking* landers and *Mariner* orbiters (**Fig. 9.19**). The color can be explained by the low level of current and past geological activity; Mars never underwent as much differentiation as the Earth or Venus, and therefore large amounts of heavy elements such as iron have remained in the crust. When combined with oxygen, iron yields a reddish material known as **iron oxide,** or rust. Thus Mars is reddish in color because it is rusty.

Seasons and Dust Storms

Mars has a substantial axial tilt relative to its orbital plane, causing the planet to have seasonal variations much like those of the Earth. The obliquity of Mars is 23°59′, very similar to the Earth's 23°27′. This might suggest a similar degree of variation from winter to summer, but the Martian seasons are a bit more complicated.

Recall that the orbit of Mars is more noncircular (i.e., its elliptical orbit is more eccentric) than is true for most other planets. The variations in the distance between Mars and the Sun are great enough to cause significant variations in the intensity of sunlight reaching Mars at different points in its orbit. The orbital eccentricity of Mars is $e = 0.0934$, and its orbital semimajor axis is $a = 1.524$ AU, which means that its perihelion distance (point of closest approach to the Sun) is

$$P = a(1 - e) = 1.524(1 - 0.0934) = 1.38 \text{ AU},$$

and its aphelion distance (greatest distance from the Sun) is

$$A = a(1 + e) = 1.524(1 + 0.0934) = 1.67 \text{ AU}.$$

The inverse square law then tells us that the intensity of sunlight at perihelion is $(1.67/1.38)^2 = 1.46$ times greater than at aphelion. This difference is large enough to have a pronounced effect on the climate.

Mars is closest to the Sun at the time of summer in the southern hemisphere and winter in the north (**Fig. 9.20**); as a result, the northern winter is mild while the southern summer is quite warm. Half a

Figure 9.18 Mars as seen from Earth. This is thought to be the finest photograph ever obtained of the red planet as seen from the Earth's surface. The image was obtained with a CCD camera during the time of the 1988 opposition of Mars, when the planet was unusually close to the Earth. (J. Lecacheux, Meudon Observatory; CCD image obtained with the 1.05-m telescope, Pic-du-Midi Observatory)

Figure 9.19 A ground-level panorama on Mars. This is the first photograph made by the *Viking 1* lander. It shows a rock-strewn plain extending in all directions. (NASA) http://universe.colorado.edu/fig/9-19.html

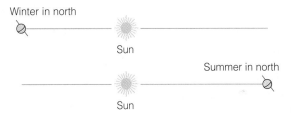

Figure 9.20 The Martian seasons. As this exaggerated view shows, in the northern hemisphere both summer and winter are moderated by the varying distance of the planet from the Sun, whereas in the southern hemisphere both seasons are enhanced. The extreme temperature fluctuations in the southern hemisphere give rise to the winds that cause the seasonal dust storms.

Martian year later, when it is summer in the north and winter in the south, Mars is at its farthest from the Sun, and the northern summer is mild while the southern hemisphere has a severe winter. Consequently, the south experiences extreme seasonal variations, while the variations in the north are only moderate.

Thus, as one season turns into the next, the change in temperature is very large and very rapid in the south. In the southern spring, the enormous temperature contrasts create immense winds as air flows from the warm region toward the cooler surroundings. Since these winds are strong enough to pick up fine dust particles from the surface, Mars undergoes dust storms on a seasonal basis, every southern spring. Sometimes these storms become so severe and widespread that they cover the entire planet **(Fig. 9.21)**. Such a storm was brewing when the first U.S. orbiter, *Mariner 9,* approached Mars in 1971; the spacecraft was not able to photograph the surface of the planet for several weeks, until the storm finally subsided.

The seasons on Mars also affect the polar ice caps, a phenomenon that was observed through

Earth-based telescopes long before any probes reached the red planet. The northern cap changes very little in size from winter to summer, but the southern cap grows very large in winter and almost disappears in summer. Spectroscopic measurements show that the caps are made of water ice and frozen carbon dioxide (known to us as dry ice). The outer surface of each cap is thought to be carbon dioxide ice, which overlies a layer of water ice.

The Story of Water on Mars

Although scientists are still uncertain whether the early Mars had lakes or oceans, the many erosion

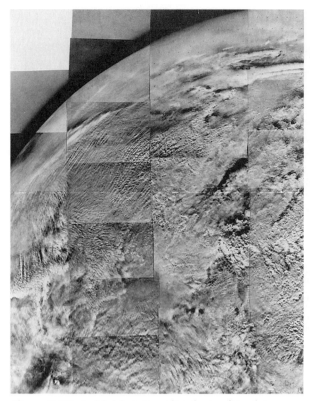

Figure 9.21 A global dust storm. This is the view of Mars that confronted the *Mariner 9* spacecraft for the first several weeks after it went into orbit around the planet.

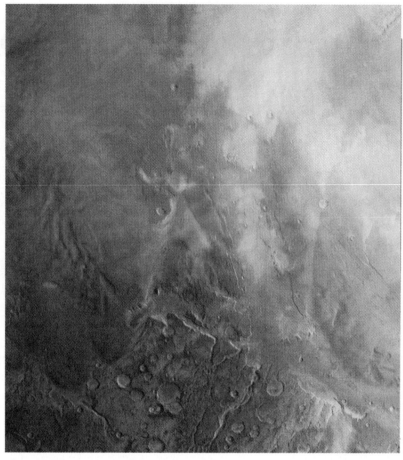

Figure 9.22 Ancient riverbeds on Mars. At the lower center are several river channels showing northward flow (upward in the figure) from the edge of a highland scarp to a lowland plains region called Amazonis Planitia. (U.S. Geological Survey; data from NASA)

channels leave little doubt that water once flowed on the planet's surface. Ancient riverbeds have been found **(Fig. 9.22),** where large volumes of water obviously flowed, eroding the surface and carrying debris downstream. Runoff channels indicate that major floods occurred as well. All of these channels are very old, lying in areas that are thought to date back some 3 billion years to the time of volcanic and tectonic activity on Mars. Since the atmosphere of Mars today could not sustain any large quantities of liquid water, we must ask how water was once able to exist there and what has happened to it.

Scientists have reached some agreement on the probable fate of at least some of the Martian water. Some of the observed flow channels emanate from curious, jumbled regions known as **chaotic terrain (Fig. 9.23).** The ground in these regions apparently has collapsed, leading to the suggestion that there once were underground ice deposits, which suddenly melted. When the ice melted, the water flowed away in a flash flood of huge proportions, and the ground left behind

collapsed, creating the chaotic terrain that is seen today. The sudden melting is thought to have been caused by heating due to volcanic activity. This indirect evidence for underground ice has led to the suggestion that a substantial portion of the original water on Mars may still reside under the surface in the form of **permafrost,** that is, a permanent layer of ice.

Apart from what may be concealed beneath the surface, today there is very little water on Mars except for some water ice in the polar caps and traces of water vapor in the atmosphere (the concentration of the vapor varies; sometimes it forms thin clouds, and on cold days it creates early morning frost). But what happened to the much larger quantities of water that once were present? Evidence in the surface structures indicates that lakes may have existed at one time, but today the atmospheric pressure is too low to allow liquid water to survive on the surface. This has led to the suggestion that the atmosphere of Mars was once much thicker (and therefore warmer due to greenhouse heating), but that changes in the climate reduced the atmospheric pressure, preventing liquid water from persisting on the surface and at the same time making the atmosphere too cold and thin to retain much water vapor. The causes of such a change in climate are unclear, but one possibility is that the lakes or oceans could have sealed their own fate by absorbing enough of the atmospheric carbon dioxide to reduce the atmospheric pressure to its modern value. Significant changes in the orbit of Mars, caused by gravitational interactions with the other planets, have also been suggested as the cause of the climatic change on Mars. An alternative scenario that is being considered seriously today is that a major impact on Mars literally blew away much of its atmosphere. The evolution of the Martian atmosphere and the fate of the planet's water supply are among the most intriguing questions facing planetary scientists today and will provide major motivation for future probes to the red planet.

Life on Mars, Past or Present

As we have seen, the possibility of life on Mars has been a recurrent theme of speculations about the red planet. Some of the speculation was fanciful, such as the myth of a civilization that built enormous canals that girdled the planet, but some was more logical, based on the knowledge that conditions under which life could form may once have existed on Mars. After all, life on the Earth is believed to have begun in the oceans; if Mars once had seas or large lakes, perhaps the same processes could have occurred there.

The most widely publicized aspect of the dual 1976 *Viking* missions was the attempt by the robot landers to detect evidence of life-forms on Mars. The robots conducted several tests, the first and most straightforward being simply to look with the television cameras for any large plants or animals. None were seen, so much more sophisticated tests were tried.

The searches for life were carried out by three different experiments on board the landers. The purpose of each experiment was to look for signs of metabolic activity in a soil sample scooped up by a mechanical arm and deposited in containers for analysis **(Fig. 9.24)**. All living organisms on Earth, even microscopic ones, alter their environment in some way just by existing. Usually, the effects involve chemical changes as the organism derives sustenance from its surroundings and ejects waste material.

Two of the three experiments, the **pyrolytic release experiment** and the **gas-exchange experiment,** showed no evidence of possible life-forms. The third, however, the **labeled release experiment,** aroused much excitement among scientists because

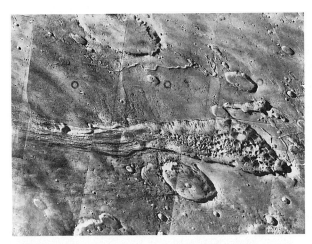

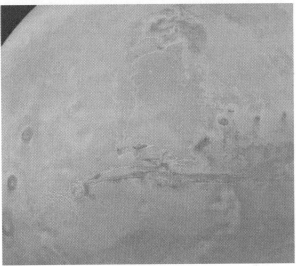

Figure 9.23 Chaotic terrain. Top: This image shows a valley floor with chaotic terrain and an ancient river channel flowing to the left. Bottom: This full-disk mosaic shows Valles Marineris, at lower center, and a region of chaotic terrain, just above Valles Marineris, at the center. River channels are seen flowing northward (up) from this chaotic region. (U.S. Geological Survey; data from NASA)

Figure 9.24 Sampling the Martian soil. Here the scoop on *Viking 2* is digging up a sample of Martian soil for one of its life-detection experiments. (NASA)

SCIENCE AND SOCIETY

Little Green Men

Over the centuries, humans have indulged in speculation about alien life-forms and distant civilizations. One of the earliest people to espouse the notion of other solar systems and other societies was Giordano Bruno, who was burned at the stake in 1600 by the Roman Catholic Inquisition for spreading his idea that the stars are distant suns, each accompanied by planets housing societies like our own.

In more modern times, such ideas are no longer regarded as so threatening—at least no one has been burned at the stake lately. While many scientists nowadays consider the presence of life elsewhere in the universe to be quite possible, if not probable, most speculations center on planets orbiting other stars. Modern knowledge of the planets in our system has discounted earlier suggestions of civilizations on Mars or Venus; now we know that Mars is barren and has a very thin atmosphere and that,

instead of a tropical oceanic climate, Venus hides a veritable hell under its clouds, with temperatures and pressures far too extreme to allow life to survive.

There have been several periods of renewed and intensified interest in the possibility that life exists elsewhere, and Mars played center stage in one such episode. It all began with the opposition of 1877, when Mars was closer to the Earth than usual (Mars was near its perihelion and Earth was near its aphelion, thus reducing the distance between the two planets). Many astronomers took advantage of the especially fine observing conditions to study Mars. The American Asaph Hall discovered the two tiny moons of Mars, and the Italian Giovanni Schiaparelli made sketches of Martian surface features, based on his visual observations as he peered at the red planet through a telescope. He drew dark, linear features, which he termed "canali," a word meaning linear features, natural or unnatural, but this was soon mistranslated as "canals," implying water-carrying channels built by a technological society.

The notion that a Martian civilization not only existed but was involved in large-scale land reclamation projects took the fancy of many people. Most notable among them was the American banker Percival Lowell, a Boston aristocrat with an abiding interest in astronomy and the wherewithal to indulge his passion for the science. Lowell was so entranced by the idea of little green men on Mars that he wrote a book (published in 1896) describing his vision of Martian society. In Lowell's analysis, this was an agricultural culture whose planet was drying up; therefore the Martians were forced to build canals to transport water from the polar ice caps to the temperate zones where crops were grown. Lowell was not the least deterred when scientists pointed out that the canals would have to be enormous (over 50 miles wide!) in order to be seen from the Earth. (Schiaparelli's sightings of linear features were later determined to be erroneous, resulting from the tendency of the human eye to connect points that are separated by small spaces.)

Lowell's enthusiasm for life on Mars never waned, although he did find another quest later in life (the search for a ninth planet, which finally succeeded

its initial results duplicated the expected effects of active life-forms. In this experiment, a nutrient solution was added to a sample of Martian soil in a closed container in the expectation that organisms in the soil would metabolize the nutrient and release waste gases in the chamber. To make the gases detectable, a small quantity of radioactive carbon (^{14}C, in which each atom consists of the usual six protons but has eight neutrons instead of the normal six) was added to the nutrient. If metabolic activity took place, then some ^{14}C should appear in gases emitted by the sample, and that is just what happened. As a check, the same experiment was performed on other samples that had been heated so that any life-forms present would have been killed, and indeed no tracer gases were emitted from those samples. The combination of activity in the normal samples and the lack of activity in the sterilized samples was consistent with

the presence of life-forms in the Martian soil.

Still the evidence was not conclusive. The other experiments failed to show positive results, and the reactions in the labeled release experiment occurred much faster than seemed likely for metabolic activity. Gases containing ^{14}C were released more quickly and in greater quantity than would ever have been possible for any earthly microorganisms. Furthermore, and perhaps more telling, a given sample would react positively only once, even though nutrient was added to some samples several times. Real life-forms should be capable of eating more than one meal. Eventually, scientists concluded that the observed activity must have been due to an unexpected type of chemical reaction between the soil and the nutrient. The reaction probably involved oxides in the soil whose chemical properties were altered by heating, so that the activity did not occur in the sterilized samples.

shortly after his death; see Chapter 7). General interest in the Martian civilization ebbed, but did not go away until the 1960s when the first close-up images showed what a desolate place Mars is. The concept of a Martian society (a hostile one this time) experienced a spectacular renewal in 1938, when Orson Welles staged his famous "War of the Worlds" radio broadcast. This was a simulated report of an invasion from Mars, featuring spacecraft landing in New Jersey and a murderous rampage across the countryside by Martians intent on capturing our planet and enslaving our populace. The broadcast was done as a series of news flashes, and many listeners, having failed to note the disclaimers, believed the invasion was really happening. Widespread panic swept through New Jersey and New York before the word got around that the "invasion" was created for a radio drama.

More serious speculations about life on Mars were based on annual changes in coloration that are seen on Mars and on the supposed detection of features due to chlorophyll in the spectrum of reflected light from Mars (this later turned out to be a misidentification). But these hopes of finding at least plants on Mars faded when the first photos from *Mariner 4* revealed nothing but craters and dust. The annual color changes were soon explained by the American astronomer Carl Sagan as being caused by seasonal wind storms, which shift the dust around on the Martian surface, causing changes in the distribution of the light and dark areas. Today the only serious hope held out for life on Mars arises from the knowledge that warmer, wetter conditions once existed there, so that microscopic life might have gained a toehold long ago. The recent discovery on Earth of possible fossil microorganisms in meteorites thought to have originated on Mars has certainly stimulated this hope.

Whether microscopic life exists on Mars or not, the quest for little green men from another planet in the solar system has died out. The once-cherished notion that the other terrestrial planets might harbor civilizations has been thoroughly debunked by reality as our observations and probes have explored these planets, and there never was much speculation of life on the gas giant planets. We are living in a time of extraordinary fascination with alien life-forms, however, even if not from one of our planets. Some of this interest may be based on little or no evidence (e.g., the current UFO and alien abduction craze), but nevertheless it indicates how intrigued people are by the possibility of life elsewhere.

There are more sober reasons to suggest that alien life-forms may exist. It is generally thought that life on Earth began from naturally occurring chemical processes, which implies that life could arise elsewhere, if the right conditions exist. It is also thought likely that planets are a natural by-product of star formation, so that finding planets orbiting other stars is probably only a matter of time (and not much time at that; recently, three stars were found to have planets orbiting them, and several intensive searches are under way). Perhaps in your lifetime, the first clues will be found that life exists out there as well, whether in the form of little green men or not.

A further test for evidence of life was carried out by a device called a **mass spectrometer,** which is capable of analyzing a sample to determine the types of atoms and molecules it contains. The mass spectrometers aboard the *Viking* landers found no evidence of organic molecules, those containing certain combinations of carbon atoms that are always found in plant or animal matter on the Earth. Hence, no doubt with some reluctance, most *Viking* scientists concluded that no evidence for life on Mars had been found.

Lack of evidence for life is not the same as evidence for the lack of life, however. After all, these experiments could test samples at only two localities on an entire planet, and furthermore, the tests were predicated on the assumption that Martian life-forms, if they exist, would be in some way similar to those on Earth. Scientists are already discussing new experiments to look for microscopic life-forms on Mars in searches to be carried out by future probes.

Even if no life exists on Mars today, the possibility remains that the planet once was the home of life-forms. We believe that life on Earth began through complex chemical reactions that took place in the early oceans. The oldest fossil evidence for life on Earth is found in rocks in western Australia that are about 3.5 billion years old. Thus life began on Earth within the first billion years after the formation of the planet. There is no obvious reason why it could not have done so on Mars as well, if the red planet underwent an early period when it had higher atmospheric pressure and liquid oceans or lakes, as is now suspected. The surface would have been warmer than it is now, due to the greenhouse effect, and we know that the surface of Mars contains the same elements from which life formed on Earth. Perhaps there were

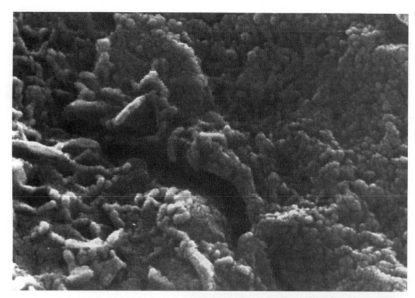

Figure 9.25 Fossils from Mars? At left is an electron microscope image of very small structures resembling bacteria, found inside of a meteorite believed to have originated on Mars. At right is a color microscopic image of the same specimen, revealing orange spots having high concetrations of organic materials which might have been deposited by bacterial life forms. (NASA) http://universe.colorado.edu/fig/9-25.html

primitive life-forms on Mars that have since died out, as the planet lost most of its atmosphere and nearly all of its water. If so, fossil remains of life-forms might be found in the Martian soil. The discovery of such remains will require detailed scrutiny of soil samples, and experiments to do this are being proposed for future missions aimed at placing landers on the Martian surface.

Even before such experiments are carried out, evidence for fossil life on Mars reached the Earth, and was recently recognized. Certain meteorites (pieces of rocky debris from space) found on Earth have been identified as coming from Mars, probably as a result of major impacts which ejected material from Mars into space. In 1996 a group of planetary astronomers anounced that in one of these Martian meteorites there are miscroscopic structures, associated with organic (carbon-bearing) materials, which appear to be fossilized life forms **(Fig. 9.25).** There is some disagreement in the scientific community about the validity of this claim, but interest in life on Mars has certainly reached a new peak, and NASA may accelerate its plans to search for fossils in Mars in the next few years.

Mercury: Close Encounters with the Sun

Mercury, the innermost of the planets, is dominated in many ways by its proximity to the Sun. As we will see, the Sun has played a major role in governing the orbital and rotational properties of Mercury, and the hot environment at the center of the solar system has largely determined the properties of the planet.

As already mentioned, Mercury is hard to observe because, from the vantage point of the Earth, the planet is never far from the Sun on the sky. Nevertheless, Earth-based observations revealed some hints of surface markings, which led to the conclusion that the planet was in synchronous rotation, always keeping the same face toward the Sun. This rotation was not unexpected, considering the enormous tidal forces that the Sun must be exerting on Mercury (see the discussion in Chapter 5).

Much more detailed observations are possible using radio wavelengths, however, and when radio telescopes were first turned toward Mercury, astronomers found some surprises. The dark side of the planet is not as cold as would be expected if the other side always pointed toward the Sun, and the planet's rotation is not synchronous. Measurements established that Mercury rotates once every 58.65 days and has an orbital period of almost 88 days. This rotation is very slow, but not as slow as would be required to keep one side facing the Sun. As a result, all portions of the surface of Mercury are exposed to direct sunlight at times, and this exposure is the reason the dark side is not as cold as had been expected.

Details of the surface structure of Mercury finally began to be obtained in 1974 and 1975 from the *Mariner 10* probe. As explained earlier in this chapter, the orbit of *Mariner 10* was adjusted so that it made three encounters with Mercury **(Fig. 9.26).** About half of the surface of the planet has been imaged, revealing that it closely resembles the Moon (see Fig. 8.21). *Mariner 10* also made magnetic field measurements, which established that Mercury has a field despite its slow rotation rate. The presence of a magnetic field provides important clues to the internal structure of the planet.

Orbit and Rotation: Spin-Orbit Coupling

Precise measurements of the rotation period of Mercury, carried out through the analysis of radar echoes bounced off the planet, showed that the rotation period of 58.65 days is precisely equal to two-thirds of the orbital period of 87.97 days. Therefore Mer-

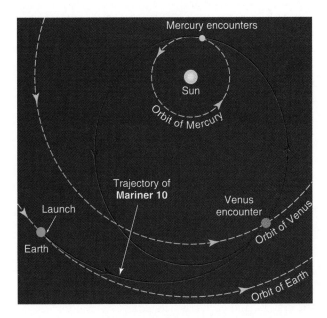

Figure 9.26 **The trajectory of *Mariner 10*.** The spacecraft flew by Venus and then encountered Mercury repeatedly because its orbital period was adjusted to be equal to twice the orbital period of Mercury.

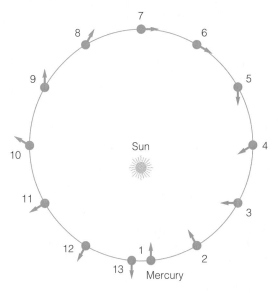

Figure 9.27 **Spin-orbit coupling.** This sketch shows how Mercury spins one-and-a-half times while completing one orbit around the Sun. At perihelion it always has the dense side facing directly toward or away from the Sun.

cury spins one-and-a-half times during each orbit around the Sun. Thus, at the same point in its orbit, Mercury has opposite sides facing the Sun on consecutive trips around **(Fig. 9.27).**

This pattern is no coincidence; clearly, the tidal force of the Sun is at work. The key is what happens at the point where Mercury is closest to the Sun, called perihelion, for the tidal force is strongest here. If one side of Mercury is a bit denser than the other, as on the Moon, then the tidal force exerted by the Sun will act to ensure that this side is aligned toward the Sun when Mercury is at perihelion. Of course, with synchronous rotation, this side would always point toward the Sun, but because Mercury's orbit is quite elongated, the urge to keep its heavy side facing the Sun is much stronger at perihelion than at other points in its orbit.

Because Mercury spins one-and-a-half times each orbit, the heavy side of the planet faces either directly toward the Sun or directly away from it each time Mercury passes through perihelion. In either case the tidal force acting on Mercury is balanced, so there is no tendency to alter the rotation further. Apparently the planet once rotated much faster, but whenever it passed through perihelion with the heavy side pointed in some random direction, tidal forces exerted a tug that tended to change its spin. This process went on until the rotation slowed to its present rate so that the tidal force is always balanced

when Mercury is closest to the Sun. If the orbit had not been so elongated, or if Mercury had been symmetric instead of having a relatively dense side, the tidal forces would have been more uniform throughout the orbit, and Mercury might be in synchronous rotation today.

In whole numbers, Mercury spins three times for every two orbits. Such a numerical relationship between the rotation period and the orbital period is called **spin-orbit coupling,** a general term applied to any situation where the rotation of a body has been modified by tidal forces so that a special relationship between the orbital and spin periods is maintained. The most common example is synchronous rotation (where the ratio is 1:1 instead of 3:2), and we will find several cases of synchronous rotation in the rest of the solar system and in the universe beyond.

The combination of orbital and rotational speeds on Mercury produces a very unusual effect for anyone who might visit the planet. At its closest approach to the Sun, when Mercury is moving most rapidly in its orbit, the orbital speed is actually greater than the rotational speed at its surface. For a short time (lasting a few hours), the Sun would appear to be moving backward in the sky, from west to east. Imagine the difficult time our ancestors would have had explaining this retrograde motion of the Sun if a similar phenomenon had occurred on the Earth!

Surface and Interior: The Effects of Heat and Collisions

We have already mentioned that Mercury has a very high density and a detectable magnetic field, both perhaps a bit surprising in view of the low mass of the planet and its slow rotation. A partial explanation lies in the environmental conditions that prevailed in the central part of the solar system at the time the planets formed. The central region of the solar nebula was very hot, which made it impossible for the more volatile elements to condense into solid form. Thus the planetesimals that formed in the innermost part of the young solar system probably contained very small quantities of volatile elements from the beginning. A more important mechanism seems necessary to explain the extreme depletion of volatile elements, however. One suggestion is that soon after its formation, Mercury was struck by a planetesimal large enough to strip away most of its mantle. This collision would have occurred after differentiation—and the formation of a metallic core—had already taken place.

The end result is that Mercury has a very low abundance of the lightweight, volatile elements and thus a very high density. In addition, the planet's high overall density means that it must have a very large core (Fig. 9.28), and this core must be at least partially molten, producing the observed magnetic field. In effect, Mercury may be viewed as a terrestrial planet that has lost most of the outer layers that the other terrestrials have retained.

We commented at the beginning of this chapter that Mercury closely resembles the Moon. Now we see that this resemblance is only skin-deep; internally, the two bodies are very different. The Moon is geologically inert, has undergone little differentiation, and has no magnetic field, whereas Mercury has a large, partially molten core and internal currents that produce a magnetic field. In composition, Mercury is much more iron-rich and lower in the volatile elements than the Moon (which, in turn, has lower abundances of volatile species than the Earth).

Despite the internal contrasts, on the surface Mercury can be mistaken for the Moon by the untrained eye (Fig. 9.29). There are craters everywhere and some smooth, dark areas that resemble maria. These features reflect some similarities in the evolutions of the surfaces of the two bodies: each has been bombarded by meteorites since its formation, with no erosion or surface geological processes to remove the craters, and each has had extensive lava flows in its past. Like the Moon, Mercury has a very old surface, which reflects the lack of crustal plate shifting and overturning despite the hot, partially molten interior.

The major differences between the surfaces of Mercury and the Moon are the lower vertical relief of the craters on Mercury (Fig. 9.30) and the global system of scarps (Fig. 9.31). The lower crater walls are due to Mercury's higher surface gravity, and the scarps are thought to be the result of crustal shrinkage as volatile gases continued to escape from the interior after the crust had hardened.

The *Mariner 10* probe was only able to map half of the surface of Mercury because the spacecraft's orbital period was equal to twice the period of Mer-

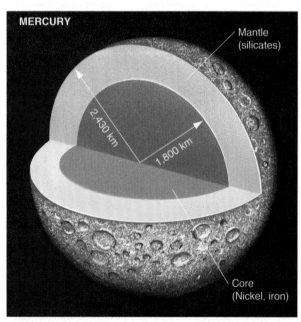

Figure 9.28 Mercury's internal structure. The presence of a magnetic field, along with the planet's relatively high density, implies the presence of a large core.

Figure 9.29 Craters on Mercury. Like the Moon, Mercury has a heavily cratered surface. Because Mercury has a greater surface gravity than the Moon, however, impact craters have lower rims and are shallower, and ejecta do not travel as far. (NASA/JPL)

cury. Due to Mercury's 3:2 spin-orbit coupling, the planet made exactly three rotations during the two orbital periods between its encounters with *Mariner 10*. Therefore every time the spacecraft came close to Mercury, the same side of the planet faced the Sun, and the other side, which was in darkness, could never be photographed.

As a result, the *Mariner 10* survey nearly missed a major surface feature. Just at the boundary between the daylight and dark sides of the planet at the time of the encounters, an enormous crater was observed **(Fig. 9.32)**. This huge circular basin must be the result of a collision with a large body some time in the distant past (the floor of the crater is covered by many, more recent craters, an indication of its great age). Interestingly, this basin lies at a position on Mercury that faces either directly toward or directly away from the Sun at perihelion, when the planet passes closest to the Sun. Thus, on every other orbit of the Sun, this crater is the hottest spot on the planet; accord-

ingly, it was named Caloris Planitia (Latin for "Plain of Heat").

The location of Caloris Planitia is probably not a coincidence. Recall that Mercury's 3:2 spin-orbit coupling is thought to have come about because one side of the planet is slightly denser than the other. If the remains of the large meteorite that created Caloris Planitia are embedded in the surface of Mercury, they may well explain why that side of the planet has

Figure 9.31 A scarp system on Mercury. An enormous scarp (cliff) runs vertically through this *Mariner 10* image, cutting across a large crater at lower center. This illustrates that the scarp formed after the period of major cratering. (NASA)

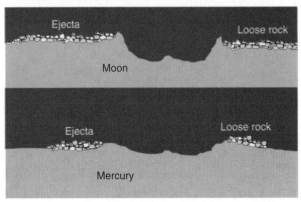

Figure 9.30 Crater formation on Mercury. Top: This is a *Mariner 10* image showing a cratered region on the surface of Mercury. Bottom: This drawing illustrates the contrasts between craters on Mercury and those on the Moon: on Mercury, the crater walls are lower and the ejecta do not travel as far due to Mercury's higher surface gravity. (Top: NASA)

Figure 9.32 Caloris Planitia. This photo shows half of the immense impact basin known as Caloris Planitia. This region is directly facing the Sun at perihelion on every other orbit. (NASA/JPL)

a higher density than the other. We may speculate that Mercury might never have fallen into its 3:2 spin-orbit coupling had it not been for the huge impact that created Caloris Planitia.

On the opposite side of Mercury from Caloris Planitia lies a region of strange, wavy surface structures, quite unlike anything else found on the terrestrial planets. This region is so unusual that geologists call it the weird terrain. These structures are thought to be the result of the collision of seismic waves that traveled around Mercury when the impact that created Caloris Planitia occurred; as these waves came together on the far side of the planet, they created stresses in the crust that forced some regions upward and others downward, leaving a record of the waves permanently frozen into the surface. Interestingly, detailed examination of the Moon's surface has revealed similar features opposite major craters, but on the Moon these areas are much less pronounced.

Summary

1. Venus, Mars, and Mercury have all been studied extensively from the Earth (radio observations being the most useful for Venus and Mercury), and all three have been visited by space probes from Earth.

2. Venus and Mars both have carbon dioxide atmospheres created by outgassing from the surface and a lack of liquid oceans to dissolve the gas; Mercury has only a very thin atmosphere.

3. The greenhouse effect is important in heating the surfaces of all the terrestrial planets that have atmospheres, but has been extreme in the case of Venus; there the thick carbon dioxide atmosphere has raised the surface temperature to 750 K.

4. Venus is tectonically active (though without continental drift) and is thought to be volcanically active and to have an internal structure much like that of the Earth. Mars underwent some volcanism and tectonic activity early in its history, but is probably not active today. Mercury has a relatively large core that must be partially molten and has had volcanic activity in the past, but the great age of the surface shows that this planet is not undergoing extensive volcanic or tectonic activity today.

5. The *Magellan* radar mapper has shown that the surface of Venus is dominated by volcanic structures and that surface stresses due to flows in the interior of the planet have created rift systems and regions of uplift; the planet has not experienced large-scale plate motions as on the Earth, however.

6. Scientists believe Venus has a dense carbon dioxide atmosphere and a high surface temperature largely because the planet's proximity to the Sun prevented it from retaining liquid oceans early in its history. The lack of seas allowed carbon dioxide to remain in the atmosphere, where it created the strong greenhouse heating that keeps the planet so hot.

7. Seasons on Mars are exaggerated by its elongated orbit; one result is annual dust storms during spring in the southern hemisphere, which sometimes involve the entire planet and shift the surface dust cover, creating seasonal changes in the planet's coloration.

8. Mars almost certainly once had abundant liquid water on its surface, as evidenced by dry channels seen today, but the amount of the water and what happened to it are not known. It may have been lost to space by evaporation and escape, or much of it may remain on Mars in the form of subsurface permafrost.

9. Although the *Viking* landers failed to find evidence for life-forms in the Martian soil, it is possible that life could have existed there if the planet had a thicker atmosphere and liquid oceans early in its history.

10. Mercury is locked into 3:2 spin-orbit coupling, spinning three times for every two orbits around the Sun. This rotation pattern was caused by Mercury's noncircular orbit and the strong tidal force due to the Sun, which acted on Mercury's nonsymmetric mass distribution to slow its rotation so that the heavy side faces either toward or away from the Sun when the planet passes through perihelion.

11. Mercury's high average density and magnetic field indicate that the planet has a very large core and a relatively low abundance of lightweight, volatile elements. The sparsity of volatile elements is due to its formation in the inner solar system, where the volatile species were not able to condense into solid form, and, more importantly, to a suspected impact with a large planetesimal, which stripped off the planet's outer layers after differentiation had occurred.

Review Questions

1. Using your knowledge of conditions on Venus and Mars, explain what the atmosphere of the Earth might be like today if life had never evolved.

2. Discuss and compare the role of the greenhouse effect on all four of the terrestrial planets.

Looking for Mercury

Even though Mercury is a very bright object, easily bright enough to be seen with the naked eye, many people never see it. The reason is that Mercury is always close to the Sun, never straying more than 28° from it (this is Mercury's angle of greatest elongation). Thus, to see Mercury, you have to look for it low in the sky near its time of greatest elongation, either just before sunrise or just after sunset, depending on where the planet is in its orbit. When Mercury is near its greatest western elongation, we see it as a morning object, rising a little before the Sun. When the planet is near its greatest eastern elongation, it sets just after the Sun, and we see it as an evening object.

The accompanying table lists the times of Mercury's greatest elongations for the coming few years. Because Mercury's synodic period is short (116 days), there are fre-

quent opportunities to see it near one of its maximum elongations (there are two per synodic period, or one every 58 days). Use this table to plan your observations; all you need to do is go outside at the appropriate time (before sunrise for a western elongation; after sunset for an eastern one) and look for Mercury low in the sky toward the diffuse glow caused by the Sun's presence just below the horizon. You must be prompt because the conditions for best viewing do not last for long.

Greatest Elongations of Mercury

Eastern Elongations	Western Elongations
April 6, 1997	May 22, 1997
August 4, 1997	September 16, 1997
November 28, 1997	January 6, 1998
March 20, 1998	May 4, 1998
July 17, 1998	August 31, 1998
November 11, 1998	December 20, 1998

3. Discuss the general structure of the surfaces of Mercury, Venus, and Mars, explaining the similarities and differences among them. In your explanation, mention the role played by tectonic activity and volcanism.

4. Compare the density and the nature (sizes, shapes) of impact craters on all four of the terrestrial planets.

5. Explain why the masses of Mercury and Venus could not be measured using the same technique that was used for the other planets.

6. Would you expect Mars, Venus, and Mercury to have charged-particle belts surrounding them similar to the Earth's Van Allen belts? Explain.

7. Discuss and compare the seasons on all four terrestrial planets.

8. Why is it thought that life might once have existed on Mars? Would you expect that the same thing might have happened on Venus? Explain.

9. Summarize the similarities and differences between Mercury and the Moon.

10. What evidence can you present to support the modern idea that planetary evolution is episodic, that is, that major changes sometimes occur quickly, rather than gradually over long periods of time? Use examples from this chapter and the preceding one on the Earth and the Moon.

Problems

1. Suppose that Mars and Mercury both are made of just two distinct kinds of material: crust, with a uniform density of 2 g/cm^3; and core, with a constant density of 10 g/cm^3. How large does the core of each planet have to be, relative to the total radius, to provide the average densities of 4 g/cm^3 and 6 g/cm^3 for Mars and Mercury, respectively? (The values for the average densities have been rounded to simplify the calculation.)

2. Using the inverse square law, calculate how much more intense sunlight is at Mercury's distance from the Sun than it is at Venus. Despite the contrast, the surface of Venus is hotter than the daylit side of Mercury. Explain why this is so.

3. The orbital eccentricity for Mercury is 0.206. Use this and the inverse square law to calculate how much more intense sunlight is on Mercury at perihelion than at aphelion.

Web Connections

The Review Questions and Problems also appear at the following URLs:
http://universe.colorado.edu/ch9/questions.html
http://universe.colorado.edu/ch9/problems.html

4. Use Wien's law (Chapter 6) to calculate the wavelength of maximum emission from Venus (surface temperature 750 K).

5. How much brighter is Venus than the Earth to an observer viewing both from the same distance? To calculate this, compare the amount of sunlight reflected from the two planets. Assume both have the same surface area, but that Venus has a higher albedo (see Appendix 8) and thus reflects a larger fraction of the incoming light from the Sun. Remember that the Sun's intensity on Venus is greater by a factor of 1.92.

6. To illustrate why the Earth's seasonal variations are not affected by the varying distance of the Earth from the Sun, calculate the relative intensity of sunlight on the Earth when it is closest to the Sun and when it is farthest away. At closest approach, the Earth is 0.983 AU from the Sun; at its farthest, it is 1.017 AU away. Compare this variation in intensity with that on Mars during its year (figures are given in the text).

7. Suppose another terrestrial planet with a mass of 0.8 Earth masses and a radius of 0.9 Earth radii were in a circular orbit 0.9 AU from the Sun. Calculate the planet's surface gravity and escape speed, and describe its likely properties (what its atmosphere, average density, degree of tectonic activity, and internal structure would be like). Support your answer by making comparisons with the properties of the actual terrestrial planets.

Additional Readings

Beatty, J. K. 1992. Mercury's Cool Surprise. *Sky & Telescope* 83(1):35.

Beatty, J. K. 1996. Life from Ancient Mars? *Sky & Telescope* 92(4):18.

Beatty, J. K., B. O'Leary, and A. Chaikin 1990. *The New Solar System*. 2nd Ed. Cambridge University Press.

Bulock, M. A. 1994. The Soil of Mars. *Mercury* 23(5):10.

Dick, S. J. 1988. Discovering the Moons of Mars. *Sky & Telescope* 76:242.

Edgett, K., P. Geissler, and K. Herkenhoff. 1993. The Sands of Mars. *Astronomy* 12(6):26.

Goldman, S. J. 1992. Venus Unveiled. *Sky & Telescope* 83(3):258.

Kargel, J. S. and R. G. Strom. 1992. Ice Ages of Mars. *Astronomy* 20(12):40.

Kasting, J. F., B. Toon, and J. B. Pollack 1988. How Climate Evolved on the Terrestrial Planets. *Scientific American* 258(2):90.

Kross, J. F. 1995. What's in a Name? *Sky & Telescope* 89(5):28.

Luhman, J. G., J. B. Pollack, and L. Colin 1994. The Pioneer Mission to Venus. *Scientific American* 270(4):90.

McKay, C. P. 1993. Did Mars Once Have Martians? *Astronomy* 21(9):26.

McSween, H. Y. 1995. Nor Any Drop to Drink. *Sky & Telescope* 90(6):18.

Robinson, C. 1995. Magellan Reveals Venus. *Astronomy* 23(2):32.

Robinson, M. 1994. Exploring Small Volcanoes on Mars. *Astronomy* 22(4):30.

Robinson, M. and W. Wadwha 1995. Messengers from Mars. *Astronomy* 23(8):44.

Stofan, E. R. 1993. The New Face of Venus. *Sky & Telescope* 86(2):22.

Strom, R. G. 1990. Mercury: The Forgotten Planet. *Sky & Telescope* 80(3):256.

Chapter 10

REMOTE-COUSINS

The Giant Planets

Chapter Web site: http://universe.colorado.edu/ch10

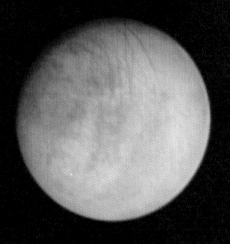

I have observed four planets, neither known or observed by any astronomers before me, that orbit around Jupiter like Venus or Mercury around the Sun.

Galileo Galilei, 1617

n moving to a discussion of the outer five planets in the solar system, we are making a major transition not only in distance, but also in style. The outer planets bear little resemblance to the terrestrial bodies, either superficially or on close examination **(Fig. 10.1)**. Furthermore, the outer planets differ markedly from the inner four in origin and evolution. Yet we will find that many of the same physical processes at work in the inner regions occur in the outer solar system. We need only understand the different starting conditions to see how these processes led to such different final results.

Overview of the Outer Planets

The four giant planets, as well as Pluto, are very different from the innermost, terrestrial planets. The giants are gaseous or fluid nearly to their cores, and the elements that compose them are the most common in the universe, but much less abundant on the terrestrial worlds. The gaseous giants, sometimes known as the **Jovian** planets because Jupiter is taken as the prototype, are cold gas balls with complex atmospheric flows driven by convection and rapid planetary rotation. Each has numerous satellites and a planetary ring system formed by swarms of orbiting debris.

Table 10.1 summarizes the basic properties of Jupiter, Saturn, Uranus, and Neptune; Pluto is omitted for now because it has very little in common with the other four. A quick glance at the table reveals how the bulk properties of the four giant planets differ from those of the terrestrials: the masses of the outer planets are far greater, their average densities are much lower, and they are much colder. Among the giant planets there is a great deal of variation. The masses range from "only" about 15 times the mass of the Earth to over 300, and the diameters range between roughly 4 and almost 11 Earth diameters. The densities are more consistent, varying between 0.7 and 1.6 g/cm³, but all are much lower than the 4 to 5.5 g/cm³ range of the rocky terrestrial bodies.

Jupiter is the giant among these giants, having a mass more than 300 times that of the Earth and a diameter nearly 11 times as great. Clearly, Jupiter was the most successful among the gaseous giants in accreting raw materials from its surroundings during the era of planetary formation. Next, in distance and in size and mass, is Saturn, whose glorious ring system has become a general symbol of celestial bodies in the minds of many. Beyond Saturn lie Uranus and Neptune, modest in size and mass compared to

Jupiter, but still giants relative to any of the terrestrial planets. These two are more alike than any other pair of planets, except possibly for the Earth and Venus.

The striking contrasts between the terrestrial and giant planets tell us that the formation processes and subsequent evolution of these planets must have been very different from the formation and evolution of the terrestrial bodies. Whereas the lightweight, volatile gases were baked out or prevented from condensing in the inner solar system, these elements were retained when the outer planets coalesced. Whereas tidal forces due to the Sun prevented the inner planets from trapping remnant gas and dust and thus forming satellites, in the outer solar system each of the giant planets was able to capture a disk of orbiting gas and ice.

Pluto followed a rather different path. Some would argue that Pluto is not a proper planet at all, but rather a big planetesimal, the largest example of a class of solar system bodies that is also represented by giant moons such as Triton of Neptune and by the many newly discovered small bodies orbiting beyond the orbit of Neptune. The origin of Pluto has been the subject of speculation, including some fanciful theories involving ejection from a former position as a satellite of one of the gas giants (usually Neptune), but currently the general consensus is that Pluto is a member of a class of primitive, icy bodies that formed in the outer solar system and have been largely unaltered since.

Exploring the Outer Solar System

Only two of the outermost five planets—Jupiter and Saturn—are easily seen by the unaided eye, so the ancient astronomers were not aware of Uranus, Neptune, or Pluto. Uranus was noticed by some observers and included in a few star charts, but was not recognized as a planet until its discovery by William Herschel in 1781. Using a fine telescope for the time, Herschel was able to discern the disk of the new body and, by tracking its motion for some time, established that it is a Sun-orbiting body located far from the center of the solar system.

The subsequent discovery of Neptune in 1846 came about because the motion of Uranus appeared to violate Newton's laws of motion and gravitation. Between 1781, when it was discovered, and 1840, Uranus wandered some 2′ away from its predicted path around the Sun. Astronomers in England (John C. Adams) and in France (Urbain Leverrier) independently reached the conclusion that another planet

Figure 10.1 The giant planets. The images of Jupiter (upper left), Saturn (upper right), Neptune (lower left), and Uranus (lower right) were all obtained at close range by the *Voyager* probes. The giant planets are very different from the terrestrial planets, being much larger and far less dense and having no solid surfaces. (NASA)

Table 10.1
Basic Properties of the Gas Giants

	Jupiter	Saturn	Uranus	Neptune
Orbital semimajor axis (AU)	5.203	9.539	19.182	30.06
Perihelion distance (AU)	4.951	9.077	18.28	29.82
Aphelion distance (AU)	5.455	10.001	20.09	30.31
Orbital period (yr)	11.86	29.46	84.01	164.79
Orbital inclination	1°18′17″	2°29′33″	0°46′23″	1°46′22″
Rotation period (hr)	9.925	10.500	17.24[a]	16.05
Tilt of axis (obliquity)	3°5′	26°44′	97°52′	28°48′
Mean diameter (relative to Earth)	10.97	9.14	3.981	3.865
Polar diameter	10.52	8.55	3.93	3.83
Equatorial diameter	11.21	9.45	4.01	3.88
Mass (relative to Earth)	317.83	95.162	14.536	17.147
Density (g/cm³)	1.326	0.687	1.318	1.638
Surface gravity (relative to Earth)	2.364	0.916	0.889	1.125
Escape speed (km/sec)	59.5	35.5	21.3	23.5
Surface temperature (K)	165	134	76	74
Albedo	0.52	0.47	0.50	0.41
Satellites	16	22	15	8

[a]The minus sign indicates that the rotation of Uranus is technically in the retrograde direction (i.e., opposite of the spins of most of the other planets), due to the fact that its north pole is tipped over so far that it points below the plane of the planet's orbit.

could be causing the discrepancy due to its gravitational pull on Uranus, and in 1846 the German astronomer Johann Galle found Neptune within 1° of the predicted position. Thus the discovery of the eighth planet turned a seeming failure of Newton's laws into a triumph and added another complex world to the known family of the Sun.

The list of planets was not extended again until 1930, when Lowell Observatory astronomer Clyde Tombaugh found Pluto after an extended search **(Fig. 10.2)**. Ironically, the motivation for the search came largely from the late Percival Lowell himself (the amateur astronomer who believed in a Martian civilization; see Chapter 9), who thought he had found discrepancies in the motions of both Uranus and Neptune that could only be explained by a ninth planet. Astronomers later determined that the discrepancies did not exist, and furthermore that Pluto is far too small to have caused any such effect. Its discovery near the predicted position was fortuitous (but not the discovery itself; Tombaugh's methodical and thorough search would surely have turned it up even in the absence of any specific predicted position).

Telescopic observations of Jupiter and Saturn are capable of revealing some of the details of their cloud tops. But Uranus, Neptune, and Pluto are all too distant for much of their surface markings to be seen from the Earth, and rather little was known about them until very recently. Jupiter displays a vivid banded appearance as well as a large reddish oval region known as the Great Red Spot **(Fig. 10.3)**, whereas Saturn has much more muted bands and a generally less colorful appearance **(Fig. 10.4)**. As if to compensate for this drab appearance, it is distinguished by a spectacular system of rings that girdle the planet. The rings show intricate structure, even when viewed from the remote vantage point of the Earth. In recent decades, both Jupiter and Saturn have been found to have many satellites. Recall from Chapter 3 that the four large moons of Jupiter were first observed by Galileo, who used the fact that they clearly do not orbit the Earth as an argument against the Earth-centered solar system.

Little about Uranus and Neptune could be determined from Earth, except for the discovery of some satellites (five for Uranus and two for Neptune) and some evidence of atmospheric clouds and banding, primarily from infrared images **(Fig. 10.5)**. From spectroscopic data, scientists learned that lightweight gases such as hydrogen and its compounds ammonia (NH_3) and methane (CH_4) dominate the outer layers of four of the giant plants (it is assumed that helium, the second most common element after hydrogen, is also abundant, but helium has few spectral lines in visible wavelengths and hence is difficult to measure directly).

Radio and infrared observations of the outer planets revealed some major phenomena that are not found on the terrestrial planets. First Jupiter and then Saturn were found to be strong emitters of radio waves. The emissions include bursts of static that are apparently caused by lightning discharges in the clouds, the normal thermal emission of any body whose temperature is greater than absolute zero (see Chapter 6), and a strange kind of continuous radio emission called **synchrotron emission.** This radia-

Figure 10.2 The discovery of Pluto. These are portions of the original photographs on which the ninth planet was discovered in 1930. The position of the planet in each is indicated by an arrow. (Lowell Observatory photograph)

Figure 10.4 **Saturn.** This image of Saturn, which was obtained by the *Hubble Space Telescope* in Earth orbit, shows much more detail than can be seen in the best ground-based images. (NASA/Space Telescope Science Institute)
http://universe.colorado.edu/fig/10-4.html

Figure 10.3 **Jupiter with Great Red Spot.** The *Voyager* spacecraft obtained this photo when it was still millions of kilometers from Jupiter. Already a vast amount of detail is evident in the atmospheric structure. (NASA) http://universe.colorado.edu/fig/10-3.html

tion, which is caused by free electrons moving very rapidly in a magnetic field, was the first clue that the giant planets are surrounded by extensive belts of charged particles. These radiation zones dwarf the Earth's Van Allen belts in both size and intensity. It was expected (and later confirmed by space probes) that Uranus and Neptune would exhibit similar phenomena, but their greater distances from the Earth made the radiation zones difficult to detect from Earth.

Infrared observations of the outer planets revealed atmospheric features, primarily those due to methane, in the cloud tops. The infrared data also provided accurate measurements of the energy emitted by the cold outer planets, whose peak radiation occurs in the infrared, according to Wien's law (Chapter 6). One major surprise was the discovery that three of the four planets (Uranus being the exception) emit considerably more energy than they receive from the Sun. This implies that the planets have some sort of internal heat source, which is either generating energy or releasing excess energy stored in the planetary interiors. The source of the extra heat is now thought to be different for each planet.

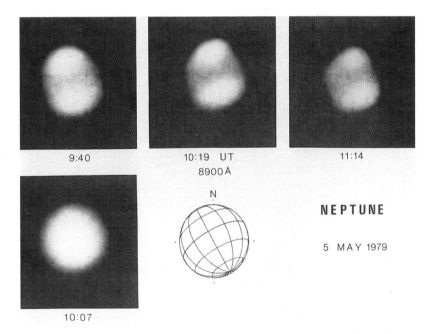

Figure 10.5 **Earth-based views of Neptune.** These images, obtained with the use of special filters that isolate the wavelengths at which methane molecules absorb light, represent the greatest amount of detail observed from the Earth. Despite the great distance to Neptune and the tiny size of its image, data such as these demonstrated the presence of atmospheric structure, even before the *Voyager 2* encounter. (NASA)

Starting in the 1970s, several successful space probes added immensely to our knowledge of the outer planets (Appendix 7). The two *Pioneer* missions *(Pioneer 10* and *Pioneer 11)* and especially the *Voyagers (Voyager 1* and *Voyager 2)* revealed that the four giant planets and their rings and moons were fantastically complex and beautiful worlds when seen close up. The *Galileo* spacecraft arrived at Jupiter in late 1995, and results from the probe dropped into the atmosphere of the giant planet as well as data from the orbiter trickled in slowly in early 1996 (a jammed antenna prevented *Galileo* from sending data to Earth efficiently). We may expect further revelations with the advent of the *Cassini* mission to Saturn (now under construction) and a planned mission to Pluto (now in its very early developmental stages). The space probes revealed that each of the gas planets is unique and intrinsically very complex. Each has more moons than previously detected and a ring system, though none is as brilliant as Saturn's rings (the rings of Uranus and Neptune were discovered through Earth-based observations, but little was known about them before *Voyager*).

General Properties of the Gas Giants

In the following sections, we will discuss and compare the important properties of the four gas giants as a group before moving on to more detailed descriptions of them individually. Pluto will be discussed later in the chapter.

The Internal Structure of the Gas Giants

As already noted, the four giant planets do not have solid surfaces, but instead are gaseous. Theoretical models of their interiors show that they probably have no surface below the visible cloud layers and, in fact, probably have little solid material anywhere inside.

Before we describe the internal structure further, it is useful to consider how astronomers can know anything at all about what goes on inside a planet that is far away and very different from the Earth and its closer neighbors. It is possible to construct mathematical models of the planets (or, for that matter, of stars or galaxies or anything else where we think the laws of physics are at work). To do this requires equations that describe the forces acting on the matter contained in the planet, heat transport from the interior outward, and conditions such as internal heat sources, rotation rate, surface temperature, overall

size and density, and chemical composition. For example, one equation would state that the inward force of gravity at any point inside the planet is balanced by an equal and opposite outward pressure force (these forces must be equal, or the planet would be collapsing or expanding, which is not observed to be happening). Another equation would describe the flow of energy outward from the interior of the planet. Others would involve additional physical processes. The goal is to write a set of equations that include all the important phenomena and then to solve these equations to find values for conditions at different positions throughout the interiors of the planets. It is possible to show that such a set of equations can have only one solution. The challenge is to find the correct equations, that is, equations that include all the important processes with sufficient accuracy.

The validity of a computed planetary model can be checked by comparing the results of the calculation with aspects that can be observed. For example, a model might predict the surface temperature and the radius of a planet, and the computed values can be compared with data obtained from observations. If the model succeeds in matching the observed surface values, then we can have some confidence that its values for the interior are also realistic. Probably the most important source of uncertainty in current models of the gas giants is the lack of detailed information on the properties of matter under conditions of extreme pressure and temperature such as exist in the deep interiors of these planets.

The results of model calculations for Jupiter, Saturn, Uranus, and Neptune show that the four share some common internal properties **(Fig. 10.6)**. Each has an outer layer of clouds above a warmer, denser region where the immense gravitational compression has forced hydrogen into its liquid state. In the two larger planets, Jupiter and Saturn, the pressure is so great in the deeper interior that hydrogen enters a new phase, known as liquid metallic hydrogen, which requires such extreme conditions that it is very difficult or impossible to create in earthly laboratories. In this state hydrogen is predicted to have a semirigid crystalline structure and to act like a liquid in some ways and like a metal in others. Below these layers of exotic forms of hydrogen, each of the giant planets is thought to have a solid core, composed chiefly of rock and metallic elements. These cores are the remainders of the original solid planetesimals from which the planets formed, with high-density material added by differentiation. This process, by which heavier elements sink to the center of a planet (discussed in Chapters 7 and 9) is much more effi-

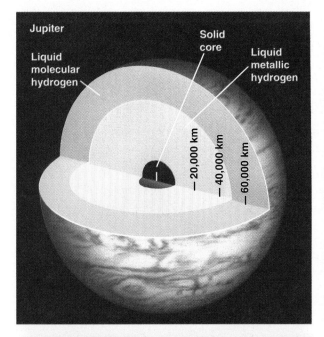

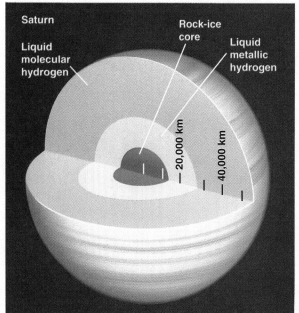

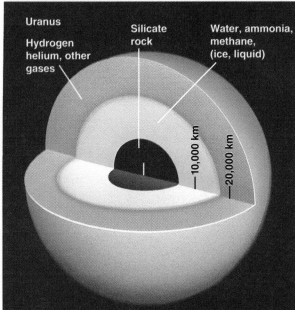

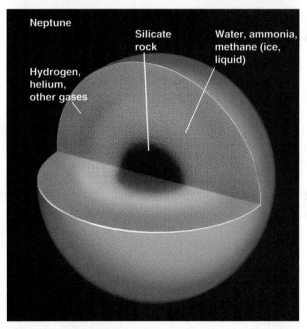

Figure 10.6 **The internal structures of Jupiter, Saturn, Uranus, and Neptune.** The two larger planets, Jupiter (top left) and Saturn (top right), have small, solid cores surrounded by a layer of liquid metallic hydrogen. Uranus (bottom left) and Neptune (bottom right) do not have liquid metallic hydrogen zones because the pressure is not sufficient, but as on Jupiter and Saturn, differentiation has caused virtually all their heavier elements to sink into the core. This model of Uranus, based on *Voyager 2* data, shows a gaseous outer zone underlain by a thick ice and liquid zone extending from about 7,800 km out to 18,000 km from the center. This zone contains about 65% of the planet's mass, while the rocky core contains roughly 24% of the mass.

cient in a body that is gaseous throughout than in a rocky body such as a terrestrial planet. Hence virtually all of the heavy, dense material in the giant planets now resides in their cores, whereas in the terrestrial worlds substantial quantities of heavy elements are still found in the crust.

Atmospheric Circulation

Heat rises from the interiors of the gaseous planets, and additional heating occurs within their atmospheric layers due to sunlight. The presence of heating sets the stage for convection to occur, creating over-

The Gas Giants and Pluto

The outer planets. Here are comparative views of a number of phenomena: the banded atmospheres of the four giants (shown to correct scale), the complex rings of Saturn and Uranus (lower center), and clouds and turbulence in the atmospheres of Jupiter (Great Red Spot, lower left) and Neptune (lower right, center).

turning motions as warm gas rises and cooler gas falls. At the same time, each of the giant planets rotates very rapidly, with days ranging in length from just under 10 hours for Jupiter to a little more than 17 hours for Uranus. Thus any horizontal motions that take place will be converted into curved or rotary patterns by a very strong Coriolis force (see Chapter 7). As a result, the giant gas planets have very active weather systems.

The rapid rotation of these planets causes what would otherwise be circular flows to stretch into planet-girdling bands, with adjacent bands moving in opposite directions relative to each other **(Fig. 10.7).** The direction depends on atmospheric pressure contrasts, as it does on the Earth and the other terrestrial planets. Gas flowing around a low-pressure center in the northern hemisphere rotates in the counterclockwise direction and is called a **cyclonic flow,** while gas flowing around a high-pressure region goes in the clockwise direction and is called an **anticyclonic flow.** On Jupiter, the dark belts are cyclonic, and within them the gas is sinking **(Fig. 10.8).** The lighter-colored zones are anticyclonic, and within them the gas is rising. The other giant planets have similar flow patterns, but they are not revealed by vivid color contrasts as they are on Jupiter.

The speed of motion increases toward the equatorial zones, as do the widths of the bands. Astronomers suspect that the widths and speeds of the moving regions vary seasonally, but so far monitoring such changes has proved difficult because the *Voyager* probes made only very brief observations as they sped by the planets, and the seasonal cycles on the outer planets are very long (equal to the orbital periods of the planets, which range from almost 12 years for Jupiter to about 165 years for Neptune). Both the *Galileo* and *Cassini* missions include orbiters that will be able to observe conditions on Jupiter and Saturn for extended periods of time. The *Hubble Space Telescope (HST)* will also provide a better opportunity to observe long-term seasonal changes on these planets, since it can resolve enough detail to detect the expected effects and will allow continuous observations over a period as long as 20 years. Interestingly, the *HST* has already shown that major storm

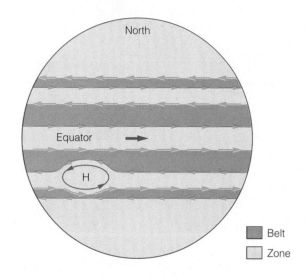

Belt

Zone

Figure 10.7 Circulation of the Jovian atmosphere. The global circulation pattern shown here indicates the location and designations of the belts and zones in Jupiter's cloud layer.

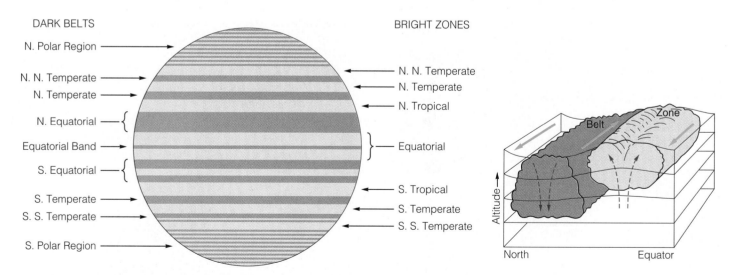

Figure 10.8 Motions in the Jovian atmosphere. These drawings indicate both the horizontal (left) and vertical (right) circulation in the clouds of Jupiter. (NASA)

systems on the giant planets can come and go on timescales of a few months: a storm arose and gradually diminished on Saturn in 1991–1992, and a new one erupted in 1994. On Neptune, a giant dark spot that had been present at the time of the *Voyager 2* encounter abruptly disappeared in late 1994 and then reappeared in 1995.

Magnetic Fields and Particle Belts

All four of the gaseous planets are surrounded by belts of charged particles. The existence of these belts indicates that each of the planets has a magnetic field, because it is the field that traps the particles. Those regions are analogous to the Earth's magnetic field and the Van Allen belts except for some differences caused by the rapid rotation of the giant planets.

Understanding why the gaseous planets with their fluid interiors have magnetic fields is not difficult. Recall from our discussions of the terrestrial planets that magnetic fields are thought to arise from electrical currents in planetary interiors, which in turn are brought about by flowing motions in a molten zone near the core. The presence of a field depends on whether a planet has an electrically conducting molten zone and on the rotation rate of the planet. The fluid nature and rapid rotation of the gaseous planets easily satisfy both conditions. Hence all four of these planets have strong magnetic fields.

The rapid rotation not only helps create the magnetic fields, but also strongly affects their shapes **(Fig. 10.9).** The rapid rotation of the giant planets forces their magnetic fields into flattened, doughnut-like shapes, rather than the more rounded shape of the Earth's field. The charged particles are therefore confined to a flattened torus (doughnut) as well, where the magnetic field is most intense. Within

these disks the intensity of the particle belt is far stronger than that of the Van Allen belts; in fact, it is strong enough to represent a serious threat to spacecraft that pass through the belts during encounters with the outer planets (charged particles can damage electronic components, for example, and some minor failures in on-board computers in the *Voyager* probes were thought to be due to charged particles).

Another factor that influences the shapes of the particle belts around the outer planets is the degree of alignment between the rotation axis and the magnetic axis of each planet. On the Earth there is a small offset between the rotational poles (the true North and South Poles) and the magnetic poles (the positions where the poles of the Earth's magnetic field lie on the surface). For the Earth the misalignment between the rotational axis and the magnetic axis is about 11°. Jupiter has a similar 11° misalignment, while Saturn's is nearly zero. Uranus and Neptune, however, have much larger misalignments, about 60° and 55°, respectively.

If the magnetic field and the rotation axis are misaligned, then as a planet spins, the magnetic field and the particle belts precess, or wobble, in time with the rotation period. Consequently, the disk of charged particles also precesses, as first one side and then the other is tipped up and down by the rotating, tilted magnetic field. Adding further complexity to the motion is the solar wind, which, as it streams past, pulls the magnetosphere out into a long tail in the downwind direction (away from the Sun). The result is that the magnetic fields and particle belts of the giant planets have rapidly precessing, disklike shapes with extended tails in the downwind direction. For Uranus and Neptune, these tails gyrate about due to the large misalignment between the magnetic equator and the direction in which the solar wind is flowing.

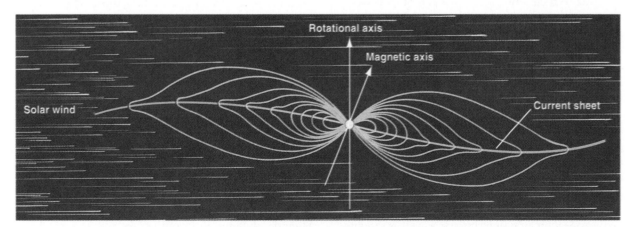

Figure 10.9 **The Jovian magnetosphere and radiation belts.** The shape of Jupiter's magnetosphere is influenced by the solar wind and by the planet's rapid rotation. The result is a sheet of ionized gas that is closely confined to the equatorial plane but wobbles as the planet's off-axis magnetic field rotates. (NASA/JPL)

The principal source of the electrons and protons that make up the particle belts is the solar wind, which provides a steady stream of charged particles from the Sun. Some particles, however, have a local source. For example, volcanic activity on Jupiter's moon Io produces free atoms that become ionized and trapped in the particle belts.

Formation and Evolution of Moons and Rings

All four of the gaseous giants have numerous satellites, ranging from 8 for Neptune to 20 for Saturn (including two discovered in 1995 by the *Hubble Space Telescope*). This is in stark contrast to the terrestrial planets, which seem to acquire moons only by accident, if at all.

In addition to having many satellites, the outer planets have ring systems, which are disks of particles (mostly ice) that orbit the planets in their equatorial planes along with the satellites. We know the rings are made of countless tiny particles because observations, first made long ago from the Earth, show that light passes through the rings. Furthermore, the rotation period of the rings is not constant, as would be expected from a rigid ring, but instead varies with distance from the parent planet in accordance with Kepler's third law. In reality, the ring particles can be viewed as a horde of very tiny moons orbiting each of the giant planets. The sizes of the particles, inferred from the manner in which they reflect light, range from microscopic to many meters in diameter. Thus each of the giant planets has a large family of orbiting bodies, with tiny particles (the rings) closest in and the larger moons farther out. Clearly, these orbiting systems must have formed as a result of some process that occurred in common for all four of the planets.

Due to their large masses and remoteness from the Sun (hence the weakness of solar tidal forces), the giant planets were able to retain disks of debris as they formed. These disks, in each case resembling a small-scale version of the solar nebula disk itself, then formed into satellites by the same kind of coalescence process that allowed planetesimals to form planets orbiting the Sun. Thus each of the giant planets ended up with several satellites, unlike the terrestrial planets, which were prevented (by tidal forces due to the Sun and by a lack of volatile elements in the inner solar system) from retaining disks and therefore have few or no natural moons.

We can understand why the disks orbiting the giant planets include ring systems as well as satellites by considering the tidal forces created by the planets themselves. Close to each of the outer planets is a region where tidal forces due to the planet are greater than gravitational forces that might hold a satellite together. In these regions the orbiting swarm of tiny particles was prevented from gathering together into large satellites. There is a well-defined limiting distance, called the **Roche limit,** inside of which a satellite cannot form because tidal forces prevent it. The location of the Roche limit depends on the mass of the parent planet and on the size (radius) of the particular satellite. A small satellite might survive where a large one could not, and we will see that some tiny moons (up to perhaps 100 km in diameter) are to be found intermingled with the outer portions of the ring systems.

Recent evidence suggests that the ring systems may occasionally be replenished with new debris resulting from the collisional destruction of a satellite. Due to their large masses and strong gravitational fields, the giant planets attract passing bodies (asteroids or comets) toward themselves, thus enhancing the probability that collisions with the existing satellites will occur. This phenomenon, called **gravitational focusing,** ensures that the moons of the giant planets will undergo relatively frequent collisions that occasionally will be energetic enough to destroy them. The destruction of a satellite injects a new swarm of particles, enriching the ring system.

Thus the rings may consist of both primordial material, left over from the time of planetary formation, and newer matter, added through the destruction of satellites. Once again, as in many other aspects of solar system formation and evolution, we find that a combination of orderly, systematic processes and random, cataclysmic events must be invoked to explain the observed phenomena.

Evidence that ring systems are occasionally renewed is largely circumstantial. It is based on the ring structures observed around Saturn, Neptune, and especially Uranus, which are unstable; that is, they cannot last for a very long time and therefore must have been created much more recently than the time of planetary formation 4.5 billion years ago. These temporary features include asymmetric (noncircular) rings of Saturn, lumpy structures in the rings of Neptune, and a broad dust ring around Uranus whose particles are gradually spiraling into the planet. Another circumstantial clue comes from the evidence of impacts on the existing satellites of the giant planets. Mimas, the innermost significant moon of Saturn, bears a crater attesting to a collision that must have very nearly destroyed the satellite. Miranda, the innermost of the five large moons of Uranus, has a very strange surface terrain, suggesting that the entire satellite might once have been fragmented and then reassembled.

Once formed, a ring system is subjected to many processes that can create intricate structure. One such process, called **orbital resonance,** creates gaps in a ring system at locations where orbiting bodies would frequently be aligned with a large satellite. The Cassini division, a large gap in the rings of Saturn, occurs at the point where ring particles would have orbital periods exactly equal to one-half the period of Mimas. A particle at this distance from Saturn would find itself lined up on the same side of the planet as Mimas on every other orbit. The frequent alignments would allow gravitational tugs from Mimas to gradually alter the particle's orbit. Over time, a gap is created at this position. Similar gaps occur where particles would have periods equaling other simple fractions, such as one-third or one-fourth, of the period of Mimas.

Another mechanism that creates structure in the rings of Saturn and the other giant planets is also gravitational in origin. **Spiral density waves,** consisting of a fixed pattern of alternating dense and rarefied regions that rotates around the parent planet, are triggered by the gravitational influence of the large, outer satellites. In any rotating fluid disk, such waves can be set up by outside forces. Another example of a spiral density wave pattern is to be found in the spiral arm structures of some galaxies (discussed in Chapter 17). In the rings of Saturn, spiral density waves show up as closely spaced series of tiny, thin "ringlets" that follow a spiral pattern. A section of a ring system where these are seen does not consist of separate, circular rings, but instead is composed of a continuous, spiral-shaped line of ring particles, similar to the grooves on a phonograph record.

A third gravitational process for creating structure in ring systems also involves the effects of satellites. By their combined gravitational forces, two closely spaced moons can confine ring particles between them **(Fig. 10.10).** The moons in this situation are called **shepherd satellites** because they act as sheepdogs, keeping their flock of ring particles in line. The shepherd satellite effect comes about because, under Kepler's laws, orbital speeds decrease with distance from the center of mass of the system. Thus the outer member of a pair of shepherd satellites moves more slowly than the ring particles between it and the inner shepherd moon, while the inner moon moves more rapidly than the ring particles. When a particle that wanders too far out passes the outer shepherd moon, it is slowed by a gravitational tug and drops back into a lower orbit. Conversely, a particle that slips into a lower orbit is

Figure 10.10 The action of shepherd satellites. This figure illustrates how a pair of small moons can keep particles trapped in orbit between them. The inner shepherd satellite overtakes the particles, accelerating them if they fall inward and thus moving them out. Similarly, the outer shepherd slows particles that move too far out, so that they fall back in. (NASA/JPL)

speeded up and boosted back into a higher orbit when the inner shepherd moon passes by. Thus the particles may wander around a bit, but the combined effect of the two moons is to keep them trapped in orbits lying between the moons. The F ring of Saturn, which has a curious, braided appearance **(Fig. 10.11),** is formed by a pair of shepherd satellites, as are several rings of Uranus.

In the following sections, we will discuss the planets individually, emphasizing their unique qualities but also noting the features they have in common. We will discuss Jupiter and Saturn together and then Uranus and Neptune because these pairs have many commonalities.

Jupiter and Saturn

The fifth and sixth planets from the Sun, Jupiter and Saturn, are also the largest. In gross properties they are in a class by themselves; Jupiter is more than 300 times more massive than the Earth, and Saturn is almost 100 times more massive (see Table 10.1). They are also very unusual in another important way,

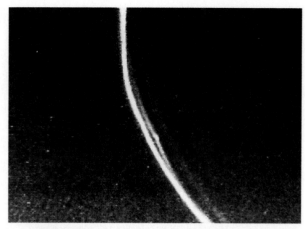

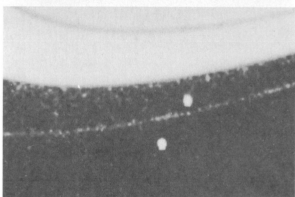

Figure 10.11 Saturn's F ring. Here we see the unusual braided appearance of this thin, outlying ring. The cause of its asymmetric shape is not well understood. The two small satellites (Pandora and Prometheus) that maintain the F ring through gravitational interaction are both visible in the lower image. (NASA)

in that their densities are far lower than the density of any terrestrial planet. Jupiter's average density is about 1.4 g/cm³ (as opposed to roughly 5.5 g/cm³ for Venus and the Earth), and Saturn's density, at about 0.7 g/cm³, is actually less than that of water. These low densities are a clear indication that the gaseous planets must have a very different composition from the terrestrials, as we have already seen. Recall that the gas giants are composed chiefly of hydrogen and its compounds, along with helium, with only very small amounts of the elements that dominate the makeup of the rocky planets.

Despite the similarities between the two largest planets, they also exhibit some notable contrasts. These will be explored in the following sections.

Atmospheres and Interiors

Both Jupiter and Saturn have light-colored bands called **zones** and darker, reddish brown bands known as **belts,** but Jupiter's bands are more vivid and col-

orful than Saturn's (see Fig. 10.1). The color of the belts on Jupiter is thought to be caused by molecular species that persist under the temperature and pressure conditions of the belts, but not under the conditions that prevail in the zones. The identity of these molecules is not known.

The *Galileo* probe penetrated several hundred kilometers into Jupiter's atmosphere, making measurements of composition and conditions until the pressure crushed it. The composition was, as expected, similar to that of the Sun, with one exception: an extreme shortage of water (as indicated by the oxygen content of the clouds). Scientists are puzzled by this result. The best explanation available initially is that the probe happened to enter the atmosphere at the location of a dry downflow, where the water had precipitated out. Such regions had been identified previously from Earth-based observations.

The belts and zones on Saturn are less prominent than those on Jupiter because Saturn's hazy upper atmosphere, subjected to a lower surface gravity than that of Jupiter, extends to a greater height and limits our view of the atmospheric bands. The speed of the flow patterns is much greater on Saturn than on Jupiter. The equatorial flows on Jupiter, for example, move at about 180 m/sec (650 km/hr), while on Saturn the corresponding winds move at 400 to 500 m/sec (up to 1,800 km/hr). In addition, the belts and zones on Saturn exhibit contrasts between the northern and southern hemispheres that may be seasonal effects, since Saturn has an obliquity of almost 27° compared with Jupiter's 3°. On Saturn the belts and zones are broader in the north, and a concentration of spots and eddies occurs at about 40° north latitude, where there is an abrupt change in flow speed at the northern edge of the equatorial belt.

The *Galileo* probe found that high wind speeds extend deeper into Jupiter's atmosphere than expected. This is causing some revision of atmospheric flow models, to incorporate the possibility that the entire gaseous envelope, not just the cloud layer, is circulating globally.

Jupiter has a large, reddish brown oval-shaped area in its southern hemisphere, known as the **Great Red Spot (Fig. 10.12).** This feature has been present for hundreds of years and was seen by Galileo and Huygens in the seventeenth century. Within the Great Red Spot, the gas is rising and rotating in a counterclockwise direction (i.e., the same as the zones in the southern hemisphere of Jupiter, except that the zones are light colored). Modern studies of the Great Red Spot indicate that it is basically a storm system, much like the great rotary storms we have on Earth, except that it is anticyclonic, unlike hurricanes and cyclones, and much longer lasting. Starting with a pair of vortices that are modified by the global wind system on Jupiter, theoretical models have succeeded in reproducing the Great Red Spot quite accurately **(Fig. 10.13)**.

Both Jupiter and Saturn display a large number of smaller, more ephemeral spots and rotary patterns (see Figs 10.1 and 10.3). These generally tend to be white, rather than dark like the Great Red Spot, but dark examples are seen on both planets as well. Occasionally, these spots can grow to large sizes. Starting in late 1990, for example, Saturn developed a "Great White Spot" that was observed by the *Hubble Space Telescope* (this spot was sufficiently prominent to be seen from Earth through modest telescopes, but much more detail is visible in the *HST* images; **Fig. 10.14**).

Models of the interiors of Jupiter and Saturn show a similar overall structure, but with Jupiter

Figure 10.12 The Great Red Spot. This close-up image was obtained in early 1996 by the *Galileo* spacecraft, which is orbiting Jupiter. Details of the rotary flow and associated turbulence are clearly visible. (NASA/JPL) http://universe.colorado.edu/fig/10-12.html

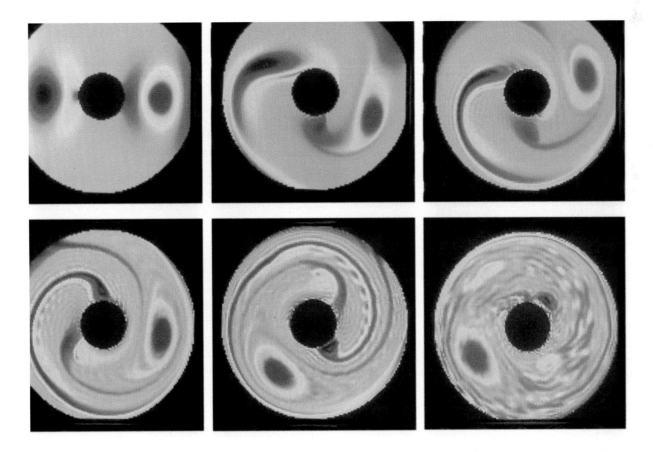

Figure 10.13 Formation of the Great Red Spot. This computer simulation, showing Jupiter from above the south pole, successfully recreates the Great Red Spot. A pair of vortices (upper left) rotating in opposite directions are modified by the strong global winds so that a clockwise (blue) vortex is destroyed while a counter-clockwise (red) vortex grows and persists. (P. Marcus and N. Socci, University of California, Berkeley and Harvard-Smithsonian Center for Astrophysics; computations done at the National Center for Atmospheric Research.)

Figure 10.14 A storm on Saturn. This *Hubble Space Telescope* image shows an elongated whitish region just above Saturn's equator. This turbulent storm system appeared quickly and then persisted for weeks. (NASA/Space Telescope Science Institute)

Figure 10.15 The Galilean satellites of Jupiter. These are the four moons discovered by Galileo, shown in correct relative size. They are Io (upper left), Europa (upper right), Ganymede (lower left), and Callisto (lower right). Ganymede is the largest satellite in the solar system. Moving from Callisto, the farthest from Jupiter, to Io, the innermost of the four, we find a wide range of geological and physical properties, as discussed in the text. (NASA) http://universe.colorado.edu/fig/10-15.html

having a much more extended liquid metallic hydrogen zone due to its greater internal pressure. As mentioned earlier, both planets emit excess infrared radiation, indicating that both are producing more heat in their interiors than can be explained by the amount of sunlight that they are receiving.

Interior models suggest that this excess heat has a different source for each planet. For Jupiter, it appears likely that this is simply heat left over from the formation of the planet, which has been able to escape only very slowly and is still trickling out today. For Saturn, however, with its lower mass and lower density, the calculations show that less heating would have occurred initially and that any primordial heat should have dissipated long ago. Hence Saturn must have an internal source that is producing heat today. Saturn has a much lower fraction of helium in its upper atmosphere than Jupiter (on Saturn the helium concentration is only 14% whereas *Galileo* measurements show that Jupiter's concentration is 24%, almost identical with that of the Sun). Therefore it has been suggested that Saturn has been undergoing continuing differentiation, as the helium atoms, twice as heavy as hydrogen molecules (H_2), have been sinking toward the planet's core, releasing gravitational potential energy. Jupiter's interior is so hot that convective motions and internal turbulence prevent this differentiation from occurring.

Tidal Forces and the Galilean Satellites of Jupiter

Among the moons of the giant planets are several that are very large (some rival or even exceed the diameter of Mercury). Jupiter has four such satellites, first observed by Galileo in 1609, while Saturn and Neptune have one each. All of these moons consist of mixtures of rock and ice and are therefore quite unlike the terrestrial planets despite their hard surfaces. The four Galilean satellites of Jupiter present a fascinating and complex series of worlds **(Fig. 10.15)** that are modified in many ways by the gravitational influence of their parent planet.

The four Galilean moons, from outermost to innermost, are Callisto, Ganymede, Europa, and Io. The detailed observations of these bodies by the *Voyager* probes revealed that they vary in spectacular fashion. The colorful images and fine detail disclosed several clear trends among these satellites (Appendix 8). For example, the numerical density of impact craters declines progressively from a very high density for Callisto, the outermost satellite, to a much lower density for Io, the innermost. This decline shows that the ages of the surfaces decrease from the outer part

Figure 10.16 An eruption on Io. This image of an eruption from a volcanic vent with a plume of ejected gas dramatically illustrates Io's present state of dynamic activity. (NASA)

Figure 10.17 Detail on the surface of Ganymede. This image, obtained by the *Galileo* orbiter in June, 1996, shows details as small as 75 m. The straight lines are ridges, probably created by crustal movement. Superimposed on these are impact craters such as the one at right center (with dark debris around it). (NASA/JPL)

of the system toward the center, with Callisto having a very old surface and Io a very young one. The observations also revealed that the densities of the moons increase with proximity to Jupiter.

Combining these observations leads to the conclusions that the outer Galilean moons contain more ice than the inner satellites (this explains the lower densities of the outer moons) and that the inner moons undergo more geological activity (this is apparent from the youth of their surfaces). Even before the first visit by a *Voyager* spacecraft in 1979, it had been suggested that this trend might be found. The reason: tidal forces acting on the inner moons were expected to cause internal heating, which in turn would cause the geological differences that were discovered. The first images showing volcanic eruptions on Io provided dramatic confirmation of this prediction **(Fig. 10.16).**

The giant planets exert very strong tidal forces on their satellites. These tides are sufficient to keep all the moons of the gas planets locked into synchronous rotation and to force them into slightly elongated shapes due to the immense stretching forces to which they are subjected. Pressing and squeezing a

body internally can cause heating, which leads in turn to internal melting and volcanic activity.

Internal heating due to tidal stress apparently extends to the outer satellites. The *Galileo* spacecraft has discovered, much to the surprise of most planetary scientists, that Ganymede has a magnetic field. This, combined with the striated surface revealed in detail by *Galileo* images, **(Fig. 10.17),** indicates that Ganymede has been tectonically active in its past, and may retain a partially fluid interior.

Tidal stresses are enhanced for the two innermost Galilean satellites. These two bodies are subjected to additional, varying forces exerted by each other. As it happens, the orbital period of Io is almost exactly half that of Europa, meaning that the two moons are regularly aligned in the same configura-

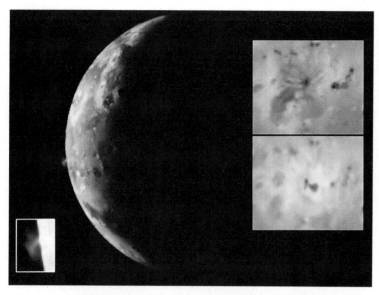

Figure 10.18 An eruption on Io observed by *Galileo*. The *Galileo* orbiter caught the Ra Patera volcanic caldera in action when it flew past Io in late June 1996, The inset at lower left shows more detail in the outburst, which sent gas as high as 100 km above Io's surface. The insets at right show the region where the eruption occurred, in 1979 as seen by *Voyager* (upper box) and in 1996 as seen by *Galileo* (lower box). Comparison of the two shows changes in the surface markings as a result of the continual eruptions. (NASA/JPL)

tion relative to Jupiter. This is another example of orbital resonance, already mentioned earlier in this chapter as the explanation of gaps in the rings of Saturn. At the times when Io and Europa are aligned, their gravitational attraction for each other adds to the tidal stresses they already feel due to Jupiter, and the two are subjected to additional internal stress. The extra tidal stress exerted on Io each time it is aligned with Europa and Jupiter has created sufficiently high temperatures in its interior to melt it. The satellite consequently is undergoing continual volcanic eruptions **(Fig. 10.18)** and has lost most of its volatile materials, which explains its high average density and low ice content. *Galileo* results, released in 1996, suggest that Io has a solid core of iron and iron sulfides. This core extends halfway to the surface of Io, and is overlain by a mantle of fluid rock, which is the source of the volcanic activity. The volcanic outpourings have continually resurfaced Io, obliterating any impact craters and giving the moon a very young surface. In addition, the gas being released by the volcanoes can escape Io's gravitational field and go into orbit around Jupiter, where it has created a doughnut-shaped ring of gas called the **Io torus (Fig. 10.19).** Particles released from the volcanoes on Io become ionized as they interact with the planetary particle belts; once ionized, these particles are forced to orbit Jupiter at the speed of rotation of the planet's magnetic field, thereby creating the torus. The 11° tilt of Jupiter's magnetic axis relative to its spin axis causes the Io torus to wobble up and down by 11° with a period of just under 10 hours, the rotational period of Jupiter.

Europa is also subjected to extra internal stress and heating due to its orbital resonance with Io, but the effects are not as severe. The surface of Europa is quite young, but no active volcanoes have been seen. It is known that Europa still contains a substantial quantity of volatiles; interestingly, models of its interior and observations of its surface features suggest that the interior of this satellite may contain large

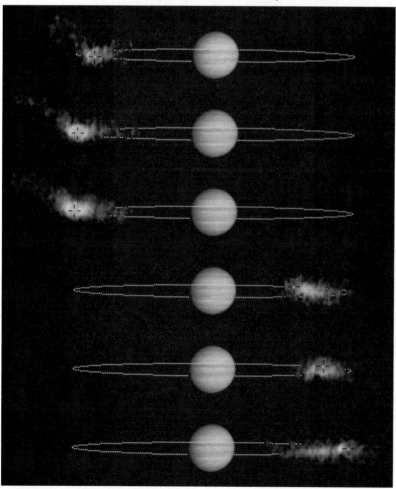

Figure 10.19 The Io torus. This remarkable series of photographs, taken through a ground-based telescope on Earth, shows the cloud of sodium atoms surrounding Io as it orbits Jupiter. The yellow color of the cloud is due to the fact that sodium atoms emit most strongly in a pair of emission lines in the yellow part of the spectrum (this is why certain types of street lights appear yellow). Once this gas, which also includes other atoms such as sulfur, escapes from Io, it becomes ionized and then is spread all the way around Io's orbit by magnetic forces. This ionized gas is not visible here, so we do not see the full extent of the torus. The sizes of Jupiter, Io (dot inside the cross-hair), and Io's orbit are all to correct relative scale. (B. A. Goldberg, G. W. Garneau, and S. K. LaVoie, JPL)

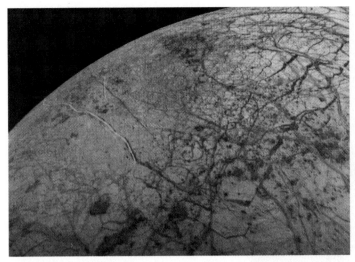

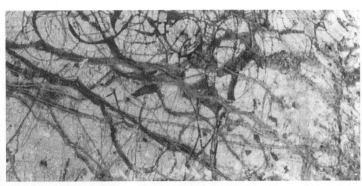

Figure 10.21 Ice floes on Europa? This *Galileo* orbiter photo, obtained in late June, 1996, shows a region of fractured ice (light material) floating on a dark medium which probably consists of slush and embedded rock. Close inspection of the shapes of the ice plates indicates that some have drifted and rotated since breaking free from their neighbors, much like the processes that occur in Arctic ice floes on the Earth. Calculations suggest that liquid water may exist of Europa from just below the surface to a depth of 100 km. The smallest details visible in this image are about 1.6 km across. (NASA/JPL)

Figure 10.20 Europa. This is the second of the Galilean satellites in progression outward from Jupiter. Its linear features and relative lack of craters indicate that it is a tectonically active moon with a young surface. (NASA)

quantities of liquid water. The surface is marked by linear features that appear to be cracks between crustal plates that are floating on the liquid water; the cracks themselves appear to be filled with ice, formed as water flowed to the surface and froze **(Fig. 10.20)**.

Recent *Galileo* images of Europa's surface **(Fig. 10.21)** show structures that closely resemble ice floes floating on the water. The fact that some of these structures have shifted position since the *Voyager* flys by more than 15 years earlier adds support to the idea that liquid water (and moderate temperatures) may prevail in the outer zones of Europa. This adds another world to the list of possible solar system sites where life could have formed.

Other Moons and the Ring of Jupiter

The small moons of Jupiter were not observed closely by the *Voyager* spacecraft, whose trajectories were optimized for making close passages by the Galilean satellites. Hence few details were known about the small moons until the arrival of the *Galileo* mission, which is to make close visits to several of the smaller satellites. We should know more about these moons soon, as the results from *Galileo* are analyzed.

The *Voyagers* did obtain images revealing that Jupiter has a single thin ring **(Fig. 10.22)**. This news came shortly after the discovery (from Earth-based observations) that Uranus has rings as well, so at that point astronomers knew that at least three of the four gas giants had rings. At the same time, rings of Neptune were suspected from Earth-based observations and were finally confirmed when *Voyager 2* reached

Figure 10.22 The geometry of the Jovian ring. Here the ring has been drawn in, showing its size relative to Jupiter. The ring's radius is about 1.8 times the radius of the planet; the ring is estimated to be no more than 30 km thick. (NASA)

the eighth planet in 1989. Thus the detection of the ring around Jupiter contributed to a growing realization that rings are the norm, rather than the exception, in the outer solar system.

Titan and the Moons of Saturn

Saturn has 20 known satellites, most of them very small and close in (several orbit within the ring sys-

Figure 10.23 **Moons of Saturn. Titan.** The largest of Saturn's moons. Titan is almost unique among the satellites in the solar system in having at atmosphere. (NASA) **Dione.** This intermediate-sized moon of Saturn has wispy, light-colored features on its surface that may be icy deposits of material that escaped from the interior and crystallized. This side of Dione, which is in synchronous rotation, always trails as the satellite orbits Saturn; the leading side is more heavily cratered. (NASA) **Enceladus.** This satellite is the shiniest object in the solar system, with an albedo of approximately 1, just like a mirror. The surface may have been coated with deposits from internal melting and outgassing. (NASA) **Iapetus.** This satellite has very unusual bright and dark areas. It appears that the dark material may be some sort of deposit. (NASA) **Rhea.** One of the seven intermediate-sized moons. Rhea shows evidence of cratering and light-colored areas that are probably composed of ice. (NASA) **Hyperion.** This satellite has a very unusual shape and is marked by several impact craters. (NASA)

tem; these are discussed in the next section). There are seven intermediate-sized moons (diameters of a few hundred kilometers) and one giant satellite, Titan **(Fig. 10.23)**.

The intermediate moons (see Appendix 8 and Fig. 10.21) are composed primarily of ice, indicated by the fact that their average densities are about 1 g/cm^3, the density of water ice. Most are heavily cratered, indicating that their surfaces are quite old,

but there are exceptions. Enceladus, for example, has a very shiny surface (albedo near 1, which means that it reflects light about as well as a mirror), which probably means that some process is continually recoating its surface with fresh ice. Some suspect that Enceladus may be volcanically active, spewing out water vapor that falls to the surface and freezes. This satellite is in orbital resonance with another of the intermediate moons, Dione, which might account for

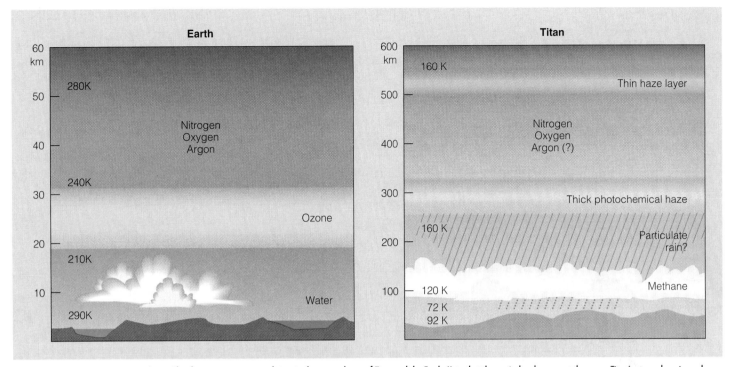

Figure 10.24 **Titan's atmosphere.** This diagram compares conditions in the atmospheres of Titan and the Earth. Note that the vertical scales are not the same; Titan's atmosphere is much deeper. (Data from J. K. Beatty, B. O'Leary, and A. Chaikin, eds. 1990. *The New Solar System.* 2nd ed. Cambridge, England: Cambridge University Press.)

the volcanic activity. Yet another satellite, Iapetus, has a large dark region on one side, evidently the result of the deposition of some kind of sooty substance of unknown origin. Large fissures and wispy, frosted regions on other moons indicate past tectonic activity, as crustal cracking and outgassing appear to have occurred. The satellite Hyperion is very asymmetric in shape, appearing as a flattened, almost disklike object.

Some of the intermediate moons have smaller bodies (sometimes called "moonlets") sharing their orbit (see Appendix 8). These **co-orbital satellites** are found either 60° ahead or 60° behind the larger satellite. The locations of these satellites coincide with a prediction made centuries ago by the French mathematician J. L. Lagrange. Working with Newton's laws of motion, Lagrange found that in a circular orbit there are stable positions that lead or trail the orbiting body by 60°. At these positions the combined effect of the central object (Saturn in this case) and the orbiting body (one of the intermediate moons, such as Dione or Tethys) acts to keep smaller bodies trapped, so that they remain in orbit at those

positions. In the next chapter, we will see that groups of asteroids are orbiting the Sun at the corresponding positions in the orbit of Jupiter.

Saturn's giant satellite, Titan, is nearly as big as Ganymede, which is the largest moon in the solar system. Titan is distinguished by its thick atmosphere (see Fig. 10.21), found by the *Voyager* probes to be composed chiefly of nitrogen. The atmospheric pressure is about 50% greater than sea-level pressure on the Earth, and the atmosphere of Titan is deeper as well **(Fig. 10.24)**; cloud layers made it impossible for the *Voyager* cameras to obtain images of the surface. Nevertheless, some information was obtained by radio instruments, and surface conditions were measured. At the surface the combination of pressure and temperature corresponds to the conditions under which methane (CH_4), which is observed in the atmosphere, can be in vapor, liquid, or solid form. Hence astronomers speculate that there may be methane lakes on Titan, perhaps with methane icebergs floating in them. Liquid nitrogen may also exist under the conditions at the surface of Titan, so nitrogen lakes or rivers are also a possibility.

Recent evidence from Earth-based radar measurements of Titan suggests that there are variations in surface smoothness that might correspond to the presence of large bodies of liquid. For a definitive answer to the question of liquid on Titan, however, we will have to await the results of the *Cassini* mission, which will include a probe that will drop into Titan's atmosphere and make measurements all the way to the surface.

We do not know why Titan has an atmosphere, while other large satellites do not (although Triton, a moon of Neptune, does have a thin atmosphere, none of the Galilean satellites of Jupiter does). That the atmosphere is composed primarily of nitrogen is also unusual; in the rest of the solar system, only the Earth and Triton have nitrogen-dominated atmospheres. In both those cases, volcanic eruptions are a leading source of nitrogen, so it is possible that Titan may be volcanically active.

The Rings of Saturn

We have already discussed most of the mechanisms thought to be responsible for the existence of ring systems in general and the maintenance of their detailed structure. Much of what we said applies to Saturn's rings, which are by far the most extensive and complex of the four ring systems of the giant planets.

Figure 10.25 The dark spokes. At the left the spokes appear bright because they scatter light in the forward direction, and they were photographed looking toward the Sun. At the right they are viewed looking away from the Sun, so they appear dark. (NASA) **http://universe.colorado.edu/fig/10-25.html**

From the Earth it appears that the rings of Saturn are relatively simple, consisting of three broad bands distinguished from each other by brightness differences and by at least one prominent gap (the Cassini division). The three bright rings were dubbed the A, B, and C rings; a fourth, dim ring (the D ring) was faintly seen, and the thin, oddly shaped F ring was discovered from *Pioneer* images. Only with the arrival of the *Voyager* missions in the late 1970s did it become apparent that the ring system was far more complex than previously suspected.

All of the gravitational mechanisms described earlier in this chapter affect the rings of Saturn. There are gaps corresponding to orbital resonances with the larger satellites (particularly Mimas), sections where spiral density waves create thin, spiral-shaped rings, and at least one ring (the F ring) that is confined and maintained by shepherd satellites.

In addition to the effects of the gravitational phenomena, there are ring structures that appear to be created by nongravitational forces. The *Voyager* images of the rings revealed curious, dark "spokes" that emanate outward from Saturn **(Fig. 10.25),** and were seen to vary with time. Careful analysis of the images shows that these spokes lie above and below the plane of the rings themselves and are thus not directly part of the ring system. This was surprising because any orbiting particles that are governed by gravitational forces should be strictly confined to the equatorial plane of Saturn, as the main rings are. The accepted explanation for the dark spokes therefore relies on nongravitational forces.

Astronomers speculate that very small ring particles could accumulate enough electrical charge (probably due to the ejection of electrons by solar ultraviolet photons) so that the force exerted by Saturn's magnetic field would be stronger than the gravitational forces acting on the particles. These very tiny particles could then be levitated above the plane of the rings by magnetic forces. The rapid rotation of the planet, along with changes in the shape of the magnetosphere caused by variations in the solar wind, could then explain the changes in the appearance of the spokes that have been seen.

Uranus and Neptune

The seventh and eighth planets from the Sun appear at first glance to be very similar to each other **(Fig. 10.26).** Both are bluish in color as seen from afar, and they are comparable in size, mass, and rotation period. Close examination, however, has revealed that they are quite distinct from each other in many ways.

Figure 10.26 **Uranus and Neptune.** These full-disk *Voyager* images show the intense blue color of Uranus (left) and Neptune (right) that is caused by methane in the atmospheres of these two planets (methane absorbs red light more efficiently than blue). Note also that atmospheric circulation and clouds are more readily seen on Neptune. (NASA) http://universe.colorado.edu/fig/10-26.html

One difference is very obvious: the tilt (obliquity) of Uranus's rotational axis is so large that the pole of the planet actually lies below the plane of its orbit **(Fig. 10.27)**. The obliquity of Uranus is 98°, meaning that the north pole points 8° below the orbital plane. This strange tilt probably came about as the result of a collision with a large planetesimal during the time when Uranus was forming. The rings and moons of Uranus lie in its equatorial plane, meaning that their orbital plane is almost perpendicular to the orbital plane of the planet as it circles the Sun. Seasons on Uranus are quite unusual; the northern hemisphere has full-time sunlight during the height of its summer, when the north pole is pointed almost directly at the Sun, and permanent darkness during winter (42 years later). On the other hand, when Uranus is at intermediate positions in its orbit so that the Sun lies over its equator, days and nights are very short, equaling half the rotation period of just over 17 hours. Neptune has a much smaller obliquity, and therefore the seasonal variations in the amount of sunlight reaching its surface are much less pronounced.

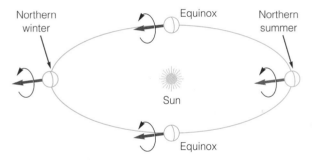

Figure 10.27 **Seasons on Uranus.** The unusual tilt of Uranus creates bizzare seasonal effects. The red arrow in each case indicates the direction of the planet's north pole, and the curved arrow, the direction of its rotation.

Atmospheres and Interiors

Like Jupiter and Saturn, Uranus and Neptune have atmospheres in which hydrogen and its compounds are dominant, and helium is also abundant. Heavy elements have sunk into the core through differentiation. The blue color of the planets is created by methane, which absorbs red light, permitting blue light from the Sun to be reflected much more efficiently **(Fig. 10.28)**. The color of Neptune is much deeper, and belts, zones, and spots are more easily seen because the atmosphere is not as deep and hazy (see Fig. 10.24). Neptune's atmosphere is also more dynamic; its more active flows are probably due to a greater temperature contrast between the poles and the equator (recall that temperature differences are one of the driving forces behind atmospheric circulation). The *Voyager 2* data on Uranus showed a surprisingly small temperature contrast from pole to equator.

Figure 10.28 **The effect of methane on the appearance of Neptune.** This is a false-color image of Neptune, designed to illustrate the effect of methane in the atmosphere. The blue color of most of the disk is natural, but the red around the edges has been enhanced to show light scattered by methane molecules. This gas absorbs red photons, causing the remaining color to be blue in regions where we see light that has passed through a region of methane absorption, such as the central portions of the disk. In regions where we see the reemitted (scattered) light from the methane, we see red. (JPL/NASA)

Figure 10.29 **The Great Dark Spot.** This *Voyager* image shows the large, dark oval feature that is very reminiscent of Jupiter's Great Red Spot, as well as a couple of smaller features. South of (below) the Great Dark Spot are a white feature, known as the "scooter" because of its rapid motion around the planet, and the "dark spot 2," a second counterclockwise storm system. Each of these features moves around the planet in the eastward direction (to the right in the view), but at different speeds, so they are not often grouped together as in this image. (JPL/NASA)

The general circulation patterns in the atmospheres of both Uranus and Neptune are similar to those of Jupiter and Saturn, with alternating belts and zones representing rising and descending currents that are forced into elongated circulatory patterns by rapid planetary rotation. Astronomers were not surprised by the similarity of Neptune's circulation pattern to those of Jupiter and Saturn, but the circulation on Uranus was unexpected because it has little or no excess internal heat to help drive convection. Furthermore, astronomers had expected that any flows on Uranus would go the other way. On Jupiter, Saturn, and Neptune, the main equatorial stream goes around the planet in the same direction as the rotation. But on Uranus, the north pole is currently pointed toward the Sun, and the pole was therefore expected to be warmer than the equator; models of atmospheric circulation show that in that case the equatorial flow should move opposite to the planet's rotation. It turns out that the models may well be correct, but that the equator on Uranus is not cooler than the pole (the temperature is nearly constant over the entire surface of the planet, with the coldest temperatures at intermediate latitudes).

Hence the question is not so much why the equatorial flow is in the normal direction, but why the equator is not colder than the pole. The answer is not known.

The many spots and storm systems discovered on Neptune when the *Voyager 2* images were returned in 1989 included a large, dark one called, naturally enough, the **Great Dark Spot (Fig. 10.29).** This was thought to be similar to the Great Red Spot on Jupiter, and it may have been, but one outstanding difference has since arisen: the Great Dark Spot disappeared in late 1994, whereas the Great Red Spot of Jupiter has been present continuously at least from the time of Galileo's observations more than 370 years ago (the Great Dark Spot on Neptune reappeared in 1995).

The upper atmosphere of Neptune is quite transparent, allowing features to be seen at great depth. For this reason the subtle contrast between the belts and zones is more readily seen on Neptune than on Uranus, whose upper atmosphere is hazy enough to obscure the flow patterns. In addition to the belts and zones, both planets also have white, wispy clouds in the upper atmosphere on occasion **(Fig. 10.30).**

Figure 10.30 **Vertical structure in the atmosphere of Neptune.** This close-up image of Neptune shows that the white "cirrus" clouds lie above other levels of the atmosphere that are observed. Here we see that the sides of the clouds that face the Sun are brighter, and that the clouds cast shadows on the lower atmosphere in the direction away from the Sun. (JPL/NASA)

These probably consist of methane ice particles and are far more common on Neptune.

Uranus and Neptune differ internally as well as externally. For example, Uranus has little or no excess heat from its interior (i.e., the amount of energy it radiates is essentially equal to the amount it receives from the Sun), whereas Neptune has a substantial amount of excess radiation. In fact, Neptune emits more excess energy relative to its mass (i.e., more energy per kilogram of mass) than either Jupiter or Saturn.

Both planets have magnetic fields, in each case strongly misaligned with the rotation axis **(Fig. 10.31)**, as described earlier in this chapter. In addition to being highly tilted relative to the rotation axes of the planets, the magnetic axes are also offset and do not pass through the centers of the planets. When this offset was first found for Uranus, it was puzzling, but was viewed as a unique case that probably could be explained by invoking some singular event. But when the same situation was found for Neptune as well, it was still puzzling, and worse, now seemed to require a more general explanation. So far, none has been found, and the curious magnetic structures of both Uranus and Neptune remain among the mysteries of the outer solar system.

Moons and Rings of Uranus and Neptune

Before the *Voyager 2* flybys, Uranus was known to have five satellites and Neptune two. The spacecraft found several additional moons; some of them are very small moonlets, but a few are of substantial size. Now 15 moons are known for Uranus and 8 for Neptune, and evidence for additional small ones has been found as well.

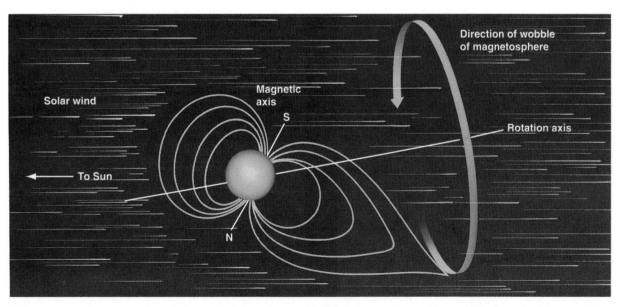

Figure 10.31 **The magnetosphere of Uranus.** Due to the combination of the high tilt of the rotation axis of Uranus and the misalignment between its rotation and magnetic axes, the magnetosphere of Uranus resembles that of a planet whose rotation axis is more normal. The tail of the magnetosphere rotates as the planet spins around its highly tilted axis.

Figure 10.32 The rings of Neptune. This *Voyager* image shows the thin, dim rings of Neptune in some detail. Here we see two distinct rings, one of which has three bright segments, which may have led to earlier suggestions that the rings are not complete circles. Other views show that an indistinct, broad region, known as the "plateau," encompasses some of the thin rings and that an inner belt of ring particles extends close to the planet. (JPL/NASA)

Each planet also has a ring system. Uranus has nine thin, dark rings that are very difficult to see from Earth (they were originally discovered from the Earth when Uranus passed in front of a distant star, and the star was seen to alternately dim and brighten as the rings passed in front of it). The rings of Neptune, suspected from Earth-based observations but not confirmed until the arrival of *Voyager 2,* are not symmetric circles; instead they are uneven and lumpy **(Fig. 10.32).** Before *Voyager,* this unevenness led to the suggestion that the rings were simply disconnected arc segments and not complete circles. Now we know that the rings are complete, but very clumpy. The cause is not understood, but the nonsymmetric nature of the rings surely indicates that dynamic processes are at work; otherwise such structures should smooth out over time.

The five relatively large satellites of Uranus **(Fig. 10.33)** have varied geological properties and present some interesting puzzles. Their diameters range from just under 500 km (Miranda) to over 1,600 km (Titania), and their densities lie between 1.3 and 1.7

g/cm³, indicating that they have substantial quantities of rocky material embedded within an icy structure. The outermost of the five is Oberon, which has a heavily cratered (hence very old) surface and appears to be inactive geologically. Moving inward, the next is Titania, which has surface fissures that may be the result of freezing and expansion in the interior; oddly enough, Titania appears to have relatively young impact craters on its surface, as though it had recently experienced a period of heavy bombardment by debris from space. By contrast, Umbriel has a very old, dark surface that may be covered by some sort of dirty deposit, whereas Ariel has a very bright, young surface that appears to have been recently coated with icy deposits, perhaps due to water vapor escaping from its interior. Finally, Miranda, the innermost of the five, has a very strange surface, covered in places by craters and in others by very unusual uplifted regions indicative of recent tectonic activity. This satellite appears to have a very unsettled geology, with very old regions coexisting with young ones on its surface. This has led to the suggestion that Miranda was shattered by a collision not long ago (geologically speaking) and has since reassembled itself, but has not had time to settle into a well-organized, differentiated state.

The strange properties of the satellites of Uranus may be related to some unusual aspects of its rings **(Fig. 10.34),** perhaps through collisions that have destroyed small moons and added material to the rings. The inner portion of the ring system consists of a broad, dark ring of dust particles (as opposed to ice). Calculations show that this ring cannot persist for a very long time without being replenished; astronomers therefore speculate that this ring either is very young or has a source of new particles to replace the ones that spiral into Uranus. Other evidence of instability among the rings of Uranus includes the fact that some of them are not perfectly circular; over the long term, gravitational forces should produce round rings. Thus the evidence suggests that the moons and rings of Uranus are changing over time. This evidence helped inspire the new view of planetary ring systems; as we discussed earlier, the rings are now thought to be ephemeral, changing entities rather than permanent, fixed structures.

The two moons of Neptune that were known before the *Voyager 2* encounter both have very unusual properties. Nereid, the smaller moon (340 km diameter), has a very elongated orbit with a period equal to about one Earth year. Triton, which is larger (2,705 km diameter), has a circular orbit but goes around Neptune in the retrograde direction. Like Saturn's Titan, this satellite has an atmosphere consisting of nitrogen and methane, but it is much thinner than Titan's atmosphere.

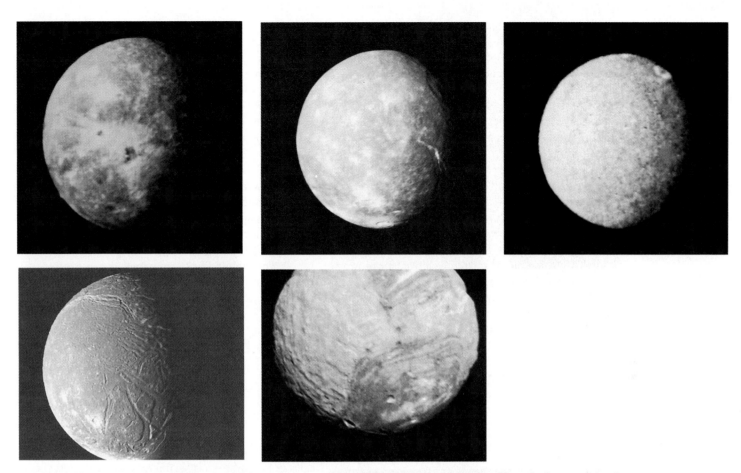

Figure 10.33 The satellites of Uranus. Starting with the outermost, the satellites are Oberon (upper left), Titania (upper middle), Umbriel (upper right), Ariel (bottom left), and Miranda (bottom right). Titania shows signs of more recent cratering and thus has a younger surface than Oberon and Umbriel, the moons on either side of it. Umbriel is the darkest of the five, an indication that its surface has probably been coated by some sort of low-reflectivity deposit. Ariel's bright surface and evidence of recent ice flows indicate that it is probably undergoing internal motions and outgassing. Miranda also shows signs of recent geological activity. Tidal stresses are insufficient to explain this satellite's youthful surface, and astronomers are looking for more exotic explanations. (NASA/JPL)

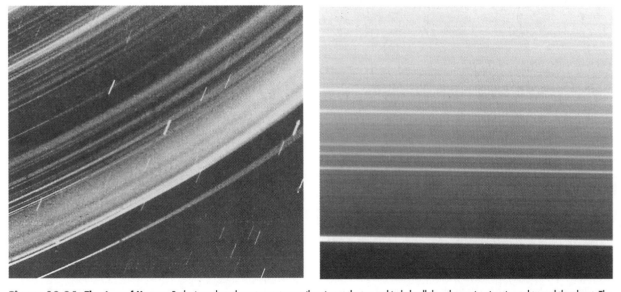

Figure 10.34 The rings of Uranus. Both views show the outermost or epsilon ring at the top and include all the other major rings inward toward the planet. The color-enhanced image brings out greater detail. (NASA/JPL)

Triton provided a wealth of surprises when the Voyager images came in **(Fig. 10.35)**. The surface could be seen clearly through the atmosphere and revealed signs of both tectonic and volcanic activity. The surface albedo is high, especially in the southern polar region, which is covered by an ice cap. Few impact craters are visible, suggesting that the surface is young. Linear fractures provide evidence of crustal plate shifting, and curious, depressed circular basins may be sites where the crust has slumped in the wake of outgassing episodes. Confirmation of current geological activity came with the discovery of several dark streaks (see Fig. 10.33), which turned out to be deposits from recent eruptions. Detailed analysis of the images showed that at least two of these regions are sites of vents that are currently active, spewing gases out of the interior and creating the dark deposits. Thus Triton was added to the short list of solar system bodies undergoing current eruptive activity. The venting on Triton appears to more closely resemble geysers than volcanoes; geysers are sites where subsurface gases are heated and forced out by expansion (as opposed to the upwelling of molten material from deep within a planet or moon). No crustal melting is involved in the outbursts on Triton; instead it is thought that subsurface pockets of frozen nitrogen are vaporized, releasing nitrogen gas, which then refreezes and is deposited on top of the methane ice of the southern polar region on Triton.

Triton has an average density of about 2 g/cm³, indicating that it is composed of about 70 percent rock and 30 percent ice. As we will see in the next section, this density is very similar to Pluto's, and it is now thought that several bodies of this type may have formed in the outer solar system. Triton's similarity to Pluto and its unusual motion have led to the suggestion that Triton may have been captured by Neptune rather than being formed in place as its satellite. It appears likely that Triton and Pluto both may have formed in the region between 20 and 30 AU from the Sun, but that Triton was then captured by Neptune while Pluto was expelled to a greater distance as a result of gravitational encounters with the major planets.

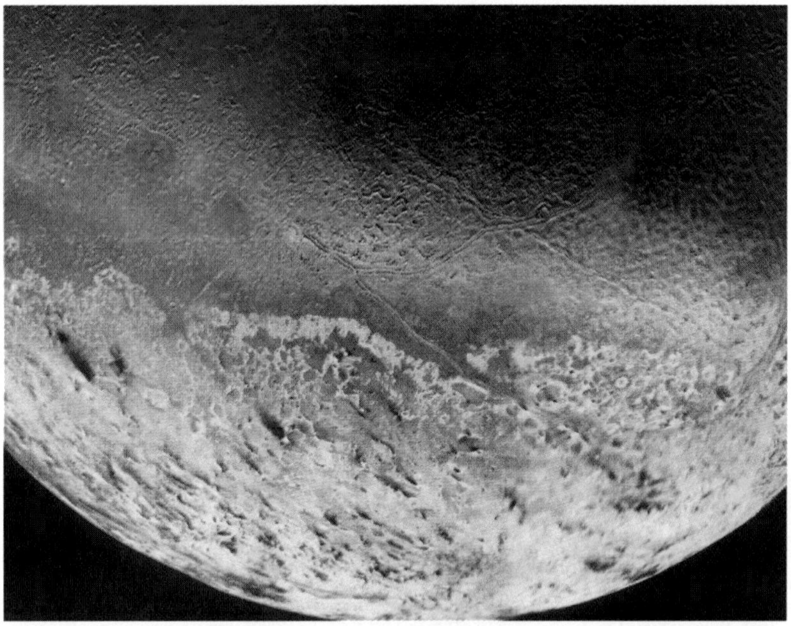

Pluto: Planetary Misfit

The little that we know about Pluto, the ninth planet, indicates that it is a nonconformist. Pluto **(Table 10.2)** simply does not fit into the systematic trends that we have found for the other planets: it has a solid surface, but is certainly not rocky throughout; therefore it does not qualify as a typical terrestrial planet. At the same time, it is not gaseous like the other cold, outer

Figure 10.35 The surface of Triton. This close-up view shows a variety of surface features that indicate that some reprocessing has taken place recently. The surface is bright, unlike old surfaces, and there are relatively few impact craters. There are also linear, grooved features suggestive of crustal plate motion. Most intriguing of all are the dark streaks near the south pole (bottom), which resemble volcanic plumes or ejecta. It is thought that these are either current or recently active geysers, which emit frozen nitrogen. If so, Triton becomes the third body in the solar system known to be volcanically active. (The Earth and Io are the other two, but recall that Venus may also have active volcanoes; see Chapter 9.) (JPL/NASA)

Table 10.2
Pluto

Orbital semimajor axis: 39.53 AU (5,900,000,000 km)
Perihelion distance: 30.22 AU
Aphelion distance: 48.53 AU
Orbital period: 247.68 years
Orbital inclination: 17°10'12"

Rotation period: 6.405 days
Tilt of axis: 122.46°

Diameter: 2,274 km (0.1785 $D_\oplus$)
Mass: 1.25×10^{22} g (0.0021 $M_\oplus$)
Density: 2.05 g/cm³
Surface gravity: 0.0675 Earth gravity
Escape speed: 1.1 km/sec
Surface temperature: 40 K
Albedo: 0.3
Satellites: 1

planets and therefore does not qualify for membership in that group either. These unique aspects of Pluto have led to some rather exotic theories about its origin, but lately scientists have begun to suspect that Pluto is not so strange after all. It may simply be the largest of a class of bodies that are actually fairly common in the outermost reaches of the solar system.

Earth-based observations have historically discovered very little about Pluto's nature, revealing only a dim point of reflected sunlight that is difficult to see without a substantial telescope. But even these limited images were enough to reveal some of the planet's major peculiarities. For example, the orbit of Pluto is very unusual, being highly tilted (by 17°) relative to the plane of the ecliptic and also highly noncircular. At its greatest distance from the Sun, Pluto is nearly 70% farther away than when it is at perihelion (when it actually moves closer to the Sun than Neptune; in fact, Pluto now is temporarily the eighth planet from the Sun, having moved inside Neptune's orbit in 1979, not to reemerge until 1999).

The *Hubble Space Telescope* provides better images **(Fig. 10.36)**, revealing some surface markings **(Fig. 10.37)**. By now a fair amount of information on Pluto has been derived through a combination of *HST* images and data provided by observations of Pluto's satellite.

Establishing Pluto's size or mass was initially very difficult, but even rough estimates showed it to be very small for a planet, the smallest of the nine by a wide margin. The mass was not measured until Pluto's satellite was discovered, but the small size made it clear that Pluto could not have enough mass to affect the orbital motions of Uranus or Neptune significantly, as some had suspected (recall the comments earlier in this chapter about Percival Lowell's original motivation to search for a ninth planet).

Spectroscopic measurements showed that Pluto has solid methane on its surface, which indicated that it has a methane atmosphere as well. In 1988, when Pluto passed in front of a background star, the star's light dimmed slowly, demonstrating that the atmosphere of the planet is quite extended. Scientists deduced that another heavier gas must also be present, which has since been identified as a mixture of carbon monoxide and nitrogen.

Because Pluto has just passed through its perihelion (closest approach to the Sun), it is speculated that the atmosphere has been enhanced by the evaporation of surface ices (see Fig. 10.35). The intensity of sunlight on Pluto is 2.8 times greater at perihelion than at aphelion (the point where a planet is farthest from the Sun). It is suspected that Pluto's atmosphere may largely freeze during the planet's passage through the remote parts of its orbit and then remain

Figure 10.36 Pluto and its satellite. The *Hubble Space Telescope* image shows the Pluto-Charon system in some detail. The size and mass of Charon, very large relative to Pluto, make this truly a double planet. The best ground-based telescopic images are barely able to resolve the two bodies, which are only about 0.6 arcseconds apart as seen from the Earth, whereas the *HST* can easily separate them. (NASA/STScI)
http://universe.colorado.edu/fig/10-36.html

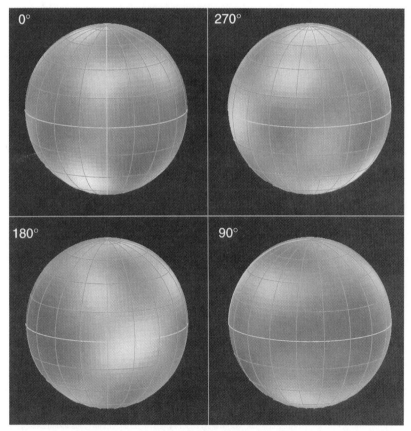

Figure 10.37 Surface detail on Pluto. These four images are computer models of surface markings on Pluto, drived from the analysis of *Hubble Space Telescope* images. The brighter areas are thought to be covered with ices such as frozen methane, while the darker regions are covered with carbon-rich deposits. The colors, which have been added artifically, are thought to realistically represent the color of Pluto's surface, with the reddish hue coming from carbon compounds. (S. Alan Stern, Southwest Research Corporation, Inc.)

entirely in frozen form except around the time of perihelion. If this is true, it is quite a fortunate coincidence that the planet happened to be approaching perihelion when it was discovered; otherwise, its atmosphere and strange seasons might not have been discovered for another 250 years!

Variations in Pluto's surface coloration indicated that its rotation period is 6.39 days, and its obliquity was estimated (with great difficulty) to be very unusual, probably over 90° (later the discovery of Pluto's satellite would enable the obliquity to be established more precisely).

In 1978 Pluto was discovered to have a satellite (see Fig. 10.34), which was named Charon. This satellite is rather large (about half the size of Pluto), making Pluto and Charon a double planet in effect, but it was difficult to detect due to its proximity to Pluto and its great distance from the Earth. Both Pluto and Charon are locked into synchronous rotation, so that each keeps the same side always facing the other. This would create an interesting effect for anyone visiting the surface of Pluto: Charon would be seen to hang in the same spot in the sky, day and night, as the background stars streamed past due to Pluto's spin. Thus Charon is analogous to the communications satellites that have been launched into synchronous orbits about the Earth, so that they always remain over the same spot above the equator.

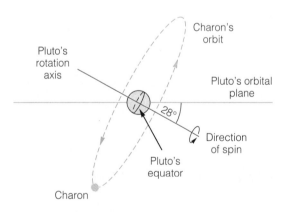

Figure 10.38 The orientation of Pluto's spin axis and Charon's orbit. Like Venus and Uranus, Pluto has its rotation axis tipped over so far that it points below the plane of the planet's orbit (which is itself tipped by an unusually large angle, 17°, relative to the ecliptic). Thus the spin of the planet is technically retrograde (i.e., backward). Charon orbits in the equatorial plane of the planet. During the late 1980s, the plane of Charon's orbit was aligned with the Pluto-Earth direction so that Pluto and Charon alternately passed in front of each other. Observations of these repeated transit events provided a wealth of information on the nature of both bodies and the transient atmosphere of Pluto.

Charon's orbital and rotation periods are both 6.39 days, the same as Pluto's rotation period.

Charon's orbital plane is highly tilted with respect to the orbital plane of Pluto **(Fig. 10.38).** By assuming that Charon orbits in the equatorial plane of Pluto (which it must, according to the laws of motion, if the two-body system is stable), astronomers were able to deduce that the pole of Pluto points about 28° below its orbital plane, meaning that the planet's obliquity is approximately 118°.

The discovery of Charon gave astronomers an opportunity to learn a great deal more about Pluto than had been possible from previous observations. First, it became possible to apply Kepler's third law to the orbit of Charon and to determine the combined masses of the two bodies. Recall from Chapter 5 that Kepler's third law, which relates the orbital period to the semimajor axis of the orbit, makes it possible to solve for the sum of the masses of the two bodies. Normally, the mass of the satellite can be ignored, and the sum can be assumed to be approximately equal to the mass of the planet alone. This is not true for the Pluto-Charon system, however, because Charon is so large compared with Pluto. Kepler's third law provides only the sum of the masses, but we can estimate the individual masses by assuming that the two bodies have similar densities; then the masses are proportional to their volumes. This calculation yields a mass for Pluto of about 0.0022 times the mass of the Earth, or about half the mass of Neptune's great moon, Triton.

Charon provides other information about Pluto as well. Recently, for a period of five years, the orbital plane of Charon was aligned with our sightline from the Earth so that the two bodies eclipsed each other repeatedly as seen from the Earth. As Pluto moves around the Sun, the angle between it, the Earth, and the Sun changes. Every half-year on Pluto (i.e., about every 124 years), the orbital plane of Charon is aligned with the direction of the Sun, which means that, as seen from the inner solar system, Pluto and Charon undergo mutual eclipses. Once again it was fortuitous that this rare event occurred not long (only 60 years or so) after the discovery of Pluto and very soon (about a dozen years) after the discovery of Charon.

The eclipses have provided astronomers with a series of opportunities to learn more about the characteristics of both bodies and their surfaces or atmospheres. When the duration of an eclipse and the orbital speeds of the two bodies are known, their sizes can be determined. For example, when Charon passes in front of Pluto as seen from the Earth, the duration of the eclipse tells us how long Charon

Science in Real Time

One of the challenges for scientists is deciding when and how to make their discoveries known. On the one hand, a great deal of caution is required because it is easy to make mistakes, but on the other hand, there is often some urgency, either because the results are thought likely to have significant impact or because the scientist fears that someone else may soon duplicate them. Thus a tension exists between competing urges, with the result that scientists are usually impatient to announce their work but at the same time are willing to slow down enough to be sure that what they are saying is correct before they make it public. The *Voyager* probes to the outer planets overturned this process by placing scientists in the very precarious, unprecedented position of having the world looking over their shoulders at the very moment of discovery—in real time.

The standard method for disseminating information about scientific discoveries is to publish in a technical journal. Publication involves a rigid procedure that includes the review of the paper by one or more other scientists, called referees. This review ensures that the paper does not contain serious blunders and helps advise the journal editor as to the significance and uniqueness of the reported research. Authors often take the additional step of circulating preliminary copies of a paper, seeking feedback from colleagues as an aid in ferreting out any problems. Even when results are of sufficient interest to be reported in the general press (newspapers and magazines), normally they are not released until after the work has passed its technical review. There are many examples of scientists too eager to become famous, who have bypassed the refereeing process and allowed their results to be publicized before being checked. Some of these escapades have resulted in serious embarrassment for the authors, when their prized discoveries later turned out to be wrong.

Imagine, then, what it must have been like for the scientists analyzing images from the *Voyager* flybys of the gas giant planets. At the time of each encounter (two each for Jupiter and Saturn; one each for Uranus and Neptune), the *Voyager* science team assembled in an auditorium at the Jet Propulsion Laboratory (JPL) in Pasadena (the NASA center responsible for the operation of planetary probes), with hordes of newscasters and reporters also present. As each wondrous new image unfolded on the huge screen, the scientists were pressed for explanation and comment *right now*. Instant interpretation was demanded by members of the news media, who were caught in a bind between the stately process of science and their story deadlines. In general, the scientists fared well, being cautious and when possible saying only what they already knew about the planets, rings, or moons being displayed on the screen. For them it was an exciting time, and the additional challenge of treading cautiously in the presence of incredible public scrutiny added to the enjoyment. It is a credit to the *Voyager* science teams that no major blunders made their way into the public consciousness, and it is a credit to the news media that in general they understood that the real-time utterings of the scientists were preliminary and not to be taken as the final word.

For the public interested in the *Voyager* results, the format of the JPL encounters was a treat. Anyone could look at the images and speculate along with the astronomers as to their meaning. Very rarely does this happen in science, but the immense popularity of the *Voyager* missions suggests that researchers should look for new opportunities to repeat the experiment of doing science in real time.

---

takes to travel a distance equal to Pluto's diameter; combining this figure with the known orbital speed of Charon tells us how far the satellite travels during an eclipse, which is the diameter of Pluto. Conversely, the diameter of Charon can be measured from the duration of its eclipses by Pluto (this technique is also used to measure diameters of double stars; see Chapter 14). Once the diameter of Pluto is known, it can be combined with the planet's mass to derive its density, which turned out to be approximately 2.0 g/cm^3, meaning that Pluto is more than half rock, the remainder being ice. Note that the density (and other general properties of Pluto) are very similar to Triton, reinforcing the notion that Pluto and Triton are close relatives, born in the same environment but now living in separate homes.

The numerical values for Charon are less certain, but apparently its density is similar to that of Pluto.

This would imply that the composition of the two bodies may also be similar, but it has become evident that at least superficially there are differences. Pluto has a much redder surface color than Charon, implying a higher abundance of organic (carbon-bearing) materials on Pluto, whereas spectroscopic data show that Charon's surface is covered with water ice. Pluto's surface has striking variations in brightness, suggesting that some processes must deposit material on the surface nonuniformly.

Apparently, the outer solar system is the home of a class of orbiting bodies that are composed of mixtures of rock and ice and have comparable sizes and masses, and Triton and Pluto are the largest examples. Recent observations (mostly done with the *Hubble Space Telescope*) have revealed a number of similar, though smaller, bodies orbiting the Sun beyond the orbit of Pluto. These will be discussed further in the next chapter.

Summary

1. The five outermost planets are all very different from the terrestrials: the four giants are large, gaseous bodies containing high abundances of lightweight elements and having many rings and moons; Pluto is more like one of the major moons of the giant planets, being composed of ice and rock.

2. Uranus was discovered accidentally, while Neptune and Pluto were found as the result of deliberate searches, based on observed irregularities in the motion of Uranus (and of Neptune, in Pluto's case; Pluto, however, turned out to be too small and distant to have caused any such irregularities, so its discovery is now regarded as fortuitous).

3. Detailed information on Jupiter, Saturn, Uranus, and Neptune was obtained through close encounters with spacecraft sent from Earth, especially the *Voyager 1* and *Voyager 2* probes. Only Pluto has yet to be observed in this fashion, and plans are now being developed to do so early in the next century.

4. The four giant planets have no solid surfaces, but instead are gaseous at high levels, with zones of liquid hydrogen below the clouds and then (in the cases of Jupiter and Saturn) liquid metallic hydrogen; all have small, solid cores of rock and metal.

5. Atmospheric circulation on the four giants is governed by convection, due to solar heating and heat rising from the interior, and rotation, which is so rapid that rotary circulation patterns are stretched into planet-girdling belts and zones.

6. All four of the giant planets have magnetic fields and charged-particle belts, created by electrons and protons trapped in the magnetic fields. The particle belts are forced into flat disklike shapes by the rapid rotation of the planets. Radio emission due to the synchrotron process occurs in these belts and is especially strong from Jupiter.

7. Three of the four giant planets emit excess radiation, in that the amount of energy radiated away (in the infrared, due to thermal emission) is greater than the amount received from the Sun. Only Uranus does not show this effect strongly. For Jupiter and possibly Neptune, the extra energy may come from heat that was created and trapped internally at the time the planets formed; for Saturn, the excess heat is thought to be created by an ongoing differentiation process.

8. The ring systems of the four giant planets lie inside the Roche limit for large satellites, implying that tidal forces prevent the formation of moons and thus maintain the disks of debris that we see as rings.

9. In general, Jupiter and Saturn are very similar, but they do offer some contrasts: Saturn has a lower average density, resulting in a deeper atmosphere that obscures its atmospheric bands; probably creates excess internal heat through differentiation (see item 7 above); and probably undergoes seasonal changes in its atmospheric circulation pattern, whereas the obliquity of Jupiter is so small that seasonal effects are minimal.

10. The four Galilean satellites of Jupiter are strongly influenced by tidal forces and by orbital resonance, with the result that geological activity decreases from the innermost of the four to the outermost. Io, which is closest to Jupiter, undergoes so much internal stress that it is in a constant state of volcanic eruption, recoating its surface and contributing particles to the Io torus. Europa, the next satellite out, appears to have extensive liquid water zones due to this heating.

11. Saturn has one giant moon (Titan), seven intermediate satellites, and a number of very small ones. Titan has a thick atmosphere of nitrogen and methane and may have liquid or solid methane on its surface. The intermediate satellites are composed largely of ice, and some show

signs of ongoing outgassing and surface recoating. The tiny moons, many of them located within the rings, are responsible for creating some of the intricate ring structure through their gravitational effects.

12. The rings of Saturn (and of the other giant planets) are shaped by gravitational forces: the tidal force due to the parent planet; spiral density waves; and shepherding forces created by small moons. In the case of Saturn's rings, electrical forces appear to play a role in creating the dark "spokes"; these are thought to be due to tiny ice particles carrying electrical charges that are acted upon by the planetary magnetic field.

13. Uranus and Neptune are very similar in gross properties, but differ in detail: Uranus has a highly tilted rotational axis, resulting in strange seasonal effects; Neptune has a more active atmosphere, with more vivid belts and zones and more spots; and the orbital and geological properties of the two planets' satellites are quite different. The rings and moons of both show indications of current or very recent dynamic processes that can renew and reshape the rings.

14. Pluto appears to have formed as a planetesimal in the region of the giant planets and was then moved to its present orbit as a result of a near-collision with one of the giant planets. Pluto has a moderate density, indicating a composition of ice and rock. Charon, Pluto's moon, has provided reliable information on Pluto's properties. Pluto and Triton appear to have formed under similar conditions, after which Triton was trapped by Neptune and Pluto was ejected into its present orbit.

Review Questions

1. How is the circulation pattern in the atmospheres of the giant planets different from the atmospheric circulation of the terrestrial planets? What would you change about the terrestrial planets if you wanted to create wind patterns in their atmospheres similar to those of the giant planets?

2. Explain why more can be learned about some properties of a giant planet from Earth-based telescopes than from probes that fly by the planet at close range (*note:* "Earth-based" here includes telescopes in orbit about the Earth as well as those on the ground). For what types of information are Earth-based observations more useful?

3. Explain how astronomers deduce the internal properties of the gaseous giant planets.

4. Explain why the giant planets all have belts of charged particles that are very intense compared with the Earth's Van Allen belts.

5. Why do the giant plants have many more satellites than the terrestrials?

6. Explain the Roche limit and its role in the formation of ring systems.

7. How is methane on Titan comparable with water on the Earth?

8. Summarize the geologies of the five major moons of Uranus, and compare their geologies with the geologies of the Galilean satellites of Jupiter.

9. What is the evidence that ring systems may not be stable, long-lasting phenomena but instead may be transient structures that change with time?

10. Explain how the discovery of Charon has helped astronomers learn about the properties of Pluto.

Problems

1. Calculate the intensity of sunlight reaching each of the four giant planets relative to the intensity of sunlight reaching the Earth. Comment on the relative importance of solar heating in driving atmospheric motions on the gas giants as compared with the Earth.

2. Demonstrate that Kepler's third law applies to the satellites of a planet by computing P^2 and a^3 for the four Galilean moons of Jupiter. You should find that the two quantities are proportional to each other; i.e., that the ratio P^2/a^3 is the same for each satellite. (Note that in doing this you are replicating Kepler in that you are neglecting the variations in the sum of the masses. Is this justifiable in this case?)

3. Io has the highest density among the Galilean satellites of Jupiter, largely because it has little or no ice in its interior, whereas the others have retained their original ice in varying proportions.

Web Connections

The Review Questions and Problems also appear at the following URLs:
http://universe.colorado.edu/ch10/questions.html
http://universe.colorado.edu/ch10/problems.html

Observing Jupiter and Saturn

ACTIVITY

The two largest of the giant planets are readily observable without sophisticated equipment, and each is situated favorably for nighttime observations at some time during each year. As the faster-moving Earth "catches up" with one of the giants, at first the planet will be up in the nighttime sky only after midnight, but as the Earth pulls even with the giant and then passes it by, the planet will become visible in the evening after sunset. You can predict when this situation will occur by noting the times of opposition for Jupiter and Saturn in the accompanying table. The best period for evening observations will be during the three months or so following opposition; but if you are an early-morning person, you can observe the planets before sunrise two or three months before opposition.

You can easily see Jupiter and Saturn without binoculars or a small telescope, but will be unable to make out any detail. In this case you can observe the motions of the planets by noting their positions relative to the fixed stars and then seeing how the positions change with time. Since Jupiter and Saturn move slowly, you will need weeks or months to notice significant changes. To do this, make a sketch each time you observe Jupiter and Saturn, showing the relative positions of a few nearby stars. In due course your sketches will begin to reveal the motions of the planets with respect to the stars. Which way do the planets appear to move? What do you find if you observe the motion through the time of opposition?

If you have access to a pair of binoculars or a small telescope, you can see some of the surface details, particularly on Jupiter, and you can see satellites as well. You should be able to see Saturn's rings, unless they happen to be aligned directly with the line of sight to the Earth; in that case their plane is so thin that they become invisible to us as we view them edge-on. For Jupiter, make a sketch of the relative positions of the four Galilean satellites, and see if you can identify them by noting their order going outward from the planet. If you take care to make these sketches as accurately as possible, and if you make a few during the course of each night you observe, you may be able to make your own estimates of the satellites' orbital periods. This will work best if you observe for several hours during each of several consecutive nights, perhaps an arduous task (but something that astronomers do routinely!). Such a strategy works best because the orbital periods of these satellites are as short as a few hours, and it is necessary to measure the position several times per orbit to see how long it takes for the satellite to go around Jupiter once.

Dates of Oppositions	
Jupiter	**Saturn**
July 4, 1996	September 26, 1996
August 9, 1997	October 10, 1997
September 16, 1998	October 23, 1998
October 23, 1999	November 6, 1999
November 28, 2000	November 19, 2000

What would the density of Io be if its volume were doubled by adding ice to it? (Ice has a density of approximately 1 g/cm³.) (*Note:* it might also be interesting to compute Io's radius in this case and compare it with the radii of the largest satellites, such as Ganymede and Titan.)

4. Calculate the position (distance from the center of Saturn) of the Cassini division, assuming it is the location where ring particles would have exactly half of the orbital period of Mimas. (*Hint:* Use Kepler's third law.)

5. Calculate the escape speed and the surface gravity on Titan, and compare your answers with the values for the Earth. Do your answers explain why Titan has an atmosphere that is thicker and deeper than that of the Earth?

6. Compare the energy absorbed by Jupiter from the Sun with the energy Jupiter emits. To find the energy absorbed, use the relation

$$E = \pi R^2 I (1 - A),$$

where πR^2 is the cross-sectional area of Jupiter (R is its radius), I is the intensity of sunlight at Jupiter's distance from the Sun, and A is the albedo of Jupiter (so the quantity $(1 - A)$ is the fraction of incident sunlight that is absorbed). The solar intensity I can be found by dividing the Sun's luminosity in watts by the area of a sphere whose radius is equal to the Sun-Jupiter distance; i.e., divide L for the Sun by $4\pi a^2$, where a is the semimajor axis of Jupiter's orbit. To find the energy emitted by Jupiter, use the Stefan-Boltzmann equation (Chapter 6). Discuss the result of your calculation.

7. The eccentricity of Pluto's orbit is 0.206. Use this to calculate the ratio of perihelion distance to aphelion distance, and hence the ratio of the intensity of sunlight on Pluto at perihelion and aphelion. Comment on the role your answer plays in the behavior of Pluto's atmosphere during its orbital cycle.

Additional Readings

Beatty, J. K. 1995. Ida and Company. *Sky & Telescope* 89(1):20.

Beatty, J. K. 1996. Into the Giant. *Sky & Telescope* 91(4):20.

Beatty, J. K. and S. J. Goldman. 1994. The Great Crash of 1994; A First Report. *Sky & Telescope* 88(4):18.

Beatty, J. K. and D. H. Levy. 1995. Crashes to Ashes: A Comet's Demise. *Sky & Telescope* 90(4):18.

Beatty, J. K., B. O'Leary, and A. Chaikin 1990. *The New Solar System*. 2nd Ed. Cambridge University Press.

Binzel, R. 1990. Pluto. *Scientific American* 262(6):50.

Burnham, R. 1994. Pluto and Charon: At the Edge of Night. *Astronomy* 22(1):40.

Esposito, L. 1988. The Changing Shape of Planetary Rings. *Astronomy* 15(9):6.

Johnson, T.V. 1995. The Galileo Mission. *Scientific American* 273(6):44.

Lunine, J. I. 1996. Neptune at 150. *Sky & Telescope* 92(3):38.

Miner, E. D. 1990. Voyager 2's Encounters with the Gas Giants. *Physics Today* 43(7):40.

O'Meara, S. J. 1994. The Great Dark Spots Jupiter. *Sky & Telescope* 88(5):30.

Sagan, C. 1995. The First New Planet. *Astronomy* 23(3):34.

Schenk, P. M. 1995. The Mountains of Io. *Astronomy* 23(1):46.

Sheehan, W. and R. Baum 1996. Neptune's Discovery 150 Years Later. *Astronomy* 24(9):42.

Tombaugh, C. W. 1979. The Search for the Ninth Planet. *Mercury* 8(1):4.

Weisman, P. and M. Segura. 1996. *Galileo* Arrives at Jupiter. *Astronomy* 24(1):36.

Chapter 11
Interplanetary Matter

ASTEROIDS, COMETS, AND METEORITES

Chapter Web site: http://universe.colorado.edu/ch11

*Meteors come from somewhere in space
that is a million times more organic than
the Earth itself.*
Lyall Watson, 1979

The planets are the dominant objects in the solar system (except, of course, for the Sun), but they are not entirely alone as they follow their regular paths through space. The abundant craters on planetary and satellite surfaces have shown us that there must have been a time when interplanetary rocks and gravel were more abundant than they are today, covering any exposed surface with impact craters. Today the rate of cratering is much lower than it once was, but some vestiges of the space debris that caused it still remain, orbiting the Sun and occasionally becoming obvious to us as they pass near the Earth or enter its atmosphere.

Overview: The Evolutionary Impact

The formation of the solar system was long considered to have been an orderly process in which each step was a natural consequence of the one before. But in our discussions of the properties of the individual planets, we came across many inexplicable phenomena, planetary puzzles that did not fit into the "evolutionary" theory of solar system formation. We found that the Earth has a satellite whose size and orbital radius cannot be explained by any orderly formation process; we learned that Venus has a slow, backward spin unlike the rotations of nearly all of the other planets; we puzzled over the enormous core and thin mantle of Mercury; and we found that Uranus is tipped over on its side. In each of these cases, we invoked a collision, usually involving a large planetesimal, to explain the observed phenomena.

On the Earth, there is evidence of many large collisions in the dim past. Despite the erosional processes that soon eliminate craters, hundreds of impact sites have been identified around the world. One of these cosmic catastrophes may have temporarily modified the global climate so drastically that it forced the extinction of the dinosaurs; other mass extinctions farther in the past may also have been triggered by impacts. Every solar system body with a hard surface is covered with impact craters. In July 1994, a fractured comet crashed into Jupiter, reminding us all of the role of impacts in solar system evolution. This event also helped to remind us that even in this late stage in solar system evolution, impacts still occur and that no planetary body, including the Earth, is immune.

The construction of the planets themselves was brought about by the collisions and mergers of smaller bodies, which in turn had built up through condensation and accretion. When the planets formed, the solar system was still aswarm with free-flying bodies of all sizes, and some planets experi-enced major collisions with large objects (accounting for the peculiarities listed above) while all were subjected to an incessant bombardment by smaller bodies from space. The earliest atmospheres of the planets may have come from accreted debris in volumes at least equal to the quantities outgassed from the planetary interiors. The original water supplies of the terrestrial planets may well have resulted from a "rainfall" of comets, rather than from the escape of water vapor from the planetary interiors.

Today the density of interplanetary bodies—thus the frequency of collisions—is far lower than in the early times of the solar system. Nevertheless, many forms of interplanetary debris remain, and our understanding of the solar system is incomplete until we have explored these remnants of the system's origin.

The most significant of the interplanetary bodies are the comets and the asteroids. Both are thought to be planetesimals that were unable to form into planets. The major distinction is that comets are icy objects that inhabit the far reaches of the solar system, far beyond the orbits of even the outermost planets (except for occasional forays into the solar neighborhood), while asteroids are rocky or metallic bodies whose normal home is within the realm of the planets. Another class of interplanetary bodies, related in a way to the comets, is made up of the icy planetesimals recently discovered to occupy a disklike zone extending from about 30 AU to well beyond the orbit of Pluto.

Meteors and meteorites are fragments of comets and asteroids that reach the Earth, or at least its atmosphere. In addition to the relatively large cometary and asteroidal bodies in interplanetary space, we also find a medium of very fine particles, called interplanetary dust, that is largely confined to the plane of the ecliptic.

Except for the dust (which is replenished by comets and asteroids), all of these forms of interplanetary debris are old, very old. They represent pristine material dating back to the time of solar system formation and thus have much to tell us about how the system formed and what it was like originally. Thus not only do interplanetary bodies influence the evolution of the entire system, but they also provide information about the past, giving us clues about how that evolution has proceeded.

The Minor Planets

As we have already seen, some of the early students of planetary motions were concerned with the distances of the planets from the Sun. Johannes Kepler, for example, was very interested in finding a mathematical relationship that would predict the positions of the planets, and he pursued a number of false leads (such as his idea that the distances could be represented by geometric solids) before discovering his third law of planetary motion. Kepler's third law can be used to predict the period of a planet, given its distance from the Sun; or its distance, if its period is known. But the law did not provide any underlying basis for explaining why planets are found only at certain distances from the Sun.

Discovery: Filling the Gap between Mars and Jupiter

In 1766 a German astronomer named J. D. Titius found a simple mathematical relationship that seemed to accomplish what Kepler had set out to do. Titius discovered that if we start with the sequence of numbers 0, 3, 6, 12, 24, 48, and 96 (obtained by doubling each one in order), then add 4 to each and divide by 10, we end up with the numbers 0.4, 0.7, 1.0, 1.6, 2.8, 5.2, and 10.0, which correspond closely to the observed planetary distances in astronomical units. A few years after Titius found this numerological device, it was popularized by another German astronomer, Johann Bode, and eventually became known as **Bode's law,** or the **Titius-Bode relation.**

The sequence of numbers dictated by Bode's law included one, 2.8 AU, where no planet was known to exist. The discovery of Uranus in 1781 and the recognition that its distance fits the sequence (the next number is $(192 + 4)/10 = 19.6$, and the semimajor axis of Uranus's orbit is 19.2 AU) aroused a great deal of interest in the "missing" planet at 2.8 AU, since Bode's law was doing so well in fitting the positions of the others.

A deliberate search for such a planet began in 1800, but the sought-after object was discovered accidentally on the night of January 1, 1801, when an Italian astronomer named Giuseppe Piazzi noticed a new object and within weeks found from its motion that it was probably a solar system body. When the orbit of the new object was calculated, its semimajor axis turned out to be 2.77 AU. Thus the object was in solar orbit between Mars and Jupiter, where Bode's

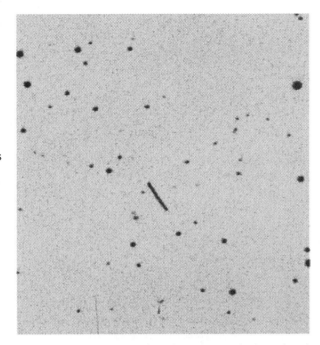

Figure 11.1 Asteroid motion. The elongated image in this photo is the trail made by an asteroid that moved relative to the fixed stars during the 20-minute exposure. Most new asteroids are discovered on photos of this type, although the first few were spotted visually. (Eleanor F. Helen, Palomar Observatory)

law had predicted that a new planet might be found. The new planet was named Ceres.

A little over a year after the discovery of Ceres, a second object was found orbiting the Sun at approximately the same distance and was named Pallas. It was clear from their faintness that both Ceres and Pallas were very small bodies and not full-scale planets. By 1807, two more of these asteroids, as they were then being called, had been found and were designated Juno and Vesta. A fifth, Astrea, was discovered in 1845, and in the next decades, vast numbers of these objects began to turn up. The process of finding them became much more efficient when photographic techniques began to be used. The orbital motion of a minor planet or asteroid causes it to leave a trail on a long-exposure photograph **(Fig. 11.1).**

Observed Properties

Today thousands of asteroids are known, with almost 5,000 of them sufficiently well observed to have had their orbits calculated and logged in catalogs. The total number is probably much higher, perhaps 100,000. The vast majority have orbits lying between the orbits of Mars and Jupiter **(Fig. 11.2).**

During the past 20 years, a number of techniques have been developed for determining the

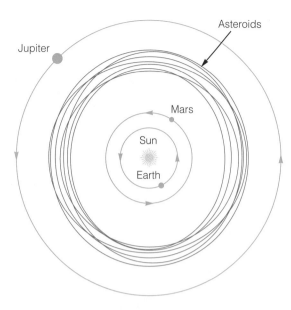

Figure 11.2 Asteroid orbits. This sketch shows typical orbital paths for the majority of asteroids, which orbit the Sun between Mars and Jupiter.

Figure 11.3 A portrait of an asteroid. This *Galileo* image of the asteroid Ida contained a surprise: a satellite. The tiny body at right, named Dactyl, orbits Ida. The existence of double asteroids had been suspected from earth-based observations, but this dramatic photo provided the proof. The asteroid itself appears to be quite typical, in that it is irregular in shape and is marked by impact craters. (NASA/JPL) http://universe.colorado.edu/fig/11-3.html

properties of asteroids. Most of these techniques involve the measurement of sunlight that is reflected from the surfaces of the asteroids. The brightness of an asteroid can reveal information on its surface reflectivity (albedo, or the fraction of light that is reflected) and on the size of the reflecting body. Most asteroids are irregular in shape, and as they rotate, their reflected brightness varies. Analysis of the variations in brightness can yield information on the shape of an asteroid and the orientation of its rotation axis.

Another technique for establishing the sizes and shapes of asteroids is the observation of stellar occultations, times when asteroids happen to pass in front of background stars. When this happens, the star is temporarily obscured; measurement of the length of time the star is invisible, coupled with the known orbital speed of the asteroid, provides a measure of the size of the asteroid. Today astronomers use extensive computer calculations to predict when known asteroids will pass in front of stars. Asteroid sizes and shapes can also be determined from radar measurements, a technique now being applied to many asteroids by using the giant Arecibo radio telescope (see Chapter 6); by performing interferometric observations (again, see Chapter 6); and by making direct observations of asteroids that pass sufficiently close to the Earth or are visited by spacecraft **(Fig. 11.3).**

The launch of the *IRAS* satellite in 1983 signaled a new era in the study of asteroid sizes. *IRAS* mapped nearly the entire sky in far-infrared wave-

lengths, where cold bodies such as asteroids glow due to their own surface temperature (see the discussion of thermal emission in Chapter 6). The Stefan-Boltzmann law relates the luminosity of a glowing object to its surface temperature and its total surface area (recall that the luminosity, or total power emitted, is proportional to the surface area and to the fourth power of the surface temperature). The infrared data from *IRAS* provided the luminosity and the surface temperature through the application of Wien's law, making it possible to solve for the surface area. Then the diameter can be computed if the shape of the asteroid is known; for simplicity, it is usually assumed to be spherical, with the result that the diameter is only a representative dimension, not an exact figure. A by-product of these studies is an estimate of the surface albedo in visible wavelengths of light. Once the total surface area is known, the albedo can be derived by comparing the amount of energy absorbed (as indicated by the infrared luminosity) with the amount reflected (as indicated by direct brightness measurements).

From these assorted techniques, astronomers have established that asteroids have a wide range of diameters **(Table 11.1),** including a few as large as several hundred kilometers. Ceres, the largest, is nearly 1,000 km in diameter, but the majority are rather small, having diameters of 100 km or less. The largest are apparently spherical, while the smaller ones are often jagged, irregular chunks of material; their irregularity is indicated by their variations in brightness as they spin. A

Table 11.1

Selected Asteroids

No.	Name	Year of Discovery	Diameter (km)	Mass (g)	Period (years)[a]	Distance (AU)[a]
1	Ceres	1801	933	1×10^{24}	4.60	2.766
2	Pallas	1802	538	3×10^{23}	4.61	2.768
3	Juno	1804	200	2×10^{22}	4.36	2.668
4	Vesta	1807	561	2×10^{23}	3.63	2.362
6	Hebe	1847	220	2×10^{22}	3.78	2.426
7	Iris	1847	200	2×10^{22}	3.68	2.386
10	Hygeia	1849	320	6×10^{22}	5.59	3.151
15	Eunomia	1851	280	4×10^{22}	4.30	2.643
16	Psyche	1852	280	4×10^{22}	5.00	2.923
51	Nemausa	1858	80	9×10^{20}	3.64	2.366
511	Davida	1903	260	3×10^{22}	5.67	3.190

[a]Periods are orbital periods; distances are orbital semimajor axes.

few asteroids are known to be binary, consisting of two objects that orbit each other as they circle the Sun, and one is known to have a satellite (see Fig. 11.3).

Asteroid compositions can be determined by analyzing visible light that is reflected from their surfaces. Atoms and molecules that are bound into solid surfaces retain their ability to absorb light only at specific wavelengths, just as free atoms or molecules do (see the discussion of spectral lines in Chapter 6). In solids, however, the mixture of different types of particles, along with their close proximity in the material, leads to much broader spectral features than the narrow lines that characterize free particles. Thus the spectrum of light reflected from an asteroid's surface has several broad regions where less light is reflected due to absorption by surface compounds. If these broad absorption features can be identified with known mineral compounds, the composition of the surface can be determined. Many of the features seen in asteroid spectra have been identified, at least as belonging to a broad class of minerals, although some remain unidentified.

When spectra of large numbers of asteroids began to be observed, researchers noticed that they tended to fall into distinct classes, based on the wavelengths and strengths of the broad absorption bands. These classes were assigned letters of the alphabet, based (where known) on the composition they indicated. Thus, the M asteroids are metallic, the C class is carbonaceous, and so on. More refined measurements have revealed many more classes than can easily be lettered according to composition, however, so the modern classification system includes many letters that have no obvious connection with the overall chemical makeup.

The C-class (carbonaceous) asteroids are the most common. About three-fourths belong to this group, with most of the rest falling into an assortment of classes containing metals and silicates. As we will see in a later section of this chapter, the meteorites are thought to be fragments of asteroids that have been fractured in collisions. Thus the compositions of meteorites provide detailed information on the composition of the asteroids. Overall the results of laboratory studies of meteorites and the inferred compositions of asteroids agree, with one major exception: very few asteroids are found with the same composition as the most common meteorites. These meteorites are chondrites, composed of rocky material with inclusions indicating that they formed under conditions of low temperature. Asteroid researchers today are trying to solve this mystery by finding the asteroids from which the chondritic meteorites originated. One suggestion is that the parent bodies are simply too small to have been detected in asteroid searches, possibly because only very small bodies could have avoided being heated during their formation during the era of planetesimal accumulation in the early solar system.

Kirkwood's Gaps: Orbital Resonances Revisited

As increasing numbers of asteroids were discovered and cataloged throughout the nineteenth century, calculations of their orbits showed remarkable gaps at certain distances from the Sun. One such gap is at 3.28 AU, and another is at 2.50 AU.

COMETS

Comet Orbit

Nucleus
Comet Halley

Tail

Nucleus

Coma

Comet West

Comet Halley

Comet Hyakutake

Comet Hale-Bopp

Ida

Gaspra

ASTEROIDS

Stones

Irons

Stony Irons

Asteroid Belt

Jupiter

Mars
Meteorites

Dactyl

Willamette
Meteorite

Hoba
Meteorite

Radiants

METEORS

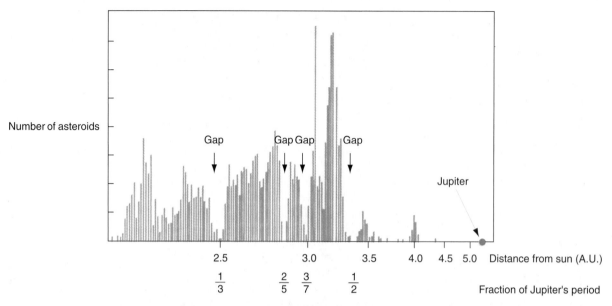

Figure 11.4 **Kirkwood's gaps.** This graph shows the distribution of asteroid orbits. Gaps appear at the distances from the Sun where asteroids would have exactly one-half, one-third, or other simple fractions of the orbital period of Jupiter.

In 1866 Daniel Kirkwood realized that these distances correspond to orbital periods that are simple fractions of the period of Jupiter **(Fig. 11.4).** The giant planet, at its distance of 5.2 AU from the Sun, takes 11.86 years to make a trip around it, whereas an asteroid at 3.28 AU, if one existed there, would have a period exactly half as long, 5.93 years. Thus, every time Jupiter made an orbit, this asteroid and the planet would be lined up in the same way, and Jupiter's gravity would subject the asteroid to regular tugs. Apparently, this effect has prevented asteroids from staying in orbits at this and other distances where the orbital periods would result in regular alignments. The gap at 2.50 AU corresponds to orbits with a period one-third that of Jupiter. These orbital resonances are exactly analogous to the ones created by the moons of Saturn, which are responsible for some of the gaps in the ring system of that planet (see Chapter 10). Since Jupiter is the most massive of the outer planets and the nearest to the asteroid belt, it has by far the greatest effect in producing gaps, but in principle the other outer planets could do the same thing.

The Origin of the Asteroids

For a long time the most natural explanation of the origin of the asteroids seemed to be the breakup of a former planet into a swarm of fragments that continued to orbit the Sun. This argument was weakened when the total mass of the asteroids was estimated and found to be much less than that of any ordinary planet (the total mass of the asteroids is estimated to

be only about 0.04% of the mass of the Earth). Another rather strong argument against the planetary remnant hypothesis is that scientists know of no reasonable way for a planet to break apart once it has formed, whereas it is easy to understand why material orbiting the Sun at the position of the asteroid belt could never have combined into a planet in the first place. It became simpler to accept the idea that the debris never was part of a planet than to explain how it first formed a planet and then broke apart.

Thus it is now considered likely that the asteroids represent material from the early solar system that never coalesced to form planets. As we have seen, the planets are thought to have formed from smaller bodies (planetesimals) that collided and merged. According to computer simulations of this process, the construction of the planets took place rather quickly and was especially rapid in the outer solar system. Thus Jupiter and Saturn accumulated large masses before the terrestrial planets had coalesced, and for a period of time, the inner solar system contained mainly small bodies while at least the two largest gaseous giant planets had already grown to nearly their present sizes. Once the two giants had formed, their gravitational forces (particularly that of Jupiter) strongly influenced the further evolution of the planetesimals near them.

Jupiter's gravity accelerated the motion of planetesimals that came near, causing these bodies to have relative speeds far too high for them to collide gently and stick together. Thus the planetesimals just inside the orbit of Jupiter, extending all the way inward to the orbit of Mars, were unable to coalesce to form a planet. Today's asteroids, therefore, are yesterday's planetesimals, and their study affords us an opportunity to learn a great deal about the early solar system.

Typical relative speeds measured today within the asteroid belt are around 5 km/sec. Not only do these speeds prevent the asteroids from forming a planet, but they also cause them to collide and occasionally to break apart. These collisions are thought to be the source of meteorites that reach the Earth, and over the billions of years since the solar system formed, they have reduced the total mass of the asteroid belt by a factor between 3 and 5 (many of the fragments escape entirely due to gravitational perturbations by Jupiter).

It is interesting that some of the asteroids are almost purely metallic, as are the meteorites that come from them. This composition implies that some of the original planetesimals were large enough to have undergone differentiation, forming metallic cores and rocky mantles and crusts. These large bodies were sufficiently heated during their formation to have melted internally, allowing the heavier metallic elements to separate out and sink to the core and harden before the planetesimals underwent the collisions that destroyed them and dispersed the fragments throughout the solar system.

The asteroids are not uniformly mixed throughout the asteroid belt, but instead tend to fall into zones where different classes of asteroids are dominant. This distribution must reflect conditions that existed in the early solar system. For example, the inner asteroid belt (around 2 AU) is dominated by E- and S-class asteroids, which are made of silicate and metallic materials; these materials, which are thought to resemble igneous rocks on the Earth, require substantial heating for their formation. In the central asteroid belt, which peaks around 3 AU, the C asteroids, consisting of carbonaceous materials, are most common, and they persist all the way out to 5 AU, near the orbit of Jupiter. In these outer regions the D and P asteroids, consisting of assorted carbonaceous and organic silicate compounds, become the most common types. These are more primitive materials, which would not have survived substantial heating.

This distribution of compositions within the asteroid belt can be understood in the context of modern theories of solar system formation. It is thought that the compounds that condensed into solid form at any given distance from the Sun were determined by the temperature at that distance. At high temperatures, for example, metallic elements can condense while the lighter elements cannot. At intermediate temperatures some carbonaceous materials can condense, while the organic silicate mixtures

can condense only at low temperatures. Thus the observed range of compositions within the asteroid belt tells us something about the distribution of temperatures as a function of distance from the Sun in the early solar system. Not surprisingly, the distribution indicates that it was hotter in the inner regions and colder in the outer portions. The rather rapid decrease in temperature from the inner portion of the asteroid belt to the outer region may have been the result of heating due to the impact of the outflowing wind from the Sun in its early formation.

Apollos and Trojans

In addition to being classified according to their compositions, some groups of asteroids have been identified on the basis of their orbital properties. The **Apollo** and **Trojan asteroids** are two such groups.

The Apollos are probably of more immediate interest to Earth-dwellers because the orbits of these bodies pass within 1 AU of the Sun, therefore raising the possibility of collision with the Earth. The impact of an asteroidal body on the Earth would be truly catastrophic.

More than 50 Apollo asteroids have been identified, and new ones are discovered from time to time (including one in 1995). There may be several hundred of them. Most are relatively small for asteroids, but this still means they are rocky or metallic bodies with dimensions of a few kilometers and masses sufficient to transfer tremendous quantities of energy to any body they might strike.

It is inevitable that eventually each of the Apollo asteroids will either collide with a planet or nearly do so and be ejected from the inner solar system as a result. On average, an Apollo asteroid may be expected to undergo such an encounter with a planet every few hundred million years. Since the solar system is 4.5 billion years old, all of the primordial Apollo asteroids should have been ejected or destroyed by collisions long ago. Therefore the Apollo objects we see today must have a more recent origin. Most likely the supply is replenished by collisions among asteroids in the main belt between Mars and Jupiter. Such a collision can disturb an asteroid's orbit so that it falls into the inner solar system. Some additional Apollo objects might be dead cometary nuclei, which would be composed of carbonaceous material left behind after all of the ices have evaporated.

The Trojan asteroids are recognized as a group because they occupy special positions in the orbit of Jupiter. Like the co-orbital satellites of Saturn, these objects have become trapped in positions, called **Lagrangian points,** which lie 60° ahead of or behind a far more massive body (Jupiter in this case; a larger satellite of Saturn in the case of the co-orbital moons). At the Lagrangian points in the orbit of Jupiter, the gravitational forces due to Jupiter and the Sun combine to trap the smaller bodies so that they are locked into position. Any wandering object that comes near will be similarly trapped, and a collection of objects has built up. As many as 200 Trojan asteroids have been identified, swarming about in small clumps at the two Lagrangian points in Jupiter's orbit.

Comets: Messengers from the Past

Among the most spectacular of all the celestial sights are the comets. Their brightly glowing heads and long, streaming tails, along with their infrequent and often unpredictable appearances, have sparked the imagination (and often the fears) of people through the ages. In antiquity, when astrological omens were taken very seriously, great import was attached to a cometary appearance **(Fig. 11.5).** Ancient descriptions of comets are numerous, and in many cases observers associated these objects with catastrophe and suffering.

Included among the teachings of Aristotle was the notion that comets were phenomena in the Earth's atmosphere. Aristotle had no good evidence for this idea and apparently adopted it because he believed that the heavens were perfect and immutable; therefore any changeable phenomena were associated with the Earth. In any case, this idea was accepted for centuries to come. In 1577 Tycho Brahe was able to prove that comets were too distant to be associated with the Earth's atmosphere because they do not exhibit any parallax when viewed from different positions on the Earth. If a comet were really located only a few kilometers or even a few hundred kilometers above the surface of the Earth, it would be seen at different positions relative to the background stars, as seen from different points on the Earth. Tycho was able to show that this was not the case and that therefore comets belonged to the realm of space.

Halley, Oort, and Cometary Orbits

A major advance in the understanding of comets was made by a contemporary and friend of Newton, Edmond Halley. Aware of the power of Newton's laws of motion and gravitation, Halley reviewed the records of cometary appearances and noted one outstanding regularity. Particularly bright comets seen in

Figure 11.5 Calamity on Earth associated with the passage of comets. This drawing is from a seventeenth-century book describing the universe. (The Granger Collection)

1531, 1607, and 1682 seemed to have similar properties, and Halley suggested that in fact all three were appearances of the same comet, orbiting the Sun with a 76-year period.

Calculations using Kepler's third law showed that for a period of 76 years, this object must have a semimajor axis of nearly 18 AU. Halley realized, therefore, that to appear as dominant in our skies as the comet does, it must have a highly elongated orbit **(Fig. 11.6),** that brings it close to the Sun at times, even though its average distance is well beyond the orbit of Saturn. Such an eccentric orbit, as a very elongated ellipse is called, had not previously been observed, even though Newton's laws clearly allowed the possibility.

Since Halley's time, searches of ancient reports of comets have revealed that Halley's comet has been making regular appearances for many centuries. The earliest records are provided by the ancient Chinese astronomers, who apparently observed its every appearance for well over 1,000 years, possibly beginning as early as the fifth century B.C.

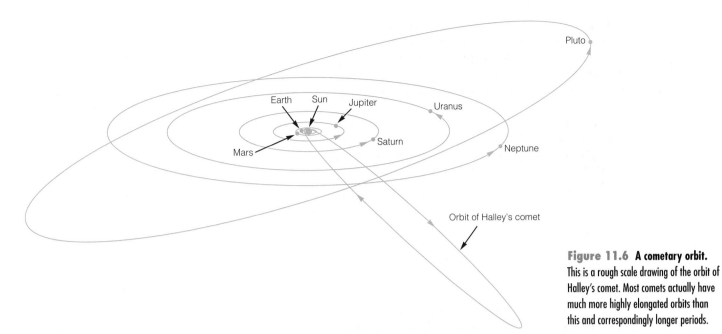

Figure 11.6 A cometary orbit. This is a rough scale drawing of the orbit of Halley's comet. Most comets actually have much more highly elongated orbits than this and correspondingly longer periods.

The most recent visit of Halley's comet was in 1985 and 1986, when it did not pass very close to the Earth, and the best views were from the Southern Hemisphere **(Fig. 11.7).** A much more spectacular appearance occurred on the comet's previous visit in 1910 **(Fig. 11.8),** when the Earth actually passed through the rarefied gases of its tail. This will happen again during the 2062 appearance of Halley. During the apparition of 1985–1986, several spacecraft were sent to encounter Halley, including the European *Giotto* mission, two Soviet *Vega* probes, and two Japanese spacecraft. In addition, two probes (the Venus-orbiting *Pioneer-Venus* mission and an interplanetary particle probe, which was renamed *Inter-*

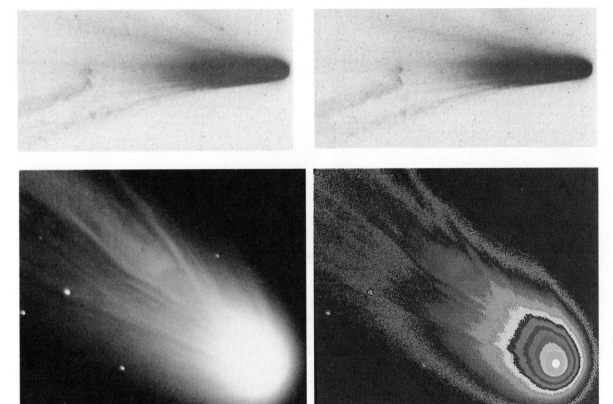

Figure 11.7 Halley's comet in 1986. These are a few of the fine photographs made from Earth during the apparition of Halley's comet in 1985 and 1986. The two negative images on top, obtained only days apart, show changes in the detailed structure of the tail, including a disconnection event, as part of the ion tail broke free of the comet (top right). The two images at the bottom were both made from the same photograph, the color-coded version being made to enhance slight brightness variations. (top: © 1986 Royal Observatory, Edinburgh; bottom: U. Fink, University of Arizona)

http://universe.colorado.edu/fig/11-7.html

Figure 11.8 **Halley's comet as it appeared in 1910.** (Palomar Observatory, California Institute of Technology)

national Cometary Explorer for the occasion), launched previously for other purposes, were able to observe Halley from space. In the end astronomers gained far more information on comets from the Halley encounters and observations than had been gathered over all the preceding centuries.

Other spectacular comets have been seen **(Fig. 11.9),** and a number have rivaled Halley's comet in brightness. Traditionally, a comet is named after its discoverer, and there are astronomers around the world who spend long hours peering at the night-time sky through telescopes, looking for a piece of immortality. Comet Hyakutake appeared in early 1996, and a comet discovered in 1995, called Comet Hale-Bopp, shows promise of being one of the brightest in several decades when it reaches the inner solar system in 1997 (**Fig. 11.10** shows it as it appeared in 1996).

When a new comet is discovered, a few observations of its position are sufficient to allow computation of its orbit. The results of many years of comet watching have shown that many comets have orbits so incredibly stretched out that their periods are measured in thousands or even millions of years. These comets, for all practical purposes, are seen only once. They return thereafter to the void of space well beyond the orbit of Pluto, there to spend millennia before visiting the inner solar system again.

Figure 11.9 **Comet Hyakutake, 1996.** This is one of the many fine photographs of Comet Hyakutake that were taken around the time of the comet's closest approach to the Earth, during late March, 1996. (R. W. Doty)

Figure 11.10 **Comet Hale-Bopp.** This is a recent photo of Comet Hale-Bopp on its approach to Sun's vicinity. The nucleus of this comet is thought to rival that of Halley's Comet in size, and there already has been a significant amount of outgassing from this first-time visitor to the inner solar system. In this image we can see several distinct rays of gas, or tails. Comet Hale-Bopp is expected to be one of the brightest comets in recent decades when it reaches its most favorable position for viewing from the Earth, in the Spring of 1996. (National Astronomical Observatory of Japan) http://universe.colorado.edu/fig/11-10.html

Table 11.2
Selected Periodic Comets

Comet[a]	Period (years)	Semimajor Axis (AU)	Year of Next Appearance
Temple (2)	5.26	3.0	1999.5
Schwassmann-Wachmann (2)	6.52	3.50	2000.8
Encke	3.30	2.21	1997.6
Wirtanen	6.65	3.55	2001.3
Reinmuth (2)	6.72	3.6	2001.2
Finlay	6.88	3.6	2002.0
Borrelly	7.00	3.67	2002.5
Whipple	7.44	3.80	2000.5
Oterma	7.89	3.96	1997.9
Schaumasse	8.18	4.05	2001.2
Wolf	8.42	4.15	2001.3
Comas Sola	8.58	4.19	2004.2
Vaisala	10.5	4.79	2002.4
Schwassmann-Wachmann (1)	16.1	6.4	1998.6
Neujmin (1)	17.9	6.8	2002.7
Crommelin	27.9	9.2	2012.6
Olbers	69	16.8	2025
Pons-Brooks	71	17.2	2025
Halley	76.1	18.0	2062.5

[a]A number in parentheses following the name of a comet indicates cases where a single observer has discovered more than one comet.

The orbits of these so-called **long-period comets** are randomly oriented. These comets do not show any preference for orbits lying in the plane of the ecliptic, in strong contrast to the planets, and about half go around the Sun in the retrograde direction. Consideration of these orbital characteristics, especially the large orbital sizes, led Dutch astronomer Jan Oort to suggest that all comets originate in a cloud of objects that surrounds the solar system. He envisioned the **Oort cloud,** as it is now called, as a spherical shell with a radius of about 100,000 AU, extending a significant fraction of the distance to the nearest star, which is almost 300,000 AU from the Sun. Modern calculations show that the Oort cloud lies primarily between 1,000 and 30,000 AU from the Sun, although its tenuous outer portions may extend as far out as Oort thought.

Occasionally a piece of debris from the Oort cloud is disturbed from its normal path, either by a collision with another object or perhaps by the gravitational tug of a nearby star, and it begins to fall inward toward the Sun. If undisturbed by other forces, a comet falling in from the Oort cloud would follow a highly elongated orbit with a period of millions of years, appearing to us as one of the long-period comets when it made its brief, incandescent passage near the Sun. In many cases, a comet is not left undisturbed, however. Instead it runs afoul of the gravitational pull of one of the giant planets—most often Jupiter. When this happens, the comet may be speeded up, causing it to escape the solar system entirely after it loops around the Sun. Or it may be slowed down, causing it to drop into a smaller orbit with a shorter period. In that case it becomes one of the numerous **short-period,** or **periodic,** comets, that reappear frequently **(Table 11.2).**

Detailed simulations of the process by which comets fall into the inner solar system from the Oort cloud show that most of the short-period comets, whose orbits tend to lie close to the plane of the planetary orbits, cannot originate in a spherical Oort cloud, however. Instead the periodic comets must arise in a region that is disk-shaped, so that they start out with orbits that are aligned with the planetary orbits. This led astronomers to deduce the existence of an inner cloud of comets, called the **Kuiper belt** after the American astronomer Gerard Kuiper, lying just outside the orbits of the outer planets. The existence of the Kuiper belt has recently been confirmed through very sensitive searches made from the ground and through the use of the *Hubble Space Telescope*. By late 1995, at least 30 objects beyond the

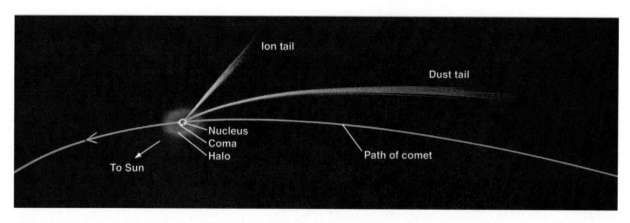

Figure 11.11 **Comet structure.** This sketch illustrates the principal features of a comet, though not necessarily to correct scale.

orbit of Pluto had been identified. *Hubble* is able to detect bodies as small as a few kilometers in size, whereas those visible to ground-based telescopes must be tens or hundreds of kilometers in diameter. Pluto and Neptune's giant moon Triton are considered by some to be the largest of the Kuiper belt objects, which collectively represent planetesimals left over from the formation of the solar system.

Thus the modern picture of cometary origins distinguishes between the long-period comets, which have very elongated, randomly oriented orbits and periods of millions of years, and short-period comets, which have smaller orbits lying close to the plane of the planets. The long-period comets arise in the inner Oort cloud, a spherical shell concentrated between 3,000 and 30,000 AU from the Sun, while the short-period comets originate in the Kuiper belt, a disk some 30 to 50 AU in radius. The total mass in the Kuiper belt is estimated to be about equal 5 percent of the mass of the Earth, while the Oort cloud is thought to contain about 10^{11} objects totaling 30 or more Earth masses. It is noteworthy that each of these collections of Sun-orbiting bodies contains far more mass than the asteroid belt.

The Structure of a Comet

A remarkably accurate picture of the nature of a comet was developed many years ago by American astronomer Fred Whipple. Whipple's model, sometimes called the "dirty snowball," envisions that, for most of its life, a comet is just a frozen chunk of icy material, probably consisting of small particles like gravel or larger boulders embedded in frozen gases. This picture was developed on the basis of observations of many comets over the years, and its basic features were verified by the close-up examination of Halley's comet in early 1986, when several space probes were able to make observations from close range.

As a comet passes through the outer reaches of its orbit far from the Sun, it does not glow, has no tail, and is not visible from Earth. As it approaches the Sun, however, it begins to warm up as it absorbs sunlight, and the added heat causes volatile gases to escape. A spherical cloud of glowing gas called the **coma** develops around the solid **nucleus (Fig. 11.11).** The principal gases found in the coma of Halley's comet **(Table 11.3)** are water vapor (H_2O; about 80% by number), carbon dioxide (CO_2; about 3.5%), and carbon monoxide (CO; most of the remaining

Table 11.3

Detected Constituents of Comets

Coma		Ion Tail	
Atoms	Molecules	Ions	Molecular Ions
Hydrogen (H)	H_2O	C^+	CO^+
Oxygen (O)	CO_2	Ca^+	CO_2^+
Sulfur (S)	C_2	O^+	H_2O^+
Carbon (C)	C_3	S^+	OH^+
Sodium (Na)	CH	Fe^+	CH^+
Iron (Fe)	CN	Na^+	CN^+
Potassium (K)	CO		N_2^+
Calcium (Ca)	CS		H_2O^+
Vanadium (V)	NH		CS_2^+
Chromium (Cr)	OH		S_2^+
Manganese (Mn)	NH_2		CS^+
Cobalt (Co)	NCN		
Nickel (Ni)	CH_3CN		
Copper (Cu)	S_2		
	HCO		
	NH_3		

16.5%). Other slightly more complex molecules such as ammonia (NH_3) and methane (CH_4) are probably present also, along with hydrogen molecules (H_2). Note that water and simple carbon compounds constitute nearly all of the mass in these ejected gases; this indicates the composition of the nucleus and helps us to understand the deduction mentioned in Chapters 8 and 9 that the terrestrial planets may have gained significant amounts of water and carbon via cometary impacts.

The observed molecular species in the coma of a comet glow by a process called **fluorescence.** The molecules absorb light from the Sun, causing them to be excited to high energy levels; then they emit visible light as they return to low energy states. A cloud of hydrogen atoms, resulting from the breakup of the molecules in the coma, extends out to great distances from the nucleus. The visible coma may be as large as 100,000 km in diameter, while the halo of hydrogen atoms may extend as much as 10 times farther from the nucleus, as demonstrated by ultraviolet observations of Halley's comet in 1986. The solid nucleus is relatively tiny, having a diameter of a few kilometers.

The *Giotto* and *Vega* probes found the nucleus of Halley to be irregularly shaped, with a long axis of about 15 km and a width ranging up to about 10 km **(Fig. 11.12).** The mass of the nucleus is estimated to be about 10^{14} kg, and its density is extremely low—0.5 to 1.0 g/cm^3—less than the density of ordinary ice. The nucleus was found to have a coating of very dark material (albedo only 0.04), probably a matrix of carbonaceous dust left behind by the evacuation of ices that evaporated previously. Hence, the dirty snowball idea has been modified to incorporate a complete outer coating of dirt, along with the traditional picture of mixed dirt and ice in the interior of the nucleus.

Another modification of the picture is that the surface of the nucleus is quite hot instead of cold and icy. The high temperature had been suspected from Earth-based infrared measurements and was confirmed by the spacecraft observations made at close range. The high temperature is the result of the blackness of the crust; its low reflectivity means that sunlight is efficiently absorbed, causing heating.

Apparently, the gas escaping from the nucleus of a comet is released in jets that can be so powerful that they alter the orbital motion of the comet. Astronomers had known for years that comets can be slowed in this way as they approach the Sun, but the close encounters with Halley's comet provided dramatic confirmation; the images revealed spectacular fountains of gas and dust erupting from beneath the crust of the nucleus (see Fig. 11.12). Apparently

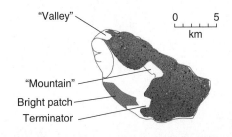

Figure 11.12 The nucleus of Halley's comet. The sketch, based on images obtained by the *Giotto* spacecraft, illustrates the size and shape, as well as some surface features, of the nucleus. The photograph is a composite of some 60 images obtained by the *Giotto* spacecraft as it flew past the nucleus of the comet. Detailed surface markings in the crust of the nucleus can be seen, as well as the bright jets of warm dust that were being ejected toward the Sun (to the left). (top: based on data from the European Space Agency; bottom: Max Planck Institute for Aeronomie, courtesy of A. Delamere, Bell Aerospace Corporation)

these gas jets are created by heat that is absorbed by the dark crust and conducted into the icy interior; there the heat causes ices to evaporate in a process called **sublimation,** in which solid ice is converted directly into gas. The released gases expand and create outward pressure, which is relieved as the gases break through the crust, forming the jets that are observed. As the nucleus rotates, jets are active only on the sunlit side, so each individual jet turns on and off as it rotates toward and then away from the Sun. Surprisingly, photos of Halley's comet in 1910 indicate that many of the jets occurred in the same locations then as they did in 1986. Apparently, the jets occur at weak points in the crust of the comet, creating craters that persist over the years between encounters with the Sun's warming radiation.

As a comet nears the Sun, the solar radiation and the solar wind force some of the gas from the coma

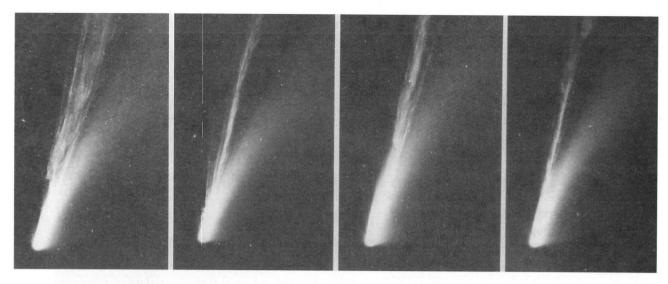

Figure 11.13 **Two tails.** These photos of Comet Mrkos (1957) show distinctly the two types of tails that characterize many comets. The ion tail points straight up in these views, while the dust tail curves to the right. (Palomar Observatory, California Institute of Technology)

to flow away from the Sun, forming the tail, which in some cases is as long as 1 AU. Often there are two distinct tails **(Fig. 11.13):** one, formed of gas from the coma, usually contains molecules that have been ionized, such as CO^+, N_2^+, CO_2^+, and CH^+; the other is formed of tiny solid particles released from the ice of the nucleus. The **ion tail,** the one formed of ionized gases, is shaped by the solar wind and therefore points almost exactly straight away from the Sun at all times. As Halley and other comets have demonstrated on many occasions, all or part of the ion tail can detach itself and dissipate and then be replaced by a new tail (see Fig. 11.7). These "disconnection events" may be related to variations in the Sun's magnetic field, which exert forces on charged particles in the ion tail.

The other tail, called the **dust tail,** usually takes on a curving shape, as the dust particles are pushed away from the Sun by the force of the light they absorb. This radiation pressure is not strong enough to force the dust particles into perfectly straight paths away from the Sun, so they follow curved trajectories due to a combination of their orbital inertia and the outward push caused by sunlight. The dust particles ejected from the nucleus are apparently quite fragile and readily break apart in collisions or when heated. Streams of gas resulting from the evaporation of icy dust particles were observed to extend as far as 50,000 km from the nucleus of Halley. Particles captured by the *Giotto* spacecraft appeared to be low-density, fluffy grains, and the farther from the nucleus they were captured, the smaller they tended to be. Infrared images obtained by the *IRAS* satellite revealed extensive trails of dust particles tracing out the orbits of several periodic comets. The quantity of

dust in these trails indicates that a typical cometary nucleus is at least half rock and dirt. In an old comet like Halley, which has made many passages near the sun where its gases were evaporated, the relative proportion of rock and dirt can be even higher.

One of the major discoveries made by the probes that encountered Halley's comet and Comet Giacobini-Zinner was that the ionized gases surrounding a comet create a trapped magnetic field. A "bow shock," analogous to a boundary of a planet's magnetosphere, builds up on the sunward side of a comet, where the solar wind encounters the trapped magnetic field and flows around it. In the direction away from the Sun, the magnetic field that is wrapped around the comet forms a sheet of trapped charged particles called a current sheet.

The gases that escape from the nucleus of a comet as it approaches the Sun are highly volatile and would not be present in the nucleus if it had ever undergone any significant heating. This confirms our expectation that comets must have formed and lived their entire lives in a very cold environment, probably never even getting as warm as 50 K before falling into orbits that bring them close to the Sun. If a comet is so easily vaporized, then once it has begun to follow a path that regularly brings it close to the Sun, its days are numbered. It may make many round trips, but eventually it will dissipate all of its volatile gases, leaving behind nothing but rocky debris. It has been estimated that Halley's comet was losing some 50 tons per second of water ice when it passed nearest the Sun in early 1986! Despite this high rate of mass loss, the comet should last at least another 100,000 years before the nucleus is entirely dissipated. There is some speculation that a comet as

Figure 11.14 **The breakup of Comet West.** This dramatic sequence shows the nucleus of Comet West fragmenting into four pieces. (New Mexico State University)

large as Halley may eventually build up a crust thick enough to halt further sublimation of its interior ices. If this is so, Halley's comet may be immortal, but it will eventually stop developing a coma and tail on its visits to the inner solar system. Several times, a comet has failed to reappear on schedule, but has been replaced by a few pieces or perhaps a swarm of fragments **(Fig. 11.14)**. In time, the remains of a dead comet are dispersed all along the orbital path, so that each time the Earth passes through this region, it encounters a vast number of tiny bits of gravel and dust, and we experience a meteor shower.

The Impact of Comet Shoemaker-Levy

In July 1994, a spectacular event occurred that brought comets into the forefront of the consciousness of both professional astronomers and the general public. An object named Comet Shoemaker-Levy 9 (or more simply Comet Shoemaker-Levy) for its discoverers (Eugene and Carolyn Shoemaker and David Levy; it was their ninth jointly discovered comet) crashed into Jupiter. This comet was a very unusual one, so unusual, in fact, that for a time after the impact questions remained about whether it might actually have been more like an asteroid in nature or perhaps a representative of a different class of outer solar system body altogether. The current consensus is that it was a comet.

Comet Shoemaker-Levy's unusual nature was noticed immediately after its discovery in 1993; rather than looking like a faint, fuzzy object as most comets do when still several AU from the Sun, this one seemed to be a stream of bodies, following each other in a common orbit **(Fig. 11.15)**. Analysis of the orbit quickly revealed the next peculiarity: Comet Shoemaker-Levy was orbiting Jupiter, not the Sun. Apparently, Jupiter's immense gravity had captured this comet into orbit. Unfortunately, by the time the comet was discovered, the orbit had been so altered that it was not possible to backtrack the comet's original trajectory and thereby deduce its origin. It was possible to backtrack its immediate past motion, however. This showed that the comet was in a highly eccentric orbit about Jupiter and that on its previous close approach to the planet, it had come very close. It was evident that Jupiter's tidal force had been responsible for breaking the comet into many fragments (more than 20 were counted) at that time.

But the most exciting result of the orbital calculations was the discovery that the comet was going to intersect Jupiter's surface on its next close passage. Thus the stage was set for the spectacular events of July 1994.

As the fateful encounter approached, images of the comet (including some from the *Hubble Space Telescope;* see Fig. 11.15) showed that the fragments

Figure 11.15 **Comet Shoemaker-Levy 9 before impact.** This *Hubble Space Telescope* image shows the string of fragments of Comet SL-9 as it approached Jupiter. Detailed analysis of photos like this, obtained over a period of time as the comet approached its fateful encounter, showed that some of the pieces themselves broke into fragments, which then slowly orbited about each other. (NASA/STScI) http://universe.colorado.edu/fig/11-15.html

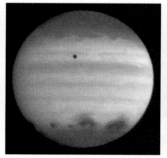

Figure 11.16 **The impact of Comet Shoemaker-Levy 9 on Jupiter.** These images include (clockwise form upper left): a combination of *Hubble Space Telescope* images, showing how the comet looked as it approached Jupiter just before the collisions began; an infrared image showing the bright impact zones of Comet SL-fragment Q; an ultraviolet HST image showing a series of dark blotches across Jupiter's southern hemisphere after most of the impacts had occurred; and an infrared image showing the spectacular hot spot created by the impact of fragment G. (NASA)

were oscillating about and gradually dispersing; there was some fear that many of them would end up missing Jupiter altogether. But this did not happen, and the string of comets began crashing into Jupiter right on schedule, on July 16, 1994. At that time, and for the ensuing several days, every telescope on Earth and in space was trained on Jupiter. The impact site on Jupiter was just out of sight as seen from Earth because the comets hit Jupiter from below and behind as seen from our direction. The *Galileo* spacecraft, on its way toward its late 1995 encounter with Jupiter, was in position to view the impacts directly, and furthermore the rapid rotation of Jupiter brought the site of each impact into view from Earth within just a few minutes. The coverage of the event was very thorough and astronomers gained a vast amount of information on the comet, on Jupiter, and on the nature of impacts **(Fig. 11.16)**.

Before the collisions, astronomers' expectations as to how dramatic the effects would be varied widely. Some thought that Jupiter would merely swallow the fragments without so much as a hiccup, while others envisioned huge impact geysers and global shock waves. One reason for all the uncertainty was that the sizes and masses of the fragments were unknown. Even the best images (those taken by the *Hubble*) failed to reveal the exact sizes of the multiple nuclei, partly because they were obscured by gas and dust that were being released due to solar heating, but also because the fragments were simply too

small to be resolved from a distance of more than 4 AU. All that could be said was that the multiple nuclei were no larger in diameter than a few kilometers, a conclusion that left a lot of room open for a wide range of mass estimates.

The optimists were proved correct: the impacts were huge. Bright flashes occurred, most spectacularly in the infrared (see Fig. 11.16), and towering clouds of ejected gas were seen. Wide-ranging wave patterns were observed in the cloud tops of Jupiter, and enormous dark blotches persisted for weeks afterward (see Fig. 11.16). Together the observations and theoretical calculations suggested that the cometary bodies did not penetrate to great depths in the Jovian atmosphere. Instead the vast energy of the impacting bodies was dissipated at high levels (within a few hundred kilometers of the cloud tops), so most of the effects on Jupiter occurred there. These effects included the deposition of enormous quantities of heat, which disrupted gases and forced them upward in columns that rose more than 1,000 km above the clouds. Spectroscopic studies of these gas columns and the resulting widening clouds of dark debris that spread from each impact site showed many lightweight elements expected to exist in both comets and the Jovian atmosphere. But the spectra also revealed a surprise: large quantities of heavy elements such as magnesium and iron, which would not be expected from either a comet or Jupiter.

It was the discovery of heavy elements that led to uncertainty about just what kind of body fell into the gaseous giant and aroused speculation that perhaps it was an asteroid. On the other hand, a pristine cometary body would contain some heavy elements, though in very small proportion, and it is possible that enough iron would have been present in a cometary nucleus to explain the spectroscopic results. The cometary hypothesis receives its strongest

Comet Watching on the WWW
In July, 1994 the Comet Shoemaker-Levy collided with Jupiter. It proved to be a spectacular event - and probably the first time that the public could see, almost in real time via the Internet, the wealth of information that was pouring in from satellites and observatories around the globe. In this web activity we'll examine images and video clips of this and other recent comet events. Our starting point is the following URL:

http://universe.colorado.edu/ch11/web.html

Impacts and You

The possibility that a major object from space may crash into the Earth may seem remote, and indeed it is—unless you take a long-term view. The chances that you will witness or be affected by such an event in your lifetime are small. It is estimated that the probability that any individual person will die due to an impact is roughly the same as that person's chances of dying in an airplane crash. Still we know that major impacts have occurred in the past, and we know that many interplanetary bodies orbiting the Sun have the potential to intersect the Earth. Given enough time, such impacts are inevitable.

Despite the small odds that it will happen to any of us, the consequences of a major impact are enormous. To some people, then, it makes as much sense to try to prepare for (and avert if possible) such an impact as it does for airlines to take all possible precautions to prevent accidents. The U.S. Congress has seriously considered funding a program proposed by a consortium of scientists to monitor the skies for potential impacting bodies (using a network of specially designed and operated telescopes) and to begin work on developing technologies for deflecting incoming bodies before they can strike the Earth. We are probably years away from the latter, but NASA has provided some seed funds to begin studying an early warning system, and astronomers at the University of Arizona and the California Institute of Technology have established search programs aimed at identifying objects in near-Earth orbits.

We can convince ourselves of the potential seriousness of an impact by a couple of methods. We can start by estimating the energy involved in a major impact. Suppose an object of spherical shape having a rocky composition (say, density is 2.5 g/cm^3 = 2,500 kg/m^3) and a radius of 1 km were to strike the Earth at the moderate interplanetary speed of 10 km/sec. The kinetic energy of a moving body is $\frac{1}{2}mv^2$ (see Chapter 5). The mass is the volume ($\frac{4}{3}\pi R^3$ for a spherical body) times the density, or 4.2×10^9 kg, yielding a kinetic energy of 5.3×10^{20} J. This is the equivalent of about 125,000 megatons of TNT, whereas the most powerful hydrogen bomb yet exploded (by the former Soviet Union) had an energy yield of about 60 megatons of TNT. Thus, even the modest object we have assumed would bring to Earth in a single moment more than 2,000 times the energy of the most powerful bomb ever exploded. If a body intersected the Earth at a crossing angle (as in the case of an asteroid in an elliptical orbit), the relative speed might be closer to the Earth's orbital speed of about 30 km/sec, and if we consider a larger asteroidal body (say, 10 km in radius, similar to one of the moons of Mars), then the energy yield is far greater. Needless to say, at either extreme of size and speed, an impact within the range considered here would be truly devastating, exacting a huge toll in lost lives and material destruction.

The second way we can realize the potential damage from a major impact is to consider the sizes of existing craters and the apparent damage caused by the impacting bodies that formed them. Craters as large as 100 km in diameter or even larger have been found on the Earth, and the scars of considerably greater impacts have been observed on other bodies such as the Moon. One particular event, which took place about 65 million years ago, carved out a large (180 km diameter) crater on the shore of the Yucatán Peninsula and lifted so much dust and debris into the atmosphere that the entire planet was covered with a layer of fallout several centimeters thick (this is the layer of iridium-rich clay marking the so-called K-T boundary, the geological stratum separating the Cretaceous and Tertiary periods). Many scientists believe that the dust in the atmosphere reduced the intensity of incoming solar radiation by so much and for so long that life-forms on our planet were devastated by indirect effects rather than by the impact itself. Many now believe that this event marked the end of the age of dinosaurs.

The impact of Comet Shoemaker-Levy on Jupiter in 1994 helped bring to focus the potential for destruction on Earth if a similar object struck here. It was in the context of the Jupiter event that a team of scientists were able to get the attention of Congress. The result was the preliminary study of an early warning system that is now under way. We are lucky to have had a chance to observe the effects of a major impact without suffering the consequences, but it remains to be seen what we will do as a result of the cost-free warning we received.

support from the apparent fragility of Shoemaker-Levy, which fell apart readily as it passed close to Jupiter on its previous orbital passage.

Whatever it was that crashed into Jupiter, the event has taught us a lot about the nature of both Jupiter and comets and has reminded us of the importance of impacts in the formation and evolution of the solar system.

Meteors and Meteorites

Occasionally, one of the countless pieces of debris floating through the solar system enters the Earth's atmosphere, creating a momentary light display as it evaporates in a flash of heat created by the friction of its passage through the air. The streak that is seen in the sky is called a **meteor (Fig. 11.17).** Most of us are familiar with this phenomenon, commonly called a shooting star, since it is often possible to see one in just a few minutes of sky-gazing on a clear night. On rare occasions an especially brilliant meteor is seen, possibly persisting for several seconds; these spectacular events are called **fireballs** or **bolides.**

The piece of solid material that causes a meteor is called a **meteoroid.** Most are very small, amounting to nothing more than tiny grains of dust or perhaps fine gravel. A few, however, are larger, solid chunks, which are responsible for the bright fireballs.

Occasionally, one of the larger meteoroids survives the arduous trip through the atmosphere and reaches the ground intact. Such an object is called a

meteorite (Fig. 11.18), and examples can be found in museums around the world. Meteorites have been the subject of intense scrutiny, for until the past 20 years or so, they were the only samples of extraterrestrial material scientists could get their hands on.

Historically, scientists scoffed at the notion that rocks could fall from the sky; that attitude changed in 1803 when a meteorite that was seen to fall near a French village was found and examined just after it fell. Such falls are rare, but they are observed occasionally and have even been known to cause damage (but so far, few injuries).

Primordial Leftovers

Meteorites are old, much older than most surface rocks on the Earth. The age of the meteorites provides us with a glimpse into the history of the solar system and therefore makes them especially interesting to scientists.

Meteorites generally can be grouped into three classes: the **stony meteorites,** which make up about 93% of all meteorite falls; the **iron meteorites,** accounting for about 6%; and the **stony-iron meteorites,** which are the rarest. These relative abundances were determined indirectly because the different types of meteorites are not equally easy to find on the ground. The majority of all meteorites found are the iron ones, although they make up only a small fraction of those that fall. The stony meteorites look so much like ordinary rocks that they are usually difficult to pick out, and some are burned up on their way through the atmosphere. A particularly produc-

Figure 11.17 A meteor. This time-exposure photogrpah caught a bright meteor as it entered the atmosphere and burned up. The maximum brightness (indicated by the wide part of the trail in the photograph) occurred when the particle reached its highest temperature and burned up as it entered. (Courtesy Hans Betlem, Dutch Meteor Society)

http://universe.colorado.edu/fig/11-17.html

Figure 11.18 A meteorite. This is a stony meteorite; the black coloring is due to heating as the object passed through the atmosphere. The light-colored spots are breaks in the fusion crust where interior material is exposed. (NASA)

tive place to search for meteorites is Antarctica, where a thick layer of ice conceals the native rock. Meteorites that fall there are relatively easy to find and are unlikely to be confused with Earth rocks.

The stony meteorites are mostly of a type called **chondrites,** so named because they contain small spherical inclusions called **chondrules (Fig. 11.19).** These are mineral deposits formed by rapid cooling, most likely at an early time in the history of the solar system, when the first solid material was condensing. A few of the stony meteorites are **carbonaceous chondrites (Fig. 11.20),** thought to be almost completely unprocessed since the solar system was formed, and therefore representative of the original stuff of which the planets were made. The primordial nature of the carbonaceous chondrites, like that of the carbonaceous asteroids mentioned in an earlier section, is deduced from their highly volatile con-tents, which indicate that they were never exposed to much heat. One particularly fascinating aspect of these meteorites is that, in at least one case, complex organic molecules called **amino acids** were found inside a carbonaceous chondrite meteorite, showing that some of the ingredients for the development of life were apparently available even before the Earth formed.

The iron meteorites have varying nickel contents and sometimes show an internal crystalline structure **(Fig. 11.21)** that indicates a rather slow cooling process in their early histories. This has important implications for their origin.

Dead Comets and Fractured Asteroids

The origins of the meteoroids that enter the Earth's atmosphere can be inferred from what we know of the properties of meteorites and of the asteroids and comets. Most meteors are caused by relatively tiny particles that do not survive their flaming entry into the Earth's atmosphere. During **meteor showers,** when meteors can be seen as frequently as several per minute, all seem to be of this type and all appear to originate from a single point on the sky called the **radiant.** This phenomenon occurs because these showers are associated with the remains of comets that have disintegrated or debris from comets still orbiting the Sun **(Table 11.4),** leaving behind a clump of gravel and dust that spreads out along the original orbit of the comet. When one of these streams of small particles enters the Earth's atmosphere, the resulting meteors appear to emanate from a single point.

The larger chunks that reach the ground as meteorites may have a different origin. It is likely that occasional collisions occur among the asteroids, sometimes with sufficient violence to destroy them and disperse the rubble that is left over throughout

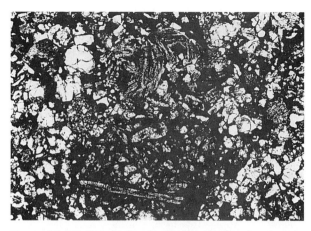

Figure 11.19 Cross section of a chondrite. This photo shows the many chondrules (light patches) embedded within the structure of this type of stony meteorite. (Griffith Observatory, Ronald A. Oriti Collection)

Figure 11.20 A carbonaceous chondrite. This example, which is not the type in which amino acids have been found, shows a large chondrule (the light-colored spot, upper center) about 5 millimeters in diameter. (Griffith Observatory, Ronald A. Oriti Collection)

Figure 11.21 Cross section of a nickel-iron meteorite. This example shows the characteristic crystalline structure indicative of a slow cooling process from a previous molten state. Such meteorites are thought to have once been parts of larger bodies that differentiated. (Griffith Observatory, Ronald A. Oriti Collection)

Table 11.4
Major Meteor Showers

Shower	Approximate Date	Associated Comet
Quandrantid	January 3	—
Lyrid	April 21	Comet 1861 I
Eta Aquarid	May 4	Halley's comet
Delta Aquarid	July 30	
Perseid	August 11	Comet 1862 III
Draconid	October 9	Comet Giacobini-Zinner
Orionid	October 20	Halley's comet
Taurid	October 31	Comet Encke
Andromedid	November 14	Comet Biela
Leonid	November 16	Comet 1866 I
Geminid	December 13	—

to be the most primitive of the meteorites, having undergone no processing in the interiors of large bodies.

From our studies of the other planets and satellites, we know that there was a time long ago when frequent impacts occurred, forming most of the craters seen today. Most of the craters on the Moon were formed more than 3 billion years ago, that is, within the first billion years or so after the formation of the solar system. Certainly, the Earth was not immune and no doubt was also subjected to heavy bombardment, but the Earth has an atmosphere, along with flowing water and glaciation, all of which combine to erase old craters in time.

A few traces are still seen, however. A very large basin under the Antarctic ice is probably an ancient impact crater, and a portion of Hudson Bay in Canada shows a circular shape thought to have a similar origin. The crater probably associated with the great impact of 65 million years ago, which is suspected of having caused the death of the dinosaurs, has been located on the shore of the Yucatán Peninsula in Mexico; the enormity of the impact is verified by a global layer of debris. A number of other suspected ancient impact craters have been found throughout the world (**Fig. 11.22** and **Table 11.5**). Although the frequency of impacts has decreased, major impacts still occur on rare occasions. The Barringer Crater **(Fig. 11.23)** near Winslow, Arizona, was formed only about 25,000 years ago, for example,

the solar system. Most meteorites are probably fragments of asteroids. The iron and the rare stony-iron meteorites apparently originated in asteroids that had undergone differentiation, while the stony ones came either from the outer portions of differentiated asteroids or from smaller bodies that never underwent differentiation at all. The chondrites probably fall into the latter category, since the chondrules reflect a rapid cooling that would be characteristic of very small bodies. This is why the chondrites are thought

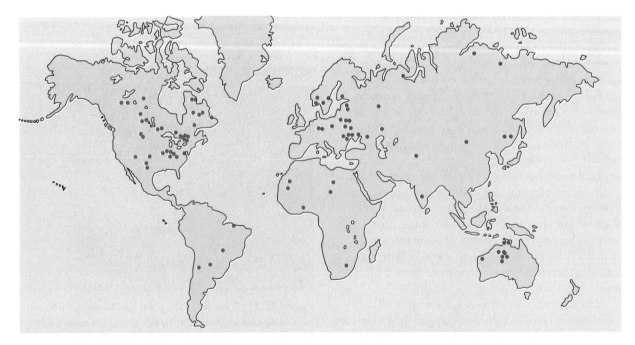

Figure 11.22 **Impact craters on the Earth.** This map shows the locations of major craters thought to have been created by impacts of massive objects. (Griffith Observatory)

Table 11.5
Some Known or Suspected Impact Craters on the Earth

Location	Diameter (km)	Location	Diameter (km)
Amirante Basin, Indian Ocean	300	Wells Creek, Tennessee	14
Yucatum Peninsula, Mexico	180	Deep Bay, Saskatchewan	13.7
Sudbury, Ontario	140	Deep Bay, Saskatchewan	12
Vredefort, Orange Free State, South Africa	100	Lake Bosumtwi, Ashanti, Ghana	10.5
Manicougan	100	Chassenon Structure, Haut-Vienne, France	10
Sierra Madera, Texas	100	Wolf Creek, Western Australia	8.5
Charlevoix Structure, Quebec	46	Brent, Ontario	3.8
Clearwater Lake West, Quebec	32	Chubb (New Quebec), Quebec	3.2
Mistastin Lake, Labrador	28	Steinem, Swabia, Germany	2.5
Gosses Bluff, Northern Territory, Australia	25	Henbury, Northern Territory, Australia	2.2
Clearwater Lake East, Quebec	22	Boxhole, Central Australia	1.8
Haughton, Northwest Territories	20	Barringer Crater, Winslow, Arizona	1.2

and the possibility exists that other large bodies could still hit the Earth. Given a long enough time, it is almost inevitable.

Microscopic Particles: Interplanetary Dust and the Interstellar Wind

The space between the planets plays host to some very tiny particles in addition to the larger ones we have just described. There is a general population of small solid particles, typically a millionths of a meter in diameter, called **interplanetary dust particles.** In addition, a very tenuous stream of gas particles flows through the solar system from interstellar space.

The presence of the dust has been known for some time from two celestial phenomena, both of which can be observed with the unaided eye, though only with difficulty. The dust particles scatter sunlight, so under the proper conditions a diffuse glow can be seen where the light from the Sun hits the dust. This is similar to seeing the beam of a searchlight stretching skyward; you see the beam only where there are small particles (either dust or water vapor) that scatter its light, so that some of it reaches your eye.

One of the phenomena created by the interplanetary dust is the **zodiacal light (Fig. 11.24),** a faintly illuminated belt of hazy light that can be seen stretching across the sky along the ecliptic on clear,

Figure 11.23 Barringer Crater near Winslow, Arizona. The impact that created this crater occurred about 25,000 years ago. (Meteor Crater Enterprises)

dark nights just after sunset or before sunrise. The second observable phenomenon created by the dust is a small bright spot seen on the ecliptic in the direction opposite the Sun. This diffuse spot, called the **gegenschein (Fig. 11.25),** is created by sunlight that is reflected straight back by the interplanetary dust, which is concentrated in the plane of the ecliptic. This is analogous to seeing a bright spot on a cloud

Figure 11.24 **The zodiacal light.** The diffuse glow of reflected sunlight from dust grains in the ecliptic plane is clearly visible in this photograph made from the summit of Mauna Kea. (W. Golisch)

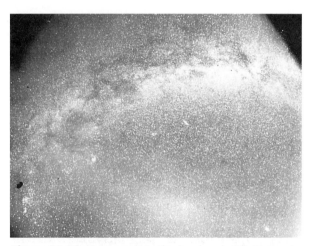

Figure 11.25 **The gegenschein.** This photo shows the Milky Way stretching across the upper portion and a diffuse concentration of light at lower center, which is the gegenschein. It is created by light reflected directly back to Earth from interplanetary dust in the direction opposite the Sun. (Photo by S. Suyama, courtesy of J. Weinberg)

Figure 11.26 **Dust bands in the outer solar system.** The IRAS satellite's infrared sensors detected emission from this spiral-shaped tail of dust, which may have been created by a collision between an asteroid and a comet. At the lower left is the Milky Way, as seen in the infrared by IRAS. (NASA)

bank or low-lying mist when you look at it with the Sun directly behind you; the bright spot is just the reflected image of the Sun and is the counterpart of the gegenschein.

The *IRAS* satellite, which mapped the sky in infrared wavelengths, observed interplanetary dust. The grains of dust are cold and emit light at infrared wavelengths, so these observations reveal much more information about the dust than observations at other wavelengths. The *IRAS* sky maps show that the zodiac glows at infrared wavelengths due to the concentration of dust in the ecliptic plane. In addition, the infrared maps revealed streaks of dust following cometary orbits and a huge spiraling ring of dust between the orbits of Mars and Jupiter **(Fig. 11.26)**. This band of dust, which extends above and below the plane of the ecliptic, is thought to have been created by a collision between asteroids or between a comet and an asteroid.

Scientists have succeeded in collecting interplanetary dust particles for direct examination **(Fig. 11.27)**. High-altitude balloons were once used to gather particles, but now new techniques have been developed. Interplanetary particles that fell into the oceans can be scooped off the seafloor, and very recently rich deposits of the particles have been found in Arctic lakes, particularly in Greenland, where they can be dredged up in vast numbers. The Earth is constantly being pelted with dust particles (which add about 8 tons per day to its mass!), and those that fall into the oceans or on Arctic ice sheets can lie undisturbed for long times. Studies of the particles show that they are probably of cometary origin, having been dispersed throughout space from the dead nuclei of old comets. These grains have a complex structure, with many segments stuck together (see Fig. 11.27). The grains are low in density and fragile, having the general characteristics attributed to grains ejected from Comet Halley (as discussed earlier in this chapter).

The interplanetary particles have an assortment of origins. Clearly, some are from the erosion of cometary nuclei. The best evidence for this is the concentration of dust along the orbital paths of comets past and present. Other grains are probably asteroidal in origin, as indicated by laboratory analysis of captured particles. The asteroidal ones tend to show signs of previous heating (through a relative lack of easily vaporized elements). And some of the particles, or at least the smaller embedded inclusions,

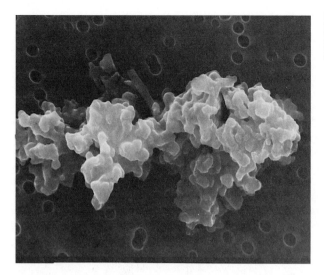

Figure 11.27 **An interplanetary dust grain.** This is a microscopic view of a tiny particle from interplanetary space. The amorphous structure is highly variable from one grain to another. (NASA photograph, courtesy of D. E. Brownlee)

are clearly interstellar in origin. The main evidence for this lies in the unusual isotopic composition that these inclusions may have; that is, they are found to have various nuclear forms of elements in ratios that are very unlike the normal pattern found in the solar system.

As we will learn in the next chapter, the space between stars in our galaxy is permeated by a rarefied medium of gas and dust. In the Sun's vicinity, the average density of the gas is far below that of any artificially created vacuum; it amounts to only about 0.1 particle per cubic centimeter (i.e., there is one atom, on average, in every volume of 10 cm³, corresponding to a cube about 1 inch on each side). Because of the motion of the Sun, the interstellar gas streams through the solar system with a velocity of about 20 km/sec. The presence of this ghostly breeze, consisting mostly of hydrogen and helium atoms and ions, was discovered in the early 1970s, when observations made from satellites revealed very faint ultraviolet emission from the hydrogen and helium atoms in the gas. The interstellar wind, tenuous as it is, has very little effect on the other components of the solar system but is nevertheless studied with some interest for what it may tell us about the interstellar medium.

Summary

1. Thousands of asteroids, or minor planets, have been discovered and cataloged; they display a variety of sizes (up to 1,000 km in diameter) and compositions (ranging from metallic ones to rocky minerals).

2. Gaps in the asteroid belt are created by orbital resonances with Jupiter, whose gravitational influence was probably also responsible for preventing the asteroids from coalescing into a planet in the first place.

3. Asteroids are classified according to their predominant surface materials, and the different classes are found to be concentrated in zones at different distances from the Sun, reflecting the range of temperatures present when the solar system formed.

4. Comets are small, icy objects that develop their characteristic comae and tails only when in the inner part of the solar system.

5. Comets may originate in the Oort cloud, some 30,000–100,000 AU from the Sun, in which case they appear as long-period comets with randomly oriented orbits when they pass through the inner solar system; or they may originate in the Kuiper belt (30–50 AU from the Sun) and appear as periodic comets, returning to the inner solar system every few years or decades and following orbits that tend to lie near the ecliptic plane.

6. When near the Sun, a comet ejects gases that glow by fluorescence. Periodic comets eventually lose all of their icy substance in this process and disintegrate into swarms of rocky debris.

7. Comets are made almost entirely of water and carbon compounds, and through impacts they have transported large quantities of these elements to the planets.

8. A comet may have two tails, one created by ionized gas and the other made of fine dust particles.

9. A meteor is a flash of light created by a meteoroid entering the Earth's atmosphere from space, and a meteorite is the solid remnant that reaches the ground in some cases.

10. Meteorites are either stony, stony-iron, or iron in composition; they are very old and thus provide information on the early solar system.

11. Most meteors are created by fine debris from comets, but most meteorites are fragments of asteroids.

Looking for Meteors

ASTRONOMICAL

ACTIVITY

If you have not seen a meteor, you should. It is not difficult, but it takes a little effort and perhaps some time. It is probably most fun to view meteors with a friend, and it is certainly more comfortable to do it during warm weather. It is possible to see a meteor at any time of year, and especially easy during the time of one of the major meteor showers. You may want to consult Table 11.4 and select a time when a shower is expected.

The procedure is simple: get outdoors, get comfortable, and look up. Obviously, you need a clear night, and it is better if it is also a dark night. So avoid full moon, and try to get away from city lights. Even without a meteor shower, the chances are still quite good that you will see a meteor in an hour or two, and often more frequently than that. Take a sky chart along and pass the time identifying constellations or finding planets. If a meteor shower is taking place, you should see meteors every few minutes or, if you are lucky, several times per minute (that is very rare, though).

The highest frequency of meteors occurs after midnight, when you are facing in the direction of the Earth's orbital motion. So you may lose some sleep if you decide to go out late at night for this. Usually, this is not necessary, though; you should succeed if you go out earlier and are patient.

---

12. Interplanetary space is permeated by fine dust particles and by an interstellar wind of hydrogen and helium atoms from the space between the stars.

Review Questions

1. How is the asteroid belt similar to the rings of Saturn?
2. Compare Bode's law with scientific theories; that is, how is it like a theory and how is it not?
3. What is the significance of the fact that some of the largest asteroids are spherical in shape? What does this tell us about their early histories? How are these asteroids related to the purely metallic ones?
4. Summarize the role of Jupiter in influencing asteroids, meteorites, and comets.
5. Discuss how the "dirty snowball" model of comets was supported or modified by the detailed close-up observations of Comet Halley in 1986.
6. What forces control the direction of a comet's tail? Why do comets tend to have two tails that are not aligned with each other?

Web Connections
The Review Questions and Problems also appear at the following URLs:
http://universe.colorado.edu/ch11/questions.html
http://universe.colorado.edu/ch11/problems.html

7. Why are carbonaceous chondrites especially important clues to the early history of the solar system?
8. Would an observer on the surface of Mercury see meteors in the nighttime sky? Might he or she find meteorites on the ground there?
9. Summarize the evidence for the existence of interplanetary dust in the solar system.
10. Summarize the ways in which the interplanetary bodies discussed in this chapter help astronomers to understand the origins of the solar system.

Problems

1. At what distances from the Sun would asteroids have one-third of the orbital period of Jupiter and one-tenth of the orbital period of Jupiter? Are there gaps in the asteroid belt at these locations? Explain.
2. Use Wien's law and the Stefan-Boltzmann law to find the surface temperature and diameter of an asteroid whose wavelength of maximum emission is 0.003 cm and whose luminosity is 2×10^{13} W.
3. What is the orbital period of an asteroid whose semimajor axis is 2.8 AU?
4. A comet has an orbital eccentricity of $e = 0.998$. Its perihelion distance (closest approach to the Sun) is 0.24 AU. Find the semimajor axis of its orbit, its aphelion distance, and its orbital period.

5. Suppose a comet in the Oort cloud has a semimajor axis of 100,000 AU. The nearest star like the Sun is Alpha Centauri, 4 light-years (about 300,000 AU) from the Sun. Compare the gravitational force on the comet due to the Sun with that due to Alpha Centauri, when the comet lies in the same direction from the Sun as Alpha Centauri.

6. Suppose Comet Halley's nucleus has a mass of 5×10^{14} kg and moves at a speed relative to the Earth of 40 km/sec. If it hits the Earth, how much kinetic energy is released? (*Hint:* Recall that the kinetic energy of a moving body is $\frac{1}{2}mv^2$, where m is the mass and v the velocity.) Compare your answer to the energy of a 1 megaton nuclear bomb; 1 megaton = 4.2×10^{15} J). Discuss the implications of your result.

Additional Readings

Aguirre, E. L. 1996. A Great Comet Visits the Earth. *Sky & Telescope* 91(6):20.

Aguire, E. L. 1996. Comet Hyakutake's Spectacular Performance. *Sky & Telescope* 92(1):22.

Baliunas, S. and S. Saar. 1992. Unfolding Mysteries of Stellar Cycles. *Astronomy* 20(5):42.

Beatty, J. K. 1991. Killer Crater on the Yucatan? *Sky & Telescope* 82(1):38.

Beatty, J. K. and S. J. Goldman. 1994. The Great Crash of 1994; A First Report. *Sky & Telescope* 88(4):18.

Binzel, R. P., M. A. Barucci, and M. Fulchignoni 1991. The Origins of the Asteroids. *Scientific American* 265(4):88.

Brandt, J. C. and R. D. Chapman. 1992. Rendezvous in Space: The Science of Comets. *Mercury* XXI(6):178.

Brophy, T. 1993. Motes in the Solar System's Eye. *Astronomy* 21(5):34.

Burnham, R. 1994. Here's Looking at Ida. *Astronomy* 22(4):38.

Burnham, R. 1994. Jupiter's Smash Hit. *Astronomy* 22(11):34.

Chapman, C. 1996. Worlds Between Worlds. *Astronomy* 23(6):46.

Chyba, C. 1993. Death from the Sky. *Astronomy* 1(12):38.

Cunningham, C. J. 1992. Giuseppe Piazzi and the "Missing Planet." *Sky & Telescope* 84(3):274.

Cunningham, C. 1992. The Captive Asteroids. *Astronomy* 20(6):40.

Durda, D. 1993. All in the Family. *Astronomy* 21(2):36.

Durda, D. 1995. Two by Two They Came. *Astronomy* 23(1):30.

Dyson, F. J. 1994. Hidden Worlds: Hunting for Distant Comets and Rogue Planets. *Sky & Telescope* 87(1):26.

Eicher, D. J. 1994. Death of a Comet. *Astronomy* 22(10):40.

Gallant, R. A. 1994. Journey to Tunguska. *Sky & Telescope* 87(6):38.

Gehrels, T. 1996. Collisions between Comets and Asteroids. *Scientific American* 274(3):54.

Grieve, R. A. F. 1990. Impact Cratering on the Earth. *Scientific American* 262(4):66.

Jokipii, J. R. and F. B. McDonald 1995. Quest for the Limits of the Heliosphere. *Scientific American* 272(4):58.

Levy, D. H. 1994. Pearls on a String. *Sky & Telescope* 86(1):38.

Levy, D. H., E. M. Shoemaker, and C. S. Shoemaker 1995. Comet Shoemaker-Levy 9 Meets Jupiter. *Scientific American* 273(2):84.

Luu, J. X. and Jewitt, D. C. 1996. The Kuiper Belt. *Scientific American* 274(5):46.

MacRobert, A. M. and J. Roth. 1996. The Planet of 51 Pegasi. *Sky & Telescope* 91(1):38.

McFadden, L.-A. and C. Chapman. 1992. Interplanetary Fugitives. *Astronomy* 20(8):30.

Morrison, D. 1992. The Spaceguard Survey: Protecting the Earth from Cosmic Impacts. *Mercury* 21(3):103.

Morrison, D. 1995. Target: Earth. *Astronomy* 23(10):34.

O'Meara, S. J. 1994. The Great Dark Spots of Jupiter. *Sky & Telescope* 88(5):30.

Robinson, M. and M. Wadwha 1995. Messengers from Mars. *Astronomy* 23(8):44.

Spratt, C. and Stephens, S. 1992. Against All Odds: Meteorites That Have Struck Home. *Mercury* 21(2):50.

Stern, A. 1992. Where Has Pluto's Family Gone? *Astronomy* 20(9):40.

Stern, A. 1994. Chiron: Interloper from the Kuiper Disk? *Astronomy* 22(8):26.

Stern, A. 1995. The Sun's Fab Four. *Astronomy* 23(6):30.

Verschuur, G. 1992. Mysterious Sungrazers. *Astronomy* 20(4):46.

Weissman, P. R. 1993. Comets at the Solar System's Edge. *Sky & Telescope* 85(1):26.

Weissman, P. 1995. Making Sense of Comet Shoemaker-Levy 9. *Astronomy* 23(5):48.

Whipple, F L. 1974. The Nature of Comets. *Scientific American* 230(2):49.

The Earth, formed out of the same debris of which the sun was born, existed at the center of a star that exploded many billions of years ago.

Isaac Asimov

Chapter 12

The Origins of Suns and Planets

Chapter Web site: http://universe.colorado.edu/ch12

he solar system is probably not unique. Everything we have learned from our studies of how the Sun and planets came into existence suggests that the same processes could have and should have occurred elsewhere, time and time again. In addition, observations tell us of sites within our galaxy where young stars appear to exist, and the data reveal evidence for at least the beginnings of planetary formation as well. Fully developed solar systems, with isolated planets orbiting a mature star, are much more difficult to detect, but recently astronomers have succeeded in a few systems where planets resembling Jupiter orbit Sun-like stars. Thus expectations run high that such planetary systems are probably quite numerous in the cosmos.

In this chapter we will explore the processes thought to give rise to suns and planets. In doing so, we will establish links between what is known of our own solar system and the general mechanism of star formation and planetary growth; these links will provide a transition from our local neighborhood to the realm of the stars.

Properties of Planetary Systems

Any successful theory of the formation of stars and planetary systems must be able to account for the general properties of such systems. But what are these properties? The only way we can answer that is to look at the example we know best: our own solar system. Thus it is useful to begin our discussion by summarizing the general features of the solar system, looking for clues to its origin. We will leave out peculiarities of individual planets that are thought to be due to random events such as collisions, because these would certainly not be duplicated elsewhere. Instead we will concentrate on the overall picture, discussing only general phenomena that we might expect to find in every system.

Perhaps the most striking physical property of the solar system is its disklike shape. All of the planets orbit the Sun in paths that lie close to the ecliptic plane (defined technically as the plane of the Earth's orbit). The orbital planes of most of the planets lie within 5° of the ecliptic, and the most deviant, Pluto (with an inclination of 17°), may be the result of a random event rather than the general formation process for the system. Surely the flattened shape of the solar system is no accident; it must be telling us something important about the way the system

formed, and we might well expect the same forces to have been at work in other systems.

The fact that all of the planets orbit the Sun in the same direction must also reflect a definite pattern. If orbital direction were determined by random events, it would be extremely unlikely to find all nine planets going around the Sun in the same direction. In addition, the great majority of all spin motions of the planets and satellites are in the same direction and must also be the result of a systematic process rather than random chance. The fact that the rotation axes of most of the planets tend to be perpendicular to their orbital planes is related to the direction of spin. Along the same vein, the orbital directions of most satellites also follow the pattern of prograde motions. When viewed from afar, our solar system would appear as a smoothly operating mechanism with virtually all of the orbital and spin motions in the same direction. We may reasonably expect that any other planetary system, if shaped by the same processes that formed our family of Sun and planets, would display similar overall uniformity.

Things become a bit more speculative when we consider the properties of the individual planets. As already mentioned, some of the peculiarities of the bodies in our system appear to be the result of singular events such as collisions, and we would not expect to see them repeated in detail elsewhere. On the other hand, the planets exhibit some systematic trends that probably do result from the overall formation process, and we might reasonably anticipate that these trends would show up in other systems. One such trend might be the tendency of the inner planets to have low abundances of the lightweight, volatile elements while the outer planets retain these species. The high temperature in the inner solar system, along with early solar activity (an outflowing wind), which is thought to have prevented the condensation or retention of the volatiles by the terrestrial planets, should operate in any other system. Similarly, the outer planets in another system would be expected to retain the lightweight gases and therefore to be much more massive than the inner planets. In short, we might reasonably expect that other solar systems would have terrestrial planets and gaseous giant planets, just as our system does.

Our theory of solar system formation is not complete enough to tell us whether the precise locations and properties of the planets came about for systematic reasons or are the result of more random processes. Thus we do not know whether exact analogs to our planets might exist; that is, whether we would always find a planet just like the Earth in location and other properties or whether we would find an exact duplicate of Jupiter. The question of whether there would be a planet at the same distance from its parent star as the Earth has some bearing on the question of whether life could be expected to develop in other planetary systems; some scientists believe that life was possible on Earth only because conditions here fell within a very narrow range that was suitable. On the other hand, the uncertainty as to whether a Jupiter-like planet might be expected has a lot to do with our chances of eventually finding planetary systems orbiting other stars because Jupiters are much easier to detect than smaller planets (this is discussed later in this chapter).

Finally, the solar system is characterized by a variety of forms of interplanetary matter, including comets, asteroids, icy planetesimals in the outer system, and interplanetary dust and gas. We should expect to find all of these in any other system because all are thought to be due to natural evolutionary processes here, rather than being the result of chance events. Indeed, astronomers have devised possible techniques for detecting evidence of cometary bodies orbiting at great distances around other stars. The expectation is that star formation is always accompanied by the formation of myriad small, icy bodies that remain after the planets have coalesced.

Many other details could be brought into this discussion, but we have included the major points that must be explained by any successful theory of the formation of stars and planetary systems. Such a theory must account for the overall disklike shape of the system; the fact that all orbital and spin motions tend to be in the same direction; the likely existence of planetary types corresponding to our terrestrial and gaseous giant planets; and the probable existence of interplanetary bodies comparable to our comets, asteroids, and icy planetesimals.

A Word About Stars

In addition to explaining planets and interplanetary bodies, a theory of solar system formation involves understanding how a star is formed, so it is useful at this point to introduce some of the basic properties of stars. Much more depth and detail will be added in the several chapters following this one.

Stars are found to have a wide range in basic physical properties such as mass, diameter, temperature, and especially luminosity (radiant power; see Chapter 6). The Sun is intermediate in these properties. It is not known whether planets form in the company of stars of all types, or whether only Sun-

like stars are born under the right conditions to also form planets. There are reasons to believe that modest stars having masses not much greater than the Sun are more likely to have planets orbiting them than are the very massive and luminous stars, so in this chapter we will tend to focus on Sun-like stars.

The masses of normal stars range from about one-tenth of the mass of the Sun up to perhaps 50 solar masses, with the less massive stars being far more numerous. Thus our theory of star formation needs to explain not only how stars of very different masses can form, but also why low-mass stars form much more readily than high-mass ones. The other basic properties of stars, such as temperature, luminosity, and diameter, all depend on mass; that is, two stars of the same mass will also be similar in these other aspects. When we speak of Sun-like stars, then, we mean stars having similar masses to the Sun, but we expect their other properties will also be similar.

Stars generally have similar chemical compositions, regardless of where their other properties fit within the range. This composition has already been mentioned in our discussions of the solar system. In brief, hydrogen is the most abundant element, making up about 73% of the mass of the Sun; helium is next, accounting for about 25%; and all the other elements together make up the remaining 2%. Within that small allotment of elements heavier than hydrogen and helium, the most common are the volatile species carbon, nitrogen, and oxygen and the metallic element iron.

A star is gaseous throughout; its structure can be understood under the assumption that it is a fluid body, held together by gravity. Furthermore, at the extremely high temperatures found inside a star, matter can exist only in gaseous form; no solid or liquid could survive. Not only is the matter gaseous throughout, but it is also highly ionized to the point where the core of a star consists only of free electrons and naked atomic nuclei. The energy that is radiated by a star arises in its core and is transmitted outward until it escapes into space in the form of a stream of photons. The energy generated at the heart of a star comes from nuclear reactions in which atomic nuclei are forced together, forming new elements and giving off immense quantities of energy in the process. As a star forms, it may glow due to gravitational compression and heating, but we do not consider it a proper star until the nuclear reactions ignite in its core.

In mentioning gravitational compression, we have invoked the single major factor that molds and creates stars and then governs their subsequent lives. The story of star formation is a story of gravity and how it can bring together diffuse matter and concentrate it into a dense, glowing ball.

Figure 12.1 A galaxy similar to the Milky Way. This spiral-shaped object consists of billions of stars arranged in a disk which has arms or streamers of luminous stars and interstellar matter. Astronomers think that the Milky Way would look like this, if we could view it from an external position. The Sun lies about two-thirds of the way out from the center of the disk. (© 1992 Anglo-Australian Telescope Board. Photo by D. Malin)

Clusters and Associations of Stars

Stars are not distributed uniformly throughout space. The universe itself is not uniform; instead, most of the visible matter is concentrated in **galaxies,** vast collections of stars that are bonded together by gravity. Our own galaxy, the Milky Way, is a giant disklike structure containing about 10^{11} stars; many other galaxies **(Fig. 12.1)** are similarly disk-shaped, while others are smooth and rounded (the general properties of galaxies are discussed in Chapter 18).

Even within a galaxy, all the stars are not distributed uniformly. Superimposed on the general backdrop of stars are concentrations called **clusters** or **associations;** these are locales where many stars are found together. The stars in a cluster are bound together by gravity, all of them orbiting a common center of mass. Clusters can range in membership from a few stars to thousands or even hundreds of thousands **(Fig. 12.2).** The smaller groupings tend to lie in the disk of the galaxy and usually have amorphous, random-appearing shapes. These are called **open clusters,** or simply **galactic clusters.** At the other extreme are the **globular clusters,** which are found outside the plane of the galactic disk; these are smoothly rounded ensembles containing more than 10^5 stars in most cases.

Within a cluster, all the stars are thought to have formed together, and therefore to be of a common age. The differences among the stars within the group are attributed to differences in mass, which lead to differences in other properties; thus the spectrum of stellar types within a cluster is basically a spectrum of stellar masses. The globular clusters are all quite old, but a wide range of ages is found among galactic clusters. Observations of clusters have much to tell us about how stars of different masses evolve once they have formed, so we will rely on clusters to help tell the story of stellar lives (see Chapter 15).

Early theories of star formation indicated that the formation of stars in clusters was much more easily accomplished than the formation of individual stars. This led astronomers to consider the possibility that all stars begin life as members of clusters, but that most had subsequently managed to escape and lead solitary lives. The Sun is certainly not part of any cluster today, but in its 4.5 billion–year lifetime it could have freed itself from the gravitational bonds of its birth group.

There is evidence that formation in clusters is indeed the norm for certain stellar types, especially the most massive and luminous stars. The hottest stars are classified as **O** and **B stars** (see Chapter 14), and virtually all of these stars are found in loosely bound groupings called **OB associations** (see Fig. 12.2). These small groupings appear not to be strongly bonded together by gravity. Instead, observations of stellar motions (via the Doppler shifts seen in stellar spectra) indicate that the groupings are usually expanding and dissipating. Because the superluminous O and B stars do not live very long, they usually do not outlive their birth clusters; as a result, we see few isolated hot, massive stars. Both the stars and the associations they form in disappear within a few million years, a very short time compared to the lifetimes of less luminous, less massive stars like the Sun.

On the one hand, then, modern theories of star formation must explain how stars can form together

Star Clusters

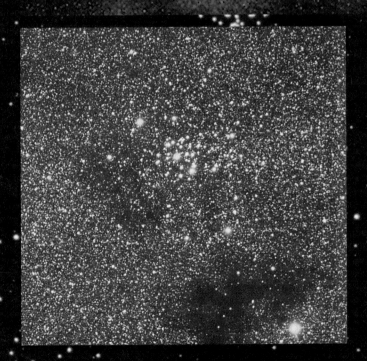

Figure 12.2 Here are four clusters of stars, illustrating some of the types that are found in our galaxy. At upper left is the well-known Pleiades cluster, a young galactic, or open, cluster; at upper right is another open cluster, known as NGC 3293; at lower left is an OB association (the grouping of blue-white stars at the upper center); and at lower right is the globular cluster 47 Tucanae. (upper left: © 1985 Anglo-Australian Telescope Board; upper right: © 1977 Anglo-Australian Telescope Board; lower left: © 1980 Anglo-Australian Telescope Board; lower right: © 1992 Anglo-Australian Telescope Board; photos by D. Malin)

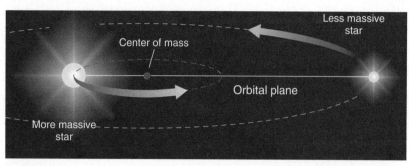

Figure 12.3 Binary star center of mass. In a double-star system, the two stars orbit about a point in space between them, which is the center of mass. The distances of the stars from the center of mass are inversely proportional to their masses, so the center of mass is closer to the more massive star. In this drawing the star at left has three times the mass of the star at right, so the star at right is three times farther from the center of mass.

in clusters—in particular, in OB associations. On the other hand, there is growing evidence that lower-mass stars can form individually without belonging to large groupings, so the theory must also account for the formation of individual stars of modest mass in isolation from others. As we will see later in this chapter, the difficulty of finding ways to form isolated low-mass stars, along with ample observational evidence that mass can be lost by young stars, has led to a radical view in which most stars form from large concentrations of matter, but then are able to eject much of it in a kind of stellar weight-loss program.

Regardless of whether they are members of clusters or not, most stars are double or multiple. A double star, called a **binary star,** consists of two members in mutual orbit about a common center of mass **(Fig. 12.3).** Binary stars are very common, although systems of three or four stars are known as well (these usually contain a binary star in orbit with another single star or another binary, so the appearance of stars in pairs seems to be the norm). As a single star, the Sun is actually in the minority among its galactic peer group. More than half of the stars in the sky are members of binary systems. Therefore our theory of star formation must include a mechanism for forming stars in pairs.

The Interstellar Medium

Stars do not arise from a vacuum but must have a source of raw material from which to form. That source of material is the diffuse medium of gas and dust that pervades the space between the stars in our galaxy **(Fig. 12.4).** The interstellar medium will be discussed in more detail in Chapter 17, but here we will review some of its major characteristics, especially in the regions where star formation is known to occur.

At first glance the interstellar medium may seem very close to a vacuum: the average density of matter in space is less than one atom per cubic centimeter, far lower than the density in even the best laboratory vacuum systems. Yet the volume of space is so large relative to the volume occupied by stars that the interstellar medium makes up a substantial fraction, some 10 to 15% by mass, of all the matter in the disk of our galaxy.

Figure 12.4 The interstellar medium. The space between the stars is filled with a very diffuse medium of gas and dust. Here we see a region where the material is denser than average, and is illuminated by the light from embedded stars, making it more easily seen on a photograph than is the case in most regions. Stars form in very dense interstellar clouds, denser and darker than the material seen here. (© 1992 Anglo-Australian Telescope Board. Photo by D. Malin)

The composition of the interstellar gas is very similar to that of the stars; it is dominated by hydrogen, then helium, and then the heavier elements in much smaller quantities. The heavy elements, particularly the metallic ones, tend to be even less abundant in the interstellar gas than in stars, because in the interstellar medium these elements are often in solid form, in the tiny particles known as **interstellar dust.** Only about 1% of the mass in interstellar space is in the form of dust, but as we shall see, the dust has important effects on star formation and on our ability to observe it.

The interstellar medium is not uniform. Instead, it is patchy, with most of the space occupied by gas and dust of very low density, far lower than the average value of one atom per cubic centimeter. Most of the mass lies in concentrated regions called **interstellar clouds,** where the densities range from 100 to 10^5 or more atoms per cubic centimeter (even the upper end of this range is about 100 trillion times more rarefied than the Earth's atmosphere and is difficult to reproduce in laboratory vacuum systems). The most concentrated clouds are dense enough to block out the light from stars lying within and behind them and therefore appear on photographs of the sky as regions that are black and devoid of stars (see Fig. 12.4).

Temperatures vary as well. The vast regions of very rarefied, ionized gas are extremely hot, having temperatures ranging from 10,000 K to nearly 1 million K. The clouds, however, are much colder, with temperatures lying between a few hundred K for the relatively diffuse clouds to less than 50 K for the denser, darker regions. In all cases, the temperature is determined by a balance of energy sources (such as ultraviolet radiation, shock waves, and rapidly moving charged particles) and energy losses (primarily in the form of radiation that escapes into the surrounding medium).

The form of matter in the interstellar medium varies, depending on the physical conditions. In the most rarefied regions, ultraviolet radiation from stars, as well as energetic subatomic particles that permeate space, ensure that the gas is ionized, consisting of free electrons and positively charged ions. In the moderately dense clouds, the gas is partially ionized, with the main elements, hydrogen and helium, in atomic form. In denser clouds the particles tend to be neutral atoms and molecules (i.e., atoms bonded together in chemical combination). The most common molecule, naturally enough, is formed of the most common element, hydrogen. In interstellar clouds, molecular hydrogen, H_2, is the most abundant form of matter. In clouds of high density, not only hydrogen but also most other elements are in molecular form. After molecular hydrogen, the most common molecular species is carbon monoxide, or CO. Most molecular species are observed at radio wavelengths, where they emit radiation in the form of emission lines.

The densest and most massive interstellar clouds, called **giant molecular clouds,** are prominent sites of star formation. These clouds occur in enormous complexes spanning many light-years and can have total masses up to 100,000 Suns or more. Within a giant molecular cloud, the density may be as high as several hundred thousand particles per cm^3, and the temperature may be as low as 10 to 20 K. Even within a single cloud or cloud complex, there can be substantial variations in conditions. While most of the material has a density averaging around a few hundred particles per cm^3, in dense knots or filaments of matter, called **dense cloud cores,** the density reaches its highest values and the temperatures are the coldest. It is within these cores that star formation occurs.

Observations of Young Stars

The fact that stars form within giant molecular clouds, which appear dark on the sky, makes the process very difficult to observe. Visible wavelengths of light are barred by the obscuring dust, and we simply cannot see far enough into the interiors of these clouds to directly observe stars in formation.

Fortunately, the same is not true of other wavelengths. The tendency of interstellar dust to block our view varies with wavelength, being strongest for the shortest wavelengths and weaker for longer wavelengths. We cannot see far into dense clouds in ultraviolet or visible wavelengths, but we have relatively clear windows into these regions in the infrared and radio portions of the spectrum. By looking through these windows, astronomers have learned a great deal about star formation.

Stars in the Process of Forming

The nearest significant giant molecular cloud complex lies in the constellation of Orion, particularly around and behind the glowing diffuse region in Orion's sword, known as the Orion nebula **(Fig. 12.5;** see also Fig. 12.4). Radio and infrared observations of the Orion dark cloud have revealed an incredibly complex region of star formation and associated activity, and much of what we know about the process is based on these observations, although similar observations of similar processes have been made for many other regions as well.

Figure 12.5 The Orion nebula. This cloud of glowing gas and dark dust-shrouded regions is visible as a faint nebulosity in the sword of Orion, with a small telescope or a pair of binoculars. The left image here is a fine, wide-field view obtained with a large telescope, ample exposure time, and sophisticated photo processing to bring out the details. The right image, obtained with an infrared telescope, shows only the central portion of the nebula, revealing many embedded stars which are detected only at infrared wavelengths, which can penetrate the dust. (upper: © 1981 Anglo-Australian Telescope Board, photo by D. Malin; lower: © 1984 Anglo-Australian Telescope Board, image by D. Allen)
http://universe.colorado.edu/fig/12-5.html

A general infrared view of the Orion complex reveals many spots that appear bright at infrared wavelengths, but are invisible at shorter wavelengths (see Fig. 12.5). These infrared bright spots are believed to be young stars in the process of formation, and hundreds of them are seen within this one giant molecular cloud alone. In addition, radio observations (at millimeter wavelengths) reveal very strong emission lines of carbon monoxide and other molecules from throughout this and other cloudy regions **(Fig. 12.6)**, tracing the distribution of the molecular gas. Young stars show up as bright infrared sources partly because their visible and ultraviolet emission cannot penetrate the clouds around them, but also because they happen to emit most of their radiation at infrared wavelengths in the first place. This is due to their relatively low (by stellar standards) temperatures, which range from a few hundred to a few thousand degrees K, depending on where the stars are in the formation process. Thus, by fortuitous good luck, the wavelengths at which young stars emit most strongly also happen to be the same wavelengths at which we have the best chance of being able to observe them

through the clouds in which they are embedded. Nature is not always so kind to astronomers.

The infrared view of young stars fits the general notion that stars form from concentrations of interstellar matter that fall in on themselves under their own gravitation. Young stars are seen primarily within the densest regions inside dark clouds, and measurements of gas motions (from Doppler shifts observed in spectral lines emitted by molecules; see Chapter 6) show that matter falls into these objects (but other, more complex motions occur as well, as described in the next section). Thus the general basis of star formation, described in Chapter 7, appears to be well founded. Nevertheless, a lot of detail remains to be understood.

Disks and Jets

Radio and infrared observations of young stars have shown that often these objects are not simple points, but instead have extended sizes and definite shapes. In particular, very often disklike shapes are revealed in both infrared and radio images. Sometimes, when there is relatively little obscuring matter, these shapes

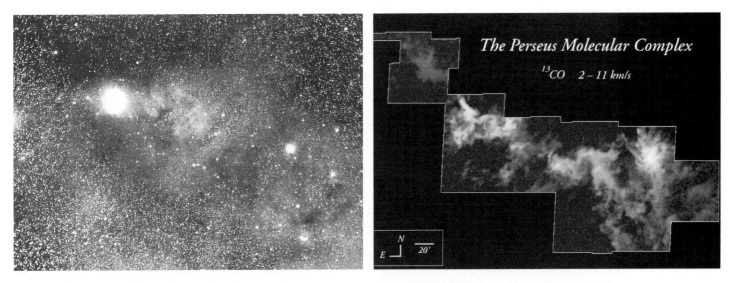

Figure 12.6 Radio-wavelength CO emission in Perseus. Above left is a photo of a region in the constellation Perseus that is suffused with interstellar clouds of gas and dust, which show up as dark regions in visible light. Above right is a map of the same region, on the same scale, revealing emission (color-coded for intensity) of carbon monoxide (the form containing the isotope ^{13}C instead of the more common ^{12}C). The radio CO data help in identifying regions of maximum cloud density, where star formation is most likely to occur. (left: Yerkes Observatory, photo by E. E. Barnard; right: CO map courtesy of D. Devine and J. Bally, University of Colorado)

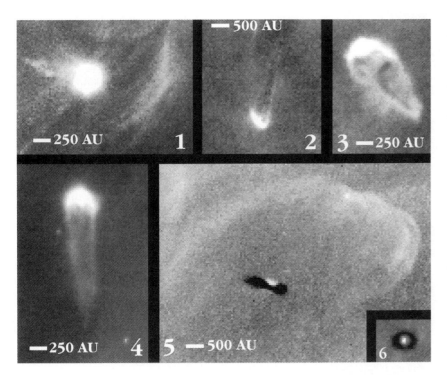

Figure 12.7 *Hubble Space Telescope* images of young stellar disks. These elongated objects, seen by *Hubble* at visible wavelengths, are images of dusty disks surrounding young stellar objects. Most have scales indicating their dimensions in AU, for comparison with the size of the solar system. The disks appear edge-on in some cases (e.g. image No. 5) and nearly face-on in others (images 1 and 6). Intense radiation from nearby very hot stars (not pictured) evaporates gas except in the "shadows" of these disks, leaving tails made up of gas that is shielded from the radiation. (NASA/STScI, courtesy of D. Devine and J. Bally, University of Colorado)

http://universe.colorado.edu/fig/12-7.html

are even seen at visible wavelengths **(Fig. 12.7).** The general existence of disks around young stars strongly confirms what we already suspected: that the disklike structure of the solar system is a natural by-product of the formation of the Sun.

The disks surrounding young stars are often far larger than the solar system, at least the part that contains planets. Evidently, some stars manage to re-

tain large disks, while others lose the outer portions or never have such extended disks in the first place. The fact that disks are detected at all at infrared wavelengths indicates that they contain high concentrations of dust either instead of planets or in addition to them. Planets observed at interstellar distances are too small to be readily seen at infrared wavelengths, but a dusty disk has so much exposed

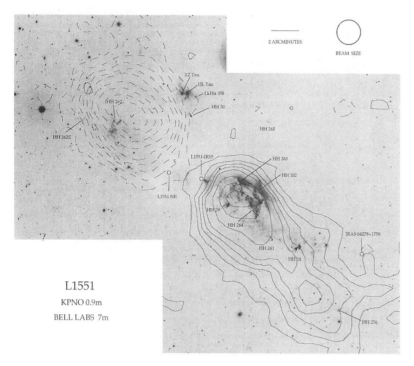

L1551

KPNO 0.9m

BELL LABS 7m

Figure 12.8 **CO map of a bipolar outflow.** This map of radio-wavelength CO emission (shown by contour lines) is superimposed on an image showing where knots of gas are emitting at the wavelength of the strong red line (H-α) of hydrogen. The CO map traces out a broad beam of outflowing gas, moving away from a young star (position indicated as L1551-IRS5) near the center of the image. Labels indicate the locations of many bright knots of gas, known as Herbig-Haro objects (thus the "HH" designations). (Courtesy D. Devine and J. Bally, University of Colorado)

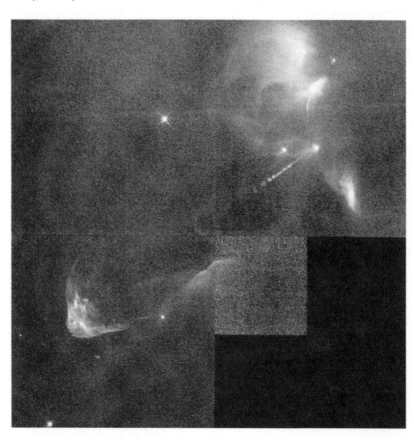

surface area that it can emit enough radiation to be seen easily, even from a great distance. If you imagine a particular amount of mass in the form of either a single, round planet or a vast swarm of tiny particles, you realize that the swarm of small particles has much more surface area exposed to space from which it can radiate, and that therefore it will radiate much more strongly than it would if the matter were all contained in a single large body.

In addition to the disks that are commonly seen, the most deeply embedded infrared sources, thought to be the youngest prestellar objects, are also often characterized by outflows of gas. These streams or jets of gas, called **molecular outflows** because they are most easily detected through their emission of spectral lines from molecules such as carbon monoxide, are usually seen on opposite sides of their associated stellar objects **(Fig. 12.8).** In these cases the outflows are sometimes called **bipolar jets.** Velocities of outflow, measured through the Doppler shifts of the molecular emission lines, are as high as several hundred kilometers per second. These jets can be observed at visible wavelengths in some cases, and detailed examination of their surroundings can reveal shock waves (bow shocks) where the jets impinge on the surrounding interstellar medium **(Fig. 12.9).** Sometimes the jets of rapidly moving gas impinge upon small knots of gas in the surrounding interstellar medium, causing them to glow **(Fig. 12.10).**

Stars in advanced stages of formation, particularly those that have reached the stage where matter is being ejected in outflows, are commonly called **T Tauri stars** after the prototype object, a variable star called T Tauri. It is thought that the Sun underwent a phase as a T Tauri star during its formation, and that the stream of rapidly moving gas it ejected at that time helped to clear the solar system of some of the leftover debris of gas and dust that permeated it.

Having summarized the observational evidence that stars form within dense interstellar clouds and the nature of some of the structures seen there, we are now ready to consider the current state of theoretical models that have been devised to explain how and why star formation occurs.

Figure 12.9 **An optical jet and its associated bow shock.** This striking image from the *Hubble Space Telescope* shows a stream of moving gas blobs (called HH34) moving down and to the left from an unseen young star at upper right in the picture. At lower left is a glowing arc of gas that is excited to emit due to the energy of impact as the leading end of the jet collides with ambient interstellar gas. (STScI/NASA, J. Hester and the WFPC2 investigation team)

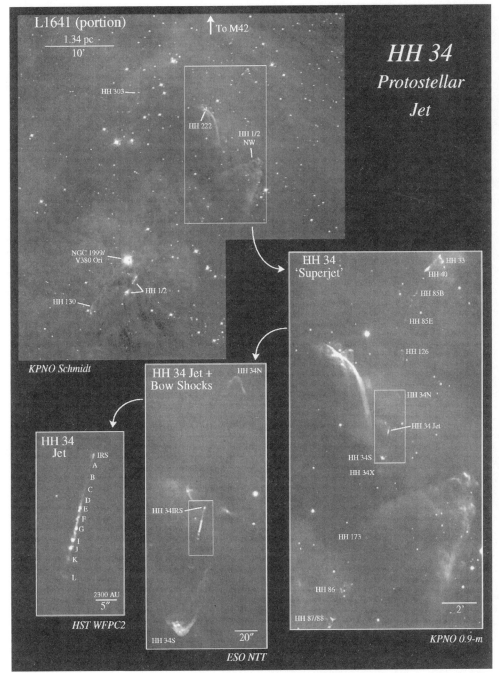

Figure 12.10 Herbig-Haro objects on many scales. This series of images shows details of the region associated with HH34, on decreasing size scales from the largest (upper left) to the smallest detectable with the *Hubble Space Telescope* (lower left). On the large scale, we see faint wisps and arcs of gas due to bow shocks where the bipolar jets are running into the surrounding interstellar gas; at lower right, we see an enlarged view, showing these arcs in more detail and revealing a number of glowing knots of gas (Herbig-Haro objects); at lower center and lower left we see details of these knots at finer and finer scales. It appears that the jets are emitted as a series of blobs, rather than being continuous streams of gas. (STScI/NASA, courtesy of J. Morse, University of Colorado)

The Modern Picture of Star Formation

Within a giant molecular cloud complex are concentrated regions called dense cloud cores. It is in these regions that stars form through gravitational collapse resulting from the forces exerted on individual gas and dust particles by each other.

Cloud Collapse

When a star forms, all the interstellar matter within a large volume of space falls together. Such a collapse would happen spontaneously in every cloud core if there were no source of pressure to counteract it. The pressure that supports most clouds against collapse comes from turbulence within the clouds and from magnetic fields that permeate the galaxy. The existence of turbulence within clouds is known from

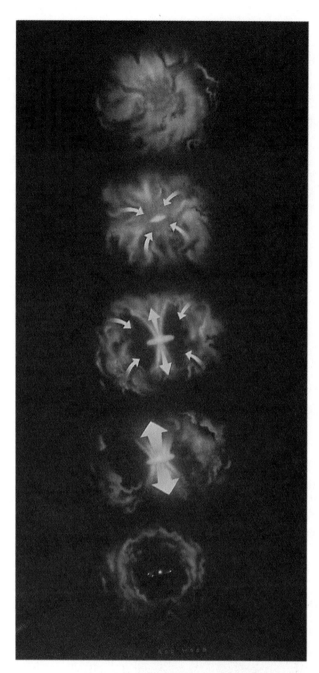

Figure 12.11 **Steps in star formation.** A rotating interstellar cloud collapses most rapidly at the center. At first infrared radiation escapes easily and carries away heat, but eventually the central condensation becomes sufficiently dense that this radiation is unable to escape, and the core becomes hot. The continued infall of gas, combined with rotation, produces a disk a bipolar outflow. The outflow, in time, carries away enough material to negate the infall, cutting off further mass gain and forming a cavity. The remaining disk either dissipates or forms planets as the outer envelope gradually disperses. (Painting by Rob Wood, in collaboration with Charles J. Lada)

the Doppler shifts of spectral lines (usually molecular lines emitted at radio wavelengths). The presence of magnetic fields is also inferred from spectral lines, because a magnetic field can split energy levels in an atom or a molecule, causing spectral lines to appear split (this is explained further in Chapter 14). A magnetic field resists being compressed, thus creating a form of pressure that helps keep interstellar clouds from collapsing.

Occasionally, however, the internal pressure in a cloud core may be overcome, either by random chance as an unusually high density of material happens to congregate in a relatively small region so that gravity becomes overwhelming, or because some outside force squeezes a section of cloud. Such a squeeze can come from an **interstellar shock wave,** a moving disturbance that may originate in a stellar explosion or a fast-moving stellar wind. Independent of any outside forces, the gradual dissipation of the cloud's internal magnetic field may also be a factor. Theoretical calculations show that the field can slowly diffuse outward toward regions of lower density, thus reducing the magnetic pressure that was partially supporting the cloud against collapse. Whatever the event that triggers collapse of a cloud, once begun, the collapse is irreversible.

A combination of theory and observation gives us a general idea of how the collapse proceeds **(Fig. 12.11).** The rate of infall is much faster at the center of the cloud than in the outer portions. Thus a dense concentration builds up relatively quickly at the center, while the atoms and dust particles at the extremities are just beginning to drift inward. The time required depends on the mass of the collapsing region; for a high-mass cloud core, the collapse may take "only" a few hundred thousand years, whereas for a lower-mass cloud core, it may take hundreds of millions of years.

As the core region becomes denser, it also becomes hotter due to compression (technically, due to the conversion of gravitational potential energy into heat energy). For a time, the heat is lost nearly as rapidly as it builds up through the emission of infrared radiation, which escapes into interstellar space. But eventually the increasing density in the core allows hydrogen molecules to form, and because molecular hydrogen is somewhat opaque to infrared radiation, heat energy is trapped in the cloud core. The temperature then rises, causing an increase in internal pressure that slows the collapse. The cloud core makes a transition from a state of rapid free-fall collapse to one in which very slow contraction occurs as heat is bottled up due to absorption of infrared radiation by the hydrogen molecules. Meanwhile, material still falling in from the outer portions of the cloud is hitting the surface of the central condensation, creating additional heating and surface shock waves. At this point the young stellar object is observable as an infrared source embedded within a dark cloud and is called a **protostar (Fig. 12.12).** The infrared luminosity of a protostar depends on its mass, which determines both its temperature and its surface area, both of which contribute to the total amount of energy released (recall the Stefan-Boltzmann law; Chapter 6).

As infall continues, the central temperature continues to rise. In time it becomes too hot for hydrogen molecules to survive, and they begin to break apart due to the energy of collisions as they move about within the cloud, running into one another. As this happens, the cloud again becomes transparent to infrared radiation, and heat energy begins to escape rapidly, as it did during the initial free-fall collapse phase. The cloud core enters a new phase of rapid contraction and becomes both smaller and hotter. But at the same time, other events occur.

In all probability, a cloud core (or any other object in space) would have had at least a slow rate of rotation before collapsing. Even if it did not, it would be likely to acquire some angular momentum (i.e., some rotation) from the outside force (shock wave) that triggered the collapse. Then, as the collapse proceeds, the cloud core spins faster and faster in order to conserve angular momentum (see Chapter 5). In the equatorial plane of the protostar, a centrifugal force develops and resists further collapse in this plane. Centrifugal force refers to the tendency of a mass to keep moving in a straight line (due to inertia), which means that in a rotating system matter tends to fly off into space. In a collapsing interstellar cloud, this tendency halts the infall of matter in the plane of the equator, where the rotational speed is greatest. The result is the formation of a disk, as the matter in the equatorial plane is prevented from further infall while the gas and dust above and below that plane continue to fall toward the center. The same kind of process leads to the formation of disks in many astronomical environments, ranging from the solar nebula to the disks of debris surrounding the outer planets to the disklike structure of galaxies.

If the angular momentum is high enough, it may help to cause the formation of a binary star. Once a disk is formed, if a large fraction of the mass lies in the disk (as opposed to the central condensation), then this matter may coalesce to form a secondary clump with sufficient mass to eventually become the second star in a binary system. After all, in our solar system we have one object (Jupiter) that contains

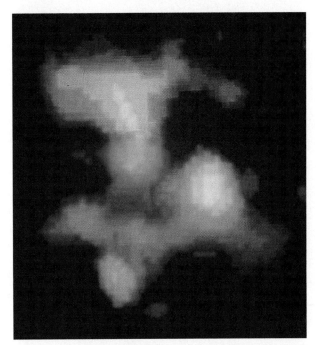

Figure 12.12 **An embedded infrared source.** In this infrared image of a region within the Orion nebula, blue indicates the location of an intense, relatively warm source that is thought to be a newborn star. (R. D. Gehrz, J. A. Hackwell, and G. Grasdalen, University of Wyoming)

more mass than all the other planets combined; if Jupiter had been larger (by a factor of 80 or so), it would have had sufficient mass to become a star. So we may think of the solar system as a failed binary star. Evidently, it is not uncommon for a disk to contain enough mass to form a binary, since a majority of all stars are members of double systems.

The formation of a disk when a cloud collapses helps explain one of the major observational features of young stars that we described in the preceding section. We see now that disk formation is a very natural process and is essentially inescapable. But what of the bipolar outflows that are also commonly seen? And how do we explain the formation of low-mass stars in isolation?

Mass Loss during Star Formation

As mentioned earlier, one of the longstanding problems of star formation theory has been how to explain the formation of low-mass stars. Calculations of the conditions under which a cloud will spontaneously collapse under its own weight show that only clouds of rather high mass will do so. This reasoning was part of the basis for the earlier belief that perhaps

Figure 12.13 **Interstellar erosion.** This spectacular image from the *Hubble Space Telescope* of the Eagle nebula illustrates one of the ways in which material in star formation regions can avoid being accreted onto young stars. The pillars of gas and dust seen here are being eroded away by the intense radiation from nearby young, hot stars. At the tips of some of the pillars are small condensations which may eventually develop into star-formation sites. http://universe.colorado.edu/fig/12-13.html

tion of the solar system was indeed triggered by a shock wave. But this evidence is controversial, and in any event it seems clear that low-mass stars can form without the help of any outside force.

Once again angular momentum conservation comes into play. The rapid spin rate that a cloud core acquires as it collapses builds up such large rotational velocities that after some point no further material can fall onto the protostar at all. Rotation alone can explain why most stars end up having small masses. The real question is how they get rid of the disk of material that has fallen in; without some form of mass ejection, these disks would far outweigh the masses of the stars they surround.

The explanation that is becoming widely accepted is that a protostar can lose mass in large quantities as it forms. The observational evidence for this lies in the detection of molecular outflows and bipolar jets, as described in a previous section, and in the vaporization of the gas surrounding young stars **(Fig. 12.13)**. The T Tauri stars mentioned earlier have been observed for some time, but only recently have they been identified as young stars in the process of ejecting matter. Somehow the infall of material onto the surface of a protostar is turned into an outflow. The exact mechanism that causes the outflow is not well understood, but many different types of stars are known to lose matter in stellar winds, so it is not altogether unreasonable that protostars should find ways of doing so as well.

The magnetic field of a protostar, which is intensified by compression as the star forms from a collapsing cloud, is probably responsible for channeling the outflow into opposing streams or jets that emanate from the poles. An active protostar consists of an equatorial disk and bipolar jets coming from the poles **(Fig. 12.14)**. The mechanism by which material feeds from the disk into the bipolar jets is related to a kind of friction that causes material to gradually spiral inward within the disk. In the absence of such friction, particles in a disk could orbit indefinitely, and the disk could last forever. But collisions between particles cause some to lose energy and fall toward the center of the disk, while at the same time new material is being added at the outer portions of the disk. At the center of the disk, the magnetic field of the protostar accelerates gas particles toward the poles and then into the bipolar jets. A disk that is gaining matter at the outside and losing it on the inside in this manner is called an **accretion disk,** and these are very common in the universe. Accretion disks are found surrounding not only protostars but also mature stars that gain matter in binary systems and exotic objects such as black holes, neutron stars, and the nuclei of active galaxies and quasars.

all stars form as members of clusters. The idea was that as a massive cloud collapses, small portions would reach sufficient densities to begin collapsing individually. In this scenario, a cloud fragments as it collapses, leading to the formation of a cluster of stars. This no doubt does happen; it is clear that clusters do form, and the cloud fragmentation model fits quite well.

But it also has become clear that stars of modest mass can form alone. We observe isolated protostars within dark clouds, and we see mature stars that are so far from companions that a prior life as a cluster member seems very unlikely. So we must explain how a cloud of only a solar mass, or even less, can collapse.

One possible explanation has already been mentioned: the compression of a cloud by a shock wave. If a small cloud is squeezed by a shock, it may reach a density high enough to simply collapse, forming a low-mass star. It has been suggested, on the basis of certain elemental abundances that may have resulted from the explosion of a nearby star, that the forma-

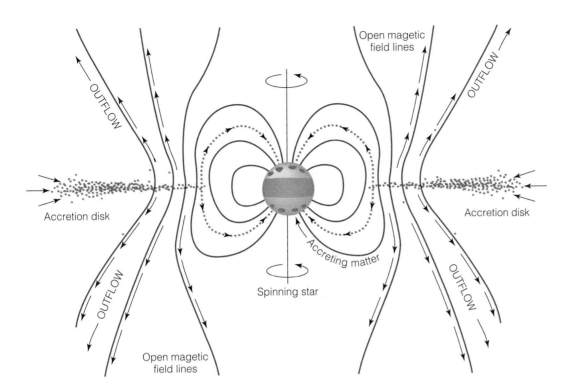

Figure 12.14 Anatomy of a protostellar disk. This is an edge-on view of a young star with a disk and bipolar jets. The accretion disk is formed due to rotation during the infall phase of star formation. As material slowly moves inward through the accretion disk, it becomes heated and ionized, reaching a point where the stellar magnetic field carries the material toward the poles, forming the bipolar outflows. (drawing courtesy of J. Bally, University of Colorado)

Bipolar jets can be very extensive, stretching for many light-years in either direction away from the protostar. Sometimes they may intersect small concentrations of interstellar gas, heating them and causing them to glow. Such knots of glowing gas have been dubbed **Herbig-Haro objects.** Like the T Tauri stars, the Herbig-Haro objects were detected long before their origin was understood; only in recent years has it been noticed that they are often found arranged in linear patterns that trace the bipolar outflows from nearby protostars.

Our story of star formation is almost complete. We have omitted only a description of the final stages of the collapse of a protostar and its ignition as a mature star, and we have not said much about how planets form.

Nuclear Ignition

Let us consider what happens to the protostar that lies at the center of the bipolar jets and the accretion disk. As the density of the protostar continues to increase, so does the internal pressure. The increased central temperature, aided by the fact that the gas again becomes opaque and traps radiant energy, slows the contraction of the protostar. As the protostar reaches a state where gravity is nearly balanced by pressure, the contraction becomes very slow. The protostar is nearly capable of living on its own as a star, except that some energy loss still occurs, and the protostar is doomed to continue shrinking slowly.

What it needs now is a source of energy to replenish what is being lost to space.

The ignition of nuclear reactions provides the needed source of energy. The principal reaction that occurs is the merging of hydrogen atoms (actually only the nuclei, which consist of single protons). This can happen only where the density and temperature are very high, that is, at the core of the protostar. High density is needed to ensure a high frequency of collisions between nuclei, while high temperature is needed to provide sufficiently high particle speeds so that the nuclei can collide with each other despite the electrical forces that try to keep them apart (atomic nuclei have positive electrical charges, and like charges repel each other).

When protons merge, they form heavier particles. Subsequent mergers build up even heavier nuclei. A stable form of matter is reached when helium nuclei (consisting of two protons and two neutrons) are formed. Thus the initial reaction phase in the core of a young star entails the conversion of hydrogen to helium.

Energy is produced in the process; each time two nuclei combine, a small fraction of their combined mass is converted into energy according to the famous equation $E = mc^2$, where E is energy, m is the converted mass, and c is the speed of light (this is discussed further in the next chapter). The resulting energy is sufficient to counteract the energy loss at the surface of the protostar, and the object enters a stable phase that will last for billions of years in some cases. Thus we say that a protostar becomes a star when its internal temperature becomes high enough for nuclear reactions to begin.

Planet Formation

During the time of slow contraction and nuclear ignition, the accretion disk is being dissipated through the bipolar jets. A stellar wind emanating in all directions apparently occurs as well, and as a result, the remaining debris within the disk is re-moved from the vicinity of the star. In addition, gas may be evaporated from the vicinity of the young star by radiation, if the star is in an environment where massive stars are producing large quantities of ultraviolet photons (see Fig. 12.13). Only large objects can survive this, so planets must form by the time the parent star reaches this stage. (This begs the question of how some mature stars can retain dusty disks, as observed. The answer is not clear, but one possibility is that these systems also contain larger bodies, whose occasional collisions produce new dust to replace what was lost. In the cases of very large disks—much larger than the solar system—the disks may simply be too extensive and massive to be completely removed by a stellar wind or by vaporization due to the in-tense radiation.)

Condensation of solid matter within an accretion disk is probably triggered by the presence of solid interstellar dust particles. As these tiny grains collide with atoms and molecules in the gas, they serve as collecting points for the deposition of additional mateial. Even at this early stage, volatile and refractory elements begin to separate, as the refractories (elements that are not easily vaporized) are more readily able to condense out of the gas onto solid surfaces. The condensation process is temperature-dependent, meaning that at the higher temperatures of the inner disk, the volatiles are even less likely to condense than in the colder outer disk.

As the dust grains build up, they gradually grow from a few micrometers to millimeters to centimeters to meters, and so on. Eventually, large bodies, hundreds of kilometers in diameter, are formed; these are the planetesimals to which we have often referred (especially in Chapters 7 and 11). Once the disk matter is concentrated in planetesimals, calculations show that the buildup of planets occurs relatively quickly. Soon the disk is converted into a family of planets with substantial quantities of interplanetary debris left over. Over the next many millions of years, this debris is gradually cleared out, either by impacting on the surfaces of the planetary bodies or by being ejected from the system due to close encounters with massive, Jupiter-like planets.

This scenario nicely accounts for the observed properties of our solar system; that is, it answers the questions we raised at the beginning of this chapter. The formation of a disk, with subsequent condensation of solid bodies and the buildup of planets, explains why the solar system has a flattened, disklike structure with all the orbital and spin motions going in the same direction. The condensation process, with its dependence on temperature, explains why the inner planets are composed primarily of heavy, refractory elements. And the residual material, the leftover debris of planetesimals, explains the existence and properties of asteroids, comets, and interplanetary dust. Thus we would expect other planetary systems to display the same general characteristics.

It is not known whether planets can form in binary star systems or, if they do form, whether they can survive. Based on what has been discussed so far, we might expect that planets could form. But if they did, they would surely have different properties than planets orbiting a single star, particularly if the two stars were close together. In that case, the radiation, the stellar wind, and the tidal force due to the second star would have significant effects on the planets. It is also likely that the presence of a massive stellar companion would create gravitational disturbances that, over time, would create chaotic, random orbits for the planets and possibly eject them from the system altogether. We probably should not consider planets of binary stars as likely homes for life because it is thought that stable conditions over long periods of time are required, and such stability would be unlikely to occur.

Brown Dwarfs

The smaller the stellar mass, the more numerous are the stars. According to calculations, the smallest star that can become hot enough to ignite nuclear reactions in its core has a mass of about 0.07 times the Sun's mass. The most massive planet in our system has a mass far smaller than this, only about 10^{-3} times the solar mass. These facts suggest that there may be a large class of objects that are intermediate between stars and planets. These objects would not have nuclear reactions, but would be warm due to the heat of formation. They would be very dim at visible wavelengths. Such objects have been dubbed **brown dwarfs.**

Realizing that brown dwarfs might be common yet difficult to detect, astronomers embarked on a laborious search. They sought brown dwarfs in binary systems, where their gravitational effects on companion stars might betray their presence, and in sensitive infrared images of star clusters, because faint sources emitting primarily at infrared wavelengths could prove to be brown dwarfs. Many faint infrared objects were found, but most were deemed to be very dim but normal stars. A few brown dwarf candidates were found, but until recently it was not clear that any of them were genuine brown dwarfs.

This changed in 1995, when several strong candidates were found, and at least two of them were demonstrated almost certainly to be legitimate brown dwarfs. One, an object in the Pleiades star cluster, is a faint point source that has a line due to lithium in its spectrum. Lithium is a very rare, light element (it comes after helium, having three protons in its nucleus), and it is readily destroyed by nuclear reactions of any kind. Thus stars do not have lithium in their spectra because reactions have destroyed it. Astronomers expect to see lithium only where no reactions have occurred. Therefore the faint object in the Pleiades cannot be a true star. It must be an object where no nuclear reactions have occurred—that is, a brown dwarf.

An even stronger case can be made for a dim object recently found orbiting a faint normal star. Infrared spectra showed the presence of methane (CH_4) in the atmosphere of this object, while orbital analysis set a limit on its mass of no more than about 50 times the mass of Jupiter, far too low to be a star. Once this object had been analyzed in infrared light, the *Hubble Space Telescope* was able to detect it directly at visible wavelengths **(Fig. 12.15).**

The number of brown dwarfs in the galaxy is still unknown, but the recent discoveries suggest that these substellar objects are probably quite numerous. Other candidates in the Pleiades have been identified, for example, and lithium data will soon be available. The number of confirmed brown dwarfs will almost certainly grow. These discoveries support the general picture of star and planet formation presented in this chapter and will play a major role in our understanding of the overall distribution of matter in our galaxy and others.

The Search for Extrasolar Planets

How do we go about trying to detect planets—small, dim objects orbiting close to distant stars? To understand the difficulty of doing this, consider that a star, even a relatively modest one like the Sun, is several billion times brighter than any planet, even one as large and reflective as Jupiter. Furthermore, a planet will appear very close to its parent star on the sky; from a distance of 10 light-years, Jupiter would be

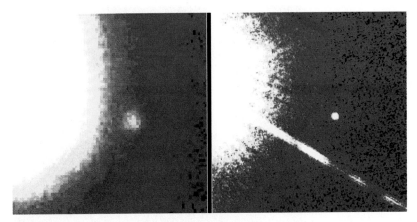

Figure 12.15 **A brown dwarf.** This *Hubble Space Telescope* image (right) shows a small, dim object close to the dwarf star Gliese 229. This image, combined with information derived from infrared observations (left), has convinced astronomers that the dim companion (dubbed Gliese 229B) is a brown dwarf, an object too low in mass to become a star, yet much larger and more massive than a planet. (STScI/NASA)
http://universe.colorado.edu/fig/12-15.html

only about 1.7 arcseconds away from the Sun. Therefore we have to decide how to go about detecting an object this close to a parent star that is several billion times brighter.

Methods for Finding Planets

The resolution of the *Hubble Space Telescope* is sufficient to separate a planet and a star, given a distance of 10 light-years or less and a planet as distant from its parent star as Jupiter is from the Sun. The problem is how to detect an object that is so dim and located very close to a star that is billions of times brighter. Some leverage can be gained by observing at infrared wavelengths, where a planet emits the bulk of its energy (this is where the peak of the planet's thermal radiation, as opposed to reflected light from its sun, is emitted; see the discussion of Wien's law in Chapter 6). Even so, the planet is dimmer than the star by a factor of millions. Still, using infrared wavelengths enhances the possibility of directly detecting a planet, and this will be one of the major goals of the *HST* after its new infrared camera is installed in early 1997.

An extrasolar planet may also be detected indirectly based on the effect of the planet on the motion of the parent star. Recall that in a planetary system, the planets and the sun orbit the center of mass of the system. A star orbited by several planets actually undergoes complex motion on a relatively small scale, and if we can somehow detect this motion, we might infer from it the presence of planets.

To see how the star's motion may be used to detect a planet, consider Jupiter and the Sun. The center of mass between them is located at the point

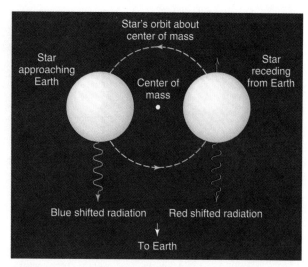

Figure 12.16 Stellar motion due to planets. A star accompanied by one or more planets will undergo small orbital motion about the center of mass of the system. Planets can be detected indirectly by measuring the motion of the parent star. One way to detect this motion is through the Doppler shift in the stellar spectrum, as the star alternately moves toward and away from the Earth.
http://universe.colorado.edu/fig/12-16.html

where the product of mass times distance from the center of mass for the two bodies is equal; that is,

$$m_S r_S = m_J r_J,$$

where m_S and m_J represent the masses of the Sun and Jupiter and r_S and r_J are their respective distances from the center of mass **(Fig. 12.16)**. Solving this for the Sun's distance from the center of mass and substituting values for the masses, we get

$$r_S = r_J(m_J/m_S) = r_J(1/1,047).$$

If we work this out, we find that $r_S = 0.00497$ AU = 7.45×10^5 km = 1.07 times the Sun's radius. Thus the Sun orbits a point just outside its own surface with a period equal to that of Jupiter, that is, 11.86 years.

The motion of the Sun would be too small to detect directly from interstellar distances (from 5 light-years' distance, the angular motion would amount to only 0.007 arcseconds), but it might be detected using indirect methods. Perhaps the best hope lies in the use of the Doppler shift. Here again we are seeking to detect the small orbital motion of the parent star, but now we try to detect this motion by looking for periodic Doppler shifts in the spectral lines of the star. As the star moves around in its tiny orbit, its spectral lines will be blueshifted as the star approaches us and redshifted when it recedes (see Fig. 12.16). For the Sun's motion about the center of mass with Jupiter, the orbital speed is about 13 m

per second. This creates only a very small Doppler shift, which is very difficult to detect. Astronomers have developed special instruments for the job, however, and shifts this small can be seen if the observations are repeated frequently.

But difficulties remain. A star can undergo small Doppler shifts in many other possible ways, and it is difficult to determine whether observed shifts are actually due to orbital motion. One way to tell is to watch a star for many years, giving it time to go through several orbits; the pattern of Doppler shifts alternating between redshifts and blueshifts should repeat itself. But this takes patience; remember that the period of the Sun-Jupiter pair is almost 12 years. To complicate matters, some stars periodically expand and contract, so that their oscillating surfaces give rise to Doppler shifts that alternate between redshifts and blueshifts, mimicking the effect of orbital motion.

Jupiters Abound!

In 1995 a Sun-like star called 51 Pegasi was found to be undergoing the kind of Doppler shift pattern that was sought, and astronomers believe that a legitimate, Jupiter-like planet has been found. But this planet has some very curious aspects. Its orbital period is only 4.15 days. Using Kepler's third law, we find from this that the orbital semimajor axis is

$$a = (P^2)^{1/3} = [(0.011 \text{ yrs})^2]^{1/3} = 0.051 \text{ AU}.$$

(Note that we assumed the mass of the parent star was equal to one solar mass, so we used the simplified form of Kepler's third law $P^2 = a^3$; see Chapter 4.) Thus this planet is orbiting very close to the star. Calculations suggest that a Jupiter could survive at such a distance, but everything we think we know about planetary formation indicates that it could not *form* there. For this reason astronomers are considering the possibility that this planet formed farther away from the star and then spiraled in, perhaps due to loss of orbital speed caused by debris in a disk around the star.

The mass of the planet was estimated from the orbital motion of the star. The Doppler shift provides a value for the star's orbital speed about the center of mass, and the speed can then be combined with the orbital period to find the circumference of the orbit and hence its radius (this is easiest if the star's orbit is assumed to be a circle, for which the circumference is $2\pi r = vP$, where v is the orbital velocity, P is the period, and r is the orbital radius; that is, the distance of the star from the center of mass; see Fig. 12.16).

Then the center-of-mass relation (above) yields the ratio of the planet's mass to that of the star. It was found that the planet orbiting 51 Pegasi has a mass near that of Jupiter.

In early 1996, several additional planetary discoveries were announced **(Table 12.1)**, in all cases involving Jupiter-like planets orbiting solar-type stars. In most of these systems, the massive planets orbit farther from their suns than in the 51 Pegasi system **(Fig. 12.17)**, but there are still significant contrasts with our own solar system. The planet orbiting 70 Virginis roams in an eccentric orbit between 0.3 and 0.8 AU from the star and has an estimated mass of 8.1 Jupiters. The planet orbiting 47 Ursae Majoris is more modest; it has a mass of about 3.5 times that of Jupiter and orbits at a distance of about 2 AU. Additional candidate systems are under study, and it seems likely that more giant planets will soon be officially confirmed.

Finding Earths

Detecting Jupiters is one thing; detecting Earths is quite another. All of the difficulties enumerated here are intensified, because terrestrial planets are smaller and dimmer than Jupiters and have smaller masses, so

Table 12.1

Planets Orbiting Other Sun-like Stars
(Extrasolar Planet and Brown Dwarf Properties)

Object	Mass	a(AU)
HD 114762 B	> 10 M_J	0.4
51 Pegasi B	> 0.47 M_J	0.051
47 Ursae Majoris B	> 2.4 M_J	2.1
70 Virginis B	> 6.6 M_J	0.45
55 ρ' Cancri		
B	> 0.78 M_J	0.11
C	> 5 M_J	≈5
Lalande 21185		
B	≈ 1.5 M_J	≈10
C	≈ 1 M_J	≈2.5
τ Bootis B	> 3.7 M_J	0.047
Upsilon Andromedae B	> 0.6 M_J	0.054

$M_\oplus$ = mass of Earth = 5.974×10^{24} kg.
M_J = mass of Jupiter = 318 $M_\oplus$.
a = semimajor axis of planet's orbit.
e = eccentricity of planet's orbit.
AU = astronomical unit (Earth–Sun distance) = 1.496×10^{11} m.
B = secondary companion (for example, 51 Pegasi B) to a star (51 Pegasi A).

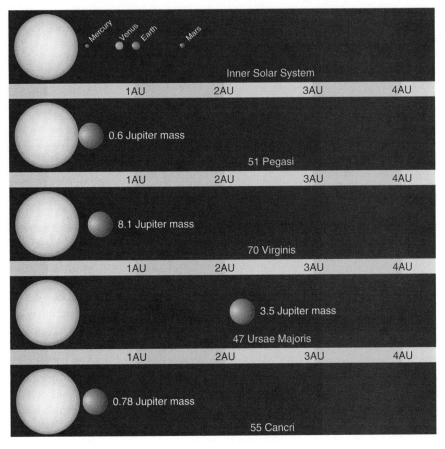

Figure 12.17 New solar systems. This drawing shows the relative sizes and distances of some of the newly-discovered planets from their parent stars. The solar system is shown for comparison.

The Chemistry of Life: Is the Earth Special?

Any planetary system that forms from interstellar gas and dust will reflect the chemical composition of its prenatal cloud. In this chapter we mentioned the composition of the cloud that gave birth to the solar system, and Appendix 9 lists the relative abundances of the elements in the Sun. You see from both sources that hydrogen and helium dominate the composition of the Sun, with all the other elements combining to total only about 2% of the mass. The solar system was built from material with this composition. Normally, we assume (as we did in this chapter) that other stars, and any planetary systems that might accompany them, were formed from material having the same composition. But this may not be strictly true, and the small differences between the solar system and other stars may be important.

A comparison between the composition of the Sun and that of other stars of similar (or younger) age in our part of the galaxy shows that the Sun is overendowed in the elements heavier than helium. In other words, the small fraction of the solar composition that is attributed to heavy elements is even smaller in other stars. The reason for this is not clear. It may be that the Sun simply represents the upper end of a range of normal values, or it may be that the solar system was infused with an extra dosage of heavy elements at the time of its formation.

There is evidence to support the latter viewpoint: ratios of different isotopes of certain elements suggest that the solar system was born with enhanced abundances of the nuclear products of a supernova explosion. It is possible that the formation of our system was triggered by a nearby supernova, which injected its nuclear debris into the collapsing cloud that was to form the Sun and planets.

The overabundance of the heavy elements is roughly a factor of two. The Sun has about twice as much iron, manganese, nickel, and the like, relative to hydrogen, as other stars typically have. This might limit the masses and sizes of the planets that can form around other stars, although that is not clear. Also unclear are the implications of the fact that carbon is among the elements that are more abundant in the solar system than elsewhere.

Carbon is the principal element on which life-forms depend. All large organic molecules have carbon as their main constituent, and it can be argued that only carbon has a sufficiently complex chemistry (diversity of chemical bonds and compounds) to provide for the needs of life-forms undergoing metabolic activities. If we take the point of view that carbon is necessary for life to exist, then we must ask how life fares in other solar systems, where there may be only half as much carbon as there is here.

Many complexities make it difficult to answer this question. It may be that local processes such as the accretion of carbon from cometary impacts are more important than the original carbon abundance in determining the quantity of carbon available for life. If so, then a planet in a system with less overall carbon than ours may still harbor life, if impacts have brought enough carbon to the planet's surface.

But if the quantity of carbon initially available in a planetary system is a decisive factor in enabling life to form, then the solar system as a home for life may be more unique than normally thought. Astronomers are fond of pointing out that most of the elements of which our bodies are made were produced in supernovae, referring to the fact that as the galaxy has aged, the heavy elements have been produced in stars and then dispersed throughout the interstellar medium by supernova explosions (and some elements are formed in the explosions themselves). Now we see that this saying may have even more truth than we thought, if the presence of life here depends critically on a supernova explosion that happened right next door as the solar system formed.

they will have less effect on the motions of their parent stars. It seems that new technologies will have to be developed before we can realistically expect to detect Earth-like planets.

One possibility, perhaps the simplest in concept, is to carefully measure the brightnesses of stars, looking for the dimming that is expected when a planet crosses in front of its parent star. The reduction in brightness caused by such an eclipse is very small, however. A planet passing in front of its star would block out an area equal to its cross section, a circle having area πR^2 where R is the planet's radius. The rest of the light coming from the star's disk would not be eclipsed. The Sun's radius is 109 times that of the Earth, so the ratio of the Earth's cross-sectional area to the area of the Sun's disk is $(1/109)^2 = 8.4 \times 10^{-5}$. This means that the Earth would block only 8.4×10^{-5} of the Sun's light as it passed in front of the Sun. To detect such events in other planetary systems, we would need to be able to detect brightness variations of only a few parts in 100,000. This is beyond the normal accuracy of stellar brightness measurements, but not hopelessly so. Instruments have been proposed to make such measurements, and perhaps soon the attempt will be made.

Note that only a small fraction of existing planetary systems could be located by this technique because it requires that the orbital plane of the planet be aligned with the line of sight toward the observer on the Earth. This would occur only through random chance, but given enough systems, at least some would happen to have the needed alignment. To find planets this way, one would have to either survey a lot of stars in hopes that some would have the needed alignment or pick systems where the orientation of the orbital plane is already known to be edge-on as seen from the Earth (and this information is normally available only for binary star systems, which may not be the best places to look for planets). Another complication is that most stars are variable in brightness at some level, so care would have to be taken to distinguish fluctuations in brightness that are intrinsic to the star from dimmings due to planetary crossings. This could be accomplished by looking for regular, repeated dimmings caused by the orbital motion of the planet. To determine whether the dimmings are periodic would require making the observations over a period of years.

Other methods for detecting Earth-like planets are under discussion but may not be feasible until farther in the future. One possibility is space-based interferometry, which would have the capability of detecting small, close-in planets directly. A space-borne interferometer would have other important applications, and such an instrument will be seriously considered as a new space observatory following the *Hubble Space Telescope* era. Another technique for seeking Earth-like planets is to look for their gravitational effects on light coming from more distant stars. A local gravitational field causes deflections in light passing by, and this **gravitational lensing** has been used as a tool for finding dim bodies (see the discussion in Chapter 16). Even a body as small as a planet causes some light deflection that may be detectable, so it has been proposed that this technique could reveal planets. The major disadvantage is that it would be difficult to know where the detected planet is between Earth and the background star whose light is deflected.

If an Earth-like planet is found, the next question will be whether it harbors life. One technique for finding out might be to perform infrared spectroscopic measurements to see whether the planet's atmosphere contains ozone. Ozone is a by-product of free oxygen in the Earth's atmosphere, and the only source of free oxygen on the Earth is plant life. Thus the detection of ozone in the atmosphere of an extrasolar planet would suggest that plant life has developed there. The technology needed to do this is probably several years in the future, however, because it requires not only separating the light of the planet from that of the parent star, but also performing spectroscopic measurements on that dim light.

The current leadership of NASA has identified the search for Earth-like planets as a major goal as we approach the next millennium, and we may expect to see some serious efforts mounted to accomplish this objective. Perhaps one of the techniques described here will prevail, or maybe some new method will be devised. In the meantime it is very exciting that other solar systems, revealed by their giant planets, are starting to be found.

Summary

1. Astronomers assume that certain general properties of our solar system are likely to appear in other planetary systems as well and therefore must be explained by any successful theory of star and planet formation. These general properties include an overall disklike structure, orbital and spin motions in the same direction, probably distinct classes of planets (e.g., terrestrial and gas giant), and the existence of interplanetary bodies.

2. Stars are gaseous bodies having a wide range of masses, luminosities, and temperatures; the Sun is near average in these properties. Their compositions are dominated by hydrogen, and they are powered by nuclear reactions in their cores. The formation and evolution of a star are governed by gravity, which holds it together and determines its internal properties.

3. Most stars are members of double or multiple systems, and many are in clusters; thus a star formation theory must explain how stars can form individually or in groups.

4. Stars form from interstellar gas and dust in an interstellar medium that is very patchy, but has a very low density on average. Stars form inside the densest of interstellar clouds, where molecular gas (molecular hydrogen and carbon monoxide mostly) dominates.

5. Young stars and stars in the process of forming are observed inside dense clouds, primarily through infrared observations of regions such as the Orion nebula.

Discovering New Planets

ASTRONOMICAL

ACTIVITY

To help you appreciate the challenges astronomers face in trying to detect planets orbiting other stars, this activity involves deriving the observable properties of a hypothetical (but realistic?) planetary system orbiting a nearby star. You will start with the actual properties of the system, and then you will calculate such observable properties as the separation between the star and its most massive planet, the relative brightness of the two, and the angular extent of the star's reflex motion as well as the Doppler shift you would expect.

Let's start with the properties of the hypothetical system. Suppose the star is identical to the Sun in all intrinsic properties such as mass and luminosity, so you can find data on the star in Appendix 4. Suppose the star is 6.84 light-years away. Now suppose there is one very massive planet orbiting at a distance of 4.83 AU and one terrestrial planet orbiting at a distance of 0.96 AU (let's say the orbits are circular). The giant planet has a mass of 1.8 times the mass of Jupiter and a radius of 1.26 times that of Jupiter, while the terrestrial planet has a mass of 1.16 Earth masses and a radius of 1.10 Earth radii (to convert these values to basic units, use data on Jupiter and the Earth from Appendix 8). The albedo of the giant planet is 0.65 and that of the terrestrial is 0.34.

First, calculate the angular separation of each planet from the parent star. To do this, you can use the small-angle approximation:

$$\theta = \frac{206,265\,a}{d},$$

where θ is the angular separation in arcseconds, a is the semimajor axis of the planet's orbit, and d is the distance to the system (important: the units of a and d must be the same).

Next, calculate the orbital periods for the two planets, using Kepler's third law. Once you have done that, you can calculate the orbital speeds of the planets. Speed is distance divided by time, so the orbital speed is

$$v = \frac{\text{orbital circumference}}{\text{orbital period}} = \frac{2\pi a}{P}.$$

You can also calculate the orbital speed for the star about the center of mass of the star-planet system. This is only worth doing for the more massive star (why do we say this?). Start by calculating where the center of mass is, remembering that the center of mass is the point between two bodies where the product of mass times distance for the two is equal. So you need to find the ratio of mass between the star and the massive planet, and this is the inverse of the ratio of their distances from the center of mass. Once you have used this to find the distance of the star from the center of mass, you can calculate the circumference of the star's orbit about the center of mass, hence the orbital speed as above.

Now you are ready to start thinking about detecting these planets. Start by comparing their angular separations

6. Young stars often are seen to have disks of gas and dust surrounding them, and many also have jets or bipolar outflows of material emanating from their polar regions.

7. The modern theory of star formation starts with the gravitational collapse of a dense cloud core, which may occur spontaneously or be triggered by an outside force such as a shock wave. As collapse proceeds, the cloud's core becomes warm, particularly once it is dense enough to absorb infrared radiation, and the core becomes observable as a protostar. As contraction proceeds, rotation of the cloud causes the formation of a disk in the equatorial plane of the newly forming star.

8. The young star may lose mass through a combination of bipolar outflows, a T Tauri wind, and evaporation of surrounding gas due to the ultraviolet radiation coming from nearby hot stars. Thus even if the collapsing cloud is initially massive, the resulting star can be as small as the Sun or even smaller.

9. The star's core eventually becomes hot enough to allow nuclear fusion reactions to begin. These reactions provide the necessary heat energy—hence pressure—to counterbalance gravity, and the star enters a long-lived stable state.

10. Planets form as tiny solid particles in the disk surrounding the young star condense and then build up in size, eventually forming planetesi-

from the parent star with the resolving power of a large telescope in space. The formula for resolving power is

$$\theta = \frac{251{,}643\lambda}{D},$$

where θ is the smallest angular separation (in arcseconds) that the telescope can detect, λ is the wavelength of observation, and D is the telescope diameter (again, λ and D have to be in the same units). Calculate the resolving power for the *Hubble Space Telescope* ($D = 2.4$ m) at wavelengths of 5500 Å (5.5×10^{-7} m) and 2.2 microns (2.2×10^{-6} m). Compare your answers with the angular separations of the two planets from the parent star. Could the *HST* resolve either planet? At either wavelength?

Now consider the relative brightnesses of the planets compared to the star. Assume that the ratio of a planet's brightness to that of the star is equal to the fraction of the star's light that the planet reflects. This fraction is given by the ratio of the planet's cross-sectional area (πR^2) to the total area that is illuminated by the star at the planet's distance from it (this is the area of a sphere whose radius is equal to a, the planet's orbital semimajor axis; i.e., $4\pi a^2$). Multiply by the albedo of the planet and you get

$$\frac{I_P}{I_S} = \frac{\pi A R^2}{4\pi a^2} = A \left| \frac{R}{2a} \right|^2,$$

where I_P/I_S is the observed ratio of the brightness of the planet compared to the star, A is the planet's albedo, R is the planet's radius, and a is its orbital semimajor axis (R and a must be in the same units).

Calculate the brightness ratio for each planet compared to the star, and comment on whether the planets could be detected.

Finally, consider the possibility of detecting the more massive planet through the reflex motion of the star as it orbits the center of mass. This may be attempted in two ways: direct detection of the angular shifting of the star's position on the sky, or by detection of the Doppler shift of the star's spectrum as it orbits the center of mass. You have already calculated the star's orbital speed about the center of mass, so you can do the Doppler shift calculation quite easily: use the star's orbital speed in the Doppler shift formula (Chapter 6) to calculate the wavelength shift of a line in the star's spectrum. Use 5000 Å as the rest wavelength and find $\Delta\lambda$, the shift. If a really good spectrograph can detect shifts as small as 0.0001 Å, can this star's motion be detected? What special conditions must be met regarding the orientation of the planetary orbit, if we are to have a chance of detecting this Doppler shift?

To find the range of angular motion the star undergoes on the sky, assume that it moves back and forth through a distance equal to twice the semimajor axis of its orbit about the center of mass (i.e., twice the distance between the star and the center of mass, already calculated above). Then use the same small-angle approximation as before, this time inserting $2a$ where we had a before, with this a referring to the *star's* semimajor axis. Is an angular shift of this magnitude detectable? Comment.

The properties we invented for this exercise seem quite reasonable for a planetary system orbiting a nearby solar-type star. By now you should have a realistic appreciation of the difficulties in detecting such systems, and especially in finding Earth-like planets beyond our own solar system.

mals, which subsequently merge to form planets. Lightweight gases cannot condense into solid form in the inner disk, but they can in the colder, outer disk, thus explaining the contrast between the dense terrestrial planets and the fluid gas giants.

11. Brown dwarfs are objects too massive and warm to be planets, but not massive enough to start nuclear reactions and become stars. The first unambiguous detections of brown dwarfs occurred in 1995, after years of searching by astronomers.

12. A planet orbiting another star may be detected directly, although this is made difficult by its dimness compared with its parent star, or it may

be detected indirectly by the measurement of the orbital motion of the parent star about the center of mass between the planet and the star. By early 1996 several Sun-like stars had been discovered to have giant planets orbiting them. Earth-like planets will be much more difficult to detect, but techniques to do so are beginning to be developed.

Review Questions

1. List properties of the solar system or individual planets that you would not necessarily expect to be duplicated in a planetary system orbiting

another star. Explain your choices.

2. What is the approximate range of masses for normal stars? In terms of mass, which stars are the most numerous, and which are the rarest?

3. What are the different kinds of stellar clusters we see in our galaxy, the Milky Way? Describe the basic properties of these different kinds of clusters.

4. Describe the physical properties of the interstellar medium (ISM), highlighting the temperatures and densities found in the different states of interstellar material. Where does new star formation occur in the ISM?

5. Summarize the modern view of star formation, with special emphasis on the formation of the protostellar disk.

6. Why is mass loss an important process in star formation? What is thought to be the primary mechanism for mass loss in protostars? Give an example of an observed phenomenon resulting from this mass loss mechanism.

7. What is the principal nuclear reaction that occurs in young stars?

8. Describe the processes whereby planets form within the protostellar disk.

9. What are brown dwarfs? Why are they so difficult to detect? What is an observational test to verify that an object is a brown dwarf and not a star?

10. Why are extrasolar planets so difficult to detect? Describe the method used to detect Jupiter-like planets orbiting solar-type stars.

Problems

1. Calculate the orbital period for a planet orbiting at a distance of 1 AU from a star having a mass equal to one-half the mass of the Sun; a 2-solar-mass star; a 10-solar-mass star; and a 50-solar-mass star.

2. Estimate the intensity of sunlight, relative to that on the Earth, for a planet orbiting at a distance of 5 AU from a star having 50 times the Sun's luminosity. Do the same for a star orbiting at 0.5 AU from a star having one-tenth of the solar luminosity.

3. If an interstellar cloud has an average density of 1,000 atoms/cm^3 (= 10^9 atoms/m^3), how much cloud volume is required to contain a mass equal to that of the Sun? If the volume is spherical, what is its radius in light-years? How much does the density have to increase as the cloud collapses to form a star like the Sun? (*Hint:* In the first part of this problem, assume that all the atoms are hydrogen in order to find the mass density in units of kg/m^3, then calculate how many m^3 are needed to contain a mass equal to that of the Sun. Once you have the volume, use the formula $V = \frac{4}{3}\pi R^3$ to find the radius, R, for the second part of the problem. For the third part, you need to compute the density of the Sun by dividing its mass by its volume.)

4. Suppose a protostar, buried inside a dense interstellar cloud, has a temperature of 1,000 K. Calculate the wavelength at which it radiates most intensely. What kind of telescope is needed to observe this protostar most efficiently?

Web Connections

The Review Questions and Problems also appear at the following URLs:
http://universe.colorado.edu/ch12/questions.html
http://universe.colorado.edu/ch12/problems.html

5. The blockage of starlight by interstellar dust grains is most effective if the wavelength of light is comparable to the diameter of the dust grains. In most clouds, the typical dust grain diameter is 5×10^{-7} m. How does this compare with the wavelength of visible light? How does it compare with the wavelength emitted most strongly by a protostar of temperature 1,000 K (see the previous problem)? What does this suggest about the best method for observing protostars inside interstellar clouds?

6. If a protostar has a temperature of 1,000 K and a radius 1,000 times the Sun's radius, what is the luminosity of the protostar?

7. If bipolar jets are being ejected by a young star at a speed of 100 km/sec, how long does it take for the star to produce jets that are 10 light-years long?

8. If the bipolar jets mentioned in Problem 7 carry away mass at a rate of 10^{15} kg/sec, how long would it take for the young star to shed a mass equal to that of the Sun?

9. What is the orbital period of the planet that has been detected orbiting the star 47 Ursae Majoris? Assume that the semimajor axis is 2 AU and the star's mass is 1.2 times the mass of the Sun.

10. Locate the center of mass between the planet orbiting 47 Ursae Majoris and its parent star; that is, how far from the center of the star is the center of mass of the system? Assume the planet has three times the mass of Jupiter and the star has 1.2 times the Sun's mass. Using the value for the period from Problem 9, what is the orbital speed of the star as it circles the center of mass? (*Hint:* See the Astronomical Activity accompanying this chapter.)

Additional Readings

Angel, J. R. P. and N. J. Woolf 1996. Searching for Life on Other Planets. *Scientific American* 274(4):60.

Black, D. C. 1996. Other Suns, Other Planets? *Sky & Telescope* 92(2):20.

Blandford, R. and A. Königl. 1993. The Disk-Jet Connection. *Sky & Telescope* 85(3):40.

Boss, A. P. 1995. Companions to Young Stars. *Scientific American* 273(4):134.

Bruning, D. 1992. Desperately Seeking Jupiters. *Astronomy* 20(7):36.

Caillault, J.-P. 1994. The New Stars of M42. *Astronomy* 22(11):40.

Henry, T. J. 1996. Brown Dwarfs Revealed—At Last! *Sky & Telescope* 91(4):24.

Kirshner, R. P. 1994. The Earth's Elements. *Scientific American* 271(4):58.

Lada, C. 1993. Deciphering the Mysteries of Stellar Origins. *Sky & Telescope* 85(5):18.

Naeye, R. 1996. Is This Planet for Real? *Astronomy* 23(3):34.

O'Dell, R. 1994. Exploring the Orion Nebula. *Sky & Telescope* 88(6):20.

Riepurth, B. and S. Heathcote. 1995. Herbig-Haro Objects and the Birth of Stars. *Sky & Telescope* 90(4):38.

Stahler, S. W. 1991. The Early Life of Stars. *Scientific American* 265(1):48.

Chapter 13
The Sun

*In the middle of
everything is the sun.*
Nicholas Copernicus,
1520

ur star, the Sun, is rather ordinary by galactic standards. Its mass and size are modest—there are stars as much as a few hundred times larger in diameter and a million times more luminous. Its temperature is also moderate, as stars go. In many respects the Sun is a very run-of-the-mill entity.

Because the Sun is an ordinary star, and because we have far more detailed knowledge of it than of other stars, it is useful to begin our discussion of stars with a chapter on the Sun. In subsequent chapters, we will see how the information on the Sun helps us to understand the nature of other stars. Conversely, we will also find that studies of other stars help us to learn more about the nature of the Sun.

The Evolutionary Context

The Sun is the central object in the solar system in more than just position. It dominates the mass of the system, containing more than 500 times the mass of all the planets and interplanetary material together. As we learned in the previous chapter, the planets and other accompanying bodies were probably formed as a by-product of the formation of the Sun. Certainly, to a distant observer, the Sun is the only member of the solar system whose existence would be obvious.

The Sun has very extensive influences on the planets. For example, it is the solar gravity that keeps the planets in orbit, preventing them from wandering off through interstellar space. The Sun's various activities, such as ejection of the solar wind, variations in energy output, and reversals of magnetic field, all have effects on the planets. In addition, through its radiation, the Sun is a significant source of energy for the planets. On Earth, virtually all forms of energy that we use (except nuclear and geothermal sources) come ultimately from the Sun, primarily through the chemical energy derived from sunlight that is stored by plants. Even fossil fuel energy, gained by burning gas and oil, is solar in origin because the energy was originally captured from sunlight and stored in plants that became fossilized.

In this chapter we will discuss the properties of the Sun as they are now, while keeping in mind the picture developed in the previous chapter, where we saw how the Sun is thought to have formed. This will help us to remember that the Sun is an evolving body: it had a beginning, it is changing with time now, and it will have an ending. Because we are so close to the Sun, we know a great deal more about its appearance and structure than we do about other stars. This knowledge will help us later when we try to understand the properties of distant suns, which we cannot hope to observe in as much detail. On the other hand, we see the Sun only as it is now, whereas by sampling many stars in different stages of their lives, we can infer how other stars live and die. This information, in turn, will be useful to us as we try to piece together the story of the Sun's evolution. Today astronomers use comparisons between the Sun and other stars to help learn more about both.

Basic Properties and Internal Structure

It is useful to begin our discussion of the Sun by considering it in the global sense. We can develop ideas about its structure and internal state from its overall properties, setting the stage for subsequent discussions of its detailed nature.

General Description of the Sun

Perhaps the most obvious attribute of the Sun is its brightness; it emits vast quantities of light. Its luminosity, or the amount of energy emitted per second, is about 4×10^{26} watts. The intensity of sunlight reaching the Earth (above the atmosphere) is about 1,354 watts/m^2. This quantity is known as the **solar constant,** although it varies slightly according to activity in the Sun's outer layers (discussed later in this chapter).

The Sun is a ball of hot gas **(Fig. 13.1).** It dwarfs all the planets in mass and radius **(Table 13.1).** Its density, on average, is 1.41 g/cm^3, not much more than that of water, but its center is so highly compressed that the density there is about 10 times greater than that of lead. The interior is gaseous rather than solid because the temperature is very high—around 15 million degrees at the center, diminishing to just under 6,000 K at the surface. At these temperatures, the gas is partially ionized in the outer layers of the Sun and completely ionized in the core, where all electrons have been stripped free from their parent atoms.

The Sun is held together by gravity. All of its constituent atoms and ions attract each other, with the net effect that the solar substance is held in a spherical shape. Gas that is hot exerts pressure on its surroundings, and this pressure, pushing outward, balances the force of gravity, which pulls the matter inward. This balance, with gravity and pressure equaling each other everywhere, is called **hydrostatic equilibrium.** How do we know that these two forces

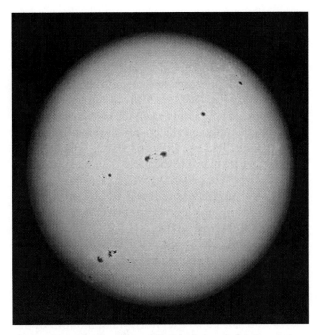

Figure 13.1 **The Sun.** This is a visible-light photograph of the Sun at a time when only a few sunspots were present. The overall yellow color reflects the surface temperature of almost 6,000 K. The spots and the darkening of the disk near the edges are discussed in the text. (National Optical Astronomy Observatories)
http://universe.colorado.edu/fig/13-1.html

are in balance? If they were not, and gravity exceeded pressure at some point, the Sun would be contracting. Or, if pressure were greater at any point, the Sun would be expanding.

Since the Sun is neither expanding nor contracting, we conclude that there must be a balance. This may seem like an extraordinary coincidence, but it is not—if the balance were disturbed, the Sun would adjust itself until a balance was reached again. Consider what would happen if the Sun were contracting. As it did so, the density and temperature would increase, driving up the pressure until equilibrium occurred. This is exactly what happened when the Sun formed. On the other hand, if the Sun were to expand, this would decrease temperature, the pressure would go down, and again a balance would be reached.

The deeper we go into the Sun, the greater the weight of the overlying layers, and the more the gas is compressed. The higher the pressure, the greater the temperature required to maintain the pressure, so we find that the pressure and temperature both increase as we approach the center. As we will soon see, it is possible to use the principle of hydrostatic equilibrium (along with a few other simple relationships) to calculate pressure and temperature conditions inside the Sun and other stars.

The composition of the Sun **(Table 13.2)** is the same as that of the primordial solar system and of

Table 13.1

The Sun

Diameter: 1,391,980 km (109.3 $D_{\oplus}$)
Mass: 1.99×10^{30} kg (332,943 $M_{\oplus}$)
Average density: 1.41 g/cm³
Surface gravity: 27.9 Earth gravities
Escape speed: 618 km/sec
Luminosity: 3.83×10^{33} erg/sec
Surface temperature: 5,780 K
Rotation period: 25.04 days (at equator)

Table 13.2

The Composition of the Sun's Photosphere

Element	Symbol	Number of Atoms[a]	Fraction of Mass
Hydrogen	H	1.0000	7.35×10^{-1}
Helium	He	8.51×10^{-2}	2.48×10^{-1}
Lithium	Li	1.55×10^{-9}	7.85×10^{-9}
Beryllium	Be	1.41×10^{-11}	9.27×10^{-9}
Boron	B	2.00×10^{-10}	1.58×10^{-9}
Carbon	C	3.72×10^{-4}	3.26×10^{-3}
Nitrogen	N	1.15×10^{-4}	1.18×10^{-3}
Oxygen	O	6.76×10^{-4}	7.88×10^{-3}
Fluorine	F	3.63×10^{-8}	5.03×10^{-7}
Neon	Ne	3.72×10^{-5}	5.47×10^{-4}
Sodium	Na	1.74×10^{-6}	2.92×10^{-5}
Magnesium	Mg	3.47×10^{-5}	6.15×10^{-4}
Aluminum	Al	2.51×10^{-6}	4.94×10^{-5}
Silicon	Si	3.55×10^{-5}	7.27×10^{-4}
Phosphorus	P	3.16×10^{-7}	7.14×10^{-6}
Sulfur	S	1.62×10^{-5}	3.79×10^{-4}
Chlorine	Cl	2.00×10^{-7}	5.17×10^{-6}
Argon	Ar	4.47×10^{-6}	1.30×10^{-4}
Potassium	K	1.12×10^{-7}	3.19×10^{-6}
Calcium	Ca	2.14×10^{-6}	6.26×10^{-5}
Scandium	Sc	1.17×10^{-9}	3.84×10^{-8}
Titanium	Ti	5.50×10^{-8}	1.92×10^{-6}
Vanadium	V	1.26×10^{-8}	4.68×10^{-7}
Chromium	Cr	5.01×10^{-7}	1.90×10^{-5}
Manganese	Mn	2.51×10^{-7}	1.01×10^{-5}
Iron	Fe	3.98×10^{-5}	1.62×10^{-3}
Cobalt	Co	3.16×10^{-8}	1.36×10^{-6}
Nickel	Ni	1.91×10^{-6}	8.18×10^{-5}
Copper	Cu	2.82×10^{-8}	1.31×10^{-6}
Zinc	Zn	2.63×10^{-8}	1.25×10^{-6}
(all others combined)			(less than 10^{-8} of total)

[a]The numbers given are relative to the number of hydrogen atoms.

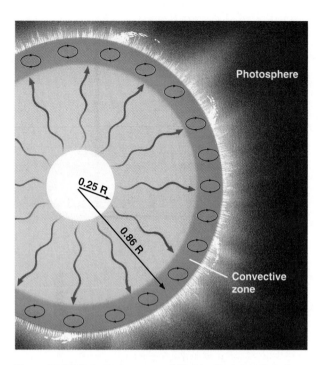

Figure 13.2 The internal structure of the Sun. This drawing shows the relative extent of the major zones within the Sun, except that the depth of the photosphere is greatly exaggerated.

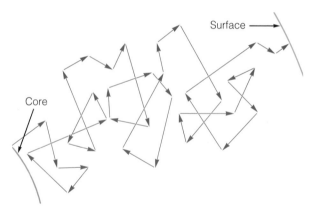

Figure 13.3 Random walk. A photon is continually being absorbed and reemitted as it travels through the Sun's interior. Because each reemission is in a random direction, the photon's progress from the core, where it is created, to the surface is very slow.

its radius **(Fig. 13.2).** Here nuclear reactions create heat and photons at X-ray and gamma-ray (γ-ray) wavelengths. This radiation eventually reaches the surface and escapes into space, but it is a laborious journey. Each photon is absorbed and reemitted many times along the way **(Fig. 13.3),** gradually losing energy. In the process, most gamma-ray photons become visible-light photons, and the energy they lose heats the surroundings. Because a photon travels only a short distance before it is absorbed and then reemitted in a random direction, progress toward the surface is very slow. Calculations indicate that it takes nearly a million years for a single photon to make its way from the solar core to the surface.

Throughout most of the solar interior, the gas is quiescent, without any major large-scale flows or currents. The energy from the core is transported by the radiation wending its slow way outward except in the layers near the surface, where convection occurs, and heat is transported by the overturning motions of the gas. As we shall see, the bubbling, boiling action in the outer portions of the Sun creates a wide variety of dynamic phenomena on the surface.

The Sun rotates, but not as a rigid body. The fact that the rotation is differential is apparent from observations of surface features such as sunspots, which reveal that, like Jupiter and Saturn, the Sun goes around faster at the equator than near the poles. The rotation period is about 25 days at the equator, 28 days at middle latitudes, and even longer near the poles. As we will see, the differential rotation probably plays an important role in governing variations in the solar magnetic field, which in turn have a lot to do with the behavior of the most prominent surface features, the sunspots.

There is evidence that the core of the Sun rotates a little more rapidly than the surface. The only clue for this is the presence of very subtle oscillations on the solar surface, which may be wave motions affected by the more rapid spin of the interior. The spin of the solar core is probably a direct result of the collapse and accelerated rotation of the interstellar cloud from which the Sun formed.

Nuclear Reactions

One of the biggest mysteries in astronomy in the early decades of this century was presented by the Sun. The problem was how to account for the tremendous amount of energy it radiates, which was particularly perplexing in view of geological evidence that the Sun has been able to produce this energy for at least 4 or 5 billion years. Some early ideas were that the Sun is simply still hot from its formation, that it is undergoing chemical reactions, or that it is

most other stars: about 73% of its mass is hydrogen, 25% is helium, and the rest is made up of traces of other elements. As we have already seen, this is similar to the chemical makeup of the outer planets and would represent the terrestrial planets as well, except that they have lost most of their volatile elements such as hydrogen and helium. It is apparent that all the components of the solar system formed together from the same material.

The ultimate source of all the Sun's energy is located in its core, within the innermost 25% or so of

gradually contracting and releasing stored gravitational energy. These mechanisms were ruled out, however, because none of them could possibly supply the energy needed to run the Sun for a long enough time.

The first hint at the solution came in the early 1900s, when Albert Einstein developed his theory of special relativity. He showed that matter and energy are equivalent, according to the famous formula $E = mc^2$, where E is the energy released in the conversion, m stands for the mass that is converted, and c for the speed of light. If mass could be converted into energy, enormous amounts of energy could be produced. Physicists began to contemplate the possibility that somehow this energy was being released inside the Sun and stars.

By the 1920s, following the pioneering work on atomic structure by Max Planck, Niels Bohr, and others, the concept of nuclear reactions began to emerge. These are transformations, like chemical reactions, except that in this case it is the subatomic particles in the nuclei of atoms that react with each other. Some reactions are **fusion reactions,** in which nuclei merge to create a larger nucleus, representing a new chemical element; and others are **fission reactions,** in which a single nucleus, usually of a heavy element with a large number of protons and neu-trons, splits into two or more smaller nuclei. In either type of reaction, energy is released as some of the matter is converted according to Einstein's formula. In the 1920s, many physicists explored these possibilities, and by the 1930s Hans Bethe had suggested a specific reaction sequence that might operate in the Sun's core.

Bethe envisioned a fusion reaction in which four hydrogen nuclei (each consisting of only a single proton) combine to form a helium nucleus, made up of two protons and two neutrons. The reaction actually occurs in several steps (see Appendix 14):

1. Two protons combine to form **deuterium,** a type of hydrogen that has a proton and a neutron in its nucleus (one of the two protons undergoing the reaction converts itself into a neutron by emitting a positively charged particle called a **positron**) and another particle called a **neutrino,** which has very unusual properties (discussed in a later section).
2. The deuterium combines with another proton to create an isotope of helium (^{3}He) consisting of two protons and one neutron.
3. Two of these ^{3}He nuclei combine, forming an ordinary helium nucleus (^{4}He, with two protons and two neutrons in the nucleus) and releasing two protons. At each step in this sequence, heat energy is imparted to the surroundings in the

form of kinetic energy of the particles that are produced, and in step 2 a gamma-ray photon is emitted as well.

The net result of this reaction, which is called the **proton-proton chain,** is that four hydrogen nuclei (protons) combine to create one helium nucleus. The end product has slightly less mass than the ingredients, 0.007 of the original amount having been converted into energy. We will learn in Chapter 15 that other kinds of nuclear reactions take place in some stars, but the vast majority produce their energy by the proton-proton chain.

It is easy to see that the proton-proton chain can produce enough energy to keep the Sun shining for billions of years. The total mass of the Sun is about 2×10^{30} kg. Only the innermost portion of this—perhaps 10%—undergoes nuclear reactions, so the Sun started out with about 2×10^{29} kg of mass available for nuclear reactions. If 0.007 of this, or 1.4 $\times 10^{27}$ kg, is converted into energy, then to find the total energy the Sun can produce in its lifetime, we multiply this mass times c^2, finding a total energy of $E = (1.4 \times 10^{27}$ kg$) \times (3 \times 10^8$ m/sec$)^2 = 1.3 \times 10^{44}$ joules. The rate at which the Sun is losing energy is its luminosity, which is about 4×10^{26} watts = 4×10^{26} joules/sec. At this rate, the Sun can last for

$$(1.3 \times 10^{44} \text{ joules})/(4 \times 10^{26} \text{ joules/sec}) = 3.2 \times 10^{17} \text{ seconds,}$$

or just about 10 billion years. Geological evidence shows that the solar system is now about 4.5 billion years old, so we can expect the Sun to keep shining for another 5 billion years or more. (In Chapter 15, we will see what happens to stars like the Sun when the nuclear fuel, hydrogen, runs out.)

Nuclear fusion reactions can take place only under conditions of extreme pressure and temperature, because of the electrical forces that normally would keep atomic nuclei from ever getting close enough together to react. Nuclei, which have positive charges, must collide at extremely high speeds to overcome the repulsion caused by their like electrical charges. The speed of particles in a gas is governed by the temperature, and only in the very center of the Sun and other stars is it hot enough (around 15 million degrees) to allow the nuclei to collide fast enough to fuse. This is why only the innermost portion of the Sun can ever undergo reactions. The high pressure in the Sun's core causes nuclei to be crowded together very densely, and this means that collisions will occur very frequently, another requirement if a high reaction rate is to occur.

SCIENCE AND SOCIETY

The Energetics of Solar Energy

The radiation incident on the Earth from the Sun can be harnessed and put to use to help satisfy our society's craving for energy. This is well known. You are probably aware of houses or office buildings that have been built to take advantage of solar energy, either passively, actively, or electrically (using devices to convert incident sunlight into current).

With passive solar heating, radiant energy is used without utilizing any special means to transport the energy throughout the house or to convert it into forms other than the heat created as sunlight is absorbed. Thus every house gets some passive solar heating. A house specially designed to take advantage of passive solar heat has large windows facing the maximum possible exposure to the Sun and dark-colored, stone or concrete walls and floors inside that heat up during the day and then reradiate their heat in the form of infrared radiation at night. In active solar heating systems, a fluid, commonly water, is allowed to absorb sunlight and is then circulated through pipes or tubes so that the heat it carries can be reradiated elsewhere in the building. The solar panels you sometimes see on the roofs of houses usually contain black pipes that absorb sunlight to heat the water that is circulating through them. Systems that convert sunlight into electricity usually make use of materials called *photovoltaics*, which release electrons when photons of light strike them (these are very closely related to the detectors used by astronomers to record the intensity of light from stars and other sources; see Chapter 6). A panel of photovoltaic cells, called a *solar array*, yields electrical current proportional to the intensity of light striking the panel over its full collecting area. Such arrays are used to power spacecraft (see, for example, pictures of the *Hubble Space Telescope*) and have even been used as the energy source in experimental cars.

How much solar energy is available, how close does it come to meeting our needs, and how can we best make use of this energy? We can answer the first part of the question by considering the solar constant, which represents the radiant energy received by the Earth at the top of the atmosphere. The value of the solar constant is 1.354 W/m^2. Atmospheric absorption (even in clear weather) reduces this by a fraction; it is probably safe to assume about half of this reaches the ground. Thus the effective solar input at the ground may be around 700 W/m^2.

A typical household may average about 1,000 W of power usage (combined total of heating plus electricity), with periods of high demand when the usage is several times higher. Let us assume that we need a total of 10,000 W at peak usage, with half for heating the house and the other half in the form of electricity to run appliances and lights. (these are reasonable numbers, but, of course, they vary depending on house size, climate, and the lifestyle of the occupants). How can we derive this needed energy from the 700 W/m^2 that are raining down on us from the Sun?

Let us address the need for heating energy first. The absorbing material in either a passive or an active solar system, usually water, rock or concrete, can have an efficiency of around 50%, meaning that about half of the incident energy is turned into heat in the absorber. In other words, the absorbing material receives a heat energy flux equal to 50% of the incident solar radiation intensity, or about 350 W/m^2. Thus, to get 5,000 W, we need $(5,000 \text{ W})/(350 \text{ W/m}^2) = 14 \text{ m}^2$ of collecting area. This seems easy enough, since this would correspond to a square surface of only about 3.7 m on a side. Indeed, it is not too hard to design a house so that it can stay warm enough through the use of solar heating alone, at least while the Sun is up. The main trick is to have a means of storing heat for nighttime and for cloudy days. This need can be addressed in active systems by using a tank to store heated water; if the tank is large enough, it is possible to store enough heat to keep the house warm overnight, though probably not for a prolonged stretch of cold cloudy weather. In a passive system, the rock or concrete absorbing material retains heat for some time, but may need to be supplemented by another source to keep the house warm all night.

What about the electrical power needs? Here the only way to use solar energy directly is to employ solar arrays, panels of photovoltaic cells. These are not very efficient; typically, only about 15% of the incident radiation can be converted into electrical energy. Thus the energy produced by a solar array is the equivalent of 15% of the incident solar radiation flux, or about 105 W/m^2. Thus, to supply the needed 5,000 W of electrical power, an array having a surface area of $(5,000 \text{ W})/(105 \text{ W/m}^2) = 48 \text{ m}^2$ would be needed. At the present time, this is not practical due to the size of the panels needed and their cost.

A great deal of research is under way to find more efficient methods of converting sunlight into other forms, for the simple reason that the amount of energy available is, in principle, more than enough for many of our needs. Areas of active work include designing more solar-efficient houses and active and passive solar heating systems and developing less expensive, more efficient photovoltaic cells. Perhaps in your lifetime it will become feasible to truly obtain all the energy you need for your daily life from the Sun.

Table 13.3

Structure of the Sun

Zone	Radius[a] (R)	(km)	Temperature (K)	Density (g/cm³)	Mass Fraction (M)[a]
Interior	0.00	0	1.6×10^7	160	0.000
	0.04	28,000	1.5×10^7	141	0.008
	0.1	70,000	1.3×10^7	89	0.07
	0.2	139,000	9.5×10^6	41	0.35
	0.3	209,000	6.7×10^6	13	0.64
	0.4	278,000	4.8×10^6	3.6	0.85
	0.5	348,000	3.4×10^6	1.0	0.94
	0.6	418,000	2.2×10^6	0.35	0.982
	0.7	487,000	1.2×10^6	0.08	0.994
	0.8	557,000	7.0×10^5	0.018	0.999
	0.9	627,000	3.1×10^5	0.0020	1.000
	0.95	661,000	1.6×10^5	0.0004	1.000
	0.99	689,000	5.2×10^4	0.0005	1.000
	0.995	692,000	3.1×10^4	0.00002	1.000
	0.999	695,300	1.4×10^4	0.0000001	1.000
Photosphere	1.000	695,990	6.4×10^3	3.5×10^{-7}	1.000
	1.000	+280	4.6×10^3	4.5×10^{-8}	1.000
Chromosphere	1.000	+320	4.6×10^3	3.1×10^{-8}	1.000
	1.001	+560	4.1×10^3	3.6×10^{-9}	1.000
(Transition)	1.002	+1,900	8.0×10^3	3.4×10^{-13}	1.000
	1.003	+2,400	4.7×10^5	4.8×10^{-15}	1.000
Corona	1.003	+2,400	5.0×10^5	1.7×10^{-15}	1.000
	1.2	+140,000	1.2×10^6	8.5×10^{-17}	1.000
	1.5	+348,000	1.7×10^6	1.4×10^{-7}	1.000
	2.0	+696,000	1.8×10^6	3.4×10^{-18}	1.000

[a]The radii are expressed in fractions of the Sun's radius $R_\odot$ as well as in kilometers. Above 695,990 km, the kilometer values are heights above the Sun's "surface," or lower boundary of the photosphere. The mass fractions are expressed in units of the Sun's mass $M_\odot$, which has the value 1.989×10^{30} kg. (Data from C. W. Allen, 1973, *Astrophysical Quantities*, London: Athlone Press.)

Modeling the Sun

Astronomers have learned a great deal about the interior of the Sun by combining observations of its surface with theoretical calculations of the internal structure. The theoretical models are constructed by writing equations to represent the known physical processes that take place in the interior, and then solving these equations for many different positions inside the Sun. For example, one equation expresses the relationship that the inward force of gravity is equal to the outward force of pressure. Another gives the rate of energy flow outward (this depends on the density and temperature at each point), and another provides the rate of energy production by nuclear reactions (which depends on temperature, density, and composition).

It can be shown that the set of equations has only one possible solution. The major uncertainty arises from the fact that some of the processes that occur inside the Sun are either too complex to be represented accurately or are not well enough known, so that approximations have to be used. Furthermore, it is possible that some physical processes, such as internal rotation or the role of the magnetic field inside the Sun, are more important than allowed for in the equations at present.

Despite these uncertainties, astronomers believe that the results of model calculations for the Sun are at least approximately correct, because they reproduce the observed surface conditions quite well. One exception is the failure to detect the predicted number of neutrinos, which has caused some consternation about the models (see the next section).

The results of one such model calculation are summarized in **Table 13.3,** which lists the temperature and density at several points inside the Sun (and extending throughout its atmosphere, which is dis-

OUR SUN

Solar Wind

Corona

Chromosphere

Solar Granulation

Sunspots

Plages

Convection Zone

Filament

Flare

Core

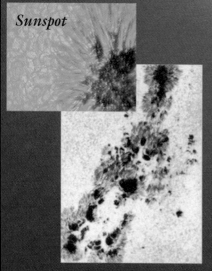

Sunspot

Solar Flare

Solar Spicules

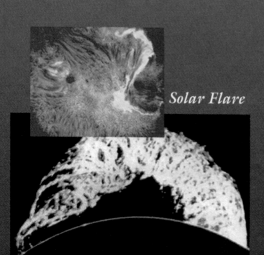

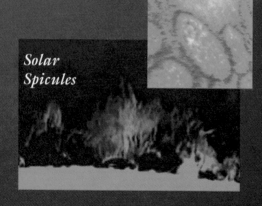

cussed in a later section). The fraction of the Sun's mass contained inside each distance from the center is listed also.

The Solar Neutrino Problem

The models of solar interior structure are, as noted above, quite convincing—they are self-consistent, and they reproduce the observed properties of the Sun. Or do they? It is generally impossible to observe what is happening inside the Sun, so we have to believe the models if we are to learn about conditions there. But there is one way to gain direct information: to detect the neutrinos that are released by nuclear reactions in the Sun's core.

Neutrinos do not interact with other matter very easily at all (technically, they can interact only through the weak nuclear force). Neutrinos pass through ordinary matter without so much as a whisper or a touch; we are all constantly inundated with neutrinos that pass right through our bodies and our surroundings, indeed right through the Earth, without pause. Thus neutrinos from the solar interior escape directly to space. If we had some way of catching and counting them, we could see whether they are being produced by the Sun at the rate expected. This would be a novel way of testing our theories of energy generation inside the Sun and other stars.

As it happens, neutrinos can interact with normal matter, though only with very low probability. Certain kinds of atoms can undergo nuclear reactions when a neutrino strikes them, and this property has been used as a means of measuring the neutrino flux from the Sun.

One of the atoms that can react with neutrinos is chlorine. Therefore an enormous quantity of chlorine, in the form of carbon tetrachloride (cleaning fluid), was placed deep in an old gold mine in South Dakota in the early 1970s and has been used ever since to detect solar neutrinos. As vast numbers of neutrinos pass through this underground reservoir, occasionally one of them will react with a chlorine atom to produce a radioactive form of argon. Every few weeks the gas emitted from the underground tank is searched for radioactive argon atoms. From the resulting count, the number of neutrinos that have passed through the tank is deduced.

The results of the Homestake Mine experiment have been baffling: only about one-third of the expected solar neutrinos have been detected. It appears that the Sun is not producing neutrinos at the predicted rate. By now several other experiments have confirmed this result, and astronomers and physicists have been considering several possible explanations. Something is wrong either with our understanding of solar nuclear reactions or with our knowledge of neutrinos. Lately, it has begun to appear that the problem is that neutrinos were not well-enough understood.

Traditional nuclear theory says that neutrinos are massless particles (like photons of light) and that they travel at the speed of light. But there have been experimental hints that perhaps neutrinos do contain some mass, which would mean that they travel at slightly less than the speed of light. Another possibility is that neutrinos may undergo spontaneous changes of state, oscillating among three possible forms (the notion that particles can change their identity is counterintuitive, but is actually common for elementary particles). If this is so, then at any given moment in time only one-third of the neutrinos from the Sun would be in the form that interacts with chlorine. This could neatly explain the deficiency of solar neutrinos.

This explanation is by no means established yet, however, and the worry over the neutrino problem continues. More experiments on the possible oscillations of neutrinos, as well as additional measurements of the quantity coming from the Sun, are under way. There is little doubt that the eventual solution to the riddle will provide us with new and fundamental knowledge of the Sun and of elementary particles.

Structure of the Solar Atmosphere

Observations of the Sun at different wavelengths of light make it clear that the outer layers are divided into several distinct zones **(Fig. 13.4).** The "surface" of the Sun that we see in visible wavelengths is the **photosphere,** with a temperature ranging between 4,000 and 6,500 K. Viewing the Sun at the wavelength of the strong line of hydrogen at 6563 Å, we see the **chromosphere,** a layer above the photosphere whose temperature is 6,000 to 10,000 K. Outside the chromosphere is the very hot, rarefied **corona,** best observed at X-ray wavelengths, whose temperature is 1 to 2 million degrees. Between the chromosphere and the corona is a thin region, called the **transition zone,** where the temperature rises rapidly. Overall, the tenuous gas within and above the photosphere is referred to as the **solar atmosphere.** Considering the solar atmosphere as a whole, we find that the temperature decreases as we move outward through the photosphere, reaching a minimum value of about 4,000 K. From there the trend reverses itself, and the temperature begins to rise as we go farther out. The chromosphere, immediately above the temperature minimum, is perhaps 2,000 km thick. Above this point the temperature rises very steeply within a few hundred kilometers to the coronal value of over a million degrees. Clearly, something is adding extra heat at these levels; shortly we will consider where this heat comes from.

First, let us discuss the photosphere, the "surface" of the Sun as we look at it in visible light. We see this level because here the density becomes great enough for the gas to be opaque, preventing us from seeing farther into the interior. The photosphere is where the Sun's absorption lines **(Fig. 13.5)** are formed, as the atoms in this relatively cool layer absorb continuous radiation coming from the hot interior.

A photograph of the photosphere reveals a cellular appearance called **granulation (Fig. 13.6)**. Doppler shift measurements tell us that the bright regions are areas where convection in the Sun's outer layers causes hot gas to rise, and the dark bordering regions are places where cooler gas is descending back into the interior. Recent detailed studies of granules and their changes over time have shown that their behavior is quite complex: granules can change appearance rapidly and can disintegrate quickly and disappear.

The temperature of the photosphere, roughly 6,000 K, is measured from the degree of ionization in the gas there. The density, roughly 10^{17} particles per cubic centimeter in the lower photosphere, was found from the degree of excitation, as described in Chapter 4. This is roughly 100 times lower than the

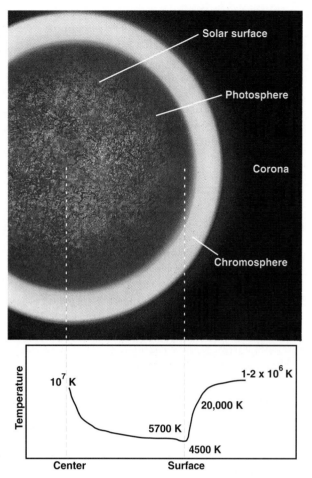

Figure 13.4 The structure of the Sun's outer layers. This diagram shows the temperatures of the convective zone, the photosphere, the chromosphere, and the corona.

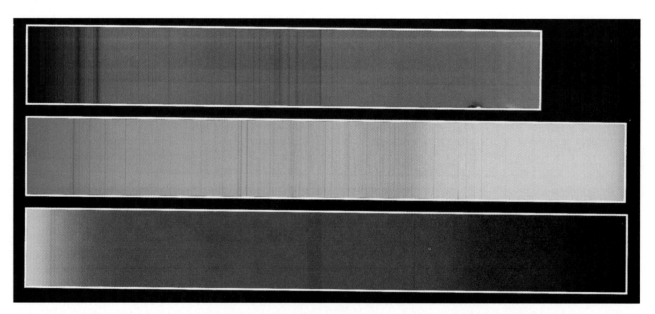

Figure 13.5 Fraunhofer lines. This is a portion of the Sun's spectrum, showing some of the dark lines cataloged by Josef Fraunhofer, a German physicist. The two lines near the far left in the upper segment are due to ionized calcium. The center segment shows several lines of the Balmer series of hydrogen, and the strongest Balmer line, H-alpha, is seen in the red portion of the lower segment. (K. Gleason, Sommers Bausch Observatory, University of Colorado)

Figure 13.6 Solar granulation. This photograph, obtained by a high-altitude balloon above much of the atmosphere's blurring effect, distinctly shows the granulation of the photosphere, which is due to convective motions in the Sun's outer layers. (Project Stratoscope, Princeton University, sponsored by the National Science Foundation.)

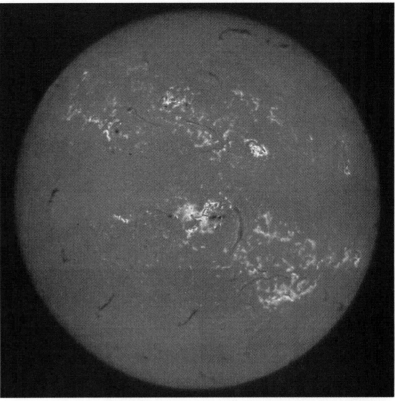

Figure 13.7 The Solar chromosphere. This is an image of the Sun, taken through a filter that allows only the wavelength of the red emission line of hydrogen to pass through. The Sun's chromosphere emits strongly at this wavelength whereas the photosphere does not; hence this image shows primarily the chromosphere and its structure. (National Optical Astronomy Observatories) http://universe.colorado.edu/fig/13-7.html

density of the Earth's atmosphere, which is about 10^{19} particles per cubic centimeter at sea level.

Most of what we know about the Sun's composition is based on the analysis of the solar absorption lines, so strictly speaking, the derived abundances represent only the photosphere. We have no reason to expect strong differences in composition at other levels, however, except for the core, where a significant amount of the original hydrogen has been converted into helium by the proton-proton reaction.

The photosphere near the edge of the Sun's disk looks darker than in the central portions. This effect, called **limb darkening** (see Fig. 13.1), occurs because we are looking obliquely at the photosphere when we look near the edge of the disk. We therefore do not see as deeply into the Sun as we do when looking near the center of the disk. Therefore, at the limb, the gas we are seeing is cooler than at the disk center and radiates less.

The chromosphere lies immediately above the temperature minimum **(Fig. 13.7).** The fact that this region forms emission lines tells us, according to Kirchhoff's laws, that the chromosphere is made of hot, rarefied gas, hotter than the photosphere behind it. When viewed through a special filter that allows only light at the wavelength of the hydrogen emission line at 6563 Å to pass through (see Fig. 13.7), the chromosphere has a distinctive cellular appearance referred to as **supergranulation,** which is similar to the photospheric granulation, but with cells some 30,000 km across instead of about 1,000 km. The chromosphere also exhibits fine-scale structure in the form of spikes of glowing gas called **spicules (Fig. 13.8)**. These come and go, probably at the whim of the magnetic forces that seem to control their motions.

The outermost layer of the Sun's atmosphere is the corona, which extends a considerable distance above the photosphere and chromosphere. Historically, the corona's existence was known only from observations during total solar eclipses, when the intense light of the photosphere is blocked by the Moon's disk, revealing a dim glow from the Sun's surroundings **(Fig. 13.9).** In modern times we can observe the corona through the use of a device built into a telescope that blocks the photospheric light or through the use of telescopes in space that observe

Figure 13.9 The corona. This photograph, obtained during a total solar eclipse, shows the type of structures commonly seen in the corona. These include giant looplike features and an overall appearance of outward streaming. This image shows the total eclipse of July 11, 1991, as seen from the Baja peninsula, in Mexico. Reddish spots along the edge of the solar disk are prominences, colored by the red emission line of ionized hydrogen. (S. Albers, G. Emerson, D. Dicicco, and D. Sime)

Figure 13.8 Spicules. This photograph shows the transient features in the chromosphere known as spicules. These spikes of glowing gas, which are apparently shaped by the Sun's magnetic field, come and go irregularly. (Sacramento Peak Observatory, National Optical Astronomy Observatories)

the Sun at wavelengths such as X rays where the corona is far brighter than the photosphere.

The corona is irregular in form, being patchy near the Sun's surface but with radial streaks at great heights suggestive of outflow from the Sun (see Fig. 13.9). The density of the coronal gas is very low, only about 10^9 particles per cubic centimeter. As we have already seen, the corona is very hot and contains highly ionized gas. The source of the energy that heats the corona to such extreme temperatures is not well understood, although a general picture has emerged, as we will see.

X-ray observations reveal the patchy structure of the corona **(Fig. 13.10).** In large regions that appear dark in an X-ray photograph of the Sun, the gas density is even lower than in the rest of the corona. These regions are called **coronal holes** and, as we shall see, are probably created and maintained by the Sun's magnetic field. The coronal holes, as well as the overall shape of the corona, vary with time **(Fig. 13.11),** showing that the corona in general is a dynamic

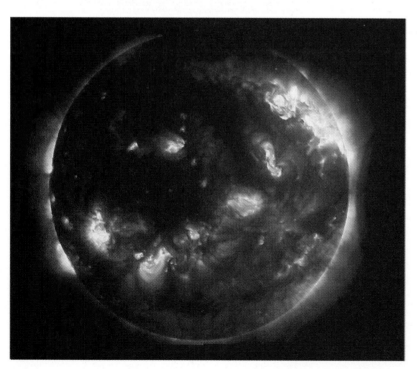

Figure 13.10   An X-ray portrait of the corona. This image was obtained by a small, rocket-borne X-ray telescope on the day of the total solar eclipse in 1991 (July 11). The bright regions are relatively dense, whereas the dark regions, called coronal holes, are less dense. All of the emission seen here originates in the corona, where the gas temperature is between 1 and 2 million degrees. (L. Golub, Harvard-Smithsonian Center for Astrophysics)

http://universe.colorado.edu/fig/13-10.html

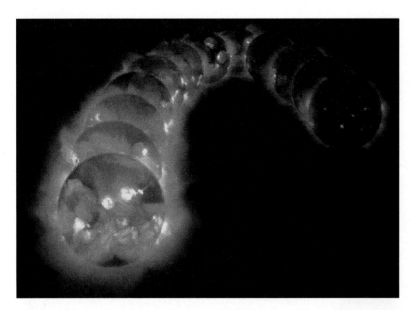

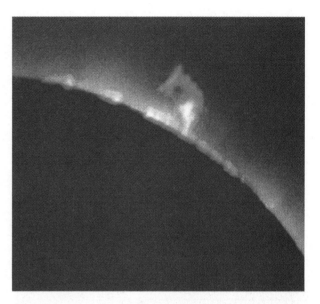

Figure 13.11 Changes in the solar corona. This series of X-ray images from the Yohkoh satellite shows the Sun's corona at 90-day intervals over a period of 4 years. We see that the coronal structure changes dramatically during the waning phases of the solar activity cycle. (Yohkoh mission of ISAS, Japan. The x-ray telescope was prepared by the Lockheed Palo Alto Research Laboratory, the National Astronomical Observatory of Japan, and the University of Tokyo with the support of NASA and ISAS)

Figure 13.12 A prominence. Here the looplike structures thought to be governed by the Sun's magnetic field can be readily seen. This photograph was obtained by *Skylab* astronauts using an ultraviolet filter. (NASA)

region. Another impressive sign of this dynamic nature is presented by the **prominences (Figs. 13.12 and 13.13),** great streamers of hot gas stretching upward from the surface of the Sun to take on an arc-shaped appearance. These are usually associated with sunspots, and both phenomena are linked to the solar activity cycle (to be discussed shortly).

The hot outer layers of the Sun have provided astronomers with a second major mystery regarding the solar energy budget. Unlike the mystery of the Sun's internal energy source, the question of the mechanism for heating the chromosphere and corona has not yet been entirely resolved. A great deal of energy is available in the form of gas motions in the convection zone just beneath the solar surface, and it is generally accepted that this energy is somehow responsible for heating the corona. It is not clear how the energy is transported to such high levels, however, although there are several ideas, each invoking some form of waves. Waves of any kind carry energy, and it has been suggested that either sound waves or magnetic waves of some kind are the agents that transfer energy from the convection zone to the corona. There is certainly enough energy in the convective motions in and below the photosphere; the problem is how this energy is transported into the higher levels.

Recent satellite observations at ultraviolet and X-ray wavelengths have shown that stars similar to the Sun in general type also have chromospheric and coronal zones, so if we can discover how the Sun operates, we will also gain a better understanding of

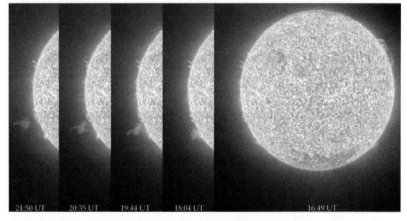

Figure 13.13 An eruptive prominence. This series of ultraviolet images from the *SOHO* satellite shows the progress of an eruptive prominence, an outburst of glowing gas from the solar surface. The motion of the gas is governed by the Sun's magnetic field, and often arc-like shapes are the result. (Courtesy of SOHO/EIT consortium. SOHO is a project of international cooperation between ESA and NASA) http://universe.colorado.edu/fig/13-13.html

how other stars work. Similarly, observations of other stars, particularly of the relationship of chromospheric and coronal activity to stellar properties such as age and rotation, can help us learn more about the Sun.

The Solar Wind

The long, streaming tail of a comet always points away from the Sun, regardless of the comet's direction of motion. The significance of this was fully real-

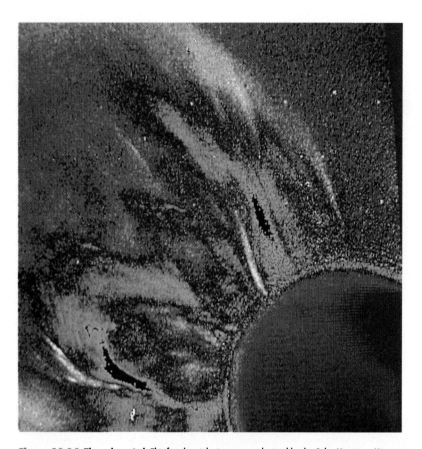

Figure 13.14 **The solar wind.** This far-ultraviolet image was obtained by the *Solar Maximum Mission* in Earth orbit. A special color separation technique was used to show the outward flow of gas. (Data from the Solar Maximum Mission, NASA and the High Altitude Observatory National Center for Atmospheric Research, sponsored by the National Science Foundation)

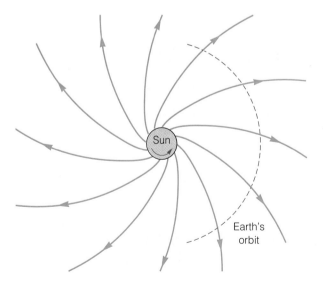

Figure 13.15 **The solar wind.** This schematic diagram illustrates how ionized gas from the Sun spirals outward through the solar system in a steady stream. The solar magnetic field creates sectors of variable density in the wind.

ized in the late 1950s, when the first U.S. satellites revealed the presence of the Earth's radiation belts and the fact that they are shaped in part by a steady flow of charged particles from the Sun. Near the Earth the solar wind reaches a speed of 300 to 400 km/sec. It is this flow of charged particles that forces cometary tails always to point away from the Sun.

The existence of this **solar wind** is evidently a natural by-product of the same heating mechanisms that produce the hot corona of the Sun. Astronomers originally thought that particles in the high-temperature region move about with such great velocities that a steady trickle escapes the Sun's gravity, flowing outward into space. Subsequently, however, X-ray observations of the Sun have shown that the situation is not that simple. The solar magnetic field governs the outward flow of charged particles. The coronal holes, mentioned earlier (see Fig. 13.11), are regions where the magnetic field lines open out into space. Charged particles such as electrons and protons, constrained by electromagnetic forces to follow the magnetic field lines, therefore escape into space primarily from the coronal holes. The speed of the solar wind is relatively low close to the Sun, but accelerates outward, quickly reaching its ultimate speed of 300 to 400 km/sec, after which it is nearly constant. The wind has nearly reached its maximum speed by the time it passes the Earth's orbit, and beyond there it flows steadily outward, persisting beyond the orbit of Pluto. At some point in the outer solar system, the wind is thought to come to an abrupt halt where it runs into an invisible and tenuous wall of matter swept up from the interstellar medium that sur-rounds the Sun. In 1995 the *Pioneer 10* spacecraft, on its way out of the solar system, apparently passed through this outer boundary of the solar wind zone, called the **heliopause,** at a distance of some 50 AU from the Sun.

Most of the direct information we have on the solar wind comes from satellite and space probe observations **(Fig. 13.14),** because the Earth's magnetosphere shields us from the wind particles. Solar wind monitors are placed on board the majority of spacecraft sent to the planets. One striking discovery has been that the density of the wind is not uniform; instead the wind flows outward from the Sun in sectors, indicating that it originates only from certain areas on the Sun's surface. As already mentioned, X-ray data have shown that the wind emanates only from the coronal holes. Because the base of the wind is rotating with the Sun, the wind sweeps out through space in a great curve, similar to the trajectory of water from a rotating law sprinkler **(Fig. 13.15).**

Occasionally, explosive activity occurring on the Sun's surface releases unusual quantities of charged particles, and some three or four days later, when this burst of ions reaches the Earth's orbit, we experience disturbances in the ionosphere that can interrupt short wave radio communications and cause auroral displays. These **magnetic storms,** as they are often called, are outward manifestations of a much more complex over-all interaction between the Sun and the Earth.

Sunspots, Solar Activity, and the Magnetic Field

Centuries ago, Chinese astronomers observed dark spots on the Sun's disk. More recently, Galileo cited spots as evidence that the Sun is not a perfect, unchanging celestial object, but instead has occasional flaws. Over the centuries since Galileo, observations of the spots, which individually may last for months, have revealed some very systematic behavior. The number of spots varies, reaching a peak approximately every 11 years, and during the interval between peak numbers of spots, their locations on the Sun change steadily from middle latitudes to a concentration near the equator. When the sunspot number is high, most of them appear in activity bands about 30° north or south of the solar equator. During the next 11 years, the spots tend to lie ever closer to the equator, and by the end of the cycle, they are nearly on it **(Fig. 13.16).** By this time, the first spots of the next cycle are already forming at middle latitudes. A plot of sunspot locations from cycle to cycle clearly shows this progression toward the equator; the plot is called a **"butterfly diagram"** from the shape of the pattern formed by the spots **(Fig. 13.17).**

A hint at the origin of the spots was found when their magnetic properties were first measured. This is accomplished by applying spectroscopic analysis to the light from the spots, taking advantage of the fact

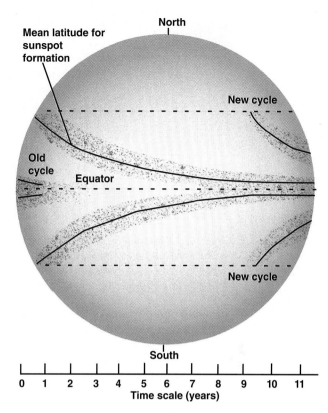

Figure 13.16 Sunspot locations. This drawing shows where sunspots appear at different times in the sunspot cycle.

that the energy levels in certain atoms are altered by the presence of a magnetic field. As a result, spectral lines formed by those levels are split into two or more distinct, closely spaced lines. The degree of line splitting, which is referred to as the **Zeeman effect,** depends on the strength of the magnetic field, so the field can be measured from afar simply by analyzing the spectral lines to see how widely split they are. This technique works for distant stars as well as for the Sun, except that for a distant star we cannot mea-

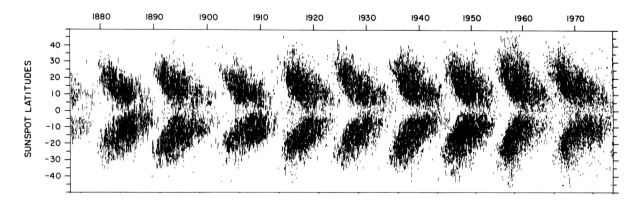

Figure 13.17 The butterfly diagram. This is a plot of the latitudes of observed sunspots through several cycles of solar activity. At the beginning of each cycle, the spots tend to appear at mid-latitudes; then, as the cycle progresses, they arise at positions closer and closer to the equator. (Prepared by the Royal Greenwich Observatory and reproduced with the permission of the Science and Engineering Council, courtesy of J. A. Eddy)

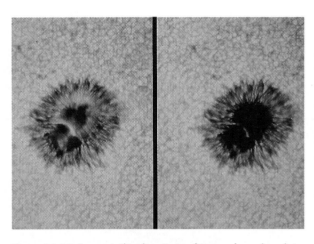

Figure 13.18 Sunspots. This telescopic view of a group of spots shows their detailed structure. They appear dark only in comparison with their much hotter surroundings. (Mt. Wilson and Las Campanas Observatories, Carnegie Institution of Washington) http://universe.colorado.edu/fig/13-18.html

Figure 13.19 Magnetic fields on the Sun. The magnetogram, or map of magnetic field strength (right), was made by measuring the splitting of spectral lines by magnetic fields (the Zeeman effect). Here colors (dark blue and yellow) are used to indicate regions of opposite magnetic polarity. Note that the polarities of east-west pairs in the southern hemisphere are reversed from those in the north. At the left is a visible-light photograph of the Sun taken at the same time, so that the correspondence between the surface magnetic field and the visible sunspots can easily be seen. (National Optical Astronomy Observatories)

sure different portions of the disk separately and what we observe represents an average field over the surface of the star.

When the first measurements of the Sun's field were made in the first decade of this century, the field was found to be especially intense in the sunspots, where it was about 1,000 times stronger than in the surrounding gas. This strong field creates a form of pressure that helps support the gas and allows it to be cooler than the surrounding gas. The magnetic field may also inhibit convection and thus prevent heated gas from below from rising up to the surface. For these reasons the spot is cooler than its surroundings and is therefore not as bright **(Fig. 13.18).** The typical temperature in a spot is about 4,000 K, compared with the roughly 6,000 K of the photosphere. Using Stefan's law, we see that the intensity of light emitted in a spot compared with the surroundings is $(4,000/6,000)^4 = 0.2$; that is, the brightness of the solar surface within a sunspot is only about one-fifth of the brightness in the photosphere outside.

When the sunspot magnetic fields were measured **(Fig. 13.19),** they were found to act like either north or south magnetic poles; that is, each spot has a specific magnetic direction associated with it. Furthermore, pairs of spots often appear together, with the two members of a pair usually having opposite magnetic polarities. During a given 11-year cycle, in every sunspot pair in the same hemisphere the magnetic polarities always have the same orientation. For example, during one 11-year cycle,

in one hemis-phere the spot to the east in each pair will generally have north magnetic polarity, while the spot to the west has south magnetic polarity. During the next cycle, all the pairs will be reversed, with the south magnetic spot to the east, and the north polar spot to the west. In the other hemisphere, these relative orientations are exactly the opposite. Between cycles, when this arrangement is reversing itself, the Sun's overall magnetic field also reverses, with the solar magnetic poles exchanging places. The Sun's magnetic field and sunspot patterns actually take 22 years to repeat themselves, so the solar cycle is really 22 years long.

Sunspot groups, often called **solar active regions (Fig. 13.20),** are the scenes of the most violent forms of solar activity, the **solar flares.** These are gigantic outbursts of charged particles, as well as visible, ultraviolet, and X-ray emissions, created when extremely hot gas spouts upward from the surface of the Sun **(Fig. 13.21).** Flares are most common during sunspot maximum, when the greatest numbers of spots are seen on the solar surface. Close examination of flare events shows that the trajectory of the ejected gas is shaped by the magnetic lines of force emanating from the spot where the flare occurred. Charged particles flow outward from a flare, some of them escaping into the solar wind. If the solar wind flowing from the site of the flare reaches the Earth, then the Earth is bathed by an extra dose of solar wind particles some three days later, with the effects on radio communications already described. The extra

Solar Activity

In this web exercise, we will search the Internet for current information on Solar activity. First, we'll find WWW sites that contain information about the sun; then we'll set out to answer a number of specific questions about solar activity. Our starting point is the following URL:

http://universe.colorado.edu/ch13/web.html

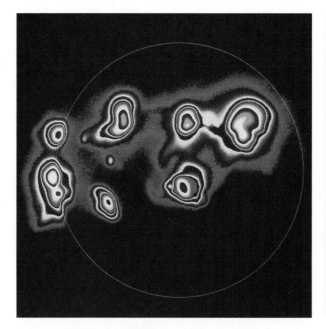
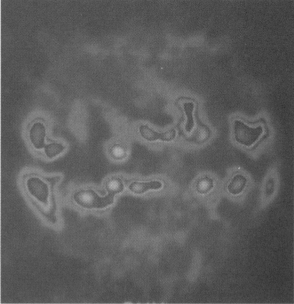

Figure 13.20 **Solar active regions.** At the left is an X-ray image and at the right is a radio map. In both, bright colors indicate the solar active regions, which emit strongly at X-ray and radio wavelengths. (left: NASA: right: National Radio Astronomy Observatory)

quantity of charged particles entering the Earth's upper atmosphere can also be responsible for unusually wide-spread and brilliant displays of aurorae. Apparently, flares occur when twisted magnetic field lines in the Sun suddenly reorganize themselves, releasing heat energy and allowing huge bursts of charged particles to escape into space.

Because of the connection between the solar activity cycle and reversals in the Sun's magnetic field, a model for the activity cycle was developed some years ago in which sunspots are envisioned as locations where tubes of magnetic field lines break through the solar surface. According to the model, these magnetic flux tubes connect the Sun's north and south poles and become twisted and kinked when the poles switch places. In the process they break through the surface in more and more places, ever closer to the equator, until the overall reversal of the field is completed and the flux tubes can become relatively smooth and straight again **(Fig. 13.22).** In this way, the flux tube model accounts for the main features of the solar activity cycle, particularly the changes in number, latitude, and magnetic field orientation of the spots during each cycle.

Although the flux tube model accounts for the general nature of sunspots and even helps in understanding what happens when the solar magnetic field reverses, a fundamental question remains: What causes the 22-year reversals of the solar magnetic

field? There must be some instability that causes the direction of flow to reverse in the interior zone where the field originates. Whatever the cause, it probably originates very deep in the solar interior, so we must look there to find the underlying explanation of the activity cycle. Perhaps new information derived from studies of solar oscillations will eventually yield the missing link.

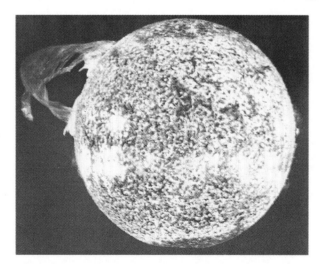

Figure 13.21 **A major flare.** The gigantic looplike structure in this ultraviolet photograph obtained from *Skylab* is one of the most energetic flares ever observed. Supergranulation in the chromosphere is easily seen over most of the disk. (NASA)

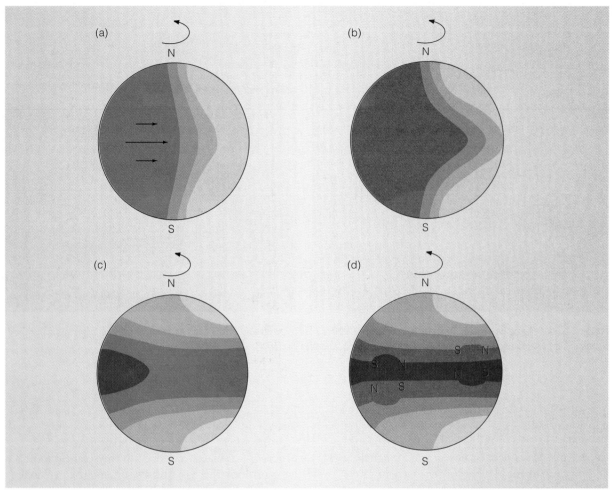

Figure 13.22 Flux tube model for solar activity. This figure depicts the model that views sunspots as locations where magnetic flux tubes break through the surface. In this scenario, after a reversal of the magnetic poles, magnetic lines connecting the Sun's poles are bent by differential rotation (a), then stretched around the Sun's interior (b and c), until they become tangled and twisted so that they break through, forming pairs of oppositely polarized sunspots (d).

There is some evidence for long-term changes in solar behavior, at intervals longer than 22 years, and these cycles will also require an explanation from deep inside the Sun, where the magnetic field is formed. The most striking known variation from the regular cycle occurred in the late 1600s, when sunspots seemed to stop altogether for more than 50 years **(Fig. 13.23)**. This period, called the **Maunder Minimum** after Edward W. Maunder, its discoverer (who also bequeathed us the butterfly diagram), was accompanied by extreme behavior of the Earth's climate, including a very cold period known in Europe as the "Little Ice Age." Only time will tell whether the Maunder Minimum was part of some very long-term cyclic behavior in the Sun. The implied link between sunspot activity and weather patterns on Earth is also quite uncertain; recent studies have suggested that a correlation may exist.

Summary

1. The Sun is a spherical, gaseous object whose temperature and density increase toward the center. From its surface, the Sun emits light whose ultimate source is nuclear reactions in the core.
2. Energy is produced in the core by nuclear fusion and is slowly transported outward by radiation, except near the surface, where energy is transported by convection.
3. The nuclear reaction that powers the Sun is the proton-proton chain, in which hydrogen is fused into helium.
4. The outer layers of the Sun govern the nature of the light it emits. These layers consist of the photosphere, with a temperature of about 6,000 K; the chromosphere, where the temperature ranges

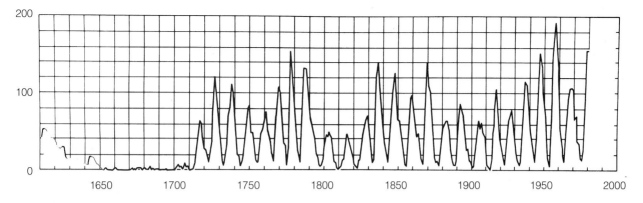

Figure 13.23 **The Sun's long-term activity.** This diagram illustrates the relative level of activity (in terms of sunspot numbers) over three centuries. It is clear that the level varies, including a span of about 50 years (1650–1700), known as the Maunder Minimum, when there was little activity. There is evidence that the Sun has long-term cycles that modulate the well-known 22-year period. (J. A. Eddy)

between 6,000 and 10,000 K; and the corona, where the temperature is over 1,000,000 K.

5. The photosphere, which is the visible surface, creates absorption lines, while the hotter chromosphere and corona create emission lines.

6. The excess heat in the outer layers is transported there from the convective zone just below the surface, but the mechanism for transporting the heat is not fully understood. It is likely to involve energy transport by the solar magnetic field.

7. The Sun emits a steady outward flow of ionized gas called the solar wind. The wind originates from coronal holes and is controlled by the solar magnetic field.

8. The sunspots occur in 11-year cycles that are part of the 22-year cycle of the Sun's magnetic field reversals. The spots are regions of intense magnetic fields where magnetic field lines from the solar interior break through the surface.

Review Questions

1. Explain in your own words why hydrostatic equilibrium causes the Sun to have a spherical shape. Do you think the rotation of the Sun might affect its shape?

2. Why do nuclear reactions occur only in the central region of the Sun?

3. Explain why the use of special filters to isolate the wavelengths of certain spectral lines allows astronomers to examine different layers of the Sun separately.

4. Why are the granules in the solar photosphere brighter than the descending gas that surrounds them?

5. Explain, in the context of Kirchhoff's rules (Chapter 6), why the chromosphere forms emission lines while the photosphere forms absorption lines.

6. Why does it take so long (about one million years) for a photon to make its way from the solar core to the surface?

7. Why do you think a sunspot group is referred to as a solar active region? What is happening in such a region at the levels of the photosphere, the chromosphere, and the corona?

8. Why do we say that the solar activity cycle is 22 years long, even though the sunspot maxima occur every 11 years?

9. Describe a research program that you could undertake to determine the effects of solar variations on the Earth's climate.

Problems

1. Using data in Appendixes 4 and 8, add up the masses of all the planets and compare the total with the mass of the Sun.

2. Use the Sun's luminosity of 3.827×10^{26} watts to calculate the value of the solar constant. Compare your answer with the value given in the text. (*Hint:* Divide the luminosity by the area of a sphere having a radius of 1 AU.)

3. How would the lifetime of the Sun be altered if its luminosity were 10 times greater than it is and its mass were twice as large?

4. The energy yield from one proton-proton chain reaction (the net conversion of four protons into a single helium nucleus) is 4.2×10^{-12} joules. If 10% of the Sun's mass undergoes nuclear

Imaging the Sun

ASTRONOMICAL

ACTIVITY

The brilliance of the Sun makes it difficult and dangerous to try to look at it directly. But with care, you can create an image of the Sun that can be viewed safely. There are at least two ways to do this: making a pinhole "camera" or using a pair of binoculars to project a solar image.

Use of binoculars is easier and has the further advantage of creating a large enough image of the Sun to allow features such as sunspots to be seen. But if binoculars are not available, it is not difficult to make a pinhole device for viewing a solar image. To do this, mount two pieces of thin, stiff cardboard onto a board so that they are parallel to each other (see the drawing). Make a small hole in one piece of cardboard; the other will be used as a screen onto which the Sun's image will be projected. The hole acts as a lens, forming an image of the Sun. The size of the image depends on the distance between the pinhole and the screen, so it is best to mount the two pieces of cardboard as far from each other as possible. Alternatively, you can use a cardboard box, with the pinhole in one end and the other end acting as the screen; the only difficulty is that unless the box is very large, the resulting image will be rather small. Whether you use two separate pieces of cardboard mounted on a board or a large box, you will need to add a shade over the screen, so that the solar image will contrast more strongly with the surrounding screen.

Using binoculars to create a solar image is very simple and requires nothing more than the binoculars and a piece of stiff paper or cardboard to act as a screen onto which the image can be projected. First, place the screen on the ground with one side propped up so that it is approximately perpendicular to the incoming sunlight (see the drawing). Then hold the binoculars about two feet above the screen, with the large end pointing toward the Sun. The binoculars will project an image of the Sun onto the screen (note: you will get two solar images if you take all the lens caps off, or you can leave the caps on one side of the binoculars and obtain a single image). You will probably have to move the binoculars around a bit until they are properly aligned so that the solar image falls onto the screen, but when you do see the image, you will find that it is large enough to allow some detail, such as major sunspots, to be seen.

You will also find that the image is very bright, perhaps too bright to be viewed without discomfort. An easy way to reduce the intensity of the image is to block out some of the light entering the binoculars; this can be done by making a hole in a small piece of cardboard and taping it over the large end of the binoculars. The intensity of the image will be reduced by the ratio of the area of the hole in your cover piece to the area of the lens when it is unblocked.

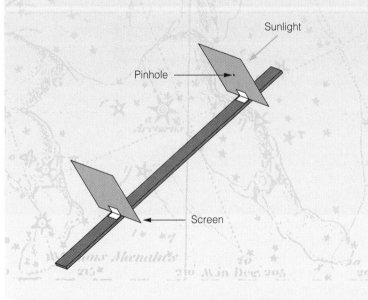

Sunlight

Pinhole

Screen

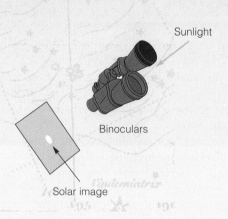

Sunlight

Binoculars

Solar image

reactions, calculate how many helium nuclei are produced (assume the Sun is pure hydrogen initially, then divide the Sun's core mass by the mass of four protons. The proton mass is given in Appendix 4). Now calculate the total energy that the Sun can produce by multiplying the number of helium nuclei formed times the energy yield per helium nucleus. Divide your answer by the solar luminosity to estimate the Sun's lifetime. How does your answer compare with the one we found in the text?

5. Suppose a gamma-ray photon of wavelength 0.1 Å is emitted at the Sun's core. On its way to the surface, it is absorbed and reemitted many times, and eventually emerges into space as a number of visible-light photons (let's say they all have a wavelength of 5500 Å). How many of these visible-light photons would be produced, assuming that half of the original energy of the gamma-ray photon was converted into visible photons?

6. Calculate the wavelengths of maximum emission for the solar photosphere (temperature 5,800 K), the chromosphere (20,000 K), and the corona (2×10^6 K). How can each of these layers best be observed?

7. If the average speed of the solar wind between the Sun and the Earth is 250 km/sec, how long does it take for a particle to reach the Earth once it is emitted by the Sun? How is your answer related to the time it takes for a solar flare to begin to affect radio communications on the Earth?

8. Suppose the temperature in a sunspot were one-half the temperature in the surrounding photosphere. How much less intense would the radiation be from the sunspot than from the photosphere?

Additional Readings

Akasofu, S.-I. 1994. The Shape of the Solar Corona. *Sky & Telescope* 88(5):24.

Bahcall, J. N. 1990. Neutrinos from the Sun: An Astronomical Puzzle. *Mercury* 19(1):53.

Dobson, A. K. and Bracher, K. 1992. Urania's heritage: Historical Introduction to Women in Astronomy. *Mercury* XXI(1):4.

Fischer, D. 1992. Closing In on the Solar Neutrino Problem. *Sky & Telescope* 84(4):378.

Foukal, P. V. 1990. The Variable Sun. *Scientific American* 262(2):34.

Golub, L. 1993. Heating the Sun's Corona. *Astronomy* 21(5):26.

Hathaway, D. H. 1995. Journey to the Heart of the Sun. *Astronomy* 23(1):38.

Kennedy, J. R. 1996. GONG: Probing the Sun's Hidden Heart. *Sky & Telescope* 92(4):20.

Lang, K. R. 1996. Unsolved Mysteries of the Sun— Part 1. *Sky & Telescope* 92(2):38.

Lang, K. R. 1996. Unsolved Mysteries of the Sun— Part 2. *Sky & Telescope* 92(3):24.

Marsden, R. G. and E. J. Smith. 1996. *Ulysses:* Solar Sojourner. *Sky & Telescope* 91(3):24.

Nesme-Ribes, E., S. L. Baliunas, and D. Sokoloff 1996. The Stellar Dynamo. *Scientific American* 275(2):46.

Peterson, C. C., M. Bruner, L. Acton, and Y. Ogawara. 1993. *Yohkoh* and the Mysterious Solar Flares. *Sky & Telescope* 86(3):20.

Wentzel, D. G. 1991. Solar Chimes: Searching for Oscillations Inside the Sun. *Mercury* 20(3):77.

Witze, A. M. 1992. The Great Stone of Ensisheim (?) 500. *Sky & Telescope* 84(5):502.

Web Connections

The Review Questions and Problems also appear at the following URLs:
http://universe.colorado.edu/ch13/questions.html
http://universe.colorado.edu/ch13/problems.html

O star,
Say something to us we can learn,
Say something! And it said, 'I burn.'
Robert Frost

Chapter 14
Properties of the Stars

Chapter Web site: http://universe.colorado.edu/ch14

rom an observational viewpoint, stars represent the fundamental form of matter in the universe. The nighttime sky is dominated by stars, and even in long-exposure photographs taken by large telescopes, it is starlight that we see. The bright nearby objects, as well as the faint, fuzzy distant galaxies, are all glowing by emitted starlight. In discussing and exploring the properties of stars, we are learning about the basic building blocks of visible matter in the universe. (Later we will learn that things might be entirely different if we delete the word **visible** from this statement, for astronomers now believe that most of the mass in the universe is dark and resides in forms not yet understood. This will be discussed in subsequent chapters.)

As in the previous chapter on the Sun, here we will confine ourselves to discussing stars as they appear to us, without speculating about how they may change as they age. But we must keep in mind that stars do change, and in the next chapter we will see how and why.

Three Ways of Looking at Stars: Positions, Magnitudes, and Spectra

Everything that we can learn about a star is contained in the light that we receive from it. Fortunately, a lot of information is there, and astronomers have learned how to extract much of it. Nearly all observations of stars fall into one of three categories, involving measurements of **position, brightness,** or **spectra.** The first two methods have rather long histories. Positional measurements date back to the first human observations of the skies, and brightness measurements (although rather crude ones) were made as long ago as the time of Hipparchus in the second century B.C. (see Chapter 3). In the following sections, each method of observation is discussed separately.

Positional Astronomy

The science of measuring star positions is called **astrometry.** Ancient astronomers used simple devices such as **quadrants** and **sextants (Fig. 14.1)** to measure angular positions on the sky, but today astronomers usually photograph the sky and carefully measure the positions of the star images on the photographic plates or electronic detectors. If several photographs of a given portion of the sky are taken and measured

separately, the results can be averaged to produce a more accurate determination than is possible from a single photograph. Today a stellar position can be measured to a precision better than 0.01 arcsecond. When the European astrometric satellite *Hipparcos* went into operation in 1990, even more accurate measurements became possible, because the fuzziness of star images caused by the Earth's atmosphere was eliminated. *Hipparcos* data are still being analyzed (one reason is that for best accuracy it is necessary to combine measurements taken over a long period of time), but it appears that typical accuracies of a few milliarcseconds (i.e., a few times 0.001 arcsecond) will be achieved.

Whenever positions are measured (for any purpose, not only in astronomy), it is necessary to state the frame of reference in which the position is defined. If you want to tell someone where you are, for example, you use your position on a map, which is a specified coordinate system or grid. In astronomy, we also use a coordinate system or grid of some kind. For measuring star positions, the most commonly used frame of reference is the **equatorial coordinate system.** In this system, the **celestial equator** (the projection of the Earth's equator onto the sky) is the basic reference for north-south positions, and a fixed point on the sky (called the **first point in Aries**) is the reference point for east-west positional measurements (details of this coordinate system are given in Appendix 11).

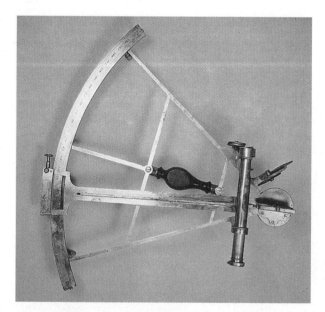

Figure 14.1 A sextant. Instruments of this type were used for centuries to measure star positions relative to each other. In modern times, sextants are used as tools for navigation. (The Granger Collection)

To make precise measurements, it is helpful not only to have a coordinate system, but also to establish precise positions of a number of stars that can be used as reference points by observers. It is not easy to measure the position of the celestial equator or the first point in Aries in order to establish the position of a particular star, so it is very helpful to have a number of reference stars available whose positions are already precisely known. Then the location of the unknown object can be measured relative to these reference stars.

In modern astrometry, a great deal of care is used to establish a grid of reference stars for this purpose. The reference stars are selected for their stability (variable stars will be discussed later), their lack of visible companions (double stars can have distorted or asymmetric images that can make positions difficult to pinpoint), and their brightness (stars that are too bright have enlarged images on photographic plates, which make their positions imprecise, and stars that are too faint are simply too difficult to measure accurately). The reference stars are also selected to provide good coverage of the sky, so that an observer wanting to establish the position of an object

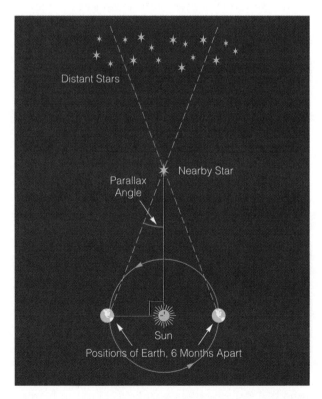

Figure 14.2 Stellar parallax. As the Earth orbits the Sun, the direction of our line of sight to a nearby star varies, so that the star appears to move back and forth against the more distant background stars. The parallax angle p is defined as one-half of the total angular motion. If p is 1 arcsecond, the distance to the star is 206,265 AU, or 1 parsec. This drawing is not to scale.

will find grid stars nearby and can easily make measurements relative to the positions of these reference stars. Every so often astronomers from around the world agree on a revised or refined reference system, and that system is then used when new positional measurements are reported.

Modern astronomers make astrometric measurements for a variety of reasons. For example, such measurements are used in analyzing stellar motions and in determining what these motions reveal about the structure of the galaxy. Astrometric data are also very important in cataloging stars with sufficient accuracy that they can be observed by other telescopes. For example, when preparing for *Hubble Space Telescope* operations, which involve much fainter objects than are commonly observed at ground-based telescopes, it was necessary to prepare an entirely new catalog of the sky, containing millions of stars. Finally, and perhaps most importantly, positional measurements can be used to measure distances to stars by making use of their parallax motions.

Recall from Chapter 2 that **stellar parallax** is the apparent shifting in a nearby star's position due to the orbital motion of the Earth **(Fig. 14.2).** Ancient and medieval astronomers were unable to detect stellar parallax, leading most of them to reject the idea that the Earth orbits the Sun. Eventually, the heliocentric theory won out, but stellar parallax remained undetected until 1838. The reason stellar parallaxes had defied earlier observers then became obvious: even for the closest stars, the maximum shift in position was only about 1 arcsecond. The stars are simply much farther away than the ancient astronomers had dreamed possible. The annual parallax motion of the star Alpha Centauri, the nearest to the Sun, is just over 1 arcsecond, comparable to the angular diameter of a dime as seen from a distance of about 2 miles—a very small angle indeed.

The successful detection of stellar parallax led to a direct means of determining distances to stars. The amount of parallax shift in a star's apparent position depends on how distant the star is; the closer it is, the bigger the shift. The parallax angle p is defined as one-half the total angular shift of a star during the course of a year (see Fig. 14.2). If you look at the long skinny triangle formed by this parallax angle p at the apex with the baseline of 1 AU at the opposite side, you can see that the more distant the star, the smaller the angle p. The two quantities p and distance d are inversely proportional; that is, $d \propto 1/p$. Astronomers have defined a unit of distance called the **parsec,** or pc (for **parallax-second**), which is the distance d to a star having a parallax angle of 1 arcsecond. In these units, the inverse proportionality becomes an equality:

$$d = 1/p.$$

Thus once the parallax angle p is measured, the distance d can be found immediately. For example, if a star has a parallax angle of 0.4 arcsecond, it is $1/0.4 = 2.5$ pc away. The closest visible star (Alpha Centauri) has a parallax angle of $p = 0.753$ arcsecond, so its distance is $d = 1/0.753 = 1.33$ pc.

In more familiar terms, a parsec is equal to 3.26 light-years, or 3.08×10^{16} m, or 206,265 AU. The use of parallax measurements to determine distances is a very powerful technique. It is the only *direct* means astronomers have for measuring how far away stars are. Unfortunately, only stars rather close to us in the galaxy have parallaxes large enough to be measured. Until recently, the smallest parallaxes that could be measured accurately were about 0.01 arcsecond, corresponding to distances of 100 pc. Our galaxy, on the other hand, is more than 30,000 pc in diameter! Thus even the advance to parallaxes as accurate as ± 0.001 arcsecond, yielding distances up to about 1,000 pc, will still probe only the nearest regions of our galaxy. Clearly, other methods of determining distance are needed if we are to probe the entire galaxy. One very powerful method will be described later in this chapter.

Stellar Brightness

The second general type of stellar observation is the measurement of the brightness of stars. This was first attempted in a systematic way over 2,000 years ago when Hipparchus established a system of brightness rankings that is still used today. Hipparchus ranked the stars in categories called **magnitudes,** from first magnitude (the brightest stars) to sixth (the faintest visible to the unaided eye). In his catalog of stars, and in most since then, these magnitudes are listed along with the star positions.

The magnitude system has since been modernized, and all astronomers use the same technique for measurement rather than relying on subjective impressions. In the mid-1800s scientists discovered that what the eye perceives as a fixed *difference* in intensity from one magnitude to the next actually corresponds to a fixed *intensity ratio* (this is called a logarithmic response). Measurements showed that a first-magnitude star is about 2.5 times brighter than a second-magnitude star; a second-magnitude star is 2.5 times brighter than a third-magnitude star; and so on **(Fig. 14.3)**. The brightness ratio between a first-magnitude star and a sixth-magnitude star was found to be nearly 100. In 1850 the system was formalized by the adoption of this ratio as exactly 100; thus the ratio corresponding to a one-magnitude difference is the fifth root of 100, or $(100)^{1/5} = 2.512$. Hence a first-magnitude star is 2.512 times brighter than a star of second magnitude, a sixth-magnitude star is $2.512 \times 2.512 = 6.3$ times fainter than a fourth-magnitude star, and so on.

Magnitudes are traditionally measured through the use of **photometers,** which are devices that produce an electric current when light strikes them. The photocell, the part of the photometer that converts light into electricity, is used in many familiar applications, such as door openers in modern buildings. The amount of electrical current produced is determined by the intensity of light, so an astronomer need only measure the current to determine the brightness of a star **(Fig. 14.4)** and hence its magnitude. In modern times, a different kind of device that allows the simultaneous measurement of many objects is more commonly used. An electronic detector called a CCD (discussed in Chapter 6) is used in place of a photometer. A CCD records a two-dimensional image, just as photographic film does, except that the data are recorded electronically by the CCD instead of as an emulsion on film. The intensity of each individual object in the field of view is recorded by the CCD, and it is a fairly simple matter of computer data analysis to determine the magnitudes of the observed objects.

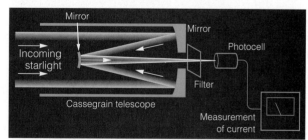

Figure 14.4 Measuring stellar magnitudes. This is a schematic illustration of photometry, the measurement of stellar brightnesses. Light from a star is focused (often through a filter that screens out all but a specific range of wavelengths) onto a photocell, a device that produces an electric current in proportion to the intensity of light. The current is measured and, by comparison with standard stars, converted into a magnitude.

Figure 14.3 Stellar magnitudes. Magnitude differences here are indicated above the line; below the line are the corresponding brightness ratios.

Once magnitudes could be measured precisely, astronomers found that stars have a continuous range of brightnesses and do not fall neatly into the various magnitude rankings. Therefore fractional magnitudes must be used; Deneb, for example, has a magnitude of 1.26 in the modern system, although it was formerly classified simply as a first-magnitude star. Each of the former categories has been found to include a range of stellar brightnesses. This is especially the case for the first-magnitude stars, some of which turned out to be as much as two magnitudes brighter than others when accurate measurements became possible. To measure these especially bright stars in the modern system, magnitudes smaller than 1 must be adopted. Sirius, for example, which is the brightest star in the sky, has a magnitude of −1.45. By using negative magnitudes for very bright objects, astronomers can extend the system to include such objects as the Moon, whose magnitude when full is about −13, and the Sun, whose magnitude is −27 (thus the Sun is about 25 magnitudes—or a factor of $2.512^{25} = 10^{10}$—brighter than Sirius). **Table 14.1** lists the magnitudes of a variety of objects.

Of course, some stars are fainter than the human eye can see, so the magnitude scale must also extend beyond sixth magnitude. With moderately large telescopes, stars as faint as fifteenth magnitude can be measured, and images obtained with modern electronic detectors at large telescopes can detect objects as faint as magnitude 31 or 32. A star of magnitude 30 is 24 magnitudes—or a factor of $2.512^{24} = 4 \times 10^{9}$—fainter than the faintest star visible to the unaided eye.

The stellar magnitudes we have discussed so far all refer to visible light. As we learned in Chapter 6, however, stars emit light over a much broader wavelength band than the eye can see. We also learned that the wavelength at which a star emits most strongly depends on its temperature. By measuring a star's brightness at two different wavelengths, it is therefore possible to learn something about its temperature. For this purpose astronomers use filters that allow only certain wavelengths of light to pass through. A star's brightness is typically measured through two such filters, one that passes yellow light and one that passes blue light, resulting in the measurement of V (for visual, or yellow) and B (for blue) magnitudes **(Fig. 14.5)**. Since a hot star emits more energy in blue wavelengths than in yellow, its B magnitude is smaller than its V magnitude. For a cool star, the situation is reversed, and the B magnitude is larger than the V magnitude (remember that the magnitude scale is backward in the sense that a lower magnitude corresponds to greater brightness).

Table 14.1	
Apparent Magnitudes of Familiar Objects	
Object	Apparent Magnitude
Sun	−26.74
100-watt bulb at 100 ft	−13.70
Moon (full)	−12.73
Venus (greatest elongation)	−4.22
Jupiter (opposition)	−2.6
Mars (opposition)	−2.02
Sirius (brightest star)	−1.45
Mercury (greatest elongation)	−0.2
Alpha Centauri (nearest star)	−0.1
Large Magellanic Cloud	+0.1
Saturn (opposition)	+0.7
Small Magellanic Cloud	+2.4
Andromeda galaxy (farthest naked-eye object)	+3.5
Brightest globular cluster (47 Tucanae)	+4.0
Orion nebula	+4
Uranus (opposition)	+5.5
Faintest object visible to naked eye	+6.0
Neptune (opposition)	+7.9
Crab nebula	+8.6
3C273 (brightest quasar)	+12.8
Pluto (opposition)	+14.9

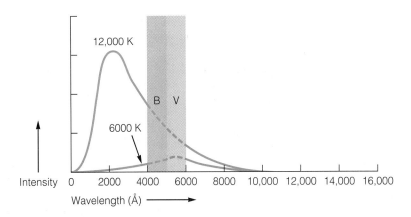

Figure 14.5 The color index. This diagram depicts continuous spectra of two stars, one much hotter than the other. The wavelength ranges over which the blue *(B)* and visual *(V)* magnitudes are measured are indicated. We see that the hot star is brighter in the *B* region than in the *V* region of the spectrum; therefore, its *B* magnitude is smaller than its *V* magnitude, and it has a negative *B* − *V* color index. The opposite is true for the cool star.

The difference between the *B* and *V* magnitudes is called the **color index.** Because the exact value of this index is a function of the temperature of a star, stellar temperatures may be estimated simply by measuring the *V* and *B* magnitudes. A very hot star might have a color index $B - V = -0.3$, while a very cool one might typically have $B - V = +1.2$.

One additional type of magnitude should be mentioned, although measuring it directly is very difficult. This is the **bolometric magnitude,** which is a measure of the intensity of light emitted by a star at all wavelengths **(Fig. 14.6).** To determine a bolometric magnitude requires ultraviolet and infrared, as well as visible observations, and must be done by using telescopes in space. This is particularly true for hot stars, which emit a large fraction of their light at ultraviolet wavelengths. For cool stars, which emit little ultraviolet but a lot of infrared radiation, bolometric magnitudes can be measured from the Earth's surface with fair accuracy. The bolometric magnitude of a star is always smaller than the visual magnitude, because more energy is included when all wavelengths are considered; greater intensity (i.e., greater brightness) means a smaller magnitude.

Often bolometric magnitudes are estimated by comparison with standard stars, whose properties are well known and assumed to represent all stars of similar type. The normal means of comparison is to measure the visual magnitude of the target star and then add a **bolometric correction,** a negative correction factor based on standard stars. Later in this chapter we will see how bolometric magnitudes are related to other properties of stars, particularly luminosity.

Measurements of Stellar Spectra

The first observations of stellar spectra, made in the mid-1800s (before scientists understood how spectral lines are formed), were accomplished using **spectroscopes,** simple devices that allow the observer to see the light from a star spread out according to wavelength, but not to record it. In the late 1800s, introduction of the technique of photographing spectra allowed a systematic study of stellar spectra to get under way. A photograph of a spectrum is called a **spectrogram (Fig. 14.7).** Spectra of many stars can be photographed at one time when a thin prism is placed in front of a telescope **(Fig. 14.8).**

Among the first astronomers to examine systematically the spectra of a large number of stars was the Roman Catholic priest Angelo Secchi, who in the 1860s cataloged hundreds of spectra using a spectroscope. He found that the appearance of the spectra varied considerably from star to star, although they were consistent in one respect: they all showed continuous spectra with absorption lines. Kirchoff's work (see Chapter 6) soon showed that this was due to the relatively cool outer layers of a star, which absorb light from the hotter interior. At first it was thought that the differing appearances of stellar spectra were caused by differences in the chemical composition of the stars, and in one early classification scheme, stars were assigned to categories based on their compositions.

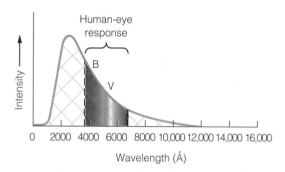

Figure 14.6 Bolometric magnitude. The bolometric magnitude includes all the energy emitted by a star at all wavelengths, not just the visible spectrum.

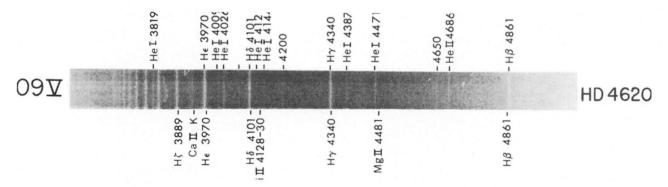

Figure 14.7 A stellar spectrum. This photograph of a stellar spectrum shows the manner in which spectra are usually displayed, as negative prints. Hence light features are absorption lines, wavelengths at which little or no light is emitted by the star (From Abt. H., W. W. Morgan, and R. Tapscott. 1968. *An Atlas of Low-Dispersion Grating Stellar Spectra.* Tucson–Kitt Peak National Observatory) http://universe.colorado.edu/fig/14-7.html

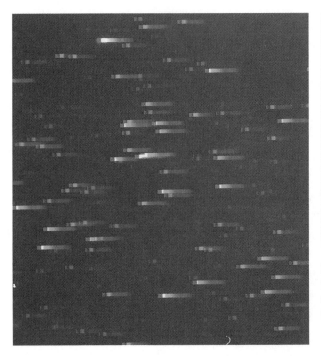

Figure 14.8 Stellar spectra. This photograph was made through a telescope with a thin prism in the light beam, so that each stellar image was dispersed into a spectrum. It is possible to measure and classify the spectra of large numbers of stars using this technique. (University of Michigan Observatories)

Figure 14.9 Annie J. Cannon. A member of the Harvard College Observatory for almost 50 years, Cannon classified the spectra of several hundred thousand stars. Today she is recognized as the founder of modern spectral classification. (Harvard College Observatory)

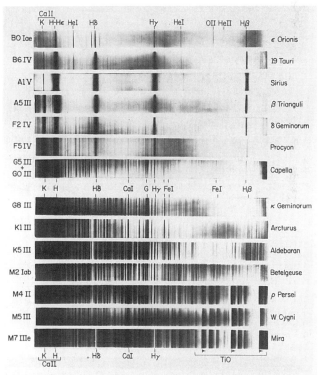

Figure 14.10 A comparison of spectra. Here we see several spectra representing different spectral classes. A few major absorption lines are identified. The spectra are arranged in order of decreasing stellar temperature (top to bottom). (University of Michigan Observatories)

The basis of the modern classification system for stellar spectra was established by a group of astronomers at Harvard, most notably Annie J. Cannon **(Fig. 14.9).** Cannon found a smooth sequence in which the pattern of strong absorption lines changed gradually from one type of spectrum to the next. Having already assigned letters of the alphabet to the various types, she placed them in the sequence **O, B, A, F, G, K, M (Fig. 14.10).**

It was later realized that the differing appearances of stellar spectra were due not to differences in chemical composition, but to differences in temperature. The hotter a star, the more highly ionized the gas in its outer layers is. The degree of ionization in turn governs the pattern of spectral lines that will form. (If this is confusing, review the discussion of ionization in Chapter 6.) Therefore Cannon's sequence of spectral types was a temperature sequence: the hottest stars are the O stars, and the coolest are the M stars. She developed a fine enough eye for subtle differences between spectra to assign subclasses to each of the major classes. In this system, the Sun is a G2 star, being intermediate between types G and K.

In the modern classification system, a few key spectral lines establish the type of a given star **(Fig. 14.11).** For the O stars, ionized helium, which requires a very high temperature for its formation, is

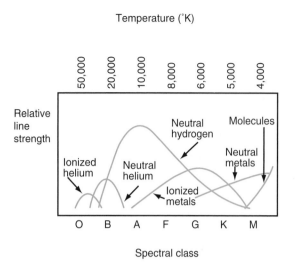

Figure 14.11 Ionization for stars of various spectral types. This diagram shows which ions appear prominently in the spectra of stars of different classes. Note that the degree of ionization evident for the hot stars (left) is much greater than for the cool ones (right); for example, helium is much more difficult to ionize than are the metallic elements. Hence an O star, which has ionized helium, is hotter than the F, G, and K stars, which have ionized metals but no ionized helium.

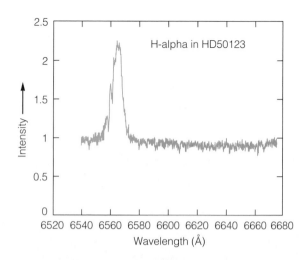

Figure 14.12 A stellar emission line. This graph shows an emission line due to hydrogen in a star belonging to a peculiar class of stars known to have hot gas surrounding them. (T. P. Snow; data obtained at the Canada-France-Hawaii Telescope).

the principal species that indicates the spectral type. For the slightly cooler B stars, it is atomic helium; for the A stars, atomic hydrogen. The F stars have strong hydrogen lines, along with lines due to certain metallic elements that are ionized once (that is, these atoms have lost just one electron). The G stars have a mixture of ionized and atomic metals, and in the K and M stars these elements are nearly all in atomic form. The cooler M stars also have strong molecular lines.

Having established this classification system around the turn of the century, Cannon cataloged nearly a quarter of a million stars, a monumental task that took about five years (although the publication of the results, called the *Henry Draper Catalog*, took place over a much longer period, during the 1920s). This catalog has been a fundamental reference for generations of astronomers, and the system of classification established by Cannon has, with some modification, been in use since its development.

Peculiar Stars

Some stars do not fit neatly into the standard spectral classes, and these are often referred to as "peculiar" stars. In most cases these stars have unusual chemical compositions, at least at the surface where the spectral lines form. Most stars have the same basic composition. (See the list of relative abundance of the elements in the Sun in Table 13.2). Others are unusual because they have emission lines in their spectra **(Fig. 14.12),** which, according to Kirchhoff's laws, means that they must be surrounded by hot, rarefied gas **(Fig. 14.13).** The so-called peculiar stars are probably quite normal but in short-lived stages of evolution, so there are not many of them around at any one time.

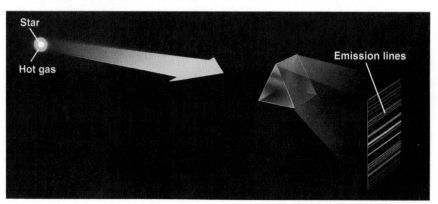

Figure 14.13 Stellar emission lines. Very few stars have emission lines in their visible-wavelength spectra (but many do at ultraviolet wavelengths; see Chapter 15). When such lines are present, it usually signifies the existence of hot gas above the surface. According to Kirchhoff's second law, a rarefied, hot gas produces emission lines.

Women in Astronomy

As in many technical fields, women are in a minority in astronomy, currently holding only about 15% of the membership in the American Astronomical Society, for example. The numbers have been increasing, and today in most American university graduate programs the fraction of women is approaching 50%. This suggests that if the trend continues, true parity may eventually be reached.

Apart from sheer numbers, there is another measure of equity for women in astronomy: the opportunities for women to advance to high positions as observatory directors or department chairs, for example. In this regard most studies show that women are still facing an uphill battle (is the chair of the department where you are taking your astronomy class a woman?). There are, however, a growing number of leaders among women in our field who serve as counterexamples. The American Astronomical Society currently has a woman president (Andrea K. Dupree of the Harvard-Smithsonian Center for Astrophysics), and the current director of the principal U.S. observatory, the National Optical Astronomy Observatories, is also a woman (Sidney Wolff, also a past president of the American Astronomical Society). The chief scientist at NASA is France Cordova, who, in that capacity, has significant influence over the direction of space science nationally. And many of today's leaders in astronomical research, in fields ranging from planetary science to cosmology, are women.

Historically, women have made many very important contributions to astronomy, even though in most cases they did so from positions of assumed inferiority.

One of the first prominent women in the field was Caroline Herschel, sister of the noted English astronomer William Herschel. Both Herschels were originally musicians from a large family in Germany. In 1772 William, already established in England as a leading astronomer, asked Caroline to become his housekeeper and assistant observer (she was allowed to make the move only after William promised to hire a maid to replace her services to their family in Germany). In due course, Caroline made many observations of her own, discovering eight comets, several new "nebulae" (now known to be galaxies), and publishing extensive catalogs of the observations she and her brother made. For these contributions to astronomy, Caroline was awarded a small annual pension as "assistant to the Astronomer Royal" and was eventually elected an honorary fellow of the Royal Astronomical Society (as a woman, she was not eligible for full membership).

Similarly, the first prominent woman in U.S. astronomy, Maria Mitchell, also rose to fame despite a lack of formal training in astronomy. The daughter of an amateur astronomer, Mitchell made several important observational discoveries, worked for nearly 20 years as a "computer" for the U.S. Naval Observatory, and was named astronomy professor and observatory director at Vassar College when that women's college was founded in 1865. Mitchell was a pioneer in establishing recognition for women in science. She was the first woman (by 95 years!) to be elected to the American Academy of Arts and Sciences, she was the first woman elected to the American Association for the Advancement of Science, and she helped found the Association for the Advancement of Women. A

small observatory named for her and dedicated to helping young women in astronomy has operated since 1889 on Nantucket Island, Mitchell's home for much of her life.

In the last quarter of the nineteenth century, the Harvard College Observatory, under the direction of Edward Pickering, hired many women to serve as assistants in measuring and analyzing the many photographic plates (images and spectra) being obtained by the observatory. The prevailing notion was that women were well suited to doing tedious measurements and calculations, though they were not thought to be as creative as men. As discussed in this chapter, however, the women at Harvard made many creative contributions to astronomy, including establishment of the spectral classification scheme for stars and, later, the development of a full astrophysical understanding of the spectra. The work of Cecilia Payne-Gaposchkin towers over many others in this field, and she was awarded the first doctorate for a woman (and the first in astronomy for anyone, regardless of gender) ever bestowed by Harvard. Eventually, Payne-Gaposchkin was deemed worthy of full status as a faculty member, and in 1956 she was named professor and department chair in astronomy at Harvard.

Comparing these historical notes with the status and role of women in astronomy today, you can see that great strides have been taken. There is still a long way to go, but the numbers and acceptance of women in the field have grown, and today there are many outstanding women who may serve as role models for young women with aspirations for careers in science. We can hope that in the future such dreamers will not be handicapped by society's attitudes toward women, as happened so many times in the past.

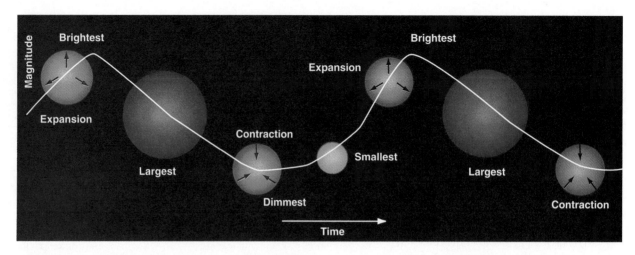

Figure 14.14 A pulsating variable star. The curved line shows how the brightness (in magnitude units) varies as the star expands and contracts. The sequence of sketches illustrates how the expansion and contraction phases are related in sequence to the variations in brightness. The surface temperature also varies. http://universe.colorado.edu/fig/14-14.html

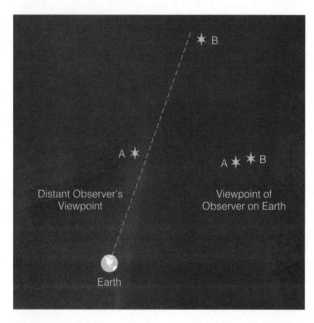

Figure 14.15 An optical double. At the left we see the line of sight from the Earth passing close to two stars (A and B) that are far apart but in nearly the same direction from Earth. At the right we see that these two stars appear very close together on the sky. This is not a true binary system.

The variable stars represent another kind of unusual star. The majority of these are stars whose brightness fluctuates regularly as they alternately expand and contract **(Fig. 14.14).** As in the case of the peculiar stars, the variables are normal stars in special stages of their lifetimes where particular combinations of atmospheric pressure and ionization conditions produce instabilities that cause the pulsations. The most widely known pulsating variable stars are **δ Cephei stars,** giant stars whose spectral type varies between F and G as they pulsate. As we will see in Chapter 18, these stars, as well as the less luminous **RR Lyrae variables,** are very useful tools in measuring distances, because their luminosities can be inferred from their periods of pulsation. Other pulsating stars include the **Mira stars** or **long-period variables,** M supergiant stars that take a year or longer to go through a complete cycle; and a variety of shorter-period variables that are not quite as regular as the δ Cephei and RR Lyrae stars. Some stars vary erratically, even explosively, and these will be discussed in Chapter 16.

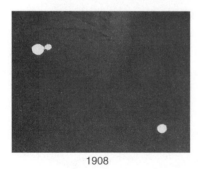

1908

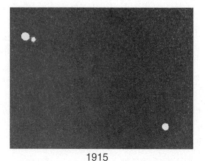

1915

1920

Figure 14.16 Binary star Krueger 60. Between 1908 and 1920, the visual binary in the upper left corner of each photograph completed about a quarter of a revolution. (Photographs by E. E. Barnard, Yerkes Observatory)

Binary stars

About half of the stars in the sky are members of double or **binary star systems,** where two stars orbit each other. All types of stars can be found in binaries, and their orbits also come in many sizes and shapes. In some systems the two stars are so close together that they are actually touching, and in others they are so far apart that it takes hundreds or thousands of years for them to complete one revolution.

Binary systems can be detected by each of the three different types of observations we have discussed **(Table 14.2).** Positional measurements, of course, tell us when two stars are very close to each other in the sky. Sometimes this proximity occurs by chance, when a nearby star happens to lie almost in front of one that is in the background **(Fig. 14.15)**; this is called an **optical double** and is not a true binary system, since the two stars do not orbit each other. A pair of stars that appear close together and are in motion about one another is called a **visual binary (Fig. 14.16).** Accurate positional measurements are needed to reveal the orbital motion, because the two stars are very close together in the sky and appear to move very slowly about each other. The two stars must be separated by many AUs to appear separately as seen from the Earth, and therefore the orbital period of a visual binary is usually many years.

It is possible for binary systems to be recognized even when only one of the two stars can be seen. If positional measurements over a long period of time reveal that a star exhibits a wobbling motion as it moves through space **(Fig. 14.17),** it may be inferred that the star is orbiting an unseen companion. Detection of binary stars in this manner is similar to the technique

Table 14.2

Types of Binary Systems

Type of System	Observational Characteristics
Visual binary	Both stars visible through telescope (usually requires photograph); change of relative position confirms orbital motion.
Astrometric binary	Positional measurements reveal orbital motion about center of mass.
Spectrum binary	Spectrum shows lines from two distinct spectral classes, revealing presence of two stars.
Spectroscopic binary	Spectral lines undergo periodic Doppler shifts, revealing orbital motion. It two sets of spectral lines are seen shifting back and forth (which occurs only if the two stars are comparable in magnitude), it is a double-lined spectroscopic binary; if only one set of lines is seen (in the more common situation where one star is much brighter than the other), it is a single-lined spectroscopic binary.
Eclipsing binary	Apparent magnitude varies periodically because of eclipses, as stars alternately pass in front of each other.

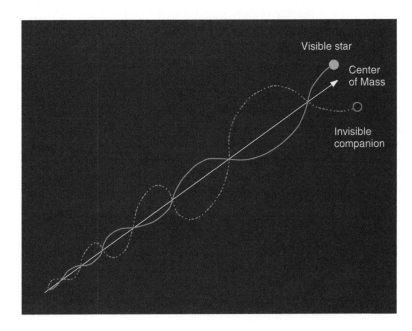

Visible star

Center of Mass

Invisible companion

Figure 14.17 An astrometric binary. Careful observation of the motion of a star across the sky in some cases reveals a curved path such as the one shown here. Such motion is caused by the presence of an unseen companion star; the visible star orbits the center of mass of the binary system as it moves across the sky.

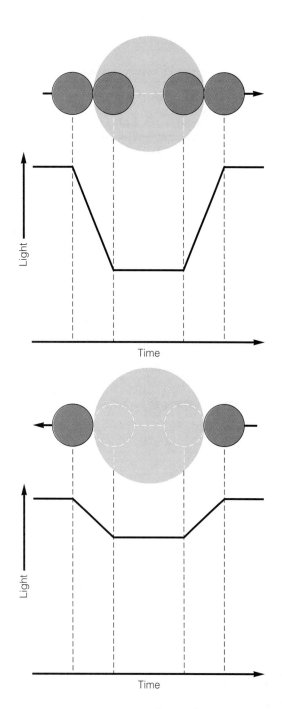

Figure 14.18 An eclipsing binary. If our line of sight happens to be aligned with the orbital plane of a double star system, the two stars will alternately eclipse each other, causing brightness variations in the total light output from the system, as shown. In the upper figure, the smaller star is passing in front of the larger one; in the lower sketch, the smaller one is passing behind. The segmented graph in each case, called the light curve, illustrates how the total brightness of the system varies as these eclipses occur.

discussed in Chapter 12 for searching for massive planets orbiting other stars. Binary stars that are detected because of variations in position are called **astrometric binaries.** Sirius, the brightest star in the sky, is an astrometric binary, having a small, dim companion (Sirius B) whose presence was discovered because of the wobbling motion observed for Sirius A.

Simple brightness measurements can also tell us when a star, which may appear to be single, is actually part of a binary system. If we happen to be aligned with the plane of the orbit, so that the two stars alternately pass in front of each other as we view the system, the observed brightness will decrease each time one star is in front of the other. This is called an **eclipsing binary (Fig. 14.18).**

Finally, spectroscopic measurements can be used to recognize binary systems even when a star appears single. Most often this is made possible by the Doppler effect, which causes the spectral lines to shift slightly in position as the stars orbit each other, alternately approaching and receding from the Earth **(Fig. 14.19).** Star pairs known by this technique to be binaries are called **spectroscopic binaries.** In many cases, one of the two stars is too dim for its light to be detected, but the system is still recognized as binary because of the shifts in the spectral lines of the brighter star. Like the astrometric method, this method is very similar to one for indirectly detecting planetary systems that we discussed in Chapter 12.

A given double star system may fall into more than one category. For example, a relatively nearby system seen edge-on could be a visual binary if both stars can be seen in the telescope; it may also be an eclipsing binary if the two stars alternately pass in front of each other; and it almost surely will be a spectroscopic binary since the motion back and forth along our line of sight is maximized when we view the orbit edge-on.

Binary stars merit our attention partly because there are so many of them, but even more importantly because of what they can tell us about the properties of individual stars. As we will see in the next section, much of our basic information on the nature of stars comes from measurements of binary systems.

Fundamental Stellar Properties

To understand how stars work, astronomers need to determine their physical properties and how they are related to each other. A number of fundamental quantities characterize a star, and in these sections we will see how the observations described previously

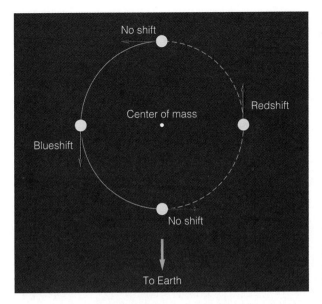

Figure 14.19 Spectroscopic binaries. As a star orbits the center of mass of a binary system, its velocity relative to the Earth varies. This produces alternating redshifts and blueshifts in its spectrum, as the star recedes from and approaches us, as shown in the left-hand diagram. At right we see spectra at a single-lined spectroscopic binary (above) and a double-lined spectographic binary (below). The three central specta above show how the lines are blue-shifted (first and third spectra) and red-shifted (middle spectra) as the star approaches and recedes. The two spectra below show how spectral lines in a double-lined system alternately separate (top) and merge (bottom).

can be used to determine those quantities. They include the luminosities, the temperatures, the radii, and, above all, the masses of stars. At several steps along the way, it is also important to know the distances to stars, and we will learn about a new, very powerful distance-determination method.

Absolute Magnitudes and Stellar Luminosities

The luminosity of a star is the total amount of energy it emits from its surface, at all wavelengths, and the basic unit of measure is the **joule per second,** or **watt** (see Chapter 5). Because the distance to a star affects its brightness, it is necessary to know the distance before the luminosity can be determined. It is also necessary to measure the brightness at all wavelengths, because luminosity refers to the *total* energy output of a star. Instead of dealing directly with basic energy units, astronomers have opted to use the stellar magnitude system that was established so long ago. Thus, to take into account the distance to a star, a quantity called the **absolute magnitude** is defined. This is the magnitude a star would have if its distance were 10 pc. In effect, when astronomers use absolute magnitudes, they are pretending to place all stars at the same standard distance **(Fig. 14.20** and **Table 14.3),** so

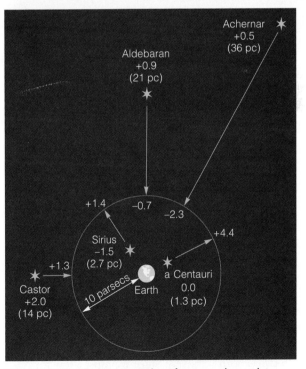

Figure 14.20 Absolute magnitudes. A few stars are shown at their correct relative distances (actual distances in parentheses). The arrows indicate the imaginary movement of these stars to a 10 pc distance from the Earth. The numbers indicate the apparent (actual distance) and absolute (10 pc distance) magnitudes.

Table 14.3
Visual Absolute Magnitudes of Selected Objects

Object	Absolute Magnitude
Typical bright quasar	−28
Brightest galaxies	−25
Milkway Way galaxy	−20.5
Andromeda galaxy	−21.1
Type I supernova (maximum brightness)	−18.8
Large Magellanic Cloud	−18.7
Type II supernova (maximum brightness)	−17
Small Magellanic Cloud	−16.7
Typical globular cluster	−8
The most luminous stars	−8
Typical nova outburst	−7.7
Vega (bright star in summer sky)	+0.5
Sirius	+1.41
Alpha Centauri (nearest star)	+4.35
Sun	+4.83
Venus (greatest elongation)	+28.2
Full moon	+31.8
100-watt bulb	+66.3

that comparisons of their brightnesses (magnitudes) amount to comparisons of their luminosities. Thus a comparison of absolute magnitudes reveals differences in luminosities, because all the effects of distance have been canceled out.

It helps to consider a specific example. Suppose a certain star is 100 pc away and has an observed magnitude of 7.3. To determine the absolute magnitude, we must find out what this star's magnitude would be if it were only 10 pc away. The inverse square law tells us that since the star would be a factor of 10 closer, it would appear a factor of $10^2 = 100$ brighter, or exactly 5 magnitudes brighter. Therefore, its absolute magnitude is $7.3 − 5 = 2.3$. The absolute magnitude can always be derived in this way from the measured or **apparent magnitude** and the distance. Other examples are given in the questions at the end of this chapter and in Appendix 13.

Let us now return to the question of luminosities. By determining the absolute magnitudes of stars, we can compare their luminosities, because the distance effect has been removed. Since we must allow for light emitted by a star at all wavelengths, the bolometric magnitude is the quantity that is used. Thus, if one star's bolometric magnitude is 5 magnitudes smaller than another's, we know that its luminosity is a factor of 100 greater. The luminosity can be ex-

pressed in terms of watts by comparing it with luminosity of a standard star that has been measured directly by measuring the intensity of light at all wavelengths and allowing for distance. Most often, however, astronomers tend to express luminosities in units of the Sun's luminosity. This has the advantage of giving some intuitive meaning to the values; it means more to say that a certain star has 29.6 times the solar luminosity than to say that it has a luminosity of 1.12×10^{28} watts. Similarly, other stellar properties are usually expressed in terms of the solar values.

Stellar luminosity values are found to have an incredibly wide range. Stars are known with luminosities as small as 10^{-4} the luminosity of the Sun and as great as 10^6 that of the Sun, a range of 10 billion from the faintest to the most luminous! The luminosity is by far the most highly variable parameter for stars; the others that we shall discuss only cover ranges of a few hundred or less from one extreme to the other.

Stellar Surface Temperatures

Earlier in this chapter, we saw that the color of a star, like its spectral class, depends on its temperature, so that the temperature can be deduced by observing either. Recall that the color index, the difference between the blue *(B)* and visual *(V)* magnitudes, is a measured quantity that indicates temperature. A negative value of $B − V$ means that the star is brighter in blue than in visual light and therefore is a hot star. A large positive value indicates a cool star. A specific correlation of color index with temperature has been developed and is used to determine temperatures from observed values of $B − V$.

More refined estimates of temperature can be made from a detailed analysis of the degree of ionization, which is done by measuring the strengths of spectral lines formed by different ions. This is basically the same as estimating the temperature from the spectral class, since in either case the point is that the strengths of spectral lines of various ions depend on how abundant those ions are, which in turn depends on the temperature. Indeed, there is a strong correlation between spectral class and temperature, and for many purposes astronomers simply use this correlation to estimate the temperatures of stars.

The temperature referred to here may be called the surface temperature, although stars do not have solid surfaces. We are really referring to the outermost layers of gas, where the absorption lines form. This region is the photosphere of a star (with the same meaning as in the Sun), and it actually has some depth, although it is very thin compared with the radius of the star.

Stellar temperatures, as we have already seen, range from about 2,000 K for the coolest M stars to 50,000 K or more for the hottest O stars.

The Hertzsprung-Russell Diagram and a New Distance Technique

We have seen that temperature and spectral class are intimately related, and we have learned how astronomers deduce the luminosities of stars. In the first decade of this century, the Danish astronomer Ejnar Hertzsprung and, independently, the American astrophysicist Henry Norris Russell **(Fig. 14.21)** began to consider how luminosity and spectral class might be related to each other. Each gathered data on stars whose luminosities (or absolute magnitudes) were known and found a close link between spectral class (temperature) and absolute magnitude (luminosity). This relationship is best seen in the diagram constructed by Russell in 1913 **(Fig. 14.22),** now called the **Hertzsprung-Russell diagram** or **H-R diagram.** In this plot of absolute magnitude (on the vertical scale) versus spectral class (on the horizontal axis), stars fall into narrowly defined regions rather than being randomly distributed **(Fig. 14.23).** A star of a given spectral class cannot have just any absolute magnitude, and vice versa.

The great majority of stars fall into a diagonal strip running from the upper left (high temperature, high luminosity) to the lower right (low temperature, low luminosity) of the H-R diagram. This strip has been given the name **main sequence.**

One group of a few stars does not fall on the main sequence but appears instead in the upper right (low temperature, high luminosity) of the diagram. Since the spectra of these stars indicate that they are relatively cool, their high luminosities cannot be due to greater temperatures than the main-sequence stars of the same type. The only way one star can be much more luminous than another of the same temperature is if it has a much greater surface area. To see this, recall the Stefan-Boltzmann law (Chapter 6), which is

$$L = 4\pi R^2 \sigma T^4,$$

where L is luminosity, R is radius, and T is surface temperature. If two stars have the same surface temperature, then the ratio of their luminosities is given by

$$\frac{L_1}{L_2} = \left(\frac{R_1}{R_2}\right)^2,$$

Figure 14.21 Henry Norris Russell. One of the leading astrophysicists of the era when a physical understanding of stars was first emerging. Russell made many major contributions in a variety of areas. (Princeton University)

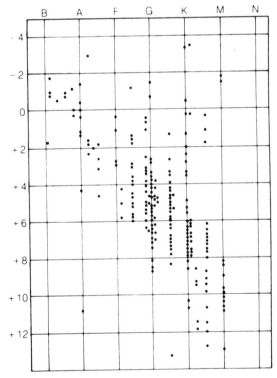

Figure 14.22 The first H-R diagram. This plot showing absolute magnitude versus spectral class was constructed in 1913 by Henry Norris Russell. (Estate of Henry Norris Russell, reprinted with permission)

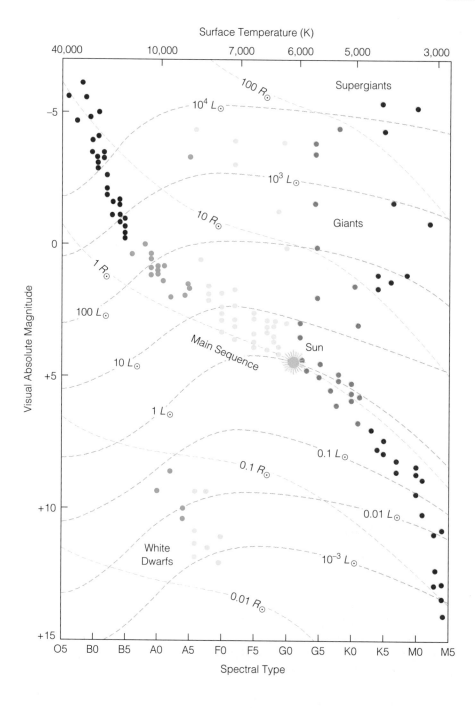

where the other terms (4π, σ and T^4) have canceled out. This can be solved for the ratio of the radii:

$$\frac{R_1}{R_2} = \sqrt{\frac{L_1}{L_2}} \, .$$

For example, if one star has 100 times the luminosity of the other, then it must have a radius 10 times larger. In this way Hertzsprung and Russell realized that the extraluminous stars sitting above the main sequence must be much larger than those on the main sequence, and they called them **giants** and **supergiants.** Antonia Maury, one of Annie Cannon's associates at Harvard, had noted earlier that stellar spectra could be classified according to the width of certain spectral lines in addition to the overall pattern of lines. No one knew how to interpret this at the time, but later, when Hertzsprung and Russell realized that some stars are much larger than others, it was discovered that the line widths were related to stellar size. A giant or supergiant star has narrower lines than a main-sequence star of the same tempera-

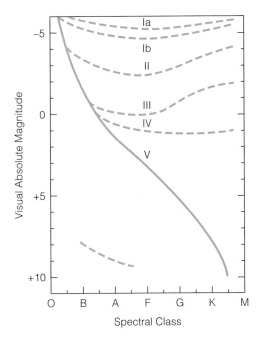

Figure 14.24 Luminosity classes. This H-R diagram shows the locations of stars of the luminosity classes described in the text. A complete spectral classification for a star usually includes a luminosity class designation.

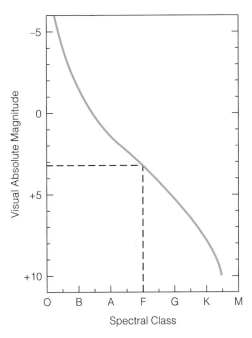

Figure 14.25 Spectroscopic parallax. Knowledge of a star's spectral class (including the luminosity class) allows the absolute magnitude, and hence the distance, to be determined. This figure shows how the absolute magnitude of an FO main-sequence star is read from the H-R diagram. The distance can then be found by comparing the absolute and apparent magnitudes of the star (as explained in the text and Appendix 13).

ture (the reason for this is that a very extended, large star has lower gas pressure in its outer layers, whereas a main-sequence star has higher pressure, which tends to broaden the lines). Thus it became possible to recognize giants and supergiants by their spectra.

The distinction among giants, supergiants, and main-sequence stars (commonly known as **dwarfs**) has been incorporated into the spectral classification system used by modern astronomers. A **luminosity class (Fig. 14.24)** has been added to the spectral type with which we are already familiar. The luminosity classes, designated by Roman numerals following the spectral type, are **I** for supergiants (this group is further subdivided into classes Ia and Ib), **II** for extreme giants, **III** for giants, **IV** for stars just a bit above the main sequence, and **V** for main-sequence stars, or dwarfs (not to be confused with white dwarfs, which are discussed later). Thus a complete spectral classification for the bright summertime star Vega, for example, is A0V, meaning that it is an A0 main-sequence star. Betelgeuse, the red supergiant in the shoulder of Orion, has the full classification M2Iab (because it is intermediate between luminosity classes Ia and Ib). It is usually possible to assign a star to the proper luminosity class by examining subtle details of its spectrum. For example, giant and especially supergiant stars have relatively low atmospheric pressures, which affect the spectral line widths and the state of ionization of certain elements.

Another group of stars (which has become known mostly since Russell first plotted the H-R diagram) does not fall into any of the standard luminosity classes but appears in the lower left (high temperature, low luminosity) corner of the diagram. Since these stars are hot but not very luminous, they must be very small, and they have been given the name **white dwarfs**. These objects have some very bizarre properties, which will be discussed in Chapter 16.

Using the H-R Diagram to Find Distances

The H-R diagram can be used to find distances to stars, even stars that are very far away **(Fig. 14.25)**. The idea is really very simple: If we know how bright a star is intrinsically (how much energy it is actually emitting from its surface), and we measure how bright it appears to be, we can determine how far away it is because we know that the difference between its intrinsic brightness and its observed brightness is due to its distance from us. The main problem lies in knowing the intrinsic brightness (luminosity) of the star, which is where the H-R diagram comes in. (Another problem is that interstellar dust can make a star appear fainter than it otherwise would, and some correction for this must be applied; see Appendix 13. For now we ignore the effects of dimming by interstellar dust.).

Once we determine the spectral class of a star, we can place it on the H-R diagram (as long as we know its luminosity class, which indicates whether it is on the main sequence or is a giant or supergiant). Once we have placed the star on the diagram, the vertical axis tells us its absolute magnitude, which is a measure of its luminosity. A comparison of the absolute magnitude with the observed apparent magnitude then amounts to the same thing as a comparison of the intrinsic and apparent brightnesses of the star, and from such a comparison the distance can be found.

The calculation of a star's distance from the difference between the apparent and absolute magnitudes can best be illustrated by considering a few simple cases. For example, if the difference $m - M$ (the apparent minus the absolute magnitude) is 5, the star appears 5 magnitudes, or a factor of 100, fainter than it would at the standard distance of 10 pc. A factor of 100 in brightness is created by a factor of 10 change in distance, so this star must be 10 times farther away than it would be if it were at 10 pc distance; therefore, it is $10 \times 10 = 100$ pc away. Similarly, a star whose apparent magnitude is 10 more than its absolute magnitude is 10,000 times fainter at its true distance than if it were 10 pc away; a factor of 10,000 in brightness corresponds to a factor of 100 in distance, so this star is 100×10 pc $= 1,000$ pc away.

The difference $m - M$ between the apparent and absolute magnitudes determines the distance, so this difference is given the special name **distance modulus.** Continuing our examples, if the distance modulus is $m - M = 15$, then the distance is 10,000 pc, and if $m - M = 20$, then the distance is 100,000 pc. For every increase of 5 in the distance modulus, the distance increases by a factor of 10. It should be obvious that if $m = M$ (that is, $m - M = 0$), the distance to the star must be 10 pc, because this is the distance that defines the absolute magnitude. **Table 14.4** summarizes the relationship between distance modulus and distance for a number of simple cases, while Appendix 13 provides formulas for calculating the distance when the distance modulus is not a simple multiple of 5.

This method is very powerful, because it can be used for very large distances. All that is needed is to be able to place a star on the H-R diagram, so that its absolute magnitude can be determined, and to measure its apparent magnitude. Because this distance-determination technique depends on placing a star on the H-R diagram by classifying its spectrum, it is called the **spectroscopic parallax** method. (Astronomers use the word *parallax* as a general term for distances, even though, technically speaking, no parallax is measured in this case.) Later we

Table 14.4	
The Distance Modulus	
m − M	**Distance**
−5	1 pc
0	10 pc
5	100 pc
10	1,000 pc (1 kpc)
15	10,000 pc (10 kpc)
20	100,000 pc (100 kpc)
25	1,000,000 pc (1 Mpc)
30	10,000,000 pc (10 Mpc)
35	100,000,000 pc (1,000 Mpc)

will see that other methods of estimating absolute magnitudes also lead to means of measuring distances in astronomy; therefore it is very important to understand how this technique works, and to be able to do simple examples.

Stellar Diameters

We have already seen that a star's position on the H-R diagram depends partly on its size, since luminosity is related to total surface area. If two stars have the same surface temperature (and therefore the same spectral class), but one is more luminous than the other, we know that it must also be larger. The Stefan-Boltzmann law (see Chapter 6) specifically relates the three quantities luminosity, temperature, and radius; use of this law allows the radius to be determined if the other two quantities are known.

Solving the Stefan-Boltzmann equation for radius R yields

$$R = \sqrt{\frac{L}{4\pi\sigma T^4}},$$

where L is the luminosity, T is the surface temperature, and σ is the Stefan-Boltzmann constant. This is easiest to use when R, L, and T are expressed in solar units, because the equation then becomes

$$R = \sqrt{\frac{L}{T^4}}$$

(this is found by writing the equation separately for the star and for the Sun and then dividing so that the constant terms cancel out; this is very similar to what we did in a preceding section, where we solved for the ratio of radii when the temperatures were equal).

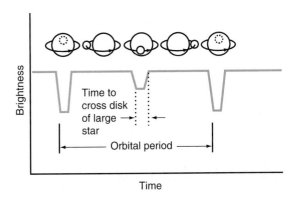

Figure 14.26 Measuring stellar diameters in an eclipsing binary. The duration of each eclipse is combined with knowledge of the orbital speeds of the stars (from the Doppler effect) to yield the diameters of the stars. (In this figure, the smaller star is hotter, making the eclipses deeper when it is obscured.)

As an example, consider a star with a luminosity of 100 $L_\odot$ and a surface temperature of 2 $T_\odot$. Its radius will be

$$R = \sqrt{\frac{100}{(2)^4}} = \sqrt{\frac{100}{16}} = 6.3 \ R_\odot.$$

Eclipsing binaries provide another means of determining stellar radius that is independent of other properties **(Fig. 14.26)**. Recall that these are double star systems in which the two stars alternately pass in front of each other as we view the orbit edge-on. An eclipsing binary is also very likely to be a spectroscopic binary, so the speeds of the two stars in the orbits can be measured from the Doppler effect. We therefore know how fast the stars are moving, and from the duration of the eclipse, we know how long it takes one to pass in front of the other; then the simple formula *distance = velocity × time* ($D = vt$) gives us the diameter of the star that is being eclipsed. Even if no information on the orbital velocity is available, the *relative* diameters of the two stars can be deduced by comparing the time they need to *enter* eclipse with the time *in* eclipse, as indicated by the sloped and flat segments of the light curve (see Fig. 14.26).

For example, suppose the time it takes one star to cross in front of the other is $4^d 13.53^h$ ($= 3.943 \times 10^5$ sec), and Doppler shift measurements tell us that the relative orbital speed of the two stars is 12.84 km/sec. The diameter of the star being eclipsed is then

$$D = vt = (12.84 \ \text{km/sec}) \times (3.943 \ 3 \ 10^5 \ \text{sec}) = 5.06 \times 10^6 \ \text{km}.$$

This is about 3.6 times the diameter of the Sun.

Eclipsing binaries provide the most direct means of measuring stellar sizes, but unfortunately they are not numerous. In most cases the radii are estimated from the luminosity and temperature as described above. In a few cases stellar radii have been measured directly by use of **interferometry,** a sophisticated technique for clarifying the image of a star by removing the blurring effects of the Earth's atmosphere (see Chapter 6). Only relatively near, large stars can be measured this way, however. This technique measures the *angular* diameter; to convert this to true diameter requires knowledge of the distance to the star.

Stars on the main sequence do not vary as widely in radius as they do in luminosity, ranging from perhaps 0.1 times the Sun's radius for the M stars at the lower right-hand end to 10 or 20 solar radii at the upper left. Of course, large variations in size occur as we go away from the main sequence, either toward the giants and supergiants, which may be 100 times the size of the Sun, or toward the white dwarfs, which are as small as 0.01 times the size of the Sun.

Binary Stars and Stellar Masses

The mass of a star is the most important of all its fundamental properties, for the mass governs most of the others. This point will be discussed at some length in the next chapter.

The only way to measure a star's mass is by observing its gravitational effect on other objects, and this is possible only in binary star systems, where the two stars hold each other in orbit by their gravitational forces. Because binary systems are common, astronomers have many opportunities to determine masses by analyzing binary orbits.

The basic idea is rather simple, although the application may be quite complex, depending on the type of binary system. We use Kepler's third law in the form derived by Newton. Remember that if the period P is measured in years, the average separation of the two stars (the semimajor axis a) is measured in astronomical units, and the masses m_1 and m_2 are in units of solar masses, then Kepler's third law is

$$(m_1 + m_2)P^2 = a^3.$$

We need only observe the period and the semimajor axis to solve for the sum of the two masses. Careful observations of the sizes of the individual orbits (actually, of the relative distances of the two stars from the center of mass; **Fig. 14.27**) also yield the ratio of the masses; when both the sum and the ratio are known, it is simple to solve for the individual masses.

Consider a case where a binary has an orbital period of 26.948 years and a semimajor axis of 14.14

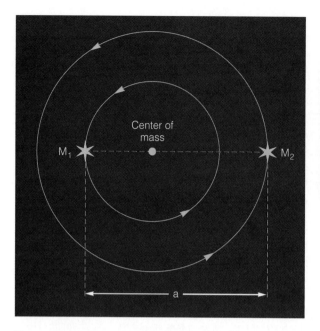

Figure 14.27 Binary star orbits. This figure illustrates the terms used in Kepler's third law. Two stars of masses m_1 and m_2 (m_1 is larger than m_2 in this case) orbit a common center of mass, each making one full orbit in period P. The semimajor axis a that appears in Kepler's third law is actually the sum of the semimajor axes of the two individual orbits about the center of mass; this sum corresponds to the average distance between the two stars.

http://universe.colorado.edu/fig/14-27.html

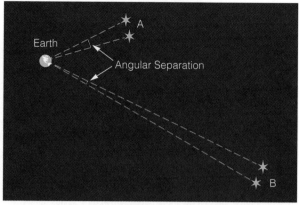

Figure 14.28 The effect of distance on measurements of binary star orbits. Here two binary systems have the same actual separation, but from Earth the nearer system (A) appears more widely separated than the more distant system (B). Therefore the distance to a binary system must be known in order to analyze the stellar orbits.

AU. Careful observations of the motions of the two stars show that the mass ratio of the two stars, derived from their relative distances from the center of mass, is 2.16. Using Kepler's third law, we find the sum of the masses:

$$m_1 + m_2 = \frac{a^3}{P^2} = \frac{(14.14)^3}{(26.948)^2} = 3.89 \ M_\odot.$$

We know that the sum of the masses is 3.89 times the Sun's mass, and that one star is 2.16 times more massive than the other. We can write

$$m_1 + 2.16m_1 = 3.89 \ M_\odot.$$

Thus $m_1 = 1.23 \ M_\odot$ and $m_2 = 2.66 \ M_\odot$.

Complications arise when some of the needed observational data are difficult to obtain. The period is almost always easy to measure with some precision, but not so the semimajor axis a. One problem is that the apparent size of the orbit is affected by our distance from the binary **(Fig. 14.28)**, so the distance must be well known if a is to be determined accurately. Another problem is that the orbital plane is inclined at a random, unknown angle to our line of sight (so the apparent size of the orbit is foreshortened by an unknown amount). In some cases it is possible to unravel these confusing effects by carefully analyzing the observations. Even in cases where this is not possible, some information about the masses of the stars can still be gained, but usually only in terms of broad ranges of possible values rather than precise answers.

The masses of stars vary along the main sequence from the least massive stars in the lower right to the most massive at the upper left. The M stars on the main sequence have masses as low as 0.1 solar mass or a bit less, while the O stars reach values as great as 60 solar masses. It is likely that stars occasionally form with even greater masses (perhaps up to 100 solar masses), but as we will see in the next chapter, such massive stars have very short lifetimes, so they are rarely encountered.

The giants, supergiants, and white dwarfs have masses comparable to those of main-sequence stars. Hence their obvious differences from main-sequence stars in other properties such as luminosity and radius have to be caused by something other than extreme or unusual masses. This is discussed in the next chapter.

Main-sequence stars exhibit a smooth progression of all stellar properties from one end to the other **(Table 14.5)**. The mass and the radius vary by similar factors, while the luminosity changes much more rapidly along the sequence. The mass of an O star is perhaps 100 times that of an M star, while the luminosity is greater by a factor of 10^8 or more. Main-sequence stars follow a so-called **mass-luminosity relation,** a numerical expression in which the luminosity of a star is proportional to an exponential power of the mass. In its simplest form, this relation states that the luminosity is proportional to the cube of the mass; i.e., $L \propto M^3$. In more precise versions of the mass-luminosity relation, the exponent (the power to which M is raised) varies somewhat along the main sequence.

Table 14.5

Properties of Main-Sequence Stars

Spectral Type	Mass ($M_\odot$)	Temperature (K)	$B - V$	Luminosity ($L_\odot$)	M_v	B.C.	Radius ($R_\odot$)
O5V	40	40,000	−0.35	5×10^5	−5.8	−4.0	18
B0V	18	28,000	−0.31	2×10^4	−4.1	−2.8	7.4
B5V	6.5	15,500	−0.16	800	−1.1	−1.5	3.8
A0V	3.2	9,900	0.00	80	+0.7	−0.4	2.5
A5V	2.1	8,500	+0.13	20	+2.0	−0.12	1.7
F0V	1.7	7,400	+0.27	6	+2.6	−0.06	1.4
F5V	1.3	6,580	+0.42	2.5	+3.4	0.00	1.2
G0V	1.1	6,030	+0.58	1.3	+4.4	−0.03	1.1
G5V	0.9	5,520	+0.70	0.8	+5.1	−0.07	0.9
K0V	0.8	4,900	+0.89	0.4	+5.9	−0.19	0.8
K5V	0.7	4,130	+1.18	0.2	+7.3	−0.60	0.7
M0V	0.5	3,480	+1.45	0.03	+9.0	−1.19	0.6
M5V	0.2	2,800	+1.63	0.008	+11.8	−2.3	0.3

Note: Masses are in units of $M_\odot$, the solar mass (whose value is 1.99×10^{30} kg); luminosities are in units of $L_\odot$, the solar luminosity (having a value of 3.83×10^{26} watts), and radii are in units of $R_\odot$, the solar radius (whose value is 6.96×10^8 m). The temperatures are *effective temperatures*, which are a measure of surface temperature. The heading $B - V$ stands for the color index, M_v stands for visual absolute magnitude, and *B.C.* stands for bolometric correction.

Other Properties

Several other properties of stars can be determined primarily by analyzing their spectra. Perhaps the most fundamental of these is the composition. It was not until the 1920s that astronomers developed methods for determining the chemical makeup of the stars. Pioneering work by the Indian physicist M. N. Saha and the American astronomer Cecilia Payne-Gaposchkin led to methods for calculating the effects of ionization due to temperature, which was necessary to correct for temperature factors in using spectral line strengths to determine the relative abundances of the elements. It is an interesting footnote that the extremely influential work of Payne-Gaposchkin, published in 1925, earned her the first Harvard doctorate to be awarded to a woman. This nicely culminated the long series of remarkable contributions by the women of the Harvard College Observatory, which started under Edward Pickering's direction in the 1890s.

We have already seen that stars are generally made of the same material in the same proportions, but it required a sophisticated analysis to show this. First, as noted, techniques had to be developed that take the effects of temperature into account, because these effects play a dominant role in controlling the strengths of lines in a stellar spectrum. Other factors, such as surface gravity and rotation, also come into play and must be incorporated into the analysis. In modern work on stellar composition, complex computer programs are used to calculate simulated spectra with different assumed compositions until a match with the observed spectrum is found **(Fig. 14.29)**.

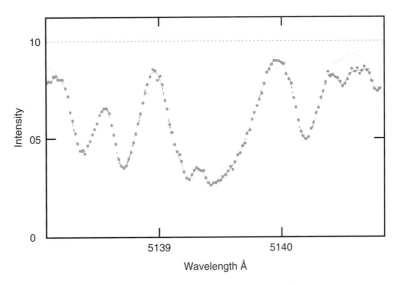

Figure 14.29 Determination of stellar composition. This diagram illustrates a modern technique for measuring the chemical composition of a star. A theoretical spectrum (dashed line) is compared with the observed spectrum (dotted and dashed line). The abundances of elements assumed present are varied in the computed spectrum until a good match is achieved. (D. L. Lambert)

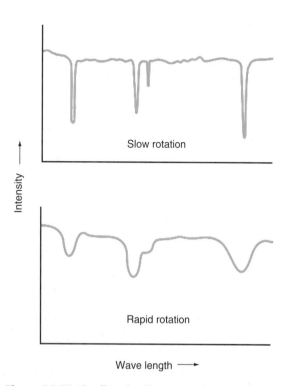

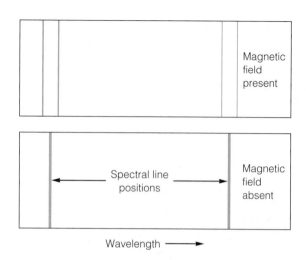

Figure 14.31 **The measurement of a magnetic field.** The presence of a magnetic field splits certain absorption lines of some elements in a process called Zeeman splitting. The amount of separation is a measure of the strength of the magnetic field.

Figure 14.30 **The effect of stellar rotation.** The Doppler effect causes spectral lines in a rapidly rotating star (bottom) to be broadened, because one edge of the star's disk is rapidly approaching us while the other edge is receding. Little broadening occurs in a slowly rotating star (top).

Another stellar property that can be learned from the spectrum is the rotational velocity at the surface, because the Doppler effect causes broadening of the spectral lines **(Fig. 14.30).** A spectral line forms in gas spread over the entire disk of a star, and in general one edge of the disk is approaching the Earth, and the other is receding. Hence some of the gas creates a blueshift, and some a redshift (the gas in the central portions of the disk has little or no shift), so the rate of rotation of the star determines the degree of broadening of the spectral lines.

A final property, whose importance is hard to assess because of the difficulty of measuring it in many cases, is the magnetic field of a star. Some spectral lines become split, under the influence of a magnetic field, because the field causes electron energy levels to be split. This phenomenon, called the **Zeeman effect,** can sometimes be measured in stars, yielding information on their surface magnetic fields **(Fig. 14.31).** This is usually difficult to do, however, and provides only approximate information. One reason this measurement is more difficult for distant stars than for the Sun is that we cannot isolate small regions on the surface of a star; it is simply too far away. Therefore we measure something representing an average field over the surface of the star, yet we expect, based on the Sun's behavior, that the field is

actually quite spotty and localized on the surfaces of other stars. Another difficulty in measuring the Zeeman effect in stellar spectra is that the line splittings are always very small, so if the star is a rapid rotator and its spectral lines are broadened by the Doppler effect, the splitting due to a magnetic field is blurred and difficult or impossible to measure.

Despite these difficulties, important information on stellar magnetic fields has been derived. For example, it has been found that magnetic fields and rotation are linked, perhaps not surprising in view of the fact that it is thought that internal rotation creates magnetic fields; and some types of stars, particularly very condensed remnants of stars that have completed their nuclear burning, have immensely strong magnetic fields.

Summary

1. There are three basic types of stellar observations: position, brightness, and spectroscopy.
2. Positional astronomy (astrometry) has developed techniques capable of measuring positions to an accuracy exceeding 0.01 arcsecond, which is sufficient to detect binary motions in some cases and stellar parallaxes for stars up to 100 pc away.
3. Stellar brightness measurements are commonly done using the stellar magnitude system, in which a difference of one magnitude corresponds to a brightness ratio of 2.512.

4. Magnitudes can be measured at different wavelengths, allowing the determination of color indices, or over all wavelengths, resulting in the determination of bolometric magnitudes.

5. Stellar spectra contain patterns of absorption lines that depend on the surface temperatures of stars and therefore can be used to assign stars to spectral classes that represent a sequence of temperatures.

6. Peculiar stars are those that do not conform to the usual spectral classes, most often because of unusual surface compositions but sometimes because they have emission lines.

7. Binary star systems can be detected by positional variations of one or both stars (astrometric binaries), brightness variations (eclipsing binaries), or periodically Doppler-shifted spectral lines (spectroscopic binaries).

8. Stellar luminosities are determined from knowledge of the distances to stars and their apparent magnitudes. The absolute magnitude, a measure of luminosity, is the magnitude a star would have if it were seen from a distance of 10 pc.

9. Stellar temperatures can be inferred from the $B - V$ color index, estimated from the spectral class, or determined from the degree of ionization in the star's outer layers.

10. The Hertzsprung-Russell diagram shows that the luminosities and temperatures of stars are intimately related and that stars that do not fall on the main sequence are either larger (red giants or supergiants) or smaller (white dwarfs) than those on the main sequence.

11. The distance to a star can be measured by determining its spectral class and then using the H-R diagram to infer its absolute magnitude, which is then compared with its apparent magnitude to yield the distance; this technique is called spectroscopic parallax.

12. Stellar diameters can be determined directly in eclipsing binaries from knowledge of the orbital speed and the duration of the eclipses.

13. Masses of stars are derived in binary systems by observing the period and the orbital semimajor axis and using Kepler's third law.

Review Questions

1. Compare the techniques used today for positional measurements in astronomy with those used by the ancient Greeks. How much more accurate are today's techniques?

2. Briefly discuss the role of stellar parallax in the development of the heliocentric theory.

3. In your own words, explain how the surface temperature of a star determines the relative strengths of the lines in the star's spectrum.

4. Explain the relationship between the luminosity of a star and its absolute magnitude. Is it the visual or the bolometric absolute magnitude that is relevent here? Explain.

5. Why is it significant that in a plot of luminosity versus surface temperature (i.e., an H-R diagram), most stars fall along the main sequence? What does this tell us about stars?

6. Can a binary system be all of the following: a visual binary, a spectroscopic binary, an eclipsing binary, and an astrometric binary? Explain what conditions would have to be met if this were to be possible.

7. Why is it difficult to directly measure a star's brightness at all wavelengths?

8. Explain why a G2 star that lies above the main sequence must have a larger radius than a G2 star that lies on the main sequence.

9. Explain the spectroscopic parallax method for finding the distances to stars. Why do you think the word *parallax* is used?

10. Summarize all of the information about a star that can be deduced from its spectrum.

Problems

1. Demonstrate why the parallax equation $d = 1/p$ is correct. To do this, consider a long, skinny triangle with a base of 1 AU and an apex (point) angle of $1''$ (i.e., the parallax angle as defined in Fig. 14.2), and use the small-angle approximation (Appendix 2). If the apex angle is $1''$, what is the ratio of the opposite side (the short one) to one of the legs (long sides) of the triangle? How does this relate to the definition and value of a parsec?

2. What is the distance (in parsecs) to the following stars: (a) a star whose parallax is $p = 0.25''$; (b) a star whose parallax is $p = 0.04''$; (c) a star whose parallax is $p = 0.005''$?

3. How much fainter is a twelfth-magnitude star than an eleventh-magnitude star? How does a third-magnitude star compare with a sixth-magnitude star? How much fainter are the faintest stars detectable to modern telescopes ($m = 31$) than the brightest stars in the sky ($m = -1$)?

4. Determine the absolute magnitude of the following stars: (1) a star 1,000 pc from the Sun with apparent magnitude $m = 12.3$; (b) a star 10,000 pc from the Sun with $m = 12.3$; (c) a star 10 pc away with $m = 6.8$; (d) a star 1 pc away with $m = 2.1$.

Observing Binary Stars

ASTRONOMICAL

ACTIVITY

As explained in the text, a visual binary is a system in which both stars are visible. In most visual binaries, however, a telescope is required to see the two stars as separate objects. Nevertheless, several visual binaries can be seen with binoculars, and one can be made out by the naked eye under good conditions. This is the pair named Mizar and Alcor, and you might enjoy having a look.

Mizar and Alcor together form one of the stars in the handle of the Big Dipper (also known as Ursa Major). The position of Mizar/Alcor is shown in the accompanying sketch. Because the Big Dipper is very close to the north celestial pole, it is up and accessible for observation all year.

Both stars are bright enough for naked-eye detection (Mizar has an apparent magnitude of 2.3; Alcor's apparent magnitude is 4.0), and their separation on the sky is large enough (about 15 arcminutes) so that both stars can be seen as distinct objects by the human eye (recall the discussion of resolving power in Chapter 4). On a clear night, test your eyes by looking at Mizar and trying to spot Alcor (North American Indians once used this as an eye test, as did several European armies). Take another look using binoculars. You will see that both stars are white in color (the spectral types are A1V for Mizar and A5V for Alcor).

Another very interesting binary system, which can be seen with the naked eye (but is better viewed through binoculars or a small telescope), is β Cygni, also known as Albireo. The two stars that form Albireo have very different spectral types (the brighter star, whose apparent magnitude is 3.1, is a K3II star;* the fainter, having apparent magnitude of 5.1, is a B8V star). Thus the brighter star has a red-orange color, while the fainter companion is blue white. The human eye tends to exaggerate color contrasts, so the colors look very different. This pair, separated by about 35 arcseconds, is a very pretty sight. The accompanying chart indicates where to look; Cygnus is up during the summer and early fall.

*The brighter member of the Albireo pair is itself a binary star; the K3II star has a B0V companion that is two magnitudes fainter.

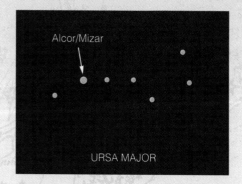

URSA MAJOR

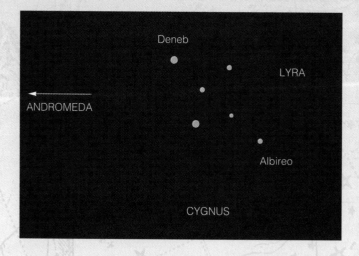

CYGNUS

5. Two stars have the same temperature, but one has an absolute magnitude $M_V = 6.54$ while the other has $M_V = -3.28$. What is the ratio of their radii?

6. A star is classified as a K4V star (i.e., a main-sequence K4 star). Its visual apparent magnitude is $+17.5$. Using the spectroscopic parallax method and the H-R diagram in Fig. 14.24, find the approximate distance to this star.

7. What is the radius (in solar units) of a star whose luminosity is 26.86 times that of the Sun and whose surface temperature is 0.65 times that of the Sun?

8. What is the sum of the masses in a binary system whose orbital period is 1 year and whose semimajor axis is 2 AU? What are the individual masses of the two stars if the center of mass between them is one-fourth of the way from the more massive star (star A) to the less massive one (star B)?

Additional Readings

Baliunas, S. and S. Saar. 1992. Unfolding Mysteries of Stellar Cycles. *Astronomy* 20(5):42.

Dobson, A. K. and K. Bracher 1992. Urania's Heritage: A Historical Introduction to Women in Astronomy. *Mercury* 21(1):4.

Hearnshaw, J. D. 1992. Origins of the Stellar Magnitude Scale. *Sky & Telescope* 84(5):494.

Hirshfeld, A. 1994. The Absolute Magnitudes of Stars. *Sky & Telescope* 88(3):35.

Kaler, J. B. 1986. The Origins of the Spectral Sequence. *Sky & Telescope* 71(2):129.

Kaler, J. B. 1986. M. Stars: Supergiants to Dwarfs. *Sky & Telescope* 71(5):450.

Kaler, J. B. 1986. The K Stars: Orange Giants and Dwarfs. *Sky & Telescope* 72(2):130.

Kaler, J. B. 1986. Cousins of Our Sun: The G Stars. *Sky & Telescope* 72(5):450.

Kaler, J. B. 1986. The Temperate F Stars. *Sky & Telescope* 73(2):131.

Kaler, J. B. 1986. White Sirian Stars: Class A. *Sky & Telescope* 73(5):491.

Kaler, J. B. 1986. The B Stars: Beacons of the Sky. *Sky & Telescope* 74(2):174.

Kaler, J. B. 1986. The Spectacular O Stars. *Sky & Telescope* 74(5):464.

Kaler, J. B. 1991. The Brightest Stars in the Galaxy. *Astronomy* 19(5):30.

Kaler, J. B. 1991. The Faintest Stars. *Astronomy* 19(8):26.

Steffey, P. C. 1992. The Truth About Star Colors. *Sky & Telescope* 84(3):266.

Terrell, D. 1992. Demon Variables. *Astronomy* 20(10):34.

Welther, B. 1984. Annie Jump Cannon: Classifier of the Stars. *Mercury* 13(1):28.

Web Connections

The Review Questions and Problems also appear at the following URLs:

http://universe.colorado.edu/ch14/questions.html
http://universe.colorado.edu/ch14/problems.html

Chapter 15

GRAVITY'S MACHINE

Stellar Structure and Evolution

When Newton saw an apple fall,
he found a mode of proving
that the earh turned round. . .
Lord Byron, 1810

Chapter Web site: http://universe.colorado.cdu/ch15

We have learned how astronomers determine the physical characteristics of stars from observations. These observations tell us a great deal about the surface properties of stars, but very little about what goes on inside them or how they evolve. To understand these secrets, we must apply the laws of physics and calculate the internal structure and evolution theoretically. If these calculations successfully reproduce the observable properties, then we can have some confidence that the calculations also accurately describe the internal conditions. We have already seen what can be learned in our discussions of the Sun in Chapter 13. In this chapter, we will rely on our previous discussions, for the same physical principles that were useful for the Sun can be applied to other stars.

Recall that the formation of stars and a bit about their internal structure and energy generation were discussed in Chapter 12. It will be useful to keep those discussions in mind as we explore the lifetimes of stars following their formation and the ignition of nuclear reactions in their cores.

Basic Stellar Structure

A star is a spherical ball of hot gas. The outer layers are partially ionized, and the interior, where the temperature and pressure are much higher, is fully ionized, so the gas there consists of bare atomic nuclei and free electrons. Continuous radiation is generated at the core of a star, and from there photons of light make their way slowly out to the surface, where they are emitted into space from a layer called the **photosphere,** just as in the Sun. Absorption lines are formed in the photosphere, so like the Sun, most stars have spectra dominated by absorption lines. The interior regions of a star are very dense because of gravitational compression, but they are still gaseous because of the high temperature. Temperatures in the cores of stars range from 10^7 to 10^8 K. Like the Sun, most stars consist primarily of hydrogen, although the abundance of hydrogen changes gradually during a star's lifetime because of nuclear reactions (recall that in the Sun's core, hydrogen is gradually being converted into helium).

The internal structure of a star, like that of the Sun, is governed by the balance between gravity and pressure that we call **hydrostatic equilibrium.** The gas particles all exert gravitational forces on each other, so that the entire star is held together by its own gravity. This force is always directed toward the center; thus the star is forced into a symmetric, spherical shape. The fact that gas is compressible explains how gravity creates a state of high density inside.

A gas that is compressed heats up, causing it to exert greater pressure on its surroundings. Thus, a star's interior is very hot and the internal pressure is high. If it were not for this pressure, gravity would cause a star to keep shrinking. A balance is struck between gravity, which is always trying to squeeze a star inward, and pressure, which pushes outward **(Fig. 15.1).** This balance, which can be stable for long periods of time only if there is a source of energy inside the star to compensate for the energy that is radiated away at the surface, plays a dominant role in determining the internal structure. Whenever the balance is disturbed, as it is during some phases of a star's lifetime, the internal structure changes.

The mass of a star determines how much it is compressed inside. The strength of gravitational compression is set by the amount of mass, and this in turn determines how much pressure is needed to balance gravity. The star will be squeezed until this pressure is reached, and then it will become stable. The pressure required to balance gravity dictates the temperature inside the star. Hence a star's mass determines the

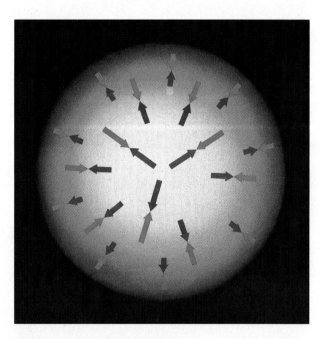

Figure 15.1 Hydrostatic equilibrium. This cutaway sketch of a star shows its spherical shape. The arrows represent the balanced forces of gravity (inward) and pressure (outward), and the shading indicates that the density increases greatly toward the center. Equilibrium is reached when the core becomes sufficiently hot to attain the pressure necessary to counterbalance gravity.

internal density and temperature as well as the overall size of the star, since radius depends on the mass and the density.

Luminosity is also governed largely by the mass. The luminosity of a star is the amount of energy per second generated inside the star that eventually reaches the surface and escapes into space, and it is determined by the temperature in the interior. As we have just seen, the temperature is set primarily by the mass. The dependence of luminosity on mass for main-sequence stars can be demonstrated by making a graph of luminosity versus mass **(Fig. 15.2)**. This is the **mass-luminosity relation,** and it was discovered observationally before it could be fully explained the-oretically.

Thus virtually all the observed properties of stars depend on their masses **(Fig. 15.3)**. This explains why the main sequence represents a smooth run of masses increasing from the lower right to the upper left in the H-R diagram: the luminosity and temperature, which define a star's place in the diagram, both depend on mass.

Of course, some stars do not fall on the main sequence, yet their masses are not different from those of main-sequence stars. This indicates that something other than mass can influence a star's properties. This other factor is the chemical composi-tion. As we shall see, this varies as a star ages. The supergiants, giants, and white dwarfs have different core chemical makeups from main-sequence stars. In the next section, we will see how these differences arise.

The amount of internal compression, and hence the temperature and luminosity of a star, depend on the average mass of the individual nuclei in the star's core. The heavier the particles that make up the gas, the more tightly they are compressed by gravity, the hotter the gas becomes, and the greater the star's luminosity. When a star is formed, it consists mostly of hydrogen, but as it ages, its core material is con-

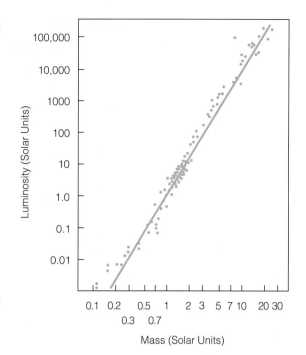

Figure 15.2 The mass-luminosity relation. This diagram, based on observed luminosities and masses for stars on the main sequence, shows how well the two quantities correlate with each other.

verted to helium at first and later possibly to other, even heavier elements. In the process, the core heats up and the star becomes more luminous.

The fact that almost all the properties of a star depend on just its mass and composition was recog-nized several decades ago and is usually referred to as the **Russell-Vogt theorem,** after the astrophysicists who first stated it. Astronomers now know that other influences, such as magnetic fields and rotation, must play roles in governing a star's properties as well. Just what these roles are is not yet well understood, however.

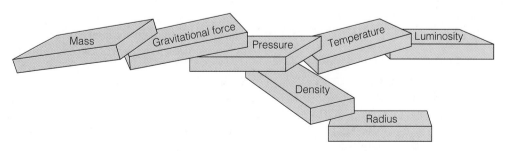

Figure 15.3 The importance of mass. A star's mass is the single quantity that governs all its other properties for a given composition. The sequence shown here is general, but the details vary for different chemical compositions.

Nuclear Reactions and Energy Transport

We have seen that the core of a star must be very hot to maintain the pressure required to counterbalance gravity. Since energy flows outward from the core, there must be some source of heat in the interior. If there were not, the star would gradually shrink.

Energy Generation

Nuclear fusion reactions provide the only source of heat capable of maintaining the required temperature over a sufficiently long period of time. A particular type of reaction was described earlier in our examination of the Sun (Chapter 13), where we discussed the solar interior. Recall that in atomic fusion reactions, nuclei of light elements such as hydrogen combine to form nuclei of heavier elements. In the process, a small fraction of the mass is converted into energy according to Einstein's famous equation $E = mc^2$. It is this energy, in the form of heat and radiation, that maintains the internal pressure in a star.

Fusion reactions can occur only under conditions of extremely high temperature and density. The beginning composition of a star is dominated by hydrogen, and ionized hydrogen consists of free protons and electrons. It is the protons that undergo the initial nuclear reactions when a star forms and joins the main sequence.

Protons carry positive electrical charges and therefore tend to stay apart because like charges repel each other. Nevertheless, protons can join together

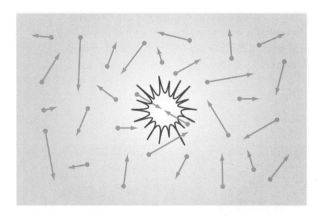

Figure 15.4 Nuclear fusion. The core of a star is a sea of rapidly moving atomic nuclei. Occasionally, a pair of these nuclei, most of which are simple protons (hydrogen nuclei), collide with sufficient velocity to merge and form a new kind of nucleus (deuterium in this case, composed of one proton and one neutron). Energy is released in the process.

to form larger nuclei because there is a force, called the **strong nuclear force,** that is stronger than the electrical force that tries to keep protons apart. But the strong nuclear force operates only over very short distances. Thus two protons can merge only if they can come close enough to each other for the strong nuclear force to overwhelm the repulsive force due to their electrical charges. One way the nuclei can get together is for them to collide at very high velocities **(Fig. 15.4).**

The high temperature of a stellar core causes nuclei to travel at very high speeds, allowing them occasionally to come close enough together to fuse. The high density of a stellar core helps also, because this density means that collisions will be frequent. Even so, only rarely do two nuclei combine in a fusion reaction; a single particle may typically collide and bounce around inside a star for millions or even billions of years before it reacts with another. There are so many particles, however, that reactions are continually occurring despite the low probability of reaction for any individual particle.

The amount of energy released in a single reaction between two particles is small—about 10^{-12} joule. A joule (defined in Chapter 5) is itself a small quantity; a 100-watt lightbulb radiates 100 joules per second. Hence a lightbulb would require 10^{14} reactions per second to keep glowing, if nuclear reactions were its energy source. A star like the Sun emits more than 10^{26} joules per second, so a tremendously large number of reactions (more than 10^{38} per second) must be occurring in its interior at all times.

In Chapter 13 we learned that the reaction that powers the Sun is the **proton-proton chain,** in which hydrogen is converted to helium, with 0.007 of the initial mass being converted to energy. The net result of this reaction is the combination of four hydrogen nuclei into one helium nucleus; two of the protons must be converted into neutrons in the process (see Chapter 13 or Appendix 14 for details of this reaction sequence). Another reaction sequence, called the **CNO cycle** (see Appendix 14), has the same net effect, but operates primarily in main-sequence stars hotter and more massive than the Sun. All stars on the main sequence produce their energy by one of these two processes that convert hydrogen into helium in their cores; the specific reaction that is dominant depends on the internal temperature, which in turn depends on the star's mass. The point on the main sequence that divides proton-proton chain stars from CNO stars corresponds roughly to spectral class F0; stars above that point produce most of their energy by the CNO cycle, and those below that point do so by the proton-proton chain.

Energy Transport

Once it is produced in the core, energy must somehow make its way outward to the stellar surface. For most stars, two different energy-transport mechanisms are at work, with one or the other dominant at different levels inside the star **(Fig. 15.5).** One of these mechanisms is **radiative transport,** meaning that the energy is carried outward by photons of light. In the core, where it is very hot, the photons are primarily gamma rays, but as they slowly move outward, being continually absorbed and reemitted on the way, they lose energy and are gradually converted to longer wavelengths. When the light emerges from the stellar surface, it is primarily in the visible-wavelength region (or the ultraviolet or infrared, if the star is very hot or very cool). Radiative transport is a very slow process, as we learned in the case of the Sun, where a single photon may take as long as a million years to make its way from the core to the surface.

The second means of energy transport inside stars is **convection,** a process already discussed at length in the chapters on planetary science and briefly in Chapter 13 on the Sun. In convection the gas is overturned as heated material rises and cooled material sinks. The same process causes warm air to rise toward the ceiling of a room and is responsible for the overturning of water in a pot that is being heated from the bottom. When conditions are right for convection to occur, it is much more efficient than radiative transport. Energy travels outward much faster in stellar convective zones than in zones where radiation is the chief means of energy transport.

Whether radiative transport or convection will be the dominant energy-transport mechanism in a star depends on the temperature structure (specifically, on how rapidly the temperature decreases with distance from the center). Model calculations show that most stars have both a convective zone and a radiative region. In stars like the Sun (i.e., stars of spectral types F, G, K, and M on the lower half of the main sequence), convection occurs in the outer layers, while radiative transport is the principal means of energy transport in the interior. For stars on the upper main sequence, the situation is reversed: convection occurs in the central core but not in the rest of the star, where radiative transport is responsible for conveying the energy to the surface.

Stellar Life Expectancies

The lifetime of a star is determined by the amount of energy it can produce in nuclear reactions and the rate at which the energy is radiated away into space

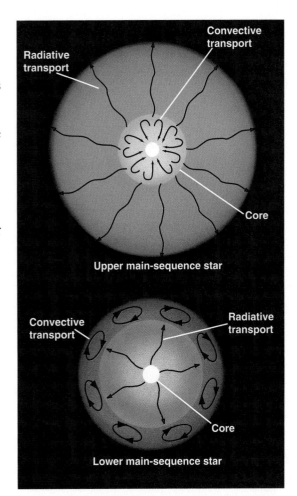

Figure 15.5 Energy transport inside stars. In a star on the upper portion of the main sequence, energy is transported by convection in the inner zone and radiation in the outer regions. The opposite is true of lower main-sequence stars.

(Fig. 15.6). When all the available nuclear fuel has been used up, the star undergoes major changes in its structure and properties and leaves the main sequence (this will be discussed in detail in the next chapter). In Chapter 13 we found that the Sun has a life expectancy of about 10 billion years, assuming that 10% of its mass undergoes reactions, and that 0.007 of that mass is converted into energy. For stars in which the CNO cycle is the dominant reaction, the calculation is made the same way, and the fraction of the initial mass converted into energy is the same. The calculation involves determining the total amount of energy available to the star during its lifetime (derived from the formula $E = mc^2$) and dividing this by the luminosity, which is the rate at which energy is expended. In mathematical terms, we can write

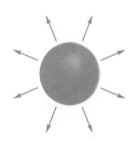

Figure 15.6 **Stellar lifetimes.** Even though the more massive star (top) has a greater supply of nuclear fuel, it loses energy so much more rapidly than the lower-mass star (bottom) that its lifetime is far shorter.

$$\tau = \frac{E}{L},$$

where τ is the lifetime of the star during its hydrogen-burning phase, E is the total energy it can produce during this time, and L is the star's luminosity. If one-tenth of the star's mass undergoes reactions (because only in the inner core is it hot enough for nuclear fusion to occur), we find

$$\tau = \frac{0.1(0.007)Mc^2}{L} = \frac{0.0007\,Mc^2}{L},$$

where M is the star's mass (note that in this we are assuming that *all* of the star's mass is hydrogen to

begin with; the error introduced by this assumption is relatively unimportant).

An easier way to do this calculation is to compare the star's lifetime with that of the Sun. If we write the above expression for a star and for the Sun and then divide the equations, the constant terms cancel out, and we find

$$\frac{\tau_\star}{\tau_\odot} = \left(\frac{M_\star}{M_\odot}\right)\left(\frac{L_\odot}{L_\star}\right),$$

where the $\star$ subscripts refer to the star and the $\odot$ subscripts refer to the Sun. If we simply express the star's mass and luminosity in solar units, the equation simplifies to

$$\frac{\tau_\star}{\tau_\odot} = \frac{M}{L}.$$

We can apply this calculation to a few stars as examples **(Table 15.1)**. A star near the top of the main sequence, for example, may have 50 times the mass of the Sun, but use its energy as much as 10^6 times more rapidly. (Recall how rapidly the luminosity of a star varies with its mass, as discussed in the last chapter.) These numbers lead to an estimated lifetime of

$$\frac{\tau_\star}{\tau_\odot} = \frac{50}{10^6} = 5 \times 10^{-5};$$

that is, this star will last 5×10^{-5} times the Sun's 10 billion–year lifetime, or only half a million years! (Actually, in such a massive star, more than 10% of the total mass can undergo reactions, and the lifetime is accordingly longer than this simple estimate; it can be as long as a million years.) By astronomical standards, such massive stars exist for only an instant before using up all their fuel and dying. This is one reason why stars of this type are very rare; only a few may be around at any given moment, regardless of how many may have formed in the past (as it happens, very massive stars also do not form very often, another reason for their rarity).

Now let us consider a lower main-sequence star; for example, an M star with mass 0.08 solar masses and luminosity 10^{-4} solar masses. Its lifetime is

$$\frac{\tau_\star}{\tau_\odot} = \frac{0.08}{10^{-4}} = 800.$$

This star will keep burning for 800 times the Sun's lifetime, or about 8×10^{12} years. This is a very long time—greater than the age of the universe. No star formed with such a low mass has had time to use up all of its fuel; all such stars that have ever been born are still on the main sequence. Partly for this reason, the vast majority of all stars in existence today are low-mass stars. (It is also true that these stars form in greater numbers, adding further to their numerical dominance.)

Heavy Element Enrichment

Clearly, nuclear reactions have an effect on the chemical composition of a star, because they change one element into another. We have stressed that all stars consist of about the same mixture of hydrogen and helium, with a trace amount of other elements, but now we find that in the core of a star, hydrogen is gradually converted into helium. While this change may not immediately affect the surface composition, it does affect the internal composition. It is this change that causes a star to evolve, because the core density is altered as hydrogen nuclei combine into the heavier helium nuclei. Furthermore, when the hydrogen runs out, the star must make major structural adjustments as its primary source of internal pressure disappears.

These adjustments are discussed later in this chapter; for now, let us consider only the change in the abundances of the elements. After the hydrogen-burning stage has ended, other nuclear reactions can take place in stars if the core temperature reaches sufficiently high levels. One such reaction is the conversion of helium into carbon by a sequence called the **triple-α reaction** (see Appendix 14; helium nuclei are called α **particles,** and in this reaction, three of these combine to form one carbon nucleus). When the triple-α reaction takes place, the composition of the star's core changes from helium to carbon.

Many additional reactions can occur in later stages if the core temperature goes even higher **(Fig. 15.7).** These reactions produce ever-heavier products, so as a star goes through successive stages of nuclear burning, elements as heavy as iron (which has 26 protons and 30 neutrons in its nucleus) replace the hydrogen that was originally predominant.

These reactions require ever-higher temperatures, because as the nuclei become larger, they acquire greater positive electrical charges and therefore repel each other more and more strongly. Therefore higher speeds are needed in order for them to come close enough together to fuse. In addition, the higher masses of these larger nuclei mean that greater energy is needed for them to reach sufficient speeds to react with each other. For both of these reasons, reactions involving the fusion of heavy nuclei require higher temperatures than those in which lightweight nuclei such as hydrogen undergo fusion. Tempera-

Table 15.1

Main-Sequence Lifetimes

Spectral Type	Hydrogen-Burning Lifetime (Years)
O5V	1×10^6
B0V	5×10^6
B5V	8×10^7
A0V	4×10^8
A5V	1×10^9
F0V	3×10^9
F5V	5×10^9
G0V	9×10^9
G5V	1×10^{10}
K0V	2×10^{10}
K5V	4×10^{10}
M0V	2×10^{11}
M5V	3×10^{11}

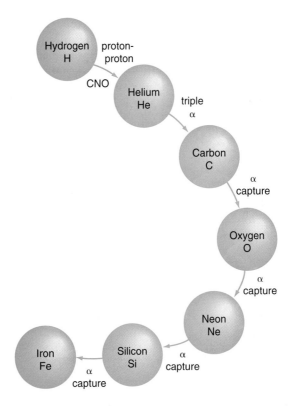

Figure 15.7 **The enrichment of heavy elements.** As a star ages, each time one nuclear fuel is exhausted, another ignites if the temperature in the core rises sufficiently (which depends on the mass of the star). The net result is an enrichment of heavy elements.

tures of hundreds of millions of degrees are needed for reactions involving the heaviest nuclei that fuse in stellar cores.

The most massive stars go through the greatest number of reaction stages, and as we have just seen, these stars live only a short time. In later chapters we will discuss the role played by these massive stars in the enrichment of chemical abundances in the galaxy.

The Outer Layers of Stars

Let us now consider other processes that may affect a star's structure and evolution. Some very important ones take place near the surface rather than deep inside. Stars have ways of losing matter, and some are shrouded in envelopes of gas and dust.

Stellar Coronae and Chromospheres

As we have seen, in most stars convection occurs at some level inside. For stars on the upper portion of the main sequence, convection takes place near the center and has no directly observable consequences. For stars on the lower portion of the main sequence, however, where convection occurs in the outer layers, there are important effects. Stars of spectral types F, G, K, and M, where surface convection is thought to occur, also are found to have **chromospheres** and **coronae.** Recall from Chapter 13 that the chromosphere of the Sun produces ultraviolet emission lines and that the corona emits X rays due to its high tem-

perature. These phenomena are often observed in other cool stars, indicating that they have chromospheres and coronae similar to those of the Sun **(Fig. 15.8).**

The presence of chromospheres and coronae seems to be linked with the presence of convection in the outer layers **(Fig. 15.9).** Stars on the lower portion of the main sequence generally have both chromospheres and coronae, indicating that somehow the kinetic energy of the turbulent motions in the convective layer is transported into higher zones, where it causes heating. As we learned in our discussion of the Sun, the precise mechanism for converting the energy of convection into heat is not well understood, but almost certainly involves magnetic fields. Observations of flare activity and ultraviolet chromospheric and coronal emission lines show that all cool stars have activity cycles similar to those of the Sun. Detailed observations of brightness variations in some stars demonstrate the presence of "starspots" similar to sunspots. Thus it seems likely that magnetic fields are important in all these stars, playing key roles in transporting energy into high levels of the atmospheres.

What of the red giants and supergiants? These stars can have surface temperatures comparable with the lower main-sequence stars, but obviously their internal structure is rather different, since they are so much larger. These stars also have convection in their outer layers, however, and indeed they also have chromospheres, although it is not certain that the extremely high temperatures characteristic of coronae are present.

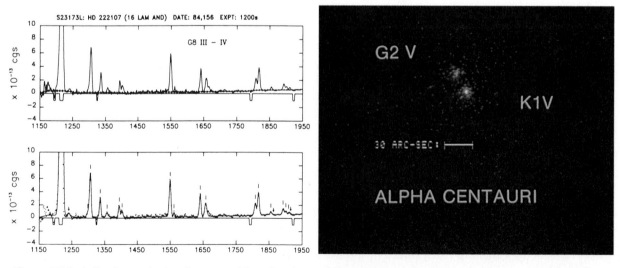

Figure 15.8 Stellar chromospheres and coronae. At left are ultraviolet spectra of the sun-like stars, showing strong emission lines at certain wavelengths, which arise in the chromospheres of these stars. At right is an x ray image of the binary star and Centauri, which consists of two cool stars. Both are x-ray emitters, from their coronal regions. (Left: NASA, courtesy J. O. Bennett; right: NASA, courtesy the Harvard-Smithsonian Center for Astrophysics)

http://universe.colorado.edu/fig/15-8.html

Stellar Winds and Mass Loss

So far we have said nothing about the possibility that the hot stars might have chromospheres or coronae. Some of the upper main-sequence stars are known to have emission lines in their spectra, indicating that they have some hot gas above their surfaces. Visible-wavelength emission lines are found in only a few extremely hot or luminous stars, however.

In the late 1960s, with the first observations of ultraviolet spectra, many hot stars (nearly all the O stars and many of the hotter B stars) were immediately found to have ultraviolet emission lines. Furthermore, there are absorption features that show enormous Doppler shifts, indicating that the stars are ejecting material at speeds as great as 3,000 km per second! The most extreme **stellar winds,** as these outflows are called **(Fig. 15.10),** occur in the O stars, but most cool giants and supergiants also have winds, usually with lower speeds.

The cause of the winds is not known, although there are indications of how the high velocities are

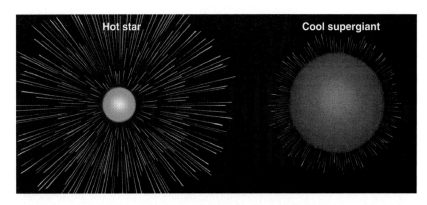

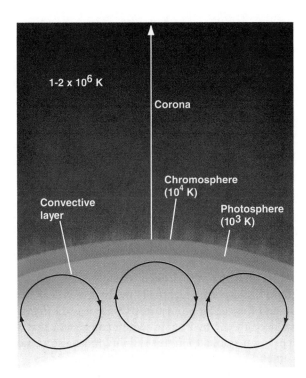

Figure 15.9 A stellar chromosphere and corona. Stars in the lower portion of the main sequence all have chromospheres and coronae, analogous to the Sun (see Chapter 11 for more details). The photosphere is the region in which the star's continuous spectrum and absorption lines are formed; the chromosphere is a thin, somewhat hotter region just above the photosphere; and the corona is a very hot, extended region outside the chromosphere. The source of heat for the chromosphere and corona is probably related to convective motions in the star's outer layers.

Figure 15.10 Stellar winds. In the drawing at the top, a luminous hot star (left) ejects gas at a very high velocity; the lengths of the arrows indicate that the gas accelerates as it moves away from the star. A luminous cool star (especially a K or M supergiant) is so extended in size that the surface gravity is very low, and material drifts away at relatively low speeds (right). In both types of stars, radiation pressure helps accelerate the gas outward. The photograph at the bottom shows a hot star surrounded by an extensive bubble of gas created by a stellar wind. (© ROE/AAT Board) universe.colorado.edu/fig/15-10.html

reached. Once gas begins to move outward, light from the star can exert sufficient force to accelerate it to high speeds. This force, called **radiation pressure,** is very weak (you certainly can't feel the breeze from a lightbulb, for example), but O and B stars are so luminous that strong acceleration of the wind is possible. For the most luminous of these stars, radiation pressure may be sufficiently strong to initiate the winds, but for the slightly less luminous O and B stars, calculations show that radiation pressure is not capable of creating the initial outflow, although it can accelerate the gas to high speed once the flow begins. Some other mechanism must start the outflow first. We do not know what this mechanism is, but recently it has been suggested that subtle pulsations might be responsible.

Whatever the cause of the winds, they have important consequences. Analysis of the ultraviolet emission lines shows that, in some cases, stars are losing matter at such a great rate that a large fraction of the initial mass may be lost during their lifetimes. An O star might lose as much as one solar mass every 100,000 years. If such a star begins life with 20 or 30 solar masses and lives a million years, it could lose 10 solar masses, a significant fraction of what it had to begin with. This loss of matter has important effects on how such a star evolves, as we shall see.

The supergiants in the upper right of the H-R diagram also lose mass through stellar winds **(Fig. 15.11),** but these winds have a distinctly different nature. The red supergiants are so large that their surface gravity is very low, and the gas in the outer

layers is not tightly bound to the star. Radiation pressure can easily push the gas outward at the relatively low speed of 10 to 20 km per second. The amount of mass lost can be just as great as in the hot stars, however, because these low-velocity winds are much denser than the high-speed winds from the O and B stars. It is thought that slow pulsations in red giant and supergiant stars may play a role in triggering mass loss, just as rapid pulsations may help create the high-speed winds in hot stars.

Because the red supergiants are relatively cool objects, the gas in their outer layers is not ionized, and even molecular species form there. In addition, small solid dust grains can condense, shrouding the star in a cloud. In extreme cases the dust cloud becomes so thick that little or no visible light escapes to the outside. The dust grains become heated, however, and emit infrared radiation, so the star can still be detected with an infrared telescope. It appears that cool giants and supergiants are major sources of interstellar dust particles (discussed briefly in Chapter 12 and treated further in Chapter 17).

Observations of Stellar Evolution

Thus far we have learned how the properties of stars are determined from a combination of observations and theory; we understand something about how astronomers know the surface properties and the internal structure of a star. We also know the essentials of how stars form, again from a combination of observation and theory. But how do we know that stars change with time, and how do we know what changes they undergo?

Once again, our knowledge comes from a combination of what we can observe and what we can deduce from the application of mathematical calculations to the laws of physics. A very simple theoretical conclusion is that if hydrogen is being converted into helium inside a star, eventually the hydrogen will run out, and this reaction must stop. Since it is the energy produced by the hydrogen-helium conversion that keeps a star stable, then this stability must also come to an end when the hydrogen is gone. Hence we realize, from purely deductive reasoning, that a star must evolve.

Our best observational information on how stars evolve comes from clusters and associations of stars. As we learned in Chapter 12, many stars in our galaxy appear in gravitationally bound groupings **(Fig. 15.12).** It is very difficult to see how clusters could form accidentally by random gatherings of preexisting stars, so astronomers believe that clusters are groupings of stars that formed together (recall also

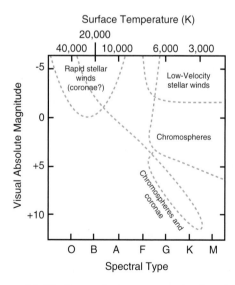

Figure 15.11 Chromospheres, coronae, and stellar winds in the H-R diagram. Here we see that most stars of type F or cooler have chromospheres, although coronae may be confined to the main sequence. Very luminous stars, hot and cool alike, have sufficiently strong winds to result in significant loss of mass.

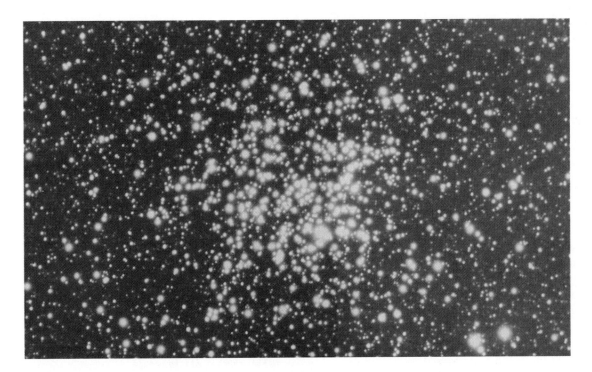

Figure 15.12 A young cluster of O and B stars. This very young cluster contains a number of blue-white O and B stars. Technically, this group of stars (NGC 6705) is too large and too cohesive to be a typical OB association; normally these contain fewer stars and are less organized in over-all structure. (© 1993 Anglo-Australian Telescope Board; photo by D. Malin.) http://universe.colorado.edu/fig/15-12.html

that there are sound theoretical reasons to expect that stars can form in clusters; again, see Chapter 12). The assumption that stars in a cluster formed together is very helpful, because it means that all of the stars within a particular cluster must have had the same initial composition and, more importantly, that they all are the same age. It is also helpful that the stars in a cluster are all at approximately the same distance from us, for this means that apparent differences in brightness correspond to actual differences in luminosity.

The fact that all stars in a cluster are of the same age and initial composition also has implications for stellar masses. As we discussed earlier in this chapter, it is mass that determines the major properties of a star. Hence the range of stellar properties within a cluster represents a range of stellar masses. This means that astronomers can learn how stars of different masses evolve by looking at clusters of different ages. But how do we know the ages of clusters? This took some time for astronomers to unravel.

If we construct an H-R diagram for the stars in a cluster, so that all the stars included have a common age and composition and lie at the same distance, some curious results are found **(Fig. 15.13).** Some clusters have "normal" H-R diagrams, with complete main sequences and very few stars not on the main sequence. But other clusters have only partial main sequences, with the upper end missing. These same

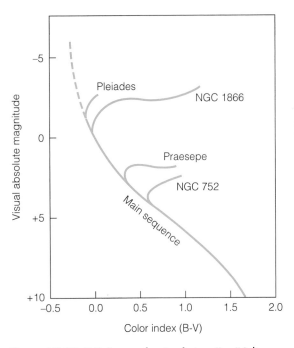

Figure 15.13 H-R diagrams for star clusters. Here H-R diagrams (using color index rather than spectral type) of several clusters are plotted on the same axes. Different symbols are used in the right-hand portion to distinguish among the giant stars in different clusters. The solid lines indicate the main-sequence turnoff points, which are related to the ages of the clusters.

clusters also contain numerous red giants and super-giants. When astronomers first began making such diagrams, they immediately thought that the differences from cluster to cluster might be an age effect, but understanding what this had to tell us about the evolution of stars took much longer. Eventually, astronomers realized that in the clusters with no stars on the upper main sequence, the stars that had been there had turned into red giants **(Fig. 15.14)**; their surface temperatures had decreased while the stars expanded in size, causing their luminosities to increase. Thus two important clues about stellar evolution came to light: (1) when a star finishes its main-sequence lifetime, it becomes a red giant; and (2) the more massive a star is, the sooner it moves off the main sequence and into red giant territory on the H-R diagram.

The variety of cluster H-R diagrams, then, represents a range of cluster ages. If a cluster has a main sequence extending nearly all the way up toward the upper left in the H-R diagram, then we know it must be a young cluster, because even the most luminous (i.e., the most massive) stars are still in their hydrogen-burning stage. If, on the other hand, a cluster has a main sequence that terminates halfway up, then it must be an old cluster, because all stars above a certain mass have had time to use up their hydrogen and become red giants. Thus the older a cluster, the lower the point where its main sequence ends, because stars of smaller and smaller mass have had time to evolve into red giants. Application of stellar evolu-

tion calculations has established a correlation between the main sequence turnoff point and age; thus we can estimate the age of a cluster of stars by plotting the positions of the individual stars on an H-R diagram and seeing where the main sequence ends.

Clusters in the disk of our galaxy have a wide variety of ages, ranging from "only" a few million years (or possibly even less) to many billions of years. The youngest clusters are the OB associations (Chapter 12), small, loosely bound groupings of very luminous, massive stars (see Fig. 15.12). The oldest clusters, which tend to lie above and below the galactic plane, are the globular clusters (see Fig. 15.12), with ages estimated as high as 11 to 14 billion years.

Observations of clusters, and especially the construction of H-R diagrams for clusters, provide us with our best observational evidence for the nature of stellar evolution. We will rely heavily on such observations in the discussions that follow.

The Life Story of a Star Like the Sun

From observations of stars in clusters and of stars in transitional phases in their lives, along with theoretical calculations, it is possible to reconstruct the full sequence of developments in the life of a star. We will also rely on the results of stellar evolution calculations. These consist of computed models for the interior structure of a star; the calculations are performed over and over with slight changes in stellar conditions at each step, taking into account the changes due to the previous step. For example, as the core hydrogen is converted into helium, at each computational step the core composition and density will be a little different than in the previous step. Thus each calculated model represents a slightly different stage in the star's life, and a sequence of such calculations provides a picture of the star's life story. The calculations, however, leave a lot of room for error, so it is important to continually check the results against what can be observed. In the following sections, we piece together the story that emerges from the combination of such calculations and observations of real stars.

The Main Sequence Life of a Star Like the Sun

When a star with the mass of the Sun arrives on the main sequence, it is similar to the Sun in all its properties. Thus it is a G2 star, with a surface temperature around 6,000 K and a luminosity of about 4×10^{26} watts. Its composition initially is about 73% hydro-

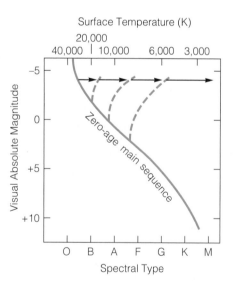

Figure 15.14 The evolution of the main sequence. As a cluster ages, the stars on the upper main sequence move to the right. The farther down the main sequence this has happened, the older the cluster.

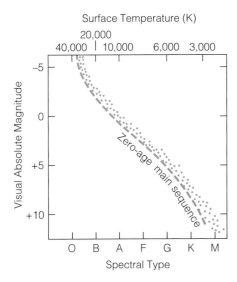

Figure 15.15 **The width of the main sequence.** As stars on the main sequence gradually convert hydrogen into helium in their interiors, the cores shrink a little and get hotter. This increases the luminosity, and the stars gradually move upward on the H-R diagram. As a result, the main sequence, consisting of stars of a variety of ages, is not a narrow strip, but has some breadth.

gen by mass, with roughly 25% helium and only about 2% other elements. It has convection in the outer layers, and it has a chromosphere and a corona. The star must change with time, because its core composition changes. **Table 15.2** summarizes the stages in the life of a star like the Sun.

In this star's core, the proton-proton chain converts hydrogen into helium. As we saw previously, this process lasts some 10 billion years for a star of the Sun's mass. During this time, the core gradually shrinks as the hydrogen nuclei are replaced by a smaller number of the heavier helium nuclei, and the internal temperature rises. This causes an increase in luminosity, and the star gradually moves up on the H-R diagram. The main sequence, therefore, is not a narrow strip but has some breadth, as stars move slowly upward on it as their luminosities increase **(Fig. 15.15).** The starting point—the lower edge of the main sequence—is called the **zero-age main sequence (ZAMS),** because this is where newly formed stars are found.

Main Sequence to Red Giant: What to Do When the Hydrogen Runs Out

When the hydrogen in the core of the star is gone, reactions there cease. By this time, however, the temperature in the zone just outside the core has nearly

---

Table 15.2

Evolution of a Star (One Solar Mass)

1. Hydrogen burning: During the 10^{10} years the star spends on the main sequence, hydrogen is converted to helium through the proton-proton chain. The luminosity gradually increases as the core becomes denser and hotter.

2. Development of degenerate core and evolution to red giant: When hydrogen in the core is used up, the core contracts until it is degenerate. A hydrogen shell source outside the core still undergoes reactions, causing the outer layers to expand and cool. The star becomes a red giant.

3. Helium flash: The core eventually gets hot enough to ignite the triple-α reaction, in which helium is converted to carbon. Because the core is degenerate and cannot expand and cool to counteract the new source of heat, the reaction rapidly takes place throughout the core. So much heat energy is added that degeneracy is quickly destroyed, and the reactions become stable.

4. Stable helium burning: Once reactions are taking place in the core again, the major energy production occurs there, and the shell source becomes less important. The outer layers contract, and the surface heats up; the star moves back to the left on the H-R diagram.

5. Second red giant stage: When the core helium is used up, the energy again comes from a shell (an inner shell, where helium is converted to carbon, and an outer shell, where hydrogen is still converted to helium). The outer layers expand again, and the star becomes a red giant for the second time.

6. Mass loss: Through a stellar wind and pulsational instabilities, the star loses its outer layers, exposing the hot core. The core continues shrinking but is never again hot enough for nuclear reactions. The star becomes a planetary nebula as the high-velocity wind from the hot core sweeps up and heats circumstellar gas left over from the red giant mass-loss phase.

7. White dwarf: The core again becomes degenerate as it shrinks. Eventually, the outer layers are gone altogether, leaving only the hot, degenerate core, which is a white dwarf.

---

reached the point at which reactions can take place, and with a little more shrinking and heating of the core, the proton-proton chain begins again in a spherical shell surrounding the core **(Fig. 15.16).** At this point, the star has an inert helium core and is producing all its energy in the hydrogen-burning shell, which steadily moves outward inside the star, eating its way through the available hydrogen.

Meanwhile the core continues to shrink and heat. This heating enhances the nuclear reactions in the shell, causing it to produce more and more energy. As the zone of energy production moves toward the surface, radiative energy builds up just outside the hydrogen-burning shell (recall that radiative energy transport is slow), and the pent-up energy forces the outer layers of the star to expand. As this occurs, the surface cools because the gas density

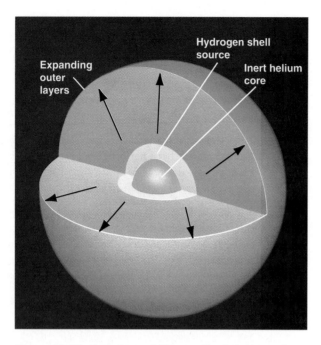

Figure 15.16 **The hydrogen shell source.** After the hydrogen in the core of a star is completely used up, the core, now composed of helium, continues to shrink and heat. This causes a layer of gas outside the core to reach the temperature required for nuclear reactions. A shell source, in which hydrogen is converted into helium, is ignited. The star's outer layers expand and cool, and the star becomes a red giant.

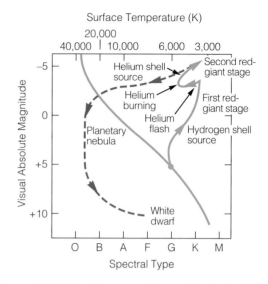

Figure 15.17 **Evolutionary track of a star like the Sun.** This H-R diagram illustrates the path a star of one solar mass is thought to follow as it completes its evolution. The dashed portion, following the second red giant stage, is less certain than the earlier stages.

decreases. Despite the decreasing surface temperature, the luminosity of the star increases because of the increased surface area, and the star moves upward on the H-R diagram **(Fig. 15.17).** It also moves to the right because the surface temperature is decreasing. The star rapidly becomes a red giant, reaching a size of 10 to 100 times its main-sequence radius **(Figs. 15.18** and **15.19).** The outer layers are constantly overturning as convection occurs to a great depth.

The helium core becomes extremely dense, but at first the temperature is not high enough for any new nuclear reactions to start. (The triple-α reaction, in which helium nuclei combine to form carbon, requires a temperature of about 100 million degrees.) As the core continues to shrink, the matter there takes on a very strange form. Electrons have a property that prevents them from being squeezed too close together, and this resistance creates a new kind of pressure, which becomes the principal force supporting the star against the inward force of gravity. The gas in the stellar core contains many free electrons, and eventually their resistance to being compressed becomes the dominant pressure that supports the core against further collapse. When this happens, the gas is said to be **degenerate.**

A degenerate gas has many unusual properties. One of them is that the pressure no longer depends

on the temperature, and vice versa. If the gas is heated further, it will not expand to compensate, as an ordinary gas would. As we will soon learn, there are important consequences if nuclear reactions start in the degenerate region of a star.

We now have a red giant star with highly expanded outer layers. Near the center is a spherical shell in which hydrogen is burning in nuclear reactions, and inside this shell is a degenerate core containing helium nuclei and degenerate electrons. The core is small, but it may contain as much as a third of the star's total mass.

Degenerate Matter and the Helium Flash

During the red giant phase, the helium core continues to be heated by the reactions going on around it. Eventually, the temperature becomes sufficiently high for the triple-α reaction to begin. When it does, the consequences are spectacular.

Ordinarily, when a gas is heated it expands, which limits how hot it can get, because an expanding gas tends to cool. In a degenerate gas, however, no such expansion occurs, because pressure does not depend on temperature in the usual way. Hence, when the reactions begin in the degenerate core of a red giant, the temperature rises quickly, but the core retains the same density and pressure. The increased temperature speeds up the reactions, producing more heat, which in turn accelerates the reactions even further. There is a rapid snowball effect, and in an in-

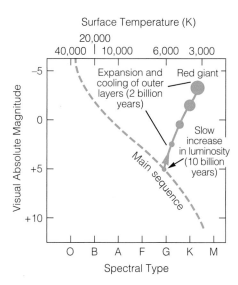

Figure 15.18 The development of a red giant. As a star's outer layers expand because of the shell hydrogen source, they cool. At the same time the surface area increases, raising the star's luminosity. The star therefore moves up and to the right on the H-R diagram.

stant (literally seconds), a large fraction of the core is involved in the reaction. This spontaneous runaway reaction is called the **helium flash.** Although it has dramatic consequences for the star's interior, calculations show that few if any effects would be immediately visible to an observer. The overall direction of the star's evolution is changed, however, as would be apparent after thousands of years.

The helium flash quickly disrupts the core, destroying its degeneracy (as temperature increases, more energy states become available to the electrons, and they can be squeezed closer together). The core then returns to a more normal state in which temperature and pressure are linked, and the triple-α reaction continues, now in a more stable, steady fashion. The outer layers of the star begin to retract as the star reverts to a more uniform internal structure, and as they do so, they become hotter. The star moves back to the left on the H-R diagram (see Fig. 15.17). Old clusters, particularly globular clusters, are found to have a number of stars in this stage of evolution, forming a sequence called the **horizontal branch** (see Fig. 15.13). This sequence is seen only in clusters old enough for stars as small as one solar mass to have evolved this far; as we learned earlier, this requires an age of 10 billion years or more.

The Final Stages of a Solar-Mass Star

In due course, the helium in the core becomes exhausted, leaving an inner core of carbon **(Fig. 15.20).** Helium still burns in a shell around the core, and

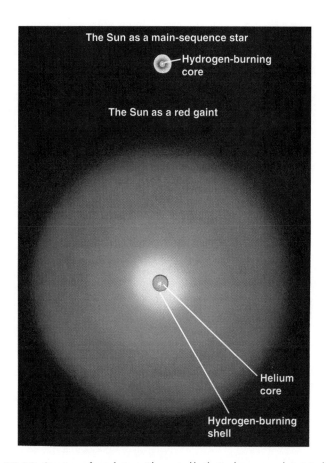

Figure 15.19 Structure of a red giant. When a star like the Sun becomes a red giant, its diameter grows much larger, but most of its mass remains concentrated in a small inner region. Energy is supplied by nuclear reactions in the hydrogen shell source. This drawing is not to scale.

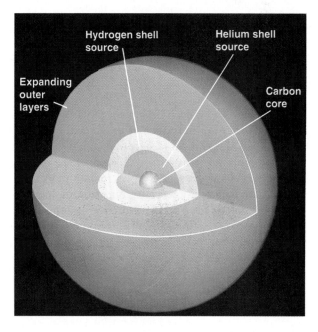

Figure 15.20 Internal structure late in a star's life. The core of this star has undergone hydrogen burning and helium burning and is composed of carbon. Hydrogen and helium shell sources are still undergoing reactions and producing energy. This is not to scale; the stellar core is actually smaller than shown here.

SCIENCE AND SOCIETY

The Quest for Controlled Fusion

The enormous energy that can be released by nuclear fusion reactions is not only the basis for the existence of stars, but may also represent a long-term solution to one of the most pressing problems facing human civilization: how to meet our ever-increasing need for energy. But harnessing fusion energy has proven to be a very complex technological task. In seeking the solution, scientists face fundamental issues of how society supports research in times of tight funding.

Humans have created nuclear fusion reactions and released immense energy in the process, but only in the instantaneous conflagration of hydrogen bombs. Fusion reactions on Earth were first achieved in a test explosion in 1952, and we are all too aware of the huge potential energy contained in the world's stockpiles of nuclear weapons. But the challenge of releasing fusion energy under controlled circumstances is far

more difficult, and efforts now spanning three decades have so far yielded only slow, torturous progress toward the ultimate goal of safe and inexpensive nuclear energy.

Why is fusion energy so desirable? First, we must be careful to distinguish between fusion and the fission reactors that have been in use since the 1950s. All of the present and past reactors have operated on fission, reactions in which large atomic nuclei (such as uranium or plutonium) are split, releasing energy in the process. Although in principle, power plants operating on fission reactions are capable of producing enough energy to run our society into the indefinite future, they also present many issues of safety and hazardous waste. The problems with fission reactors include the potential for runaway reactions (meltdown), accidental release of radioactive gases, and the production of radioactive by-products with long half-lives. In recent years, society has been turning away from fission. A significant fraction of the world's energy is still produced by atomic power plants, but this fraction is not growing and may decline as old

reactors are deactivated.

The issue of atomic power production using fission reactors is complex and involves many trade-offs between the need for energy, the perceived safety of the power plants, the risks associated with other forms of energy (such as the atmospheric pollution and resource depletion associated with fossil fuels), and other factors. Society has yet to develop a coherent response to these issues.

In contrast, controlled fusion power would avoid nearly all of these problems. There is no chance of runaway reactions, because the reactions can work only if intense pressure and temperature are maintained; a failure of either would simply stop the reaction. There are few dangerous by-products and virtually none that are radioactive (the major danger comes from neutrons produced during the reactions, but these can be stopped by shielding). Furthermore hydrogen, the raw material required for fusion, is available in virtually infinite supply, so there is no danger of ever running out of fuel. Advocates of controlled fusion point out that a wholesale switch to this energy source would not only provide cheap, plentiful energy, but would also greatly reduce

there could even be an active hydrogen-burning shell farther out in the star at the same time. The star expands and cools, becoming a red giant for the second time. This second red giant stage is short-lived, however, lasting perhaps 1 million years. A degenerate core again develops, but without another dramatic flare-up in its interior. In fact, no more nuclear reactions ever occur in the core of this low-mass star, because the temperature never reaches the extremely high levels needed to cause heavy elements like carbon to react.

The star in its second red giant stage may be viewed as having two distinct zones: (1) the relatively dense interior, consisting of a carbon core inside a helium-burning shell, which in turn is surrounded by a hydrogen-burning shell; and (2) the very extended, diffuse outer layers. The inner portion may contain

up to 70% of the mass, but occupies only a small fraction of the star's volume. As the nuclear fuel in the interior runs out, the core shrinks. In the meantime, the outer layers expand in a slow stellar wind, similar to the wind the star ejected during its earlier red giant stage.

The hot stellar core soon develops a rapid stellar wind, more akin to the winds of O and B stars than to the slow winds of red giants and supergiants. The gas flowing outward from the hot core collides with the gas remaining from the extended outer envelope of the star, sweeps up this leftover gas, and creates a shell of gas around the dying star. This shell contains gas from both red giant wind phases, as well as gas being ejected at high speed by the hot core. As the gaseous shell is heated by the energy of the high-speed wind crashing into it and also by ultraviolet

our dependence on fossil fuels, thus reducing the associated problems of atmospheric pollution and depletion of global coal and oil reserves.

So what is the problem? Why haven't we already made the switch to fusion energy? The major difficulties are technological. Creating the immense temperatures and pressures required for fusion reactions to occur is a very tough challenge. After all, what we have to do is create on Earth conditions similar to those in the core of the Sun, but without the Sun's enormous mass to bring about the required compression gravitationally.

The essence of the challenge is to find a way to compress hydrogen fuel sufficiently so that it will react. Two general methods have been explored in fusion programs under way in the United States and abroad. One, pioneered in the former Soviet Union and also emphasized in the U.S. fusion program, is to use magnetic fields to confine and compress a plasma (a gas consisting of free electrons and nuclei). A magnetic field exerts a force on charged particles, so in theory it is possible to use very strong magnetic fields to squeeze a plasma enough to make it react. In the largest such effort to date, the U.S. pro-

ject located at Princeton University has succeeded in producing energy by fusion, but only for very brief moments of time. Recent tests have temporarily reached the "break even" point, where the energy produced exceeds the energy put in (it takes huge quantities of energy to create the heated, confined plasma in the first place). While promising, this development came only after many years of expensive, laborious work, and no one can be sure how many more years it will take to develop the technology to the point where sustained reactions can be maintained.

The second method being exploited is to focus an array of powerful lasers on a tiny pellet of fuel, relying on the radiation of the lasers to create the conditions necessary for fusion to occur. As the fuel pellet is vaporized almost instantaneously, the required conditions can occur in the resulting plasma for a brief moment. If enough lasers are used and they repeatedly fire bursts of radiation at the fuel, scientists believe that they can achieve sustained energy production that produces more energy than is required to run the lasers. Experiments of this type have been conducted at the Livermore National Laboratory in California, and indeed short-lived fusion

reactions have been created. But again, it is difficult to predict how long it will take to reach the point where reactions can be sustained over a long term.

The quest for fusion has proven to be a test of our mettle as a society. It is expensive, and few results have been achieved after some three decades of effort. On the other hand, the potential rewards are nearly limitless, and few would question the worth if success eventually comes. But political reality makes it difficult to sustain funding, and the U.S. fusion program has taken severe cuts, with a drop of no less than 33% in federal funding in the 1996 budget. (This cut applies to the "civilian" fusion research program, which is supported by the Department of Energy, which emphasizes magnetic confinement techniques. Fusion research that is sponsored by the Department of Defense, including the laster technology being developed at Livermore, is not affected by these cuts.)

The outlook for the fusion effort is uncertain. Today the societal issues rival the technological problems as obstacles to progress, and it is clear that we will have to be patient if sustained fusion is to be the long-term solution to our energy needs.

radiation from the stellar core, it begins to glow. The result is one of the most beautiful and striking of all astronomical objects, the **planetary nebula (Fig. 15.21).** Named for their resemblance to the planets Uranus and Neptune, some of these nebulae are bright enough and nearby enough to be familiar objects to owners of small telescopes.

The star that remains at the center of a planetary nebula lies far to the left on the H-R diagram, having a surface temperature (as high as 200,000 K) that is often much hotter than even the O and B stars on the main sequence (see Fig. 15.17). In essence, such a star is the naked core of the original star.

Some nuclear reactions may still be going on in a shell inside the star, but they do not last much longer, since the fuel in the core is depleted. The density, already very high, increases as the stellar rem-

nant condenses under the force of gravity. A larger and larger fraction of the star's interior becomes degenerate. As the core shrinks, it stays hot, but its luminosity decreases because of its diminishing surface area. The star's position on the H-R diagram lowers as its luminosity goes down (see Fig. 15.17). Eventually, the pressure of the degenerate electron gas stops the collapse, and the star becomes stable again. But now it is a very small object, so dim that it lies well below the main sequence in the lower left-hand corner of the H-R diagram. It is smaller than any normal star, and it is called a **white dwarf.**

A white dwarf is a bizarre object in many ways. It is supported by degenerate electrons whose peculiar properties govern its internal structure. It has a mass as great as the Sun (or even slightly greater), yet it is about the size of the Earth. This means that its den-

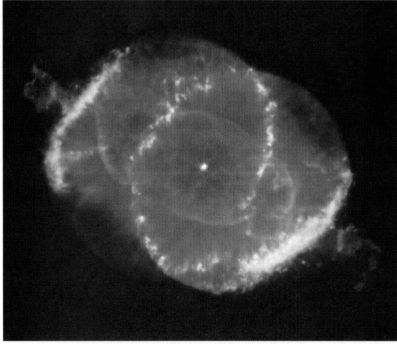

Figure 15.21 Planetary nebulae. These two images show some of the wide range of shapes and colors presented by planetary nebulae. At top is a ground-based telescopic image of the Helix nebula (NGC 7293) and below is a *Hubble Space Telescope* image of NGC 6543. The colors in each case represent emission by different gases in the nebulae. (Top: © 1979 Anglo-Australian Telescope Board, photo by David Malin; bottom: NASA/STScI) http://universe.colorado.edu/fig/15-21.html

sity is incredibly high—roughly a million times that of water. A cubic centimeter of white dwarf material would weigh a ton at the surface of the Earth!

A white dwarf does not do much. It undergoes no further nuclear reactions, so it does not evolve further. It gradually cools as it radiates away its internal heat energy, so it slowly becomes dimmer and dimmer. Under certain circumstances, a white dwarf may be stirred into nuclear action again (these are described in the next chapter), but failing that, the white dwarf stage is the end of the lifetime of a star like the Sun.

The Evolution of Massive Stars

As in solar-mass stars, the life of a more massive star is governed by the eternal battle against gravity. A star can exist only if it finds a way to counteract gravity; that is, it must find a source of pressure that can balance the weight of the stellar matter and prevent collapse. We will see that the massive stars often ultimately lose this battle, whereas the smaller-mass stars like the Sun may win it by virtue of degenerate electron gas pressure, which can last virtually forever.

The Main Sequence Phase

Stars more massive than the Sun follow the same general course of events as the Sun, but with significant differences. The major similarity is that the majority of the star's lifetime is spent on the main sequence, while nuclear reactions in the core convert hydrogen into helium. These reactions produce the energy necessary to balance gravity and supply the luminosity of the star. The principal contrasts with the evolution of a less massive star are that everything happens more rapidly in the massive star, and many more stages of nuclear reactions can occur. Further, because the core region of a massive star is convective, so that overturning motions extend well out of the core, more of the star's hydrogen gets mixed into the inner region where nuclear reactions can take place. Thus a greater fraction of the material in a massive star becomes involved in the reactions.

Even though the first nuclear-burning stage is the same as in the Sun, the details are different. Where the Sun produces most of its energy via the proton-proton chain, the massive star derives more of its energy from the CNO cycle. For stars at least twice as massive as the Sun, the higher internal temperature ensures that the CNO cycle is more efficient and is thus the dominant reaction (the steps in the CNO cycle are outlined in Appendix 14).

The luminosity and surface temperature of a hydrogen-burning star—hence its position on the main sequence—are determined by its initial mass. A star of 5 solar masses would have a surface temperature of approximately 12,000 K and be spectral type B8, for example, and a star of 10 solar masses would have a surface temperature of about 20,000 K and be spectral type B2. A star of 20 solar masses would have a surface temperature of roughly 30,000 K and would be an O9 star.

The core temperature inside a star governs the rate of nuclear reactions, which is the reason that massive stars evolve more rapidly than lower-mass stars. Reaction rates are very sensitive to temperature, so a modest increase in temperature will cause an enormous increase in energy output. As we learned earlier in this chapter, the mass of a star controls the internal temperature (through hydrostatic equilibrium), so the more massive stars have higher temperatures and thus higher reaction rates, higher luminosities, and shorter lifetimes. A 5-solar-mass star will exhaust its hydrogen supply in about 100 million years (about 1% of the Sun's lifetime), whereas a 10-solar-mass star will run out of hydrogen in much less time, perhaps only 1 million years.

Post-Main Sequence: Multiple Reaction Stages

The high core temperatures of massive stars have another important impact on the reactions: they allow further reaction stages to occur. After each form of nuclear fuel is depleted, the core contracts and the temperature increases due to compression. The temperature soon reaches a sufficiently high value that a new reaction stage can begin. In the massive stars, helium will combine by the triple-α reaction to form carbon (while hydrogen continues to burn in a shell outside the core), and when a significant amount of carbon is present, it may react with helium to form oxygen through a process called an **α-capture reaction.** In α-capture reactions, helium nuclei (which are called α particles) are added to larger nuclei, increasing the mass of the nucleus by addition of 2 protons and 2 neutrons and changing its identity. Carbon, with 6 protons and 6 neutrons, becomes oxygen, with 8 protons and 8 neutrons. Oxygen can then add another α particle to become neon, with 10 protons and 10 neutrons, and so on. At sufficiently high temperatures, even more complex reactions can occur, including carbon mergers with other carbon nuclei, oxygen with oxygen, and even silicon with silicon. In another sort of reaction that becomes important, nuclei capture free neutrons, thereby building up the heaviest of elements. The

core of a massive star becomes a soup of vastly energetic nuclei flying about, merging through all sorts of reactions, and building up abundances of heavier and heavier elements.

After each reaction stage, when a new reaction starts in the core, the previous one continues in a shell outside the core. Thus, at one point, the star may have an oxygen core where neon is being produced, surrounded by a carbon shell where oxygen is forming; outside the shell is a helium zone where carbon is being made, and this zone is surrounded in turn by a hydrogen-burning shell where helium is forming. At a later time, the same star will have even more nuclear-burning layers, as additional reactions start in the core.. The star acquires an onionlike structure in which each layer has a different composition **(Fig. 15.22)**. Ultimately, a core made of iron will develop, and this signals the end of the orderly progression of reaction stages, for reasons described shortly.

As the successive reaction stages occur in the star's core, its position on the H-R diagram changes in a fashion similar to that of a solar-mass star as it goes from main sequence to red giant. A massive star also makes this transition, although no helium flash occurs because the core is too hot to become degenerate. Soon after the star becomes a red giant, helium burning begins in the core, causing the interior structure to readjust itself and reverse the red giant phase temporarily. The surface contracts and heats up, and the star turns to the left and down in the H-R diagram **(Fig. 15.23)**. When the helium is used up, a second red giant phase follows, as the energy is again supplied primarily by a shell source and the outer layers expand and cool for a second time. Each time one form of fuel is depleted, a new red giant phase begins, and each time a new fuel ignites, the star contracts and moves back to the left and down in the H-R diagram. The number of these loops a star will undergo depends on how much mass it has, because this governs the internal temperature the star will ultimately reach, and this in turn determines how many reaction stages can occur.

The situation is complicated by the fact that a massive star loses mass nearly continuously throughout its lifetime, starting with a rapid stellar wind while the star is on the main sequence, and followed by many slow red giant wind phases. The continual stellar winds alter the internal conditions and prolong the lifetime of the star, while ejecting a substantial fraction of the original mass. This makes it difficult to

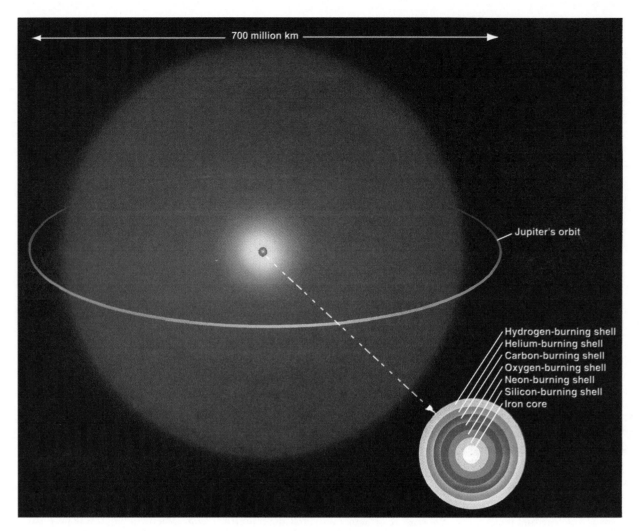

Figure 15.22 A massive star at the end of its life. This star has undergone all possible reaction stages in its core. The core consists of multiple layer of different composition, including six current or former shell sources and an inert iron core. The star is about to undergo collapse and supernova explosion.

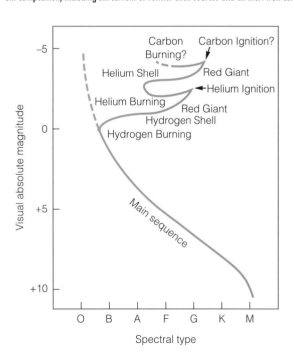

predict exactly what the lifetime of a given star will be or how many reaction stages it will undergo. The final fate of a massive star is determined by the outcome of a sort of race between two competing factors: on the one hand, the succession of nuclear reaction stages, which is prolonged by high mass; and on the other hand, the loss of mass, which acts to curtail the succession of reactions.

An Explosive Final Stage

We have seen that a star like the Sun will become a white dwarf when all possible nuclear reactions have

Figure 15.23 (left) The evolution of a massive star. This evolutionary track in the H-R diagram shows several steps followed by a star in the range of 5–10 solar masses. Each time the nuclear fuel in the core is exhausted, a shell burning source becomes dominant, causing the outer layers to expand and moving the star into the red giant region. When a new core source ignites, the star moves back to the left. The more massive the star, the greater the number of burning stages.

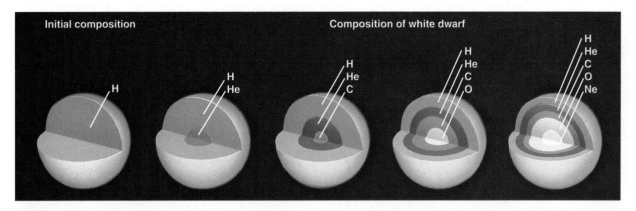

Figure 15.24 **White dwarf composition.** The more nuclear burning stages a star goes through, the greater the variety of elements produced in its interior. The white dwarf that results, therefore, has a composition that depends on the number of reaction stages, and this, in turn, depends on the mass of the star.

run their course. The star becomes stable, with the pressure of degenerate electron gas supporting it against gravity, and it remains a white dwarf perpetually (except under certain rare circumstances, described in the next chapter). Some more massive stars also become white dwarfs, although their compositions are different from the white dwarfs produced by low-mass stars **(Fig. 15.24)**. If nuclear reactions proceed through helium burning, as is expected for the Sun, then the resulting white dwarf will be made of carbon; if the next nuclear stage occurs before burnout, then the white dwarf remnant will be made of oxygen.

For stars above a certain final mass, however, a white dwarf becomes an unattainable endpoint, because there is a firm limit on how much mass degenerate electron gas pressure can support. A stable white dwarf cannot exist if its mass is above this so-called **Chandrasekhar limit,** which is about 1.4 solar masses. Above this mass, the inward force of gravity overwhelms the outward pressure force due to electron degeneracy. It is the final mass of the star that counts, so a star that starts life with more than 1.4 solar masses can become a white dwarf if it loses enough matter along the way, through stellar winds. Calculations are somewhat uncertain, largely because rates of mass loss are poorly known, but it now appears that stars with initial masses as high as 6 to 8 solar masses may lose enough material to become white dwarfs. Stars that begin life with more than 6 to 8 solar masses apparently cannot lose enough matter to become white dwarfs, and something else must happen to them.

Stars whose initial masses range from 8 to about 20 solar masses probably end up in another very compact state, even smaller and denser than a white dwarf. Many of them reach this state by violently expelling their outer layers in a **supernova explosion (Fig. 15.25).** The cause of a supernova is related to the composition of the star's core. If reactions proceed all the way to the point where the core is made of iron, no further energy-producing reactions can occur stably. Elements heavier than iron can combine, but only if additional energy is available. These reactions require more energy than they create and are said to be **endothermic** (reactions that produce more energy than they use are **exothermic**). When the iron core of a massive star reaches a temperature of more than 10^9 K and a density of around 10^{10} g/cm^3, endothermic reactions begin. These reactions remove heat energy from the gas, thus reducing the pressure and allowing the core to collapse. It is as if a sudden vacuum formed at the heart of the star, and the core collapses very rapidly in free fall.

Within a few ten-thousandths of a second, the density becomes so high that electrons and protons are forced together, forming neutrons in **inverse β-decay** reactions **(Fig. 15.26).** In the process, a vast quantity of **neutrinos,** which are tiny, chargeless, possibly massless, particles, are released. The neutrinos escape directly into space, carrying away enormous amounts of energy from the core of the star and further enhancing the core collapse.

The core is rapidly converted into a sea of neutrons, which are squeezed closer and closer together. But, like electrons, neutrons have a limiting density and resist further compression after a certain point. The neutron gas becomes degenerate, and the pressure of this degenerate neutron gas becomes sufficient to support the stellar core against further collapse. At this point the core, which had been in free-fall collapse, suddenly becomes incredibly rigid and hard, and material still falling in now crashes onto its surface. A violent rebound, or shock, wave is formed and moves rapidly outward through the star. This shock may become stalled as it runs into the large infalling mass of the outer layers of the star, but within a second or less, the enormous burst of neu-

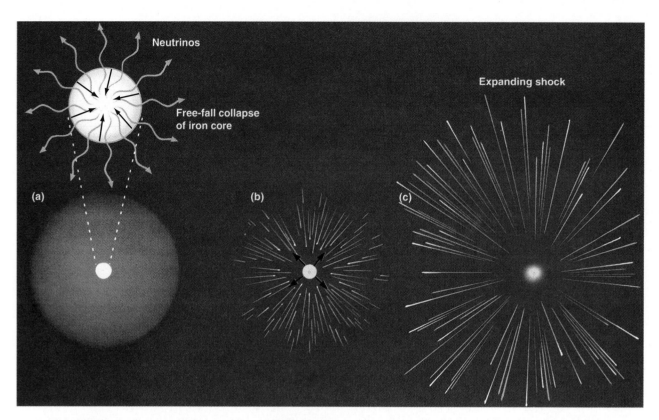

Figure 15.25 **Type II supernova.** The tiny iron core at the star's center undergoes rapid collapse because the only nuclear reactions that can occur there remove heat energy (i.e., they are endothermic). The compression converts the core to pure neutrons, releasing copious quantities of neutrinos and becoming rigid in the process (a). The outer layer of the star falls in onto the neutron star core, and rebounds (b). This creates a shock wave which travels outward (c) and, with the help of the outflowing neutrinos, blows off the star's outer layers in a supernova explosion, leaving behind a neutron star remnant.

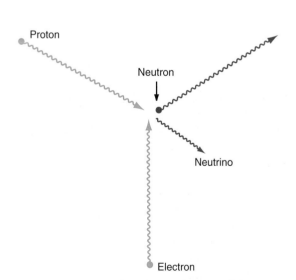

Figure 15.26 **Inverse beta decay.** In this process, which occurs at a significant rate only under conditions of very high pressure, a proton and an electron are forced together, forming a neutron and a neutrino. Energy is absorbed in the process, rather than being produced.

trinos coming from the core pushes the shock on its way again. The force is so great that the entire outer portion of the star is blown away into space.

Observationally, such an event is spectacular in the extreme. The star temporarily rivals an entire galaxy in luminosity, and spectroscopic measurements (using the Doppler effect) reveal outward velocities of tens of thousands of kilometers per second. Strong emission lines of hydrogen are seen (due to the fact that the outermost portion of the star never underwent nuclear reactions), and the object is called a **Type II supernova** (we will learn what a Type I supernova is in the next chapter).

The core that is left behind by the collapse and rebound of the outer layers is made almost entirely of neutrons and is incredibly dense. A mass between 1 and 2 solar masses is compressed into an object about 10 km in diameter. This object, called a **neutron star,** is as dense as 10^{14} to 10^{15} g/cm^3, comparable to the density of an atomic nucleus. Needless to say, matter in this form has very bizarre properties that cannot be reproduced on the Earth. For years after the theoretical possibility of neutron stars was recognized, there was no observational evidence that they exist, but this has changed dramatically, as discussed in the next chapter. It now appears that neutron stars are quite common in our galaxy.

The Most Massive Stars

What of the stars that start life with more than 20 solar masses? These stars go through the same succession of nuclear reactions as the moderate-mass stars, although they do this so rapidly that their outer layers do not have sufficient time to respond to all the internal changes. As a result, these stars may not actually go through the red giant part of the H-R diagram. In a matter of several hundred thousand to a million years, all the possible reactions are completed and an iron core is formed. In the meantime, the star stays in the upper portion of the H-R diagram, remaining so hot and luminous at all times that a rapid stellar wind is constant. This wind removes mass so effectively that the inner regions of the star are exposed as the outer layers are stripped off, and the star enters a class of peculiar emission-line objects known as **Wolf-Rayet stars.** These stars are identifiable by strong emission lines, usually dominated by either carbon (if triple-α burning layers are exposed) or nitrogen and oxygen (if the CNO-burning region is brought to the surface). The evolutionary status of the W-R stars, as they are commonly designated, was not understood until it was realized that massive stars undergo such rapid mass loss that the winds can expose the inner regions that were previously thought to be forever hidden from view.

The final stages in the life of a very massive star are more uncertain than for the other mass ranges we have discussed. Calculations show that the formation of an iron core should always be followed by collapse and at least the temporary formation of a neutron star core. The subsequent rebound shock can be stalled by the high mass of the overlying layers, however, so no supernova explosion occurs. As additional material falls onto the neutron star, it soon surpasses the maximum mass (between 2 and 3 solar masses) that can be supported by degenerate neutron gas pressure, and the neutron star suddenly collapses.

This time nothing can stop the collapse, and gravity wins this battle. No form of pressure can halt the infall, and it continues without limit. In less than one-thousandth of a second, the star becomes a **black hole.** The name derives from the fact that as the core shrinks and its surface gravity increases, a point is reached where even photons of light are unable to escape. After that point, we have no means of observing the remains of the star directly or communicating with it in any way.

The mass limit dividing stars that become neutron stars from those that become black holes is not accurately known and may not be a firm limit anyway; other factors such as the rotation speed of the stellar core may also be important. We can see that the rate of mass loss during the star's "normal" lifetime is an important factor, because this affects the final mass the star has when it reaches the iron-core stage. In our discussion we have used 20 solar masses as the approximate dividing line between stars that become neutron stars and those that form black holes, but this should not be regarded as firm. Some astronomers think that more likely an initial mass of 30 to 40 solar masses is required to make a black hole.

Explosive Nuclear Reactions and the Heaviest Elements

Our discussions have stressed that iron is the endpoint of the succession of nuclear reaction stages in stellar cores, because further reactions are endothermic and precipitate the collapse of the core. The main sequence of reactions does indeed stop with iron and so does not produce elements that are heavier. Indeed, the composition of matter in our galaxy—and in other galaxies with similarly advanced states of stellar processing—has a high abundance of iron (you may recall that in our discussion of star formation in Chapter 12, we noted that iron is the most common element in the interstellar gas and dust after hydrogen, helium, carbon, nitrogen, and oxygen; now you have an idea why this is so).

It is actually misleading to say that no elements heavier than iron are formed during the stable nuclear burning stages in the life of a massive star. The neutron-capture reactions that were mentioned earlier in this chapter do build up some quantities of elements beyond iron by adding neutrons to nuclei. These processes are called **neutron-capture reactions.** More specifically, the reactions that occur in stable stellar cores are called **slow-process,** or **s-process,** neutron captures. The reason for this is that there is sufficient time between neutron captures for the nuclei to undergo radioactive decay, resulting in the further conversion of the nuclei to new forms. The decay that occurs as a neutron converts itself spontaneously into a proton and an electron is called a **β-decay.** The electron escapes the nucleus and leaves the proton behind, thus increasing the atomic number by one.

Another form of neutron-capture reaction occurs during the brief, incandescent moments of a supernova explosion. Vast quantities of free neutrons are

Figure 15.27 **The region of Supernova 1987A before and after the explosion.** UPPER PANEL: At the left is a photograph made before the supernova occurred; the photograph at the right was made afterward. At this point, the supernova was a fourth-magnitude object, on its way to an eventual peak brightness of magnitude 2.9. LOWER PANEL: This photograph shows both the supernova (lower right) and the Tarantula nebula (upper left). The large size of its image compared with normal stars is due to blurring on the photographic emulsion caused by the brilliance of the supernova; it was actually a point source, no bigger than the images of other stars. (© upper: 1987 REO/AAT Board; lower: David Dunlap Observatory) http://universe.colorado.edu/fig/15-27.html

flying around during the collapse and shock expansion, and these can merge with nuclei to rapidly build up heavier nuclei. The collisions, which occur too frequently for decay to take place between captures, are called **rapid-process,** or **r-process,** neutron-capture reactions. The result is a buildup of neutron-rich nuclei, in contrast to the proton-rich elements that are created by s-process reactions. Thus the supernova process itself is responsible for the present-day abundances of many of the heavy elements known to us on Earth. It is fascinating to think of the history of the atoms of which we and our surroundings are made.

Supernovae: The Event of 1987

Supernovae are rare events, because very massive stars are rare. It is estimated that supernovae occur at a rate of one every 40 to 60 years in our galaxy, but most of them either take place very far from us or are hidden within the vast complexes of dark clouds where stars form. Supernovae bright enough to be seen with the naked eye are very unusual indeed: before 1987 the most recent one had been Kepler's supernova, which was observed in 1604 (amazingly, another major one, dubbed Tycho's supernova, had occurred only 32 years earlier). Several supernovae per year are usually seen in other galaxies scattered around the sky, but these are invariably too faint to be observed except with large telescopes. After nearly 400 years without a "local" supernova, one finally occurred in 1987, not in our own galaxy, but in the satellite galaxy to the Milky Way known as the Large Magellanic Cloud **(Figs. 15.27** and **15.28).**

Supernova 1987A, as it became known because it was the first one observed in 1987, reached a peak apparent magnitude of about 2.9, making it readily visible to the naked eye (so long as that eye was taken to the southern hemisphere; the Large Magellanic Cloud lies about 70° south of the celestial equator). Astronomers around the world were instantly at work observing the event from both ground-based telescopes (at such sites as Cerro Tololo in Chile and the Anglo-Australian Observatory; see Chapter 6) and space-based instruments. Unfortunately, a series of delays meant that the *Hubble Space Telescope* was still on the ground at the time, but several other spacecraft were in operation, providing ultraviolet, X-ray, and infrared data.

The presence of strong hydrogen emission lines soon demonstrated that this was a Type II supernova, thought to be the result of the iron-core collapse of a massive star, as just described. But there were some peculiarities: the maximum luminosity was lower than expected for such an explosion, and the precursor star, once it was identified from existing

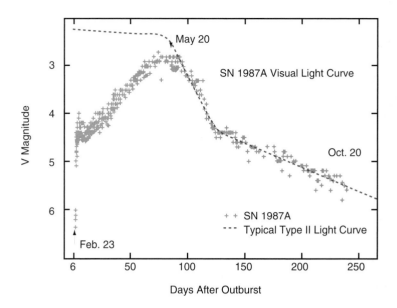

Figure 15.28 **The light curve of Supernova 1987A.** This graph shows the variation in magnitude of Supernova 1987A from its discovery until late fall 1987. The early part of the light curve is very unusual compared with other known supernovae, but once this one reached its peak, it began to behave very much like the light curves of other Type II supernovae. The reason for the difference in the early period is that the star that exploded was smaller than is thought typical, so it had less surface area to emit light. The later part of the light curve is consistent with energy production from the decay of radioactive nickel (^{56}Ni and then ^{57}Co). (J. Doggett and R. Fesen, University of Colorado)

photographs of the region, turned out to be a blue supergiant rather than a red one, which was contrary to standard theory. These oddities were eventually traced back to differences in chemical composition between our galaxy and the Large Magellanic Cloud, which is relatively deficient in heavy elements. The lower quantities of heavy elements in stars there cause some subtle differences in their structure and evolution, as compared with Milky Way stars. Models of Supernova 1987A were calculated, and the best match to the observations was found to be a star with a final mass of 20 solar masses and a diameter about 43 times that of the Sun. The relatively low luminosity was attributed to the smaller size of this star, as compared with a red supergiant, which meant that there was less surface area available from which light could be radiated.

One of the most exciting observations of Supernova 1987A had nothing to do with light, but instead consisted of the detection of the flood of neutrinos that formed as the core collapsed (recall the inverse β-decay reactions that were described earlier in this chapter). The neutrinos, which are very difficult to detect because of their tendency to pass right through normal matter without interacting in any way, were "seen" by two different experiments. The neutrinos arrived at the Earth a few hours before the supernova could be seen in visible light. This difference in timing came about because the neutrinos were able to escape directly from the interior of the dying star during collapse, while the flash of visible light was not emitted until later, when the rebound shock broke through the stellar surface. This timing

difference helped verify the radius of the exploding star, because theory gives us firm numbers on the speed with which a shock can travel through the interior of a star.

The number and energy of the neutrinos were nicely consistent with theoretical models of what should be produced in an iron-core collapse supernova, and of course this was gratifying to the theorists. It is noteworthy that the majority of the energy produced in a Type II supernova comes out in the form of neutrinos: the visible light contains only about 10^{-4} of the neutrino energy, and the mechanical (kinetic) energy of the expanding gas layers is about 10^{-2} (1%) of the total. Thus what we see of a supernova represents only a tiny portion of the total energy involved.

The visible radiation came from different processes at different stages during the outburst. Initially, the light was simple thermal radiation, as the surface of the star was strongly heated as it was expanded away by the outgoing shock front. As the expanding envelope of gas cooled, the energy of radioactive decay soon became dominant and was the chief source of energy for the next several months. The nuclear reactions during the explosion produced substantial quantities of radioactive nickel, which decays over time to cobalt and then to iron. The decay stages produce heat energy, and this was enough to make the expanding gas cloud glow. Verification of the decay process came from gamma-ray observations (from high-altitude balloons), which revealed emission lines due to the decay reactions (an atomic nucleus has energy levels similar to electron

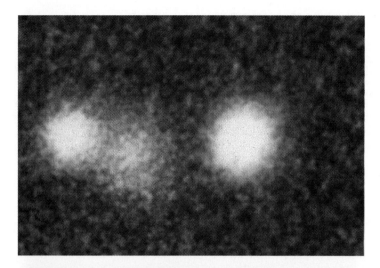

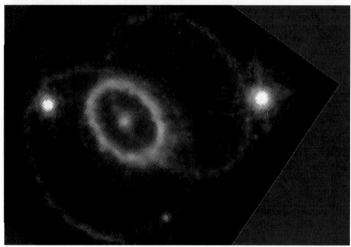

Figure 15.29 **The ring around Supernova 1987A.** The upper photo shows a ground-based image of the remnant of SN 1987A (the reddish blob in the left center). The lower image, from the Hubble Space Telescope, reveals that the supernova remnant consists of an intricate pattern of loops or rings of gas, glowing due to the radiation received when the star exploded. Astronomers think that these rings of gas were left over from an earlier phase when the star ejected matter in a red-giant wind. (upper: © 1992 Anglo-Australian Telescope Board, photo by David Malin; lower: NASA/STScI) http://universe.colorado.edu/fig/ 15-29.html

mass, they might account for the "dark matter" that dominates the universe (this will be discussed in Chapters 17, 18, and 20). But the limit implied by the Supernova 1987A observations apparently rules out neutrinos as the sole source of the missing mass in the universe.

It is considered virtually certain that a neutron star formed during Supernova 1987A; not only is this the expectation of the model calculations, but no other explanation for the observed burst of neutrinos is known (recall that the neutrinos are thought to come from the inverse β-decay reaction as the neutron star forms). So far, however, no direct detection of the neutron star has occurred, and this is becoming problematic for astronomers, who have expected that the neutron star would be detected, initially at X-ray wavelengths, as the expanding gas cloud cools and thins. The conflict with expectations is not serious at this point, but could become so soon if the neutron star continues to defy detection. Already there is some speculation that the neutron star accreted enough mass to resume its collapse and become a black hole. Supernova 1987A will continue to be watched for years to come. Already some very interesting manifestations of its interaction with surrounding matter have been seen. *Hubble Space Telescope* observations have revealed a complex series of rings around the original stellar location; these are thought to be preexisting gas shells from earlier mass-loss stages, being heated by radiation from the supernova blast **(Fig. 15.29).** Study of the form of these gas rings will provide useful information on the process of mass loss by pre-supernova stars. Within a decade or so, a more physical interaction with these rings of gas is expected to occur, when the expanding outer layers of the supernova itself crash into them.

Stellar Evolution in Binary Systems

We return to the main thread of this chapter, the evolution of stars, by considering what might happen in a double star, or binary, system. As noted previously, about half the stars in the sky are binary. In most cases, the two stars are well separated and have little or no effect on each other; each follows its own evolutionary path as though in isolation. In some cases, however, the two stars have profound effects on each other.

If either star is massive enough to have a strong stellar wind during the main-sequence phase, then matter can be transferred to the companion star **(Fig. 15.30).** Even if no wind is present initially, one will

energy levels, except that the energies are so high that the photons emitted or absorbed lie at gamma-ray wavelengths instead of visible wavelengths).

The neutrino observations produced an interesting spin-off. Based on the arrival times of neutrinos of different energies, it was possible to place firm limits on the mass of the neutrino itself. This question had been somewhat controversial, because despite the original theoretical expectation that these particles are massless, at least one experiment had suggested that they do have some mass. The implications were potentially very important: having mass might help explain the low rate of detection of neutrinos from the Sun; furthermore, if neutrinos have

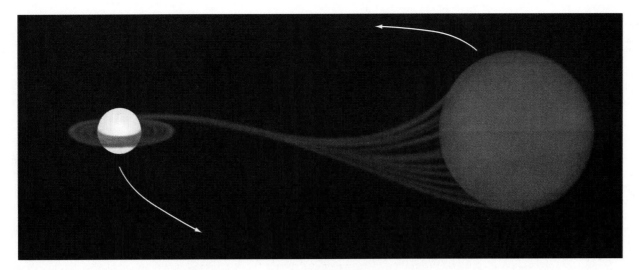

Figure 15.30 Mass transfer in a binary system. When two stars in a binary system are close together, material from one may be transferred to the other. There is a point in space between the two stars at which material would feel equal gravitational forces from each star; if one star has a wind or expands as a red giant so that its outer layers reach this point, then matter can flow from one star to the other. This can alter the evolution of both.

likely develop during the red giant phase, and mass transfer can occur then. If this happens, the evolutions of both stars may be affected. The lifetime of the mass-losing star will be extended, as we have already discussed; the evolution of this object will not necessarily be any different than if it were a single star, unless it later accretes matter from the same companion to whom it is now donating mass. As for the recipient of the mass transfer, its evolution will be accelerated as its mass increases.

Mass transfer is most likely to occur in close binary systems, where it is easier for the matter to reach the point in space between the two stars where the gravitational forces are equal. Once material has reached this point, it can flow across into orbit around the other star. Normally, the matter has enough angular momentum that it cannot fall directly onto the companion star, but instead forms a disk (recall our discussions of accretion disks in Chapter 12). Gas within the disk gradually loses orbital energy due to collisions between particles and drifts inward, eventually falling onto the surface of the star. In this manner the companion to the mass-losing star can gain mass gradually, although it is believed that in some circumstances instabilities can develop in the disk that result in large clumps of matter falling onto the star more quickly.

The realization that mass transfer can occur in close binary systems was stimulated in part by observations of Algol, the famous "demon star," a well-known, bright, eclipsing binary. Analysis of the orbit of this system provided accurate information on the masses of the two stars; one is a B main-sequence star, and the other is a G red giant. The B star has more mass than the red giant, which led to a conundrum: How could the less massive of the two stars have the more advanced evolution? We know full well that massive stars evolve more rapidly than less massive stars, so surely the B star should have become a red giant first. The answer, we now recognize, is that the G giant was originally the more massive star and thus was the first to evolve off the main sequence. But once it became a red giant, it developed a stellar wind and transferred mass back to its companion. The companion, originally a lower main-sequence star, has now gained enough mass to become a B star. In due course, when this star becomes a red giant, another mass transfer may occur, and the ratio of masses may again be reversed. Many other systems are observed in which one or more stages of mass transfer have occurred, muddying our normally clear picture of stellar evolution.

When one of the two stars in a binary system has evolved all the way to its final endpoint as a white dwarf, neutron star, or black hole, then spectacular effects will occur if mass is then transferred from the companion star to the stellar remnant. These effects will be discussed in the next chapter.

Summary

1. A star is gaseous throughout, with temperature and density increasing toward the center. The mass of a star, through its control of hydrostatic equilibrium (the balance between gravity and pressure), governs most of the basic properties of the star.

Building a Model Star

ASTRONOMICAL

ACTIVITY

Table 13.3 provides information on interior conditions in the Sun, which can be used to develop an intuitive understanding of the internal structure of a typical star. To do this, you will need three sheets of graph paper and a ruler. Preferably, the graph paper should have relatively heavy lines marking 1 cm squares and lighter grid lines marking 1 mm squares as well.

Make three graphs, each with the relative radius of the Sun on the horizontal axis, from 0 at the left to 1.0 at the right. On the vertical axis of the first graph, plot the values for temperature; on the second graph, plot the values for density; and on the third, plot the values for mass fraction. All of these values are given in Table 13.3. Note that you should choose your horizontal and vertical scales before beginning to plot the points for each graph. If your graph paper has 1 cm squares, a convenient horizontal scale would be 1 cm = 0.1 $R_\odot$, so that your horizontal axis will be 10 cm long (plot only the points out to R = 1.0 $R_\odot$; ignore the remaining values, which represent the atmosphere of the Sun rather than the interior). For your vertical scales, note the maximum value you will have to plot, and find a value per centimeter that will provide a convenient vertical height in the range of 15 to 20 cm. For example, for the temperature plot, you could choose a scale of 1 cm = 1.0×10^6 K, since the maximum value of 1.6×10^7 K is 16 times greater than this, providing a vertical height of 16 cm. Similarly, you could choose a scale of 1 cm = 10 g/cm³ for density and a scale of 1 cm = 0.05 $M_\odot$ for mass fraction.

Once you have made points on the graphs for each of the values in the table, connect the points in each graph with a smooth line. This line shows how the value of each plotted quantity varies with radius from the center of the Sun to its surface. Right away you will find some interesting results. For example, you will see that the temperature peaks sharply toward the center of the Sun, as does the density, and that the mass fraction grows quickly as you move outward from the center. On all three graphs you may have difficulty distinguishing the values in the outer portion of the Sun, because they are so low relative to the central value (for temperature and density) or are very close to 1.0 throughout the outer portion (for mass fraction). For this reason graphs like this are often plotted in logarithmic units, which can more easily show points covering a very wide range of values.

To help you visualize some of the important aspects of the internal structure of a star, use your graphs to answer some questions. First, as we already noted, the temperature drops off very quickly with distance from the Sun's center. To illustrate this, find the relative radius where the temperature has dropped to one-half its central value, to one-tenth of the central value, and to one-hundredth of the central value. In similar fashion, use your other graphs to find where the density drops off to one-half, one-tenth, and one-hundredth of its central value; and where the mass fraction reaches one-half, three-fourths, and nine-tenths of the total mass of the Sun.

How would these results differ if the Sun were uniform throughout its interior with constant density and temperature? Indicate on the same graphs how the curves would look in that case. To do this, you will have to choose values for the temperature and the density; the average between the central and surface values will do. For mass fraction, note that the fraction will vary from 0 at the center to 1.0 at the surface, but the graph will not be a simple straight line, because the mass inside a spherical volume of constant density grows with the cube of the radius. Calculating the exact shape of this curve may be a bit complicated but you may be able to deduce its general nature with some simple experiments of plotting mass fraction versus R, where the mass fraction grows with the cube of R.

Now that you have done this exercise for the Sun, think about what a similar model for a red giant star might be like. Based on the discussions in this chapter, how do you think the graphs of temperature, density, and mass fraction for a red giant might differ from the graphs you have made for the Sun?

2. Nuclear fusion reactions occur in the core of a star, where the temperature and density are high enough to allow nuclei to collide with sufficient velocity and frequency. In most stars (those on the main sequence), hydrogen is converted into helium by reactions in the core.

3. Following the hydrogen-burning phase, stars may undergo a series of additional reaction stages, requiring a higher core temperature and creating more massive nuclei at each step.

4. The energy inside a star is transported by convection or by radiation. On the lower main sequence (including the Sun), radiative transport dominates in the inner parts of the star, while convection operates in the outer layers. On the upper main sequence, convection is

dominant in the core, while radiative transport dominates in the outer layers.

5. The lifetime of a star depends on its mass and how rapidly it uses up its nuclear fuel; the lifetime decreases dramatically from the lower main sequence to the upper main sequence.

6. Many stars have surface phenomena, such as stellar winds, chromospheres, and coronae, which may involve the loss of mass. Hot, luminous stars usually have high-speed stellar winds, while cool stars usually have chromospheres and sometimes also coronae. Very luminous, cool stars have extensive low-speed winds and sometimes form dust particles in their outer layers.

7. The evolutionary development of stars is deduced from observations of star clusters, where the stars all have the same age and initial composition, making it possible to trace the differing evolutions of stars having different masses, and from theoretical calculations of stellar structure models in which changes that occur with time are taken into account.

8. A star spends about 90% of its lifetime on the main sequence, converting hydrogen to helium in its core, and then alters its internal structure as the internal equilibrium is disturbed.

9. A star like the Sun will expand to become a red giant when its core hydrogen runs out and its energy is produced by a shell hydrogen source. When helium begins to fuse into carbon, the star contracts again, but never starts a new reaction stage after that, as the core shrinks while the outer layers escape and the star becomes a white dwarf.

10. Stars more massive than the Sun undergo more nuclear reaction phases but live shorter lives. Those starting with masses of 8 solar masses or less are likely to lose enough matter to end up as white dwarfs, while more massive stars will either explode as supernovae and become neutron stars or will collapse to form black holes.

11. Supernova 1987A was the first naked-eye supernova in nearly 400 years and has helped astronomers to test models of the evolution of massive stars.

12. In some binary systems, mass transfer between the stars affects the evolution of one or both of the stars.

Review Questions

1. How do we know that a star must be in a state of balance between gravity and pressure? What would happen if this balance were disturbed?

2. Explain why the more massive a star is, the shorter its lifetime.

3. Describe the role of convection in the structure and evolution of a star like the Sun.

4. Summarize how a star on the upper main sequence (an O or B star) and one on the lower main sequence (K or M) differ in structure, energy generation and transport, and outer layers.

5. Summarize the importance of star clusters in studies of stellar evolution.

6. If a cluster is observed to have O stars in it, how do we know that it must be a young cluster? Is it possible for the same cluster to also contain M stars? Explain.

7. Explain why the main sequence is a broad strip rather than a thin line. Where do you think the Sun's position lies within this broad strip? Explain.

8. Explain how white dwarfs of similar mass can have very different chemical compositions.

9. Why is each successive nuclear reaction stage undergone by a massive star shorter than the previous stage?

10. The relative abundances of elements in the galaxy show a steady decline with increasing atomic mass. That is, hydrogen is most abundant, helium is next, and so on. But iron is *more* abundant than elements close to it in weight. Given what you have learned in this chapter, explain why this is so.

Problems

1. Estimate the hydrogen-burning lifetime of a star of 15 solar masses and luminosity 10,000 times that of the Sun, and of a star of 0.2 solar masses and luminosity 0.008 that of the Sun.

2. If a star has an initial mass of 20 solar masses and a lifetime of 5 million years, how much of its mass will it lose if the rate of mass loss is one solar mass every 500,000 years?

3. How much will the Sun's luminosity increase when it becomes a red giant, if its radius increases by a factor of 20 and its surface temperature drops to half its present value?

4. Suppose a 5-solar-mass star whose main-sequence radius is twice the Sun's radius becomes a red supergiant star with a radius 200 times the Sun's radius. By how much would this star's surface gravity and escape velocity decrease? Discuss the

importance of your answer in view of the stellar winds from red supergiants.

5. What is the wavelength of maximum emission for the central star of a planetary nebula, whose surface temperature is 100,000 K? What kind of telescope would be most appropriate for observing such a star?

6. When the Sun becomes a white dwarf, its radius will shrink to about 0.01 times its present radius, but its mass will remain approximately the same. Calculate what its density, surface gravity, and escape speed will be when it is a white dwarf.

7. Suppose a neutron star has a radius that is 1/1,500 the radius of a white dwarf having half the mass of the neutron star. Compare the average densities of the two.

8. A supernova may have an absolute magnitude as small (bright) as $M = -19$. How much more luminous is such a supernova than the Sun, whose absolute magnitude is approximately $M = +5$?

9. The neutrinos from Supernova 1987A arrived at Earth about 4 hours before the visible light. The neutrinos and the light traveled through space at the same speed, but the neutrinos were emitted 4 hours earlier than the light. The time delay equals the time it took for the shock wave inside the exploding star to make its way from the core to the surface of the star. If the shock moved at a speed of 2,000 km/sec, what was the radius of the star before it exploded? Express your answer in units of the Sun's radius.

Additional Readings

Boss, A. P. 1991. The Genesis of Binary Stars. *Astronomy* 19(6):34.

Eicher, D. J. 1994. Ashes to Ashes and Dust to Dust. *Astronomy* 22(5):32.

Furth, H. P. 1995. Fusion. *Scientific American* 273(3):174.

Hayes, J. C. and A. Burrows. 1995. A New Dimension to Supernovae. *Sky & Telescope* 90(2):30.

Hoagland, M. 1995. Solar Energy. *Scientific American* 273(3):170.

Kaler, J. B. 1993. Giants in the Sky: The Fate of the Sun. *Mercury* XXII(2):34.

Kaler, J. B. 1994. Hypergiants. *Astronomy* 22(3):32.

Kwok, S. 1996. A Modern View of Planetary Nebulae. *Sky & Telescope* 92(1):38.

Marschall, L. A. 1994. Cosmic Chameleon: The Supernova in M81. *Astronomy* 22(2):40.

Marschall, L. A. and K. Brecher 1992. Will Supernova 1987A Shine Again? *Astronomy* 20(2):30.

Naeye, R. 1993. Supernova 1987A Revisited. *Sky & Telescope* 85(2):31.

Nesme-Ribes, E., S. L. Baliunas, and D. Sokoloff 1996. The Stellar Dynamo. *Scientific American* 275(2):46.

Soker, N. 1992. Planetary Nebulae. *Scientific American* 266(5):78.

Stahler, S. W. 1991. The Early Life of Stars. *Scientific American* 265(1):48.

Talcott, R. 1996. Unwinding the Helix. *Astronomy* 24(7):44.

Van Buren, D. 1993. Bubbles in the Sky. *Astronomy* 21(1):46.

Web Connections

The Review Questions and Problems also appear at the following URLs:
http://universe.colorado.edu/ch15/questions.html
http://universe.colorado.edu/ch15/problems.html

Chapter 16
Stellar Remnants

Chapter Web site: http://universe.colorado.edu/ch16

*What may only appear to be a
white puff of smoke in space
is actually the violent end of one
stellar system and the possible
beginnings of another.*

P. Scoma, 1968

ur story of stellar evolution is almost complete. We have seen how the properties of stars are measured, and we have learned how they form, live, and die. The only thing we have not considered is what is left of a star after its life cycle is completed.

As we saw in the previous chapter, the form of remnant that remains at the end of a star's life depends on the mass the star had when it finally ran out of nuclear fuel. Three possibilities were mentioned: white dwarf, neutron star, and black hole. Another form of remnant is the cloud of gas left behind in the wake of a supernova explosion. Whether accompanied by a neutron star, a black hole, or no trace at all of the original star, supernova remnants have much to tell us about stellar deaths and will be included in this chapter along with the compact remnants of the stars themselves.

White Dwarfs, Black Dwarfs

A star whose final mass is less than 1.4 solar masses will end its life as a white dwarf. The star may never have exceeded this mass limit, or it may have started life with more than 1.4 solar masses but shed enough matter to finish under the limit. White dwarfs are small, dim objects, but they are so numerous that many are found close enough to us to be observed directly. In contrast, the other forms of stellar remnants usually can be detected only through their indirect effects on matter that may approach them.

> **Table 16.1**
>
> **Properties of a Typical White Dwarf**
>
> Mass: $1.0\ M_\odot$
> Surface temperature: 10,000 K
> Diameter: $0.008\ D_\odot$ ($1.0\ D_\oplus$)
> Density: $5 \times 10^5\ g/cm^3$
> Surface gravity: 1.3×10^5 Earth gravities
> Luminosity: 2×10^{31} erg/sec ($0.005\ L_\odot$)
> Visual absolute magnitude: 11

Properties of White Dwarfs

Table 16.1 summarizes the properties of a typical white dwarf. An isolated white dwarf is not a very exciting object from an observational point of view. It is rather dim **(Fig. 16.1)**, and its spectrum is nearly featureless, except for a few very broad absorption lines created by the limited range of chemical elements it contains **(Fig. 16.2).** The great width of the lines is caused by the immense pressure in the star's outer atmosphere. Pressure tends to smear out the atomic energy levels, because of collisions between atoms. In effect, the energy levels become broadened, and this results in broadened spectral lines when transitions of electrons occur between the levels. The pressure is so high because a white dwarf has a very high surface gravity. Recall from Chapter 5 that surface gravity is proportional to the mass of an

Figure 16.1 Sirius B. The dim companion to Sirius (to the left of the large image of Sirius A) was the first white dwarf to be discovered; analysis of its mass, temperature, and luminosity led to the realization that it is very small and dense. (Lick Observatory) http://universe.colorado.edu/fig/16-1.html

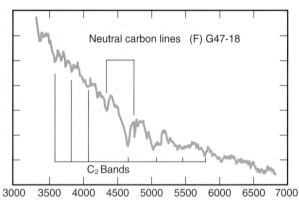

Figure 16.2 A white dwarf spectrum. Here we see that white dwarf spectra have few strong features, and that the spectral lines tend to be rather broad. This white dwarf has a large abundance of carbon as a result of the triple-α reaction. (G. A. Wegner)

object and inversely proportional to the radius. A white dwarf has about the same radius as the Earth, but the mass of the Sun, which is some 330,000 times the Earth's mass, so we can estimate the surface gravity on a white dwarf as about 330,000 Earth gravities! This is the reason the gas at the surface is highly compressed and under great pressure.

A white dwarf can also have a very strong magnetic field, and this has further effects on spectral lines formed in its atmosphere. If the star had a magnetic field before its contraction to the white dwarf state, not only is that field preserved, but it is intensified in the process of contraction. As we have seen previously in Chapters 13 and 14, a magnetic field causes certain spectral lines to be split into two or more parts. In a white dwarf, where the lines are already very broad, this splitting (called the **Zeeman effect**) usually just makes the lines appear even broader.

The lines in the spectrum of a white dwarf are also shifted, always toward longer wavelengths. This shifting is not a Doppler shift caused by motion of the star. If it were, it would be telling us that somehow all the white dwarfs in the sky are running away from us at very high speeds. Instead this is a **gravitational redshift.** One prediction of Einstein's theory of general relativity is that photons of light should be affected by gravitational fields. In effect, Einstein predicted that photons of light should lose energy in escaping a gravitational field and thus be shifted to longer wavelengths (i.e., lower energies). White dwarfs, with their immensely strong surface gravities, provide a confirmation of this prediction. In later sections, we will see even more extreme examples of gravitational redshifts.

Left to its own devices, a white dwarf will simply cool off, eventually becoming so cold that it no longer emits visible light. At this point, it cannot be seen and simply persists as a burnt-out cinder that may be referred to as a **black dwarf.** It will remain as a source of infrared radiation that may be detected long after it has faded from view at visible wavelengths, but eventually it will become so cold that it will be undetectable even at infrared wavelengths. The cooling process takes a long time, however; several billion years may go by before the star becomes cold and dark. This may seem surprising, since the white dwarf has no source of energy and stays hot only as long as it can retain the heat that it contained when it was formed.

A white dwarf takes so long to cool because heat is trapped inside; it simply cannot get out quickly. One reason for this is the small surface area of a white dwarf, which means that there is not much surface from which to radiate energy away. The second,

and more important, reason is that the thin layer of ordinary gas at the surface of a white dwarf blocks radiation from escaping. This thin layer is **opaque,** meaning that it absorbs photons from the interior that are trying to get out. Eventually, after being absorbed and reemitted many times, a photon gets through, but the net effect is that energy is bottled up in this thin layer of gas, which therefore acts as a thermal blanket.

Inside the white dwarf, energy is transported very efficiently by a process called **conduction,** which is more efficient than radiative transport in this case. Conduction refers to the transport of heat energy by fast-moving electrons, and in the degenerate electron gas of a white dwarf interior, electrons can travel very freely, more freely than photons of radiation. The conduction process is so efficient that the entire interior has a uniform temperature. The temperature cannot vary from place to place, because the freewheeling electrons ensure that heat is uniformly distributed throughout.

The degenerate electron gas fills nearly the entire volume of a white dwarf's interior **(Fig. 16.3)**. The atmosphere of ordinary gas, so important in trapping heat energy inside the star, is only about 50 km thick (remember, the white dwarf itself is about the size of the Earth, with a radius of some 5,000 to 6,000 km). Of course, along with the degenerate electron gas,

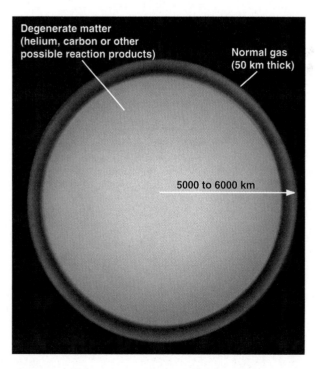

Figure 16.3 The internal structure of a white dwarf. The star's interior is degenerate and would cool very rapidly, except that the outer layer of normal gas acts as a very effective insulator, trapping radiation so that heat escapes only very slowly.

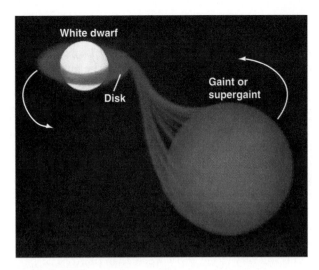

Figure 16.4 Mass transfer to a white dwarf. Matter is tranferred onto a white dwarf by a red giant companion. Theoretical calculations show that the new material will orbit the white dwarf in a disk but will eventually become unstable and fall down onto the white dwarf. When it does, nuclear reactions flare up, creating a nova outburst.

White Dwarfs, Novae, and Supernovae

Ordinarily, a white dwarf cools off, dims, and is never seen again. In certain circumstances, however, it can be resurrected briefly for another role in the cosmic drama. Some very spectacular phenomena are attributed to white dwarfs in binary systems.

In the previous chapter, we discussed the evolution of stars in close binary systems, where mass can transfer from one star to the other. Consider such a system, in which the star that was originally more massive has evolved far enough to become a white dwarf. Now the companion begins to transfer matter to the white dwarf **(Fig. 16.4),** probably because the companion has expanded to become a red giant. The white dwarf may respond to the addition of the new material in a variety of ways, depending on its composition and on the rate at which mass is added. All of the possible responses lead, sooner or later, to an explosive outburst, or to many repeated outbursts, because of the degenerate nature of the white dwarf.

the interior is populated with atomic nuclei, whose composition depends on the nuclear processing that took place while the star was "alive." If the star never progressed beyond the hydrogen-burning stage, the interior may be composed primarily of helium; if the star reached the stage where the triple-α reaction occurred but no farther, the interior may be mostly carbon; or if the original star was sufficiently massive to have undergone several reaction stages, the interior may be some heavier element such as oxygen or neon. In any case, however, the mass of the white dwarf must be less than the limit of 1.4 solar masses.

If the transferred gas reaches the white dwarf with a high velocity, so that heating occurs because of the energy of impact on the white dwarf's surface, there may be an instant nuclear explosion, enhanced by the fact that the degenerate dwarf material cannot expand and cool when heated (recall the discussion of the helium flash in the previous chapter). In other cases, mass may accrete onto the white dwarf for some time, perhaps decades or centuries, before the new material becomes hot enough to undergo nuclear reactions. The intensity of the outburst depends on how much matter is involved. If enough material is consumed in this explosion, the formerly very dim

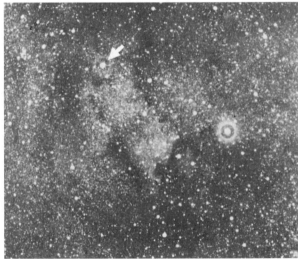

Figure 16.5 Nova Cygni 1975. This was one of the brightest novae in recent years. The photograph at the left shows the region without the nova; at the right is the same region near the time of peak brightness, when the apparent magnitude of the object was near $+1$. (E. E. Barnard Observatory, photograph by G. Emerson)

white dwarf may become more luminous than any ordinary star **(Fig. 16.5).** The resulting outburst is called a **nova.** Its luminosity may be 20,000 to 600,000 times that of the Sun (corresponding to an absolute magnitude in the range −6 to −9). A star that becomes a nova may do so repeatedly, with the outbursts usually separated by decades.

In some binary systems, rather small amounts of matter reach the white dwarf at regular intervals, creating relatively minor flare-ups. Such systems are referred to in general as **cataclysmic variables,** but there are a number of specific types, each named after a particular prototype. The outbursts can occur as frequently as several times a day, but once every few days is typical.

Material that transfers from one star to the other in a binary system cannot fall directly onto the second star, because of the relative orbital motions of the two stars; that is, because of the high angular momentum of the transferred gas. Instead an **accretion disk** is formed, as described in Chapters 12 and 15. From this accretion disk, material can then slowly descend onto the star at a rate determined by a number of factors such as the rate of mass gain by the disk and the presence of various instabilities that can develop in the disk's motion.

If material trickles onto the white dwarf at a slow rate, so that the impact does not cause much heating, the white dwarf can gradually gain mass without the infalling material causing any explosions. In this way, a white dwarf can approach the 1.4-solar-mass limit beyond which degenerate electron gas pressure can no longer support it. As a white dwarf gains mass, it becomes smaller, another peculiar property of degenerate matter. Thus the white dwarf increases in density. What happens as the white dwarf approaches the mass limit depends on its composition; it may, more or less quietly, suddenly contract, becoming either a neutron star or a black hole. If its composition is carbon, however, theory shows that nuclear reactions will begin instead, and the degeneracy will cause these reactions to very rapidly consume much of the white dwarf's mass, creating an immense explosion. This is another mechanism for producing a supernova explosion.

In the previous chapter, we discussed supernovae caused by the collapse and rebound of massive stars at the end of their nuclear lifetimes; these are called **Type II supernovae.** The original basis for classifying supernovae as Type I or II was according to whether lines due to hydrogen appear in the spectrum. If hydrogen is seen, the supernova is a Type II; if not, it is **Type I.** The massive stars we discussed in the previous chapter have outer layers of normal gas, meaning gas composed primarily of hydrogen, so

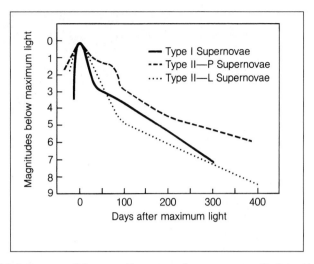

Figure 16.6 Supernova light curves. The two types of supernovae are readily distinguished by their spectra and their light curves, which show marked differences. Type I supernovae usually have no hydrogen lines in their spectra and are thought to be explosions of carbon white dwarfs that have accreted matter. Type II supernovae have hydrogen lines in their spectra and are the explosions that occur in massive stars when all nuclear fuel is gone. Peak brightnesses for the two types are slightly different, and, as seen here, the rates of decline are also quite different. Type II supernovae have been subclassified into types P and L. (J. Doggett and D. R. Branch)

strong spectral lines of hydrogen naturally occur when the star explodes.

Now we see that supernovae can also be caused by the addition of mass to a white dwarf, which has little or no normal, hydrogen-dominated gas in its outer layers. The original hydrogen in the core was converted to other elements, and the outer atmosphere is so thin that it also contains very little hydrogen. Thus a supernova created by an exploding white dwarf is usually classified as a Type I supernova. The understanding of the mechanisms for the two types of supernovae came after the original classification scheme had been established, but it is interesting to note that the simple classification based on surface composition reflected a fundamental difference in the processes that create the two types of explosions. Not all superficial classifications devised by astronomers end up being so meaningful.

Supernovae created in both ways are generally similar in maximum brightness and in the length of time they require to become dim again. But they differ in many details, including the types of spectral lines observed and the shape of the **light curve,** a graph showing the brightness variations with time **(Fig. 16.6).**

White dwarfs, then, are most prominent observationally in binary systems in which mass transfer from a normal companion takes place. Later in this chapter, we will see that other forms of stellar remnants are also best observed in these circumstances.

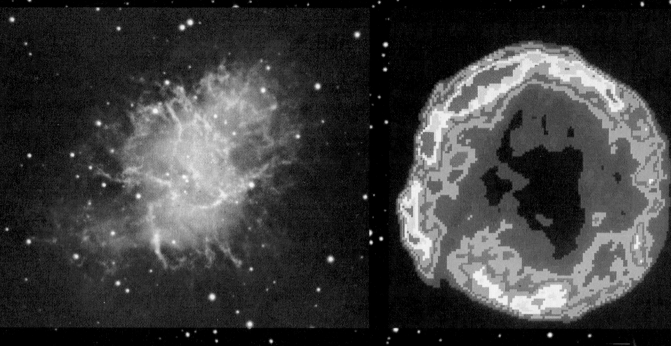

Figure 16.7 **Supernova remnants.** At the upper left is the Crab nebula, the remnant of a star that was seen to explode in A.D. 1054. AT the upper right is a radio image of the remnant of Tycho's supernova. Below is a photograph of part of the Vela supernova remnant, showing its wispy, filamentary structure. (Top left: Lick Observatory photograph; top right: National Radio Astronomy Observatory; bottom: © 1992 AAT Board; photo by D. Malin) http://universe.colorado.edu/fig/16-7.html

Supernova Remnants

We have learned that stars can explode as supernovae in two different ways, in some cases leaving a stellar remnant (a neutron star or black hole). In all cases another form of matter will be left over as well—an expanding gaseous cloud called a **supernova remnant.**

Several supernova remnants are known to astronomers **(Table 16.2).** A few, like the prominent Crab nebula **(Fig. 16.7)**, are detectable in visible light, but many are most easily observed at radio wavelengths. This is partly because visible light is affected by the interstellar dust that can hide our view of a remnant, whereas radio waves are largely unaffected. More importantly, however, as a supernova remnant ages, its emission in visible light fades out long before it stops emitting radio waves. Thus there is an extended time period (10,000 to 20,000 years) when the remnant emits radio radiation and little else. The radio emission is not thermal radiation (i.e., due to the temperature of the gas), but is produced by a different mechanism, called **synchrotron emission.**

Synchrotron emission, which derives its name from the particle accelerators in which it is produced on Earth, occurs when electrons move rapidly through a magnetic field. The electrons must have speeds near that of light; as they travel through the magnetic field, they are forced to move in a spiraling path, and they emit photons as they do so **(Fig. 16.8).** The emission occurs over a very broad range of wavelengths, including some visible, ultraviolet, and even X-ray radiation **(Fig. 16.9**, next page), but these other wavelengths are often not as easily detected as radio, because the radio emission is the strongest and most persistent. Synchrotron radiation can be distinguished from thermal radiation in two ways: (1) the intensity of synchrotron radiation varies smoothly over a wide range of wavelengths, rather than peaking strongly at a particular wavelength; and (2) synchrotron radiation is usually polarized (see Chapter 6).

Supernova remnants that are detected at visible wavelengths usually glow in the light of several strong emission lines—most notably, the bright H-α line of atomic hydrogen at a wavelength of 6563 Å, in the red portion of the spectrum. These lines often show large Doppler shifts, indicating that the gas is still moving rapidly as the entire remnant expands outward from the site of the explosion that gave it birth. Often a filamentary structure is seen, suggestive of turbulence, but probably modified in shape by magnetic fields.

Table 16.2
Prominent Supernova Remnants

Remnant	Age (yr)	Distance* (l.y.)
Vela X	10,000	1,600
Cygnus Loop	20,000	2,300
Lupus	975	4,000
IC 443	60,000	4,900
Crab nebula	936	6,520
Puppis A	4,000	7,500
Cassiopeia A	200	9,130
Tycho's supernova	400	7,800
Kepler's supernova	370	14,340

*Distance data are from a modern re-analysis by J. Saken, R. A. Fesen, and M. J. Shull, 1991. *Astrophys. J., Suppl.*

Today remnants are visible at the locations of several famous historical supernova explosions. The Crab nebula is the most prominent, having been created in the supernova observed by Chinese astronomers in A.D. 1054. Other supernovae seen by Tycho Brahe (in 1572) and by Kepler and Galileo (1604) also left detectable remnants, although neither is as bright in visible wavelengths as the Crab. Apparently, a supernova remnant can persist for 10,000 years or more before becoming too dissipated to be recognizable any longer. One of the best-studied remnants, Cassiopeia A (see Fig. 16.9), is apparently only about 300 years old, but the supernova that created it was quite dim and not widely

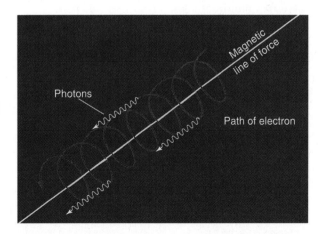

Figure 16.8 The synchrotron process. Rapidly moving electrons emit photons as they spiral along magnetic field lines. The radiation that results is polarized, and its spectrum is continuous but lacks the peaked shape of a thermal spectrum. The electrons must be moving very rapidly (at speeds near that of light), so whenever synchrotron radiation is detected, a source of large quantities of energy must be present.

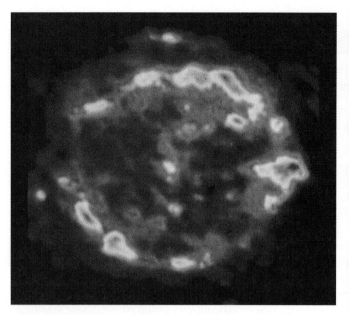

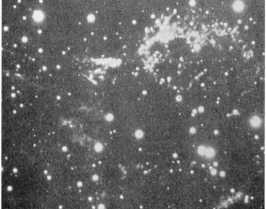

Figure 16.9 Emission from a supernova remnant. These are three different images of the same supernova remnant, Cassiopeia A. At the left is a radio map, at the upper right an X-ray image, and at the lower right a visible-light image. The remnant is less obvious in visible light because there are many other, brighter objects in the field of view. (Left: National Radio Astronomy Observatory; top right: Harvard-Smithsonian Center for Astrophysics; bottom right: K. Kamper and S. van den Bergh)

observed. This supernova may have been obscured by interstellar dust, or it may have been an unusually dim one intrinsically.

The energy of a supernova explosion is immense—a hundred times greater than the total amount of radiant energy the Sun will emit over its entire lifetime. While most of this energy is emitted and escapes in the form of neutrinos, some goes into radiation (which also largely escapes), and some takes the form of mass motions as the remnant expands. The kinetic energy of the gas motions heats the expanding gas to very high temperatures as it plows its way outward through the surrounding interstellar medium. The entire process has a profound effect on the interstellar gas and dust that permeates the galaxy, as we shall see in Chapter 17.

Neutron Stars

We have referred to a neutron star as a stellar remnant composed entirely of neutrons that are in a degenerate state, similar to that of the electrons in a white dwarf. The pressure created by the degenerate neutron gas is greater than that of a degenerate electron gas, so a star slightly too massive to become a white dwarf can be supported by the neutrons. Again, however, there is a limit on the mass one of these objects can have. This limit, which depends on other factors such as the rate of rotation, is between 2 and 3 solar masses. Thus neutron stars come from a class of stars whose masses at the end of their nuclear lifetimes are between 1.4 and about 2 or 3 solar masses. In most cases neutron stars are remnants of supernova explosions (of Type II) and are thus the end products of stars that were originally quite massive.

The structure of a neutron star **(Table 16.3** and **Fig. 16.10)** is even more extreme than that of a white dwarf. All of the mass is compressed into an even smaller volume (now the radius is about 10 km), the density is comparable to that of an atomic nucleus (i.e., in the range of 10^{14} to 10^{15} g/cm^3), and the gravitational field at the surface is immensely strong (a simple scaling argument, such as we made earlier for a white dwarf, suggests that the surface gravity on a neutron star of 2 solar masses is roughly a million times the Earth's surface gravity!). The layer of normal gas that constitutes the atmosphere is only cen-

Table 16.3
Properties of a Typical Neutron Star
Mass: 1.5 $M_\odot$
Radius: 10 km
Density: 10^{14} g/cm^3
Surface gravity: 7×10^9 Earth gravities
Magnetic field: 10^{12} Earth's field
Rotation period: 0.001–4 sec

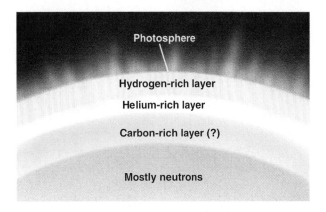

Figure 16.10 The structure of a neutron star. As this diagram shows, the outer regions of a neutron star may consist of thin layers of various elements that were produced by nuclear reactions during the star's lifetime. These outer layers are thought to have a rigid crystalline structure because of the intense gravitational field of the neutron star.

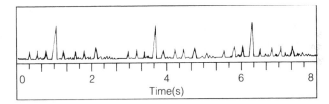

Figure 16.11 The radio emission from a pulsar. Here, plotted on a time scale, is the radio intensity from a pulsar. The radio emission is weak or nonexistent except for a very brief flash once each cycle (in some cases, there is a weaker flash between each adjacent pair of strong flashes).

timeters thick, and beneath it may be zones of different chemical composition resulting from previous shell-burning episodes. Each zone is only a few meters thick at the most. Inside these surface layers is the incredibly dense neutron core, which takes the form of a crystalline lattice. The temperature throughout is very high, but because the surface area is so small, a neutron star is very dim indeed. The compression of the star's original magnetic field may result in its having a magnetic field that is much more intense than even that of a white dwarf. Neutron stars generally cannot be observed directly because they are so dim, but the immense gravitational and magnetic fields associated with them can produce effects that are quite dramatic and readily observed.

Pulsars: Cosmic Clocks

The properties of neutron stars were predicted theoretically several decades ago, but until 1967 astronomers did not expect that they could be observed because of their low luminosities. In that year, however, British radio astronomers Jocelyn Bell and Antony Hewish discovered curious pulses of radiation during a routine survey of radio sources. These pulses, coming from a single point on the sky, were so perfectly regular that at first it was not known how any natural process could cause them. Soon additional sources of pulses were found (it became easier to spot them once the first one was found, because astronomers knew what to look for; special data analysis techniques were needed to locate the rapidly pulsing sources). The suggestion that the pulses might come from interstellar beacons left by a space-faring civilization (the so-called *little green men*, or *LGM*, hypothesis) aroused some initial excitement, but then astronomers immediately began to search for a more natural explanation. The name **pulsar** was adopted for these mysterious sources.

The most rapid among the pulsars discovered initially—the one located in the Crab nebula—flashes 30 times a second, and several others are now known that repeat even more rapidly than the Crab pulsar. In contrast, the slowest known pulsars have cycle times of more than a second. In every case the pulsar is only "on" for a small fraction of each cycle **(Fig. 16.11).**

The quest for a natural explanation of the pulsars quickly turned toward compact stars of some type. It was well known that a variety of stars pulsate regularly, alternately expanding and contracting, but none were known to pulsate so rapidly. Theoretical studies showed that such quick variations as those exhibited by the pulsars should occur only in very dense objects, denser even than white dwarfs. This led astronomers to think of neutron stars, bizarre objects known only in obscure theoretical studies until then. Enough was known from the properties of pulsating objects, however, to rule out rapid expansion and contraction of neutron stars as the cause of the observed pulsations—a neutron star should vibrate even more rapidly than the observed pulsars! A second possibility—that the rapid periods were produced by *rotation* of the objects—was considered. Like the physical contraction and expansion

STELLAR REMNANTS

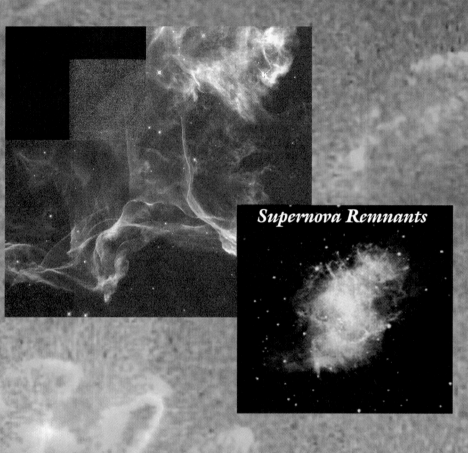

Supernova Remnants

STELLAR NURSERIES

Planetary Nebula

Diffuse Nebula

Interstellar Nebula. Stars form from the interstellar medium and, in aging and dying, they return material to the interstellar medium. At the top we see the results of mass-loss processes; at the bottom, we see steps on the way toward star formation.

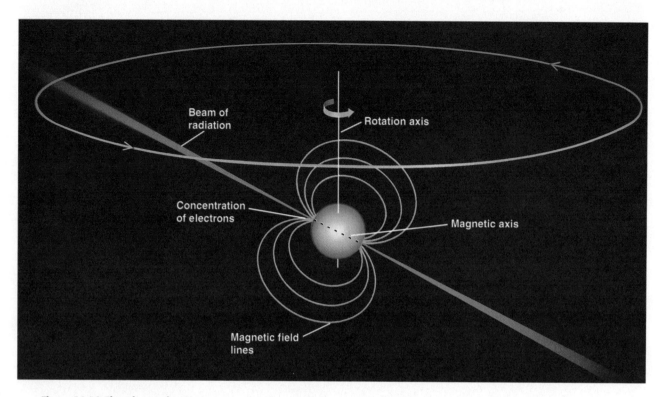

Figure 16.12 **The pulsar mechanism.** Here we see a rapidly rotating neutron star, with its magnetic axis out of alignment with its rotation axis. Synchrotron radiation is emitted in narrow beams from above the magnetic poles, where charged particles, constrained to move along the magnetic field lines, are concentrated. These beams sweep the sky as the star rotates, and if the Earth happens to lie in a direction covered by one of the beams, we observe the star as a pulsar.

hypothesis, the rotation hypothesis ruled out ordinary stars and white dwarfs. To rotate several times per second, a normal star or white dwarf would have to have a surface speed in excess of the speed of light, a physical impossibility. The object would be torn apart by rotational forces before it approached such a rotational speed. A neutron star, on the other hand, could rotate many times per second and remain intact.

When the Crab pulsar was discovered, a great deal of additional information became available. Astronomers found that the pulse rate of this object was very gradually slowing, something that could best be explained if the pulses were linked to the rotation of the object. The rotation could slow as the pulsar gave up some of its energy to its surroundings. This suggestion soon received a major boost with the discovery that the amount of energy being lost by the Crab pulsar if it were a spinning neutron star matched the amount of energy needed to explain the emission observed to be coming from the Crab nebula. The source of energy that powers the synchrotron emission from the nebula had been a persistent mystery. Now it was suggested that the slowdown of the rotating neutron star could provide the necessary energy, either by magnetic forces exerted on the surrounding ionized gas or by transfer of energy from the pulsar to the surrounding material

through the emission of radio waves. This coincidence settled the matter for most astronomers: the pulsars were rapidly rotating neutron stars.

The remaining question was how the rotation created the pulses. Evidently, a pulsar acts like a lighthouse, with a beam of radiation sweeping through space as it spins. But why should a rotating neutron star emit beams from just one or two points on its surface?

The most probable explanation has to do with the strong magnetic fields that neutron stars are likely to have. If a neutron star has a strong field, then electrons from the surrounding gas are forced to follow the lines of the field, hitting the surface only at the magnetic poles. The result is an intense beam of electrons traveling along the magnetic field, but especially concentrated near the magnetic poles of the neutron star, where the field lines are crowded together. The rapidly moving electrons emit synchrotron radiation as they travel along the field, creating narrow beams of radiation from both magnetic poles of the star. If the magnetic axis of the star is not aligned with the rotation axis **(Fig. 16.12),** these beams will sweep across the sky in a conical pattern as the star rotates. If the Earth happens to lie in the direction intersected by one of these beams, then we see a flash of radiation every time the beam sweeps by us.

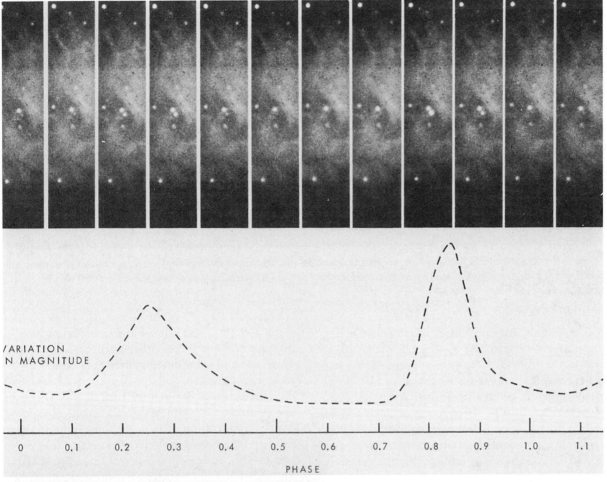

VARIATION
IN MAGNITUDE

0   0.1   0.2   0.3   0.4   0.5   0.6   0.7   0.8   0.9   1.0   1.1

PHASE

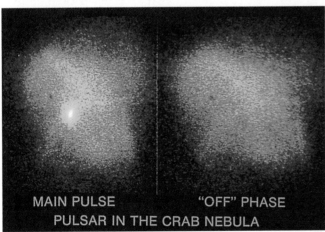

MAIN PULSE   "OFF" PHASE
PULSAR IN THE CRAB NEBULA

Figure 16.13 Visible and X-ray flashes from the Crab pulsar. At the top is a sequence of visible-light photographs accompanied by a light curve, showing how the pulsar in the Crab nebula flashes on and off during its 0.03-second cycle. Below is a pair of X-ray images, showing the pulsar "on" and "off" in different parts of its cycle. (Top: National Optical Astronomy Observatories; bottom: Harvard-Smithsonian Center for Astrophysics) http://universe.colorado.edu/fig/16-13.html

Synchrotron radiation is normally emitted over a very broad range of wavelengths, suggesting that pulsars may "pulse" in parts of the spectrum other than the radio. Indeed, this is the case: the Crab pulsar has been identified as a visible-light and X-ray pulsar **(Fig. 16.13).** Since special conditions (i.e., non-alignment of the magnetic and rotation axes, and a beam that crosses the direction toward the Earth) are required for a neutron star to be seen from Earth as a pulsar, it follows that there should be many neutron stars that do not show up as radio pulsars. In the next section, we will see how some of these neutron stars are detected.

Neutron Stars in Binary Systems

Earlier we made a general statement that stellar remnants are often most easily observed when they are in

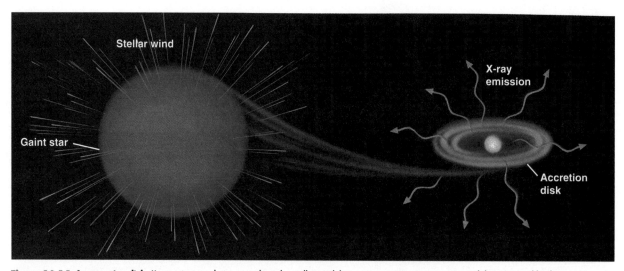

Figure 16.14 An accretion disk. Here a giant star, losing mass through a stellar wind, has a neutron star companion. Material that is trapped by the neutron star's gravitational field swirls around it in a disk, which is so hot from compression that it glows at X-ray wavelengths. A nearly identical situation can arise when the compact companion is a black hole.

binary systems. We have already seen that a white dwarf in a binary can flare up violently if it receives new matter from its companion star. A neutron star reacts similarly in the same circumstances. It might seem surprising at first that neutron stars can be in binary systems, since the formation of a neutron star involves the violence of a supernova explosion. But evidently it is possible for a supernova to occur without disrupting the binary and sending the two stars on separate paths.

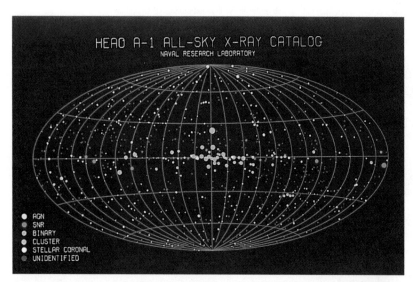

Figure 16.15 An X-ray map. This shows the locations in our glaxay of X-ray sources cataloged by the *High Energy Astronomical Observatory 1*, an orbiting X-ray satellite that operated in the late 1970s. Most of the sources here are binary systems in which one member is a compact stellar remnant such as a neutron star of a black hole. The concentration of sources along the galactic equator indicates that the massive stars that can end up as neutron stars or black holes tend to form, evolve, and die close to the plane of our disk-like galaxy (see the discussion in Chapter 17). (Smithsonian Astrophysical Observatory)

If a neutron star is in a binary system in which the companion object is either a hot star with a rapid wind or a cool giant losing matter because of its active chromosphere and low surface gravity, some of the ejected mass will reach the surface of the neutron star. As in the case of mass transfer involving white dwarfs, conservation of angular momentum prevents the material from falling directly onto the neutron star; instead, an accretion disk is formed **(Fig. 16.14)**. The individual gas particles orbit the neutron star like microscopic planets and fall inward only as they lose energy in collisions with other particles. The accretion disk acts as a reservoir of material, slowly feeding it inward toward the neutron star.

The disk is very hot because of the immense gravitational field close to the neutron star. As gas in the accretion disk slowly falls in closer to the neutron star, gravitational potential energy is converted into heat. The gas in the disk reaches temperatures of several million degrees, hot enough to emit X rays. Hence a neutron star in a mass-exchange binary system is likely to be an X-ray source, and a number of such systems have been found by X-ray telescopes **(Fig. 16.15)**. It is often obvious that an X-ray source belongs to a binary system, because the X rays are periodically eclipsed by the companion star. Eclipses are likely in such binaries because the two stars are close together (mass exchange would not occur unless it were a close binary) and because the mass-losing companion is likely to be a large star, either an upper main-sequence star or a supergiant.

The so-called binary X-ray sources (a few are listed in **Table 16.4**) are among the strongest X-ray-emitting objects known. All of them appear to be emitting their X rays from accretion disks surround-

Table 16.4
Selected Binary X-Ray Sources

Source	Period (Days)	Mass of System ($M_\odot$)	Nature of Stars
Cygnus X-1	5.6	40	O9Ib supergiant; probable black hole
Centaurus X-3	2.09	20–25	B0Ib-III giant; pulsar (neutron star)
Small Magellanic Cloud X-1	3.89	15–25	B0Ib supergiant; pulsar (neutron star)
Hercules X-1	8.95	20–30	B0.5Ib supergiant; probable neutron star
Vela X-1	1.70	2–5	HZ Hercules (A star); pulsar (neutron star)
V616 Monocerotis	0.32	11	K dwarf; probable black hole (companion mass $\simeq 8\ M_\odot$)
Large Magellanic Cloud X-3	1.70	15	B3 main-sequence star; probable black hole (mass $\simeq 9\ M_\odot$)
Large Magellanic Cloud X-1	4.2	25–40	O giant; probable black hole (mass $\simeq 10\ M_\odot$)
V 404 Cygni	6.47	8–15	K dwarf; probable black hole
Nova Muscae 1991	0.43	4–6	K dwarf; probable black hole
Nova Ophinchi 1977	0.70	74.1	K star; probable black hole
J.422 + 32	0.21	4.5	M dwarf; probable black hole
J1655 − 40	2.61	4–5.2	F-G star; probable black hole
GS2000 + 25	0.34	5.3–8.2	K dwarf; probable black hole

ing compact stellar remnants that are receiving mass from a stellar companion. Most likely, many of them are neutron stars, although it is probable that some are black holes, which would similarly produce X-ray emission as material fell inward (this is discussed a bit later). Some are known definitely to be neutron stars because they are also pulsars.

Neutron stars can also emit X rays in a slightly different way, which also occurs in mass-exchange binaries. Having passed through the accretion disk, material that reaches the surface of a neutron star can become hot enough to undergo nuclear reactions. As fresh material containing hydrogen builds up on the surface of the neutron star, the incredible gravitational compression is sufficient to heat it to such a high temperature that eventually it begins to react. This causes a sudden flare-up at X-ray wavelengths, as the hydrogen is quickly consumed in the reactions. For a brief period of time (only a few seconds), the neutron star is very bright **(Fig. 16.16)**. Then it quickly dies down and does not flare up again until more hydrogen-rich material accumulates on its surface.

Systems where these repeated and sporadic outbursts occur are known as X-ray burst sources, or simply **bursters.** These are similar in some ways to novae, but one important difference is that the flare-up of a burster occurs much more rapidly than that of a nova. Another is that most of the emission occurs only in very energetic X rays, so these outbursts do not show up prominently in visible light.

The Most Rapid Pulsars

In the past few years, astronomers have found a few pulsars with astonishingly short periods of pulsation, even by the standards of rotating neutron stars. Rather than emitting a few times per second, these objects pulse hundreds of times per second (the current record-holder flashes 885 times per second!). Most pulsars slow with age, so ordinarily the pulsars that spin and emit the most rapidly are the youngest. But no young pulsar was known to have a period as short as a few thousandths of a second (i.e., a few milliseconds), so astronomers thought it unlikely that these **millisecond pulsars** were simply very young neutron stars that had not yet slowed down. Further evidence of this came from the fact that the millisec-

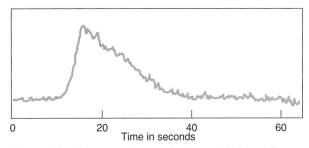

Figure 16.16 The X-ray light curve of a burster. This schematic illustration of the X-ray intensity from a burster shows how rapidly the emission flares up and then drops off. In analogy with the nova process involving white dwarfs, these outbursts are thought to be caused by material falling onto the surface of a neutron star and igniting brief episodes of nuclear reactions.

The Stuff We Are Made Of

You may have read or heard references to the notion that we are stardust—that we are made of elements that were created in the nuclear reactions of stars, long ago and far away. This is true, and you are now in good position to appreciate just what this means. Nuclear reactions in stars are the source of virtually every atom of every element except for hydrogen, most helium, and tiny quantities of other lightweight elements such as lithium, beryllium, and boron.

The key role of stellar nuclear reactions in element production was not established until the 1950s, nearly two decades after the discovery of the reactions that power the Sun and most stars. In the late 1940s, the Russian-American astrophysicist George Gamow considered a different source for the elements and found it wanting. Gamow was interested in the growing evidence that the universe had undergone an early phase where it was very hot and dense, and he wondered whether nuclear reactions might have taken place then that could have created all the elements. He conducted theoretical studies of conditions in the early universe and the reactions

that would have occurred, eventually finding that only hydrogen, some helium, and trace quantities of other very light elements could have been formed before the universe became too cool for reactions to continue.

In 1957, the team of Geoffrey and Margaret Burbidge, Willie Fowler, and Fred Hoyle published an epochal paper, in which a full theory of stellar nucleosynthesis was developed ("nucleosynthesis" refers to the formation, or synthesis, of elements by nuclear reactions). The four authors, all of them British except for Fowler, worked together in the United States at the California Institute of Technology. The paper, known to this day as "B^2FH," successfully explained the observed composition of the galaxy as a whole, a monumental achievement. Since that time, many further advances have been made, particularly in the understanding of reactions that occur explosively during supernova outbursts and in the refinement of our detailed knowledge of reaction rates, but the overall picture has not changed significantly.

Look at the table of elemental abundances in Appendix 9. This table represents the Sun specifically, but the Sun formed from interstellar material that came from the debris of previous gener-

ations of stars. Thus the solar composition represents the makeup of the galaxy, at least in a general way (but see the comments in the Science and Society box in Chapter 12). You see that hydrogen is the most abundant element by far, followed by helium; these two elements constitute roughly 98% of the mass. Following hydrogen and helium; there are very low abundances of the next few elements and then a major peak in quantities of carbon, nitrogen, and oxygen. Then there is a sharp decline, followed by a gradual rise in abundance until another peak is reached for iron. After that it is all downhill, with ever-decreasing quantities of elements with increasing mass. It was this general picture that was explained by B^2FH.

The predominance of hydrogen and helium arises from the hot, dense conditions in the early universe, as Gamow had shown. The very low quantities of the next three elements (lithium, beryllium, and boron) come about because there are no reactions that produce these elements under stable conditions in either the early universe or stellar interiors. There is a gap, and then carbon is very abundant. Carbon is the key element on which living organisms are based, and it is interesting to know that only by the sheerest of coincidences does carbon exist at all. Carbon is produced by the triple-α reaction, in which

ond pulsars were not found near the sites of recent supernova explosions. Quite to the contrary, most of them were found among the oldest regions in the galaxy, in globular clusters, where there are no massive, young stars that could explode in supernovae. Supernovae simply do not occur in globular clusters, so young pulsars should not exist there either. So how could a pulsar that might be many billions of years old be spinning hundreds of times per second?

It was noticed that many of the millisecond pulsars are in binary systems, where mass transfer could occur. This observation led to the suggestion that in these binaries the neutron stars had gained mass from

their companions through an accretion disk. The angular momentum that would be transferred along with the mass could cause the neutron stars to "spin up," increasing their spin rates to the very high values that are observed. To visualize this process, imagine that matter swirls onto the surface of the neutron star at a speed higher than the rotation rate, and that the impact of the matter hitting the surface pushes the star in the direction of rotation, speeding it up.

In some systems, the neutron star is even helping its companion to lose mass. The neutron star with its accretion disk is a source of intense X-ray emission. This emission heats the surface of the nearby com-

three helium nuclei fuse, but actually this reaction occurs in two very closely spaced steps. Its success depends critically on the existence of an excited energy state in the carbon nucleus that happens to have just the right energy to allow the merger to occur. If the carbon nucleus did not have this exact energy state, the triple-α reaction could not take place, and the universe would consist only of hydrogen and helium. We would not be here.

Once you have carbon, the other elements follow more easily. Nitrogen and oxygen are created in the CNO cycle reactions that occur in massive stars. Thus the other major elements needed for life to exist are formed in massive stars. Once the CNO cycle has occurred, you have not only hydrogen and carbon, but also nitrogen and oxygen, the main elements in many biochemical reactions.

The next several elements are formed in subsequent reaction steps, again in massive stars (remember that low-mass stars never get beyond the hydrogen-burning or helium-burning stage). As long as a stellar core still has some helium left, α-capture reactions can take place, adding helium nuclei to lighter elements. Additional intermediate-weight nuclei are formed when the stellar core becomes hot enough for carbon-carbon fusion to occur, followed by oxygen-oxygen fusion, and ultimately silicon burning, in which silicon nuclei merge with each other. Eventually, the core of a massive star produces iron, and

when this is done, the star has reached the end of its rope. By this time virtually all of the elements required for life have been produced, including the minerals and other moderately heavy species that are required in many important biological processes. Life-forms make very little use of elements beyond iron, so as far as life is concerned, perhaps the nucleosynthesis sequence could have stopped once stars made iron.

But this is not so, because in addition to forming the necessary elements, there must be a mechanism for distributing them and making them available for future generations of stars and planets and people. For this we need stellar winds and supernova explosions. Stellar winds from red supergiants release large quantities of CNO products into the interstellar medium. Some remain in gaseous form, and some condense into interstellar particles (we are stardust, remember?). As the galaxy rotates and as stellar explosions stir up the interstellar medium, these stellar by-products become mixed together and spread throughout space, so when a new star and planetary system form, these raw materials for life are already available. In effect, we need supernovae to stir the pot in which life simmers.

Supernova explosions play another key role in addition to helping to spread CNO products throughout the galaxy. They also form new elements in very rapid nuclear fusion reactions that take place during the fiery moments of their outbursts. Most of these are relatively

heavy elements that do not end up playing a major biological role, with one very important exception: the formation of iron. While it is true that iron is created in the core of a massive star, little of this iron escapes into the interstellar medium during a supernova explosion. Instead most of the iron in the core is compressed during the final collapse of the star and converted to neutrons or lost forever into the depths of a black hole. But in layers of the star outside the core, rapid fusion reactions produce more iron, and most of this escapes. Furthermore, Type I supernova, in whch a carbon white dwarf explodes (discussed in this chapter), produces large quantities of iron. Most of the iron in our galaxy and in our bodies comes from a combination of the two types of supernovae, with the majority produced by white dwarf explosions. It is fair to say that our industrial society, from manufacturing plants to automobile scrap yards, is based on the debris of exploding white dwarf stars.

Life—and all of the comforts of our society—came about ultimately because of processes discussed in this chapter. Perhaps you enrolled in this class thinking astronomy has little relevence to your daily life, but now you know that in a very real sense, astronomy is the *explanation* of your daily life.

panion star, causing gas to evaporate from it and escape in a form of stellar wind. In turn, some of the escaping matter falls into the accretion disk surrounding the neutron star, further enhancing its X-ray emission and at the same time helping the neutron star to increase its spin rate. One such system has been found where the companion to the pulsar has lost so much mass that it is nearly gone. This neutron star has been dubbed the "black widow" pulsar, because it is literally eating its companion.

The process of creating a millisecond pulsar is nearly identical to the mechanism for forming an X-ray burster, described earlier. It is possible that

the bursters are neutron stars on their way to becoming millisecond pulsars.

Black Holes: Gravity's Final Victory

In the last chapter, we saw what happens to a massive star at the end of its life: it collapses on itself, and there is no barrier—neither electron degeneracy nor neutron degeneracy—that can stop it. In the eternal battle of gravity against pressure, gravity wins in this case, while losing out to pressure when a white dwarf or neutron star forms.

Of all the exotic objects that can result from the final stages of stellar evolution, a black hole is the most bizarre. It is possible to visualize a white dwarf or a neutron star; these are physical objects, having material substance, and even though they have some very unusual physical properties, at least they are stable objects. But a black hole is a different story. It is inherently unstable in the sense that there is no balance between gravity and pressure, and the severe gravitational distortion of space and time in the near vicinity of a black hole is not easily visualized or understood at an intuitive level. Furthermore, we cannot hope to observe what goes on once the star has shrunk beyond a certain point. For these reasons, black holes are among the most fascinating of objects to scientists and nonscientists alike.

We will start trying to understand the properties of black holes by considering what happens as a massive star collapses. The gravitational effects are described by general relativity, and without the full mathematical framework of that discipline, we will not be able to demonstrate many of the concepts here. Suffice it to say, though, that observations are fully consistent with what the theory says, at least as far as the observations can go.

While we discuss the effects of the incredibly strong gravitational fields near the black hole, it is very important to keep in mind that there are no unusual effects at substantial distances from it. At any point outside the star's original surface, the gravitational field is unchanged by the black hole and acts just as it did when the star was still stable before collapse. There is a common misconception that black holes reach out and suck in material from their surroundings, clearing out large volumes of space around them, but in fact the unusual effects occur only very close to the black hole at points well inside the original star. Planets, spaceships, binary star companions, or other objects would be able to orbit the black hole or pass by it just as they would have when it was an ordinary star, as long as they do not venture in too close.

In order to understand something of what happens close to the collapsing star, we must have some idea of what general relativity has to say about the effects of mass and gravity on space, and on the behavior of photons.

Brief Comments on General Relativity, Gravity, and Photons

Einstein postulated that accelerations caused by changing motion and those caused by gravitational fields are equivalent. This is a basic tenet of general relativity. One consequence is that to an observer being accelerated at a very high rate, photons of light would appear to follow curved paths. The classic "thought experiment" that helps us to see why this is so involves a person inside an enclosed box with no windows **(Fig. 16.17)**. If the box is sitting on the Earth's surface, the person inside "feels" an acceleration of 1 gravity, and his weight feels normal. If that same box were in space and accelerating upward (with respect to the person) at a rate of 1 gravity, the person inside would feel exactly the same. For the person inside the box, there would be no observational or experimental difference between the two situations.

Now think about the person in the box that is being accelerated through space at 1 gravity, and let him have a flashlight. If he turns on the flashlight so that it emits photons perpendicular to the direction of acceleration of the box, the photons will lag be-;hind as the box continues to accelerate. In effect, the photons will fall toward the floor of the box because they no longer follow the motion of the box once they have left the flashlight.

Figure 16.17 General relativity. The person in the spaceship has no experimental or intuitive means of distinguishing whether he is motionless on the surface of the Earth or in space accelerating at a rate equivalent to 1 Earth gravity. The implication of this is that the acceleration due to motion and that due to gravity are equivalent, which in turn implies that space is curved in the presence of a gravitational field.

If we accept the postulate that there is no observational or experimental difference between the box that is accelerating through space at 1 gravity and the one sitting on the Earth's surface, then photons from the flashlight held by the person on the Earth had better also fall toward the floor. Thus gravity deflects photons of light. This was a basic prediction of Einstein's theory, and it has been verified many times by observations and experiments. The first confirmation came in 1919 with the observation that the Sun's gravity deflects light from a distant star seen just adjacent to the Sun's disk, so that the light passed very close to the solar surface on its way to the Earth (this observation had to be made during a total solar eclipse, so that stars near the limb of the Sun could be seen).

In short, photons are affected by gravity, just as particles of matter are. Einstein thought in terms of curvature of space itself as a way to understand how photons could follow curved paths near massive objects. This curvature can be visualized by thinking of an elastic surface on which massive bodies sit: the surface will sag, creating a dimple, at the locations of the objects **(Fig. 16.18)**. If this surface represents the curvature of space near massive objects, then you can visualize how photons would be forced to curve if they came near the dimples created by massive objects.

Photons near a Collapsing Star

Not only are photon paths curved near massive objects, but photons escaping from a massive object lose energy in the process and are therefore redshifted in wavelength. This effect is very small for normal stars (although the minute gravitational red-

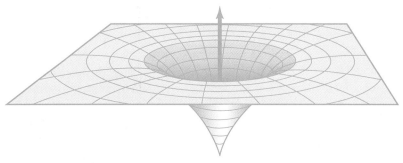

Figure 16.18 Geometry of space near a black hole. Einstein's theory of general relativity may be interpreted in terms of a curvature of space in the presence of a gravitational field. Here we see a representation of how this curvature varies near a black hole (this is only a representation, not a realistic illustration, because the curvature of space near a black hole is actually a distortion of three-dimensional space, not two-dimensional space as shown here).

shift of photons leaving the Sun's surface has been measured), but it can become very large if the gravitational field is sufficiently strong. Earlier in this chapter we discussed the gravitational redshifts of photons escaping from white dwarfs; now we will learn about the effect on photons leaving the surface of a very massive star that is collapsing to a very small size.

Let us consider a photon emitted from the surface of a collapsing star **(Fig. 16.19)**. If the photon is emitted at any angle away from the vertical, its path will be bent over farther. If the gravitational field is strong enough, the photon's path may be bent over so far that the photon falls back onto the stellar surface. Photons emitted straight upward follow a straight path, but they lose energy to the gravitational field, causing their wavelengths to be redshifted (those emitted at an angle will also lose some

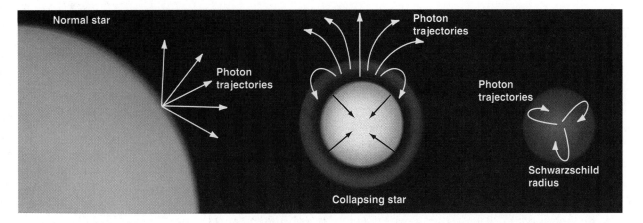

Figure 16.19 Photon trajectories from a collapsing star. Light escapes in essentially straight lines in all directions from a normal star (left), whose gravitational field is not sufficient to cause large deflections. At an intermediate stage of collapse (center), photons emitted in a cone nearly perpendicular to the surface can escape, while others cannot. Those emitted at just the right angle go into orbit around the star, while those emitted at greater angles fall back onto the stellar surface. After collapse has proceeded to within the Schwarzschild radius, no photons can escape.

Table 16.5
Schwarzschild Radii for Various Objects

Object	Schwarzschild Radius
Sun	3 km
Earth	0.9 cm
150 lb person	1×10^{-23} cm
Jupiter	2.9 m
Star of 50 $M_\odot$	150 km
Typical globular cluster	5×10^9 km
Nucleus of a galaxy	$10^6 - 10^8$ km ($10^{-7} - 10^{-5}$ pc)
Massive cluster of galaxies	10^{15} km (100 pc)
The universe	10^{26} km (10^{13} pc)

energy and will therefore be redshifted a bit). When the gravitational field has become strong enough, even a photon emitted straight up loses all of its energy and cannot escape.

The radius of the star at the time when its gravitational field becomes strong enough to trap photons is called the **Schwarzschild radius,** after the German astrophysicist who first calculated its properties some 60 years ago. The Schwarzschild radius depends only on the mass of the star that has collapsed. One way to see this is to consider that the photons cannot escape because the escape speed exceeds the speed of light (this is technically not a perfect representation of what happens, but it leads to the correct result). From Chapter 5 we know that the escape speed for a particle traveling upward from a body of mass M and radius R is

$$v_{esc} = \sqrt{\frac{2GM}{R}}.$$

If we set this equal to c, the speed of light, and solve for R, we get

$$R = \frac{2GM}{c^2}.$$

This is the expression for the Schwarzschild radius. Note that it depends only on the mass of the star; all other terms are constants. For a star of 10 solar masses, the value comes out to 30 km. A star of twice that mass would have twice the Schwarzschild radius, and so on **(Table 16.5).** It is easy to calculate the Schwarzschild radius of a star by remembering that it is 3 km times the mass in solar units. For example, a 20-solar-mass

star has a Schwarzschild radius of 20×3 km = 60 km. Every object has a Schwarzschild radius; for the Sun, it is, of course, 3 km, and for the Earth, it is about 9 mm. We normally do not think of such low-mass objects becoming black holes, but they could, if they were somehow compressed to a small enough size.

Properties of a Black Hole

When a collapsing star has shrunk inside its Schwarzschild radius, it is said to have crossed its **event horizon,** because an outside observer cannot see it or anything that happens to it after that point. We can have no hope of ever seeing what happens inside the event horizon, but since no force is known that could stop the collapse, we assume that it continues. The mass approaches a state of **singularity,** meaning that it is concentrated in an infinitesimally small point.

A singularity is a mathematical concept, not a physical one, and for the time being, science cannot describe the state of the former star's matter. One reason is that the fields of quantum mechanics and gravity have not yet been merged; that is, the behavior of gravity on the smallest size scales is not yet understood. But there has been no lack of speculation about the behavior of matter inside a black hole. One of the more popular concepts is the **wormhole,** a passage through spacetime to some other location (perhaps in another universe); at the other end of a wormhole is a **white hole,** a location where matter that entered the black hole reappears. These ideas have formed the central theme of many a science fiction story. Because they can be described mathematically, scientists do not rule out the possibility that such phenomena exist.

Although what happens to the matter that falls into a singularity is a subject for speculation, we can be more concrete about what happens just outside the event horizon. From the outsider's point of view, the rate of time is relativistically distorted by the extreme gravitational field and acceleration in such a way that the collapse would seem to slow gradually, taking an infinite time to reach the event horizon. This slowdown, however, is most significant in the last moments before the star's disappearance, and by then the star would be essentially invisible anyway, as most escaping photons would be redshifted into the infrared or beyond. The star would seem to disappear rather quickly, despite the stretching of the collapse in time caused by relativistic effects. To an observer unfortunate enough to be falling into the black hole, the fall would go very quickly; the rate of time would not be distorted by relativistic effects in the same way it would for the distant observer. The infalling observer would suffer serious discomfort, however, and would not survive even to the point of crossing the event horizon. Tidal forces would be immense as the observer fell in near the black hole and soon would be strong enough to tear any material substance apart (science fiction stories about spaceships using black holes for transportation between universes seem to neglect this problem).

Mathematically, the external properties of a black hole can be described completely by three quantities: its mass, its electrical charge, and its spin. The mass, of course, is determined by the amount of matter that collapsed to form the hole, plus any additional material that may have fallen in later. The electrical charge, similarly, depends on the charge of the material from which the black hole formed; if the material contained more protons than electrons, for example, the black hole would end up with a net positive charge. Because particles with opposite charges attract each other, and those with like charges repel, it is thought that electrical forces would maintain a fairly even mixture of particles during and after the formation of the black hole, so that the overall charge would be nearly zero. To illustrate this, imagine that a black hole was formed with a net negative charge and that afterward it was surrounded by ion-ized gas (as it likely would be, with some of the matter from the original star still drifting inward). Then the negative charge of the hole would repel addi-tional electrons, preventing them from falling in, while protons would be accelerated inward. In a short time, enough protons would be gobbled up to neutralize the negative charge.

The spin of a black hole is not so easily dismissed, however. It stands to reason that if the star were spinning before its collapse—and most likely it would have been—then conservation of angular momentum would cause the rotation to speed up greatly as the star shrank. A high spin rate actually shrinks the event horizon, allowing an outside observer to see closer in to the singularity residing at the center. It is even possible mathematically—although it presents a physical dilemma—to have suf-ficient spin that there is no event horizon and what-ever is in the center is exposed to view. A **naked singularity,** as this has been dubbed, would not pro-duce the usual gravitational effects of a black hole, and it would be possible to blunder into one without any forewarning (so perhaps the space-faring science fiction hero should use naked singularities for trans-portation—but how to find them?).

For the most part, it is assumed that the spin rate is never so large that a naked singularity can form, and, in fact, black hole properties are usually speci-fied by mass alone, neglecting the effects of charge and spin. Thus the assumption is that our main hope

of detecting a black hole is by its gravitational effects, determined entirely by the mass.

Before turning to a discussion of how to find a black hole, we should mention that some black holes may be formed by processes other than the collapse of individual massive stars. These include supermas-sive black holes, which apparently exist in the cores of some galaxies, where they formed when thousands or millions of stellar masses coalesced in the center. These other possibilities will be discussed later; for now, we turn to the hunt for stellar black holes.

Do Black Holes Exist?

The mass that goes into a black hole during its for-mation still exists there, hiding inside the event hori-zon. Even though no light can escape, the gravita-tional effects persist. The gravitational force of the star is exactly the same as it was before the collapse, except at points so close that they would have been inside the original star.

Our best chance of detecting a black hole, then, is to search for an invisible object whose mass is too great and whose volume is too small for it to be any-thing else. Even if we do find such a thing, the evi-dence is really only circumstantial, because our con-clusion that the object is a black hole relies on the theory that says neither a white dwarf nor a neutron star can survive if the mass is sufficiently great.

The best method we have for deter-mining the mass of an object is when it is in orbit around a companion star, where Kepler's third law can be applied to find the masses. The search for stellar black holes, then, leads us to examine binary sys-tems, looking for invisible, but massive, objects.

This search has been facilitated by another property of black holes: if they find new material to pull into themselves, under most circumstances (i.e., in mass-exchange binaries) the trapped matter forms an accretion disk, just as we described in the case of a neutron star in a mass-transfer binary. We have already seen that such a disk becomes so hot that it emits X rays. Thus some of the binary X-ray sources probably contain black holes rather than neu-tron stars. The best way to distinguish between the two possible types of remnants in these systems is to determine the mass of the invisible companion, to see whether it is too great to be a neutron star.

The determination of the masses in a binary sys-tem is difficult, if not impossible, if only one of the two stars can be seen, because the information on

WEB activity

Black Holes and Neutron Stars
In this web activity we will take a 'Virtual Trip' to study Black Holes and Neutron Stars. We will examine scien-tifically accurate simulations of what it would look like to travel to a black hole or a neutron star. Our starting point is the following URL:
http://universe.colorado.e du/ch16/web.html

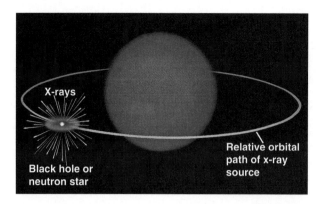

Figure 16.20 Eclipses of an X-ray source in a binary system. Because the separation of the two stars in a mass-transfer system is always very small, and because the mass-losing star is usually very large (a red giant or a hot giant or supergiant with a stellar wind), eclipses occur easily. If the mass-receiving star is a neutron star or black hole, and therefore an X-ray emitter, the X rays are likely to be eclipsed periodically by the companion. Hence many such systems are eclipsing X-ray binaries.

Figure 16.21 Cygnus X-1. This X-ray image obtained by the orbiting *Einstein Observatory* shows the intense source of X rays at the location of a dim, hot star that is thought to be a normal star that is losing mass to its invisible black hole companion. (Harvard-Smithsonian Center for Astrophysics)

http://universe.colorado.edu/fig/16-21.html

their speeds and on the inclination angle of the orbit is incomplete. The fact that many X-ray binaries are eclipsing systems (at least the X-ray source is eclipsed; **Fig. 16.20**) helps to determine the tilt of the orbit in some cases. The orbital period is usually easily determined from the eclipse frequency as well, but the orbital velocities, needed to deduce the sizes of the orbits, are often very difficult or impossible to measure. The unseen star, of course, emits little or no light, so there is no hope of measuring Doppler shifts in its spectral lines. The normal companion star is often a hot giant or supergiant, and these stars usually have very broad spectral lines whose shifts cannot be measured accurately. Nevertheless, sometimes enough information can be derived to at least place limits on the mass of the invisible companion. In one such case, the X-ray binary called Cygnus X-1 **(Fig. 16.21)**, the collapsed object appears to contain at least 8 solar masses. This object is one of several leading black hole candidates (see Table 16.4).

Although definitive proof for stellar black holes is still lacking, the circumstantial evidence in favor of their existence is strong. Many of the candidates listed in Table 16.4 are now considered to be airtight cases. It has become more complex and difficult to find other explanations for them (such as assuming there are other, dim stars in the system to help explain the observed orbital motions) than to accept that they are black holes. To believe that black holes do not form, one has to invent some way of preventing collapse and also some way of explaining the observational evidence for invisible massive objects in some binary systems. It is far less complicated to accept that black holes do form and exist in these binaries, because doing so requires the fewest unproven assumptions. The principle called Occam's razor, a guiding philosophy for scientists, states that the explanation of any phenomenon that requires the fewest arbitrary assumptions is most likely to be the correct one.

Most astronomers today have adopted the concept of black holes and not only believe that they exist, but consider them an integral part of our universe, playing numerous important roles in the evolution of stars and of the universe.

Summary

1. A white dwarf gradually cools off, taking billions of years to become a cold cinder.
2. If new matter falls onto the surface of a white dwarf—for example, in a binary system in which the companion star loses mass—then the white

dwarf may become a cataclysmic variable if the matter arrives in small amounts with high energy, or it may become a nova if matter builds up slowly to the point where a larger quantity is involved in the explosion. If a star exceeds the white dwarf mass limit, it may become a neutron star or black hole, or it may explode as a Type I supernova.

3. Massive stars are likely to explode as Type II supernovae when all possible nuclear reaction stages have ceased. The supernova explosion creates an expanding cloud of hot, chemically enriched gas known as a supernova remnant.

4. In some cases, a remnant of less than 2 to 3 solar masses is left behind in the form of a neutron star, consisting of degenerate neutron gas.

5. A neutron star is too dim to be seen directly in most cases but may be observed as a pulsar (depending on the alignment of its magnetic and rotation axes, and our line of sight), and in a close binary system, it may become a source of X-ray emission.

6. Some neutron stars that receive new material in clumps flare up occasionally as X-ray sources called bursters.

7. If the final core mass of a star exceeds 2 or 3 solar masses, it will become a black hole at the end of its nuclear reaction lifetime.

8. The immensely strong gravitational field near a black hole traps photons of light, rendering the black hole invisible.

9. A black hole may be detected by its gravitational influence on a binary companion or by the X rays it emits if new matter falls in, as in a close binary system in which mass transfer takes place. A black hole binary X-ray source can be detected only by analysis of the orbits to determine the mass of the unseen object.

Review Questions

1. Why is the final mass of a star, rather than the initial mass, the factor that determines what form of remnant the star will leave? How can the final mass be different from the initial mass?

2. Explain why white dwarfs, with no source of energy, can stay hot for as long as a billion years.

3. Describe the differences between a nova, a Type I supernova, and a Type II supernova.

4. Compare the properties of a white dwarf with those of a neutron star. Include a discussion of the form of pressure that counteracts gravity in each object.

5. How are neutron stars detected from the Earth?

6. Explain why the pulses from pulsars were finally judged to be the result of rapid rotation of neutron stars, rather than rapid expansion and contraction or rotation of some other type of star. What role did the Crab nebula and its pulsar play in this discovery?

7. Both neutron stars and black holes can be X-ray sources in binary systems where mass exchange occurs. How can astronomers tell which kind of remnant is present in a given binary X-ray source?

8. Explain why some massive stars become black holes, while others become neutron stars or blow up completely, leaving no remnant.

9. Why would a neutron star or a black hole be expected to have a very rapid rotation rate?

10. Summarize the role of mass transfer in binary systems in both the formation and the detection of stellar remnants.

Problems

1. Suppose mass is added to the surface of a white dwarf and undergoes nuclear reactions, creating a nova as it releases energy according to $E = mc^2$. How much energy is released if the mass that undergoes reactions is 10^{-4} solar masses? If the reacting mass in a Type I supernova is 1 solar mass, how much energy does it release? (In both cases, assume that 1% of the reacting mass is converted into energy.)

2. Compare the average density, the surface gravity, and the escape speed for a neutron star (mass of 2 solar masses and radius 10^{-4} times the Sun's radius) with the same properties of the Sun.

3. If the rotation period of the Crab pulsar is 0.033 sec and its radius is 10 km, what is the speed of a point on its surface at the equator? What fraction of the speed of light is this?

4. Apart from the effects created by its particle beams, a neutron star also emits radiation from its surface because of its high temperature (i.e., it emits thermal radiation, as discussed in Chapter 6). Suppose a neutron star has a surface temperature of 1 million degrees. What is its wavelength of maximum emission? What kind of telescope would be needed to try to detect this radiation?

5. The surface area of a neutron star is so small that it has a very low luminosity despite its high temperature. To show this, calculate the luminosity for a neutron star with surface temperature of 10^6 K and radius of 10 km, and compare your answer

Black Hole or Neutron Star?

ASTRONOMICAL

ACTIVITY

Suppose an eclipsing X-ray binary is discovered. Observations of its X-ray intensity produce a "light curve" as shown in the accompanying figure. The visible star (the companion to the X-ray source) is found to have spectral type B5V. Observations of its radial velocity produce the velocity curve shown in the figure. From this information and a few assumptions (described below), you can determine the masses of the two stars and decide whether the X-ray source is a black hole or a neutron star.

To do this, you may need to review some of the discussion of binary stars in Chapter 14. You will need to use Kepler's third law in the form $(m_1 + m_2)P^2 = a^3$, where m_1 and m_2 are the masses of the two stars in units of the Sun's mass, P is the orbital period in years, and a is the semimajor axis of the orbit in AU.

To apply Kepler's third law, you need to determine the period P and the semimajor axis a, and you must convert the period to units of years and the semimajor axis to units of AU. You can find the period in days from either the light curve or the velocity curve; to convert to years, divide the period in days by 365.25.

To find a, you will have to use the information given on the orbital speed of the visible star, along with its period. The velocity curve shows how the observed radial velocity of the star varies as the star orbits the center of mass. The radial velocity reaches its extreme values when the star is moving directly toward or away from the Earth; the greatest positive value of the radial velocity occurs when the star is moving directly away from the Earth, and the greatest negative value of the radial velocity occurs when the star is moving directly toward the Earth. Thus, by finding these extreme values for radial velocity, you will find v, the orbital speed of the star.

The star travels a distance equal to the circumference of its orbit in a time equal to its orbital period. Since distance equals speed times time, you can write the equation

$$2\pi a = vP,$$

where $2\pi a$ is the circumference of the circular orbit having radius a, and vP is the distance the star travels in time P at speed v. Now you can solve this equation for a and substitute the values for v (from the velocity curve) and P (from the light curve—but you must convert P to seconds, since v is in units of km/sec). The value of a that you find will be

in units of kilometers, but for Kepler's third law you need a in AU. To convert, divide your value in kilometers by the number of kilometers in one AU (see Appendix 2).

Once you have both a in AU and P in years, you can solve Kepler's third law for the sum of the masses of the two stars. Next, estimate the mass of the B5V star using standard tables of mass versus spectral type (see Table 14.4), and subtract. The remainder is the mass of the companion object, which is the source of the X rays. Finally, use the information given in this chapter to decide whether the mass you found falls into the range for neutron stars, or whether the object is so massive that it must be a black hole.

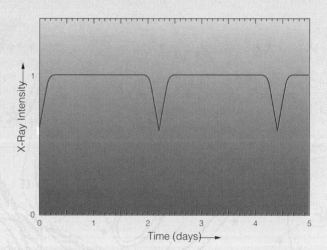

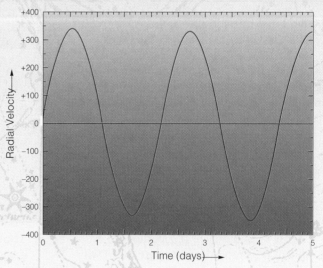

with the luminosity of the Sun. How much more energy per square meter of surface area does the neutron star emit?

6. Calculate the Schwarzschild radius for (a) the Sun; (b) a star of 20 solar masses; and (c) the Milky Way galaxy (mass approximately 10^{42} kg).

7. Suppose an X-ray binary system is found in which the visible star is a red giant of a spectral type thought to have a mass of about 12 solar masses. The orbital period is 3.65 days, and the semimajor axis is 0.12 AU. Calculate the sum of the masses from Kepler's third law, then try to decide whether the compact companion star is a neutron star or a black hole.

Additional Readings

Abramowicz, M. A. 1993. Black Holes and the Centrifugal Force Paradox. *Scientific American* 268(3):74.

Backer, D. C. and Kulkarni, S. 1990. A New Class of Pulsars. *Physics Today* 43(3):26.

Bailyn, C. 1991. Problems with Pulsars. *Mercury* 20(2):55.

Bernstein, J. 1996. The Reluctant Father of Black Holes. *Scientific American* 274(6):80.

Cannizzo, J. K. and R. H. Kaitchuk, 1992. Accretion Disks in Interacting Binary Systems. *Scientific American* 266(1):92.

Charles, P. A. and R. M. Wagner 1996. Black Holes in Binary Stars: Weighing the Evidence. *Sky & Telescope* 91(5):38.

Crosswell, K. 1992. The Best Black Hole in the Galaxy. *Astronomy* 20(3):30.

Graham-Smith, F. 1990. Pulsars Today. *Sky & Telescope* 80(3):240.

Kaler, J. 1991 The Smallest Stars in the Universe. *Astronomy* 19(11):50.

Kaler, J. 1994. Stellar Oddballs. *Astronomy* 22(9):50.

Kaspi, V. M. 1995. Millisecond Pulsars: Timekeepers of the Cosmos. *Sky & Telescope* 89(4):18.

Nather, R. E. and D. E. Winget. 1992. Taking the Pulse of White Dwarfs. *Sky & Telescope* 83(3):374.

Overbye, D. 1991. God's Turnstile: The Work of John Wheeler and Stephen Hawking. *Mercury* 20(4):98.

Piran, T. 1995. Binary Neutron Stars. *Scientific American* 272(5):52.

Starrfield, S. and S. N. Shore 1994. Nova Cygnis 1992: The Nova of the Century. *Sky & Telescope* 87(2):20.

Starrfield, S. and S. N. Shore 1995. The Birth and Death of Nova V1974 Cygni. *Scientific American* 272(1):76.

Terrell, D. 1992. Demon Variables. *Astronomy* 20(10):34.

Trimble, V. and S. Parker 1995. Meet the Milky Way. *Sky & Telescope* 89(1):26.

Van den Heuvel, E. P. J. and J. van Paradijs 1993. X-Ray Binaries. *Scientific American* 269(5):64.

Webbink, R. F. 1989. Cataclysmic Variable Stars. *American Scientist* 77:248.

Wheeler, J. C. 1996. Wormholes and Spacetime Paradoxes. *Astronomy* 23(2):52.

Web Connections

The Review Questions and Problems also appear at the following URLs:
http://universe.colorado.edu/ch16/questions.html
http://universe.colorado.edu/ch16/problems.html

*Along the Milky Way
stand the places of the illustrious gods;
the common people of the skies lie apart
on either side.*

Bulfinch's *Mythology*

Chapter 17
The Milky Way Galaxy

Chapter Web site: http://universe.colorado.edu/ch17

e have discussed stars as individuals, as though they exist in a vacuum, isolated from the rest of the universe. For the purposes of analyzing the structure and evolution of stars, this is a suitable approach, but if we want to understand the full stellar ecology, the properties of individuals must be discussed in the context of the larger environment.

Even a casual glance at the nighttime sky shows that the stars tend to be grouped rather than randomly distributed. The most extensive concentration is the Milky Way **(Fig. 17.1),** a diffuse band of light stretching from horizon to horizon, but clearly visible only in areas well away from city lights. To the ancients the Milky Way was merely a cloud. Galileo, with his primitive telescope, recognized it as a region with a great concentration of stars. In modern times we know the Milky Way as a **galaxy**—a great pinwheel of billions of stars to which the Sun belongs **(Table 17.1).** The hazy streak across our sky is a cross-sectional, or edge-on, view of the galaxy from our vantage point within its disk.

Figure 17.1 The Milky Way. This illustration (top) shows the cross-sectional view of our galaxy that we see from Earth. The bottom image is an infrared view of the Milky Way, obtained by the IRAS satellite. (Top: Lund Observatory; bottom: NASA)

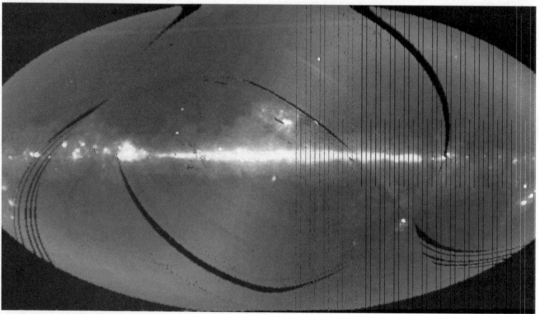

Table 17.1
Properties of the Milky Way

Mass: 1.0×10^{11} M$_\odot$ (interior to Sun's position)
Diameter of disk: 30 kpc
Diameter of central bulge: 10 kpc
Sun's distance from center: 8.5 kpc
Diameter of halo: 100 kpc (very uncertain)
Thickness of disk: 1 kpc (at Sun's position)
Number of stars: 4×10^{11}
Typical density of stars: 0.1 stars/pc^3 (solar neighborhood)
Average density of interstellar matter: 10^{-24} g/cm^3 (roughly equal to one hydrogen atom/cm^3)
Luminosity: 2×10^{10} L$_\odot$
Absolute magnitude: -20.5
Orbital period at Sun's distance: 2.4×10^8 years

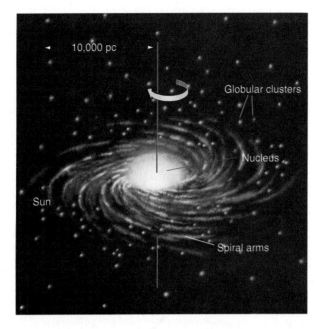

Figure 17.2 The structure of our galaxy. This sketch illustrates the modern view of the Milky Way.

The overall structure of the Milky Way resembles a phonograph record with a large central bulge, the center of which is called the **nucleus (Figs. 17.2** and **17.3).** Nearly all the visible light is emitted by stars in the plane of the **disk,** although the galaxy also has a **halo,** a distribution of stars and star clusters centered on the nucleus but extending well above and below the disk. The most prominent objects in the halo are the **globular clusters**—very old, dense clusters of stars characterized by their distinctly spherical shape.

To describe the size of the Milky Way, we must use a new unit of distance. In Chapter 14 we discussed stellar distances in terms of parsecs and found that the nearest star is about 1.3 parsecs (pc) from the Sun. When we expand to the scale of the galaxy, we speak in terms of **kiloparsecs (kpc),** or thousands of parsecs. The visible disk of the Milky Way is roughly 30 kpc in diameter, and the disk is a few hundred parsecs thick. Light from one edge of the galaxy takes about 100,000 years to travel across to the far edge.

As we can see with a little thought, even as vast and diverse a system as a galaxy must evolve with time. We have learned that stars change, as they use up nuclear fuel in their lifelong battle against gravity. All stars create new elements in their cores, forming first helium and then successively heavier elements. Most stars have relatively low masses and keep their new material to themselves as they finish their lives and become white dwarfs. But the most massive stars are extravagant spendthrifts, spewing out significant portions of their matter in stellar winds and life-ending explosions. As it happens, these stars are also the ones that form the greatest variety of elements during their nuclear-burning lifetimes. So the most massive stars, rare though they are, have the greatest

Figure 17.3 A spiral galaxy similar to the Milky Way. This galaxy (NGC 2997) probably resembles our own, as seen from afar. (© 1980 AAT Board; photo by D. Malin)
http://universe.colorado.edu/fig/17-3.html

impact on galactic evolution, gradually enriching the galaxy with enhanced abundances of all the elements, including the heaviest ones.

The galaxy, therefore, must change. As its interstellar gas is enriched with heavy elements, the raw materials for new generations of stars are altered. Stars formed today (or even at the time of the Sun's formation) start life with a greater abundance of heavy elements than did the first generation of stars in the galaxy's distant past. In time, this enrichment will become greater and greater, thus changing the nature of the stars and of the galaxy itself.

Other changes have taken place as well, and of course the galaxy itself had to form somehow in the first place. We will discuss the formation and evolution of the galaxy later in the chapter, but for now we need to keep in mind that the galaxy is an evolving entity, and that there is a rich interplay between stars and the interstellar medium from which they form and to which they return processed material as they live and die.

A New Scale of Distances

Our galaxy is so large that we must discuss new methods of measuring distance. We cannot rely on stellar parallax for measuring distances beyond a few hundred parsecs, and the method of spectroscopic parallax, in which a star's luminosity is estimated on the basis of its spectral type (see Chapter 14), is both uncertain and limited to relatively nearby stars (up to

a few thousand parsecs). Here we will discuss two additional methods for finding distances, one of them related to spectroscopic parallax and the other one quite different.

Main-Sequence Fitting

In Chapter 15 we learned that plotting H-R diagrams of clusters of stars can be very useful, particularly for studying how stars of different masses evolve. Another advantage derives from the fact that the stars in a cluster all lie at approximately the same distance from us. A typical galactic cluster may be only a few parsecs in diameter, even though it is many hundreds or thousands of parsecs away. Hence the differences in apparent brightness among stars within a cluster reflect real differences in luminosity.

If we can determine the luminosity (or absolute magnitude) of an object, we can always find its distance by measuring its apparent brightness (or magnitude), because we know that the inverse square law dictates how far away an object of a given luminosity must be to appear as faint as it is. We used this principle in the method of spectroscopic parallax (Chapter 14), and we can use it again when comparing cluster H-R diagrams.

When the H-R diagram is plotted for a cluster, it is normally done with apparent magnitude, instead of absolute magnitude, on the vertical axis **(Fig. 17.4)**, because the absolute magnitudes of the stars are usually unknown. Using apparent magnitude preserves the correct distribution of stars in the diagram and shows the shape and extent of the main sequence,

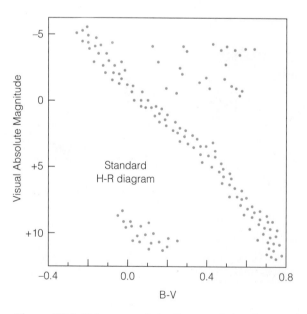

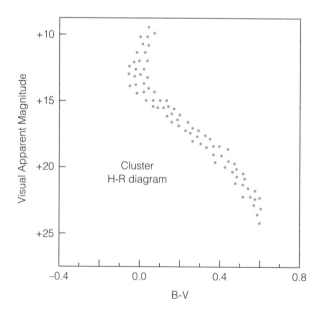

Figure 17.4 Main-sequence fitting. This is a standard H-R diagram, with a cluster H-R diagram next to it using apparent magnitude on the vertical axis. A comparison of the two diagrams leads to an estimate of the distance to the cluster.

since this depends only on the *relative* magnitudes of the stars within the cluster. It is possible to determine the absolute magnitude scale for the cluster H-R diagram by fitting the main sequence to the "standard" H-R diagram. Figure 17.4 illustrates this process by showing a cluster H-R diagram, with an apparent magnitude scale, next to a standard diagram having an absolute magnitude scale. By comparing these two diagrams, you can see that if the cluster diagram is placed on top of the standard one, the two magnitude scales would be superimposed, and you could determine the absolute magnitude scale for the cluster diagram. In the process, you would find the difference between apparent and absolute magnitudes for the stars in the cluster; that is, you would find the distance modulus for the cluster (see Chapter 14). The distance modulus is directly related to the distance, so you would now know how far away the cluster is.

This method of finding distances is called **main sequence fitting.** It is identical to the stellar parallax method in most respects; the only difference is that instead of using spectral types of individual stars to estimate absolute magnitudes, here you use the collective positions of stars in the cluster H-R diagram to determine the absolute magnitude scale. Main-sequence fitting is more accurate and more powerful than spectroscopic parallax for two reasons. For one thing, because main-sequence fitting is based on the collective behavior of a number of stars, it eliminates uncertainties due to possible individual peculiarities. In addition, it can be applied to greater distances due to the greater ease with which distant clusters, as opposed to individual stars, can be identified and observed.

Variable Stars and the Period-Luminosity Relation

We have stressed that in any situation where we can find both the apparent and the absolute magnitude of an object, we can find its distance. Both the spectroscopic parallax and cluster main-sequence fitting techniques depend on this concept. Here we discuss another way of estimating luminosities, and hence distances, to certain types of stars. In Chapter 14 we briefly mentioned variable stars, including those that pulsate regularly **(Table 17.2).** These stars physically expand and contract, changing in brightness as they do so. One of the first such stars to be discovered was δ Cephei, which is sufficiently bright to be seen with the unaided eye. Following the discovery of δ Cephei in the mid-1700s, other similar stars were found, and these as a class became known as **Cepheid variables,** named after the prototype. Another type of pulsating variable, the **RR Lyrae stars,** were found to exist primarily in globular clusters and to have pulsation periods of less than a day, whereas the Cepheids have periods ranging from 1 to 100 days.

In 1912, through observations of variable stars in the Magellanic Clouds (two small, relatively nearby galaxies), American astronomer Henrietta Leavitt discovered that there is a definite relationship between the pulsation period of these stars and their luminosities. For the Cepheids, the period increases with increasing luminosity (or decreasing absolute magnitude), while the RR Lyrae stars all have about the same luminosity. This correlation, called the **period-luminosity relation,** has a profound implication: When a star is recognized as a variable belonging to one of these classes, its absolute magnitude

Table 17.2
Types of Pulsating Variables

Type of Variable	Spectral Type	Absolute Period (Days)	Change of Magnitude	Magnitude
δ Cephei (Type I)	F, G supergiants	3–50	−2 to −6	0.1–2.0
W Virginis (Type II Cepheids)	F, G supergiants	5–30	0 to −4	0.1–2.0
RR Lyrae	A, F giants	0.4–1	0.5 to 1.2	0.6–1.3
RV Tauri	G, K giants	30–150	−2 to −3	Up to 3
δ Scuti	F giants	0.1–0.2	2	0.1
β Cephei (β Canis Majoris)	B giants	0.1–0.2	−3 to −5	0.1
Long-period variables (Mira)	M supergiants	80–600	+2 to −3	3–7

can be determined simply by measuring its period and using the established period-luminosity relation **(Fig. 17.5)**. Once the absolute magnitude is known, the distance can be found by comparing the absolute and apparent magnitudes. (Average values must be used, since the magnitudes vary with the pulsations.)

Because Cepheids are giant stars, they are very luminous and can be observed at great distances. This makes them very powerful tools for measuring distances beyond those reached by other techniques we have discussed.

The Properties of the Galaxy

Finding out what sort of galaxy we live in has proved to be a difficult task, one that is not completed yet. We see only the view from the interior, with no possibility of getting outside and seeing the external appearance. It has taken centuries to reach the understanding of the Milky Way's structure that was outlined in the introduction to this chapter, and even today we find large uncertainties and many surprises as new bits of information are gleaned laboriously from indirect evidence. The story of the discovery of the galaxy is an interesting one.

Discovering the Galaxy: Its Size and Shape and the Location of the Sun

The idea that our Sun belongs to a disk-shaped collection of vast numbers of stars dates all the way back to 1734, when the Englishman Thomas Wright proposed that the Sun and other stars orbited a central point, and that the stars were arranged in either a

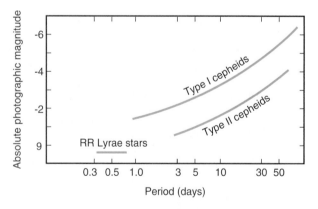

Figure 17.5 The period-luminosity relation for variable stars. This diagram shows how the pulsation periods for Cepheid and RR Lyrae variables are related to their absolute magnitudes. Initially, astronomers did not recognize that there are two types of Cepheids with somewhat different relationships, and this led to some early confusion about distance scales.

shell or a ring (later interpreted by the influential German philosopher Immanuel Kant as a disk) rotating about a common center. Soon thereafter the German Johann Lambert published his theory that the Sun and its neighbors belonged to a lens-shaped system, which was part of a hierarchy of ever-larger such systems. Lambert's work influenced the English astronomer William Herschel, who was well known for his discovery of Uranus and his advances in telescope making and the study of stars in general. Herschel studied the problem of the Milky Way's structure and concluded that it is a disklike system including the Sun, but later, due to the many irregularities in the distribution of stars, he became less certain about this conclusion.

Even though the idea of a disk-shaped system became accepted early on, many significant questions remained. Because the solar system is located within the disk, we have no easy way of getting a clear view of where we are in relation to the rest of the galaxy. All we see is a band of stars across the sky, which tells us that we are near the plane of the disk. It is not so easy to determine where we are with respect to the edge and center of the disk or to determine the size and the detailed shape of the disk.

Early in this century no one knew to what extent our galaxy is permeated with an interstellar medium of gas and dust in interstellar space, so the extent to which interstellar obscuration affects our view of distant stars was unknown. The possibility that our view is affected by interstellar material was considered, but because there was no direct evidence that the effect was significant, it was largely ignored. Among those who derived the shape and size of the galaxy and our place within it, the most influential was the Dutch astronomer Jacobus Kapteyn, who in the early years of this century used counts of stars of differing magnitude as a means of tracing the galaxy. Kapteyn found that the density of stars in space appeared to fall off with distance in all directions, though more slowly in the plane of the disk. From this he was able to develop a map of the galaxy that showed it to be a fat disk with the Sun very near the center.

But other measurement techniques soon showed that this picture was not correct. One of these techniques, which was employed by Harlow Shapley **(Fig. 17.6)**, made use of the globular clusters, the spherical star clusters found outside the confines of the galactic disk. These clusters tend to contain RR Lyrae variables, so Shapley could determine their distances and hence their locations with respect to the Sun and the disk of the Milky Way. He found that the globular clusters are distributed in a spherical arrangement centered on a point several thousand parsecs from the Sun **(Fig. 17.7)**, and he argued that this point must

Figure 17.6 Harlow Shapley. Shapley's work on the distances to globular clusters was a key step in the determination of the size of the Milky Way. Shapely also played a prominent role in the discovery of galaxies outside our own. Much of his observational work was done before 1920, when he was a staff member at the Mount Wilson Observatory. He later became director of the Harvard College Observatory. (Harvard College Observatory) http://universe.colorado.edu/fig/17-6.html

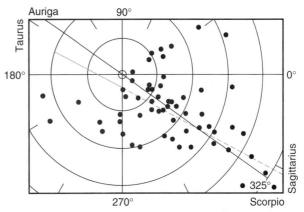

Figure 17.7 Shapley's measurements of globular clusters. This is one of Shapley's original figures illustrating the distribution of globular clusters in the galaxy. This is a projection down onto the galactic plane. The Sun is at the point where the straight lines intersect, and each circle centered on that point represents an increase in distance of 10,000 pc. Note that the distribution of globulars is centered some 10 to 20 kpc from the Sun. (Estate of H. Shapley, reprinted with permission.)

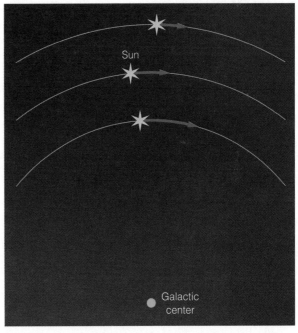

Figure 17.8 Stellar motions near the Sun. Stars just inside the Sun's orbit move faster than the Sun, whereas those farther out move more slowly. Analysis of the relative speeds (as inferred from measurements of Doppler shifts) and distances of stars like these led to the realization that the Sun and stars near it are orbiting a distant galactic center.

represent the center of the galaxy. It would not make physical sense for the globular clusters to be concentrated around any location other than the center of the entire galactic system.

Shapley's conclusion, first published in 1917, was not widely accepted initially, but other supporting evidence was found in the 1920s. Two scientists, Jan Oort of Holland (already mentioned in connection with the origin of comets; see Chapter 11) and Bertil Lindblad of Sweden, carried out careful studies of the motions of stars in the vicinity of the Sun. They found that these motions could best be understood if the Sun and the stars around it were assumed to be orbiting a distant point; that is, they found systematic, small velocity differences between stars, similar to those between runners on a track who are in the inside and outside lanes **(Fig. 17.8).** It appeared from these studies that the Sun is following a more-or-less circular path about a point several thousand parsecs away, which indicates that the center of the galaxy is located at that distant center of rotation.

This supported Shapley's view of the galaxy, although the Sun's distance from the center was still uncertain.

In 1930, by finding that distant star clusters appeared farther away than they should, the Swiss-born American astronomer Robert Trumpler was able to show that the galaxy is permeated with a medium of diffuse dust, which makes all stars appear fainter than they should. In other words, because of the dust, star brightnesses fall off with distance faster than expected from the inverse square law. The discovery of the general interstellar dust medium laid to rest the apparent conflict between the findings of Kapteyn and those of Shapley, Oort, and Lindblad, and the only major question remaining was the true size of the galaxy. Determining this required refinement of the period-luminosity relation for variable stars, which was confused for awhile by the initial failure to recognize that there are two types of Cepheids (see Fig. 17.5).

Eventually, a consensus was reached that the Sun is about 10 kpc from the center of a disklike galaxy whose total diameter is about 30 kpc. More recently, evidence has emerged that indicates the Sun is a little closer to the center, perhaps only 7 or 8 kpc out (the current "official" distance, used by astronomers through international agreement, is 8.5 kpc).

We see that the problem of determining where we are in our galaxy is not necessarily completely solved even today. There are also significant uncertainties about the shape of the galaxy and its extent beyond the Sun's position. Regarding the shape,

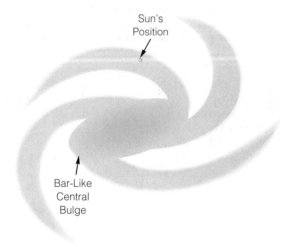

Figure 17.9 Elongated central bulge of the Milky Way. This sketch illustrates the recent finding that our galaxy may belong to the barred spirals, galaxies with elongated, or bar-like, central regions. Evidence for this comes from gas motions in the inner portion of the galaxy. (L. Blitz, University of Maryland, and D. Speigel, Princeton University)

recent analyses of the motions of interstellar gas clouds indicate that the central region of the galaxy may be slightly elongated **(Fig. 17.9).** This would place our galaxy in the category of **barred spirals,** though it is apparently not an extreme case (the classification of galaxies by shape is discussed in the next chapter). The question of the size of the galaxy is complicated by huge uncertainties about the extent and density of the halo of the galaxy and is discussed in a later section of this chapter.

Galactic Rotation and Stellar Motions

An important result of the work of Lindblad and Oort was an understanding of the overall motions in the galaxy. Oort's analysis was especially useful in this regard; he showed not only that the Sun and stars near it are orbiting the distant galactic center, but also that the rotation of the galaxy is differential. This means that the stars follow their own orbits at their own speeds, rather than moving together as though part of a fixed structure. Thus the disk of the galaxy is not rigid, but instead acts like a fluid, made of individual particles (stars) moving independently.

If the galaxy were a rigid disk, then the orbital speed would increase with increasing distance from the galactic center. Think about a record on a turntable. In order for the entire disk to rotate as one body, the outer edge, which has to travel much farther than the inner portions in order to make one revolution, must travel correspondingly faster. On the other hand, if the galaxy were made of individual stars that had no effect on each other, so that each one simply orbited the center as though all the mass were concentrated there, then we would expect the orbital speeds to decrease with distance from the center. To see this, recall Kepler's laws of planetary motion; the third law in particular, which relates orbital period to orbital radius (semimajor axis), shows that orbital speeds decrease with distance from the center (we know, for example, that each planet in succession outward from the Sun moves more slowly than the ones closer in).

So what do we actually find when we examine the orbital speeds of stars in the Milky Way? We can measure the rotation curve of the galaxy **(Fig. 17.10);** this is a graph showing how the orbital speed varies with distance from the center. What we find is an intermediate case; we see neither an increase nor a decrease in orbital speed as we go outward. This tells us that the galaxy is not a rigid disk, but on the other hand, the individual stars do not completely ignore the effects of other stars around them, so they do not act as completely free agents either.

The Milky Way Galaxy

Dark matter

Unmapped region

Halo

Our Sun

Orion Spur

Globular clusters

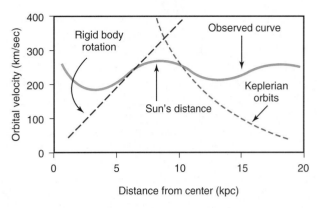

Figure 17.10 The rotation curve for the Milky Way. This diagram shows how the stellar orbital velocities vary with distance from the center of the galaxy. The fact that the curve does not simply drop off to lower and lower velocities beyond the Sun's orbit, as it would if all the mass of the galaxy were concentrated at its center, indicates that there is a lot of mass in the outer portions of the galaxy. (Data from M. Fich)

The rotation curve cannot be measured closer to the center than about 1 kpc and has been measured no farther out than about 16 kpc. It is likely that there is a small inner disk, extending to perhaps 700 pc, where rigid body rotation does occur, and that orbital speeds rise quickly from zero at the very center to the steady level of above 200 km/sec, which extends as far out as the speeds can be measured.

The Sun's orbital speed is approximately 220 km/sec, and its orbital radius is about 8.5 kpc. Using these numbers to estimate the orbital period (by dividing the circumference of the orbit by the orbital speed, assuming the orbit to be a circle) yields a period of about 240 million years. Thus, in the Sun's 4.5-billion-year lifetime, it has orbited the galaxy some 18 to 20 times.

Individual stellar motions do not necessarily follow precise circular orbits about the galactic center. Thus far we have described the overall picture that emerges from looking at the composite motions of large numbers of stars. If we look at the individual trees instead of the forest, we find that each star in the great disk has its own particular motion, which may deviate slightly from the ideal circular orbit. These individual motions are comparable to the paths of cars on a freeway, where the overall direction of motion is uniform, but a bit of lane-changing occurs here and there.

The Sun has a velocity of about 20 km per second with respect to a circular orbit, in a direction about 45° from the galactic center and slightly out of the plane of the disk. Most stars near the Sun have comparable departures from perfect circular orbits, amounting to only minor deviations from the overall orbital velocity of about 220 km per second. Later in this chapter, we will mention the **high-velocity stars,** which deviate strongly from the circular orbits followed by most stars in the Sun's vicinity.

The Mass of the Galaxy

Once the true size of the Milky Way was determined, it became possible to estimate its total mass. This could be achieved by measuring the star density in the vicinity of the Sun and then assuming that the entire galaxy has about the same average density. But a much simpler and more accurate technique is possible if we assume that the stars in our region of the galaxy approximately obey Kepler's laws (for this purpose this is a reasonable assumption).

Kepler's third law, in the more complete form developed by Newton, expresses a relationship among the period, the size of the orbit, and the sum of the masses of two objects in orbit about each other:

$$(m_1 + m_2)P^2 = a^3,$$

where m_1 and m_2 are the two masses in units of the Sun's mass, P is the orbital period in years, and a is the semimajor axis in AU. If we consider the Sun to be one of the two objects and the galaxy itself to be the other, we can use this equation to determine the mass of the galaxy. As we have already seen, the orbital period of the Sun is roughly 240 million years: $P = 2.4 \times 10^8$ years. The orbit is nearly circular, with the galactic nucleus at the center, so the semimajor axis is approximately equal to the orbital radius; that is, $a = 8.5$ kpc $= 1.8 \times 10^9$ AU. Now we can solve Kepler's third law for the sum of the masses:

$$m_1 + m_2 = a^3/P^2 = 1.0 \times 10^{11} \text{ solar masses.}$$

Since the mass of the Sun (1 solar mass) is inconsequential compared with this total, we can say that the mass of the galaxy itself is about 1.0×10^{11} solar masses. The Sun is slightly above average in terms of mass, so we conclude that the total number of stars in the galaxy must be a few hundred billion.

This method gives us only the mass inside the orbit of the Sun. Any mass that lies farther out has no effect on the motion of the Sun and its neighbors and therefore is neglected. Astronomers once assumed that not much mass was to be found beyond the Sun's orbit, since we live quite far from the huge central concentration of stars in the galactic disk. But, as noted above, the velocities of stars and interstellar gas do not fall off at greater distances from the center. Instead orbital speeds remain as high as the Sun's at least as far out as 15 kpc. This indicates that a great deal of mass—possibly as much as 90% of the mass in the entire galaxy—lies at great distances from the center.

The Mass in the Halo: The Dark Matter Problem

The discovery that the halo must contain most of the galaxy's mass created a dilemma for astronomers: What is this mass? The visible disk of the galaxy appears to become much less dense with increasing distance from the center, and the halo appears to be devoid of stars, with the exception of the globular clusters and a scattering of isolated stars. The discrepancy has become so troublesome that today astronomers consider the "dark matter" problem to be one of the most vexing of all. Furthermore, it is not limited to our galaxy. As we will see in chapters to come, several lines of evidence indicate that the majority of the mass in the universe is also in some dark, invisible form. Perhaps the problem of the unseen mass in the outer Milky Way is related; if so, solving the problem on the "local" scale of our own galaxy may help solve one of the great mysteries of the universe at large.

Many candidates for the dark matter in our galactic halo have been proposed, but so far none has proved satisfactory. One obvious possibility would be that the halo contains huge numbers of very faint, low-mass normal stars, but recent searches with the *Hubble Space Telescope* have shown that this is not the case. It has also been established that the quantity of interstellar gas and dust in the halo is insufficient to account for the mass. Another suggestion is that the halo contains large numbers of stellar remnants, such as white dwarfs, black holes, or neutron stars, or possibly substellar objects known as brown dwarfs (see Chapter 12), but the jury is still out on these ideas. Recently, however, a novel observational search has

shed new light on dim stellar objects as a major component of the dark matter in the galactic halo.

Instead of looking for dim, compact objects directly, which would be a virtually hopeless endeavor, astronomers have searched for their gravitational effects on light coming from more distant sources. Because photons of light are affected by gravitational forces (as discussed in the previous chapter), their directions of travel can be altered as they pass near a massive object. If light from a very distant source passes close by an intervening object, the light can be brought to a focus by the gravitational effect of the foreground body **(Fig. 17.11)**. If the background source, foreground object, and observer are aligned precisely enough, the observer looking at the background object will see an increase in brightness as the object's light is focused by the intervening body. Thus the presence of the body in-between can be inferred from the brightening of the background source. This effect is called **gravitational lensing,** and it occurs in the realm of distant galaxies (as seen in Chapters 19 and 21). But it can also occur for objects as small as a star or even smaller, if the alignment is sufficiently precise; in this case the effect is referred to as gravitational **microlensing.**

Normally, the chances of finding such an exact alignment are small, but if enough background sources are considered, then by random chance some coincidences ought to occur. This has led at least two different groups of astronomers to search stars in the Magellanic Clouds, which are concentrations of billions of stars, for occasional flare-ups that could be due to gravitational lensing by objects more nearby, in our own galaxy's halo. Photographs of large regions, each containing millions of Magellanic Cloud stars, are repeated at frequent intervals and then compared (by computer—it would be impractical for a person to do this) to find stars that increased in brightness due to chance alignments with

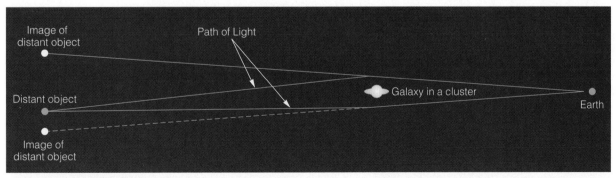

Figure 17.11 Gravitational lensing by a cluster of galaxies. This diagram illustrates how a galaxy in a cluster can create multiple images of a more distant object, such as a galaxy or quasar (see chapter 19). The mass of the foreground object creates a curvature in space, so that photons travelling from the more distant object are diverted, forming distorted or multiple images.

foreground objects. This should be a way to estimate the number of dim objects in the halo of our galaxy.

Enough microlensing events have now been recorded to allow some statistical conclusions to be reached. The masses of the lensing objects can be estimated from the duration of the brightening, and it appears that a significant number of objects with masses around 0.3 to 0.5 solar masses exist in the halo. This suggests that many white dwarfs are present because ordinary red dwarfs of this mass range (lower main-sequence stars) would have been detected directly. Recent estimates indicate that as much as half of the mass needed to explain the halo may be present in the form of white dwarf stars. This is an encouraging result in view of the difficulty astronomers have had in finding other forms of matter to explain the mass of the halo.

The Interstellar Medium

We have already discussed the interstellar medium, first when considering the formation of stars and planetary systems (Chapter 12), and more recently when learning how the size and shape of the Milky Way were established (this chapter). Here we will discuss the interstellar material in the context of its role in the galaxy, both in the distribution of matter and in its importance for galactic evolution.

Astronomers have known for many decades that there is material in space between the stars. We have mentioned the haze of interstellar dust that permeates interstellar space, and photographs show very obvious concentrations of dark or glowing clouds here and there in the galaxy **(Figs. 17.12** and **17.13)**. These clouds are regions of relatively high density, whereas the lower-density material filling most of the volume of the galaxy is virtually transparent. In all, the interstellar gas and dust constitute some 10 to 15% of the mass of the galactic disk and play very important roles in its evolution. Stars form from interstellar clouds, as we have seen, and late in life many stars return some of their substance to the interstellar medium. Consequently, the interstellar medium is studied in detail by astronomers interested in learning more about the lives and deaths of stars and about the galaxy itself.

Figure 17.12 **An emission nebula.** This dense cloud of interstellar gas and dust is being heated by hot stars embedded within it, which cause the gas to glow. Much of the emission is due to hydrogen atoms, which produce strong emission at 6563 Å, accounting for the red color seen here. (© 1984 ROE/AAT Board)
http://universe.colorado.edu/fig/17-12.html

Figure 17.13 Dark clouds. These dark, patchy regions are dense, cold interstellar clouds, the kind of regions where stars can form. The interior temperatures may be as low as 20 K, and the density roughly 10^5 particles per cubic centimeter, a high vacuum by earthly standards, but quite dense for the interstellar medium. The dark cloud in the lower photo, colored by gaseous emission and reddening due to dust absoption, is called a cometary globule because of its shape. (upper © 1984 Anglo-Australian Telescope Board; lower: © 1992 AAT Board; photos by D. Malin) http://universe.colorado.edu/fig/17-13.html

Table 17.3
Interstellar Medium Conditions

Component	Temperature (K)	Density (per cm³)	State of Gas	How Observed
Coronal gas (intercloud)	10^5–10^6	10^{-4}	Highly ionized	X-ray emission, UV absorption lines
Warm intercloud gas	1,000	0.01	Partially ionized	21 cm emission, UV absorption lines
Diffuse clouds	50–150	1–1,000	Hydrogen in atomic form, others ionized	Visible, UV absorption lines, 21 cm emission
Dark clouds	20–50	10^4–10^6	Molecular	Radio emission lines, IR emission and absorption lines

The interstellar medium is a mixture of gas and dust, with a wide range of physical conditions **(Table 17.3)**. The interstellar dust makes distant stars appear dimmer than they otherwise would and complicates distance determinations based on apparent magnitude measurements. The tendency of the dust to make stars appear dimmer is called **interstellar extinction.** Because the particles tend to block out short-wavelength light more effectively than the longer wavelengths, red light penetrates the interstellar medium more easily than blue light **(Fig. 17.14)**. Thus distant stars appear not only dimmer but also redder than they otherwise would. This trend of increasing extinction with decreasing wavelength continues throughout the spectrum, so that the extinction is very severe at ultraviolet wavelengths, but is minimal in the infrared portion of the spectrum **(Fig. 17.15)**. Thus it is virtually impossible to observe the inside of even a moderately dense interstellar cloud with an ultraviolet telescope, but we can see deep inside the densest clouds at infrared and radio wavelengths. Infrared observations are a partic-ularly effective technique for observing stars in the process of formation, as discussed in Chapter 12.

Interstellar grains are formed in the material ejected by various kinds of aged or dying stars. Red giant and supergiant stars, for example, are known to form grains in their outer atmospheres, and grain formation also takes place in planetary nebulae, novae, and supernova outbursts. Grains form through condensation, which occurs whenever the proper combination of pressure and temperature prevails. The formation of dust grains in the atmosphere of a red giant is analogous to the formation of frost on the lawn on a cool, humid morning.

Most (about 99%) of the mass in interstellar space is in the form of gas particles rather than grains. The gas is observed by means of absorption lines formed in the spectra of distant stars **(Fig. 17.16)**, mostly in the ultraviolet portion of the spectrum, or by means of various kinds of emission lines. Dense clouds that are heated by nearby hot stars glow by producing emission lines, primarily of hydrogen **(Fig. 17.17)**. The strongest emission line of hydrogen lies at 6563 Å in

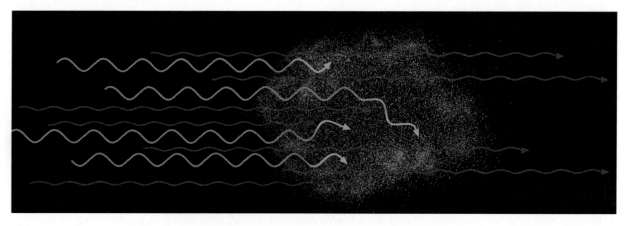

Figure 17.14 Scattering of light by interstellar grains. The grains tend to absorb or deflect blue light more efficiently than red, so a star seen through interstellar material appears red.

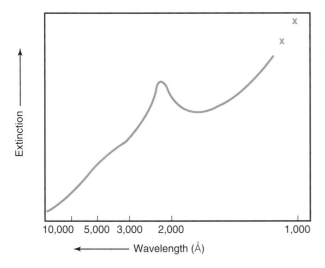

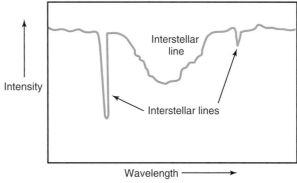

Figure 17.15 Interstellar extinction. The extinction of starlight by inter-stellar dust is greater at short wavelengths, especially in the ultraviolet, and decreases toward longer wavelengths. It is very small in the infrared and virtually nonexistent at radio wavelengths.

Figure 17.16 Interstellar absorption lines. Gas atoms and ions in space absorb photons from distant stars, creating absorption lines in the spectra of the stars. Here we see some interstellar absorption lines at ultraviolet wavelengths, where most of these lines are found. The interstellar lines can be distinguished from the star's own spectral lines because they are narrower, usually have a different Doppler shift (because the star and the cloud producing the line move at different velocities), and represent different states of ionization.

the red portion of the spectrum, so color photographs of these **emission nebulae** or **H II regions** show a vivid red appearance (see Fig. 17.12).

Interstellar gas also produces radio emission lines. Again, hydrogen atoms play an important role, emitting at a wavelength of 21 centimeters. Since hydrogen is the most abundant element, observations of the **21-centimeter emission line** are very useful for determining the structure of the galaxy (discussed in the next section). Molecules in the densest interstellar clouds also form radio emission lines at wavelengths ranging from a few millimeters to several centimeters.

Molecules, like atoms, have definite energy states, and photons are emitted or absorbed when the molecules change energy states. The energy levels that produce lines in the radio portion of the spectrum are related to the rotational energy of a molecule. Being an extended object, a molecule can rotate, and the energy of rotation can have only certain values, just as the energy level of an electron orbiting an atom can have only certain values. Thus a spinning molecule can slow its spin only by making a sudden drop to a lower energy state, and in doing so, it emits a photon whose energy corresponds to the loss of rotational energy of the molecule. Rotational energy levels are very closely spaced, so photons corresponding to changes in rotational energy have low energies and long wavelengths, usually a few millimeters or centimeters (i.e. radio wavelengths). Each kind of molecule has its own unique set of rotational energy states and, therefore, its own unique spectrum of radio emission or absorption lines.

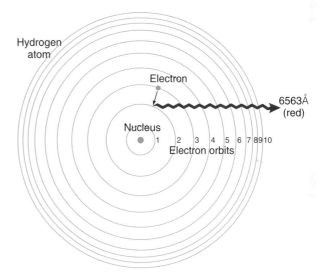

Figure 17.17 Hydrogen lines from an emission nebula. In heated interstellar gas, electrons and protons combine to form hydrogen atoms, usually with the electron initially in an excited state. It then drops down to the lowest state, emitting photons at each step. The strongest emission at visible wavelengths corresponds to the jump from level 3 to level 2 and has a wavelength of 6563 Å. This is why emission nebulae appear red. (The energy level spacings are not drawn to scale.)

Over 100 different molecular species have been identified in dense, "dark" interstellar clouds (see Appendix 15), and more are being found frequently as radio telescope technology improves. By far the most abundant molecule is hydrogen (H_2), followed by carbon monoxide (CO). Ironically, hydrogen molecules do not have strong radio emission lines, and their presence must be inferred indirectly except in more rarefied clouds, where they can be observed through ultraviolet absorption lines. The less dense interstellar clouds have few molecules but are composed of a mixture of atoms and ions. Ionization in interstellar space occurs primarily through the absorption of ultraviolet photons rather than through collisions between atoms, as in the case of a star's interior or atmosphere.

The range of physical conditions in interstellar space is enormous. The densest interstellar clouds have densities of perhaps 10^4 to 10^6 particles per cubic centimeter. (Compare this with the Earth's atmospheric density of about 2×10^{19} particles per cubic centimeter!) These clouds are very cold, with temperatures typically between 20 and 50 K. The less dense "diffuse clouds" have densities of 1 to 1,000

particles per cubic centimeter and temperatures of 50 to 150 K. Most of the volume of interstellar space is filled by even more tenuous material, with a density as low as 10^{-4} particles per cubic centimeter and a temperature as high as 100,000 to 1 million K!

The enormously energized hot intercloud medium is heated by the blast waves from supernova explosions and the rapid outflows from hot stars with winds. The space between the obvious clouds is a violent place filled with superheated gas. The clouds themselves are often in motion, pushed around by the same forces that provide the energy for the intercloud heating. The same blast waves that stir up the interstellar gas and dust within the galactic disk also sometimes eject material out of the plane into the halo. Galactic "fountains" of gas rising to great distances above and below the plane have been inferred to exist, primarily from radio-wavelength observations and the use of the Doppler effect to derive the velocities of the gas clouds. In the next section, we will learn more about how the interstellar material is distributed in the galaxy.

Spiral Structure

So far we have spoken of the galactic disk as though it were a uniform, featureless object, but we know that this picture is not completely accurate. The Milky Way is a spiral galaxy, and if we could see it face-on, we would see the characteristic pinwheel shape normally found in galaxies of this type **(Fig. 17.18)**.

A common misconception about the spiral structure in the Milky Way (and other spiral galaxies) is that there are few stars between the visible spiral arms. In reality, the density of stars between the arms is nearly the same as it is in the arms. The most luminous stars, however—the young, hot O and B stars—tend to be found almost exclusively in the arms. This is because the interstellar gas and dust tend to be concentrated in the arms, so young stars tend to be concentrated there also. Because these are the brightest stars, their presence in the arms makes the spiral structure stand out.

The fact that we live in a spiral galaxy was not easily discovered, again because we are located within it and see only a cross-sectional view. It was not until 1951 that investigations of the distribution of luminous stars revealed traces of spiral structure in the Milky Way, and even that technique was limited to a small portion of the galaxy. The obscuration caused by interstellar material limits our view of even the brightest stars to a local region, about 1,000 pc from the Sun at most.

Figure 17.18 **Spiral structure.** Like our galaxy, this galaxy has prominent spiral arms, because hot, luminous young stars tend to be concentrated in the arms. This is the galaxy NGC 300, in Sculptor. (© 1992 AAT Board; photo by D. Malin)

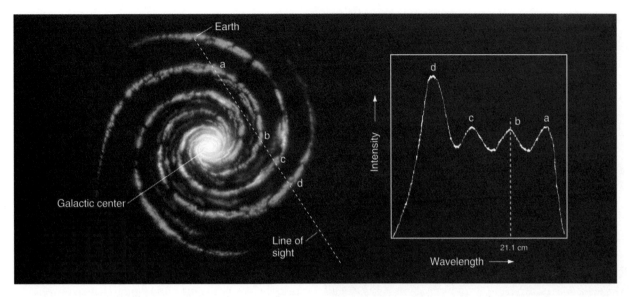

Figure 17.19 21-centimeter observations of spiral arms. At the left is a schematic diagram of the galaxy, showing the direction in which a radio telescope might be pointed to record a 21-centimeter emission-line profile like the one at the right. The 21-centimeter line has many components; each corresponds to a distinct spiral arm, and each is at a wavelength reflecting the Doppler shift between the velocity of that arm and the Earth's velocity.

Tracing the Arms Using the 21-centimeter Line of Hydrogen

A major advance in measuring the structure of the galaxy also occurred in 1951, when the 21-centimeter radio emission of interstellar hydrogen atoms was first detected. It had been predicted that hydrogen should emit radiation at this wavelength, and the Americans Ed M. Purcell and H. I. Ewen, using a specially built radio telescope, were the first to detect this emission.

One great advantage of using the hydrogen 21-centimeter emission for measuring galactic structure is its ability to penetrate the interstellar medium to great distances. Whereas light from the brightest stars can be detected at distances of at most one or two thousand parsecs, clouds of atomic hydrogen in space can be "seen" by radio telescopes from all the way across the galaxy, at distances of several thousand parsecs. Because hydrogen gas is the principal component of the interstellar medium and tends to be concentrated along the spiral arms, observations of the 21-centimeter radiation can be used to trace the spiral structure throughout the entire Milky Way **(Fig. 17.19).**

When a radio telescope is pointed in a given direction in the plane of the galaxy, it receives 21-centimeter emission from each segment of spiral arm in that direction. Because of differential rotation, each arm has a distinct velocity from the others, which are either closer to the center or farther out. Therefore, instead of a single emission peak at the laboratory wavelength of exactly 21.1 centimeters, what we see is a cluster of emission lines near this wavelength but separated from each other by the Doppler effect. By combining measurements such as these with Oort's mathematical analysis of differential rotation, it was possible to reconstruct the spiral pattern of the entire galaxy. Radio emission from the carbon monoxide (CO) molecule has also been used to map the distribution and velocities of relatively dense interstellar clouds, providing further information on galactic rotation and spiral structure.

The spiral pattern of the Milky Way is much more complex than in some galaxies, where two arms elegantly spiral out from the nucleus. Instead the Milky Way consists of bits and pieces of a large number of arms, giving it a definite overall spiral form, but not a smoothly coherent one. It has recently been discovered that the central region of the galaxy has an elongated, barlike structure (as mentioned earlier in this chapter) resembling galaxies in a class known as the barred spirals (these will be discussed in Chapter 18). About half of all spiral galaxies are barred spirals, so it is not surprising that the Milky Way may belong to this class.

The Maintenance of Spiral Arms

Maintaining spiral arms, which is a distinct question from forming them in the first place, is a complex business and is not well understood in all aspects. To see why some mechanism is needed, consider the fact

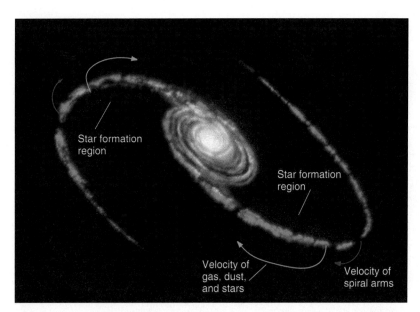

Figure 17.20 **The density wave theory.** This drawing depicts the manner in which circular orbits are deformed into slightly elliptical ones by an outside gravitational force. The nested elliptical orbits are aligned in such a way that there are density enhancements in a spiral pattern. This pattern rotates at a steady rate and does not wind up more tightly.

that the galaxy has rotated more than 50 times since it formed, yet the spiral arms have not wrapped up tightly around the nucleus, as they would have if they were simply streamers of gas that trailed the rotation.

The first theory to successfully explain the persistence of spiral arms appeared in 1960 and is still being refined. The essence of this theory is that the large-scale organization of the galaxy is imposed on it by wave motions. We have understood waves as oscillatory motions created by disturbances, and we know that waves can be transmitted through a medium over long distances while individual particles in the medium move very little. In the case of water waves, for example, a floating object simply bobs up and down as a wave passes by, whereas the wave itself may travel a great distance. Here we are distinguishing between the wave and the medium through which it moves.

The waves that apparently govern the spiral structure in our galaxy are not transverse waves like those in water, but compressional waves, similar to sound waves and to certain seismic waves (the P waves; see Chapter 8). In this case the wave pattern consists of alternating regions of high and low density. When a compressional wave passes through a medium, the individual particles vibrate back and forth along the direction of the wave motion. In contrast to water waves, there is no motion perpendicular to the direction of wave motion.

The waves that are thought to explain the spiral structure in galaxies like the Milky Way are called **spi-**

ral density waves. These consist of a spiral-shaped wave pattern centered on the galactic nucleus, creating a pinwheel shape of alternating dense and relatively empty regions **(Fig. 17.20)**. The density waves have more effect on the interstellar medium than on stars, so the spiral arms are characterized primarily by concentrations of gas and dust. The accumulation of interstellar material in the arms leads, in turn, to concentrations of young stars, because star formation is enhanced in regions where the interstellar material is compressed (see Fig. 17.20).

We have encountered spiral density waves in one other circumstance: the ring system of Saturn (Chapter 10). We learned that in portions of the ring system, the thin "ringlets" actually consist of a tightly wound spiral structure. In that case gravitational perturbations due to Saturn's satellites were thought to cause the development of the spiral density waves.

The same question arises in connection with the spiral density waves on the galactic scale: What causes them to form? What would make a rotating, uniform disk change its appearance and develop spiral density waves in the first place? Studies have shown that almost any disturbance to such a disk will create a spiral structure, at least temporarily. One common cause is thought to be gravitational forces imposed by neighboring galaxies. In our case these forces could be supplied by the Magellanic Clouds, which are actually satellites of the Milky Way. We already know that the gravitational forces due to these galaxies have created a distortion in the shape of the Milky Way's disk, and it appears quite likely that they also are the cause of the spiral density waves that formed and maintain the spiral arms. Another mechanism for creating spiral density waves, which does not depend on nearby neighbors, arises in barred spirals. The asymmetric shape of the central bulge in such a galaxy can create the disturbance needed to initiate spiral density waves. Since our galaxy possibly has a weak barred structure, this provides another mechanism that may have contributed to the formation of the spiral arms.

Regardless of how the spiral density wave structure was established, the properties of the resulting wave pattern are the same. It is important to understand that the wave pattern rotates as a fixed structure, not differentially. The stars and interstellar material follow their own orbits about the galaxy independently of the rotation speed of the spiral density wave pattern; hence the stars and interstellar material can overtake and pass through a spiral density wave, or a wave can overtake and pass the stars and gas, depending on their relative speeds at any given distance from the galactic center. Material passing through a spiral arm is slowed, so a density concentration accumulates there. This accumulation is especially effective for gaseous material, and this is why the interstellar gas and dust in particular highlight the spiral structure of the galaxy. There is a minor buildup of stars in the arms as well, but as noted earlier, the stars in the galactic disk are actually distributed almost uniformly.

As we mentioned earlier, the reason spiral arms stand out so clearly in a distant view of a spiral galaxy is that the brightest stars, the massive O and B stars, tend to be concentrated in the arms. Stars of these types are so short-lived that they go through their entire life cycles before they have time to move out of the spiral arms in which they form. Because these stars are very luminous, the spiral arms contribute a major quantity of light, even though they do not actually represent significant concentrations of mass. The Sun happens to be in a spiral arm now, but this arm is not the one the Sun formed in. Recall that the Sun has orbited the galaxy several times since it was formed. The Sun is moving faster than the spiral pattern and has passed through spiral arms several times as it caught up and overtook them. That we are in a spiral arm at this point in the lifetime of the solar system is purely coincidental.

In its simplest form, a spiral density wave pattern is double; that is, there are just two spiral arms emanating from opposite sides of the nucleus. Some galaxies have this simple spiral structure, but many, including the Milky Way, are more complicated. The density wave theory allows the possibility of more arms if the waves have shorter "wavelength," or distance between them, but the rather chaotic, fragmented appearance of our galaxy's spiral structure may require another explanation.

It is possible that in some spiral galaxies the structure is not governed by spiral density waves at all. The more chaotic and disorganized spirals, possibly including the Milky Way, consist of many fragments of spiral arms and have no simple overall spiral pattern. The fragmentary spiral arms in these galaxies may be gigantic regions of star formation that have been stretched into arclike shapes by the differential rotation of the galaxy **(Fig. 17.21)**. If so, then the spiral structure in such galaxies must be regarded as a constantly changing and evolving pattern rather than the long-standing pattern that is maintained in other galaxies by density waves.

The Galactic Center: Where the Action Is

It is impossible to observe the central regions of our galaxy at visible wavelengths because of obscuration

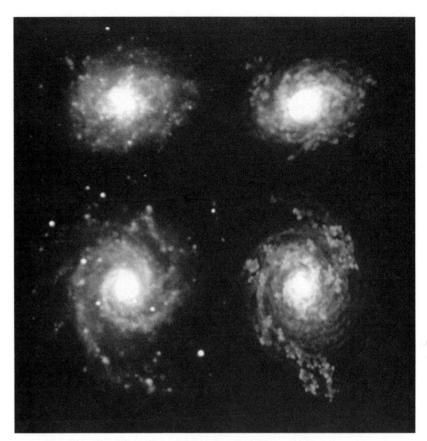

Figure 17.21 Spiral arms created by star formation regions. An alternative mechanism to density waves for creating spiral arms is sequential star formation, which creates regions of bright nebulae and concentrations of young stars, which are then stretched into arc-like structures by the rotation of the host galaxy. The two galaxies at left in this panel are images of real galaxies; at right are the corresponding model galaxies, based on the sequential star-formation model. Some astronomers have suggested that the Milky Way's spiral structure may have been created in this way, but current models favor overlapping spiral density waves instead. (Courtesy Philip Seiden, IBM Thomas Watson Research Center)

by the intervening interstellar gas and dust **(Fig. 17.22)**. From Shapley's work on the distribution of globular clusters, as well as Oort's analysis of galactic rotation, it was established by the 1920s that the center of the galaxy lies in the direction of the constellation Sagittarius. There we find immense concentrations of interstellar matter and a great concentration of stars. It is one of the richest regions of the sky to photograph, although the best views are seen only from the Southern Hemisphere.

The central region of the galaxy has different structures when examined on different scales. On the largest scale, the structure is dominated by a roughly spherical bulge consisting of old stars. This bulge is large enough (its radius is several hundred parsecs) that at some points we can see its outer regions, above and below the plane of the intervening disk (see Fig. 17.22). As we will see in the next chapter, this bulge probably represents the innermost, highly concentrated region of the galactic halo.

Figure 17.22 The galactic center. The dark regions are dust clouds in nearby spiral arms. Because of obscuration, photographs such as this do not reveal the true galactic center, but only nearby stars and interstellar matter in the plane of the disk and the outer portions of the central bulge of the galaxy. (© 1980 ROE/AAT Board) http://universe.colorado.edu/fig/17-22.html

At the center of the bulge lies the nucleus of the galaxy. This region cannot be observed at visible wavelengths, but can be probed using infrared and radio telescopes, because the interstellar gas and dust do not block these wavelengths. Here again we see different structures when we look at different size scales. The use of interferometry in radio observations (see Chapter 6) allows us to observe details as small as a tiny fraction of a parsec, even though the galactic center is about 8,500 pc away from the Earth.

Radio data long ago revealed a central complex of hot gas, which is called Sagittarius A (Sgr A). This was the first indication of highly energetic activity at the galaxy's core. Detailed observations subsequently showed that Sgr A consists of several smaller sources of emission, including one called Sgr A East, which has since been identified as a probable supernova remnant, and one called Sgr A West, which appears to be very close to the true center of the Milky Way. But Sgr A West itself consists of at least two distinct, tiny sources. One, called Sgr A*,

The Milky Way at Different Wavelengths
In this web exercise we will locate and analyze images of our Galaxy, the Milky Way, at several different wavelengths—examining in detail the information that each different wavelength regime provides. Our starting point is the following URL:
http://universe.colorado.edu/ch17/web.html

has a diameter of less than 0.001 parsec (that is, less than 100 AU, or roughly the size of the solar system). The other component of Sgr A West, lying very near to Sgr A*, is a bright infrared source, which has been identified as a dense cluster of recently formed, massive stars. Deciding which of these two objects is the true center of the galaxy has been controversial, but most astronomers believe that Sgr A* is the more likely one. This, then, appears to be the point about which the entire galaxy rotates.

The nature of Sgr A* is not known. Studies of stellar motions in the vicinity indicate an increase in orbital velocity with decreasing distance from Sgr A; from these measurements it is possible to estimate the mass located at the central point. It is difficult to extend these observations close enough to the center to be certain of the mass, but the best indications are that a mass of about 1 million suns is located there within a very small volume. While some astronomers argue that this mass could exist in the form of a very dense cluster of stars, it is difficult to see how such a cluster could be present in such a small volume, and many believe that the object is a massive black hole. As we will see in later chapters (especially Chapter 20), many galaxies show evidence of having similar or more extreme activity at their nuclei, and in many cases it has been concluded that a supermassive black hole is the probable source of the energy.

Just outside the tiny, energetic source that represents the galactic center lies a fantastic "mini-spiral" structure, which shows up in both radio and infrared observations **(Fig. 17.23).** The gas in this region is ionized, but the source of the heating is not well understood. The mini-spiral is not part of the overall spiral structure of the galaxy. Its size scale is very much smaller (its extent is only about 2 pcs, as compared with thousands of parsecs for the spiral arms), and the mini-spiral is tilted with respect to the plane of the galactic disk. Surrounding the mini-spiral is a thin shell of gas that appears to form a boundary between the hot, ionized gas in the central region and the colder, denser gas lying outside (this shell appears as a red glowing boundary in Figure 17.23).

On a somewhat larger size scale, another very interesting structure is seen **(Fig. 17.24).** High-resolution radio maps reveal a large, arclike structure projecting out of the plane of the galactic disk. This structure consists of filamentary arcs, which lead astronomers to suspect that they are supported by magnetic forces, although it is not understood how and why the galactic magnetic field would project nearly perpendicular to the plane of the disk.

The energetic activity at the core of the galaxy is reflected in structures much farther out as well. At a distance of about 3 kpc from the center lies a spiral

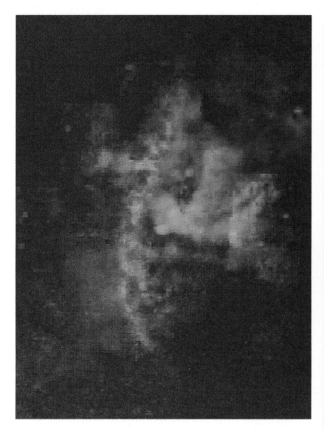

Figure 17.23 The central portion of the galactic nucleus. This false-color infrared image shows the "mini-spiral" in blue, ionized gas in green, and the dense, hot shell in red. The diameter of the shell is only about 5 pc. (AAT Board; courtesy of D. A. Allen)

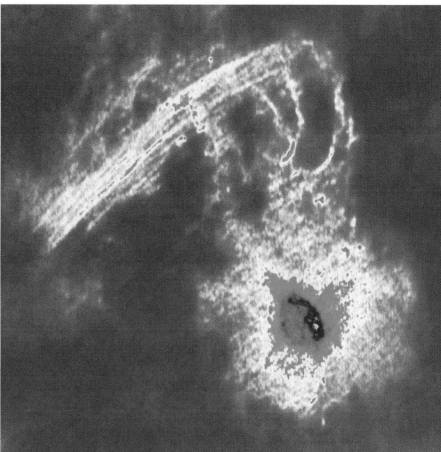

Figure 17.24 Wispy structure outside the nucleus. This image, made at a wavelength of 6 centimeters (where thermal continuum emission is measured, rather than a spectral feature such as the 21-centimeter line), shows an arc of gas extending some 200 pc above the plane of the galactic disk. (F. Yusef-Zadeh, M. Morris, and D. Chance)

arm that is moving outward at a speed of about 50 km/sec. This expansion, measured through the Doppler shift of the 21-centimeter emission line as well as absorption lines of the molecule CO, may be the result of a long-ago explosive event at the galactic center. Even farther out, some 5 or 6 kpc from the center, is a ringlike structure encircling the entire galaxy, which contains a high concentration of dense clouds. It is very difficult to get a clear picture of this "6-kiloparsec ring," but one possible interpretation is that it is formed of swept-up material from an early outburst at the galactic center.

The overall view we get of the nucleus of the Milky Way is that it is a region of great turmoil, a place where star formation has taken place recently, in contrast with the rest of the central bulge, and a place where violent events may have occurred in the past. In addition, at the very center is a mysterious, very compact energy source that might be a million-solar-mass black hole. As we will see in later chapters, it is possible that all of these forms of energetic activity are related to even more spectacular happenings at the cores of many other galaxies.

The Evolution of the Galaxy

At the beginning of this chapter, we argued that a galaxy must change with time, simply because the stars it contains change. As they form, live, and die, stars inject altered material into the interstellar medium, and this material forms the substance of future generations of stars. Now that we have surveyed the structure and overall properties of the galaxy, we are in position to consider the way in which it formed and the manner in which it is changing with time.

Stellar Populations and Elemental Gradients

A good starting point is to look at observational evidence that indicates that the galaxy is indeed evolving and changing with time. The best evidence for this—and the basis for our modern picture of Milky Way formation and evolution—comes from the study of properties of stars in distinctly different regions of the galaxy.

We have already seen that most of the young stars in the galaxy tend to lie along the spiral arms. Indeed, it is the brilliance of the hot, massive O and B stars and their associated H II regions that delineates the arms, making them stand out from the rest of the galactic disk. We have also noted that there are few young stars and relatively little interstellar material in the nuclear bulge of the galaxy, as well as in the globular clusters and the halo in general. These findings, along with studies of the dynamics (motions) of stars in different parts of the galaxy, have led astronomers to think of the galaxy as though it had two distinct components: a flattened disk including the spiral arms, and a "spheroidal" component, consisting of the central bulge and the halo. The disk component has overall rotation, while the spheroidal component does not; instead the stars there orbit the center of the galaxy in randomly oriented paths.

Careful scrutiny has revealed a number of distinctions between the stars inhabiting the two distinct substructures of the galaxy. Analysis of the chemical compositions of the stars has shown that those in the halo and central bulge tend to have relatively low abundances of heavy elements such as metals, while stars in the vicinity of the Sun (and in the disk and spiral arms in general) contain these species in greater quantity. Nearly all stars are dominated by hydrogen, of course, but the heavy elements are present in even smaller traces in stars in the spheroidal component than in the disk stars. The relative abundance of heavy elements in a halo star may typically be a factor of 100 below that found in the Sun; if iron, for example, is 10^{-5} as abundant as hydrogen in the Sun, it may be only 10^{-7} as abundant as hydrogen in a halo star.

The evidence for two distinct groupings of stars in our galaxy originally arose from observations of another galaxy, our neighboring spiral M31 (the Andromeda galaxy; **Fig. 17.25**). In the 1940s it was noticed that the stars in the disk and central bulge of this galaxy have distinctly different collective colors: blue-white, hot stars are all confined to the disk (and concentrated in the spiral arms) while the stars in the central bulge have a yellowish color, indicating a lack of hot, young stars. This observation led to the idea that stars in a galaxy like ours can be classified into distinct groupings, called **populations.** Stars in the central bulge were designated **Population II** stars, while those in the disk and spiral arms were **Population I** stars. These ideas were soon adapted to the Milky Way, which seems to resemble M31 in general. The Sun is often taken as representative of Population I stars in our galaxy.

Further distinctions between the two groups were found as well **(Table 17.4).** Most of these differences reflect the differing conditions at the times the stars in the two populations formed. Population II stars, which include the membership of the globular clusters, clearly constitute an older population than Population I stars, which encompass not only the Sun and its neighbors, but also the very young O and B stars found in spiral arms. Thus the distribution of the two stellar populations provides important clues to the history of the galaxy, which we will exploit later in this chapter.

Almost every time an attempt is made to classify astronomical objects into distinct groups, the boundaries between them turn out to be indistinct. There are usually intermediate objects whose properties fall between categories, and the stellar populations are no exception. For example, the stars in the disk of the galaxy that do not fall strictly within the spiral arms have properties intermediate between those of Population I and Population II and are often called simply the **disk population** (most recently, astronomers have spoken of a "thin disk" and a "thick disk"; the latter is somewhat older and may represent an exten-

Figure 17.25 The Andromeda galaxy. This great spiral galaxy is thought to resemble the MIlky Way in many respects: it is similar in structure and similarly large (both are among the largest of spiral galaxies).In this color image shows the yellow-white central bulge and the blue-white outer regions, which reflect the contrasting populations of stars residing in these different regions. (University of Oregon)

http://universe.colorado.edu/fig/17-25.html

Table 17.4

Stellar Populations

	Population I	Population II
Age	Young to intermediate	Old
Location	Disk, spiral arms	Halo, central bulge
Composition	Normal metals	Low in heavy elements
Constituents	Disk, arm stars	Stars low in heavy elements
	O, B stars	High-velocity stars
	Interstellar matter	Globular clusters
	Type I Cepheids	Type II Cepheids
	Type II supernovae	RR Lyrae stars
	The Sun	Type I supernovae
		Planetary nebulae

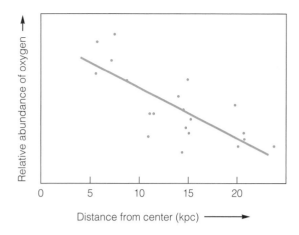

Figure 17.26 An abundance gradient. This diagram shows how the abundance of oxygen (relative to hydrogen) varies with distance from the center of the Andromeda galaxy. More stellar generations have lived and died in the dense regions near the center, so nuclear processing is further advanced in the central region than it is farther out. (Data from W. P. Blair, R. P. Kirschner, and R. A. Chevalier, 1982, *Astrophysical Journal* 254:50)

sion of the central bulge). Furthermore, there are gradations within the two principal population groups, and we therefore speak of "extreme" or "intermediate" Population I or Population II objects. It is much more accurate to view the stellar populations as a sequence of stellar properties represented by very low abundances of heavy elements at one end and Sun-like abundances at the other.

Within the disk and spheroidal structures of the galaxy, there are systematic variations in stellar composition. Changes that occur gradually over substantial distances are referred to as **gradients,** and the variations of stellar composition from one part of the galaxy to another are therefore referred to as **abundance gradients.** There is a gradient of increasing heavy element abundances from the halo to the disk **(Fig. 17.26).** Within the disk itself, a similar gradient extends from the outer to the inner portions; thus the central regions of the disk have relatively high abundances of heavy elements.

The distribution of the elements within the galaxy, measured through the distributions of Population I and II stars and also through observations of elemental gradients, will help us greatly later when we try to piece together a chronology of galactic evolution. Another set of data that will help comes from studies of the motions of stars in the different regions of the galaxy. The disk component, the flat structure to which the Sun belongs, rotates about the center, while the spheroidal component, consisting of the bulge and halo, has no overall rotation. Instead stars there orbit the center in randomly oriented planes, and the orbits tend to be highly elongated (eccentric) ellipses, resembling those of comets that have fallen into the inner solar system from the Oort cloud (Chapter 11).

Most of the stars near the Sun belong to Population I; they are disk stars orbiting the galactic center in approximately circular paths, like the Sun. There are also a few Population II stars near the Sun, and they are distinguished by several properties in addition to their compositions. Most easily recognized are their motions, which depart drastically from those of Population I stars. Because Population II stars from the halo have randomly oriented orbits, those that pass near the Sun generally come through the disk on paths that lie nearly perpendicular to the Sun's motion in the plane of the disk **(Fig. 17.27).** The Sun and the other Population I stars in its vicinity move in their orbits at speeds around 220 km/sec, so there is a large velocity difference between the Sun and a Population II halo star that is passing through in a perpendicular direction. Therefore, from our perspective here on the Earth, these Population II objects are classified as **high-velocity stars.**

Historically, high-velocity stars were discovered and recognized as a distinct class even before the differences between Population I and Population II were enumerated. It was only later that these objects were equated with Population II stars following randomly oriented orbits in the galactic halo.

Stellar Cycles and Chemical Enrichment

The fact that stars in different parts of the Milky Way have distinctly different chemical makeups is readily explained in terms of stellar evolution. Recall (from discussions in Chapter 15) that as a star lives its life, nuclear reactions gradually convert light elements into heavier ones. The first step occurs while the star is on the main sequence, when hydrogen nuclei deep in its interior are being fused into helium. Depending on the mass of the star, later stages may include the fusion of helium into carbon and possibly the formation of even heavier elements by the addition of further helium nuclei. The most massive stars, which undergo the greatest number of reaction stages, also form heavy elements in the fiery instants of their deaths in supernova explosions.

All of these processes work in the same direction: they act together to gradually enrich the heavy element abundances in the galaxy. Material is cycled back and forth between stars and the interstellar medium, and with each passing generation, a greater supply of heavy elements is available. As a result, stars formed where a lot of stellar cycling has occurred in the past are born with higher quantities of heavy elements than those formed where little previous cycling has occurred. Stars formed before much cycling occurred therefore tend to have low abundances of

heavy elements. This explains why old stars belong to Population II; they formed out of material that had not yet been chemically enriched.

Population I stars, such as the Sun, condensed from interstellar material that had previously been processed in stellar interiors and in supernova explosions. Therefore, as a rule, they formed at moderate to recent times in the history of the galaxy and are not as ancient as Population II objects.

Elemental abundances can vary with location as well as with age. As noted in the preceding section, the disk component of the galaxy has generally greater abundances of the heavy elements than the spheroidal component, especially the halo. This clearly implies that more cycles of star formation and star death have occurred in the disk, and of course this is verified by the fact that only in the disk (i.e., the spiral arms) do we see significant star formation taking place today.

There is also a distinct abundance gradient within the disk of the galaxy, with the stars nearer the

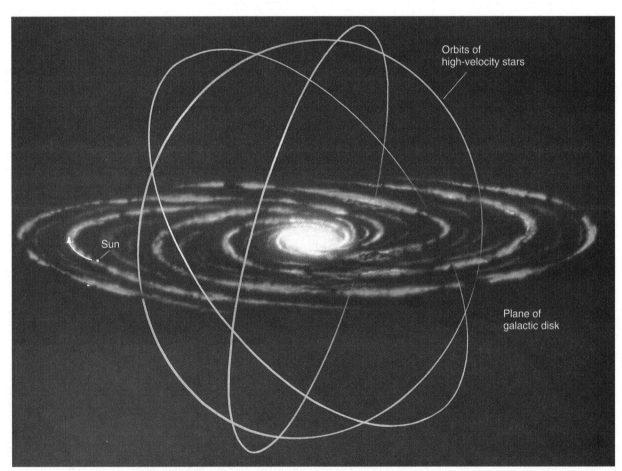

Figure 17.27 The orbits of high-velocity stars. These stars follow orbits that intersect the plane of the galaxy. When such a star passes near the Sun, it has a high velocity relative to us, because of the Sun's rapid motion along its own orbit.

Astronomy and Lasers

It is difficult to think of a technological device that has had more impact on our society than the laser. Lasers were invented in the 1960s, and today they are *everywhere*. It is interesting to know that Mother Nature was making lasers for billions of years before humans discovered the concept.

You may not fully appreciate all of the applications of lasers that affect your daily life. You may go to the occasional laser light show or notice that your instructor uses a laser pointer in the classroom, but did you realize that lasers are integral parts of CD players? Or that they are used in many medical applications, including delicate surgery? Or that holograms, whose colorfully irridescent play of light fascinates everyone, are made with lasers? In addition to these consumer-oriented applications, lasers have an enormous variety of technical uses, including measurement techniques that have revolutionized mapping and surveying; atomic and molecular research, where lasers are used to probe the behavior of matter on the smallest scales; precision atomic clocks, where the frequency of laser light is used as a timing reference; the development of laser weapons with potentially far-reaching effects; data storage on optical disks and CD-ROMs; and even a new class of devices that will replace radar guns in the enforcement of automobile speed limits.

In addition to all of the laser-based technology on Earth, astronomers have discovered naturally occurring lasers in space. One might even argue that if lasers had not been developed when they were, the discovery of natural lasers would inevitably have led to their invention here on Earth soon enough anyway.

To understand the operating principle of a laser, you need to recall how atoms emit and absorb photons of light (discussed in Chapter 6). An atom has electrons that can occupy distinct energy levels corresponding to different orbits about the nucleus. Photons are emitted or absorbed when electrons either drop from a high level to a lower one (emitting a photon) or jump from a low level to a higher one (absorbing a photon). In most natural situations, the electrons will be found in the lowest possible energy states, but in some circumstances (due to collisions between atoms or the absorption of light), some of the electrons can reside temporarily in elevated, or excited, energy states.

Some atoms have special excited states, called metastable states, where an electron can stay for a long time instead of immediately dropping down to a lower state. In a large collection of atoms where there is an energy source to excite the electrons, at any time a number will be sitting in the metastable state. The operation of a laser depends on the existence of these long-lived excited states.

When there is an appropriate metastable state and a source of energy, there will be an overpopulation of electrons in that state; that is, there will be a kind of bottleneck in which electrons dropping from higher levels will get stuck temporarily. This is called a population inversion. The operation of a laser depends on having such an inversion, so that there will always be electrons in the excited state. The excited electrons can then be induced to give up their extra energy by emitting photons through a process called stimulated emission. This occurs when a photon with the same energy as the energy of the excited electron state strikes an atom. The incoming photon causes the electron to release its energy in the form of another photon, which will have precisely the same wavelength as the incident photon. Thus one photon becomes two. Both travel in the same direction, and both are in phase with each other, meaning that the peaks and troughs of the wave oscillations match.

Now imagine a volume with many atoms whose electrons lie in the same excited state. If a single photon enters and stimulates the emission of another, then the two can stimulate the emission of yet others, and in a nanosecond you have a flood of photons all in phase and all having the same precise wavelength. In effect the energy of the initial photon has been magnified many times over, and a beam of light emerges that is very tightly collimated and very intense.

If you are thinking carefully about this, you might be wondering how energy is conserved in this situation. A single photon triggers the emission of many more, and the resulting beam has far more energy than the original incoming photon. But recall that energy has to be injected into the collection of atoms in the first place, creating the population inversion. In other words, you must first "pump" the atoms with energy, so that their electrons are excited to the higher level. When an incoming photon then triggers the emission of many others, what it is doing is releasing the energy that was previously inserted when the electrons were excited.

Thus a laser is a device in which a collection of atoms is pumped to an excited state; then entering photons stimulate the emission of many more, producing an intense beam of light in which all of the photons have precisely the same wavelength and are in phase with each other. The name laser stands for "Light Amplification by Stimulated Emission of Radiation."

In practice, a laser consists of a chamber filled with a gas, liquid, or solid material that has an appropriate excited state. The atoms are usually pumped to the excited state by radiation, but other energy injection methods can be used. The photons that trigger the stimulated emission are emitted naturally, when some of the excited electrons drop spontaneously to the lower level. Once a few photons are emitted in this way,

Continued

they stimulate the emission of an avalanche of additional photons. The laser can operate continuously if the pumping mechanism is operated continuously, so that newly excited electrons are always available. Most lasers have semireflective mirrors at either end of the emitting chamber, which (1) allow the photons to bounce back and forth several times before escaping, so that the number of stimulated emissions is increased, and (2) keep all of the photons traveling back and forth along precisely the same direction (by having the mirrors precisely parallel to each other).

The principle of the laser was discovered by American physicist Charles Townes in 1951. Instead of considering visible wavelengths, Townes envisioned a mechanism for stimulating the emission of short-wavelength radio photons. He later turned his idea into reality, inventing the **maser**—"Microwave Amplification by Stimulated Emission of Radiation." Shortly afterward, Townes (along with colleague Arthur Schawlow) patented the concept of the laser as well and is recognized today as its inventor (Townes, who has devoted most of his career to astrophysics, is today a Professor Emeritus at the University of California, Berkeley, and is still quite active in astronomical research).

A maser operates in the same way as a laser, except that the energy levels involved are not electron energy levels in atoms, but rotational energy levels in molecules. A molecule can vibrate and rotate in addition to having electrons that reside in specific energy states, and the vibrational and rotational energy is quantized in a similar fashion to electron energy states. This means that the molecule can exist only in discrete, specific states of vibrational and rotational energy. It also means that photons of radiation can be absorbed or emitted by a molecule when it makes transitions between these states. The radio emission lines used to detect molecules inside dark interstellar clouds are produced when molecules, excited by collisions, release rotational energy as they drop from an excited state to a lower one. The structure of rotational levels that molecules can occupy gives rise to the possibility that molecules can be masers; that is, there can be metastable rotational levels in which a population inversion can be created. If this occurs, then a powerful beam of radiation can be created by stimulated emission. This is the maser.

In 1963, astronomer Harold Weaver and several colleagues discovered a mysteriously powerful radio emission line coming from the Orion nebula. The wavelength corresponded to well-known emission from the simple molecule OH (the hydroxyl radical), but the intensity was so much greater than expected that initially Weaver and his colleagues thought the emission must be due to some other unidentified species, which they dubbed "mysterium." Soon they realized that it was OH, and that the maser process had amplified the intensity of the line. The first natural maser had been discovered. Radiation from stars and especially from embedded infrared sources (such as protostars) within the Orion nebula is pumping OH molecules to an excited rotational state, from which they can be stimulated to emit on the laser principle. Today many additional molecular masers have been discovered in dense interstellar clouds and in the material surrounding red supergiant stars, where there is ample infrared radiation to pump the molecules to excited states. Even planetary atmospheres give rise to masers: infrared emission from carbon dioxide (CO_2) amplified by the maser effect has been found in the atmospheres of both Venus and Mars, with solar radiation providing the pumping.

Very recently, evidence for natural lasers as well as masers has been found. The *Hubble Space Telescope* has detected unexpectedly strong emission due to ionized iron coming from a massive, unstable star called η Carinae ("eta" Carinae). This star is undergoing a massive stellar wind phase and, at the same time, is a powerful source of ultraviolet radiation, thus providing both of the ingredients needed for a laser: a source of pumping radiation, and a volume containing atoms (ions in this case) that have excited metastable states that can be overpopulated by the radiation. The intensity of the ionized iron emission line cannot readily be explained by any other process, and astronomers are becoming convinced that a natural laser has been found. It is interesting to realize that nature is capable of creating such high-tech devices as lasers and masers.

center having higher abundances of heavy elements than those farther out. This distribution reflects enhanced stellar cycling near the center, so even within the disk structure of the galaxy, there has been more stellar processing near the center than out near the edge. In this case the reason is not necessarily an age difference, but rather a density contrast: the density of interstellar material is higher in the inner disk, and therefore stars have formed more readily and more often there.

The relative abundances of the elements follow a pattern that is explained by the stages in stellar evolution that we have already discussed. After hydrogen and helium, much of which was already present when the first stars began to form, the next most abundant elements are carbon, nitrogen, and oxygen, which are produced in the triple-α and CNO reaction cycles, which many stars undergo. Beyond these elements, abundances generally fall off with increasing nuclear mass (i.e., with increasing atomic number or mass number). A relative peak in abundance, however, occurs at iron (26 protons, 30 neutrons in its most common form).

For this reason, iron and its nearby relatives are called the **iron peak elements.** The explanation for this peak is seemingly not hard to find, since iron formation is the end-point of nuclear reactions in massive stars. But in reality the situation is not quite so simple because much of the iron formed in a stellar core is then converted into neutrons and locked up in a neutron star, or lost from the galactic ecology forever when the core collapses to form a black hole. The iron peak is actually the result of nuclear reactions that occur during supernova explosions. We learned in Chapter 15 about Type II supernovae, which are created when the core of a massive star collapses and rebounds, and in Chapter 16 we discussed Type I supernovae, which take place when a carbon white dwarf gains enough new matter to undergo total destruction in a nuclear explosion. Both types of supernovae create iron, but calculations show that more of the galactic iron supply comes from the Type I events. Thus it is fair to say that most of the iron in your body, or in your car, or in the neighborhood scrap yard, came originally from exploding white dwarf stars.

Beyond the iron peak, elemental abundances continue to drop off with increasing nuclear mass. The most massive elements—the heavy metals and related species—are therefore the most rare both in the galaxy and on the Earth. Thus the most treasured elements, such as gold and platinum, are valuable not only for their unique chemical properties but also for their scarcity.

Galactic History

We are now in a position to tie together all the diverse information on the nature of the Milky Way and to develop a picture of its formation and evolution. Many important aspects are still uncertain, however, and some of these will be pointed out.

The pertinent facts that must be explained include the size and shape of the galaxy, its rotation, the distribution of interstellar material, elemental abundance gradients, and the dichotomy between Population I and Population II stars in terms of composition, distribution, and motions. The task of fitting all the pieces of the puzzle together is made easier by the fact that we can reconstruct the time sequence—we know that stars with high abundances of heavy elements were probably formed more recently than those with lower abundances.

The oldest objects in the galaxy are in the halo, which is dominated by the globular clusters but contains some isolated, dim, red stars as well. Estimates based on the main-sequence turnoff in the H-R diagrams for globular clusters indicate ages as high as 14 to 16 billion years, and we conclude that the age of the galaxy itself is comparable (but there are some interesting hints that this picture may be more complicated, as discussed in the next section).

The spheroidal shape of the halo and galactic bulge components, centered on the nucleus of the galaxy, suggests that when these structures formed, the galaxy itself was round. Evidently, the progenitor of the Milky Way was a gigantic spherical gas cloud, consisting almost exclusively of hydrogen and helium. Very early, perhaps even before the cloud began to contract **(Fig. 17.28)**, the first stars and globular clusters formed in regions where localized condensations occurred. These stars contained few heavy elements and were distributed throughout a spherical volume, with randomly oriented orbits about the galactic center. Eventually, the entire cloud began to fall in on itself, and as it did, star formation continued to occur, so that many stars were born with motions directed toward the galactic center. These stars assumed highly eccentric orbits, accounting for the motions observed today in Population II stars.

Apparently, the pregalactic cloud was originally rotating, for we know that as the collapse proceeded, a disklike shape resulted. Rotation forced this to happen, just as it did in the contracting cloud that was to form the solar system (see Chapters 7 and 12). Rotational forces slowed the contraction in the equatorial plane, but not in the polar regions, so material continued to fall in there. The result was a highly flattened disk. Stars that had already formed before the disk took shape retained the orbits in which they were born; stars are unaffected by the fluid forces (e.g., viscosity) that caused the gas to continue collapsing to form a disk.

Stars that were formed after the disk had developed differed from their predecessors in at least two respects: they contained greater abundances of heavy elements, because by this time some stellar recycling had occurred, enriching the interstellar gas; and they were born with circular orbits lying in the plane of the disk. These are the primary traits of Population I stars.

Since the time of the formation of the disk, additional, but relatively gradual, changes have occurred. Further generations of stars have lived and died, continuing the chemical enrichment process, particularly in the inner region of the disk, and creating the chemical abundance gradient mentioned earlier. Apparently, the enrichment process was once more rapid than it is today, because the present rate of star

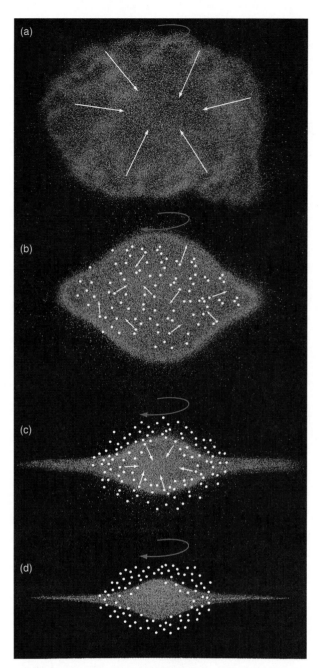

Figure 17.28 Formation of the galaxy. The pregalactic cloud, rotating slowly, begins to collapse (a). The stars formed before and during the collapse have a spherical distribution and noncircular, randomly oriented orbits (b). The collapse leads to a disk with a large central bulge, surrounded by a spherical halo of old stars and clusters (c). The disk flattens further and eventually forms spiral arms (d). Stars in the disk have relatively high heavy element abundances because they formed from material that had been through stellar nuclear processing.

formation is too slow to have built up the quantities of heavy elements that are observed in Population I stars. At some time in the past, probably when the disk was forming, there must have been a period of intense star formation, during which the abundances of heavy elements in the galaxy jumped from almost none to nearly the present level. A large fraction of the galactic mass must have been cooked in stellar interiors and returned to space in a brief episode of stellar cycling whose intensity has not been matched since.

Apparently, massive stars were created at a greater rate during this early phase of intense star formation than they are today. Studies of star formation have shown that isolated dark clouds tend to form low-mass stars, while large star-formation regions, characterized by intense ultraviolet radiation and warm temperatures, tend to form stars of greater mass. Thus, when intense star formation occurred early in the history of the galaxy, many massive stars were born; these stars quickly produced heavy elements and dispersed them into the interstellar medium through supernova explosions.

If large numbers of massive stars did form early in the galaxy's history, then today there may be many compact stellar remnants such as neutron stars and black holes. The possibility that these exotic objects might account for the dark matter in the galactic halo was discussed earlier in this chapter.

We have now recounted almost all the events that led up to the present-day Milky Way, except for the formation of the spiral arms. It is not known when this took place, but it was probably soon after the formation of the disk itself. We see few gas-rich, disk-shaped galaxies without spiral structure, which leads to the conclusion that a disk galaxy containing large quantities of interstellar matter does not exist long in an armless state.

It is difficult to guess exactly what the future of the galaxy will be. The interstellar gas and dust are being consumed by star formation, but only gradually, so the general appearance of the galaxy may not change substantially for many billions of years. Perhaps in the very distant future, when most of the ingredients for new stars have been used up, our galaxy will become an armless disk containing only old stars and stellar remnants.

The Age of the Galaxy

The scenario we have described for the formation of the galaxy is thought to be generally correct, but some substantial uncertainties remain. Most of these have to do with the time scale on which the galaxy formed and the duration of certain stages along the way.

Estimates of the age of the galaxy using different techniques exhibit some discrepancies. We have already mentioned the use of globular cluster ages as an indicator of the age of the galaxy. The general idea is that the galaxy is probably comparable in age to the oldest systems within it—the globular clusters. This assumption would not be correct, however, if some delay occurred between the time the galaxy formed and the time the globular clusters developed. Recently, we have received some hint that we do not fully understand the ages of the globular clusters.

The age of a cluster is determined from the main-sequence turnoff point, as described in Chapter 15. We did not emphasize, however, that while this method provides reliable *relative* ages, it relies on stellar evolution models to yield actual ages in years. The accuracy of the model calculations is always a bit uncertain, especially in cases of old clusters whose chemical composition is deficient in heavy elements.

Recent analyses of globular clusters in our galaxy reveal that they are probably not all the same age. Some appear to be as "young" as 11 billion years, while others are as old as 16 billion years. If the globular clusters really have such a wide range of ages, the formation of the galaxy was probably not as simple as outlined in the previous section. There must have been a rather extended period of time (perhaps 5 billion years) before the galaxy completed its contraction to a disklike shape. During this time the galaxy must have retained the roughly spherical shape of the halo, while the globular clusters formed here and there at different times.

It is not easy to explain why the galaxy would have undergone such a drawn-out early phase prior to collapsing to form a disk. One suggestion is that the galaxy formed from the conglomeration of several smaller systems of stars, which themselves formed earlier (and at different times, accounting for the spread in ages now observed). In this picture, our galaxy was created by the merger of a number of subgalaxies, systems of stars much like globular clusters.

Thus, rather than the entire galaxy forming from a single cloud of gas that collapsed, each of these subsystems would have formed separately from the collapse of a cloud. As we will see in the coming chapters, there is growing evidence that galaxies in general formed in this way. In that case, the earliest stellar systems in the universe were smaller than galaxies, with galaxies forming later as the result of mergers of these smaller systems.

The age of the Milky Way can be estimated in at least two other ways, both of them independent of the globular cluster method. One method uses white dwarfs as age indicators. Through a statistical analysis of the luminosities of white dwarfs scattered around our part of the galaxy, it is possible to determine the age of the oldest (this depends on knowledge of the time required for white dwarfs to cool to the point of being too dim to be observed). This method yields a rather low estimate of the age of the galaxy—around 9 billion years. The significance is a bit unclear, however, since the white dwarfs near enough to be observed all lie in the local portion of the galactic disk, which is certainly not as old as the oldest portions of the galaxy.

The other method of age estimation also depends in part on theoretical calculations. From knowledge of the rates of the nuclear reactions that produce heavy elements, it is possible to estimate how old the galaxy must be in order to contain the quantities of these elements that exist today. This method involves substantial uncertainties and consequently does not provide very precise answers. It has been shown, however, that the current abundances of heavy elements are consistent with an age for the galactic disk of 11 to 15 billion years, in keeping with the age estimated from the oldest globular clusters.

It is not understood why the white dwarf and globular cluster methods of age determination provide such different answers. Certainly, the picture we have developed of the formation and evolution of our galaxy is subject to further revision as this question is resolved and the wide range of globular cluster ages is further explored.

Recently, the age of the galaxy has emerged as one of the most critical of all questions facing astronomers today. New evidence on the age of the universe, based on *Hubble Space Telescope* data, appears to suggest that the age of the universe might be less than the 14 to 16 billion years assigned to some globular clusters. If these *HST* results are confirmed, then astronomers have to face the fact that either their estimates of globular cluster ages or their understanding of the evolution of the universe as a whole is seriously in error.

Summary

1. The Milky Way is a spiral galaxy, consisting of a disk with a central nucleus and a spherical halo where the globular clusters reside. The disk is about 30 kpc in diameter.

2. Distances within the Milky Way can be determined by main-sequence fitting and by use of the period-luminosity relation for variable stars.

3. Star counts seemed to indicate that the Sun was in the densest part of the Milky Way, but mea-

surements of the distribution of globular clusters and analysis of stellar motions show that the Sun is very far (about 8.5 kpc) from the center of the galaxy. The discrepancy was resolved when it was discovered that interstellar extinction affects the star counts.

4. The mass of the galaxy is estimated by applying Kepler's third law to the orbits of stars, but the result is accurate only for the disk. The rotation curve shows that orbital velocities do not decrease with distance from the center, thus indicating that an enormous quantity of mass in some unseen form resides in the outermost portions of the galaxy.

5. Between 10 and 15% of the mass of the galactic disk is in the form of interstellar matter, including both dust particles and gas. This medium ranges from dark, cold clouds, where the gas is molecular and stars form, to more diffuse clouds and to a very hot intercloud medium with temperatures as high as 1 million K. The interstellar material is stirred up by the supernovae, which account for the very hot regions and for motions of interstellar clouds.

6. The spiral structure of our galaxy is most easily and directly measured through radio observations of the 21-centimeter line of hydrogen atoms in space. The spiral pattern is complex with many segments of spiral arms.

7. A variety of evidence indicates that chaotic, energetic activity is associated with the central core of our galaxy. The data show that a compact, massive object is at the center, and the best explanation is that it is a massive black hole.

8. Observations of our own galaxy and similar ones show that heavy elements are more abundant in the central bulge and disk than in the halo. These differences led to the categorization of stars into populations, which are indicators of relative age.

9. Regions with high abundances of heavy elements are thought to have undergone many cycles of enrichment through the formation, evolution, and explosion of massive stars, whereas regions where heavy elements are relatively scarce are thought to have ceased forming new generations of stars long ago.

10. Because the halo is dominated by old stars, it is thought that the galaxy initially was a spherical object. The disk formed later, as material enriched due to stellar processing collapsed gravitationally and was forced into a disk by rotation. The wide range of ages among halo clusters suggests that the original spherical pro-

togalaxy may have built up from the merging of many smaller systems of stars.

Review Questions

1. Compare the reasoning of Shapley in deciding that the globular clusters must be concentrated around the galactic center with that of Aristarchus who, more than 2,000 years earlier, deduced that the Sun is at the center of the solar system.

2. What are the advantages of using pulsating variable stars as distance indicators, compared with other distance methods we have discussed?

3. Why was it so difficult for astronomers to determine the size and shape of the galaxy, as well as our location in it? How is this difficulty related to the fact that the probable barlike structure of the central bulge of the galaxy was not discovered until 1990?

4. Explain the assumptions made in determining the galactic mass from Kepler's third law, and discuss the validity of those assumptions. What does this have to do with the dark matter problem in the galactic halo?

5. How do you think astronomers can distinguish between a cool star that is intrinsically red and a hotter star that appears just as red because of interstellar dust?

6. Explain why the 21-centimeter line of atomic hydrogen is a powerful tool for mapping the structure of the galaxy.

7. Discuss the two alternative mechanisms for the formation and maintenance of spiral arms. Which appears to fit our galaxy better?

8. Summarize the evidence that a massive black hole might exist at the core of the galaxy.

9. Given the great age of the globular clusters, do you think their chemical composition might differ from that of younger stars? Explain.

10. Explain why the evolution of the most massive stars, rather than the far more common lower-mass stars, has been the main contributor to the enrichment of heavy elements in the galaxy.

Problems

1. A Type I Cepheid variable is observed to have a pulsation period of 10 days. If its average apparent visual magnitude is +12, estimate its distance. (*Hint:* You will have to estimate the average visual

Stellar Statistics

ASTRONOMICAL

ACTIVITY

Taking a detailed and complete census of all the stars in the galaxy is impossible, but getting a good idea of the relative numbers of stars of different types is not so difficult. One way to do this is to count the stars in a representative volume of space and determine the relative numbers of different types. Then, by assuming that the galaxy as a whole has a similar distribution of types, you have learned something about the large-scale statistics of stars in the galaxy. This technique is very similar to the polling methods used to determine national opinions and preferences; a small sample of people is questioned, and then the results are applied to the overall population of the nation. You can use information in Appendix 10 to carry out a limited census of stellar types

in the galaxy. The second table in Appendix 10 is a list of nearby stars; it is a complete (or nearly so) compilation of all stars within 5 pc of the Sun. This is an extremely small volume of space to use as representative of the galaxy as a whole, but even so you will find some meaningful trends.

Draw an H-R diagram for the stars in the table, placing spectral type on the horizontal axis and absolute visual magnitude on the vertical (note that your scale will have to extend as far as +17 to include stars fainter than those shown on the H-R diagrams in the text). Omit white dwarfs and stars for which the spectral type is not given.

What does your diagram tell you about the relative numbers of different types of stars in the galaxy? Look at the first list in Appendix 10, which provides data on the 50 brightest stars in the sky. Would the H-R diagram for this sample of stars look the same as the one you just drew for the nearest stars? Why would the list of brightest stars not be a representative sample of the relative numbers of stars of different types in the galaxy?

absolute magnitude from Figure 17.5 and then use the distance modulus, as discussed in Chapter 14.)

2. A cluster of stars is observed and its H-R diagram is constructed. Comparison of the main sequence with that in the standard H-R diagram shows that the cluster main sequence is 25 magnitudes fainter than the main sequence in the standard diagram; that is, the difference between the visual apparent magnitude and the visual absolute magnitude for the cluster is 25 magnitudes. How far away is the cluster? Could it lie within the disk of the galaxy, or is it too far away?

3. Suppose that interstellar obscuration makes a star appear five magnitudes fainter than it would otherwise. If the star's apparent magnitude is +17, what would the apparent magnitude be if the star were not obscured by dust? If the star's absolute magnitude is +2, how far away is it? Compare your answer with what you would find if you did not correct the apparent magnitude for the effects of dust.

4. If the precise rest wavelength of the 21 cm emission line is 21.102 cm, what is the velocity of a cloud whose 21 cm line is observed to have a wavelength of 21.117 cm? Is this cloud moving toward the Earth or away from it?

5. Suppose the orbital speed for a star orbiting the galaxy at a distance of 4 kpc from the center is 100 km/sec. If you use Kepler's third law as we did in the text, what value would you find for the central mass of the galaxy? Compare your answer with the value found using the Sun's orbital distance and speed, and comment on the implications for the structure and rotation of the galaxy. (*Hint:* To find the orbital period for the star, you need to divide the circumference of its orbit, which is $2\pi a$, by the star's orbital speed.)

6. If there is an object at the galactic center with a mass equal to 1 million solar masses, what is the orbital period for a star 10 pc from this object? If this star's orbit is circular, what is its orbital speed? Comment on the relationship of your result to observations suggesting the presence of a massive black hole at the galactic core.

7. If the mass of the galaxy is 1.5×10^{11} solar masses, what is the orbital period for a globular cluster whose semimajor axis is 20 kpc? About how many times has this cluster orbited the galaxy since the galaxy formed?

8. To appreciate what the nighttime sky might be like if the Earth orbited a globular cluster star, calculate the apparent magnitudes of the first five

stars in the list of the brightest stars (Appendix 10) if each star were only 0.1 pc away (the average distance between stars in a globular cluster).

Additional Readings

Binney, J. 1995. The Evolution of Our Galaxy. *Sky & Telescope* 89(3):20.

Chevalier, R. A. and C. L. Sarazin 1987. Hot Gas in the Universe. *American Scientist* 75:609.

Crosswell, K. 1992. Galactic Archeology. *Astronomy* 20(7):28.

Crosswell, K. 1995. What Lies at the Heart of the Milky Way? *Astronomy* 23(5):32.

Crosswell, K. 1996. The Dark Side of the Galaxy. *Astronomy* 24(10):40.

Elitzur, M. 1995. Masers in the Sky. *Scientific American* 272(2):68.

Gingerich, O. and B. Welther 1985. Harlow Shapley and the Cepheids. *Sky & Telescope* 70(6):540.

Goldsmith, D. and N. Cohen 1991. The Great Molecule Factory in Orion. *Mercury* 20(5):148.

Jayawardhana, R. 1995. Destination: Galactic Center. *Sky & Telescope* 89(6):26.

Knapp, G. 1995. The Stuff Between the Stars. *Sky & Telescope* 89(5):20.

Lomberg, J. 1993. A Portrait of Our Galaxy. *Sky & Telescope* 86(6):38.

Mateo, M. 1994. Searching for Dark Matter. *Sky & Telescope* 87(1):20.

O'Dell, R. 1994. Exploring the Orion Nebula. *Sky & Telescope* 88(6):20.

Oort, J. 1992. Exploring the Nuclei of Galaxies (Including Our Own). *Mercury* 21(2):57.

Townes, C. and R. Genzel 1990. What Is Happening in the Center of Our Galaxy? *Scientific American* 262(4):46.

Tucker, W. 1994. A Brightening Star Reveals Dark Matter. *Astronomy* 22(8):40.

Van Buren, D. 1993. Bubbles in the Sky. *Astronomy* 21(1):46.

Van Den Bergh, S. and J. E. Hesser 1993. How the Milky Way Formed. *Scientific American* 268(1):72.

Verschuur, G. L. 1992. Interstellar Molecules. *Sky & Telescope* 83(3):379.

White, R. E. 1991. Globular Clusters: Fads and Fallacies. *Sky & Telescope* 81(1):24.

Wynn-Williams, G. 1993. Bubbles, Onions, and Sheets: The Diffuse Interstellar Medium. *Mercury* XXII(1):2.

Web Connections

The Review Questions and Problems also appear at the following URLs:
http://universe.colorado.edu/ch17/questions.html
http://universe.colorado.edu/ch17/problems.html

Galaxies are like cities of stars
suspended forever in the blackness of night.

Denise Wolf, 1986

Chapter 18
The Universe of Galaxies

We have spoken of our galaxy as one of many, a single member of a vast population that fills the universe. Given all that we have learned about the ordinary position of human beings in the cosmos, this is no surprise. Having traced our painful progression from the geocentric view to the realization that we occupy an insignificant planet orbiting an ordinary star in an obscure corner of the galaxy, we should be surprised if we found that our galaxy held any kind of unique status in the larger environment of the universe as a whole. It doesn't.

The Mystery of the Nebulae

By the beginning of this century, astronomers were well aware of the numerous "nebulae" scattered around the sky, but were uncertain of their nature. Some, of course, were glowing gas clouds in our own galaxy, but others, particularly those with spiral shapes, did not fit this picture so well. Spectroscopic measurements showed that their light had an ab-

sorption-line spectrum closely resembling the spectra of stars, yet individual stars could not be discerned. Many astronomers had a natural predisposition to think that the nebulae were distant galaxies like our own, but demonstrating this was not easy.

The great observatory on Mt. Wilson, overlooking Pasadena, California, provided most of the essential data used in the quest to learn about the nebulae, and thus most of the main players were American astronomers. Mount Wilson opened in the early years of this century with a 1.5 m (60-inch) telescope, which was joined by a fine 2.5 m (100-inch) instrument starting in 1919 (the Mt. Wilson 2.5 m telescope was to be the largest in the world for some 38 years, until the great 5 m [200-inch] instrument at Mt. Palomar, near San Diego, was completed). The 2.5 m telescope was destined to aid in many crucial discoveries about our place in the universe, none of them more significant than the question of whether our galaxy stood alone or was one of many.

Among those supporting the "island universe" idea, that the nebulae were distant galaxies, was Heber D. Curtis, an American astronomer who observed novae in a few of the nebulae and interpreted their faintness as an indication of their great distances (but he underestimated the distances, not realizing that what he had seen were supernovae; which are much more luminous than ordinary novae). Chief among those opposing this viewpoint was Harlow Shapley, whose contributions to the understanding of the size and structure of the Milky Way we discussed in the last chapter. Shapley's size estimate for our galaxy was, we now know, far too large, and this allowed him to treat at least some of the nebulae as objects contained within our own star system. He was encouraged in this belief by measurements, later found to be incorrect, of rotation in some of the spiral nebulae, which were published by Adrian van Maanen. Measurements of photographs taken a few years apart seemed to show definite rotation, something that could not have been detected in any object as large as a galaxy in times so short. Shapley also argued that the distribution of the nebulae on the sky indicated a local position: the nebulae tended to be seen only above and below the plane of the Milky Way, supporting the view that they were somehow distributed about the plane of our galaxy. Now we

Figure 18.1 Edwin Hubble. Hubble's discovery of Cepheid variables in the Andromeda nebula led to the unambiguous conclusion that this object lies well beyond the limits of the Milky Way and must therefore be a separate galaxy. Hubble later made important discoveries about the properties of galaxies and what they tell us about the universe as a whole (see Chapter 19). (Niels Bohr Library, California Institute of Technology)

http://universe.colorado.edu/fig/18-1.html

realize that interstellar extinction, caused by dust in the disk of our galaxy, accounts for this distribution because it makes distant galaxies impossible to observe close to our galactic plane. But Shapley's work came a decade before the discovery of the diffuse dust medium.

In 1920 Curtis and Shapley conducted a debate before the National Academy of Sciences, aimed at airing the issues and helping to determine the nature of the nebulae. The first part of the debate was devoted to the size scale of the Milky Way, which was crucial to Shapley's case. Subsequently, the nature of the nebulae was discussed. No clear-cut "winner" emerged, but the debate did help to shape thinking and identify new lines of research. The erroneous measurements of the galaxy's size by Shapley, and of rotation in some of the nebulae by van Maanen, delayed acceptance of the island universe idea. But in 1924 the matter was settled unequivocally by the American Edwin Hubble **(Fig. 18.1).** Hubble obtained a series of fine photographs of M31, the Andromeda nebula, and was able to identify individual stars, including some Cepheid variables. Use of the period-luminosity relation established a distance for M31 that was far too great to fit within the Milky Way, even if Shapley's overestimate of our galaxy's size were accepted.

Hubble's great discovery changed forever our perception of our place in the cosmos. It was truly revolutionary to think that despite the vastness of our galaxy of hundreds of billions of suns, it was only one of a multitude of star systems equally as vast.

The Basic Properties of Galaxies

Even though galaxies are ensembles of stars rather than individual objects, they have systematic properties that can be measured and categorized. The ultimate goal of doing so is to learn how galaxies form and evolve and what role they play in the structure and composition of the universe at large.

The First Step: Classification by Shape

Having established that the nebulae were truly extragalactic objects, Hubble began a systematic study of their properties. The most obvious basis for classifying the nebulae was to do it according to shape. Hubble did this, designating the spheroidal nebulae as **elliptical galaxies (Fig. 18.2),** distinct from the **spiral galaxies (Fig. 18.3).** Within each of the two general types, Hubble established subcategories on the basis of less dramatic gradations in appearance **(Table 18.1).** The ellipticals displayed varying degrees of flattening

Figure 18.2 An elliptical galaxy. These smooth, featureless galaxies are probably more common than spirals. (National Optical Astronomy Observatories)
http://universe.colorado.edu/fig/18-2.html

Figure 18.3 Spiral galaxies. Here are four typical examples of spiral galaxies. At the top left is NGC 3627; at the top right is NGC 6946 (these designations refer to the Messier Catalog, see Appendix 17); at the bottom left is NGC 3623; at the bottom right is NGC 4321 (these designations are from the *New General Catalog*). (top left, bottom left: © 1992 AAT Board, photos by D. Malin; top right: U.S. Naval Observatory; bottom right: J. D. Wray, University of Texas) http://universe.colorado.edu/fig/18-3.html

Table 18.1

Types of Galaxies

Type	Designation	Characteristics
Elliptical	E0–E7	Spheroidal shape; subtype determined by the expression $10(1 - b/a)$, where a is the long axis and b is the short axis of the galaxy.
Dwarf elliptical	dw E	Spheroidal shape; very low mass and luminosity.
S0	S0	Disklike galaxies with no spiral structure.
Spiral	Sa–Sc	Disklike galaxies with spiral arms; subtype determined by relative size of nucleus and openness of spiral structure; Sa refers to large nucleus and tightly wound arms, whereas Sc refers to small nucleus and open spiral arms.
Barred spiral	SBa–SBc	Spiral galaxies with elongated, barlike nuclei; subtypes determined the same way as in spiral galaxies.
Irregular I	Ir I	Disklike galaxies with evidence of spiral structure, but not well organized.
Irregular II	Ir II	Galaxies that do not fit into any of the other types; some Ir II galaxies have been found to be normal spirals heavily obscured by interstellar gas and dust.

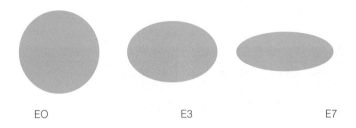

EO   E3   E7

Figure 18.4 The shapes of ellipticals. The numerical designation (following the letter E) is given by the formula $(1 - b/a) \times 10$, where b is the short axis and a is the long axis of the galaxy image. Here are three examples: the EO galaxy has $a = b$ (that is, it is circular); the E3 galaxy has $a = 1.4b$; and the E7 has $a = 3.3b$. The E7 galaxy is the most highly elongated of the ellipticals.

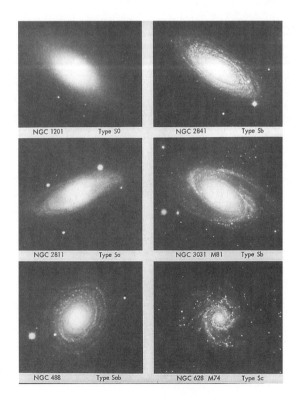

Figure 18.5 An assortment of spirals. This sequence shows spiral galaxies of several subclasses. (Palomar Observatory, California Institute of Technology)

and were sorted out according to the ratio of the long axis to the short axis **(Fig. 18.4)**, with designations from E0 (round) to E7 (the most flattened). The number following the letter E is determined by the formula $10(1 - b/a)$, where a is the long axis and b is the short axis, as measured on a photograph.

Hubble based his classification of the spirals on the tightness of the arms, the compactness of the nucleus, and the presence or absence of an elongated central bulge **(Fig. 18.5)**. Hubble found that about half of all spirals have an elongated central region, with the spiral arms emanating from the ends of the central bar-like structure. He called these galaxies **barred spirals (Fig. 18.6)**. Subclasses of both barred and normal spirals were based on the tightness of the spiral arms and the relative size of the central region, with categories ranging from Sa or SBa (tight spiral, large nucleus) to Sc or SBc (open arms, small nucleus).

Following Hubble's original work on galaxy classification, an intermediate class of **S0 galaxies** has been recognized. These appear to have a disk shape, but no trace of spiral arms.

The Milky Way is probably an Sb in Hubble's system—that is, intermediate in both characteristics —although it is difficult to be sure, because we cannot get an outsider's view of what our galaxy looks like. Recall from the previous chapter that recent evidence shows that the Milky Way might be a barred spiral. It is not clear, however, whether the central bulge is elongated enough to change the Milky Way's classification to SBb.

Hubble arranged the types of galaxies in an organization chart that has become known as the "tuning-fork" diagram **(Fig. 18.7)**. Because there are two types of spirals, Hubble chose not to force all the

Figure 18.6 Barred spirals. Above we see photos of two examples of this group, which makes up about half of all spiral galaxies. (Left. © 1977 Anglo-Australian Telescope Board; right: © 1992 Anglo-Australian Telescope Board; bottom: Palomar Observatory, California Institute of Technology)

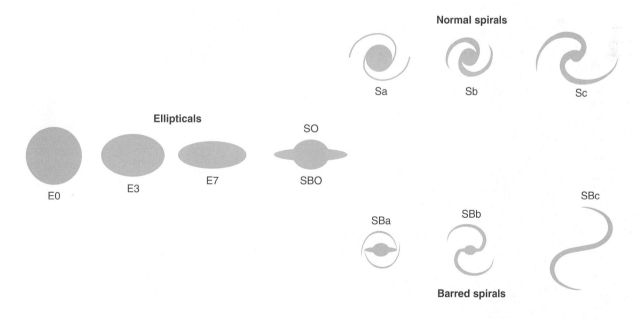

Figure 18.7 The tuning-fork diagram. This is the traditional manner of displaying the galaxy types, as originally devised by Hubble. For quite some time this was thought to be an evolutionary sequence, although the imagined direction of evolution was reversed at least once. Now we know that in the normal course of events galaxies do not evolve from one type to another.

GALAXIES

These galaxy types are arranged according to the classical "tuning-fork" diagram. This was once thought to represent an evolutionary sequence, but it is now recognized as merely a convenient way to display galaxy types.

EO

E4

E7

Sa

SBa

Figure 18.8 Irregular galaxies. Not all galaxies fit the standard classifications. At the left is M82, a well-known example of a galaxy with an unusual appearance. For a time, it was thought that the nucleus of this galaxy was exploding, but more recent analysis indicates that the odd appearance is due instead to a very extensive region of interstellar gas and dust surrounding the center of a spiral galaxy that is undergoing an episode of intense star formation. At the right is another galaxy shrouded in gas and dust, the starburst irregular galaxy NGC1313. (left: Lick Observatory photograph: right: © 1992 AAT Board; photo by D. Malin)
http://universe.colorado.edu/fig/18-8.html

types into a single sequence but to split them into two branches. For a time astronomers thought that this diagram represented an evolutionary sequence, but when later studies showed that all types of galaxies contain old stars, it became clear that one type of galaxy does not evolve into another. The tuning-fork diagram is not an age sequence after all, and the differences in galactic type must be explained in some other way.

Most of the galaxies listed in catalogs are spirals, which are about evenly divided between normal and barred spirals. Only about 20 to 30% of the listed galaxies are ellipticals, and a comparable number are S0 galaxies. The remaining few percent are called **irregular galaxies** because they do not fit into the normal classification scheme **Fig. 18.8.** Because there are probably many small, dim, elliptical galaxies that are usually not sufficiently prominent to appear in catalogs, it seems likely that ellipticals actually outnumber spirals in the universe. This certainly is the case in dense clusters of galaxies, as we will see. **Table 18.2** lists a few promi-

nent galaxies whose properties indicate the typical values for galaxies of different types.

Although the irregular galaxies are, by definition, misfits, some systematic characteristics have been found even in their case. Most have a hint of spiral structure, although they lack a clear overall pattern; these have been designated as **Type I irregulars.** The rest, a small minority, simply do not conform in any way to the normal standards and are classified as **Type II irregulars** or peculiar galaxies (for example, see Fig. 18.8). The Magellanic Clouds are both Type I irregulars.

Recently, based largely on infrared observations, many Type II irregular galaxies have been found to be embedded in so much gas and dust that they are invisible except when observed at infrared wavelengths. Because very active star formation is taking place in these galaxies, as a class they are called **starburst galaxies** (see Fig. 18.8).

Finding the Distances to Galaxies

To probe the physical nature of the galaxies, we must first determine their distances. Without this information, such fundamental parameters as masses and luminosities cannot be deduced.

Classification of Galaxies
In this web activity we'll review the basic classification scheme of galaxies. After working through a review of the classification system, we'll set out and search the internet to find and classify additional galaxies. Our starting point is the following URL:
http://universe.colorado.edu/ch18/web.html

Table 18.2
Selected Bright Galaxies

Galaxy	Type	Angular Diameter (arcmin)	Diameter (kpc)	Distance (kpc)	Apparent Magnitude[a]	Absolute Magnitude[a]	Mass ($M_\odot$)
Large Magellanic Cloud	Ir I	460	7	55	0.1	−18.7	10^{10}
Small Magellanic Cloud	Ir I	150	3	63	2.4	−16.7	2×10^9
Andromeda (M31)	Sb	100	16	700	3.5	−21.1	3×10^{11}
M33	Sc	35	6	730	5.7	−18.8	1×10^{10}
M81	Sb	20	16	3,200	6.9	−20.9	2×10^{11}
Centaurus A	EOp	14	15	4,400	7	−20	2×10^{11}
Sculptor system	dw E	30	1	85	7	−12	3×10^6
Fornax system	dw E	40	2	170	7	−13	2×10^7
M83	SBc	10	12	3,200	7.2	−20.6	—
Pinwheel (M101)	Sc	20	23	3,800	7.5	−20.3	2×10^{11}
Sombrero (M104)	Sa	6	8	1,200	8.1	−22	5×10^{11}
M106	Sb	15	17	4,000	8.2	−20.1	1×10^{11}
M94	Sb	7	10	4,500	8.2	−20.4	1×10^{11}
M82	Ir II	8	7	3,000	8.2	−19.6	3×10^9
M32	E2	5	1	700	8.2	−16.3	3×10^9
Whirlpool (M51)	Sc	9	9	3,800	8.4	−19.7	8×10^{10}
Virgo A (M87)	E1	4	13	13,000	8.7	−21.7	4×10^{12}

[a]The apparent and absolute magnitudes represent the light from the entire galaxy. Because the light from a galaxy comes from a large area in the sky, a galaxy of a given apparent magnitude is not as easily seen by the eye as a star of the same magnitude.

We have already mentioned one technique for distance determination that can be applied to some galaxies: the use of Cepheid variables. These stars are sufficiently luminous to be identified as far away as a few million parsecs (**megaparsecs,** abbreviated *Mpc*), which is sufficient to reach the Andromeda nebula and several other neighbors of the Milky Way. It is not adequate for probing the distances of most galaxies, however, so other techniques had to be developed.

Recall from our discussions of stellar distance determinations (Chapter 14) that we can always find the distance to an object if we know both its apparent and its absolute magnitude. This is what we do when we use the spectroscopic parallax method, the main-sequence fitting technique, and even the Cepheid variable period-luminosity relation. A general term for any object whose absolute magnitude is known from its observed characteristics is **standard candle,** and an assortment of these are used in extending the distance scale to faraway galaxies.

The most luminous stars are the red and blue supergiants, which occupy the extreme upper regions of the H-R diagram. These can be seen at much greater distances than Cepheid variables and therefore are important links to distant galaxies. The absolute magnitudes of these stars are inferred from their spectral classes, just as in the spectroscopic parallax technique, although these stars are so rare that there is substantial uncertainty in assuming that they conform to a standard relationship between spectral class and luminosity. In a variation on this technique, it is simply assumed that there is a fundamental limit on how luminous a star can be, and that in a collection of stars as large as a galaxy, there will always be at least one star at this limit. Then, to determine the distance to a galaxy, it is necessary only to measure the apparent magnitude of the brightest star in it and to assume that its absolute magnitude is at the limit, which is about $M = -8$. This technique extends the distance scale by about a factor of 10 beyond what is possible using Cepheid variables—that is, to 10 or more megaparsecs.

Other standard candles come into play at greater distances. These include supernovae **(Fig. 18.9),** which at peak brightness always reach approximately the same absolute magnitude. (Care must be exercised in making measurements, however, because different types of supernovae differ in absolute magnitude, and questions have been raised about how well the absolute magnitudes are known for the different types.) Supernovae can be observed at distances of hundreds of megaparsecs and are therefore very useful distance indicators; the major drawback is that distances can

Figure 18.9 A supernova in a distant galaxy. Before the distinction between novae and supernovae became clear, these flare-ups (arrow) contributed to the controversy over the nature of the nebulae. Now that these occurrences in galaxies are known to be supernova explosions, comparisons of their apparent and assumed absolute magnitudes provide distance estimates. (Palomar Observatory, California Institute of Technology)

be measured only to galaxies where supernovae happen to be observed. We will see that the use of supernovae as distance indicators is playing a major role in a current controversy about the size and age of the universe.

Another technique called the **Tully-Fisher method,** in which the luminosity of a galaxy is in-

ferred from its mass, shows great promise for spiral galaxies. The mass is not measured directly. Instead astronomers measure the rotational velocity of the galaxy, which depends on its mass. The rotational velocity is estimated from the width of the 21-centimeter radio emission line from hydrogen clouds in the galaxy. The Doppler effect broadens the line according to the maximum spread of velocities from one side of the galaxy's disk to the other **(Fig. 18.10).** Thus the luminosity (hence the absolute magnitude) of a spiral galaxy can be inferred from the width of its 21-centimeter emission line, and its distance can then be found by measuring the apparent magnitude and comparing it with the inferred absolute magnitude. This technique works best when infrared magnitudes are used because infrared light is relatively unaffected by interstellar dust within the galaxy being observed. The Tully-Fisher method is now being used in efforts to improve the intergalactic distance scale, which has very important implications for our understanding of the universe as a whole. A similar method has been developed for elliptical galaxies.

It is important to keep in mind how uncertain all galactic distance determinations are. To assume that all objects in a given class have identical basic properties such as luminosity is always a risky business, especially when such assumptions are applied to objects as distant as external galaxies or as rare as the brightest star in a galaxy. Furthermore, each method as we go farther out in the universe depends on the ones that come before, so an error in one method (say, the

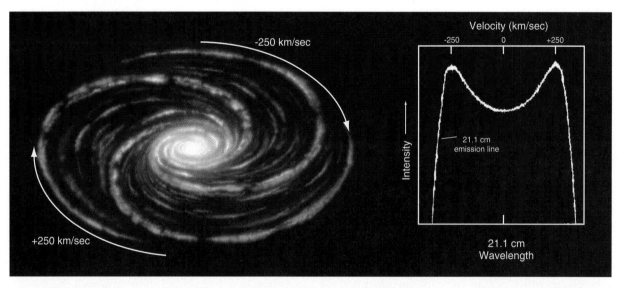

Figure 18.10 Rotation velocity and the width of the 21-centimeter line. This shows how the rotation velocity of a spiral galaxy is related to the width of the 21-centimeter emission line from hydrogen gas in the galaxy. Because of the Doppler effect, 21-centimeter photons from one side of the galaxy are shifted toward shorter wavelengths, those from the other side are shifted toward longer wavelengths, and those from the central regions are not shifted. The result is a 21-centimeter line that is broadened according to how rapidly the galaxy is rotating. Because the rotation speed is governed by the central mass of a galaxy, the width of the line is related to the mass of the galaxy and can be used to estimate its absolute magnitude for distance determination.

Cepheid variable period-luminosity relationship) leads directly to systematic errors in the others. We have few options other than the existing methods, however, so we must simply recognize the inherent limitations in accuracy and take them into account. The uncertainties in distance determinations carry over to our measurements of other properties of galaxies.

The Expansion of the Universe: A New Yardstick

Edwin Hubble followed his proof of the extragalactic nature of the nebulae with another great discovery. This one was built on the work of several others as well as his own observations, and it had an even more profound impact on our understanding of the cosmos and our place in it than did the earlier demonstration that our galaxy is not alone. Hubble's new announcement got right to the heart of what makes the universe tick, what it was like in the past, and what may come of it in the future.

The American astronomer Vesto Slipher, working at the Lowell Observatory in Arizona, had measured spectra of many nebulae in the years prior to 1920 (i.e., before the nebulae were recognized to be galaxies). The spectra of the nebulae generally looked like spectra of moderately cool stars, with absorption lines (usually much broader than in spectra of single stars) of singly ionized metals and of hydrogen (review the stellar spectral classification discussion in

Chapter 14 if this does not sound familiar). Slipher found a curious phenomenon: the spectral lines of galaxies were nearly always shifted toward the red, compared with laboratory wavelengths. If these shifts were due to the Doppler effect, then all nebulae must be moving away from the Earth.

Subsequently, Hubble and his colleague Milton Humason gathered spectra of additional nebulae (known to be galaxies by then), and eventually Hubble made a graph showing redshift versus distance **(Fig. 18.11)**. This graph showed a truly revolutionary trend: the more distant a galaxy was, the faster it was moving away from the Earth. This suggested that the universe is expanding uniformly.

Hubble announced his results in 1929, and studies of the universe have been profoundly influenced by it ever since. He and Humason spent the next two decades gathering additional measurements of distances and redshifts, expanding the graph farther and farther **(Fig. 18.12)**. Today the relationship between distance and redshift is known as the **Hubble law**.

Mathematically, the Hubble law can be expressed very simply: $v = H_0 d$, where v is the speed of recession of a galaxy at distance d, and H_0 is the **Hubble**

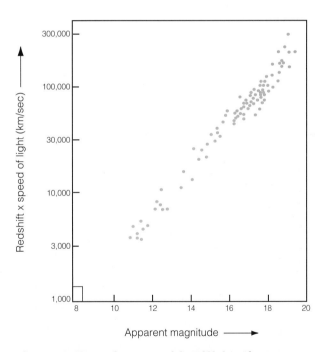

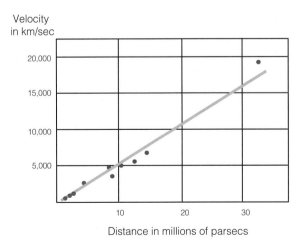

Figure 18.11 The Hubble law. This early diagram prepared by E. Hubble and M. Humason shows the relationship between galaxy velocity and distance. The slope of the relation has since been altered as data for more galaxies have been added and the measurement of distances has improved, but the general appearance of this figure is the same today. (Estate of E. Hubble and M. Humason. Reprinted with permission)

Figure 18.12 A modern version of the Hubble law. Often, apparent magnitude is used on the horizontal axis instead of distance, because the two quantities are related, particularly if the diagram is limited to galaxies of the same type, as this one is. The small rectangle at the lower left indicates the extent of the relationship as Hubble first discovered it; today many more, much dimmer galaxies have been included. The most rapid galaxies are not actually traveling at the speed of light as this diagram implies, because relativistic corrections have not been applied (see the discussion in the next chapter). (Adapted from J. Silk, 1980, *The Big Bang* [San Francisco: W. H. Freeman])

constant. In the graph of velocity versus distance, H_0 is the slope, or the proportionality constant. If we can measure the value of H_0, we can use it to determine speed when distance is known, or distance when speed is known.

The fact that there is a straight-line relationship between redshift and distance means that we can use the Hubble diagram to find distances. All we need in order to find the distance to a galaxy is the Hubble diagram and a spectrum from which the redshift can be measured. Distances found in this manner are called **cosmological distances,** because they are based on the cosmological interpretation of the diagram, namely, that the universe is expanding and that the observed redshifts are due to this expansion.

To use the Hubble diagram to find distances, all that is needed is to establish the correct slope of the straight-line relationship in the diagram; that is, to determine the value of H_0. Once H_0 is known, then the Hubble relation can be solved for distance *d*:

$$d = v/H_0.$$

This provides a new and powerful method for finding distances to the most faraway objects in the universe, provided that *v* can be measured from the redshift, that the value of H_0 is known, and assuming that the redshift is cosmological.

But finding the value of H_0 with accuracy has proved to be a very difficult problem. The difficulty is that distances to galaxies have to be measured by some other technique in order to determine the slope of the diagram; that is, you have to be able to draw the diagram accurately before you can use it to find accurate distances to other galaxies. But the farther out we go, the more inaccurate the distance techniques become, as we have seen in the previous section.

Another difficulty arises from the fact that the redshift-distance relation is not a sharp line, but a fuzzy line that includes a lot of scatter. This is due to a combination of uncertainty in determining the distances and the fact that individual galaxies can have local motions that are not part of the universal expansion. Within a cluster of galaxies, for example, the individual members may move with speeds of a few hundred kilometers per second, and these motions can be in any direction. The Doppler shifts due to local motions are added to the shift due to the expansion of the universe, thus creating a range of speeds for galaxies at a specific distance from us.

The current estimates for the value of H_0 range between 45 and nearly 90 km/sec/Mpc. For now we will take a value of 75 km/sec/Mpc, which means that the velocities of galaxies increase by 75 km/sec

for every million parsecs of distance. Thus, for example, a galaxy having a velocity (determined from its redshift) of 150 km/sec would be at a distance of 2 Mpc.

The use of the Hubble diagram to find distances to faraway galaxies is extremely useful, but also extremely uncertain. Thus astronomers tend to be very careful, either discussing only *relative* velocities or being sure to state what value of H_0 they are assuming. In Chapter 20 we will explore further the ramifications of universal expansion and the drive to find the correct value of H_0, which has become something of a Holy Grail to modern astronomers.

Weighing the Galaxies

The masses of nearby galaxies can be measured in the same manner as the mass of our own galaxy: by applying Kepler's third law to the orbital motions of stars or gas clouds in the outer portions. All that is required is to measure the orbital velocity at some point well out from the center and to determine how far from the center that point is (which in turn requires knowledge of the distance to the galaxy). Then Kepler's third law leads to

$$M = \frac{a^3}{P^2},$$

where M is the mass of the galaxy (in solar masses), and a and P are the semimajor axis (in AU) and the period (in years), respectively of the orbiting material at the observed point. (See the more complete discussion in Chapter 17 to remind yourself how this equation was developed.) Note that the orbital period is far too long—typically hundreds of millions of years—to be observed directly. The period must be estimated by knowing the orbital speed and the distance from the center of the galaxy, and then computing how long it takes for a star to complete one orbit.

One of the difficulties with this technique is that both the distance and the orientation of the galaxy must be known before the true orbital velocity and the semimajor axis can be determined. The orientation can usually be deduced for a spiral galaxy, since it has a disk shape whose tilt can be seen, and the distance can be estimated using one of the methods just outlined. In most cases the orbital velocities are measured at several points within a galaxy from the center out as far as possible, and the data are plotted as a rotation curve **(Fig. 18.13),** which is simply a diagram showing how orbital velocity varies with distance from the center. The most effective method of obtaining velocity data on the outer portions of a spiral galaxy is to measure the 21-centimeter emission from hydrogen, which can be detected at greater distances from the center than visible light from stars can be measured **(Fig. 18.14).**

The technique of measuring rotation curves can best be applied to spiral galaxies, where there is a disk with stars and interstellar gas orbiting in a coherent fashion. Elliptical galaxies have no such clear-cut overall motion, and a slightly different technique must be used. The individual stellar orbits are randomly oriented in the outer portions of an elliptical galaxy, so that there is a significant range of velocities within any portion of the galaxy's volume. This range of velocities, called the **velocity dispersion,** is greatest near the center of the galaxy, where the stars move fastest in their orbits, and smaller in the outer regions. Furthermore, the greater the mass of the galaxy, the greater the velocity dispersion at any distance from the center, so a measurement of this parameter can lead to an estimate of the mass of a galaxy. The velocity dispersion is deduced from the widths of spectral lines formed by groups of stars in different

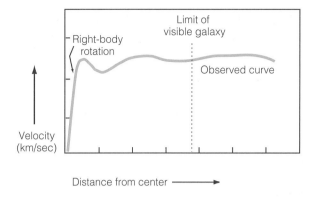

Figure 18.13 A schematic rotation curve for a galaxy. As this diagram illustrates, in most spiral galaxies, the rotation velocity does not drop off in the outer regions, as it would if most of the mass were concentrated at the center of the galaxy. The data on rotation speed in the outermost regions come from radio observations.

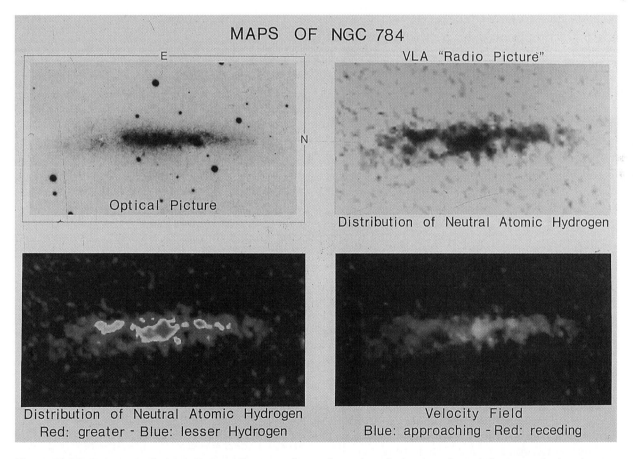

Figure 18.14 Measuring a galactic rotation curve. This sequence of images illustrates how radio observations of atomic hydrogen are used to derive the rotation curve for a spiral galaxy. At the upper left is a visible-light photograph of the nearly edge-on galaxy; at the upper right is a radio map obtained in the 21-centimeter line of hydrogen. In the two lower panels, the radio data have been color-coded, on the left to indicate relative intensities of the hydrogen emission, and on the right to indicate relative line-of-sight velocities. In this last panel we can see that one side of the galaxy (red) is receding and the other side (blue) is approaching. Comparison of the velocities leads to a determination of the rotational speed as a function of distance from the galactic center. (National Radio Astronomy Observatory)

portions of a galaxy **(Fig. 18.15):** the greater the internal motion within the region observed, the greater the widths of the spectral lines, due to the Doppler effect (caused by the fact that some stars are moving toward the Earth and others away from it).

Kepler's third law can sometimes be used in a different way entirely to determine galactic masses. Here and there in the cosmos are double galaxies, orbiting each other exactly like stars in binary systems. In these cases, Kepler's third law can be applied to derive an estimate of the combined mass of the two galaxies. The uncertainties are even more severe than in the case of a double star, however, because the orbital period of a pair of galaxies is measured in hundreds of millions of years. Thus the usual problems of not knowing the orbital inclination or the distance to the system are compounded by inaccuracies in estimating the orbital period, something that is usually well known for a double star. Still, this technique is useful and has one major advantage: it takes into account *all* the mass of a galaxy, including whatever part of it is in the outer portions, beyond the reach of the standard rotation curve or velocity dispersion measurements. Interestingly, galactic masses estimated from double systems are generally much larger than those based on measurements of internal motions within galaxies, possibly indicating that most galaxies have extensive halos containing

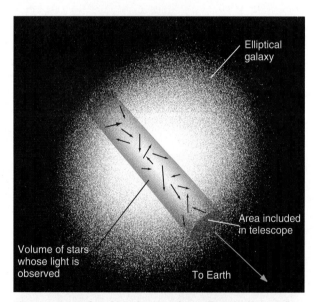

Figure 18.15 The velocity dispersion for an elliptical galaxy. An elliptical galaxy has no easily measured overall rotation, so the mass cannot be estimated from a rotation curve. Instead the average velocities of stars at a known distance from the center are used. Light from a small area of the galaxy's image is measured spectroscopically. The random motions of the stars in the observed part of the galaxy broaden the spectral lines by the Doppler effect, and the amount of broadening is a measure of the average velocity of the stars in the observed region. At a given distance from the center of the galaxy, the higher the average velocity, the greater the mass contained inside that distance.

large quantities of matter. As noted in Chapter 17, independent evidence indicates that our own galaxy has a massive halo, containing perhaps as much as 90% of the total mass.

Luminosities of Galaxies and the Mass-to-Light Ratio

Once the distance to a galaxy has been established, its luminosity and size can be deduced directly from the apparent magnitude and apparent diameter. Both quantities are found to vary over wide ranges, with luminosities as low as 10^6 and as high as 10^{12} times that of the Sun, and diameters ranging from about 1 to 100 kiloparsecs (kpc).

Elliptical galaxies generally display a wider range of luminosities and sizes than the spirals. Spirals tend to be more uniform, with luminosities usually between 10^{10} and 10^{12} solar luminosities and diameters between 10 and 100 kpc. Ellipticals display a much broader range of properties. The smallest elliptical galaxies are called **dwarf ellipticals.** These may be very common, but they are too dim to be seen at great distances, so we can only say for sure that there are many of them near our own galaxy and the Andromeda galaxy.

Galaxies of a given type seem to be fairly uniform in other properties, just as stars of a given spectral type are the same in other ways. One quantity often used by astronomers to characterize galaxies is the **mass-to-light ratio (*M/L*),** which is simply the mass of a galaxy divided by its luminosity, with both measured in solar units. Values of the mass-to-light ratio typically range from 5 to 200. Any value larger than 1 means that the galaxy emits less light per solar mass than the Sun; that is, the galaxy is dominated in mass by stars that are dimmer than the Sun or by dark matter. Even the smaller mass-to-light ratios that are observed for galaxies are much larger than 1; a value of *M/L* = 50, for example, means that 50 solar masses are required to produce the luminosity of one Sun. Interestingly, counts of stars in the Sun's part of our galaxy indicate a mass-to-light ratio near 1, whereas we expect (based on observations of similar galaxies) that it is much larger than 1 for our galaxy as a whole. This indicates that significant mass in galaxies like ours must be in forms that are very difficult to see, perhaps not in the form of normal stars at all. It is possible that the invisible mass resides in the halos of galaxies in some unseen form. Recall from Chapter 17 that this appears to be true of our own galaxy. The whole question of the nature of the unseen "dark matter" in galaxies is currently an area of intense interest in astronomy.

Elliptical galaxies tend to have the largest mass-to-light ratios, consistent with our earlier statement that these galaxies are relatively deficient in hot, luminous stars. Spirals, on the other hand, which contain some of these stars, have lower mass-to-light values. It is worth stressing again, however, that even the low mass-to-light ratios for these galaxies are much greater than 1, indicating that they too are dominated in mass by very dim stars or dark matter (including, of course, interstellar gas and dust, but this usually accounts for less than 25% of the mass of spiral galaxies). Hence a spiral galaxy, with all its glorious bright blue disk stars, actually has far more dim red ones.

The colors of galaxies can also be measured by using filters to determine the brightness at different wavelengths. In general the spirals are not as red as the ellipticals, again indicating that the latter galaxies contain a higher percentage of cool, red stars. There are also color variations within galaxies: in a spiral, for example, the central bulge is usually redder than the outer portions of the disk where most of the hot, young stars reside.

Both the mass-to-light ratios and the colors of galaxies are indicators of the relative content of Population I and Population II stars. Recall that Population I stars tend to be younger and include all the bright, blue O and B stars. By contrast, Population II objects are old and include only cool, relatively dim main-sequence stars and red giants. Therefore a red overall color, along with a high mass-to-light ratio, implies that Population II stars are dominant, while a low mass-to-light ratio and a bluer color means that some Population I stars are included. Thus elliptical galaxies seem to contain almost entirely Population II objects, while spirals contain a mixture of the two populations.

This dichotomy between the two types of galaxies is also found when the interstellar matter content of spirals and ellipticals is compared. Photographs of spirals, especially those seen edge-on, often clearly show the presence of dark dust clouds, and face-on views typically reveal a number of bright H II regions. Neither shows up on photographs of elliptical galaxies. Radio observations of the 21-centimeter line of hydrogen bear this out; emission is usually present in the spectra of spirals but is weak or absent in the spectra of ellipticals.

The Distribution of Galaxies in the Universe

If galaxies represent the majority of matter in the universe, then we can learn about the overall structure of the cosmos by tracking the galaxies. Just as

stars do not populate the galaxy uniformly, so too the galaxies tend to be scattered about in a patchy distribution. They are grouped on various distance scales, up to sizes that approach that of the universe itself.

Although galaxies may be considered the largest single objects in the universe (if indeed an assemblage of stars orbiting a common center can be viewed as a single object), there are yet larger scales on which matter is organized. Galaxies tend to be located in clusters **(Fig. 18.16)** rather than being distributed uniformly throughout the cosmos, and these in turn have an uneven distribution, with concentrations of clusters referred to as **superclusters.**

Clusters of galaxies range in membership from a few (perhaps half a dozen) to many hundreds or even thousands (Appendix 16). Just as stars in a cluster orbit a common center of mass, so galaxies in a group or cluster are gravitationally bound together and follow orbital paths about the center of mass of the system. Most large, rich clusters have relatively smooth, rounded overall shapes, whereas smaller groupings such as the one to which the Milky Way belongs are often less regular in shape, with a more-or-less random distribution of members.

Figure 18.16 A cluster of galaxies. Several galaxies are visible in this photograph of a portion of a large cluster of galaxies, called the Fornax cluster. (© 1984 Royal Observatory, Edinburgh)

The LOCAL GROUP

Figure 18.17 These are the galaxies of the Local Group, arranged to represent their actual physical relationships to the Milky Way http://universe.colorado.edu/fig/18-17.html

27

6

26

5

4

2

1

12 13

7 8

2

1 3

6

5 4

10 9

11

17

18

19

20

22

23

25 24

31

30

29

(1) *Milky Way*
(2) *Draco*
(3) *Ursa Minor*
(4) *SMG*
(5) *LMG*
(6) *Carina*
(7) *Sextans*
(8) *Ursa Major*
(9) *Pegasus*
(10) *Sculptor*
(11) *Fornax*
(12) *Leo I*
(13) *Leo II*
(14) *Maffei*
(15) *NGC 185*
(16) *NGC 147*
(17) *NGC 205*
(18) *M32*
(19) *Andromeda I*
(20) *Andromeda II*
(21) *Andromeda (M31)*
(22) *M33*
(23) *LGS 3*
(24) *IC 1613*
(25) *NGC 6822*
(26) *Sextans A*
(27) *Leo A*
(28) *IC 10*
(29) *DDO 210*
(30) *Wolf-Lundmark-Melotte*
(31) *IC 5152*

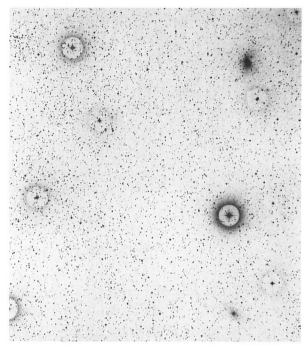

Figure 18.18 Maffei 1 and 2. This infrared photograph reveals two large galaxies (the fuzzy images, upper right and lower right) that for a while were considered possible members of the Local Group. More extensive analysis has shown, however, that they probably are not. (The circular marks are overexposed images of closer stars.) (H. Spinrad)

Figure 18.19 A new member of the Local Group? The spiral galaxy seen faintly in the background was first detected by radio observations, and then by a visible wavelength telescope. Analysis of the distance suggests that this galaxy lies outside of the Local Group. (Harry Ferguson, STScI)

The clustering of galaxies is part of the basic framework of the universe, which consists of a hierarchy of structures on ever-increasing size scales, up to the superclusters. Thus when discussing the arrangement of galaxies into groups and clusters, we are beginning to raise questions related to cosmology, the study of the universe as a whole.

The Local Group

The Milky Way belongs to a small group of galaxies known as the Local Group. This group consists of about 30 members arranged in a random distribution **(Fig. 18.17, preceding pages).** Despite the relative proximity of these galaxies to us, it has been difficult to ascertain their properties in some cases because of obscuration by our own galactic disk. We cannot even say with certainty how many members the Local Group has.

Among the member galaxies are three spirals, two of which—the Andromeda galaxy (see Fig. 18.17) and the Milky Way—are rather large and luminous. These are probably the brightest and most massive galaxies in the entire cluster. The Local Group is about 800 kpc in diameter and has a roughly disklike overall shape, with the Milky Way located a little off-center. Beyond the outermost portions of the Local Group, few conspicuous galaxies are found for a distance of some 1,100 kpc.

Two large galaxies discovered in the 1970s through infrared observations **(Fig. 18.18)** were thought for a while to be members of the Local Group. In 1994 another large spiral galaxy was discovered, like the first two, lying behind obscuring dust in the disk of the Milky Way **(Fig. 18.19).** It appears that all three may be beyond the limits of the Local Group. If these recently discovered large galaxies had proven to be members of the cluster, they would have significantly altered our reckoning of the size and mass of the group.

Most of the other members are ellipticals, many of them dwarf ellipticals (see Fig. 18.17). There may be additional members of this type undetected so far because of their faintness. Four irregular galaxies (see Fig. 18.17), two of them the Large and Small Magellanic Clouds, are also found within the Local Group. In addition there are globular clusters, which are probably distant members of our galaxy, but they lie so far away from the main body of the Milky Way that they appear isolated. A list of known members of the Local Group appears in Appendix 16.

The Magellanic Clouds and the Andromeda galaxy (also known commonly as M31, its designation in the widely cited Messier Catalog; see Appendix 17) have been particularly well studied because of their proximity and prominence and because of what they can tell us about galactic evolution and stellar processing. The Large and Small Magellanic Clouds appear to the unaided eye as fuzzy patches, most easily visible on dark, moonless nights. They lie near the south celestial pole and can therefore be seen only from the Southern Hemisphere. Their name originated from the fact that the first Europeans to see them were Ferdinand Magellan and his crew, who made the first voyage around the world in the early sixteenth century.

The Magellanic Clouds are considered to be satellites of the Milky Way. They have smaller masses and follow orbits about it, taking several hundred million years to make each circuit. Lying between 50 and 65 kpc from the Sun, both are Type I irregulars. They contain substantial quantities of interstellar matter and are quite obviously the sites of active star formation, with many bright nebulae and clusters of hot, young stars. Measurements of the colors of these galaxies, and of the spectra of some of their brighter stars, indicate that their heavy element abundances are 5 to 10 times lower than those of Population I stars in our galaxy. This seems to indicate that the Magellanic Clouds have not undergone as much stellar cycling and recycling as the Milky Way. These and other Type I irregular galaxies may generally be viewed as galaxies in extended adolescence that have not yet settled down into mature disks. In the case of the Magellanic Clouds, the Milky Way's tidal forces are probably the reason for the unrest.

The great spiral M31, the Andromeda galaxy, is the most distant object visible to the unaided eye, lying some 700 kpc from our position in the Milky Way. All that the eye can see is a fuzzy patch of light, even if a telescope is used, but when a time-exposure photograph is taken, the awesome disk stands out (although the spiral arms are still difficult to see because we view the disk from an oblique angle). The Andromeda galaxy is so large, extending over several degrees across the sky, that full portraits can be obtained only by using relatively wide-angle telescope optics. (Most large telescopes have extremely narrow fields of view.)

The Andromeda galaxy is probably very similar to our own galaxy and thus has taught us quite a bit about the nature of the Milky Way. The two stellar populations were first discovered through studies of stars in Andromeda, and the effects of stellar processing on the chemical makeup of different portions of the galaxy are better determined for Andromeda than for our galaxy. The study of this galaxy and other members of the Local Group has been very important in the development of distance-determination techniques for more distant galaxies and clusters.

Rich Clusters and Galactic Cannibalism

Many clusters of galaxies are much larger than the Local Group and contain hundreds or even thousands of members **(Fig. 18.20)**. The density of galaxies in such a cluster is relatively high, and therefore there are numerous close encounters between galaxies as they follow their individual orbits about the center. When two galaxies pass close by each other, they exert mutual gravitational forces that can have profound effects. These close encounters apparently were very frequent during the era when the rich clusters formed, which accounts for the smooth overall distribution of galaxies within the clusters, with the greatest density toward the center. By contrast, small groups or clusters rarely reach this state and retain a less regular appearance.

The central regions of large, rich clusters have higher concentrations of elliptical and S0 galaxies **(Fig. 18.21)** than are found in small groups or isolated galaxies. Near the center of a rich cluster, some 90% of the galaxies may be ellipticals or S0s, whereas about 60% of noncluster galaxies are spirals. This contrast is probably a direct result of the frequent near-collisions between galaxies in dense clusters. When two galaxies have a close encounter, the tidal forces they exert on each other stretch and distort them. Under some circumstances any interstellar

Figure 18.20 A portion of a rich cluster of galaxies. Most of the objects in this photograph are galaxies. This cluster lies in the constellation Hercules. (Palomar Observatory, California Institute of Technology)

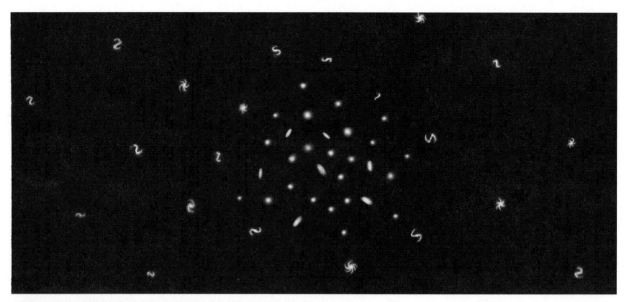

Figure 18.21 Galaxy types in a rich cluster. In dense, highly populous clusters of galaxies, nearly all the members in the central portion are ellipticals, and a giant, dominant elliptical is often found at the center. Only in the outer portion are many spirals seen.

matter in them can be pulled out and dispersed **(Fig. 18.22).** The outer regions, such as the halos, can also be stripped away. The net effect is analogous to what happens to rocks in a tumbler: the galaxies are gradually ground down into smooth remnants. A spiral galaxy subjected to these cosmic upheavals may assume the form of an elliptical. In time, most of the spirals in a cluster, particularly those in the dense central region, may be converted into ellipticals and S0s.

The frequent gravitational encounters between galaxies during cluster formation have another interesting effect: they cause a buildup of galaxies right at the center of the cluster. When two galaxies orbiting within a cluster come together, one always gains

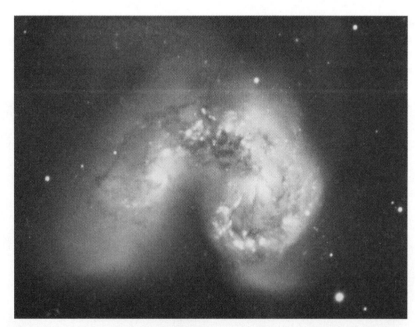

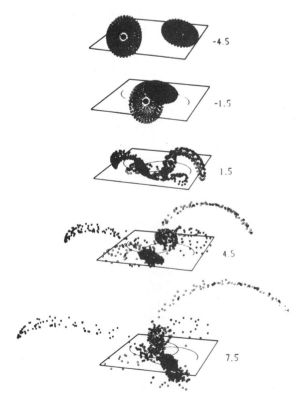

Figure 18.22 A collision between galaxies. The photo above shows a pair of galaxies undergoing a near-collision. At the right is a computer simulation of their interaction, showing how the present appearance was created. Astronomers believe this type of interaction between galaxies is responsible in some cases for the removal of interstellar matter, converting spiral galaxies to elliptical or S0 galaxies. (© 1992 AAT Board; photo by D. Malin; computer simulation from A. Toomre and J. Toomre) http://universe.colorado.edu/fig/18-22.html

energy and moves to a larger orbit, while the other (the more massive of the two) loses energy and drops closer to the center. Thus a gradual sifting process—like the differentiation that occurs inside a planet as the heavy elements sink toward the core—gradually builds up a dense central conglomeration of galaxies at the heart of the cluster. There these galaxies may actually merge, the end result being a single gigantic elliptical galaxy **(Figs. 18.23** and **18.24),** which continues to grow larger as new galaxies fall in.

Another distinctive characteristic of some rich clusters of galaxies is the existence of a very hot gaseous medium filling the spaces between the galaxies. This intracluster gas was discovered through X-ray observations **(Fig. 18.25),** which showed that the temperature of the gas is as high as a hundred million degrees, much hotter even than the highly ionized gas in the interstellar medium in our galaxy. This observation raised the possibility that the general intergalactic void is filled with such gas, although if it is, the gas outside clusters must be cooler than the intracluster gas, or it would have been detected by its X-ray emission.

A different type of X-ray measurement indicates that the hot intracluster gas originates in the galaxies themselves, rather than entering the cluster from the intergalactic void. Spectroscopic measurements made

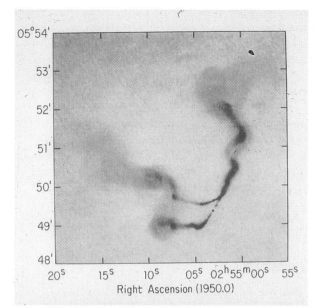

Figure 18.24 A radio image of a galactic merger. This is a radio map of a giant elliptical galaxy with two nuclei (the bright points at lower center, each with a pair of wispy gaseous jets emanating from it; the jets are discussed in Chapter 19). Evidently, this galaxy is in the process of forming from the merger of two galaxies whose centers have not yet completely merged. (National Radio Astronomy Observatory)

Figure 18.23 A giant elliptical galaxy. This is M87, a well-known example of a dominant central galaxy in a rich cluster. This galaxy is discussed further in Chapter 20. (© 1992 AAT Board; photo by D. Malin)

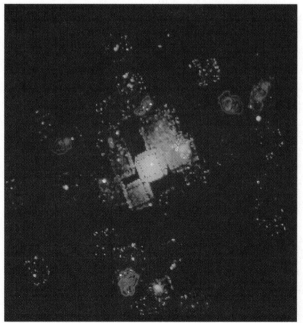

Figure 18.25 Intracluster gas in the Virgo cluster. This is an X-ray image from the *Einstein Observatory*, showing emission from the entire central portion of this cluster of galaxies. The X rays are being emitted by very hot gas that fills the space between the galaxies. The X-ray emission is shown in blue; radio images of galaxies in the cluster are in red. (X-ray data from the Harvard-Smithsonian Center for Astrophysics; radio data and image processing by the National Radio Astronomy Observatory)

at X-ray wavelengths have revealed that the gas contains iron, a heavy element, in nearly the same quantity (relative to hydrogen) as the Sun and other Population I stars. Such a high abundance of iron could only have been produced in nuclear reactions inside stars. Therefore this intracluster gas must once have been involved in part of the cosmic recycling that goes on in galaxies, as stars gradually enrich matter with heavy elements before returning it to space. How the gas was expelled into the regions between galaxies is not clear, but it may have been swept out during near-collisions between galaxies, or it may have been ejected in galactic winds created by the cumulative effect of supernova explosions and stellar winds.

Cluster Masses

There are two distinct methods for measuring the masses of clusters of galaxies, and both are quite uncertain. This is a crucial problem, as we shall see in Chapter 20, because of the importance of knowing how much mass the universe contains.

The simpler and more straightforward of the two methods is to estimate the masses of the individual galaxies in a cluster, using techniques described earlier in this chapter, and add them up. In many cases, particularly for distant clusters where it is impossible to measure the rotation curves or velocity dispersions of individual galaxies, the only way to estimate a cluster's mass is to measure its brightness and then use a standard mass-to-light ratio to derive the mass. This is inaccurate, however, because it depends on knowing the cluster's distance from us and because it assumes that the galaxies adhere to the usual mass-to-light ratios for their types. It also neglects any matter in the cluster that may lie between the galaxies.

The second method **(Fig. 18.26)** is similar to the velocity dispersion technique used to estimate masses of elliptical galaxies. The mass of a cluster is estimated from the orbital speeds of galaxies in its outer portions; the faster they move, the greater the mass of the cluster. An advantage of this method is that it measures all the mass of the cluster, whether it is in galaxies or between them, but a disadvantage is that the necessary velocity measurements, particularly for a very distant cluster, are difficult. Furthermore, the technique is valid only if the galaxies are in stable orbits about the cluster; if the cluster is expanding or if some of the galaxies are not really gravitationally bound to it, the results will be incorrect. In rich clusters at least, the smooth overall shape and distribution give the appearance of a bound system, and this technique is probably valid. It may not be valid for small clusters such as the Local Group. This method always leads to an estimated cluster mass that is much greater than that derived by adding up the masses of the visible galaxies, suggesting that much of the matter is in some invisible form.

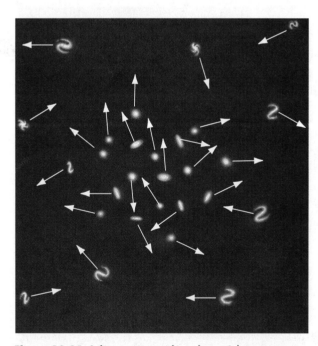

Figure 18.26 Galaxy motions within a cluster. Galaxies move randomly within their parent cluster. In a large cluster, where the overall distribution of galaxies is uniform, analysis of the average galaxy velocities is used to estimate the total mass of the cluster, just as the velocity dispersion method is used to measure the masses of elliptical galaxies.

Gravitational Lenses and Dark Matter in Clusters

In our discussion of the derivation of masses of clusters of galaxies, we found that there seems to be much more mass present in many clusters than can be accounted for by the visible galaxies in those clusters. This implies that more mass is present in these clusters than we can see in the form of stars and galaxies. The invisible matter inferred to be present has been dubbed **dark matter.** Whatever this dark matter is, it has profound implications for the overall distribution of mass in the universe and plays a key role in determining the ultimate fate of the universe.

Very recent observations of clusters of galaxies have provided new insights into the distribution of the dark matter, while leaving its form and origin as mysterious as ever. Several clusters have been found with huge glowing arcs or rings of light in and around them **(Fig. 18.27).** These so-called **luminous blue arcs** are truly gigantic if they lie at the same distance from us as the clusters in which they are seen.

They appear to be millions of light-years in length, comparable in size to the largest structures associated with individual galaxies.

Astronomers now think that the luminous arcs are not material structures at all, but images of galaxies farther away that have been extended, magnified, and distorted due to the effect of gravity on photons of light. As we learned in Chapter 16, light is deflected when passing near a very massive object because of the curvature of spacetime created by the intense gravitational field of the massive object. In Chapter 17 we learned that individual stars can act as **gravitational lenses,** focusing the light of background objects and enhancing their brightness. Now we find that clusters of galaxies, with their enormous masses, can also act as gravitational lenses.

The diagram in **Fig. 18.28** illustrates how a gravitational lens is formed. If the source of light, the observer, and the massive object between them are perfectly aligned, the image formed by a gravitational lens will be a circle. If the alignment is not quite perfect, the image can consist of one or more arcs of light or of multiple images. Several years ago, both members of a closely spaced pair of very distant objects called **quasars** (discussed in the next chapter) were observed to have virtually identical properties, and it was deduced that they were two images of the same object, formed by the gravitational lensing effect of a galaxy between the Earth and the quasar. Since then many other examples of multiple images of quasars have been found (e.g., **Fig. 18.29**). Now it is thought that the luminous arcs seen around some galaxy clusters are also the result of gravitational lenses.

If a distant galaxy lies behind a massive cluster of galaxies, then depending on the precision of the alignment as seen from the Earth, the gravitational field of the cluster can form one or more arclike, magnified images of the background galaxy. From the size and shape of the image, it is possible to calculate the mass contained in the cluster of galaxies, because the properties of the image depend on the strength of the gravitational field that forms the lens. Thus we have a new method of determining the masses of galaxy clusters, one that is quite independent of the usual assumptions that must be made about the orbits of individual galaxies within the cluster.

This new method yields results that correspond very well with those obtained by other methods. The masses of clusters are still found to be much larger than the sum of the masses of the visible galaxies. Thus this new tool is helping to confirm the growing belief among astronomers that the universe contains

Figure 18.27 A luminous arc. The blue, arclike structure shown here is gigantic in scale, dwarfing the individual galaxies in this cluster. (NASA/STScI) http://universe.colorado.edu/fig/18-27.html

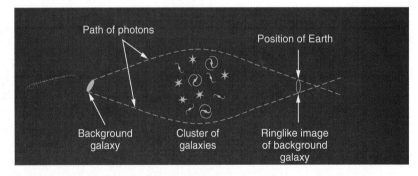

Figure 18.28 Gravitational lens creating luminous arcs. The gravitational field of a foreground cluster of galaxies can deflect the light from a distant object, creating a ringlike image if the alignment is perfect as shown here or arclike images if the alignment is not quite perfect.

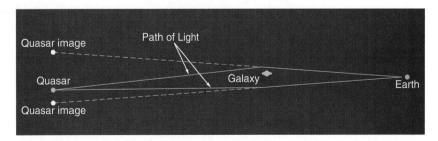

Figure 18.29 A double quasar image. This diagram illustrates how an intervening galaxy can act as a gravitational lens forming two images of a distant quasar.

Figure 18.30 **Dark matter in a cluster of galaxies.** It is possible to determine the distribution of dark matter in a cluster of galaxies by analyzing the luminous arcs created by the gravitational lensing effect of the dark matter. The left panel shows the optical image of a cluster of galaxies; close examination reveals a number of short, blue arcs. The right panel shows, on the same scale, the distribution of dark matter required to produce the blue arcs gravitationally. The dark matter is concentrated toward the center of the cluster but has a smoother distribution than the galaxies. (A. Tyson, Bell Laboratories)

vast quantities of invisible, dark matter. At the same time, it is also providing new and dramatic confirmation of one of the predictions made by Einstein's theory of general relativity.

Through the analysis of the luminous arcs observed in and around a cluster, it has actually been possible to determine the distribution of the dark matter in the cluster **(Fig. 18.30).** In some clusters, a large number of luminous arcs are seen. Using the equations that describe the deflection of light by a gravitational field, astronomers have been able to deduce the mass distribution that would be needed to create the observed pattern of arcs. In this way, they have actually been able to map the dark matter in several clusters of galaxies. The maps show that the dark matter traces the distribution of visible galaxies in each cluster, but without much fine-scale variation from place to place. So whatever the dark matter is, it appears that, at least in some situations, it tends to be associated with the glowing matter that makes up the visible stars and galaxies in the universe.

Superclusters, Voids, and Walls

We turn now to consider the overall organization of the matter in the universe. We noted earlier that clusters of galaxies may represent the largest scale on which matter in the universe is clumped, but that there is some evidence for a higher-order organiza-tion, even in the case of the Local Group **(Fig. 18.31).** It appears that clusters of galaxies are themselves concentrated in certain regions, commonly referred to as superclusters.

The reality of superclusters was difficult to demonstrate, because it is a very difficult observational job to map out the distribution of faint galaxies on the sky in such a way that coverage is complete, at least to some limit in brightness. The longstanding debate over whether superclusters exist eventually faded from the spotlight as evidence for the reality of groupings of clusters accumulated, and by the 1970s the notion was generally accepted. But then the significance of the groupings became an issue, because it could be shown that a certain amount of superclustering would occur over time due to random encounters and gravitational attraction. One could conclude from this that the superclusters had nothing to do with any fundamental large-scale structure in the universe; that is, that they did not show any early tendency for the universe to be lumpy. In this view, the overall distribution of galaxies and of clusters of galaxies was basically uniform, with only the concentrations that would be expected from random gatherings that developed after galaxies and clusters had formed.

But in the past decade an entirely new view of the superclusters has emerged. Using large telescopes and lots of observing time, astronomers have

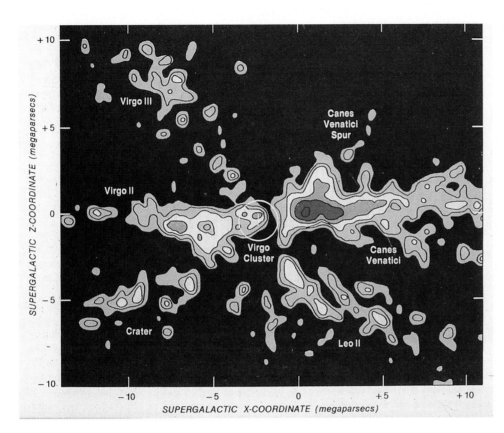

Figure 18.31 **The Local Supercluster.** This is an artist's conception of the cluster of galaxy clusters to which the Local Group belongs. The Local Group is at position (0,0) in this diagram. (Illustration from *Sky and Telescope Magazine* by Rob Hess. © 1982 by the Sky Publishing Corporation; also with the permission of R. B. Tully)

attempted to map the locations of galaxies in three dimensions over large portions of the sky. From these studies, a surprising picture of the structure of the universe itself has emerged. Not only are clusters of galaxies grouped into larger structures, but the superclusters themselves are not uniformly distributed through the universe. Instead they form a filamentary network, giving the universe a very nonhomogeneous structure **(Fig. 18.32).** Enormous sheets or strings of galaxies are seen, with relatively empty regions called voids between them. Typical sizes of the voids and of the long filaments of galaxy clusters are 100 Mpc or more. Sizes on this scale are becoming significant compared to the universe as a whole.

The uneven, almost cell-like structure that has been revealed cannot be the result of random gravitational groupings that arose after galaxies formed. Instead this structure indicates that somehow the early universe became lumpy, perhaps even before the time of galaxy formation. These findings present both an answer and a puzzle for astronomers: they help to answer the question of how and when matter first formed dense regions that could form into galaxies and clusters, but they raise the larger question of how and when the universe became lumpy in the first place. This will be a central issue in our discussion of universal evolution in Chapter 20.

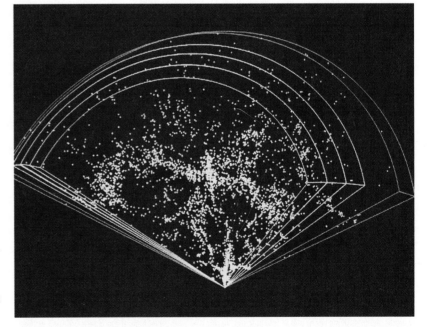

Figure 18.32 **Arclike superclusters.** This shows the distribution of galaxies in a region of the sky that has recently been mapped completely, showing that superclusters are arranged in curved filamentary structures. Such groupings provide evidence that there is a large-scale structure in the universe and that matter is not distributed uniformly. (M. Geller, J. Huchra, M. Kurtz, and V. de Lapparent, Harvard-Smithsonian Center for Astrophysics)

Figure 18.33 The *Hubble* Deep field image. This remarkable photo was obtained by the *Hubble Space Telescope* in late 1995, and represents many hours of continuous observations of a single small region of sky, chosen because it appears nearly empty of objects in photographs taken through ground-based telescopes. Thus most of the objects seen here were previously undetected. This wealth of galaxies, some very distant and therefore seen as they were when the universe was very young, is providing astronomers with unprecedented information about the formation and evolution of galaxies. (NASA/STScI) http://universe.colorado.edu/fig/18-33.html

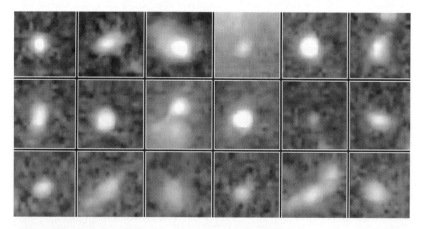

Figure 18.34 Embryronic Galaxies? These objects, selected from *Hubble Space Telescope* images because of their enormous redshifts, appear to be young galaxies or sub-galactic stellar systems yet to merge into galaxies. Because of light-travel time, these objects are seen as they were about 11-billion years ago, when the universe was perhaps only ⅙ of its present age. (Rogier Windhorst and Sam Pascarelle (Arizona State University) and NASA/STScI)

Galaxies at the Edge

It is not so clear now as it was a few years ago that there are no galaxies in the voids. Recent studies have revealed significant numbers of faint galaxies in some of the regions thought to have been devoid of galaxies. The voids are still significant because even if faint galaxies are present, the lack of luminous ones is still indicative of an uneven distribution of matter and of galaxies.

Very recent *Hubble Space Telescope* images have changed our perception of the overall density of galaxies in the universe and at the same time have provided us with new information about the nature of galaxies at the earliest times. A small spot of sky was chosen in which few objects were visible to the limiting magnitude of existing sky images and maps. One of the *HST*'s fine cameras was focused on that spot, making repeated exposures over many hours. The result was stunning: some 1,500 objects were seen, almost all of them too faint to have appeared in previous images **(Fig. 18.33)**. The objects appear to be very young galaxies, seen as they were when the universe was only a small fraction of its present age. Closer examination of other distant objects **(Fig. 18.34)**, seen as they were 11 billion years ago, suggests that today's galaxies formed when smaller systems merged, much like the process recently suggested for the formation of the Milky Way (Chapter 17).

Their implication for our present discussion of the distribution and density of galaxies in the universe is itself quite remarkable. The new *HST* image suggests that there may be five times more galaxies in the universe than previously thought. This will have important implications for our understanding of the mass density of the universe and for its evolution.

Summary

1. The true nature of the nebulae was not established until the mid-1920s, when Edwin Hubble identified Cepheid variables in the Andromeda nebula and used them to show that its distance was so great that it had to be a galaxy comparable to the Milky Way.

2. Galaxies are categorized by shape in two general classes: spirals and ellipticals. There are also many S0 galaxies (disk-shaped, but with no spiral structure) and irregular galaxies.

3. Distances to galaxies are measured using a variety of standard candles, such as Cepheid variables, extremely luminous stars, supernovae, and galaxies of standard types. One useful tech-

A Great Debate Revisited

In 1920 a famous debate on the scale of the universe took place before the U.S. National Academy of Sciences. This was the Shapley-Curtis debate, often described as a debate on the nature of the nebulae. While this was one of the most famous scientific debates of modern times, it was neither as unique nor as decisive as it is often said to have been. Scientists engage in debates all the time. You could even say that debate is at the heart of what scientists do, because very often they disagree or at least differ in their interpretations, and discussion and comparison of ideas are often the best way to make progress toward finding the best explanation.

In its most common form, scientific debate is an extremely informal affair that may involve only two or three people discussing different views of some recent results while having lunch together. On a slightly more organized level, the seminar or colloquium is a common form of debate as well. Here someone makes a presentation of recent research, and the audience, usually consisting of experts, may respond with questions or, sometimes, challenges. Researchers find this a very helpful way to test ideas before committing them to publication. If you are a student at a research university, you will probably find notices of seminars and colloquia on a wide variety of topics. You might enjoy sitting in on a few to get a flavor of the nature of the debate that takes place.

A very similar process occurs at meetings and conferences, where scientists gather together to compare results and ideas. The presentations are scrutinized by audiences who are usually not reluc-tant to raise questions or offer differing interpretations. But the formal debate is a rarity in science today and probably never was as common as one might think from such events as the Shapley-Curtis encounter in April 1920. That "debate" was actually a pair of presentations intended for a nontechnical (and non-astronomical) audience at the National Academy's annual spring meeting in Washington. The remarks were very different in technical level, and together the two speakers took only about 80 minutes. This was hardly enough time for Shapley and Curtis to argue and counterargue in defense of their opposing viewpoints. Following the April meeting, each speaker published his discussion, providing far more technical detail than had been possible during the spoken remarks, and the pair of papers that resulted really constitutes the basis of the "debate" as it is viewed historically.

The issue at the heart of the Shapley-Curtis disgreement was the size scale of the galaxy. This was an essential part of the very closely related issue of the nature of the nebulae, because most of the arguments over whether the nebulae were separate galaxies or objects within the Milky Way hinged on just how big the Milky Way is. As we saw in this chapter, Shapley had overestimated the size of the galaxy and therefore thought that the nebulae were contained within it, while Curtis argued for a smaller galaxy that was one of many. No clear winner was declared at the time, although the general consensus appears to have been that Curtis backed up his point of view with better evidence.

In April 1996, a new debate took place before the National Academy of Sciences. Just as in the Shapley-Curtis debate 76 years earlier, the topic was the scale of the universe, and the location of the debate was once again the auditorium of the Smithsonian's Museum of Natural History in Washington, D.C.

This time the antagonists were Gustav Tammann, a Swiss astronomer affiliated with a research group led by Alan Sandage of the Mount Wilson Observatory in Pasadena, California, and Sidney Van den Bergh of the Dominion Astrophysical Observatory, Victoria, British Columbia. The debate concerned the value of the Hubble constant H_0. As we have seen in this chapter, the expansion rate of the universe, embodied in this constant, determines the distance scale to the most faraway galaxies. In a very real sense, a debate over the value of H_0 is a debate over the size of the universe.

Why is there enough difficulty in measuring H_0 to justify a debate? And was this debate really an important step toward the eventual resolution of the dispute? The answer to the first question was hinted at in this chapter, where we discussed the use of standard candles to establish distances to galaxies, which are used in turn to determine the expansion rate. Two large teams of astronomers are using *Hubble Space Telescope* observations to develop standard candles so that they can find the value of H_0. One team, led by Sandage, is relying exclusively on supernovae as their distance indicators, while the other team, led by Wendy Freedman (also of the Mount Wilson Observatory, which was Shapley's home institution at the time of the 1920 debate), is using several different standard candles. Each team's methods ultimately depend on finding Cepheid variables in distant galaxies and applying the period-luminosity relation to establish their distances and hence the distances to other objects, which can then be used as standard candles.

The two teams have been gradually converging, and within a short time, their values of H_0 may well overlap. The latest estimates, as of mid-1996, are $H_0 = 68-78$ km/sec/Mpc from the Freedman group, while Sandage and his co-workers are finding $H_0 = 57$ km/sec/Mpc; both groups consider these results preliminary, however. Over the past three years, the two teams have

Continued

been coming closer to agreement, so there is reason to hope that we will soon have a consensus value for the expansion rate, and hence the size scale, of the universe.

The debate of 1996 had mostly symbolic value. In these days of instantaneous communications, astronomers circulate their latest results electonically, and the majority of the debating takes places in the forums of lunchrooms, electronic mail, and informal gatherings at seminars and conferences. The debate before the National Academy was a nice commemoration of the 1920 event, and it helped to inform the general public about the state of affairs today. But it was not the first, the only, or even the most meaningful of the many encounters the two H_0 teams will have before the question of the size scale of the universe is settled.

---

nique uses a correlation between the width of the 21-centimeter line and luminosity to determine absolute magnitude and distance.

4. The expansion of the universe, announced by Hubble in 1929, can be used to estimate distances to galaxies from their spectral redshifts, if the value of the Hubble constant H_0 is known.

5. Masses of spiral galaxies are determined by applying Kepler's third law to the outer portions, where the orbital velocity and period are determined from a rotation curve. The internal velocity dispersion is used to determine the masses of elliptical galaxies. Galactic masses can also be determined by applying Kepler's third law to binary galaxies.

6. While all galaxies are dominated by Population II stars, spirals tend to contain a greater proportion of Population I stars, have substantial quantities of interstellar matter, and generally seem to be in a state of continuous evolution and stellar cycling. Ellipticals, on the other hand, have few or no Population I stars, contain little or no cold interstellar matter, and generally do not seem to have active stellar cycling at the present time.

7. Many galaxies are members of clusters rather than being randomly distributed throughout the universe. Clusters range from small, amorphous groups, such as the Local Group to which the Milky Way belongs, to very large groups with thousands of members.

8. The dominance of elliptical galaxies in rich clusters is probably the result of the conversion of spirals into ellipticals by collisions and tidal interactions with other galaxies. Many rich clusters have a giant elliptical galaxy at the center that probably formed from the merger of several galaxies that settled there as a result of collisions.

9. Masses of clusters can be determined by summing the masses of individual galaxies, by analyzing the internal velocity dispersion of the galaxies in the clusters, or by observing and mapping luminous blue arcs, which are gravitationally lensed images of background objects. These methods all agree that the majority of mass associated with clusters of galaxies is invisible, or dark, matter.

10. Clusters of galaxies tend to be grouped into aggregates called superclusters, but these may be random concentrations rather than fundamental inhomogeneities of the universe. Superclusters appear to be sheetlike, with an overall distribution resembling a series of huge arclike structures, and this may tell us something about the early distribution of matter in the universe.

Review Questions

1. Why was it difficult for astronomers of the early 1900s to accept the idea that the nebulae might be distant galaxies, rather than clouds within our own galaxy?

2. How do we know that the tuning-fork diagram does not represent an evolutionary sequence?

3. Review the techniques for finding distances to galaxies, and explain why each method depends on techniques developed previously for finding distances to closer objects.

4. Why has it been difficult for astronomers to pinpoint the value of the Hubble constant H_0?

Web Connections

The Review Questions and Problems also appear at the following URLs:
http://universe.colorado.edu/ch18/questions.html
http://universe.colorado.edu/ch18/problems.html

Finding the Andromeda Galaxy

ASTRONOMICAL

ACTIVITY

To see the Andromeda galaxy with the naked eye is an awesome experience and not too difficult to do. You will need a very clear, dark night with no Moon (i.e., the Moon should be at third quarter or new moon phase, if you want to view Andromeda during the early evening). And you will have better luck if you can get out in the country, away from city lights.

The constellation Andromeda is up during the evening in late summer and early autumn. The accompanying diagram shows where the galaxy is. To begin, find the large-scale pattern of Pegasus; this is a kite-shaped arrangement of bright stars. Then follow the diagram to the grouping near the Andromeda galaxy, and look for a faint, fuzzy patch of light at the indicated position.

The galaxy is large and dim and will not reveal its spiral structure or any other details to the unaided eye (only a time-exposure photograph taken through a telescope can do that). Nevertheless, the galaxy is an awesome sight, particularly when you consider that you are looking beyond the confines of our home galaxy, and that your eye is receiving photons of light that have been traveling for 2 million years. The Andromeda galaxy is about a thousand times more distant than any of the stars you can see and about six times farther away than the other neighbor galax-

ies that are naked-eye objects, the Magellanic Clouds (which are visible only from the Southern Hemisphere). To view the Andromeda galaxy is to look as far away (and as far back in time) as the human eye can see.

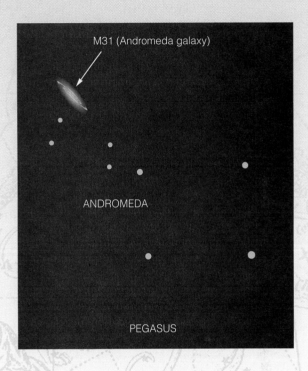

Problems

5. The mass-to-light ratio for spiral galaxies is usually much larger than 1, and for elliptical galaxies it is larger yet. What does this tell us about the types of stars that emit most of the light from such galaxies?

6. Elliptical galaxies are dominated by Population II stars, whereas spiral galaxies contain the younger Population I stars. Why are spiral galaxies thought to be as old as elliptical galaxies?

7. Explain why the relative number of galaxies of different types may not be the same in a rich cluster of galaxies as it is among galaxies not belonging to such clusters.

8. Why would it be surprising if the Local Group included a giant elliptical galaxy among its members?

9. Summarize the evidence for the presence of dark matter in clusters of galaxies.

10. Why is it thought that the intracluster gas observed in rich clusters of galaxies has been expelled from the galaxies in the cluster rather than being truly intergalactic gas?

1. To help understand the difficulty of observing individual stars in other galaxies, calculate the apparent magnitude of a very luminous supergiant having visual absolute magnitude $M_V = -8$, if this star were in a galaxy 100,000 pc away (i.e., about twice as far as the Magellanic

Clouds). How faint would this star be if it were 1,000,000 pc (a little farther than the Andromeda galaxy) away?

2. Suppose an elliptical galaxy is measured to be twice as long as it is wide. What type of elliptical is it?

3. A Type I supernova is observed in a distant galaxy. The absolute magnitude of such a supernova is assumed to be -19. If the supernova has an apparent magnitude of $+24$ (near the limit that can be observed with the largest telescopes), how far away is the galaxy in which it lies?

4. If the value of the Hubble constant is $H_0 = 50$ km/sec/Mpc, how far away is a galaxy whose ionized calcium line (rest wavelength of 3933.67 Å) is observed at a wavelength of 3978.72 Å? What would your answer be if $H_0 = 100$ km/sec/Mpc?

5. Suppose that the rotation curve for a spiral galaxy shows that at a distance of 15 kpc from the galactic center, the rotation speed is 150 km/sec. Assuming that Kepler's third law can be applied, what is the mass of this galaxy? (*Hint:* You will have to find the period for stars at 15 kpc from the center by using the relationship *time = distance/speed*, where the distance is the orbital circumference of $2\pi a$, a being the orbital radius.)

6. If a dwarf elliptical galaxy has an absolute magnitude of $M = -15$ and could be detected with an apparent magnitude as faint as $m = +20$, how far away can this type of galaxy be found? Compare this distance with the diameter of the Local Group and with the distance to the Virgo cluster of galaxies, a moderately large cluster 15 Mpc away.

7. How long does it take for light to reach us from (a) the Large Magellanic Cloud, whose distance is 55 kpc; (b) the Andromeda galaxy, at its distance of 700 kpc; (c) the Virgo cluster of galaxies, lying about 19 Mpc away; and (d) a cluster that is 5×10^9 pc away? Discuss the implications of your answers: How do the light-travel times compare with the age of the solar system, and what do they suggest about observations of the evolution of the universe?

Additional Readings

Barnes, J. L. Hernquist, and F. Schweizer 1991. What Happens When Galaxies Collide. *Scientific American* 265(2):40.

Cowen, R. 1994. The Debut of Galaxies. *Astronomy* 22(12):44.

Crosswell, K. 1996. The Dark Side of the Galaxy. *Astronomy* 24(10):40.

Dressler, A. 1993. Galaxies Far Away and Long Ago. *Sky & Telescope* 85(4):22.

Eicher, D. J. 1995. Galaxy Time Machine. *Astronomy* 23(4):44.

Elmegreen, D. M. and B. Elmegreen. 1993. What Puts the Spiral in Spiral Galaxies? *Astronomy* 21(9):34.

Gallagher, J. and J. Keppel. 1994. Seven Mysteries of Galaxies. *Astronomy* 22(3):38.

Geller, M. J. and J. P. Huchra 1991. Mapping the Universe. *Sky & Telescope* 82(2):134.

Hodge, P. 1993. The Extragalactic Distance Scale: Agreement at Last? *Sky & Telescope* 86(4):16.

Hodge, P. 1993. The Andromeda Galaxy. *Mercury* 22(4):98.

Hoge, P. 1994. Our New! Improved! Cluster of Galaxies. *Astronomy* 22(2):26.

Impey, C. 1996. Ghost Galaxies of the Cosmos. *Astronomy* 23(6):40.

Keel, W. C. 1993. The Real Astrophysical Zoo: Colliding Galaxies. *Mercury* XXII(2):44.

Lake, G. 1992. Understanding the Hubble Sequence. *Sky & Telescope* 83(4):515.

Lake, G. 1992. Cosmology of the Local Group. *Sky & Telescope* 84(6):613.

Oort, J. 1992. Exploring the Nuclei of Galaxies (Including Our Own). *Mercury* 21(2):57.

Phillips, S. 1993. Counting to the Edge of the Universe. *Astronomy* 21(4):38.

Schramm, D. N. 1991. The Origin of Cosmic Structure. *Sky & Telescope* 82(2):140.

Steiman-Cameron, T. 1993. A Peculiar Twist. *Astronomy* 21(6):36.

Stephens, S. 1993. What Happens When Galaxies Collide? *Mercury* XXII(3):78.

Sulentic, J. 1992. Odd Couples. *Astronomy* 20(11):36.

Taken as a whole, the universe is absurd.
Walter Savage Landor, 1824

Chapter Web site: http://universe.colorado.edu/ch19

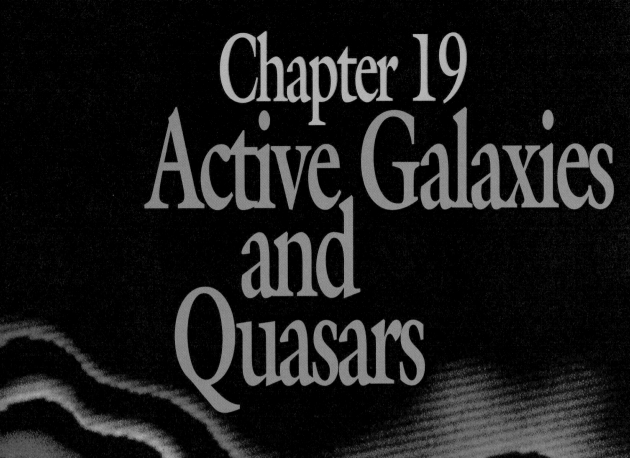

Chapter 19
Active Galaxies and Quasars

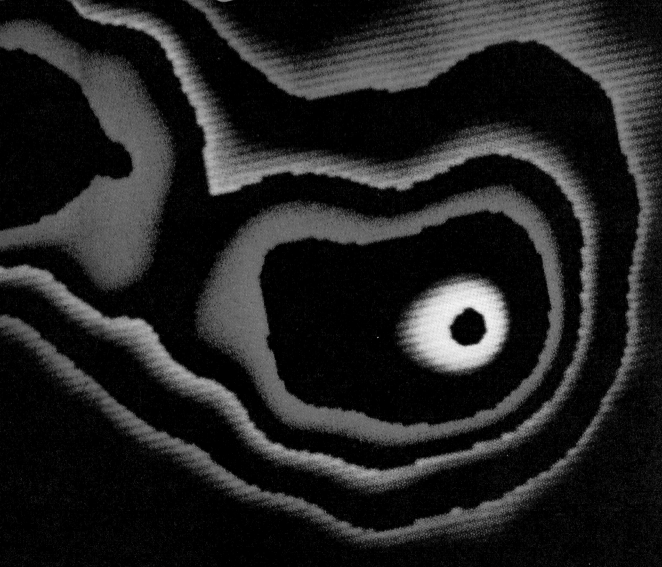

n discussing the characteristics of galaxies in the preceding chapter, we overlooked a variety of objects that have unusual traits. As in many other situations in astronomy, the so-called peculiar objects, once understood, have quite a bit to tell us about the more normal ones. For example, we will find that the unusual galaxies, like peculiar stars, often represent transitional phases in the evolution of normal members of their class. Thus we will learn a lot about the origins and evolution of galaxies by studying these unusual objects. We will also learn quite a bit about the universe, especially its early times.

Active Galaxies

Galaxies with unusual shapes or excess emission of energy in one portion of the spectrum or another have been recognized almost since the time galaxies were first discovered. Early on, astronomers found some galaxies with unusual shapes, and by the 1950s several galaxies were discovered to be immensely powerful sources of radio waves while others were found to have unusually bright visible emission from their nuclear regions. More recently, galaxies emitting most of their radiation at infrared wavelengths have been found, and many have been shown to emit X rays from their cores. The ability of today's giant ground-based telescopes to produce high-resolution images of faint objects, along with the fine images from the *Hubble Space Telescope*, has revolutionized our view of the most distant and unusual galaxies.

The Radio Galaxies

Other than the Sun, most of the first astronomical sources of radio emission to be discovered are galaxies. Most of these, when examined optically, have turned out to be large ellipticals, often with some unusual structure **(Fig. 19.1).** The first of these objects

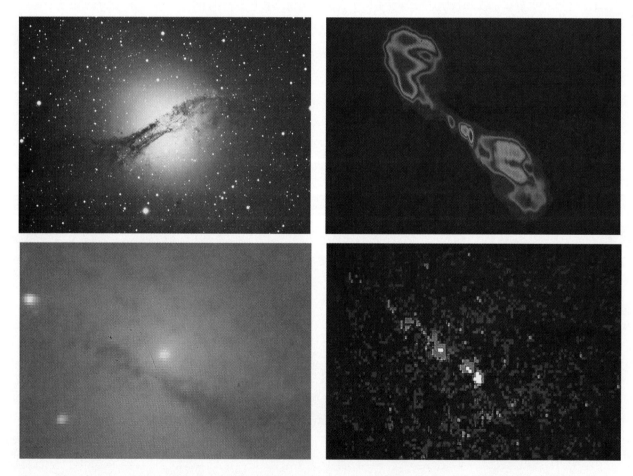

Figure 19.1 **Centaurus A.** This is a giant elliptical radio galaxy, showing a dense lane of interstellar gas and dust across the central region. At the upper left is a visible-light photograph; at the lower left is an infrared image showing the galactic core; at the upper right is a radio image; and at the lower right is an X-ray image. Note that the radio and X-ray emission comes from the lobes above and below the visible galaxy. (Upper left © 1980 AAT Board, photo by D. Malin; lower left: © 1992 AAT Board; image provided by D. A. Allen; upper right: NRAO; lower right: Harvard-Smithsonian Center for Astrophysics) http://universe.colorado.edu/fig/19-1.html

to be detected, and one of the brightest at radio wavelengths, is called Cygnus A (named under a preliminary cataloging system in which the ranking radio sources in a constellation are listed alphabetically). An image of this galaxy shows a strange double appearance, and after sufficient refinement of radio observing techniques (see the discussion of interferometry in Chapter 6), it was eventually found that the radio emission comes from two locations on opposite sides of the visible galaxy and well separated from it **(Figs. 19.2** and **19.3).** High-resolution observations with the *Very Large Array* have revealed remarkable detail in some radio galaxies **(Fig. 19.4).**

These radio-bright galaxies are called **radio galaxies.** Ordinary galaxies also emit radio radiation, but the term *radio galaxy* is applied only to the special cases in which the radio intensity is many times greater than the norm and is comparable to, or even greater than, the energy emitted in visible wavelengths by the stars in a galaxy. In these galaxies most of the emitted energy comes from some nonstellar source. The core of the Milky Way is a strong radio source as viewed from the Earth, but it is not in a league with the true radio galaxies. Typical properties of radio galaxies are listed in **Table 19.1.**

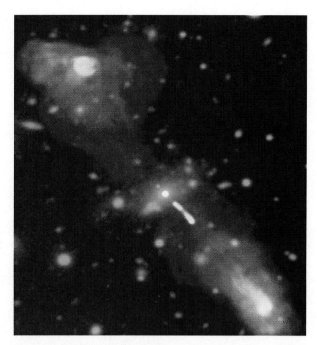

Figure 19.2 An image of a radio galaxy. This is the object 3C219. The visible galaxy is the small blue spot at the center; the enormous radio lobes are shown in red, orange, and yellow. Near the center is a well-defined jet of ionized gas, apparently flowing outward from the core of the galaxy. (National Radio Astronomy Observatory) http://universe.colorado.edu/fig/19-2.html

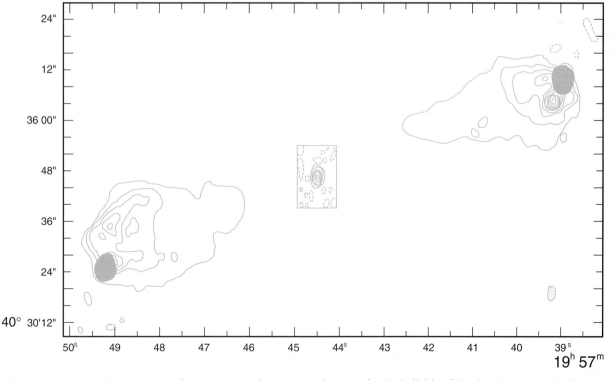

Figure 19.3 A map of a radio galaxy. The intensity contours here represent radio emission from the double lobes of the radio galaxy Cygnus A. The blurred image at the center illustrates the size of the visible galaxy compared with the gigantic radio lobes. (Mullard Radio Astronomy Observatory, University of Cambridge)

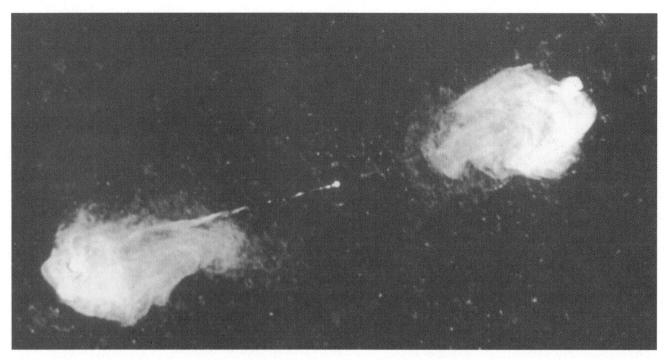

Figure 19.4 **Details of radio lobes.** This image of Cygnus A was obtained with the *Very Large Array* and shows fine detail suggesting turbulence and flows in the lobes. (National Radio Astronomy Observatory)

Table 19.1
Properties of Radio Galaxies

Galaxy type: Most often elliptical or giant elliptical.

Radio luminosity compared with optical luminosity: 0.1 to 10.

Radio source shape: Either double-lobed or compact central source; often jets are seen that emit throughout the spectrum.

Variability: Intensity variations in times as short as days may occur in the compact radio sources.

Nature of spectrum: Usually synchrotron or inverse Compton spectrum, usually polarized.

Although the double-lobed structure is common among elliptical radio galaxies, the visual appearance varies quite a bit. We have already noted the double appearance of Cygnus A; another bright source, the giant elliptical Centaurus A (the object shown in Fig. 19.1), appears to have a dense band of interstellar matter bisecting it, and it has not one, but two, pairs of radio lobes, one much farther out from the visible galaxy than the other. Other giant elliptical radio galaxies present other kinds of strange appearances. One of the most famous is M87, also known as Virgo A, which has a jet protruding from one side that is aligned with one of the radio lobes **(Fig. 19.5).** Careful examination shows that this jet contains a series of blobs or knots that appear to have been ejected from the core of the galaxy in sequence.

While jets detectable at visible wavelengths are rare, many galaxies have been discovered to have radio-emitting jets. These include galaxies already known to be peculiar, such as Centaurus A (see Fig. 19.1) and at least one spiral galaxy. In some cases, double jets are seen on opposite sides of a galaxy **(Fig. 19.6).** Even when jets are not seen, they may well be present, because it seems likely that the jets are responsible for the double radio lobes commonly seen in radio galaxies. The visibility of jets probably depends on the energy of the moving streams of gas and on our viewing angle, for reasons mentioned below.

The size scale of the radio galaxies can be enormous. In some cases, the radio lobes extend as far as 5 million pc from the central galaxy. Recall that the diameter of a large galaxy is only about 2% of this distance, and that the Andromeda galaxy is less than 1 million pc from our position in the Milky Way. Thus a giant radio galaxy, including its lobes, may be comparable in size to the entire Local Group of galaxies!

Measurements of the radio spectra show that these galaxies emit by the synchrotron process (already mentioned in Chapter 16), which requires a magnetic field and a supply of rapidly moving electrons. The electrons are forced to follow spiral paths around the magnetic field lines, and as they do, they emit radiation over a broad range of wavelengths.

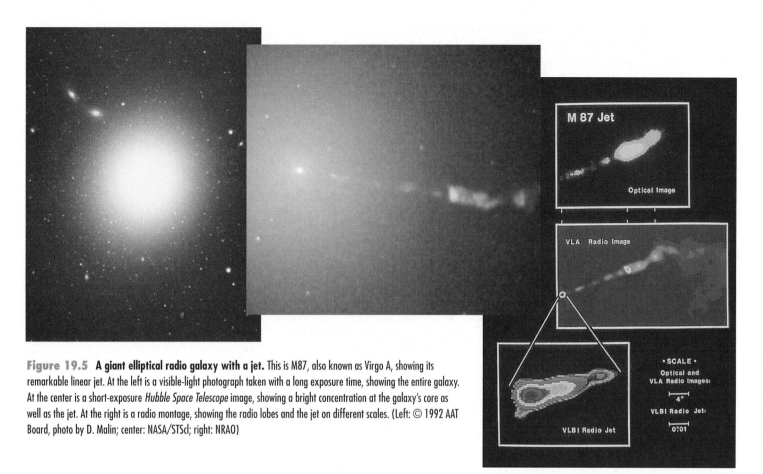

Figure 19.5 **A giant elliptical radio galaxy with a jet.** This is M87, also known as Virgo A, showing its remarkable linear jet. At the left is a visible-light photograph taken with a long exposure time, showing the entire galaxy. At the center is a short-exposure *Hubble Space Telescope* image, showing a bright concentration at the galaxy's core as well as the jet. At the right is a radio montage, showing the radio lobes and the jet on different scales. (Left: © 1992 AAT Board, photo by D. Malin; center: NASA/STScI; right: NRAO)

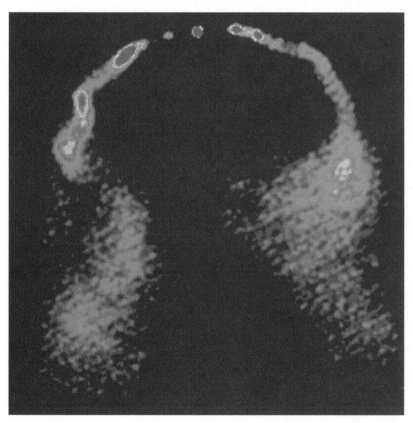

Figure 19.6 **Dual radio jets.** While visible jets are rare, radio images often reveal jetlike structures. Here is a radio image of a galaxy with a pair of jets emanating from opposite sides and curving away from the galaxy, apparently because the galaxy is moving through an intergalactic gas medium. (National Radio Astronomy Observatory)

The characteristic signature of synchrotron radiation, in contrast with the thermal radiation from hot objects such as stars, is a sloped spectrum with no pronounced peak at any particular wavelength. We say that this emission is **nonthermal,** because it is not created by the temperature of the source as normal stellar emission is. The radiation from a nonthermal, or synchrotron, source is also polarized (see Chapter 6). The shape of the synchrotron spectrum, along with its polarization, enables astronomers to distinguish it from thermal emission.

The visibility of the jets depends on how energetic they are and on the angle from which we view them. If the jets are **relativistic,** meaning that the gas is moving at a significant fraction of the speed of light, then most of their radiation is emitted in the forward direction. Therefore the galaxies in which we see jets (either optically or in radio wavelengths) are thought to be cases where the jets are relativistic and are nearly aligned with our line of sight. Naturally, these would be only a fraction of all galaxies that have jets, thus explaining why no jets are seen in many radio galaxies, even though they have the characteristic dual radio lobes. The emission from a jet is not perfectly aligned with the magnetic field, but instead spreads out in a conical shape, so that we see jets over a small range of angles. If the alignment with our line of sight had to be perfect, we would see a far smaller number of galaxies with jets.

It is fun to think for a minute about the implications of the presence of jets and radio lobes. First and foremost, we are led to suspect that something incredible must be going on in the core of a radio galaxy, something that is capable of producing highly energetic jets of ionized gas over hundreds of millions of years. The alignment of the jets with the radio lobes indicates that the jets created the lobes and continue to feed them. Yet the jets would require hundreds of millions of years to produce the mass in the lobes. The energy needed to produce that much high-velocity gas over that long a time is stupendous—comparable to the entire luminous output of the host galaxy over its lifetime. Another remarkable feature is the stability of the direction of the jets over that long time interval. If it took hundreds of millions of years to produce the side lobes, and if these lobes today are still aligned with the jets, then the whole system must have retained exactly the same orientation in space for all that time. About the only way this can happen is in a rotating system that is free of significant outside forces. In that case, the orientation of the rotational axis remains stable; this suggests that the jets emanate along the rotational axis. These considerations lead us to conclude that the ultimate source of the jets and radio lobes lies at the core of a radio galaxy, that immense quantities of energy must be generated there, and that rotation must be a feature of the system. We will return to the question of how all of this comes about shortly.

Seyfert Galaxies and Active Galactic Nuclei

Although the strongest radio emitters among galaxies are the giant ellipticals, they are by no means the only galaxies with evidence for highly energetic activity in their cores. Some spiral galaxies also display such behavior. We have seen that even the Milky Way is not immune; in Chapter 17 we summarized the evidence for the existence of a compact, massive object at its center. Some other spiral galaxies have much more pronounced violence in their nuclei. Many of these spirals are called **Seyfert galaxies (Fig. 19.7** and **Table 19.2),** after Carl Seyfert, an American astronomer who discovered and cataloged many of them in the 1930s.

Seyfert galaxies look like ordinary spirals, except that the nucleus is unusually bright and blue, rather than the yellow of most normal spiral galaxy nuclei. About 10% of them are radio emitters, with spectra indicating that the synchrotron process is at work. The radio emission usually comes from the nucleus rather than from double lobes. The spectrum of the visible light from a Seyfert nucleus typically shows emission lines, which are usually not seen in normal galaxies and do not come from stellar radiation. Thus there must be extensive regions of hot interstellar gas in the central regions of these galaxies. In all cases, high degrees of ionization are indicated, and high-velocity gas motions are usually found as well. Some Seyferts have broad emission lines indicating speeds of several thousand kilometers per second in the gas that produces the lines, whereas for others the lines are narrower. It is now thought that the difference between the broad-line and narrow-line Seyfert nuclei is due to the different angles from which we view disklike structures. When seen nearly edge-on, a rotating disk has higher velocity along the line of sight than when viewed face-on. Therefore, if the gas in a Seyfert nucleus is in a rotating disk, we will see higher Doppler shifts, and hence broader lines, when we happen to view the disk edge-on.

The energy coming from the nucleus of a Seyfert galaxy can literally outshine the rest of the galaxy, and a short-exposure photograph may reveal only the nucleus. The rest of the galaxy is visible only on

longer-exposure photographs. Other galaxies are also seen to have energetic activity in their cores, and today astronomers tend to use the term **active galactic nuclei** (usually abbreviated simply **AGN**) to describe all galaxies with such activity. The study and understanding of AGNs has been a primary area of modern astronomical research. It is difficult not to be fascinated by the notion that at the center of an otherwise ordinary galaxy is something that is capable of generating more energy than an entire galaxy, yet is so small it cannot be directly observed. We will return to the question of how these prodigious amounts of energy are produced after we examine another class of objects with some obvious similarities to both the radio galaxies and the AGNs.

BL Lac Objects

An object called BL Lacertae puzzled astronomers for many years. Originally thought to be a variable star (hence its name, which comes from a catalog of variable stars in which stars are identified by pairs of letters and the constellation name), this object has a very nonstellar spectrum. Instead of the normal absorption-line spectrum, BL Lac has only continuous emission, with no spectral lines, having the characteristic shape of the synchrotron spectrum. BL Lac is also a very strong radio source, further evidence that it is not a normal star. The brightness is variable in times of a few days.

Eventually, long-exposure photographs revealed that BL Lac is surrounded by a faint fuzzy cloud. This fuzz seemed completely featureless, and attempts to measure its spectrum were thwarted by the difficulty of isolating its faint light from the much brighter core emission of the object, which drowned out the faint light from the fuzz. Finally, in the 1970s, astronomers succeeded in obtaining a spectrum of the fuzz and found that it resembled a normal galactic spectrum, as would be seen in an elliptical galaxy. Thus BL Lac joined the ranks of galaxies with very energetic nuclei.

Many other BL Lac objects have now been found. They are invariably strong radio emitters, always by the synchrotron process, and they are usually variable. They are usually found embedded in the cores of elliptical galaxies. The objects range in luminosity, some of them being brighter than any normal galaxy. Some are so bright that they are given the name **blazars.**

There is reason to believe that BL Lac objects emit bipolar jets of energetic particles, just as many radio galaxies do, but that the BL Lac jets appear bright only when seen nearly on-axis; that is, when the beam is directed almost straight toward us. This would be consistent with the idea, already men-

Table 19.2
Properties of Seyfert Galaxies

Spiral galaxies.
Luminosities comparable to brightest normal spirals.
Bright, compact blue nuclei.
Nuclei show emission lines of highly ionized gas.
About 10% are radio sources.
Most are variable in times of days or weeks.
Emission from the nucleus is synchrotron radiation.
Some have radio jets.

Figure 19.7 A Seyfert galaxy. This is the galaxy NGC 1566, a spiral with a small, bright blue nucleus that is characterized by emission lines. The high-energy sources in the nuclei of these galaxies are thought to be related to other active galacic nuclei (AGN's) discussed in this chapter. (© 1987 Anglo-Australian Telescope Board, photo by David Malin) http://universe.colorado.edu/fig/19-7.html

tioned, that some radio galaxies probably have unseen jets, made invisible by the angle between the direction of the beam and our line of sight. Perhaps the main difference between the BL Lac objects and radio galaxies or spiral galaxy AGNs is that the BL Lac objects are radio galaxies that happen to emit their jets directly toward and away from the Earth, enhancing the brightness of the jet relative to the underlying galaxy.

Energy Sources for Active Galaxies: Black Holes and Accretion Disks Revisited

By this time you can probably guess what the source of energy in the active galaxies might be. We are faced with a need to explain enormous amounts of energy coming from very small volumes of space, and this naturally leads to the suggestion that black holes might be involved. This idea came up almost immediately when the enormity of the energy requirement became clear, but only recently has it been confirmed to the satisfaction of most astronomers.

What is the evidence that the source of energy is small? The mere fact that the bright cores of active galaxies such as Seyferts appear small in images is not enough, because at the distances typical of these galaxies even a rather large core would look small (i.e., would have a small angular size). It would be possible to pack an immensely bright, dense cluster of stars within the observed bright regions, for example. Of course, the *Hubble Space Telescope* has improved our view quite a bit, as we shall see.

An early clue to the small size of the energy-emitting regions came from their time variability. Some AGNs undergo variations in brightness in times as short as a few days or even less than a day. It

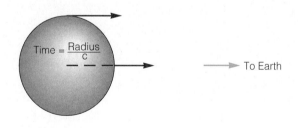

Figure 19.8 **The implication of the time variability for the size of the emitting object.** This sketch illustrates why an object cannot appear to vary in less time than it takes for light to travel across it. As a simple case, this spherical object is assumed to change its luminosity instantaneously. On Earth we observe the first hint of this change when light from the nearest part of the object reaches us, but we continue to see the brightness changing gradually as light from more distant portions reaches us.

is easy to see that the size of a region that varies cannot be larger than the distance light travels in the time of the shortest variations **(Fig. 19.8).** If the entire object were to brighten in an instant, we would see the change occur gradually, as light from the near side reached us first and light from the far side reached us last. The duration of the brightening as we saw it would be comparable to the light-travel time across the object. Thus, if we see the core of a galaxy change brightness in a few days or hours, then the distance across the emitting region must be less than a few light-days or light-hours. Generally speaking, this evidence limits the diameters of AGNs to sizes comparable to that of the solar system. Hence we are faced with explaining how an object no bigger than our planetary system can produce energy comparable to many billions of suns. Black holes definitely come to mind.

The energy observed to come from an AGN cannot be coming directly from the black hole. After all, the basic property of a black hole is that it is black; no energy can escape from within its event horizon. Therefore the emitted energy must come from material outside the event horizon. This situation should sound familiar, because we have already addressed it in discussing observations of stellar black holes (Chapter 16). Matter close to a black hole is likely to have angular momentum, meaning that it is orbiting around the black hole rather than falling straight in. The swirling, infalling material forms an accretion disk, a thin rotating disk of gas. But in the case of AGNs, the disk is far larger and more massive. It is heated by gravitational compression to the point where it glows at X-ray wavelengths, and its heat energizes gas in the vicinity so that it emits strongly at the wavelengths of spectral lines characteristic of ionized gas. Thus, when we see an AGN, we presumably are observing the glow from the accretion disk and the gas around it. Several galaxies with active nuclei are now suspected of housing supermassive black holes **(Table 19.3).**

So far our evidence is entirely circumstantial. We surmise that black holes are present because enormous amounts of energy are being emitted from small regions in galactic cores, and we use our knowledge of accretion disks in stellar binary systems thought to contain black holes to develop a similar accretion-disk model for the galactic nuclei. But to nail down the black hole hypothesis, we need to determine the masses of the central objects.

How do we measure these masses? Here astronomers use essentially the same techniques described in the previous chapter for measuring masses of galaxies, but now these techniques have to be extended to points very close to the galactic center. The key is to measure orbital speeds for stars and make use of the

Table 19.3

Black Hole Candidates

(in order of increasing distance from Earth)

Galaxy		Galaxy Constellation	Galaxy Type	Distance (millions of light years)	Black-hole mass (Sun)[b]
Milky Way core		Sagittarius	Sbc	0.028[c]	2×10^6
MGC 221	M32	Andromeda	E2	2.3	3×10^6
NC 224	M31	Andromeda	Sb	2.3	3×10^7
NGC 4258	M106	Canes Venatici	Sbc	24	4×10^7
NGC 3115		Sextans	S0	27	2×10^9
NGC 4594	M104	Virgo	Sa	30	5×10^8
NGC 3377		Leo	E5	32	8×10^7
NGC 4486	M87	Virgo	E0	50	$3 \times 10^9 0$
NGC 4261		Virgo	E2	89	$1 \times 10^9 50'$

fact that at any given distance from the center, the speed depends on how much mass lies interior to that distance. This was true of both the rotation-curve and the velocity-dispersion techniques we discussed previously. Now the goal is to measure Doppler shifts at positions very close to the core of a galaxy. But visible-wavelength observations made from the ground are limited in resolution, and it is not possible to measure orbital speeds close enough to the center to rule out some more normal form of matter.

Radio observations can overcome this problem, however. Using interferometry (see Chapter 6), astronomers can improve on the angular resolution of visible images. Radio telescopes cannot detect stars, but they can measure the orbital speeds of gas clouds containing molecules that emit at radio wavelengths. Using this technique, a group of radio astronomers in 1995 attained resolution 50 times sharper than that provided by *Hubble* images and were able to measure orbital speeds as close as 0.2 pc from the center of an active galaxy **(Fig. 19.9)**. The orbital speed there was so

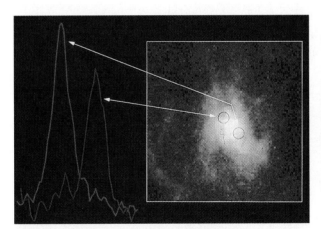

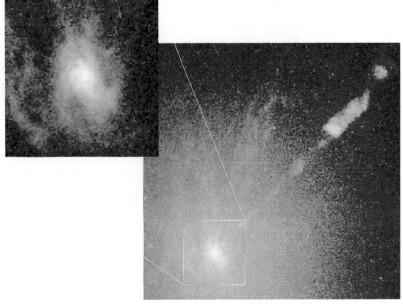

Figure 19.9 High-velocity gas in the heart of M87. The giant elliptical galaxy M87, known for its jet of relativistic gas (see Fig. 19.5), is seen in Hubble Space Telescope images to harbor a small, rotating disk of gas at its core (right). This disk (enlarged at center) displays high velocities as seen through Doppler shift measurements of emission lines (left). The high orbital speed of the gas in the disk, combined with its proximity to the center, indicates that the mass lurking there is too great to be anything but a supermassive black hole. (NASA/STScI) http://universe.colorado.edu/fig/19-9.html

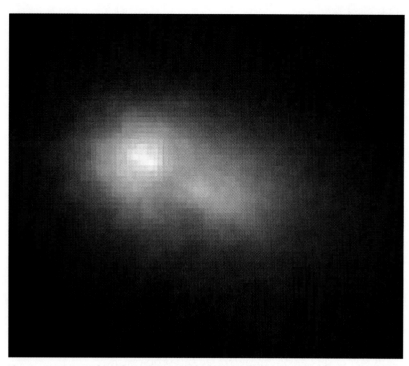

Figure 19.10 **Evidence for a supermassive black hole in the Andromeda galaxy.** This *Hubble Speace Telescope* image reveals that the Andromeda galaxy has a double core, suggesting that two stellar systems are in the process of merging. In addition, rapid orbital motions are observed surrounding the core region, indicating the presence of an enormous unseen mass. (NASA/STScI)

high that the central mass was estimated to be almost 1 billion solar masses, compelling evidence indeed for a black hole. There is virtually no way to avoid the conclusion that a black hole is at the center of this galaxy; the density of matter in the core region is simply too high to be supported by any conceivable physical force. This evidence, along with recent *Hubble* images, has persuaded most astronomers that supermassive black holes are indeed the energy sources in active galaxies. Some recent *Hubble* images have even revealed disk-like structures in the inner regions of some galaxies, although these disks are probably too big to be the accretion disks themselves. Even our neighboring spiral, the Andromeda galaxy, shows evidence for a rapidly-rotating inner disk and a double nucleus **(Fig. 19.10)** suggesting that its core has formed a black hole.

The similarities between the accretion disks in galactic nuclei and those associated with stars in formation (Chapter 12) and stellar remnants in binary systems (Chapter 16) are uncanny. Then we spoke of stellar-sized disks; now we speak of systems having millions to hundreds of millions of solar masses. Yet the physical processes by which the disks form and emit energetic radiation are nearly identical.

The resemblance becomes even more striking when we consider the jets that emanate from many AGNs. Recall that young stars often have bipolar outflows fed by jets of high-velocity ionized gas, streaming outward along their rotational axes and presumably confined by magnetic fields. The enormous dual radio-emitting lobes and the jets that feed

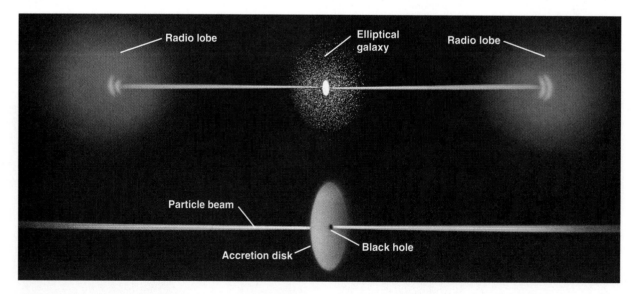

Figure 19.11 **A geometrical model for a quasar, a Seyfert galaxy, or a radio galaxy.** All of these objects have certain features in common that fit the picture shown here. A central object (most likely a supermassive black hole with an accretion disk) ejects opposing jets of energetic charged particles. These jets produce a synchrotron radiation and build up double radio-emitting lobes on either side of the central object. These lobes extend well beyond the confines of the galaxy in which the object is embedded.

them in radio galaxies can be described in the same terms, although the details of the mechanisms that produce the narrow beams are probably different. A sketch of the general model for a galactic black hole with its accretion disk and jets **(Fig. 19.11)** is almost interchangeable, except for size scales, with a sketch of the disk and jets from a young stellar object.

We surmise that as a supermassive accretion disk forms in the heart of a galaxy, the galaxy's magnetic field is squeezed and intensified as it is carried with the matter that is being compressed. Magnetic lines of force emanating outward along the rotational axis of the disk transport charged particles outward at relativistic speeds, creating synchrotron emission and filling enormous radio lobes on opposing sides of the galaxy. The similarity in the formation, structure, and physics of accretion disks and bipolar jets in objects ranging from low-mass stars to high-mass stellar remnants to galaxies is one of the most intriguing discoveries in astrophysics in recent years.

The Quasars

More than 35 years ago astronomers discovered a new class of objects, having the appearance of stars but with properties totally unlike anything previously observed. The story begins in 1960, when spectra were obtained of two starlike, bluish objects that had been found to be sources of radio emission **(Fig. 19.12)**. No radio stars were known, and astronomers were very interested in these two objects, called 3C 48 and 3C 273 (their designations in a catalog of radio sources that had recently been compiled by the radio observatory of Cambridge University in England). The objects were called **quasars** (short for **quasi-stellar radio sources**), or **quasi-stellar objects (QSOs),** because of their starlike appearance.

The spectrum of 3C 48 was obtained and found to consist of emission lines whose atomic identities were unknown. Some three years later, the spectrum of 3C 273 was also found to consist of emission lines, but in this case they were recognized as the strong Balmer lines of hydrogen, shifted toward far longer wavelengths than those measured in the laboratory. The lines in the spectrum of 3C 48 were then identified and also found to show very large shifts in wavelength. If these shifts are due to the Doppler effect, then tremendous speeds are implied. The redshift in the spectrum of 3C 273 is 16%, indicating that this object is moving away from us at 16% of the speed of light, or 48,000 km/sec! In 3C 48, the shift was even greater—about 37%—corresponding to a speed of 111,000 km/sec.

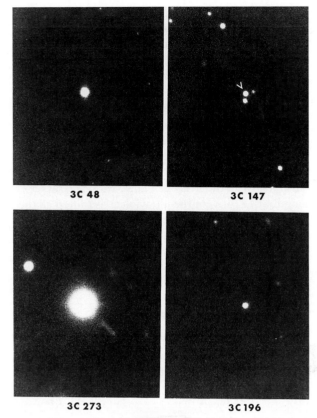

Figure 19.12 **Quasi-stellar objects.** These starlike objects are quasars, whose spectra are quite unlike those of normal stars. As explained in the text, these objects are probably more distant, and therefore more luminous, than normal galaxies. (Palomar Observatory, California Institute of Technology)
http://universe.colorado.edu/fig/19-12.html

The Relativistic Doppler Effect

The Doppler shift formula given in Chapter 6 applies only when the speed is a small fraction of the speed of light. A more complete form of the Doppler shift expression, which must be used when speeds are higher than a few thousand kilometers per second, is given by

$$\Delta\lambda/\lambda = z = \sqrt{\frac{(1 + \frac{v}{c})}{(1 - \frac{v}{c})}} - 1,$$

where the symbol z is commonly used to represent $\Delta\lambda/\lambda$. If we solve this for the velocity v, we get

$$v = c\left[\frac{(z+1)^2 - 1}{(z+1)^2 + 1}\right].$$

Using this formula, we find that 3C 273 is actually moving away from us at a speed of 44,000 km/sec, while 3C 48's speed is 91,000 km/sec.

Since the early 1960s, thousands of additional quasars have been discovered (the current total is over 5000). Only a small fraction are radio sources, and many differ from the first two in other ways. Invariably, though, the quasars have highly redshifted emission lines, and in very many cases, weak absorption lines are also seen. In many quasars the redshift is so huge that spectral lines with rest wavelengths in the ultraviolet portion of the spectrum are shifted all the way into the visible region. The grand champion today is a quasar with a redshift of 485%, so that the strongest of all the hydrogen lines, with a rest wavelength of 1216 Å (in the ultraviolet), is shifted all the way to 7113 Å (almost in the infrared). This quasar is not traveling away from us at 4.85 times the speed of light, however; use of the relativistic Doppler shift formula yields a speed of 94.3% of the speed of light—still an enormous speed!

The Origin of the Redshifts

To understand the physical properties of the quasars, we must first discover the reason for the high redshifts. We have already tacitly adopted the most obvious explanation—that they are a result of the Doppler effect in objects that are moving very rapidly—but we must still ascertain the nature of the motions. Furthermore, at least one alternative explanation has been suggested that has nothing to do with motions at all.

One consequence of Einstein's theory of general relativity, verified by experiment, is that light can be redshifted by a gravitational field. Photons struggling to escape an intense field lose some of their energy in the process, and as this happens, their wavelengths are shifted toward the red. We have already been exposed to this concept in Chapter 16, when we discussed the behavior of light near black holes, which have such strong gravitational fields that no light can escape at all. For a while it was considered possible that quasars are stationary objects sufficiently massive and compact to have large gravitational redshifts.

This suggestion has been largely ruled out for two reasons. One is that there is no known way for an object to be compressed enough to produce such strong gravitational redshifts without falling in on itself and becoming a black hole. If enough matter were squeezed into such a small volume that its gravity produced redshifts as large as those in quasars, no known force could prevent this object from collapsing further. A neutron star does not have as large a gravitational redshift as those found in quasars, and we already know that the only possible object with a stronger gravitational field than that of a neutron star is a black hole.

The second objection is that even if such a massive, yet compact object could exist, its spectral lines would be very much broader than those observed in quasar spectra. The high pressure would distort electron energy levels so that the spectral lines would be smeared out (as in a white dwarf, but more extreme), and light emitted from slightly different levels in the object would have different gravitational redshifts, again causing spectral lines to be broadened.

Others have suggested even more exotic explanations for the redshifts of quasars. These explanations generally wander into the realm of unproven (and in some cases unprovable) new physical laws. For example, it has been suggested that over the vast distances of the universe as a whole, the values for basic physical constants, such as the quantum values that govern the relationship between photon energy and wavelength, might change. It is possible to hypothesize new physics that would explain the high redshifts without requiring high velocities or even very great distances for the quasars. Most astronomers, however, find these ideas unlikely and, worse, untestable. It is generally accepted that the explanation requiring the fewest unproven assumptions or new physical laws is the one most likely to be correct (recall our many references to Occam's razor), and the assumption that quasar redshifts are due to the Doppler effect certainly requires fewer unproven assumptions or new physical laws than the alternatives.

For these reasons, we are led to accept the Doppler shift explanation for the redshifts. The problem then is to explain how such large velocities can arise. One possibility is that the quasars are relatively nearby (by intergalactic standards) and are simply moving away from us at very high speeds, perhaps as the result of some explosive event. This "local" explanation requires some care. To accept it, we must explain why no quasars have ever been found to have blueshifts. In other words, if quasars are nearby objects moving very rapidly, it is not easy to see why they are all receding from us, with none approaching. It might be possible to argue that they originated in a nearby explosion (at the galactic center, perhaps) so long ago that any that happened to be aimed toward us have had time to pass by and are now receding. There are serious difficulties with this picture, though, primarily in the amount of energy that would be required to get all these objects moving at the observed velocities—an amount that would dwarf the total light output of the galaxy over its entire lifetime.

The Cosmological Hypothesis

We are left with the explanation that we have already alluded to: that the quasar redshifts are due to

Astronomical Brand Names

Do you own a Quasar television? Do you use Comet cleanser in your kitchen? Is your car a Pulsar or a Saturn? These are just a few examples of astronomy's many legacies to our society. Product names and terms that evoke astronomical images seem to be favorites in the marketplace and in the language of the superlative.

It could be fun to try to assemble a list of astronomical brand names. Almost certainly there are more astronomical brand names for automobiles than for any other products on the market. We have not only Pulsar and Saturn, but also Comet, Galaxy, Nova, Eclipse, Vega, and Mercury, not to mention the Japanese Subaru, named after the star cluster that we call the Pleiades. Can you think of others?

A famous passenger airliner of the 1950s was the Constellation. A cargo plane is called the Hercules, which is the name of a constellation, but it may have been named after the mythological Greek strongman rather than the star pattern. The aerospace industry has used several other names that are both constellation names and figures in Greek legends, such as the Apollo, Orion, and Saturn launch vehicles. The aerospace world has also given us the term *astronaut*, for those who travel into space. As a side benefit we get the immortal expression "You don't have to be a rocket scientist," which somehow has come to mean that things are not as complicated as they seem.

Don't forget about food. We have Milky Way and Mars candy bars, for example, and a candy called Starbursts. In the sports and entertainment industry, we have the Astrodome and Astroturf. *Aster* is the Latin word for star and is also the name of a flower. Aster is also the base of the word asterisk, the name

of the ubiquitous symbol for a footnote. The word *lunatic*, derived from the Latin word for the Moon (luna), refers to someone who is a little unbalanced mentally and comes from the superstition that people get a little crazy at the time of the full moon.

Astronomical terms have crept into our language in many other ways. We speak of the astronomical odds against winning the lottery, and we call performers and athletes stars when they excel—that is, when they give stellar performances. A rapid rise to the top of one's profession is not just fast, it is meteoric. A newcomer may burst onto the scene like a nova or, in more extreme cases, a supernova. Records in sports are not just broken, they are eclipsed. And the term *black hole* has found innovative applications, such as the place where your money goes every month right after you get paid.

If you had any doubt about the importance of astronomy in your everyday life, or if anyone asks you what use astronomy is, now you have some truly cosmic answers.

Doppler shifts in objects so distant that the expansion of the universe can explain their enormous velocities. In this view the quasars are said to be at "cosmological" distances, meaning that they obey Hubble's relation between distance and velocity, just as galaxies do. If we adopt this assumption and assume that H_0 = 75 km/sec/Mpc, we find a distance to 3C 273 of

$$d = \frac{v}{H} = \frac{44{,}000 \text{ km/sec}}{75 \text{ km/sec/Mpc}} = 586 \text{ Mpc.}$$

Similarly, 3C 48 is just over 1,200 Mpc away. (These distances are based on the assumption that the rate of expansion of the universe has been constant, something that is probably not strictly true.)

There is a growing body of evidence supporting the cosmological interpretation of quasar redshifts. Perhaps the best evidence that the quasars are at cosmological distances is the fact that some have been found in clusters of galaxies with the same redshift as

the galaxies. Making this observation was extremely difficult, because at the distances at which the quasars exist, the much fainter galaxies are very hard to see, even with the largest telescopes. But today many such associations between quasars and ordinary galaxies have been found, and the commonality of their redshifts makes it very difficult to accept any explanation other than the cosmological one (unless one decides to claim that even the normal galaxies in these clusters have redshifts due to some other cause).

Additional support comes from the cases where quasars are found to be embedded within ordinary galaxies, a phenomenon that is discussed in a later section. For now we can note that in these cases, the redshift of the quasar matches that of the host galaxy, further demonstrating that the redshift is cosmological and not due to some peculiar property of the quasar itself.

The BL Lac objects help fill out the picture as well. They do not have spectral lines whose redshift can be measured, but their host galaxies do, as we have seen. The redshifts of the galaxies appear completely consistent with the usual cosmological interpretation. And the BL Lac objects—the central, energetic cores of these galaxies—are very similar to quasars. They are very luminous, rivaling quasars in energy output in some cases, and their variability and synchrotron spectra also resemble those of quasars. If BL Lac objects are close relatives of quasars, then this helps to cement the relationship between quasars and galaxies and supports the cosmological interpretation of quasar redshifts.

Look-Back Time and the Evolving Universe

The mere fact that the redshifts of quasars are best understood as being due to the expansion of the universe brings in some very important consequences. One of them has a lot to tell us about the evolution of the universe itself.

It is important to note that all quasars have substantial redshifts, until recently always much greater than those of normal galaxies (although now very faint galaxies, or perhaps smaller stellar systems leading to galaxy formation, are being found with redshifts as large as any quasar). The fact that all quasars have large redshifts, combined with the cosmological interpretation of the redshifts, tells us that all quasars are very distant.

Consider the length of time required for light from a quasar to reach the Earth. If the distance is measured in the tens or hundreds of millions of light-years, or even billions of light-years, then the light we receive has been traveling for tens of millions, hundreds of millions, or even billions of years. This is called the **look-back time.** Look-back time is always a consideration in astronomy. Even the most nearby star is seen as it was a few years ago, and phenomena on galactic or intergalactic distance scales are seen as they were hundreds, thousands, or millions of years ago. But these timescales are very short compared with the age of the universe, so we normally do not worry about whether we are viewing things that may no longer exist. But when we consider objects as far away as the quasars, we must allow for the possibility that we are seeing back to a time when the universe was different than it is today.

We see only very distant quasars. This tells us that quasars existed only long ago. Thus quasars represent a phenomenon that occurred in the early universe, but does not exist today. Therefore whatever the explanation of quasars turns out to be, it will prove to be something that does not (or cannot) happen today. The fact that galaxies are being found

with redshifts comparable to those of quasars tells us that galaxies already existed at the early times when quasars were common.

The existence of quasars only in the distant past of the universe is one form of evidence that the universe must be evolving—that is, changing with time. If there were quasars long ago but not today, then the ancient universe was different from the modern universe. This notion was resisted by astronomers who preferred cosmologies in which the universe is eternal, having no beginning and no end. Some of these astronomers have attempted to find alternative explanations for the quasar redshifts because they were unhappy with the implication that the universe has changed with time, but so far no satisfactory alternative explanation of the redshifts has been proposed.

Interestingly, even before the quasars were discovered, the radio galaxies had already offered evidence for an evolving universe, because these galaxies were also more common in earlier times. This was inferred not from redshifts, but from the fact that there are proportionally more faint radio galaxies than bright ones. Statistically speaking, this implies that more radio galaxies exist at great distances than in the nearby universe. As we have just argued with respect to the quasars, if there are more radio galaxies at great distances, that means there were more of them at early times, and that the universe has changed.

Quasar Luminosities

Another profound implication of the vast distances to quasars is their vast energy outputs. Quasars billions of light-years away can appear as bright as nearby stars (typical visual magnitudes for quasars are in the range from $m = 14$ to $m = 18$; see Chapter 14 for a review of magnitudes). An object billions of light-years away, but as bright as a local star, must be very luminous indeed. Conversion of apparent magnitudes for quasars to absolute magnitudes reveals values much smaller (hence brighter) than are ever seen in any ordinary galaxy.

The brightest galaxies have luminosities of about 10^{12} times the solar luminosity, or about 10^{38} watts. The Milky Way, a rather substantial spiral galaxy, has a luminosity of about 10^{11} suns, or roughly 10^{37} watts. Bright quasars, on the other hand, are found to have luminosities as high as 10^{15} suns, or 10^{41} watts. These are by far the most luminous objects in the known universe.

The related objects—AGNs, radio galaxies, and BL Lac objects—have lesser luminosities than quasars, but not by much. The brightest BL Lac objects in particular are almost as luminous as qua-sars, and all of these active objects are very bright compared with normal galaxies. As we will see, this similarity is just one of many that lead us to believe that the sources of energy in all these objects may be related.

The Properties of Quasars

Before we discuss the possible origin of the quasars, it is useful to explore their properties a bit further. Many quasars are being studied with care, primarily by means of spectroscopic observations. Within the past 20 years or so, the development of space-based observatories has added ultraviolet, infrared, and X-ray data to the information gathered at visible wavelengths, thus providing a fairly complete view of the characteristics of these objects **(Table 19.4)**.

The blue color that characterized the first quasars discovered is a general property of all quasars (except for extremely redshifted ones, where the blue light has been shifted all the way into the red part of the spectrum), but the radio emission is not. Only a small fraction of quasars are radio sources, contrary to the early impression that they all are. This type of misunderstanding is known as a **selection effect,** because the detection of new quasars depended on the incorrect (or incomplete) assumption that they were all radio sources. Thus only those that are radio emitters were found. More recent searches for quasars, based on their peculiar colors or on their strong X-ray emission, have turned up many new finds, and astronomers now realize that only about 10% are quasars are radio sources. The radio emission, and usually the continuous radiation of visible light as well, shows the characteristic synchrotron spectrum. Apparently, all quasars emit X rays, however, again by the synchrotron process, so X-ray surveys have proved to be a better method of finding quasars.

In some cases, photographs of quasars reveal some evidence of structure instead of a single point of light. The most notable of these is 3C 273, which played such a key role in the initial discovery of quasars. This object shows a linear jet extending from one side **(Fig. 19.13)**, closely resembling the jet emanating from the giant radio galaxy M87 (see Fig. 19.5). This similarity in structure suggests that similar physical processes are occurring in these two rather different objects. It is quite possible that other quasars have similar structures that we cannot see due to their great distances from us (recall that 3C 273 is one of the nearest quasars).

Many quasars vary in brightness, usually over periods of several days to months or years (although, in at least some cases, variations are seen over times

of just a few hours). This variability is very important, for it provides information on the size of the region in the quasar that is emitting the light. As noted earlier, variations in brightness in times of days means that the emitting object must be no larger than light-days across. This means that we are now contemplating objects that are more luminous than any galaxy, yet are still comparable in size to our solar system. These properties obviously place stringent limitations on the nature of the emitting objects.

The spectra of quasars have already been described in broad outline, but they contain a great deal of detail as well **(Fig. 19.14)**. All quasars have emission lines, generally of common elements such as hydrogen, helium, and often carbon, nitrogen, and oxygen

Table 19.4
Properties of Quasars
Characterized by large redshifts.
Spectra dominated by emission lines of highly ionized gas.
Optical luminosities 100 to 1,000 times those of normal galaxies.[a]
About 10% are radio sources.
Appear as compact, blue objects.
Many are variable in times of days or weeks.
Emission due to synchrotron radiation.
Some have radio or optical jets.
[a]Assuming that quasars are at cosmological distances.

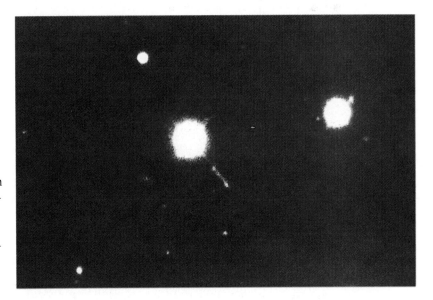

Figure 19.13 A quasar with a jet. This is 3C 273, one of the first two quasars discovered. This visible-light photograph reveals a linear jet very much like those seen in many radio galaxies. The radio structure of quasars is usually double-lobed, also similar to the structure found in radio galaxies. (National Optical Astronomy Observatories)

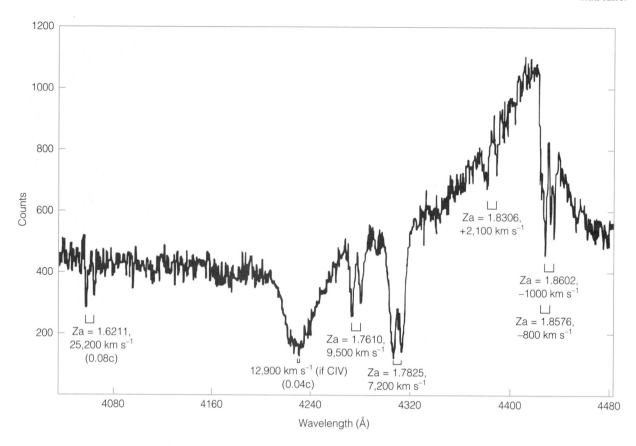

Figure 19.14 A quasar spectrum. This spectrum shows several features typical of quasar spectra: a broad emission line centered near 4400 Å, absorption near 4240 Å, and narrow absorption lines at several wavelengths. All features seen here are due to carbon that has lost three electrons (designated CIV). Redshifts are indicated (Za indicates the redshift of an absorption-line system). (Data from R. J. Weymann, R. F. Carswell, and M. G. Smith, 1981, *Annual Reviews of Astronomy and Astrophysics*, 19:41.)

(some of the latter elements only have strong lines in the ultraviolet and therefore are best observed when the redshift is sufficiently large to move these lines into the visible portion of the spectrum). In many cases, the emission lines are very broad, showing that the gas that forms them has internal motions of thousands of kilometers per second. The degree of ionization tells us that the gas is subjected to an intense radiation field, which continually ionizes the gas by the absorption of energetic photons.

Quasar Absorption Lines

Many quasars show absorption lines in their spectra in addition to the emission features. The absorption lines are nearly always seen at a smaller redshift than the emission lines, indicating that the gas that creates the absorption is not moving away from us as rapidly as the quasars themselves. Many quasars have multiple absorption redshifts; that is, they have several distinct sets of absorption lines, each with its own red-

shift, and each therefore representing a distinct velocity. This shows that there are several absorbing clouds in the line of sight.

These absorbing clouds are thought to lie in the foreground, somewhere along the line of sight between the Earth and the background quasar. Their distances from us are thought to be cosmological, meaning that the redshifts are imparted to them by the expansion of the universe. The fact that the clouds virtually always have lower redshifts than the quasars behind them is consistent with this idea and explains why they can lie between us and the quasars. When a quasar has more than one set of absorption lines at different redshifts, there are multiple clouds in the line of sight at different distances.

What are these intervening gas clouds? They seem to fall into two categories: absorption-line systems that include lines of heavy elements, and systems that contain only lines of hydrogen (and possibly helium). The **metallic-line systems,** as the former are called, are thought to be halos of normal galaxies that happen to lie along the lines of sight toward quasars. As we have learned from studying our own galaxy, a galaxy can be surrounded by a very extended volume of gas. Because galaxies have heavy elements as a result of stellar evolution, we would expect to find some heavy elements in the gas of a galactic halo. The enormous extent of the halos allows them to be seen projected against the images of distant quasars even though the visible galaxy may lie quite a distance off to the side of the quasar's position on the sky **(Fig. 19.15).** Thus the study of the metallic-line systems can tell us a great deal about galactic halos: their numbers, their sizes, and their compositions.

The absorption-line systems containing only hydrogen lines are sometimes called **Lyman-α forest systems.** These systems often consist of just one spectral line, the Lyman-α line of atomic hydrogen, repeated over and over at many different redshifts (the word *forest* is applied because these lines may be so numerous that they almost blend together). The Lyman-α line has a rest wavelength in the far-ultraviolet, at 1216 Å, so quasars having Lyman-α forest lines must have sufficient redshifts (greater than about 3) to shift this line into the visible portion of the spectrum. The lack of metallic (i.e., heavy element) lines at the same redshifts indicates that these clouds contain few or no heavy elements. Instead we infer that they are made up almost solely of hydrogen and helium. This is the composition we would expect in a cloud of primordial material that has never formed any stars. Thus the Lyman-α forest systems, particularly those seen at high redshifts (meaning they are seen as they were at the time when galaxies were still forming), probably represent clouds of gas that are yet to form into stars and galaxies. Some of these metal-free clouds have survived to more modern times (i.e., they have low redshifts) and may be cosmic leftovers that will never form stars.

Thus the quasar absorption lines have much to tell us about galaxies along the way toward the quasars or about intergalactic gas clouds left over from the earliest times in the history of the universe, but they are thought to be otherwise unrelated to the quasars themselves, which are just background light sources against which we can detect the absorption lines. We return now to the question of what quasars are.

Galaxies in Infancy?

A variety of theories, some of them rather fanciful, have been proposed to explain the quasars. One idea has gradually become widely accepted, however, and

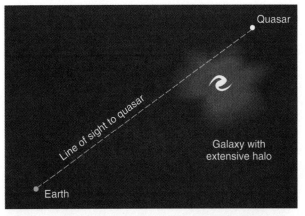

Figure 19.15 A quasar seen through a galactic halo. The gas surrounding the intervening galaxy forms absorption lines. Because the quasar and the galaxy are both receding from Earth at cosmological speeds, the redshift of the galactic halo absorption lines is less than the redshift of the quasar emission lines.

we will restrict ourselves to discussing that hypothesis, keeping in mind that there are other suggestions.

The prevailing interpretation of the nature of quasars was inspired by the fact that they existed only in the long-ago past, by their similarities to the BL Lac objects (which clearly are galaxies), and by their resemblance to the nuclei of Seyfert galaxies. The resemblance to Seyfert nuclei is striking: both are blue; both are radio sources in about 10% of the cases; both vary in brightness on similar timescales; and both have very similar emission-line spectra. Seyferts lack the complex absorption lines often seen in the spectra of quasars, but this would be expected if the quasar absorption lines are formed in the halos of intervening galaxies, because Seyfert galaxies are not far enough away that there are likely to be many other galaxies along their lines of sight. The nuclei of Seyferts also differ from quasars in the amount of energy they emit, being considerably less luminous.

The picture that is developing is that quasars are young galaxies, with some sort of youthful activity taking place in their centers. In this view, the Seyfert galaxies are related to quasars, and in some cases may be their descendants, which are still showing activity in their nuclei but with diminished intensity because today less gas is being fed into them through their accretion disks.

The notion that quasars may be infant galaxies is supported by some rather direct evidence. As noted earlier, some quasars have been found associated with clusters of galaxies, showing that they can be physically located in the same region of space at the same time. In addition, very recent *Hubble Space Telescope* images **(Fig. 19.16)** show that quasars always reside at

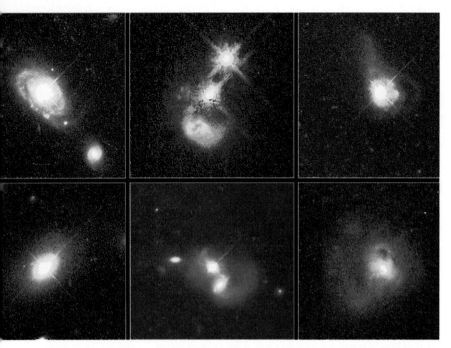

Figure 19.16 Quasar "host" galaxies. Each one of these galaxies imaged with the *Hubble Space Telescope* has a quasar at its core. Some are normal galaxies, while others appear to be galaxies undergoing collisions or mergers. (NASA/STScI)

the cores of galaxies, and that these host galaxies often show signs of turmoil. A short-exposure photograph of a Seyfert galaxy **(Fig. 19.17)** shows only the nucleus, which is very much brighter than the surrounding galaxy. The picture emerging from all this evidence is that quasars may form in young galaxies that are colliding or merging, and that modern galaxies (including the Seyferts) are their descendants. It has been suggested that the Milky Way may even harbor a defunct quasar at its core.

Although we can make a strong case that quasars are young galaxies, we are still far from understanding their origins. Yet their origins are clearly relevant to understanding the source of their immense energies.

Quasar Formation and Evolution

Another observational tendency of quasars may help us to shed some light on how they form and why they shine so brightly. Recent surveys aimed at finding all the quasars in specific regions of the sky show that they tend to be grouped, rather than being randomly distributed. In other words, quasars tend to appear in clusters. Based on what we have already learned about the interactions among galaxies in clusters, this leads us to suspect that similar interac-

tions may have something to do with forming quasars. If clusters of quasars represent clusters of objects in the early universe, then the formation of quasars might have something to do with the collisions and mergers of these young objects. Thus the suggestion is that quasars are young galaxies, and that the energetic phenomena associated with quasars may be related to interactions among these pre-galactic objects.

If two young galaxies, still full of interstellar gas and dust, should collide and merge, inevitably matter would settle toward the center of the resultant object. Calculations of the dynamics (motions) of the gas in such situations show that it is possible for this infalling matter to become so concentrated that it forms a black hole and an accretion disk. Gravity takes over at the core of the cannibalistic object, and as additional collisions and mergers occur, the black hole gains more and more mass, becoming more and more luminous as its accretion disk grows. Some merged objects would gain more mass than others, producing active young galaxies with varying degrees of activity and luminosity. Those objects we see today (but really very long ago) as full-fledged quasars would be the ones that gained the most mass. The AGNs would be either galaxies that never were as active as quasars or else former quasars whose energy output has declined as the supply of mass to feed their black holes diminished.

In this picture, the source of the energy in a quasar is basically the same as in the active galaxies already discussed. A supermassive black hole heats a surrounding accretion disk to the point where it emits X rays. The presence of jets in at least a few quasars also fits this picture, as the magnetic field associated with the collapsed matter would channel ionized gas outward along the rotational axis.

This view of quasar formation and evolution has proved to be very satisfying to astronomers, because it accounts for all of the observed properties of the objects and furthermore provides a unifying picture of quasars and galaxies, all nicely tied in with what is known of the history of the universe. For example, this scenario helps to explain why quasars are seen only at early times in the universe, while the AGNs and radio galaxies are more recent, and vestigial traces of intense activity are found in modern-day galaxies such as the Milky Way.

As a large galaxy formed and built up a black hole at its core, for an extended time new matter would have spiraled in, feeding the accretion disk. But as the interstellar gas and dust in the central region of the galaxy were gradually consumed, a cavity of relatively low density would have been cre-

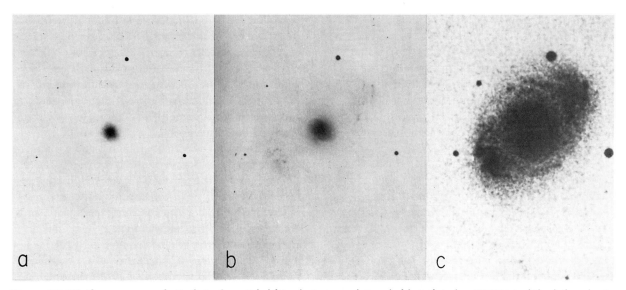

Figure 19.17 Three exposures of a Seyfert galaxy. At the left is a short-exposure photograph of the Seyfert galaxy NGC 4151, in which only the nucleus is seen, resembling the image of a star. The center and right-hand images are longer exposures of the same object, revealing more of its fainter, outer structure. This sequence illustrates how a quasar, which is even more luminous than a Seyfert galaxy nucleus, can appear as a starlike image even though it is embedded in a galaxy. (Photographs from the Mt. Wilson Observatory, Carnegie Institution of Washington, composition by W. W. Morgan)

ated, thus cutting off the supply of new mass to the accretion disk. As a result, new material would have eventually stopped feeding into the jets of rapidly moving ions streaming out along the polar axes, while at the same time the emission from the accretion disk itself would have been reduced. The quasar that was born with the galaxy would gradually diminish in power, first reaching the luminosity levels typical of radio galaxies and AGN and then perhaps diminishing further so that today there may be no external sign of its energetic past.

We have seen (in Chapter 17) that our own galaxy may have a supermassive black hole at its core. This raises the possibility that the Milky Way once harbored a quasar, but it is not at all clear that the Milky Way was ever as powerful as a quasar (the mass proposed for our "local" supermassive black hole is far smaller than that found in some AGNs). It is likely that a wide variety of activity occurred during the era of galaxy formation. Probably not all galaxies underwent a quasar stage, and it is possible that many of today's normal galaxies never had active nuclei.

While astronomers may feel that the essential nature of the quasars is now understood, a key gap in our knowledge of the early history of galaxies remains. What is still missing is a complete model for the initiation of galaxy formation. How and when did matter become congregated in lumps in the early universe, leading in turn to star formation and the development of galaxies? How did the galaxies form and acquire their shapes? We can gain some information on these questions by studying the most distant —hence the youngest—galaxies.

Galaxies in Infancy: Formation of Stellar Systems

As we will find in the next chapter, one of the premier puzzles facing astronomers today is the origin of galaxies. This question arises at two levels: First, how did the universe become clumpy enough to form galaxies and clusters of galaxies? Second, by what process did the galaxies form? The first question will be discussed in Chapter 20. The second can be addressed here in the light of startling new images, most of them obtained by the *Hubble Space Telescope*.

Traditional Ideas of Galaxy Formation

All theories of galaxy formation begin with a vast cloud of intergalactic gas, which collapses on itself gravitationally, forming stars in the process. The eventual result is either a spiral or an elliptical galaxy. Which you get may be determined by the density and the rotation of the initial cloud. The denser it is, the more quickly the gas is consumed by star formation, and the more likely the end result will be a galaxy having little or no leftover interstellar matter; that is, an elliptical galaxy. The role of rotation competes with the role of density; the more rapid the rotation, the more likely the galaxy is to be flattened into a disk. Thus a low-density cloud with relatively high angular momentum (i.e., relatively rapid spin) is likely to produce a disk galaxy, while a dense, slowly rotating cloud is more likely to yield an elliptical galaxy.

In many cases, the disk galaxies subsequently become spirals through perturbations caused by outside forces such as gravitational effects of nearby neighbor galaxies. The development of barred central regions in about half of all spirals is thought to be related to the amount of mass in the halo. A massive halo can suppress the formation of an elongated central bulge, which otherwise will form due to natural instabilities in a round disk.

The picture just described seems to account nicely for the general properties of observed galaxies. Or at least it would, if observations supported the idea that the galaxies we see today are what formed when the universe was young. Instead we are finding more and more evidence that galaxies have changed in many ways and that their distribution and form have been very dynamic over the lifetime of the universe.

Galaxies in Turmoil

We have already noted that in rich clusters of galaxies a preponderance of ellipticals is found near the central regions, with a greater percentage of spiral galaxies lying farther out. This distribution, along with other evidence for galactic mergers, suggests that the galaxies in such clusters change form over time. In the central regions, disks and spirals are eroded away into ellipticals. Galactic cannibalism also plays a role. Lurking at the centers of many rich clusters are giant ellipticals, often radio galaxies with active nuclei, which appear to have grown to their huge sizes by eating galaxies that have wandered too close and fallen in. In this chapter we have discussed how such a process can create the supermassive black holes that energize the cores of active galaxies. The evidence is overwhelming that in at least some environments galaxies can and do change with time.

Another line of evidence supports this view as well. It is indicated by two types of data: one comes from evolutionary studies of our own galaxy; the other, from recent infrared observations of distant galaxies.

Some time ago it was calculated that the Milky Way could not have produced its present-day abundances of heavy elements if star formation and evolution had always occurred at the rate they do today. The rate of **astration,** as the processing and enrichment of chemical elements by stellar evolution is called, is too slow to have produced the observed quantities of carbon, nitrogen, oxygen, and the iron-peak elements over the lifetime of the galaxy. This suggests either that these elements were already present when the galaxy formed, or that the rate of astration must have been much higher at some time in the past. Thus it was suggested years ago that perhaps our galaxy experienced an era of very intense star formation and evolution at some point in the past and in that brief period of time built up substantial quantities of the heavy elements.

Supporting evidence for this idea comes from infrared maps of the sky, particularly those obtained by the *IRAS* satellite in the early 1980s. Many previously unknown infrared sources were found that coincided in position with faint, peculiar galaxies or sometimes had no optical counterpart at all. The distribution of these sources on the sky suggested that all were extragalactic. They were identified as a new class of previously unsuspected infrared galaxies.

When infrared spectroscopy of these objects was subsequently done and their redshifts measured, most turned out to be quite distant. Thus these are young galaxies, seen as they were billions of years ago. They must be telling us something about how galaxies form and evolve.

The most logical explanation for the huge infrared luminosities of these galaxies is that they are so immersed in interstellar dust that visible light from the stars within cannot escape without being absorbed by dust grains. The grains in turn warm up to the point where they release their stored heat energy as infrared radiation. This hypothesis is confirmed by the presence in the spectra of these galaxies of infrared features due to solid dust, such as emission due to silicates (a common component of dust in our galaxy). The infrared galaxies are analogous to huge versions of the dark dust clouds that we see in the Milky Way; they are opaque to visible light but glow brilliantly at infrared wavelengths.

But why should young galaxies be so shrouded in dust? The accepted explanation is that these galaxies are undergoing extreme episodes of star formation, creating huge numbers of massive, brilliant, short-lived stars out of interstellar gas and dust. Bottled up behind the dusty veil surrounding such a galaxy are swarms of massive stars, creating heavy elements and returning those elements to the interstellar medium as they blow off their outer layers in rapid stellar winds and then explode in supernovae. A galaxy cannot sustain this level of activity for long, and in due course it slows its rate of star formation and becomes a visible galaxy as the dust is consumed by star formation. A galaxy going through this intense period of star formation and astration is called a **starburst galaxy (Fig. 19.18).** It is possible that our own Milky Way once experienced such an episode, creating at that time most of the heavy elements we find today.

Looking Back: Infant Galaxies in Stress

Perhaps the best way to see how galaxies form is to take a look at some of them in the process. We can do this because a telescope is a time machine. The farther away we look, the farther back in history we are viewing the universe because of light-travel time. The problem has always been that very faraway galaxies are very dim. This, combined with the blurring caused by the Earth's atmosphere, has prevented ground-based telescopes from providing very sharp pictures of galaxies at distances of billions of light-years.

The *Hubble Space Telescope* has changed that, providing unprecedented resolution in images of extremely faint objects. Images of groups and clusters of galaxies obtained in 1994 show that young galaxies, particularly those of disklike structure, have very unsettled appearances **(Fig. 19.19).** The spirals and disks appear to be not yet fully formed when the universe was roughly one-third of its present age, while ellipticals already appear to have evolved to their present

Figure 19.18 A starburst galaxy. This irregular galaxy (NGC 1313) is dominated by interstellar gas and dust, and is host to widespread regions where star formation is occurring. Because of the high level of star-forming activity, galaxies like this are dubbed starburst galaxies. Some are so filled with gas and dust that the only outward sign of them is the infrared glow from the dust, which completely obscures the galaxy in visible wavelengths. (© Anglo-Australian Telescope Board. Photo by D. Malin)

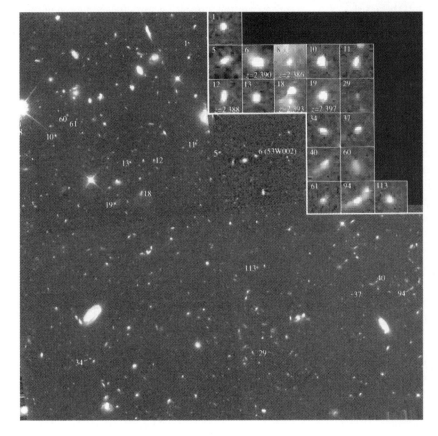

Figure 19.19 Embryonic galaxies. Many recent observations, made with large ground-based telescopes as well as the *Hubble Space Telescope,* have revealed very distant, irregular objects that are thought to be very young galaxies or, like the objects featured here, smaller systems on their way to merging and forming galaxies. The objects in this *HST* image are too small to be full-fledged galaxies and tend to be grouped, supporting the "bottom up" model for galaxy formation, as discussed in the text. (*HST* image provided by S. Pascarelle, R. Windhorst, W. C. Keel, and S. C. Odewahn; research sponsored by NASA/STScI)

shapes. These findings upset a couple of long-standing beliefs of astronomers who study galaxy formation: they showed that all types of galaxies did not necessarily form at the same time and suggested that at least for some types of galaxies, formation was a more recent event than had been suspected.

In early 1996 a new *Hubble* project revealed similar results, but pushed the picture out to greater distances and back to earlier times. This project, already mentioned in Chapter 18, involved pointing the *Hubble* at a nearly blank spot of sky for several hours to build up an image of the faintest objects ever observed. Hordes of distant, young galaxies showed up, increasing estimates of the density of galaxies in the universe and also providing a better glimpse of the nature of galaxies in the early universe. Again, a strange assortment of shapes and colors was seen, indicating that galaxy formation was not a singular event but a process that lasted for billions of years as protogalaxies settled only gradually into the shapes we see today. The weirdly distorted shapes of these young galaxies show that they are experiencing gravitational stress as they interact with neighbors, and their colors indicate that they are undergoing widely varying rates of star formation.

At the same time (early 1996), astronomers using ground-based telescopes have begun to identify very distant, very immature galaxies. With redshifts indicating that they are being seen as they were when the universe was only 15% of its present age, these primordial galaxies show signs of intense star formation and generally elliptical overall shapes. These results, combined with the deep-sky image obtained by the *Hubble*, are helping to create a consistent picture of galaxy formation in which the first systems formed were small and elliptical. These systems apparently then merged to form larger galaxies, including disks and spirals.

Studies of the objects seen at the earliest times in the history of the universe will continue, as the *Hubble* program is extended to other regions of the sky, and as new supergiant telescopes on the ground, equipped with active optics to compensate for atmospheric distortion, push the frontier farther and farther back in time.

Summary

1. Radio galaxies emit vast amounts of energy in the radio portion of the spectrum. These galaxies, often giant ellipticals, commonly show structural peculiarities. The radio emission is synchrotron radiation and in most cases is produced from lobes of hot gas on opposite sides of the visible galaxy.

2. Many radio galaxies have jets of ionized gas that appear to be the source of the side lobes and thus suggest that activity at the galactic core is responsible. The jets are visible only if their particles are moving at relativistic speeds.

3. Seyfert galaxies are spiral galaxies with compact, bright blue nuclei that produce emission lines characteristic of high temperatures and rapid motions.

4. BL Lac objects are active nuclei of elliptical galaxies that appear so bright to us because their jets are pointed in our direction.

5. Radio galaxies, Seyfert galaxies, and BL Lac objects are generally categorized as active galactic nuclei (AGNs) because each has a source of immense energy at its core. This energy source is almost certainly a supermassive black hole with an accretion disk and jets emanating along the rotational axis.

6. Several point sources of radio emission found in the early 1960s looked like blue stars and were therefore called quasi-stellar objects, or quasars.

7. Quasars have emission-line spectra with very large redshifts, corresponding to speeds that are a significant fraction of the speed of light. If these redshifts are cosmological, then quasars are on the frontier of the observable universe, are extremely luminous, and are seen as they were billions of years ago.

8. Many quasars have absorption lines at a variety of redshifts (always smaller than the redshift of the emission lines), which are probably created by intervening galactic halos (having spectral lines of heavy elements) or by primordial intergalactic gas clouds (having lines of hydrogen and helium only).

9. The most satisfactory explanation of the quasars is that they are the cores of very young galaxies. This interpretation is supported by the fact that they are seen only at great distances (that is, they existed only long ago), and that they resemble the nuclei of Seyfert galaxies. In several cases, photographs have revealed galaxies surrounding quasars, supporting this suggestion. The core energy source is the same as in an AGN, that is, a supermassive black hole.

10. Observations of very distant, and hence young, galaxies reveal that galaxy formation was an extended process, starting with ellipsoidal systems, which then merged to form disks and spirals. Young galaxies (particularly spirals) show weirdly distorted shapes, indicating that they had not yet settled into their final form.

Finding Quasar Redshifts

ASTRONOMICAL

ACTIVITY

The accompanying figure is a sketch of the spectrum of a quasar, showing both emission lines and absorption lines. Also given is a table listing spectral lines (either emission or absorption) that might appear in the spectrum of a typical quasar. You can use this list to identify features in the spectrum and determine its redshift (*Hint:* the lines represent more than one redshift; that is, this quasar has an emission redshift and at least two distinct absorption redshifts).

First, you will need to identify lines belonging to each redshift system, and then use the wavelengths you measure from the spectrum, along with the rest wavelengths given in the table, to determine the redshift $\Delta\lambda/\lambda$. You will have to identify the lines by pattern recognition, because the wavelengths are shifted so far that they are not even close to the laboratory wavelengths. A useful starting point is to assume that the 1216 Å line of hydrogen (H I) is normally the strongest line in each redshift system; note also that this line has the shortest wavelength of the lines typically seen. Once you have tentatively identified an H I line, compute its redshift and then see whether other lines from the list appear at approximately the same redshift.

Spectral Lines of a Typical Quasar	
Atom or Ion	**Rest Wavelength (Å)**
H I	1216
C III	1909
C IV	1549
N V	1240
MG II	2800
SI II	1260
	1304
	1526
SI IV	1398

If you find several coincidences, then you have identified a redshift system. Note that to make things a little more complicated (but more realistic), not all of the lines in the table appear in each of the redshift systems in the quasar spectrum.

Once you have found the redshifts, you can determine the velocities of the quasar and of each absorbing cloud (using the relativistic Doppler shift formula); then you can use the Hubble law to find their distances (indicate what value of the Hubble constant H_0 you chose).

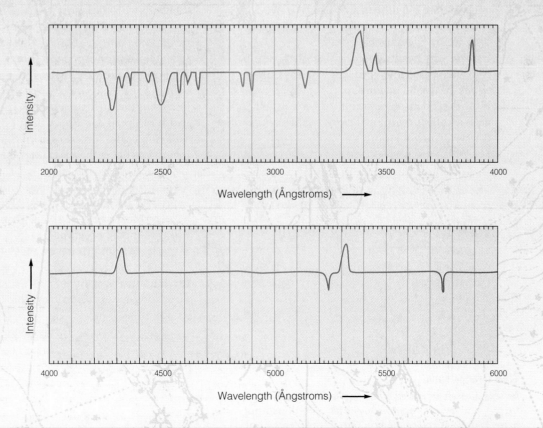

Review Questions

1. Summarize the properties of radio galaxies, pointing out the ways in which they differ from normal galaxies.

2. Why do astronomers suggest that massive black holes may be responsible for the vast amounts of energy being emitted from the cores of radio galaxies, Seyfert galaxies, and quasars?

3. Why do we see jets in some radio galaxies but not in others, even though nearly all of them have double radio lobes?

4. Explain, in your own words and using your own sketch, why the timescale of variations in light from a source such as a quasar places a limit on the diameter of the emitting region.

5. Discuss the similarities and differences between the accretion disks and bipolar outflows in young stars and the accretion disks and jets in quasars and AGNs.

6. What is the significance of the fact that quasars all have high redshifts?

7. Why are quasar absorption lines always observed at smaller redshifts than the emission lines? Would this necessarily be true if the redshifts of the emission lines were not cosmological?

8. If quasars are young galaxies, how does this help demonstrate the difficulty of using faraway galaxies as standard candles in estimating large distances in the universe?

9. Explain why nearly all of the first quasars to be discovered are radio emitters, whereas only about 10% of all quasars are.

10. How do recent observations of very distant galaxies affect traditional views of the process of galaxy formation?

Problems

1. Consider a galaxy that is 1 Mpc away. Your task is to determine how close to the center of this galaxy the *Hubble Space Telescope* can measure stellar velocities in the quest to determine whether a supermassive black hole is present. The diffraction limit for the *HST* at visible wavelengths is about 0.05 arcseconds. Use the small-angle approximation (Appendix 2) to determine how close to the center of this galaxy the *HST* could make measurements. Compare your answer to the size of the solar system (i.e., express it in AU).

2. If the central black hole in a galaxy that is 1 Mpc away has a mass of 2×10^8 solar masses, what is the Schwarzschild radius of the black hole? Could an object this size be resolved by the *Hubble Space Telescope*? (*Hint:* Review the information on black holes in Chapter 16.)

3. If the orbital speed in a galaxy is 100 km/sec at a distance of 5 pc from the galactic center, what is the central mass of this galaxy?

4. Suppose a quasar is observed to have a visual apparent magnitude of $m = +17.68$. Its redshift indicates that its distance is 1,000 Mpc. What is its visual absolute magnitude? How does this compare with the visual absolute magnitude of a galaxy like the Milky Way? (*Hint:* You will need to remind yourself about the distance modulus in Chapter 14 or in Appendix 13.)

5. Suppose that the red hydrogen line (rest wavelength 6563 Å) is observed as a broad emission line from the nucleus of a Seyfert galaxy. The width of the line is 200 Å (that is, it extends for 100 Å in either direction from the rest wavelength). What is the maximum speed of the gas in the region where this line is emitted?

6. The Lyman-α line of hydrogen has a rest wavelength (in the ultraviolet) of 1216 Å. If a quasar has a redshift of $z = 3.46$, at what wavelength will this line be observed?

7. A quasar is found to have a redshift of $z = 2.45$. What is its recession velocity? How far away is it if the Hubble constant is $H_0 = 75$ km/sec/Mpc? How long has the light from this quasar been on its way to us?

Web Connections

The Review Questions and Problems also appear at the following URLs:
http://universe.colorado.edu/ch19/questions.html
http://universe.colorado.edu/ch19/problems.html

Additional Readings

Barnes, J. L. Hernquist, and F. Schweizer. 1991. What Happens When Galaxies Collide. *Scientific American* 265(2):40.

Blandford, R. and A. Königl. 1993. The Disk-Jet Connection. *Sky & Telescope* 85(3):40.

Burbidge, G. and A. Hewitt 1994. A Catalog of Quasars Near and Far. *Sky & Telescope* 88(6):32.

Courvoisier, T. J.-L. and E. I. Robinson 1991. The Quasar 3C273. *Scientific American* 264(6):50.

Crosswell, K. 1993. Have Astronomers Solved the Quasar Enigma? *Astronomy* 21(2):28.

Djorgoski, S. G. 1995. Fires at Cosmic Dawn. *Astronomy* 23(9):36.

Finkbeiner, A. 1992. Active Galactic Nuclei: Sorting Out the Mess. *Sky Telescope* 84(2):138.

Ford, H. and Z. I. Tsvetanov. 1996. Massive Black Holes in the Hearts of Galaxies. *Sky & Telescope* 91(6):28

Miley, G. K. and K. C. Chambers. 1993. The Most Distant Radio Galaxies. *Scientific American* 268(6):54.

Preston, R. 1988. Beacons in Time: Maarten Schmidt and the Discovery of Quasars. *Mercury* 17(1):2.

Veilleux, S., G. Cecil, and J. Bland-Hawthorne. 1996. Colossal Galactic Explosions. *Scientific American* 274(1):98.

Man said to the universe, "I exist."
"However," replied the universe,
"that fact has not created in me a sense of obligation."
Stephen Crane 1889

Chapter 20
Cosmology

THE PAST, PRESENT, AND FUTURE OF THE UNIVERSE

Chapter Web site: http://universe.colorado.edu/ch20

iven the time and opportunity, humans through the ages have devoted themselves to speculation on the grandest scale of all. Poets, philosophers, and theologians have approached the question of the origin and future of the universe in countless ways. Scientists, too, have investigated these questions, using the framework of physical laws. In doing so, they bring together all their accumulated knowledge of the many parts that make up the whole of the universe: the ways in which stars form, evolve, and die; the nature of galaxies and their formation and evolution; the distribution of matter in the universe. Putting these pieces together with the basic physics of atoms and subatomic (elementary) particles, astronomers and physicists have been able to construct a general picture of an evolving universe, in which the overall properties were determined by physical conditions during the very brief moments at the beginning of time, when the universal expansion began. Many questions remain beyond the reach of known physics, but even these issues are gradually submitting to the intense scrutiny of the scientists who are probing the frontiers of cosmology.

Technically, cosmology is the study of the universe as it now appears, and that is really what the preceding 19 chapters have all been about. The study of the origin of the universe is **cosmogony,** a word that applies to the big bang as well as to the earlier theories on the origin of the solar system, once thought to be the entire universe. In practice, the general subject of the nature of the universe and its evolution is lumped under the heading of **cosmology,** and it is on a pursuit of this subject that we now embark.

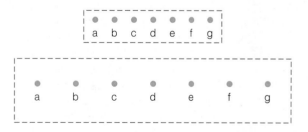

Figure 20.1 Velocity of expansion as a function of distance. At the top is a row of galaxies and at the bottom is the same row 1 billion years later, when the universe (represented by the rectangular box) has expanded so that the distance between adjacent galaxies had doubled from 1 to 2 Mpc. From the viewpoint of an observer in any galaxy in the row, the recession velocity of its nearest neighbor is 1 Mpc per billion years; of its next nearest neighbor, 2 Mpc per billion years; of the next, 3 Mpc per billion years; and so on. For any pair of galaxies, the velocity is proportional to the distance in all cases.

Historically, the theory of cosmology has often gotten ahead of the observations. From the earliest cosmologies of the Greeks to the speculations of Immanuel Kant and the Marquis de Laplace, from Albert Einstein's general relativity to the modern inflationary theory, the models have often been constructed in the absence (or ignorance) of observational constraint. Here we will take things a bit out of chronological order and begin with an overview of the observed properties of the universe at large. We will follow that with a discussion of theoretical models and how the observations support or constrain these models.

Observing the Universe: The Big Picture

A few general properties of the universe must be accounted for in any reasonable cosmological theory. We have already been introduced to some of them, so here we will include only a brief review. Others will be new to us and will be explained in more detail.

The Implications of the Expanding Universe

In Chapter 18 we described the discovery of the universal expansion by Edwin Hubble, and in that chapter as well as the next, we used the redshifts caused by the expansion as tools for finding distances to normal galaxies as well as to quasars and other exotic objects at the fringes of the universe. But we did not discuss the nature of the expansion or its implications for the past and future of the universe.

The first point to consider is the meaning of the phrase "uniform expansion." We may be tempted to think that because all galaxies are seen to recede from us, we must be located at the center of the universe. But if we think about what the universe would look like from another point, we will see that an observer anywhere would see the same thing that we do: all galaxies would appear to recede. In a uniformly expanding environment, all points are spreading apart from each other, so from the reference frame of any one point all others appear to recede. There is no center, or to put it another way, the center is everywhere.

Uniform expansion means that at any given time the entire universe is expanding at a constant rate everywhere. In this situation any object will appear to recede from any observer with a speed that depends on their separation; that is, the farther away a galaxy is, the faster it appears to be moving, no matter where in the universe the observer is. To help see this, consider **Figure 20.1,** where the changes in distance between several galaxies are indicated for different times. We see there that the separation between any pair of galaxies grows in proportion to the initial separation, which means that the speed of recession is proportional to distance. This is what Hubble found from his observations of galaxy redshifts.

Perhaps the most profound implication of the expansion of the universe is that the universe must have had a beginning. Rather than a universe that is infinite in age, that has existed forever and will continue endlessly into the future, we now must contemplate a cosmos that did not always exist and may, by inference, not always exist in the future.

How do we reach this conclusion? Consider this: if all the galaxies in the universe are moving away from each other, they were closer together in the past. If we backtrack far enough, we see that there must have been a time when all the galaxies were together in one place. The view that the universe was once gathered in a single point and has expanded since then and therefore has a finite age is commonly called the **big bang** cosmology. The term was introduced by the British astrophysicist Fred Hoyle during a radio interview in the late 1950s, and it has stuck (ironically, Hoyle himself has resisted accepting the big bang model of the universe, preferring models in which the universe is infinite in age and does not evolve). In a recent contest, astronomers could not find a name more suitable, even though to some the name *big bang* is abhorrent. One reason the term is inappropriate is its suggestion that something explosive happened, and its implication that this explosion occurred at some location within the universe. Difficult though it may be, we have to think instead of the fabric of space itself as expanding; there is no "someplace" that blew up, no vacuum into which matter has expanded since.

By observing the speed of separation (given by the Hubble constant), we can calculate how long ago all the galaxies were gathered together. This is a measure of the age of the universe, because there must have been a **singularity** at the point in time when all the matter in the universe was concentrated in a single point, and ordinary matter in the form of atoms, stars, and galaxies could not have existed before then.

How do we find the age of the universe? To get an approximate answer, we can use the simplest of mathematical relationships: distance equals speed times time, or $d = vt$. This can be rewritten for time as $t = d/v$. From the Hubble law we have the relationship $v = H_0 d$. If we substitute this for v in the preceding relation, we get

$$t = \frac{d}{H_0 d} = \frac{1}{H_0} \ .$$

The age of the universe is simply the inverse of the Hubble constant! Now we see clearly why modern astronomers are making such a huge effort to find the correct value of H_0.

Let us try the calculation for a representative value for H_0, say, 75 km/sec/Mpc (current estimates range from 45 to 90 km/sec/Mpc). To get an answer in normal units of time, we must convert H_0 to standard units. There are 3.09×10^{19} km in a megaparsec, so our value for H_0 of 75 km/sec/Mpc becomes 2.42×10^{-18}/sec (the distance units have canceled out, and we are left with inverse time as the units). Now solving for the time yields $t = 1/2.42 \times 10^{-18}$/sec $= 4.13 \times 10^{17}$ sec, or 13 billion years. Thus this simple consideration of an equally simple observation yields a parameter of enormous consequence: the age of the cosmos itself!

Unfortunately, there are some complications. A correction factor is required, which depends on the density of matter in the universe. For the cosmological model that is the currently most popular, the correction factor is 2/3; that is, the age is $2/3 H_0$, yielding a value of about 9 billion years in the example above. Other cosmological models require different factors; if the density is smaller, as some scientists argue, the factor is closer to 1 and thus yields a greater age.

A second complication arises from our tacit assumption that the rate of expansion has been constant throughout the history of the universe. But this is unlikely: surely, the self-gravitation of the matter in the universe has acted to slow the expansion since it began. Therefore the expansion rate must have been faster in the past, which would mean that the universe has reached its present size in less time than we have calculated by assuming the rate to be constant. Thus our result is an upper limit on the age. The fact that the Hubble constant has probably varied with time is the reason we use the notation H_0; the subscript 0 indicates that this is the value *now*, as opposed to some time in the past.

If we could measure the distances to galaxies far enough away (by an independent method), we might begin to see the effect of a higher value of the Hubble constant long ago. The slope would be steeper for the most distant galaxies **(Fig. 20.2).** Attempts to measure this upturn have so far been very imprecise, but the quest is worthwhile because of what it would tell us about the expansion rate long ago, and what

Satellites & Cosmology
In this web activity we will use the WWW to learn how astronomical satellites have played an important role in our understanding of cosmology. Our starting point is the following URL:
http://universe.colorado.edu/ch20/web.html

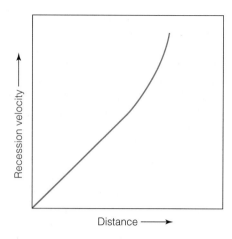

Figure 20.2 **The early expansion rate.** Because the universal expansion was more rapid at early times than it is today, there should be an upturn in the Hubble diagram for the most distant galaxies. This upturn, shown schematically here, is seen for galaxies that are so distant that the observed recession velocities represent early times in the universal expansion.

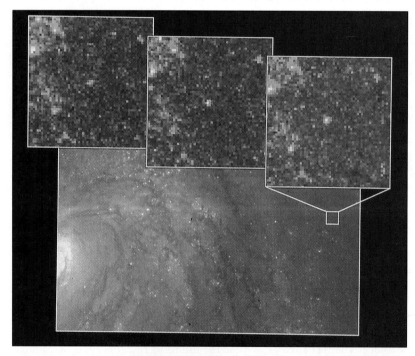

Figure 20.3 **A Cepheid variable in a distant galaxy.** This series of *Hubble Space Telescope* images shows how progress is being made toward determining the value of the Hubble constant H_0. The galaxy M 100 lies too far away for ground-based telescopes to resolve individual stars in it. But the Wide Field/Planetary Camera (WFPC2) aboard the *HST* can do so, and the upper three panels seen here show brightness fluctuations in a Cepheid variable star in the galaxy. Use of the period-luminosity relation for such stars establishes an accurate distance to the galaxy, which is then used to calibrate other distance indicators and refine the distance scale to more faraway galaxies. (NASA/STScI) http://universe.colorado.edu/fig/20-3.html

that in turn would tell us about the rate of slowdown (deceleration) of the expansion. As we will see later in the chapter, the deceleration rate is an important clue to the future of the universe.

A second difficulty with using H_0 to determine the age of the universe is that the value of H_0 is not well established despite decades of intense effort by astronomers. To determine H_0 requires independent measurements of the distances to galaxies that are very far away, so that the relationship between speed and distance can be calibrated for large distances. This is very difficult, as all of our distance-determination methods become more and more uncertain as we observe galaxies more and more remote from us (see the discussion in Chapter 18).

Currently, there is a major controversy among astronomers about this, and the *Hubble Space Telescope* is right in the middle of it. At least two different groups of astronomers are using the *Hubble* to find H_0 by locating Cepheid variables in galaxies as far away as possible **(Fig. 20.3)** and using the period-luminosity relationship to find their distances (as described in Chapter 17). These distances can then be combined with redshift information to determine H_0. But even the *Hubble* cannot detect individual Cepheid variables at truly large cosmological distances, so other methods have to be used to extend the measurements out far enough to get a good value for H_0. One way to do this is to use the Cepheids to calibrate other standard candles such as supernovae, which can then be observed in galaxies much farther away. The two groups doing this have used variations on the same basic method, and so far they do not agree on the answer. One finds a value of H_0 near 57 km/sec/Mpc; the other finds a value more like 73 km/sec/Mpc. The work is continuing, and we can hope that agreement will be reached soon.

In the meantime, the larger value that is being quoted has some problematic implications. If H_0 is more than about 60 km/sec/Mpc and the density of the universe has the higher value preferred by many cosmologists (so that the 2/3 correction factor is needed), then the age of the universe becomes uncomfortably small. Considerations of stellar evolution and the aging of stellar clusters (particularly, the globular clusters; see Chapter 17) lead to age estimates for our galaxy of 12 to 16 billion years, yet a value larger than 60 km/sec/Mpc for H_0 implies an age for the universe that is *younger* than this! Clearly, something is wrong with the measurement of the ages of the globular clusters, the measurement of H_0, or our assumptions about the relationship between H_0 and age. It will be interesting to watch this controversy for the next few years as astronomers try to get it all sorted out.

The Cosmic Background Radiation

The observation that the universe is expanding has another potent implication. At very early times, when the matter of the universe was squeezed together in one place, the physical conditions must have been extreme. It would have been very hot and very dense for a time, until the expansion allowed the universe to cool as the density decreased. The big bang models hold that the universe began in such a state.

In the 1940s a group of physicists led by the Russian-American George Gamow investigated the possibility that conditions during the early moments of the big bang might have been sufficient to allow nuclear fusion reactions to take place. They wanted to know whether the heavy elements we see in today's universe might have been created by reactions at the very beginning of the expansion. They found that the answer was no (largely due to the shortness of the time when it was hot and dense enough for reactions), but in the process Gamow and his colleagues realized that the early universe must have been filled with thermal radiation.

Remember from Chapter 6 that any object glows due to its temperature, and that the wavelength of peak radiation is inversely proportional to the temperature. At very early ages (much less than a second), the temperature of the universe would have been in the billions of kelvins, and the radiation would have been gamma rays. As the expansion proceeded and the universe cooled, the radiation would have shifted to longer wavelengths. Eventually, the radiation would have stopped interacting with the matter, but as the universal expansion continued, the wavelength would have continued to shift to longer and longer values. From our point of view today, this radiation, if it remains, will have been cosmologically redshifted to much longer wavelengths.

Gamow and his co-workers estimated the modern value of the radiation's temperature to be 5 K. They did not think much about detecting the radiation, whose peak wavelength according to Wien's law would have been near 0.06 cm (in the short-wavelength radio spectrum), and their prediction of its existence was largely forgotten.

Then, in the early 1960s, a group of physicists at Princeton unwittingly reproduced Gamow's work, but with a twist: they set out to actually find the remnant radiation that was left over from the early expansion. This group, headed by Robert Dicke and James Peebles, was immediately upstaged by two radio astronomers at the nearby Bell Laboratories. These two, Arno Penzias and Robert Wilson **(Fig. 20.4)**, were testing a new microwave receiver and having difficulty tracking the source of a persistent static that

they were picking up. Word reached Penzias and Wilson that Dicke's group (specifically, Peebles) had predicted the existence and temperature of relic radiation from the early expansion of the universe, and that the Princeton physicists were building an apparatus to detect it. The two groups exchanged information on their discoveries and published the results in a pair of papers in 1965. Penzias and Wilson were later awarded the Nobel Prize for detecting the cosmic radiation, while Dicke and Peebles were left out.

The radiation was found to represent a temperature of about 2.7 K, a bit cooler than predicted by either Gamow or Peebles, but nevertheless essentially consistent with expectations. The peak emission occurs near 0.1 cm (i.e., 1 mm), in the microwave portion of the radio spectrum. The temperature value has since been refined, by an orbiting observatory called the *Cosmic Background Explorer*, or *COBE*, which was able to observe the full spectrum (part of which is blocked by the atmosphere) and derive a value of 2.726 K **(Fig. 20.5)**.

The detection of the **cosmic background radiation**, as it is now commonly called, was a seminal event in astronomy (other terms for the radiation are the **3 degree background** and the **microwave background**). Even more than the expansion of the universe, this discovery provided a direct window into the earliest times of creation. But before it could be accepted as such, certain qualifications had to be met. These were that the spectrum should be perfectly

Figure 20.4 The discoverers of the universal background. Robert Wilson, left, and Arno Penzias, in front of the horn-reflector antenna at Bell Laboratories with which they found a persistent noise that turned out to be the remnant radiation from the big bang. (Courtesy of AT&T)

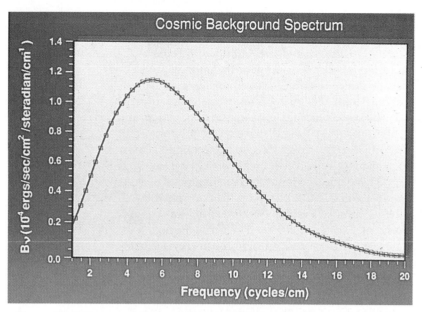

Figure 20.5 **The measured background spectrum.** This shows the very accurate spectrum of the background radiation, as measured by the *Cosmic Background Explorer.* The solid line represents a purely thermal spectrum for an object having a temperature of 2.730 K; the dots represent actual measured intensities. (NASA/COBE Science Working Group) http://universe.colorado.edu/fig/20-5.html

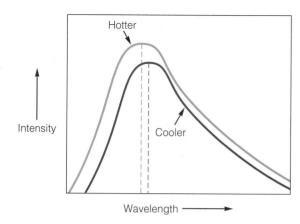

Figure 20.6 **The effect of a Doppler shift on the cosmic background radiation.** If the observer on the Earth has a velocity with respect to the radiation, this velocity shifts the peak of the spectrum a little and affects the temperature that the observer deduces from the measurements. Motion of the Earth results in a daily cycle of tiny fluctuations in the observed temperature, known as the 24-hour anisotropy.

thermal (i.e., that it should follow exactly the shape expected for blackbody radiation; see Chapter 6 and Fig. 20.5) and that it should come from all directions in space uniformly (i.e., that it be **isotropic**). Both tests have been passed with flying colors, but it took some 25 years for astronomers to be able to say so definitively (*COBE* was the ultimate key). Today there are few dissenters from the interpretation of the radiation as the remnant of the early expansion of the universe, and the big bang picture is widely accepted.

The background radiation can be used as a tool for measuring motions relative to the universal expansion. In a very real sense, the radiation provides a fundamental frame of reference for us. If a galaxy has local motion (for example, its orbital motion within its cluster), this will result in a slight Doppler shift of the background radiation as seen from that galaxy **(Fig. 20.6).** The Earth moves locally as a result of several motions: its orbit around the Sun, the Sun's orbit around the galaxy, the galaxy's motion within the Local Group, and any motion the Local Group might have within the local supercluster. The sum of all these components is an overall motion of the Earth at a speed of a few hundred kilometers per second relative to the frame of reference of the background radiation, and it shows up in the form of a contrast between the temperature of the radiation in opposite directions **(Fig. 20.7).** In the direction toward which we are moving, the radiation appears a little

hotter than its actual temperature of 2.726 K, and in the opposite direction it appears a little cooler. This **dipole anisotropy** can be analyzed to reveal our motion relative to the expansion and to tell us something about the large-scale motion of our cluster of galaxies. What it shows is that the Local Group and other clusters nearby are all streaming toward a distant point beyond the great Virgo cluster of galaxies. No obvious concentration of mass exists there that might be pulling us in that direction gravitationally, so the unseen mysterious object has been dubbed the **Great Attractor.**

Thus the cosmic background radiation tells us not only about the structure of the universe in the distant past, but also provides information about the distribution of unseen matter in the present. As we have seen, there is ample additional evidence that the universe contains large quantities of invisible mass.

The Evidence for Dark Matter

Several lines of argument favor the conclusion that as much as 90% of the mass in our galaxy is in some form other than normal stars (see Chapter 17). On scales beyond our galaxy, one of the most direct lines of evidence for large quantities of dark matter comes from clusters of galaxies, where measurements of gravitational effects invariably show the presence of much more mass than can be accounted for by the

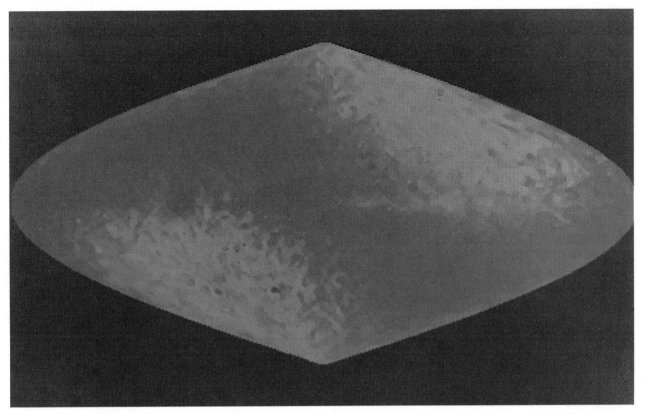

Figure 20.7 The 24-hour anisotropy as measured by *COBE*. This global sky map shows slightly different background temperatures in opposite directions as different colors. (NASA/COBE Science Working Group)

visible galaxies in the cluster (see Chapter 18). On scales beyond clusters of galaxies, the evidence for dark matter becomes more indirect. Other than the local streaming motion of our galaxy and many others toward the Great Attractor, there is little opportunity to analyze motions of clusters of galaxies to look for anomalous effects that might require the presence of large amounts of unseen mass. The composition of the Great Attractor itself is still unknown, but recently some have argued that it may be an ordinary (but very massive) supercluster containing the usual allotment of luminous galaxies. On the other hand, some have suggested that it is indeed a concentration of dark matter.

All the lines of evidence seem to point toward the same conclusion: more than 90% of the mass of the universe is in some form that cannot be seen. Astronomers and physicists today are devoting considerable effort to determining the nature of this "missing mass," which is not missing but merely invisible.

One often-discussed possibility is that neutrinos, the elusive subatomic particles produced in nuclear reactions, have mass. Recall from Chapter 13 that standard theory says that these particles, which permeate space, traveling freely through matter and vacuum alike, are massless. This assertion may not be correct, however, for recent experiments suggest that neutrinos may contain minuscule quantities of mass after all. If so, neutrinos could account for much of the dark matter, because huge quantities of neutrinos were produced during the early big bang expansion.

Other kinds of elementary particles have been postulated to exist on the basis of elementary particle theory. These particles have not been detected experimentally (the energies required far exceed those possible in any existing particle accelerators), but it is possible that they are sufficiently numerous and massive to explain the dark matter. We must await new generations of particle accelerators to find out how significant such particles are.

Large Scale Structure in the Universe

We have already seen (Chapter 18) that galaxies are arranged in sheetlike superclusters on distance scales of tens to hundreds of megaparsecs. This suggests that the universe is not uniform on the largest scales, but questions remain. Are the observed structures the result of random aggregations of galaxies in a universe that is uniform (homogeneous) overall? Or are the structures the result of order that was imposed on the universe from the beginning, from times even earlier than the era of galaxy formation? These are questions of profound importance, because they get right to the heart of the issue of how the universe evolved from the earliest times.

It turns out that both viewpoints may be correct. Numerical simulations starting with a random distribution of galaxies show that gravitational encounters among them will gradually create clusters of galaxies and clusters of clusters of galaxies, at least up to a point (size scales around 10 Mpc). On the other hand, calculations of gravitational collapse of matter in the early universe show that very much larger structures might have been the first to form, and that they could have spontaneously fragmented and formed clusters and superclusters as they broke up. Thus there is evidence for both the "bottom-up" and the "top-down" theories of the evolution of galaxies and structure in the universe.

The theoretical models for the "top-down" formation depend critically on the initial conditions that are assumed, and these depend in turn on the nature of the dark matter, since most of the mass of the universe is in invisible form. The physical state and dynamics of the dark matter control how the universe first became clumpy. The large structures would have formed first if the dark matter is "hot," referring to whether the particles of which it is made are moving at relativistic speeds or not. In most hot dark matter

models, the mass consists of neutrinos, which move at the speed of light. Cold dark matter would be in the form of something more massive and hence slower moving than neutrinos; this could be some form of subatomic particle (as yet undiscovered) or even ordinary matter in the form of dust, rocks, dim compact stars, or subluminous gas clouds. Small structures would have formed first in the cold dark matter scenario. Since we do not yet know what the dark matter is, we do not know which picture is correct.

In the meantime, there is ample evidence (discussed in Chapters 17 and 19) that galaxies formed from the merging of smaller systems. This evidence, along with the statistical analysis of random clustering, shows that the bottom-up picture may well explain structures up to the size scale of clusters of galaxies, if not superclusters. Although bottom-up formation seems to fit the cold dark matter picture, it leaves a nagging problem: the bottom-up model cannot explain very large structure if it exists, and the enormous walls and voids of galaxies (see Chapter 18) argue that it does.

Confirmation of very large structure in the universe came in spectacular fashion, provided by the *COBE* observations **(Fig. 20.8).** Subtle variations in the intensity of the cosmic microwave emission (measured in the form of very slight temperature variations) revealed patchiness in the radiation background. This patchiness could only have formed at very early times, *before* the radiation had separated itself from matter during the expansion of the young

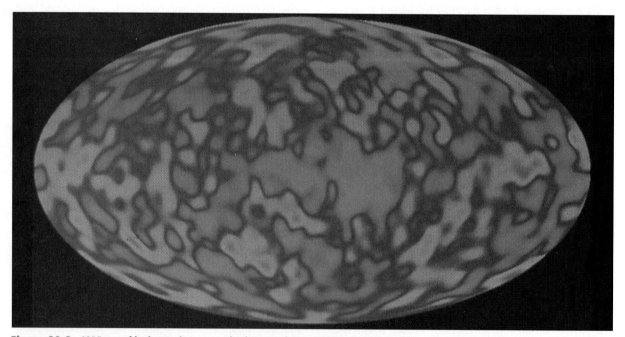

Figure 20.8 *COBE* **map of background structure.** This shows, in galactic coordinates, the subtle variations in background radiation temperature that are interpreted as evidence for density variations (i.e. lumpiness) in the early universe. (NASA) http://universe.colorado.edu/fig/20-8.html

universe. Here lay virtual proof that the universe became lumpy very early on, before galaxies had begun to form. Thus the observations tell us that both the top-down and bottom-up pictures are valid, that some elements of the hot dark matter models must merge with the cold dark matter picture that explains the smaller structures so well. Recent computer simulations of the formation of structure in the universe have confirmed that a combination of preexisting structure (top-down model) and gravitational grouping of smaller systems into larger ones can very accurately reproduce the observed distribution of galaxies. This combination of seemingly opposed mechanisms has been likened to the synthesis that took place when the wave and particle theories of light were merged into the photon model (see Chapter 6).

Keeping in mind the observational constraints provided by the universal expansion, the cosmic background radiation, and the large-scale structure of the universe, we are ready to consider the theories of cosmology that have been advanced to explain these phenomena.

Modeling the Cosmos

To even begin to study the universe as a whole, we must make certain assumptions. These can, in principle, be tested, although it is not clear whether this can be done practically. One of the most important assumptions we make is that the known laws of physics can be applied to the universe as a whole. This is a well-justified assumption, which has been verified in many ways (for example, observations of spectral lines in very distant objects such as quasars show that atomic properties and constants had the same values at early times and different locations in the universe as they do here and now).

A central rule traditionally set forth by cosmologists is that the universe must look the same at all points within it. This does not mean that the appearance of the heavens should be literally identical everywhere; it means that the general structure, the density and distribution of galaxies and clusters of galaxies **(Fig. 20.9)** should be constant. This assumption states that the universe is **homogeneous.**

A related, but slightly different assumption has to do with the appearance of the universe when viewed in different directions. The assumed property in this case is **isotropy;** that is, the universe must look the same to an observer no matter in which direction the telescope points (isotropy was already mentioned in connection with the background radiation). The assumption of isotropy does not imply

identical constellations or patterns of galaxies and stars in all directions, just comparable ones.

These two assumptions have their philosophical roots in the hard lesson that astronomers have learned over the centuries: the Earth does not occupy a privileged place in the cosmos. In more general terms, we assume that no other location is privileged either. All of our observational experience, as well as the success of the theory of relativity (which rests entirely on this foundation), tells us that there are no preferred reference frames or locations in the universe. Thus cosmologists feel justified in adopting the principles of homogeneity and isotropy as basic conditions of the universe. We find that we can make considerable progress in modeling the universe by adopting these postulates.

Einstein's Relativity: Mathematical Description of the Universe

The most powerful mathematical tool for describing the universe was developed by Albert Einstein **(Fig.**

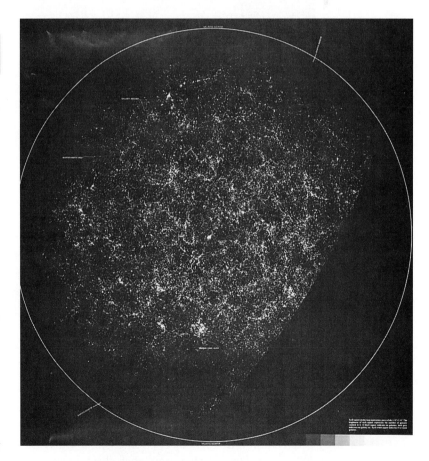

Figure 20.9 The cosmological principle. The universe is assumed to be homogeneous and isotropic, meaning that it looks the same to all observers in all directions. Here is a segment of the universe, filled with galaxies. Their distribution, though not identical from place to place, is similar throughout. (© 1978, 1979 by The Co-Evolution Quarterly, Box 428, Sausalito, California)

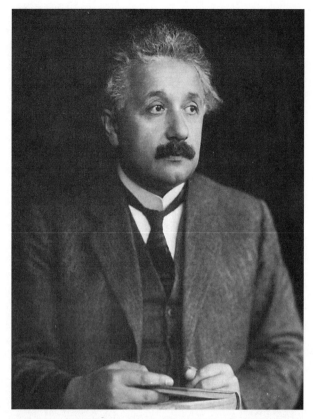

Figure 20.10 Albert Einstein. Among Einstein's great contributions was the development of a mathematical formalism to describe the interaction of gravity, matter, and energy in the universe with space-time. This framework, general relativity, has withstood all observational and experimental tests applied to it so far. (The Granger Collection) http://universe.colorado.edu/fig/20-10.html

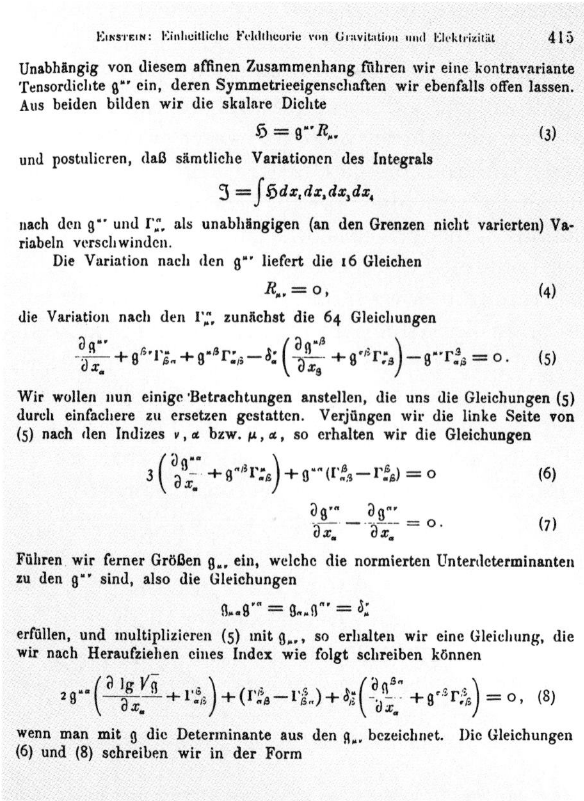

EINSTEIN: Einheitliche Feldtheorie von Gravitation und Elektrizität 415

Unabhängig von diesem affinen Zusammenhang führen wir eine kontravariante Tensordichte $\mathfrak{g}^{\mu\nu}$ ein, deren Symmetrieeigenschaften wir ebenfalls offen lassen. Aus beiden bilden wir die skalare Dichte

$$\mathfrak{H} = \mathfrak{g}^{\mu\nu} R_{\mu\nu} \qquad (3)$$

und postulieren, daß sämtliche Variationen des Integrals

$$\mathfrak{J} = \int \mathfrak{H}\, dx_1 dx_2 dx_3 dx_4$$

nach den $\mathfrak{g}^{\mu\nu}$ und $\Gamma^{\alpha}_{\mu\nu}$ als unabhängigen (an den Grenzen nicht variierten) Variabeln verschwinden.

Die Variation nach den $\mathfrak{g}^{\mu\nu}$ liefert die 16 Gleichen

$$R_{\mu\nu} = 0, \qquad (4)$$

die Variation nach den $\Gamma^{\alpha}_{\mu\nu}$ zunächst die 64 Gleichungen

$$\frac{\partial \mathfrak{g}^{\mu\nu}}{\partial x_\alpha} + \mathfrak{g}^{\beta\nu}\Gamma^{\mu}_{\beta\alpha} + \mathfrak{g}^{\mu\beta}\Gamma^{\nu}_{\alpha\beta} - \delta^{\nu}_{\alpha}\left(\frac{\partial \mathfrak{g}^{\mu\beta}}{\partial x_\beta} + \mathfrak{g}^{\tau\beta}\Gamma^{\mu}_{\tau\beta}\right) - \mathfrak{g}^{\mu\nu}\Gamma^{\beta}_{\alpha\beta} = 0. \qquad (5)$$

Wir wollen nun einige Betrachtungen anstellen, die uns die Gleichungen (5) durch einfachere zu ersetzen gestatten. Verjüngen wir die linke Seite von (5) nach den Indizes ν, α bzw. μ, α, so erhalten wir die Gleichungen

$$3\left(\frac{\partial \mathfrak{g}^{\mu\alpha}}{\partial x_\alpha} + \mathfrak{g}^{\alpha\beta}\Gamma^{\mu}_{\alpha\beta}\right) + \mathfrak{g}^{\mu\alpha}(\Gamma^{\beta}_{\alpha\beta} - \Gamma^{\beta}_{\alpha\beta}) = 0 \qquad (6)$$

$$\frac{\partial \mathfrak{g}^{\nu\alpha}}{\partial x_\alpha} - \frac{\partial \mathfrak{g}^{\alpha\nu}}{\partial x_\alpha} = 0. \qquad (7)$$

Führen wir ferner Größen $\mathfrak{g}_{\mu\nu}$ ein, welche die normierten Unterdeterminanten zu den $\mathfrak{g}^{\mu\nu}$ sind, also die Gleichungen

$$\mathfrak{g}_{\mu\alpha}\mathfrak{g}^{\nu\alpha} = \mathfrak{g}_{\alpha\mu}\mathfrak{g}^{\alpha\nu} = \delta^{\nu}_{\mu}$$

erfüllen, und multiplizieren (5) mit $\mathfrak{g}_{\mu\nu}$, so erhalten wir eine Gleichung, die wir nach Heraufziehen eines Index wie folgt schreiben können

$$2\mathfrak{g}^{\mu\alpha}\left(\frac{\partial \lg \sqrt{\mathfrak{g}}}{\partial x_\alpha} + \Gamma^{\beta}_{\alpha\beta}\right) + (\Gamma^{\beta}_{\alpha\beta} - \Gamma^{\beta}_{\beta\alpha}) + \delta^{\mu}_{\beta}\left(\frac{\partial \mathfrak{g}^{\beta\alpha}}{\partial x_\alpha} + \mathfrak{g}^{\tau\beta}\Gamma^{\beta}_{\tau\beta}\right) = 0, \qquad (8)$$

wenn man mit $\mathfrak{g}$ die Determinante aus den $\mathfrak{g}_{\mu\nu}$ bezeichnet. Die Gleichungen (6) und (8) schreiben wir in der Form

Figure 20.11 The field equations. Here are portions of the equations that describe the physical state of the universe. To a large extent, the science of cosmology involves finding solutions to these equations. (The Granger Collection)

20.10). His theory of general relativity represents the properties of matter and its relationship to gravitational fields. Within the context of his theory, Einstein developed a set of relations, called the **field equations (Fig. 20.11),** that express in mathematical terms the interaction of matter, radiation, and gravitational forces in the universe. Although alternatives to general relativity have been developed and their consequences explored, most research in cosmology today involves finding solutions to Einstein's field equations and testing these solutions with observational data.

As we learned in Chapter 16, the basic premise of general relativity is that acceleration created by a gravitational field is indistinguishable from acceleration due to a changing rate or direction of motion. One consequence of the equivalence of gravity and acceleration is that an object passing near a source of gravitational field (that is, any other object having mass) undergoes acceleration and therefore follows a curved path. In a universe containing matter, this means that all trajectories of moving objects are curved, and it is often said that space itself is curved. The degree of curvature is especially high close to massive objects **(Fig. 20.12),** but the universe also has an overall curvature as a result of its total mass content. The solutions to the field equations specify, among other things, the degree and type of curvature. We will return to this point shortly.

Einstein's solution to the equations, developed in 1917, had a serious flaw (in his view): it did not allow for a static, nonexpanding universe. In what he later admitted was the biggest mistake he ever made, Einstein added an arbitrary term called the **cosmological constant** to the field equations, primarily for the purpose of allowing the universe to be stationary, neither expanding nor contracting.

Others developed different solutions, always by making certain assumptions about the universe. Some of these assumptions were necessary to simplify the field equations so that they could be solved. In 1917, for example, Willem de Sitter developed a solution that corresponded to an empty universe, one with no matter in it.

By the 1920s, solutions for an expanding universe were found, primarily by the Soviet physicist Aleksander Friedman and later by the Belgian Georges

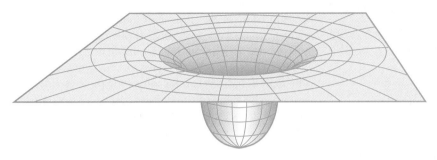

Figure 20.12 Curvature of space near a massive star. This curved surface represents the shape of space very close to a massive star. The best way to envision what this curvature of space actually means is to imagine photons of light as marbles rolling on a surface of this shape.

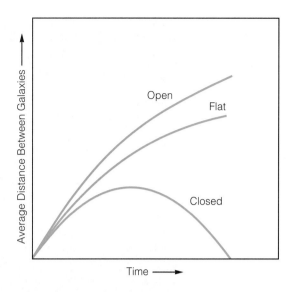

Figure 20.13 The three possible fates of the universe. This diagram shows the manner in which the average distance between galaxies will change with time for the open, flat, and closed universe. In the first case, galaxies will continue to separate forever, although the rate of separation will slow. In the second case, the rate of separation will, in an infinite time, slow to a halt but will not reverse. In the third case, the galaxies eventually begin to approach each other, and the universe returns to a single point.

LeMaître, who went so far as to propose an origin for the universe in a hot, dense state from which it has been expanding ever since. This was the true beginning of the hot big bang idea, and it was developed some three years before Hubble's observational discovery of the expansion. Of course, once scientists found that the universe is not static but is actually in a dynamic state, the original need for Einstein's cosmological constant disappeared. Nevertheless, modern cosmologists usually include it in the field equations, but its value is normally assumed to be zero.

The general relativistic field equations allow for three possibilities regarding the curvature of the universe and three possible futures **(Fig. 20.13).** A central question of modern studies of cosmology involves deciding which possibility is correct. One is referred to as "negative curvature." An analogy to this is a saddle-shaped surface **(Fig. 20.14),** which is curved everywhere, has no boundaries, and is infinite in extent. This type of curvature corresponds to what is called an **open universe,** a solution to the field equations in which the expansion continues forever, never stopping. (It does slow down, however, because the gravitational pull of all the matter in it tends to hold back the expansion.)

A second possibility is that the curvature is "positive," corresponding to the surface of a sphere **(Fig. 20.15),** which is curved everywhere, has no boundaries, but is finite in extent. This possible solution to the field equations is called the **closed universe,** and it implies that the expansion will eventually be halted by gravitational forces and will reverse itself, leading to a contraction back to a single point.

The third and last possibility is that the universe is a **flat universe,** with no curvature. In this possibility, the momentum of the outward expansion is precisely balanced by the inward gravitational pull of the matter in the universe, so that the expansion will eventually come to a stop but will not reverse itself.

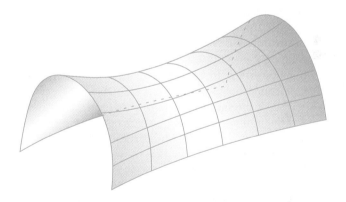

Figure 20.14 A saddle surface. This is a representation of the geometry of an open universe, which has negative curvature, is infinite in extent, has no center in space, and has no boundaries.

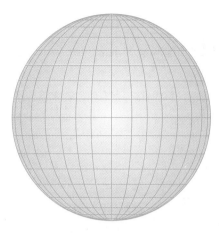

Figure 20.15 **The surface of a sphere.** This surface represents a closed universe, which has positive curvature and a finite extent but no boundaries and no center.

This balanced state requires a perfect coincidence between the energy of expansion and the inward gravitational pull of the combined total mass of the universe and may therefore seem very unlikely. Recently, however, models have been developed in which such a balance is expected.

Open, Closed, or Flat: The Observational Evidence

Today substantial intellectual and technological resources are being devoted to the question of the fate of the universe. Theorists are at work developing and refining the solutions to the field equations or seeking alternatives to general relativity that might provide equally or more valid representations. Observers are busily attempting to test the theoretical possibilities by finding situations in which competing theories should lead to different observational consequences. This is a difficult job, because most of the differences show up only on the largest scales, and therefore detecting them requires observations of the farthest reaches of the universe.

Within the context of general relativity, the premier question traditionally has been whether the universe is open or closed. Historically, two general observational approaches have been used in answering this: determining whether there is enough matter in the universe to produce sufficient gravitational attraction to close it, or measuring the rate of deceleration of the expansion to see whether it is slowing rapidly enough to eventually stop and reverse itself.

The total mass in the universe is the quantity that determines whether or not gravity will halt the expansion. The field equations express the mass content of the universe in terms of the density, the amount of mass per cubic centimeter. This is convenient for observers, because it is obviously simpler to measure the density in our vicinity than to try to observe the total mass everywhere in the universe. The field equations can be solved for the **critical density,** the value of the density that would produce an exact balance between expansion and gravitational attraction (the density corresponding to a flat universe). If the actual density is greater than the critical density, then there is sufficient mass to close the universe, and the expansion will stop. The critical density depends on the value of Hubble's constant H_0; for a value near 75 km/sec/Mpc, it is calculated to be roughly 10^{-29} g/cm^3, or about 3 protons per cubic meter, a very low value by any earthly standards.

The parameter that astronomers seek is usually expressed in terms of the ratio of the actual density to the critical density. The symbol Ω_0 ("omega nought") usually represents this ratio. A value of $\Omega_0 = 1$ represents a flat universe, with smaller values corresponding to an open one and a value larger than 1 implying a closed universe. (A value of $\Omega_0 = 1$, now preferred in some models of the universe, yields the correction factor of $2/3$ in calculating the age of the universe from $1/H_0$, as discussed earlier in this chapter.) Along with H_0, Ω_0 is one of the key numbers that determine the fate of the universe in standard big bang cosmology.

The most straightforward way to measure the density of the universe is to simply count galaxies in some randomly selected volume of space, add up their masses, and divide the total by the volume. Care must be taken to choose a very large sample volume, so that clumpiness from clusters of galaxies is not important. This technique yields values for the density between 10^{-31} and 10^{-30} g/cm^3, only a small percentage of the critical density, implying that Ω_0 is less than 1.

It may seem that we have already answered the question, and that the universe is open, but this method may overlook substantial quantities of mass, as indicated by the ample evidence for dark matter. It is noteworthy and perhaps very significant that the quantity of dark matter suspected to be present in the universe is very nearly what is needed to meet the critical density. The evidence indicates that the unseen matter constitutes about 10 times what is visible, and this would mean that, to a fair approximation at least, $\Omega_0 = 1$. There is no reason in the standard big bang cosmology for this to be so. The universe could have been strongly unbalanced in favor of continued expansion or on the side of a reversal of the expansion, yet it is balanced or nearly so. Many cosmologists regard this balance as very significant, and as we will see, it has stimulated the development of models in which a flat universe is expected.

Another approach to answering the question of whether the universe is open or closed is also being pursued vigorously. In this case the objective is to compare the present expansion rate with what it was early in the history of the big bang to see how much slowing, or deceleration, has occurred **(Fig. 20.16).** The expansion has certainly slowed; the question is how much. If it has only decelerated a little, then we infer that is not going to slow down enough to stop and reverse itself; but if it has decelerated quite a bit, then we conclude that the expansion is coming to a halt, and that the universe is closed.

The present evidence on deceleration is considered ambiguous by most astronomers. In effect, to measure the deceleration requires determining the curvature at the extreme upper right-hand end of the Hubble diagram (see Fig. 20.16), which is very difficult in view of the many uncertainties in determining the distances to the most faraway galaxies.

An indirect technique that has recently been employed to measure the deceleration avoids many of the difficulties of observing extremely distant galaxies. Some light elements were created in the early stages of the expansion, and the amounts that were produced depend on the rate of expansion during the first moments of the big bang. The strategy, therefore, is to measure the abundance of some element that was produced in the big bang and then to derive the deceleration from that measurement.

The most abundant element (other than hydrogen) produced in the big bang is helium, so if we could measure how much helium was created, this might be a good indicator of the deceleration. One problem with helium, however, is that it is also produced in stellar interiors, so it is difficult to determine how much of what we see in the universe today is really left over from the big bang. Another problem with using helium as a probe of the early universal expansion rate is that the quantity produced is not expected to vary much with different expansion rates; that is, the primordial helium abundance, even if we knew it well, is not a very sensitive indicator of the early expansion rate.

Better candidates are ^{7}Li, an isotope of the element lithium, and deuterium, the form of hydrogen that has one proton and one neutron in the nucleus. These species were also produced in the big bang and, as far as is currently known, are not made in any other way. To date, most efforts have been concentrated on measuring the present-day abundance of deuterium, although in principle ^{7}Li can be mea-

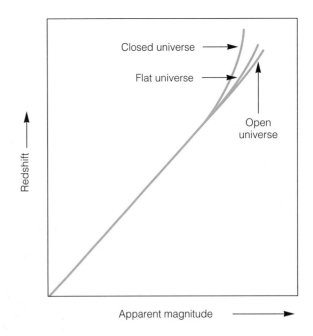

Figure 20.16 **The effect of deceleration in the Hubble diagram.** This figure shows the relationship between velocity of recession and apparent magnitude (an indicator of distance) for the three possible cases: closed (upper curve), flat (middle), or open (lower). Present data seem to favor the open universe, although there is substantial uncertainty, largely because of problems in applying standard candle techniques to galaxies so far away that they are being seen as they were at a young age. The distinction is also made difficult by the subtlety of the difference between the shapes of the curves.

sured as well. While deuterium can be destroyed by nuclear processing in stars, it is not produced in that way. (Even if it were, it would not survive the high temperatures of stellar interiors without undergoing further reactions.) The present deuterium abundance in the universe should therefore represent an upper limit on the quantity created in the big bang. Direct measurements of the amount of deuterium in space are possible through observations of ultraviolet absorption lines of interstellar deuterium atoms **(Fig. 20.17).** The abundance of deuterium found is sufficiently high that it implies a low density, in turn pointing to a small amount of deceleration.

This result appears to have received support from recent ground-based observations. Even though the principal spectral lines of deuterium are in the ultraviolet portion of the spectrum, which can only be observed from space, in very distant galaxies and quasars the redshifts are high enough to shift these lines into the visible portion of the spectrum. Recent observations obtained at the *Keck Telescope* appear to show strong deuterium lines in the absorption-line spectrum of a quasar, suggesting that the deuterium abundance in intervening gas clouds is high, but the result only places a lower limit on the deceleration

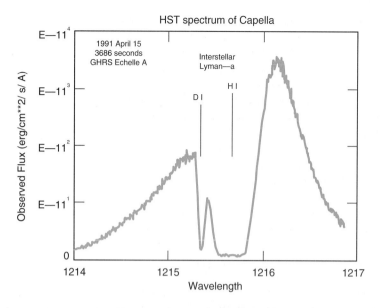

HST spectrum of Capella

1991 April 15
3686 seconds
GHRS Echelle A

Interstellar
Lyman—a

D I   H I

Figure 20.17 *An ultraviolet absorption line of interstellar deuterium.* The abundance of deuterium in space is determined from the analysis of absorption lines it forms in the spectra of background stars. This is a portion of a spectrum obtained by the *Hubble Space Telescope*, showing a broad ultraviolet emission line of hydrogen, formed in the chromosphere of the red giant star Capella. There are two interstellar absorption lines superimposed on the broad emission; the stronger one, at the right, is due to hydrogen atoms along the line of sight to Capella; the weaker one, at the left, is due to deuterium atoms. Analysis of the strengths and shapes of these absorption lines yields the relative abundance of deuterium and hydrogen in space. (NASA/STScI; data provided by J. L. Linsky)

rate because of problems with line blending in the observed spectrum. The uncertainties in both the ultraviolet and the visible-wavelength results render the deceleration studies inconclusive for now. For example, the ultraviolet measurements of deuterium absorption lines refer only to rather nearby interstellar material, so the "local" region is used as a probe of a property of the universe as a whole. Because these observations seem to indicate an uneven distribution of interstellar deuterium throughout our part of our own galaxy, it seems possible that there are local sources of deuterium, and that these measurements do not necessarily indicate the universal abundance. A new space observatory, the *Far Ultraviolet Spectroscopic Explorer* (*FUSE*), will measure the deuterium abundance in many more regions, including intergalactic gas clouds. *FUSE* will be launched in late 1998.

A Model of a Balanced Universe

It is very significant that all the observational techniques used thus far to discover whether the universe is open or closed have pointed toward a fairly close

balance; that is, the data do not point to a universe that is closed or open by a wide margin. Whatever the correct answer is, we can say that the universe is close to being flat. This is very significant, for such a close balance is not necessarily expected in the big bang theory and must be regarded in this theory as a coincidence. There are modified forms of the big bang model, however, in which a flat or nearly flat universe is expected.

The Inflationary Universe

The big bang cosmology is widely accepted, but some nagging problems remain. One is the improbability that a universe that started from a singularity should turn out to be as symmetric as the observed universe. It is unlikely that the universe would appear as homogeneous and isotropic as observations tell us it is. A great deal of effort has been put into testing this property of the universe, because any departures from homogeneity would provide information on imbalances or asymmetries in the early epochs of the expansion. It is also surprising that the universe is so closely balanced between open and closed.

A new kind of big bang model has been developed in which the close balance and the homogeneity of the universe are easily understood. Calculations have shown that, at a certain point very early in the history of the universe, when the temperature was around 10^{27} K, conditions were right for small regions to separate themselves from the rest of the universe and then to expand very rapidly to much larger sizes. A reasonable analogy would be the creation of bubbles in a liquid, with the bubbles then growing much larger almost instantaneously. In this cosmological model, each "bubble" becomes a universe in its own right, with no possibility of communication with other bubbles.

The major advantage of this so-called **inflationary universe** model is that the universe born of a tiny cell in the early expansion would be expected to remain symmetric and homogeneous as it grew. The reasons for this are somewhat abstract but have to do with the fact that the various portions of the tiny region that was to expand into the present universe were so close together originally that they were in a form of equilibrium, with little or no variation in physical conditions being possible. In a larger region, such as in the standard big bang cosmology, no such equilibrium would have existed, and there would be no reason to expect such uniformity throughout the resulting universe.

Another advantage of the inflationary model is that in such a universe, the expansion would be expected to go on forever but would continue to slow, approaching a stationary state; that is, a flat universe is the natural result of rapid expansion from a tiny bubble as envisioned in this model. As we have seen, observations tell us that our universe is nearly flat and may be precisely flat.

The inflationary universe theory has reached a high degree of acceptance among cosmologists. This does not mean that the big bang model is wrong, but rather that it is a subordinate part of a much bigger picture, much as Newton's concept of gravity is now seen as a subset of Einstein's more complete theory, general relativity.

How It All Started

One of the most pleasing aspects of the inflationary model is also one that is very difficult to understand intuitively. In the standard big bang, we are forced to accept that the state of the universe as it began to expand was a given, something that just was. In other words, the big bang theory can only explain what happened after the beginning. In the inflationary model, the entire universe (and the infinite number of others that are possible) was literally created out of a vacuum. This means that no untidy, inexplicable initial conditions have to be invoked.

In a vacuum, energy may be present even though mass is not, and quantum mechanics tells us that particles can appear out of nothing. These occurrences may be viewed as random events, reflecting the fact that on the smallest of scales the fabric of the universe undergoes fluctuations. The spontaneous appearance of particles out of a vacuum has been observed in laboratory experiments. Inflation theory postulates that if the energy of the vacuum is high enough, a strange state called **false vacuum** is created, in which a peculiar form of "negative" pressure acts as a repulsive force, causing the local region of space-time to expand if matter is present. Another strange property of false vacuum is that during expansion the density remains constant, which means that additional matter is created out of the energy of the vacuum. In a very brief moment (about 10^{-30} seconds), the early universe would have expanded in size by a factor of 10^{25} or more and would have grown enormously in mass. This was the inflation phase already mentioned, which forms the basis of the inflationary model of the universe. The cooling that occurred during the expansion eventually (that is, after 10^{-30} seconds) would have broken the false vacuum, and the universe would then have been transformed into the more normal form of space we are familiar with, containing matter and having enough residual energy to power the expansion that has been proceeding ever since.

In the inflationary theory, then, the origin of our universe was a random quantum mechanical event, something now believed to occur naturally and inevitably, requiring nothing more than a vacuum with sufficiently high energy. It is now being speculated that perhaps new universes can begin within existing ones, that generation after generation of new bubbles of false vacuum occur randomly and expand. In this view it seems likely that our universe was born of another earlier one; it would be pretentious to think that, of all the possible universes, ours was the first.

If a new universe formed from a random fluctuation in our own, there might be no way for us to know it, because the expansion could take place in different dimensions of space-time from those that we occupy and can sense. In effect, a tiny bit of space-time, if given enough energy, could pinch off and disappear from our universe, spawning a whole new universe that we cannot perceive. An enormous concentration of energy in a tiny space would be required for this to happen, so it is a very low-probability event. But in a sufficiently large universe, even a low-probability event might occur somewhere at some time.

As already noted, once the false vacuum was broken and the universe reverted to normal vacuum filled with radiation and matter, the rest of the expansion is described very well by the standard big bang theory. Hence, the long legacy of big bang cosmologies still provides a useful description of the evolution of our universe. This description is summarized in the next section.

The History of Everything

It is impossible to give a physical description of the universe at the precise moment that the expansion (or the inflation of our "bubble") began; it is physical nonsense to deal with infinitely high temperature and density. We can, however, calculate the conditions immediately (10^{-43} seconds, in the standard models) after the expansion started, and at any later time. Many of the most interesting events in the early history of the universe, and the ones currently under the most active investigation, occurred very early, before even a ten-thousandth of a second had passed. Under the conditions of density and temperature that existed then, matter and the forces that act on it were quite different from anything we can experience, even in the most advanced laboratory experiments. Even the familiar subatomic particles such as protons and neutrons could not exist, but were preceded in existence by their constituent particles.

SCIENCE AND SOCIETY

What Is a Theory, Really?

The notion that some scientific finding or explanation is a "theory" has been consistently misunderstood by the general public. The news and print media sometimes go so far as to imply that if something is "only" a theory, it is merely one of a number of ideas, all of which may have equal validity. Therefore some people think that to say that a concept or explanation is a theory is a criticism, carrying with it a negative implication. This idea seems to come up especially frequently in discussions of the evolution of life and of cosmology.

It is important to understand that a theory is often nearly as firm and incontrovertible as a "fact." It is not correct to assume that a theory is merely a suggestion, an idea that someone came up with in the absence of supporting evidence. It is true that many theories start out in that fashion, initially without evidence, but normally we would refer to such a suggestion as a "hypothesis." In many instances when a scientist refers to the theory of something, it is a well-established construction that has a long history of meeting the challenges to it. This is true of evolution theory, and it is true of big bang cosmology.

A theory is the best explanation that science can offer for a particular phenomenon, and if it has withstood the tests of observation or experiment and of making successful predictions, it becomes widely accepted. The big bang theory falls into this category, as does evolution theory. In this chapter we have seen much of the evidence favoring the big bang theory, including its simple explanation of observations, such as the expansion of the universe, and its predictive power, as demonstrated by the discovery of the microwave background radiation. In the next chapter, the theory of evolution will be discussed, without as much detail, but in that case as well, the supporting evidence is quite overwhelming. No competing theory for the big bang or for evolution has been as capable of explaining phenomena or making predictions.

Of course, both the big bang and evolution are subject to change if the evidence demands it. This has already happened to both in many instances. Currently, for example, the inflationary form of the big bang theory is widely favored because it explains some details of the expanding universe that earlier forms of big bang theory could not. The modification of the original theory to accommodate inflation did not invalidate the big bang; rather, it lent the big bang more weight precisely because it showed that the theory was general and broad enough to accommodate new evidence. Similarly, when the *COBE* satellite detected ripples in the background radiation, some in the media described this discovery as possibly invalidating the big bang theory because the theory did not explain these ripples; in fact, quite the opposite was true. Now cosmologists are working with modifications of the big bang, invoking asymmetries or structure early in the expansion, which again shows the

---

In a sense it is utterly amazing that we presume we can know anything about the early universe. But what we know about the behavior of matter under conditions of immense temperature and pressure, much of it learned from particle accelerators here on Earth, shows that matter becomes surprisingly simple in its properties and its interactions under those conditions. In many ways it is easier to model the early universe, and to verify the models with experiments and observations, than it is to model and observe the interior of a planet or the structure of a galaxy. This and the ability to compare the predictions of our models with the observed universe of today and of long ago (through the microwave background; the existence of helium, deuterium, and ^{7}Li formed in the early big bang; and the look-back time when viewing distant objects) give us confidence that we can learn something valid and useful about the early universe.

The Radiation Era

Current particle physics theory holds that the most fundamental particles are **leptons,** which include electrons and positrons, and **quarks.** Modern theory also provides a basis for believing that all four of the fundamental forces in nature may really be manifestations of the same phenomenon. So far it has been possible to show that three of the fundamental forces (the electromagnetic force and the weak and strong nuclear forces) are manifestations of the same basic interaction (the electromagnetic and weak nuclear forces have been observed in laboratory experiments to be combined under the appropriate conditions, though the addition of the strong nuclear force under more extreme conditions has been demonstrated only theoretically so far). Under physical conditions that we are used to, these forces behave very differently, but when the density and temperature are

power of the basic model.

Even in the ongoing debate over the age of the universe and the apparent discrepancy between the expansion time and the ages attributed to the oldest star clusters, few scientists are arguing that big bang theory should be thrown out. Instead people are discussing what modifications might be needed (is the density of the universe actually lower than critical, so that the age is greater than in the case where Ω is less than 1? Or is the value of the cosmological constant, first introduced by Einstein, not zero after all?). These are minor modifications, which, if required, will not alter the basic structure of the big bang theory.

Why do we cling to a theory in the face of new evidence that requires modifications? At what point are we merely propping up a failed paradigm? When do we abandon ship altogether and seek a wholly new model? The answer is this: only when a better theory is found. Some cosmologists favor non–big bang theories, but they cannot explain or satisfy, in as simple and elegant a fashion, all of the observational and predictive tests that the big bang cosmology has met. Similarly, opponents of evolution theory find many points to criticize, but

they offer nothing to replace evolution except far more complicated scenarios (along with a liberal dose of pure faith).

The principle of Occam's razor enters into the development and acceptance of a theory. This is the idea, mentioned several times in the text, that nature usually chooses the explanation that requires the fewest complications or the smallest number of unsubstantiated assumptions. One of the theories offered as an alternative to the big bang in recent years was the *steady state theory*, the idea that the universe has no beginning and no end, but has always existed exactly as it is now. To sustain this idea in the face of the observed universal expansion, the proponents of the steady state theory had to postulate that new matter is continually being formed in intergalactic space to offset the density decrease that would otherwise take place as the galaxies moved apart. When confronted with the discovery of the cosmic background radiation, these same scientists invented an otherwise unsuspected population of distant point sources that would be so numerous that their summed radiation would emulate the observed background. (Observations of the homogeneity of the radiation and

its perfect agreement with the spectrum of a black body, as shown especially by *COBE*, have now made this position totally untenable. Steady state theorists have been in retreat, although some are still at work devising new ways of accommodating the observations under the steady state paradigm.) These attempts to save the steady state model have violated Occam's razor because they require the adoption of many unsubstantiated assumptions. For this reason, most astronomers prefer the big bang theory.

From your studies of astronomy and of the nature of science, you should be in a better position to analyze what it means when someone says that something is "only" a theory. You should think about whether this is really a criticism, and whether the critic is offering anything that is better. And, of course, you should always be prepared, as every scientist should be, to drop a weak explanation or theory in favor of a better one.

very high, as they were early in the expansion of the universe, the forces are indistinguishable. Some aspects of the theory that show this have been confirmed by laboratory experiments. The theoretical framework connecting the three forces is called **Grand Unified Theory (GUT** for short). This name may be somewhat exaggerated, because the fourth force, gravity, has not yet been shown to be unified with the other three, and some theories of gravity indicate that it may not be.

The unification of the other three forces implies that, in the very early moments following the beginning of the expansion, until 10^{-35} seconds had passed, the universe contained only leptons and quarks and related particles and radiation, and the only forces operating in it were gravity and the unified force that was later to become recognizable as the electromagnetic, strong, and weak forces.

As the universe expanded and cooled, the three forces sequentially separated and became distinguish-

able from each other. Meanwhile a rich stew of particles was brewing, because particles can appear spontaneously from photons of radiation, and in its earliest times the universe was filled with very high energy photons. Recall that mass and energy are interchangeable according to the special relativity relation $E = mc^2$; particles (mass) can appear from photons (energy) if sufficient energy is present in the photons. This was the case for a brief time during the early expansion of the universe. As long as the energy contained in radiation exceeded that embodied by the particles that had formed (as calculated from $E = mc^2$), the radiation field dominated the structure of the universe. We refer to this early time as the **radiation era.**

When particles appear spontaneously from photons, they do so in pairs with opposing properties. Thus, instead of a single electron, an electron and its oppositely charged counterpart, the **positron,** would appear. Protons appeared paired with **antiprotons,** particles identical in mass but having opposite charges. For each type of particle, there is an antiparticle, its opposite in key properties. (In some cases of neutral particles, the antiparticle and the normal particle are the same.) Pair production of particles from energetic photons (or from high-energy particle collisions) has been observed in particle accelerators.

Once a particle-antiparticle pair formed, each of the two particles moved freely through the young universe, until by random chance it encountered its antiparticle. When this happened, the two particles annihilated each other, converting their combined mass into energy and creating a new photon. Thus an equilibrium existed in which energy and matter were continually being converted back and forth, as photons created particle pairs and particle pairs created photons.

This process could have continued forever if not for the expansion and cooling of the universe. The cooling meant that the energy of the photons decreased steadily, as the wavelength of maximum emission shifted according to Wien's law (Chapter 6). Soon the photon energies were insufficient to produce particle pairs, and pair production ceased. This occurred at different times for particles of different masses; thus as the universe expanded, each type of particle passed a threshold after which that particle was no longer produced.

Because particles and their antiparticles were both created, eventually they found each other and annihilated. If particles and their antiparticles had formed in exactly equal numbers, then in due time all would have annihilated, and there would be no matter in the universe today. The fact that there is matter tells us that particles and antiparticles must have been produced in slightly unequal numbers. The current understanding of how this came to be is that at the time when the four forces were separating from each other, there were very slight asymmetries that favored normal particles over their antiparticles. Theorists estimate that normal particles had to outnumber antiparticles by only one part in 10^{16} to leave all the matter in the universe that we see today. It is thought possible that in other universes that may exist (see the preceding section), different physical laws could prevail, and perhaps antimatter could dominate over normal matter.

The First Nuclear Age

Let us now consider an epoch in the early universe when matter was beginning to take on familiar forms, and the four fundamental forces were already acting as four distinct forces. In the inflationary theory, this followed the phase of rapid expansion, which was finished by 10^{-30} seconds after the beginning. The temperature dropped to 10 billion degrees (10^{10} K) by 1.09 seconds after the start, and by then protons and neutrons were appearing **(Fig. 20.18)**. At this point, most of the energy of the universe was still in the form of radiation. Within a few minutes, conditions became better suited for nuclear reactions to take

Figure 20.18 Element formation in the big bang. This diagram illustrates the relative abundances of various light elements formed by nuclear reactions during the early stages of the big bang expansion. The relative abundances are shown as functions of time and temperature. (Data from R. V. Wagoner, 1973, *The Astrophysical Journal* 179:343)

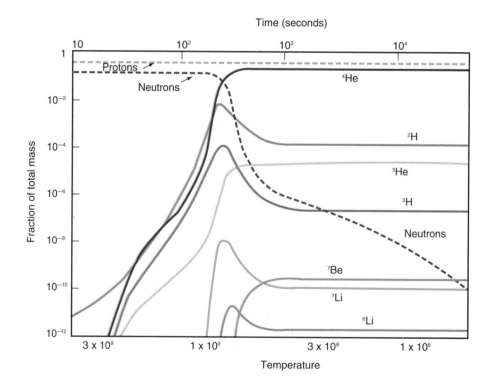

place efficiently, and the most active stage of element creation began (see Fig. 20.18). The principal products, in addition to the helium and deuterium mentioned in an earlier section, were tritium (another form of hydrogen, with one proton and two neutrons in the nucleus) and the elements lithium (three protons and four neutrons) and to a minor extent beryllium (four protons and five neutrons). Nearly all the available neutrons combined with protons to form helium nuclei, a process that was complete within 4 minutes of the beginning of the expansion. At this point, with some 22 to 28% of the mass in the form of helium, the reactions were essentially over, except for some production of lithium and beryllium over the next half-hour (see Fig. 20.18).

The rapidity with which the universe expanded and cooled allowed little time for further reactions. There was another problem as well: no stable nucleus exists with a nuclear mass number of 8 (i.e., total number of protons plus neutrons). Thus there was no way to build up heavier species than ^{7}Li and some short-lived ^{7}Be (beryllium), so nuclear production in the big bang was halted. Virtually all of the heavier elements we find in today's universe were formed by stars, either in stable reactions in stellar cores or in supernova explosions. (Stars surmount the mass 8 gap by virtue of the triple-α reaction, in which three helium nuclei collide virtually simultaneously; see Chapter 15).

The Matter-Dominated Universe

The expansion and cooling of the early universe continued, but nothing significant happened for a long time after the nuclear reactions stopped. Eventually (at least 500,000 years after the beginning), the density of the radiation had dropped sufficiently that its energy was less than that contained in the mass (that is, the energy derived from $E = mc^2$ became greater than that contained in the radiation). At that point, it is said, the universe became **matter-dominated** rather than **radiation-dominated.**

The matter and radiation continued to interact, however, because free electrons scatter photons of light very efficiently. The strong interplay between matter and radiation finally ended nearly a million years after the start of the big bang, when the temperature became low enough to allow electrons and protons to combine into hydrogen atoms. The atoms still absorbed and reemitted light, but much less effectively, because they could do so only at a few specific wavelengths. From this time on, the matter and the radiation went separate ways. The radiation simply continued to cool as the universe expanded, reaching its present temperature of 2.7 K some 10 to 16 billion years after the expansion began.

The scenario just described does an admirable job of explaining the early stages of the big bang in the context of today's universe. Both the composition of the universe and the present-day background radiation are nicely consistent with this model, as are the homogeneity and near-flatness of the observed universe. The major element that is unexplained is the large-scale structure.

As we have seen, it is not difficult to explain the presence of galaxies, and even of clustering up to a certain scale, as the result of random aggregations in a cold dark matter (bottom-up) universe. But the enormous sheets or filaments of galaxies could not have formed in this way. These imply instead that some structure must have been imposed on the universe from very early times, even before galaxy formation was complete.

The confirmation that primordial structure existed came from the *COBE* results described earlier in this chapter. Very subtle variations (measured in *millionths* of kelvins!) appear in the map of radiation intensity over the sky, corresponding to fluctuations in the density of matter at the time when the radiation decoupled from matter. The origin of these fluctuations is not clear, but models of the early expansion have been developed in which a mixture of hot dark matter and cold dark matter can account for them. As we discussed earlier in this chapter, hot dark matter is capable of forming large structures while cold dark matter forms small ones. This is another way of saying that both the hot dark matter (i.e., top-down) and cold dark matter (bottom-up) models may be correct.

We have reached a fascinating conclusion, namely, that dark matter has governed the properties of the universe we live in. The structure that we see when we map the galaxies is nothing but the tip of the iceberg. In a very real sense, all of the ordinary matter that we are familiar with means nothing, and what really counts is what is missing, the 90% of the mass out there whose nature we cannot fathom. This is a sobering note on which to end our discussion of the formation and evolution of the universe.

What Next? The Future of the Universe

Any discussion of the future of the universe is obviously a speculative venture. We cannot even answer, with absolute certainty, the basic question of whether it is open or closed. Nevertheless, just as theorists have attempted to describe the early stages of the universe, as discussed in the last section, efforts have been made to calculate future conditions in the universe. The uncertainties are much greater in describing the future, however, so the following discussion should be regarded as very speculative.

If the universe is open or flat (as in the inflationary model), then it will have no definite end; it will just gradually run down. The radiation background will continue to decline in temperature, approaching absolute zero. As stellar processing continues in galaxies, the fraction of matter that is in the form of heavy elements will continue to grow, and the supply of hydrogen, the basic nuclear fuel, will diminish. It is predicted that all the hydrogen will be gone by about 10^{14} years after the birth of the universe, so the universe in which stars dominate has now lived approximately one ten-thousandth of its lifetime. The recycling process between stars and the interstellar medium will continue until this time, but gradually the matter will become locked up in black holes, neutron stars, white dwarfs, and black dwarfs. Dead and dying stars will continue to interact gravitationally, eventually colliding often enough in their wanderings that all planets will be lost (by about 10^{17} years) and galaxies will dissipate as their constituent stars are lost to intergalactic space (by about 10^{18} years).

At later times, new physical processes will take over. Given enough time, even black holes can evaporate through a low-probability quantum mechanical effect that allows particles to appear spontaneously in space outside the event horizon. This process, called **quantum mechanical tunneling,** operates on the quantum mechanical principle that the location of a particle is described by a probability function. Thus the position is not precisely known at any given time, and a particle has at least a small probability of being somewhere else (this is probably one of the most counterintuitive ideas in quantum mechanics, but it has been verified many times in experiments). Given enough time, the mass in a black hole can tunnel out, and this will lead eventually to the evaporation of all black holes in the universe. The final stage that has been foreseen occurs at an age of 10^{100} years, when sufficient time has passed for all black holes to evaporate, leaving nothing but a sea of positrons, electrons, and radiation.

If the universe is closed, then someday, perhaps some 50 billion years from now, the expansion will stop and will be replaced by contraction. The deterioration described above will still take place until the final moments when the universe once again becomes hot and compressed, entering a new singularity. In some views, purely conjectural and without possibility of verification, such a contraction would be followed by a new big bang, and the universe would be reborn. This concept of an oscillating universe, pleasing to the minds of many, will not occur

unless the present weight of the evidence favoring an open or flat universe is found to be in error.

Summary

1. Observational constraints on the nature of the universe as a whole are provided by the expansion of the universe, the cosmic background radiation, the evidence for dark matter, and the large-scale structure of the universe.

2. The expansion of the universe is uniform, meaning that the speed of recession between any two points is proportional to their separation, and that there is no preferred location, that is, no center.

3. The age of the universe can be estimated from the time it would have taken for the present separation between galaxies to develop at the present rate of expansion. This depends critically on the Hubble constant H_0 (the age is given by $t = 1/H_0$, modified by a correction factor depending on the mass density of the universe), which is so poorly known at present that age estimates range from 9 to about 15 billion years.

4. The universe is filled with thermal radiation leftover from the earliest times of the expansion. Today this radiation, called the cosmic background radiation, has a temperature of about 2.7 K and is observed in the microwave portion of the radio spectrum.

5. Very slight temperature differences (Doppler shifts) in the radiation as seen in opposing directions reveal the motion of the Earth relative to the reference frame of the expanding universe and show that there is a concentration of unseen mass in the direction of the Virgo cluster of galaxies.

6. Several lines of evidence from observations indicate that some 90% of the mass in the universe is invisible; the unseen mass is referred to as dark matter.

7. Very large-scale structure in the universe, as indicated by subtle variations in the cosmic background radiation, show that large structures formed very early in the history of the universe. At the same time, it has become clear that galaxies and clusters of galaxies probably formed from the gravitational clustering of smaller systems. Therefore modern models for the formation of structure in the universe combine the "top-down" and "bottom-up" theories.

8. Modern cosmological theories are usually based on general relativity, whose field equations describe the state of matter and energy in the

Measuring the Hubble Constant

ASTRONOMICAL

ACTIVITY

The accompanying table provides you with data that can be used to determine the value of the Hubble constant. The table lists the measured wavelength, for several galaxies, of the spectral line of ionized calcium (designated Ca II), whose rest wavelength is 3933.663 Å. Also given is the observed apparent magnitude of each galaxy, which can be used to derive its distance if the absolute magnitude is known (see Chapter 14 or Appendix 13 for a reminder of how to do this). Assume that all the galaxies are type Sb, and that all have the same absolute magnitude, $M = -21.0$.

To derive the value of the Hubble constant, you will need to compute the recession velocity of each galaxy, using the Doppler shift formula. You will have to use the relativistic formula for the Doppler shift if the redshift $\Delta\lambda/\lambda$ is greater than about 0.01 (i.e., if the velocity is as large as 1% of the speed of light). The relativistic Doppler shift formula is given in Chapter 19. Once you have found the distances and the velocities of all the galaxies, convert the distances to units of megaparsecs, and make a graph (on graph paper with uniform-size squares in each direction) showing distance on the horizontal scale and recession velocity on the vertical scale (see Fig. 18.12).

The points representing the individual galaxies should form an approximately straight line running diagonally from the lower left toward the upper right. Using a ruler, draw the straight line that seems to fit best through the points (the points will tend to scatter a little around this line; this reflects observational uncertainty and local motions of galaxies). The slope of this line is the Hubble constant. To measure the slope, pick a point on the line, preferably near its upper end, and read off the values for distance (from the horizontal axis) and recession velocity (from the vertical axis) corresponding to that point. Then divide the recession velocity by the distance; the result will

be the value of H_0, in units of km/sec/Mpc (note that for a straight line, you will get the same value no matter where you pick your point).

How does your value of H_0 compare with modern values as discussed in the text? Compute the age of the universe on the basis of your value (see the text). What do you think might be the major uncertainties in the value of H_0 derived by this method?

Galaxy	m	Wavelength (Å)
1	11.12	3961.032
2	10.70	3954.567
3	13.60	4023.457
4	13.40	4013.540
5	6.39	3936.418
6	12.96	4004.984
7	13.20	4010.998
8	12.27	3979.163
9	11.94	3975.318
10	10.05	3951.800
11	13.25	4005.384
12	12.39	3983.278
13	13.45	4018.496
14	11.64	3970.550
15	12.67	3987.264
16	13.32	4007.522
17	13.64	4019.836
18	12.79	3992.052
19	8.60	3939.568
20	12.21	3983.942
21	13.82	4033.938
22	13.17	4005.251
23	10.55	3955.754
24	9.19	3946.402
25	13.82	4030.038
26	12.91	3997.113
27	10.15	3949.429
28	11.83	3969.021
29	12.55	3993.383
30	13.49	4016.084

universe. The models show three possible fates for the universe: continued expansion forever (open universe); eventual reversal of the expansion (closed universe); and a perfect balance, meaning that the expansion will slow but never stop (flat universe).

9. Observational evidence, based on the matter content of the universe (i.e., whether the density is greater or less than the critical density) and on the rate of deceleration of the expan-

sion, indicates that the universe is close to being flat.

10. A flat universe is expected in the inflationary theory, a form of the big bang model, which holds that the universe underwent a period of extremely rapid expansion at very early times (less than 10^{-30} seconds after the beginning). In this model, the initial expansion could have been triggered by a random energy fluctuation in a vacuum, and it suggests that other

universes could have formed in the same manner.

11. Particle pair formation from energetic photons, followed by a brief period when nuclear fusion reactions occurred, created an initial universal composition of roughly 75 to 80% hydrogen and nearly all the rest helium, with only traces of the light elements lithium and beryllium. Subsequently, when atoms formed (i.e., when electrons and protons became bound into hydrogen atoms), the radiation filling the early universe became decoupled from the matter and has continued to cool, being observed today as the cosmic microwave background.

12. The universe will gradually dissipate in the distant future as the expansion continues, eventually leaving nothing but elementary particles and very cold radiation. If the universe is closed and therefore collapses back to a point in the far future, then conditions of high temperature and density will be repeated.

Review Questions

1. In your own words, explain how all galaxies can be receding from our own galaxy, even though our galaxy is not at the center of the universe.

2. How is the value of the Hubble constant measured? Why is it difficult to obtain a precise value?

3. Why is the discovery of the microwave background radiation considered to be strong evidence in favor of the big bang theory?

4. Is the background radiation homogeneous and isotropic? Explain.

5. It has been suggested (as discussed in Chapter 19) that many galaxies have very massive black holes in their cores. If so, would this change observational estimates of the mass density of the universe and therefore help determine whether the universe is open or closed? Explain.

6. Why is the deuterium abundance a better indicator than helium of the early expansion rate of the universe?

7. Discuss the assumptions that the universe is homogeneous and isotropic in the context of the Copernican (Sun-centered) hypothesis and Hubble's discovery that the Milky Way is just one of many galaxies in the universe.

8. Explain in your own words why general relativity implies that space is curved near massive objects.

9. Explain why the big bang cosmology is widely accepted by astronomers.

10. Why were elements heavier than beryllium not produced during the big bang?

Problems

1. Calculate the age of the universe if the value of H_0 is 55 km/sec/Mpc and again if it is 80 km/sec/Mpc. Discuss the implications of both results, as compared with the age of globular clusters. (Assume that $\Omega_0 = 1$, so that the correction factor of 2/3 is needed).

2. Suppose the temperature of the background radiation were 250 K instead of 2.7 K. What would be the wavelength of maximum radiation intensity? What kind of telescope would be required to observe it?

3. The interstellar medium in the disk of our galaxy has an average density of about one hydrogen atom for every 10 cm³ of volume. Convert this density into units of g/cm³, assuming that the mass of a hydrogen atom is 1.7×10^{-24} g. How does your answer compare with the critical density for closing the universe? What implications does this have for whether the universe is open or closed?

4. Suppose that, on average, the universe contains one galaxy like ours (mass about 10^{11} solar masses) for every volume of 10 Mpc³. What is the average density of the universe, and how does it compare with the critical density? (*Hint:* A volume of 10 Mpc³ corresponds to a cube that is about 2.2 Mpc on each side. Convert this to centimeters so that you can convert the volume of the cube to cubic centimeters for comparison with the critical density.)

5. Suppose a positron and an electron combine, annihilating each other and producing a photon containing the total rest mass energy of the particle pair. What is the wavelength of the photon? (To do this, you must first calculate the rest mass energy from $E = mc^2$ and then use this energy in the formula $E = hc/\lambda$ to find λ, the wavelength of the photons. The mass m is the total mass of the positron and the electron. The electron mass, which is equal to the positron mass, is given in Appendix 4.) What kind of telescope would be needed to detect such photons?

6. If the universe has a temperature of 10^{10} K, is it hot enough to produce electron-positron pairs? (To find out, find the wavelength of maximum emission from Wien's law and compare it with

the wavelength calculated in problem 5 above. If the wavelength of maximum emission is shorter than the wavelength representing the energy of an electron-positron pair, then the universe is hot enough to produce such pairs from photons.)

7. What is the threshold temperature for the universe below which it stopped producing proton-antiproton pairs; that is, at what temperature did the wavelength of maximum emission become too great for the photons to equal the rest mass energy of a proton-antiproton pair? (To do this, calculate the wavelength corresponding to the rest mass of the particle pair, using the proton mass from Appendix 4, and then substitute the resulting value for the wavelength into Wien's law to find the temperature at which photons of this wavelength are emitted.)

Additional Readings

Abbott, L. 1988. The Mystery of the Cosmological Constant. *Scientific American* 258(5):106.

Allégre, C. J. and S. H. Schneider 1994. The Evolution of the Earth. *Scientific American* 271(4):66.

Bruning, D. 1996. Stellar Graveyard. *Astronomy* 24(2):44.

Brush, S. G. 1992. How Cosmology Became a Science. *Scientific American* 267(2):62.

Burstein, D. and P. L. Manly 1993. Cosmic Tug of War. *Astronomy* 21(7):40.

Cowen, R. 1994. The Debut of Galaxies. *Astronomy* 22(12):44.

Crosswell, K. 1996. White Dwarfs Confront the Universe. *Astronomy* 23(5):42.

Davies, P. 1992. The First One Second of the Universe. *Mercury* 21(3):82.

Deutsch, D. 1994. The Quantum Physics of Time Travel. *Scientific American* 270(3):68.

Eicher, D. J. 1995. Galaxy Time Machine. *Astronomy* 23(4):44.

Frank, A. 1996. A River in the Universe. *Astronomy* 24(8):44.

Freedman, W. 1992. The Expansion Rate and the Size of the Universe. *Scientific American* 267(5):54.

Geller, M. J. and J. P. Huchra 1991. Mapping the Universe. *Sky & Telescope* 82(2):134.

Haisch, B. M. and A. Rueda. 1996. A Quantum Broom Sweeps Clean. *Mercury* 25(2):12.

Halliwell, J. J. 1991. Quantum Cosmology and the Creation of the Universe. *Scientific American* 265(6):76.

Hodge, P. 1993. The Extragalactic Distance Scale: Agreement at Last? *Sky & Telescope* 86(4):16.

Horgan, J. 1994. Particle Metaphysics. *Scientific American* 270(2):96.

Jayawardhana, R. 1993. The Age Paradox. *Astronomy* 21(6):38.

Kaku, M. 1996. What Happened Before the Big Bang? *Astronomy* 23(5):34.

Kanipe, J. 1996. Dark Matter and the Fate of the Universe. *Astronomy* 24(10):34.

Kinney, A. L. 1996. Fourteen Billion Years Young. *Mercury* 25(2):29.

Kirshner, R. P. 1994. The Earth's Elements. *Scientific American* 271(4):58.

Lake, G. 1992. Cosmology of the Local Group. *Sky & Telescope* 84(6):613.

Linde, A. 1994. The Self-Reproducing Inflationary Universe. *Scientific American* 271(5):48.

Monda, R. 1992. Shedding Light on Dark Matter. *Astronomy* 2(2):44.

Nather, R. E. and D. E. Winget 1992. Taking the Pulse of White Dwarfs. *Sky & Telescope* 83(3):374.

Odenwald, S. 1991. Einstein's Fudge Factor. *Sky & Telescope* 81(4):362.

Odenwald, S. 1996. Space-Time: The Final Frontier. *Sky & Telescope* 91(2):24.

Parker, B. 1988. The Cosmic Cookbook: The Discovery of How the Elements Came to Be. *Mercury* 17:171.

Peebles, P. J. E., D. N. Schramm, E. L. Turner, and R. G. Kron 1994. The Evolution of the Universe. *Scientific American* 271(4):52.

Roth, J. and J. R. Primask. 1996. Cosmology: All Sewn Up or Coming Apart at the Seams? *Sky & Telescope* 91(1):20.

Sadoulet, B. and J. W. Cronin 1992. Subatomic Astronomy. *Sky & Telescope* 83(1):25.

Schramm, D. N. 1991. The Origin of Cosmic Structure. *Sky & Telescope* 82(2):140.

Schramm, D. N. 1994. Dark Matter and Cosmic Structure. *Sky & Telescope* 88(4):28.

Spergel, D. N. and Turok, N. G. 1992. Textures and Cosmic Structure. *Scientific American* 266(3):52.

Talcott, R. 1992. COBE's Big Bang. *Astronomy* 10(8):42.

Tyson, A. 1992. Mapping Dark Matter With Gravitational Lenses. *Physics Today* 45(6):24.

Web Connections

The Review Questions and Problems also appear at the following URLs:

http://universe.colorado.edu/ch21/questions.html
http://universe.colorado.edu/ch21/problems.html

Chapter 21
The Chances of Companionship

The universe is in the business of making life.
Cyril Ponnamperuma, 1958

Chapter Web site: http://universe.colorado.edu/ch21

We have attempted to answer all of the fundamental questions about the physical universe that can be treated scientifically. Having done this, we know our place in the cosmos: we know something of its scale and age, and we realize how insignificant our habitat is. In this chapter we contemplate whether we as living creatures are unique in the universe, or whether even that distinction must be shared. Nothing that we have learned so far leads us to rule out the possibility that other life-forms, some of them intelligent, may exist. We believe that the Earth and the other planets in our solar system are a natural by-product of the formation of the Sun, and we have evidence that some of the essential ingredients for life were present on the Earth from the time it formed.

Furthermore, we have the spectacular news that fossil bacteria-like organisms appear to be present in meteoritic material from Mars. This discovery, if validated by further studies, is one of the most important ever made. If life existed on Mars, then in future

Figure 21.1 Charles Darwin. Scientific inquiries by Darwin led to an understanding of evolution, one of the most profound concepts of human intellectual development. (The Granger Collection)

http://universe.colorado.edu/fig/21-1.html

textbooks the kind of speculative discussion you will find in this chapter will be replaced by more objective analysis. The entire subject of life in the universe, of the chances for companionship in the cosmos, will become an area of new and legitimate scholarly study rather than fiction.

Science cannot yet tell the full story of how life began, however, and we have not found incontrovertible evidence that it actually does exist elsewhere. For now the mystery remains.

Life on Earth

We start our discussion of possible extraterrestrial life by discussing the origins of the only life we know. Besides giving us some insight into the processes thought to have been at work on the Earth, this will help us later, when we are speculating about whether the same processes have occurred elsewhere.

Life's Origin: The Panspermia Hypothesis

Before the time of Charles Darwin **(Fig. 21.1)** in the mid-1800s, the view was widely held that life could arise spontaneously from nonliving matter. Darwin's work in the study of evolution, showing how species develop gradually as a result of environmental pressures, helped make such an idea seem improbable.

An alternative to the spontaneous formation of life was proposed in 1907 by the Swedish chemist Svante Arrhenius, who suggested that life on Earth was introduced billions of years ago from space, arriving originally in the form of microscopic spores that float through the cosmos, landing here and there to act as seeds for new biological systems. This idea, called the **panspermia hypothesis,** cannot be ruled out, but several arguments make it seem unlikely. Such spores would take a very long time to permeate the galaxy, and their density in space would have to be very high for one or more of them to reach the Earth by chance. More importantly, it seems very unlikely that the spores could survive the hazards of space, such as ultraviolet light and cosmic rays.

The panspermia hypothesis has some modern-day advocates, although they appear to be in the minority at present. One group, led by the widely regarded British astrophysicist Fred Hoyle, argues that organic materials including bacteria exist on dust grains, and that the introduction of these grains onto the Earth's surface not only brought life to our planet in the first place, but even today causes earthly epidemics of flu and other illnesses. Other arguments for panspermia are that there has not been enough time in Earth's history for random collections of atoms to form the complex molecules of life, or that there was no viable mechanism on the early Earth to produce large molecules such as RNA and DNA.

Even if the panspermia concept is correct, the question of the ultimate origin of life remains, although it is transferred to some other location. In view of what is known today about the evolution of life and early conditions on the Earth, scientists generally agree that life arose through natural processes occurring here and was not introduced from elsewhere.

Earthly Origins of Life

At the earliest times after the Earth had coalesced from the debris orbiting the young Sun, few organic materials could have survived on the surfaces of any terrestrial planets. The Earth was largely molten for a period, and the intense radiation and wind from the proto-Sun would have dispatched most of the volatile elements to the outer system anyway.

The first atmosphere of the Earth must have been acquired from impacts of cometary bodies. These fragments of icy debris condensed in the outer solar system (see Chapter 11), and then were perturbed into larger orbits due to the gravitational effects of the giant planets, primarily Jupiter. Subsequently, further perturbations brought comets into the inner system occasionally, and some intersected the surfaces of the terrestrial planets, depositing their hydrogen, carbon, oxygen, nitrogen, and other essential ingredients. It is interesting to realize that had Jupiter not formed early in the history of the solar system, the entire mechanism for transporting cometary material to the Earth might not have functioned, and life here might not have been possible.

As noted in Chapter 11, even such complex species as **amino acids** may have been present in the solar system before the Earth formed; these molecules form the basis of proteins, which are the fundamental substance of living matter. Traces of them have been found in some meteorites **(Fig. 21.2)**, and we know that meteorites are very old, representing the first solid material in the solar system. We also know that many kinds of complex molecules, including organic (carbon-bearing) molecules, exist inside dense interstellar clouds (see Chapter 12 and Appendix 15), and we may speculate that perhaps amino acids may also have formed in these regions (a tentative identification of interstellar radio emission from glycine, the simplest amino acid, has recently been announced). But as noted above, if any of these materials came to the Earth, it must have been after our planet had cooled sufficiently for them to survive.

Figure 21.2 **A meteorite containing primordial amino acids.** This is a section of the Murchison meteorite, which fell in Australia. Amino acids found in the carbonaceous chondrite were apparently present in it when it fell. (Photo by John Fields, the Trustees, the Australian Museum)

It is believed that the early atmosphere of the Earth was composed in part of hydrogen and hydrogen-bearing molecules such as ammonia (NH_3) and methane (CH_4), as well as water (H_2O). Therefore, the first organisms must have developed in the presence of these ingredients. In the 1950s, scientists began to perform experiments in which they attempted to reproduce the conditions of the early Earth. The starting point of these experiments was to place water in containers filled with the type of atmosphere just described. Water was introduced because it is apparent that the Earth had oceans from very early times, and because it is thought that life started in the oceans, where the liquid environment provided a medium in which complex chemical reactions could take place. Reactions occur much more slowly in solids and in gases—in solids because the atoms are not free to move about easily and interact, and in gases because the density is low, and particles encounter each other much less frequently, thus keeping reaction rates slow. Water is the most stable and abundant liquid that can form from the common elements thought to be present when the Earth was young.

The first of these experiments **(Fig. 21.3)** was performed in 1953 by the American scientists Harold Urey and Stanley Miller, who concocted a mixture of methane, ammonia, water, and hydrogen and exposed it to electrical discharges, a possible source of energy on the primitive Earth (ultraviolet light from the Sun is another, but it was more difficult to work

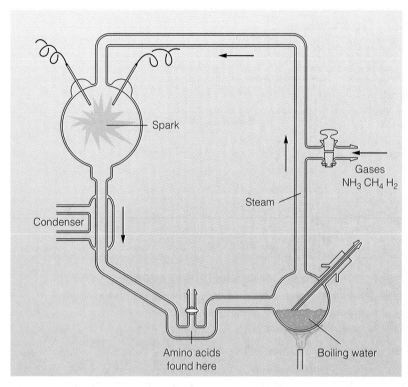

Figure 21.3 Simulating the atmosphere of early Earth. Urey and Miller constructed this apparatus to reproduce conditions on the primitive Earth, in hopes of learning how life-forms could have developed.

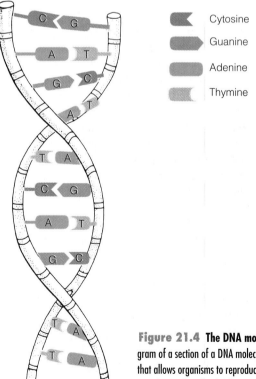

Figure 21.4 The DNA molecule. This is a schematic diagram of a section of a DNA molecule. DNA carries the genetic code that allows organisms to reproduce themselves. A critical question in understanding the development of life on Earth is to learn how DNA arose. (© West Publishing Co.)

with in the lab.) After a week, the mixture turned dark brown, and Urey and Miller analyzed its composition and found large quantities of amino acids. Other experimenters later showed that exposure to ultraviolet light produced the same results. These experiments demonstrated that at least some of the precursors of life probably existed in the primitive oceans almost immediately after the Earth had cooled enough to support liquid water. Other similar experiments have produced more complex molecules, including sugars and larger fragments of proteins.

The direction things took once amino acids and other organic molecules existed is not so clear. Somehow these building blocks had to combine to form **ribonucleic acid (RNA)** and **deoxyribonucleic acid (DNA; Fig. 21.4).** These very complex molecules carry the genetic codes that allow living creatures to reproduce. Experiments have successfully produced molecules that are fragments of RNA and DNA from conditions like those that prevailed on the early Earth, but not the complete forms required. Maybe it is simply a matter of time; if such experiments could be performed for years or millennia, perhaps the vital forms of these proteins would appear. The major hurdle to forming RNA and DNA is arranging the very long sequences of amino acids in the proper order and orientation. Some suggestions have been made as to how this might have happened (for example, it has been proposed that the correct alignment could have occurred in deposits along crystalline structures in material such as quartz or certain clays), but it remains a major unknown. How nature took the step from amino acids and other fragments to the first self-replicating complex organic molecules such as RNA remains one of the areas of greatest uncertainty in our present understanding of how life began.

The Evolution of Life

Fossil records **(Fig. 21.5)** tell us that the first microorganisms appeared more than 3.5 billion years ago, when the Earth was barely 1 billion years old. Following their appearance, the evolution of increasingly complex species seems to have followed naturally **(Table 21.1).** At first the development was very slow, only reaching the level of simple plants such as algae 1 billion years later. Increasingly elaborate multicellular plant forms followed and gradually altered the Earth's atmosphere by introducing free oxygen. Meanwhile, the gases hydrogen and helium, light and fast-moving enough to escape the Earth's gravity, essentially disappeared. Nitrogen, always present from outgassing and volcanic activity, became more predominant through the decay of dead organisms. By about 1 or 2 billion years ago, the Earth's atmosphere had reached its present composition.

The first broad proliferation of animal life occurred about 600 million years ago, and the great reptiles arose some 350 million years later. The dinosaurs died out after about 200 million years, and mammals came to dominance about 65 million years ago. Our primitive ancestors appeared only in the last 3 or 4 million years. Once the development of intelligence had provided the ability to control the environment, the entire world became our ecological home, and our physical evolution essentially stopped. It remains to be seen whether future ecological pressures will lead to future evolution of the human species.

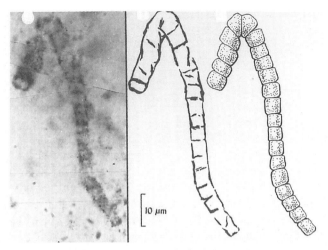

Figure 21.5 A fossil microorganism. This fossil alga is evidence that primitive life-forms existed on the Earth billions of years ago. (Photo courtesy of J. William Schopf, University of California, Los Angeles)

Table 21.1
Steps in the Evolution of Life on Earth

Era	Period	Age (yr)	Biological Developments
Archeozoic		3.5×10^9	
	Archean		No life forms
Proteozoic		$1.5–3.5 \times 10^9$	
	Algonkin		Radiolaria; marine algae
Paleozoic		$0.25–1.5 \times 10^9$	
	Cambrian		Marine faunas, primitive vegetation
	Ordovician		Fishlike vertebrates
	Silurian		Air-breathing invertebrates, first land plants
	Devonian		Fishes and amphibians; primitive ferns
	Mississippian		Ancient sharks, mosses, ferns
	Pennsylvanian		Amphibians, insects
	Permian		Primitive reptiles, mosses, ferns
Mesozoic		$70–250 \times 10^6$	
	Triassic		Reptiles, dinosaurs, ferns
	Jurassic		Toothed birds, primitive mammals, palms
	Cretaceous		Decline of dinosaurs, modern insects, birds, snakes
Cenozoic		$1.5–70 \times 10^6$	
	Paleocene		Large land mammals
	Eocene		Primates, first horse, modern plants
	Oligocene		Larger horse, modern plants
	Miocene		Proliferation of mammals, larger horse, modern plants
	Pliocene		Grassland mammals, earliest hominids
Modern		1.5×10^6 to present	
	Pleistocene		Hominids, modern-size horse, modern plants
	Holocene		*Homo sapiens*, present-day vegetation

Could Life Develop Elsewhere?

The scenario just described, if at all accurate, seemingly should occur almost inevitably, given the proper conditions. If this is so, then the question of whether life exists elsewhere amounts to asking whether the conditions that existed on the primitive Earth could have arisen elsewhere. It is clear that no other planet in the solar system could have provided an environment exactly like that on the early Earth, although conditions on the early Mars might have been suitable. The recent discovery of possible fossil organisms in meteoritic material from Mars greatly enhances this possibility. Mars is the only planet where life-forms have been sought so far **(Fig. 21.6)**, and perhaps they have been found **(Fig. 21.7)**. If this conclusion is borne out by further analysis, then we will have much stronger grounds for speculating about life in other planetary systems. Given the billions of stars in the galaxy and the vast number that are very similar to the Sun, it seems highly probable that the proper conditions must have been reproduced many times in the history of the galaxy.

Assumptions about Life's Requirements

So far we have worked from the tacit assumption that life on other planets, if it exists, must be similar to life on Earth, and we have considered only the question of whether other Earth-like environments may exist.

We may question the premise that life could have developed only in the form that we are familiar with, however. Here, obviously, we must indulge in speculation, having no examples of other types of life at hand for examination.

Life as we know it is based on carbon-bearing molecules, and some have argued that only carbon is capable of combining chemically with other common elements in a sufficiently wide variety of ways to produce the complexity of molecules thought to be necessary. Carbon has a uniquely complex chemistry, having more ways to bond with neighboring atoms (of carbon and other species) than any other common element. It is difficult to invoke exotic life-forms based on elements that are rare on Earth, because the composition of the galaxy is quite uniform everywhere: an element that is rare here will likely be rare elsewhere also. This may seem to rule out life-forms based on anything other than carbon. However, it has been pointed out that another common element, silicon, also has a very complex chemistry and therefore might provide a basis for a radically different type of life. But silicon has some disadvantages compared to carbon: its chemistry is not really as complex, and there is no liquid solvent for silicon compounds that is likely to exist under moderate conditions; that is, water has no counterpart that would play the same role for silicon-based life.

Another assumption that might be subject to question is that water is a necessary medium for carbon-based life. As noted earlier, it is the most abun-

Figure 21.6 Searching for life in the solar system. Mars was long thought to be the most likely home for extraterrestrial life in the solar system. Here we see a *Viking* lander in a simulated Martian environment. The *Viking* missions reached Mars in 1976 but found no evidence for life-forms. (NASA)
http://universe.colorado.edu/fig/21-6.html

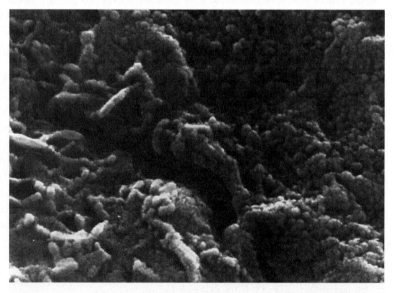

Figure 21.7 Fossil life from Mars? This meteorite, found in Antarctica, is almost certainly from Mars, having been blasted out of the surface by an impacting body. Carbon-rich deposits in the rock, along with microscopic structures resembling earthly bacteria, have convinced some scientists that fossil organisms from Mars are present. (NASA) http://universe.colorado.edu/fig/21-7.html

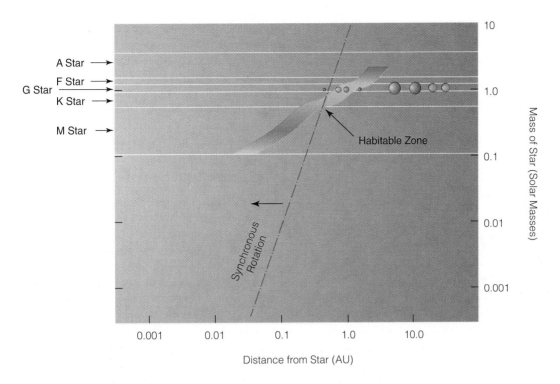

Figure 21.8 Habitable zones. This figure illustrates how the location and width of the habitable zone depends on stellar type. The habitable zone is defined as the region around a star where liquid water could exist on a planetary surface. (Data from Kasting, J. F., Whitmire, D. P., and Reynolds, R. T. 1993, *Icarus,* 101, 108–128)

dant liquid that can form under the temperature and pressure conditions of the early Earth, and it is thought that only a liquid medium could support the required level of chemical activity. Mobility was needed to form the complex molecules of life in the first place, and water still plays an essential role in the internal transport of nutrients and waste products. Other liquids can exist under other conditions, however, and it is interesting to consider whether life-forms of a wholly different type (but probably still carbon-based) might arise in oceans of strange composition. One might speculate, for example, about what goes on in the lakes of liquid methane or liquid nitrogen that are thought possibly to exist on Titan, the mysterious giant satellite of Saturn.

The Habitable Zone

Although speculating about whether life can form from exotic ingredients under physical conditions wholly unlike those here on Earth is interesting, it is probably unnecessary. It seems likely that many other Earth-like environments exist. This is an assumption that can be addressed, at least partially removing it from the realm of speculation.

We have referred to "Earth-like" environments without saying what we really mean. In the study of prospects for life to develop in other planetary sys-

tems, the term **habitable zone** has been developed. This refers to the region around a star where the surface conditions on a terrestrial planet would fall within the range where water is liquid, thus allowing life to form. The location and extent of the habitable zone depend on the type of star **(Fig. 21.8)** and also on the assumptions that are made about atmospheric composition and greenhouse heating. The general result is that the zone is close to the parent star and small for cool stars and farther away and broader for hot stars. Thus a very hot star (say, an O star) would have a very broad habitable zone located very far from the star, while a cooler star (an M star) would have a very thin zone close to the star.

We can use the habitable zone, along with some other considerations, to decide which kinds of stars would be most likely to have planets where life could develop. You might think those with the largest habitable zones, the hot stars, would be the winners, but there is another important consideration: the life expectancy of the star. As we learned in Chapter 15, an O star might live only a million years or so, while an A star (to pick a more moderate example) might live for a few hundred million years. Given that it took about a billion years for the first primitive life-forms to develop on Earth, we can probably rule out hot stars as hosts for life-bearing planets (these stars present another problem anyway: they do not appear to form disks and preplanetary systems and hence may not form planets as readily as cooler, lower-mass stars do). On the basis of stellar life expectancy, scientists generally rule out stars much hotter than the Sun, a G star.

At the cool end of the stellar classification sequence, we have the K and M stars. As seen in Figure 21.7, these stars have very small habitable zones, which would lower the chances that a planet might form in just the right location. Another potential problem is that the radiation from such a star would be very deficient in energetic photons and would arise instead primarily at infrared wavelengths. If visible or even ultraviolet radiation as an energy source is a necessary ingredient for life to form and evolve, then these low-mass, cool stars may not be viable. And they have a bad habit, particularly the M stars, that might make life very problematic on any planet close enough to be in the habitable zone. All cool main-sequence stars experience flare activity, much like that of the Sun, but it is more frequent and far more energetic than the normal stellar luminosity. When a solar flare occurs, we hardly notice it on Earth, but if we lived on a planet in the habitable zone around an M dwarf, a flare event could easily be fatal. An additional problem with these very cool stars is that any planet in the habitable zone would be so close to its parent star that it would be tidally locked into synchronous rotation, keeping one side permanently facing its sun. This would create a very uneven distribution of light and heat over the surface of the planet, possibly diminishing the chances that life could form and persist. For all of these reasons, the planets of K and M stars are unlikely places to look for life.

We are left with the stars very similar to the Sun as the best candidates for life's formation. Solar-type stars (i.e., G stars) have moderately large habitable zones (although there is some disagreement as to just how large they are); they have long lifetimes; and the habitable zones are far enough away from the stars to minimize the effects of flares and other disagreeable behavior. But even if we confine ourselves to solar-type stars, there are still vast quantities of them in our galaxy alone. There should be many habitable planets.

Strategies for Getting in Touch

In view of the limitations that prohibit faster-than-light travel, it is exceedingly unlikely that we will ever be able to visit other solar systems, seeking out life-forms that may live there. We will continue to explore our own system, so there is a reasonable chance that if life exists on any of the other planets of our Sun, we will someday discover it. It seems, however, that our best hope of finding other intelligent races in the galaxy will be to make long-range contact with them through radio or light signals. Since this requires both a transmitter and a receiver, we can hope to contact only civilizations as advanced as ours, with the capability of constructing the necessary devices for interstellar communication.

The Drake Equation

A mathematical exercise in probabilities has been used for some years as a means of assessing, as objectively as possible, the chances for making contact with an extraterrestrial civilization. The aim is to separate the question into several distinct steps, each of which can then be treated independently. The underlying assumption is that the number of technological civilizations in our galaxy today with the capability for interstellar communication is the product of the number of planets that exist with appropriate conditions, the probability that life developed on those planets, the probability that such life has developed intelligence that gave rise to a technological civilization, and, finally, the likelihood that the civilization has not killed itself off through evolution or catastrophe.

Mathematically, the so-called **Drake equation,** (after Frank Drake, who has been its best-known advocate) is written

$$N = R_* f_p\, n_e\, f_l f_i f_c L,$$

where N is the number of technological civilizations currently in existence in our galaxy, R_* is the number of stars of appropriate spectral type formed per year in the galaxy, f_p is the fraction of these that have planets, n_e is the number of Earth-like planets per star, f_l is the fraction of these on which life arises, f_i is the fraction of those planets on which intelligence has developed, f_c is the fraction of planets with intelligence on which a technological civilization has evolved to the point at which interstellar communications would be possible, and, finally, L is the average lifetime of such a civilization.

By expressing the number of civilizations in this way, we can isolate factors about which we can make educated guesses from those about which we are more ignorant. It is an interesting exercise to go through the terms in the equation one by one to see what conclusions we reach under various assumptions. People who do this have to make sheer guesses for some of the terms, and the result is a variety of answers ranging from very optimistic to very pessimistic. In the following, we will adopt middle-of-the-road numbers for most of the unknown terms.

The Appeal of E.T.

The notion that there may be life "out there," especially intelligent life, has been popular for centuries. There are probably many reasons for people's readiness to believe in extraterrestrial life, ranging from observations of seemingly inexplicable but real phenomena to deeply hidden psychological needs. And there is always the possibility that aliens do exist and have visited the Earth, although good evidence for that is lacking.

The interest in aliens takes many forms. As long ago as the late sixteenth century, Giordano Bruno preached that the stars were other suns, each with orbiting planets that housed civilizations (Bruno was burned at the stake for this in 1600). Much more recently, starting with a pilot's observation of unusual clouds near Mt. Rainier in 1947, there have been occasional spates of "UFO" sightings. For well over a century, science fiction, some of it stimulated by the putative Martians of Percival Lowell and Orson Welles, has invoked alien creatures and civilizations of many descriptions. Today's youth have grown up in an environment where UFO sightings and unexplained phenomena are standard television fare, and where, according to some claims, as much as *10% of the U.S. population* has been abducted, probed, experimented upon, and then released by aliens (but the abductees have suppressed the memory!).

The "evidence" for frequent visits by aliens is weak, if not entirely absent. No physical artifacts of a landing by alien spacecraft have ever been recovered, and furthermore well over 90% of all UFO sightings have been explained quite readily by conventional effects. One could easily conclude that humans have an underlying need to believe in higher powers (not to mention government conspiracies), and that this need to believe often overrides critical thinking.

It is not appropriate here to delve into the possible psychological motivations people might have for believing in close encounters with aliens, although there is a widespread literature on the subject that you may wish to explore. But it is timely to consider some of the physical constraints on alien visitors.

The laws of physics dictate that a spacecraft cannot travel faster than the speed of light. Therefore, allowing for acceleration time, it must take decades or centuries for a ship to travel the distance from even the nearest star to the Earth (recall that the nearest star is over 4 light-years away). Furthermore, the energy required for such a journey is immense; it has been estimated that the energy needed to boost a colonizing spaceship from Earth to a nearby star is enough to power the entire United States at its present rate of consumption for one hundred years! Is it likely that Congress would ever approve such an expenditure? This enormous energy requirement would be faced by any civilization planning interstellar travel and cannot be lessened by advanced technology.

Apart from the fascinating question of why so many people seem to be putting out a welcome mat for aliens, we can think about the impact on society if aliens were indisputably detected. As discussed in this chapter, the first contact, when and if it occurs, will very likely come in the form of detected radio signals. This could happen in our lifetime, as the evidence for the existence of planetary systems spurs on the search for signals. What will you think when a message is received from the stars? How will society react?

Surely, the impact would rival or exceed that of any past broadenings of human vision. The discovery that other life-forms exist, particularly other intelligent, technological forms, would likely have greater ramifications for religion, science, sociology, and society than previous scientific breakthroughs. To date the discovery that the Earth is not the center of the universe and that our galaxy is just one of a vast number are considered by some to be the two most significant discoveries of all time. The discovery that we are not alone in the universe would represent the next logical step in this progression. Others liken the news of alien civilizations, should it come, to earlier cultural encounters such as the infusion of Greek civilization into western Europe (by way of the Arabs) in the twelfth and thirteenth centuries, or the interaction between European explorers and native populations in the Americas, Africa, and Southeast Asia.

To a large degree, the impact will probably depend on the form of the contact. If a weak radio signal is picked up that is indisputably from an alien civilization but contains no other information (such as an incidental transmission), then the impact might be less significant than if a deliberate message containing information is received. Imagine the reaction if we received a message describing a galactic civilization, or new laws of physics, or perhaps a threat to take over the Earth!

Given the time and energy constraints imposed on any civilization thinking about interstellar travel, and given the lack (so far) of physical evidence, does it make sense to accept the notion that UFOs are sightseeing or spying spacecraft from another star system or that an alien civilization would expend the required time and energy routinely and often?

The first two factors, $R_\star$ and f_p, are, in principle, quantities that can be known with some certainty from observations. The rate of star formation in the equation refers to stars similar to the Sun. As discussed in the preceding section, there are good reasons to exclude stars much hotter or cooler than the Sun as hosts to potential life-supporting planets. Taking these considerations into account, some estimate that up to 10 suitable stars form in our galaxy per year; for the sake of discussion, we will be more cautious and adopt $R_\star = 1/\text{year}$. This may even be a little high for the present epoch in galactic history, but the star-formation rate was surely much higher early in the lifetime of the galaxy, and low-mass stars of solar type formed that long ago are still in their prime. Thus the adoption of an average formation rate of one Sun-like star per year is probably reasonable.

From what we know of the formation of our solar system, the formation of planets seems almost inevitable, except perhaps in double- or multiple-star systems. For the sake of discussion, let us assume that $f_p = 1$; that is, that all stars of solar type have planets. Observations have already begun to prove what has long been assumed: planets are indeed formed around solar-type stars. At this writing (mid-1996), nearly 10 stars have been found to have Jupiter-class planets, and others are suspected. This does not prove that Earth-like planets form also, but it lends us a basis for supposing so.

If we define "Earth-like" planets as those lying within the habitable zone, it is difficult to assess how many there are. We do not know enough about the mechanism that causes planets to form where they do. And the newly discovered systems (Chapter 12) show that not all planetary systems are organized like our own. So we can only guess what fraction of planetary systems contain an Earth-like planet. If we take the pessimistic point of view, as some do, that the habitable zone is actually very thin, then the value of n_e may be as small as 10^{-6}. Others, more optimistic, think the value is close to 1; that is, nearly every planetary system orbiting a Sun-like star has at least one planet in the habitable zone. Let us be moderate and assume $n_e = 0.1$; that is, in 1 out of 10 planetary systems around solar-type stars, there is a planet within the temperature zone where life can arise.

Now we get to the really speculative terms in the equation. We have no way of estimating how likely it is that life should begin, given the right conditions. From the seeming naturalness of its development on Earth, it can be argued that life would always begin if given the chance. Let us be optimistic here and agree with this, adopting $f_l = 1$.

If we are satisfied that life probably has formed naturally in many places in the universe, we can address a related and, to many, a more important question: Given the existence of life, how likely is it that intelligence will follow? Here we have no means of answering, except to reiterate that as far as we know, the evolution of our species on Earth was the natural product of environmental pressures. This is another area of little direct knowledge and lots of speculation. For now we take $f_i = 1$.

At this point, it is instructive to put the values adopted so far into the equation. We find:

$$N = (1/\text{year})(1)(0.1)(1)(1)(1)L = 0.1L.$$

Having taken our chances and guessed at the values for all the other terms, we now face a critical question: How long can a technological civilization last? Ours has been sufficiently advanced to send and receive interstellar radio signals for only about 70 years, and our society is sufficiently unstable to lead some pessimists to think that we will not last many more decades. If we take this viewpoint and adopt 100 years as the average lifetime, then we find

$$N = 10,$$

meaning that we should expect the total number of technological civilizations present in the galaxy at one moment to be very small, about 10. If this number is correct, then the average distance between these outposts of civilization is nearly 10,000 light-years **(Fig. 21.9)**. The time it would take for communications to travel between civilizations would be very much longer than their lifetimes, and we would have no hope of establishing a dialogue with anyone out

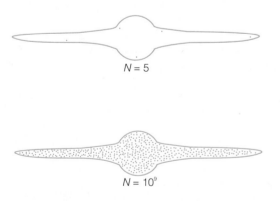

Figure 21.9 Possible values of N. The number of technological civilizations in the galaxy *(N)* may be very small (upper), in which case the average distance between them is very large; or *N* may be large, so that the distance between civilizations is relatively small (lower). The chances for communication with alien races are much higher in the latter case, where the distances are only a few tens of light-years.

there. If this estimate is correct, it is not surprising that we haven't heard from anybody yet.

We can be more optimistic, though, and assume that a technological civilization solves its internal problems and lives much longer than 100 years. Extremely optimistic people would argue that a civilization would be immortal; it would colonize star systems other than its own, so that it would be immune to any local crises such as planetary wars or suns expanding to become red giants. In that case, allowing a few billion years for the development of such civilizations, we can set $L = 10^{10}$ years (i.e., nearly equal to the age of the galaxy), and we find

$$N = 10^9 \, ,$$

where the average distance between civilizations is only about 15 light-years (see Fig. 21.8), coincidentally comparable to the distance our own radio signals have traveled **(Fig. 21.10)** since the early days of radio and television. If this estimate of N is correct, we could be hearing from somebody very soon.

We have presented two extreme views of the likelihood that other civilizations exist in our part of the galaxy. As we mentioned earlier, opinions among scientists who seriously study this question vary across this large range. Those who favor the optimistic viewpoint advocate making deliberate attempts to seek out other civilizations.

The Strategy for Searching

The probability arguments outlined in the preceding section are amusing and perhaps somewhat instructive, but obviously not very accurate. There are entirely too many unknowns in the equation for us to develop a reliable estimate of the chances for galactic companionship. Perhaps we will not know for certain what the answer is until we actually make contact with another civilization.

The problem of developing an experiment to search for interstellar signals or to send them is that we do not know the ground rules. There are an infinite number of ways in which a distant civilization might choose to communicate, and we cannot search for all of them. We must try to make our coverage of the possibilities as broad as possible and make reasoned guesses as to the best methods to use.

In view of the power that is transmitted by radio signals and the relative lack of natural noise in the galaxy in that part of the spectrum, it has normally been assumed that radio communications are most likely to succeed, although other techniques have been tried **(Fig. 21.11)**. Several searches for extraterrestrial radio communications have been carried out, starting with *Project Ozma* in the 1960s, in which a large radio telescope at the U.S. National Radio Astronomy Observatory was used to search for signals from the directions of nearby Sun-like stars.

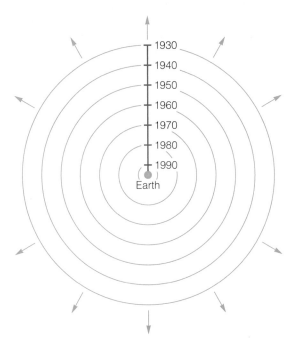

Figure 21.10 Earth's message to the cosmos. As our entertainment and communications broadcasts travel out into space, they provide a history of our culture for anyone who may be receiving the signals. At the present time, the growing sphere that is filled with our broadcasts has a radius of over 50 light-years.

Figure 21.11 Another message from Earth. This recording of a message from Earth is traveling beyond the solar system aboard the *Voyager* spacecraft. Only an advanced race of beings would have the technological skills necessary to learn how to listen and to decode the message of peace that it contains. (NASA)

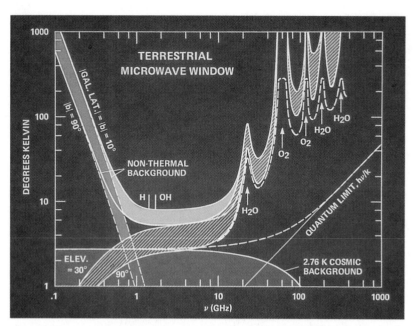

Figure 21.12 **The galactic noise spectrum.** This graph illustrates the relative intensities of various sources of natural background noise, from both astronomical sources and the Earth's atmosphere. As seen here, the region between the lines of hydrogen and the hydroxy radical (OH) has the lowest background noise. For this reason, the so-called water hole has been proposed as a likely spectral region in which interstellar communications might be carried out. (NASA/SETI Institute)

One of the problems faced by anyone wanting to search for extraterrestrial radio communications is obtaining observing time on a radio telescope. Most major radio facilities are already in high demand by astronomers for a wide range of research projects that are more likely to produce significant results; thus it is difficult to persuade the proposal review panels that large blocks of time should be devoted to the search for extraterrestrial civilizations. This problem has generally prevented any large-scale, systematic searches from being conducted, with a couple of exceptions. One project called *Serendip* has worked in a parasitic fashion by attaching a special receiver to radio telescopes carrying out routine astronomical research. The extra receiver searches for intelligent signals at selected frequencies in whatever direction the telescope happens to be pointed. Thus the search does not cover any planned pattern, but simply takes advantage of the "free" observing time in the expectation that at least some of the time the telescope will be pointed toward interesting stars.

Most of the early searches were not only limited in available telescope time, but were also confined to a narrow wavelength band within the radio spectrum. This meant that difficult choices had to be made. Starting with the assumption (or hope) that some

civilizations might be trying to send signals deliberately, some guesswork led to the hypothesis that the signals might be sent at or near the same wavelengths most often observed by radio astronomers. Accordingly, many of the early searches have concentrated on the 21 cm wavelength of atomic hydrogen. The reasoning is that alien scientists would realize the significance of this wavelength and might therefore use it in attempts to send signals that would be noticed by distant colleagues. Another good argument for searching at this wavelength is that many of the world's radio telescopes are already equipped with receivers designed to operate at 21 cm.

Another wavelength that has been considered, but tried only in a limited search, lies between the 21 cm line of hydrogen and a series of lines due to the hydroxyl radical (OH), lying between 17.4 and 18.6 cm. This region is dubbed the "water hole" because both hydrogen and OH are constituents of water. An advantage to searching there is that this region lies just in the wavelength range where natural galactic noise is minimized **(Fig. 21.12).**

Recent advances in computer technology have made it possible to scan a wide range of frequencies and then analyze the data to look for signals at any frequency within the observed range. The quantity of data that must be processed is huge, so the computer programs that do the work are very complex. But it has been possible to automate the process, even including instructions for the computer to send an electronic (E-mail) message to the human on duty if a signal is found! The development of these techniques has largely eliminated the need to try to guess the wavelength at which aliens might be communicating. We simply look for messages at all wavelengths in the microwave portion of the spectrum (still deemed the most likely spectral band due to the low energies involved and the relatively noise-free galactic environment).

A few searches using this new technology have already been conducted. A dedicated search for extraterrestrial civilizations called *META* has been under way for some time at Harvard University. There a small radio telescope has been equipped with a special receiver and a sophisticated computer data-analysis system, so that a wide range of frequencies can be searched. The telescope is devoted completely to the search, and it is expected that thousands of nearby stars will be scrutinized for nonnatural radio signals in years to come. Interestingly, major funding for this project has been provided by movie producer Steven Spielberg, who has been responsible for a number of films about contact with extraterrestrial civilizations.

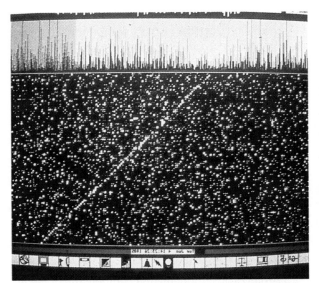

Figure 21.13 A successful detection of an extraterrestrial signal. The clearly visible bright streak across this display of radio intensities represents the signal received from the *Pioneer 10* spacecraft, over 3.5 billion km from the Earth and emitting with a power of only about 1 watt. Each row of dots in the lower-central position of this display represents an individual spectrum, taken at successive times 2 seconds apart. The reason the signal from *Pioneer 10* appears as a slanted line is that the Earth's velocity relative to the spacecraft was changing due to the Earth's orbital motion, so that the signal was Doppler-shifted to a slightly different frequency in each successive spectrum (NASA/SETI Institute)

Figure 21.14 A radio telescope used in the search for life. This is the 64-m Parkes radio telescope, a part of the Australia Telescope, located in New South Wales, Australia. This instrument was used during 1995 and 1996 for the southern-hemisphere portion of the targeted search (pointed observations of selected Sun-like stars) as part of the privately-funded *Project Phoenix*. (Courtesy CSIRO and the Australia Telescope)

In 1993 the U.S. space agency, NASA, initiated a project called the *High Resolution Microwave Survey (HRMS),* which was to use a combination of dedicated telescopes and shared time on other facilities to carry out a search for over a million stars, covering a very broad range in frequencies. The program called for a combination of an all-sky survey over a wide range of frequencies and a smaller, more intensely focused survey of a selected number of solar-type stars that were considered to be especially good candidates. The pointed observations would provide much more sensitive data over a broader spectral region than the all-sky survey. Early tests showed that the detection technique was quite powerful, as the weak signal from the *Pioneer 11* spacecraft, by then well beyond the orbit of Pluto, was detected easily **(Fig. 21.13)**.

The *HRMS* was funded by Congress after years of struggle on the part of its backers (the search for extraterrestrials is an easy target for politicians wanting to show that they can be frugal with the taxpayers' money), but after a year of operations the funding was canceled. The untiring supporters of the *HRMS* have managed to keep the pointed survey going (under the new name *Phoenix*) with donated funds and hope to continue for a few years, achieving at

least a substantial portion of the original goals. As of mid-1996, *Project Phoenix* had completed a several-months-long search using radio telescopes in Australia **(Fig. 21.14)** (with no detections so far).

One pessimistic possibility is that we are all listening, and no one is sending signals. In that case, we can only hope to pick up accidental emissions, such as entertainment broadcasts on radio or television, and these would be much weaker and more difficult to detect. Deliberately sent signals can be detected over much greater distances than accidental transmissions, because more power would be put into a deliberate signal, and it could be directed specifically toward candidate stars, whereas our accidental radio and television signals are broadcast indiscriminately in all directions. Therefore it makes a big difference whether someone out there is trying to send a message or not.

Humans have attempted to send messages. In 1974, a message from Earth was sent from the giant Arecibo radio telescope **(Fig. 21.15)** toward the globular cluster M13, about 25,000 light-years away. The globular cluster was chosen because it contains hundreds of thousands of stars, many of which are similar to the Sun in spectral type, and all of them would be within the beam of the radio telescope's transmission.

Figure 21.15 **The giant Arecibo radio dish.** This is the largest single radio antenna in the world. The 300-meter dish is built into a natural bowl in the mountains of Puerto Rico. Since the telescope cannot be pointed in arbitrary directions, it must rely on the Earth's rotation to scan the sky. (Arecibo Observatory, part of the National Astronomy and Ionosphere center operated by Cornell University under contract with the National Science Foundation) http://universe.colorado.edu/fig/21-15.html

The message consists of a stream of numbers that, when arranged into a two-dimensional array, form a pattern that illustrates schematically such things as the structure of DNA, the form of the human body, Earth's population, and the location of the Earth in the solar system. If this message is received and understood by anyone in M13, it will be some 25,000 years from now, and any answer that they may send back will arrive here about 50,000 years from now.

It is very intriguing that several of the ongoing searches have detected signals that appear to fit the criteria for possible communications from extraterrestrial civilizations. But so far none has been seen twice, even in cases where a repeat observation was performed within minutes of the initial detection.

Until such a signal is seen repeatedly, it is not possible to conclude that extraterrestrial communications have been detected. But scientists around the world are watching with anticipation as the attempts to detect signals from extraterrestrial civilizations continue, knowing that success will be perhaps the most significant discovery of all time.

Summary

1. Although some scientists have suggested that life started on Earth spontaneously or by primitive spores from space, most scientists today accept the theory that life began through natural, evolutionary processes.
2. Amino acids, fundamental components of living organisms, are formed readily in experiments designed to simulate early conditions on the Earth.
3. The steps that led to the development of the necessary forms of RNA and DNA are not yet fully understood and probably occurred over a very long period of time.
4. Fossil evidence provides a record of the evolutionary steps leading from the first primitive lifeforms to modern life.
5. The conditions that prevailed on the early Earth have probably been duplicated on other planets in the galaxy, and possibly for a time earlier in the history of Mars in our own solar system.
6. It is often assumed that only Earth-like life could develop, because carbon is nearly unique in the complexity of its chemistry, but some have suggested that at least one other element (silicon) may have the necessary properties.
7. The habitable zone, the region surrounding a star where a terrestrial planet could sustain liquid water, depends on stellar spectral type. For a combination of reasons having to do with the size and location of the habitable zone and the lifetime and evolutionary behavior of the parent star, Sun-like stars appear to be the most likely to support life-bearing planets.
8. Estimates of the number N of technological civilizations now in the galaxy can be made, with great uncertainty, based on what is known of the formation rate of Sun-like stars, what is guessed for the probability that such stars have planets with the proper conditions, and the probability that life, leading to intelligence and technology, will develop on these planets. Estimates range from $N = 1$ to $N = 10^9$.

Constructing an Alien

It is clear that the forms life has taken here on Earth have been influenced very strongly by local conditions. Our eyes can see only the portion of the electromagnetic spectrum that penetrates the atmosphere; our skeletal systems are of appropriate size and strength to allow motion under one Earth gravity; we breathe the gases that are most common in the Earth's atmosphere; and we derive our energy from food sources that are abundant on the Earth. All of these characteristics of humans and other earthly species would be very different if the Earth itself were very different.

We have shown in this chapter that Earth-like conditions do not exist anywhere else in the solar system (although conditions similar to the early Earth might once have prevailed on Mars), so it is not likely that the solar system harbors any other life-forms similar to us. But this does not rule out the possibility that life of a rather different kind than we are familiar with may exist elsewhere in the solar system.

To stimulate your thinking about this, try writing a description of a possible life-form that might exist on another planet (or a satellite or an interplanetary body) in the solar system. Try to think of realistic ways in which this creature would get food (and what kind of food it might use), and think about how its body chemistry might work. Such factors as the physical state of matter in its environment will come into play; for example, if you choose to invoke a life-form that lives in the Sun's outer layers, you should explain how it remains in solid form even though all matter in that environment is vaporized; conversely, if you choose a very cold environment, you need to think about how the alien's body might transport nutrients internally when everything around it is frozen solid.

You might want to think about motion and how the creature could travel around to catch food or pursue changing environments as the seasons change. What kind of sensory organs might be useful?

Perhaps in doing this exercise, you will hit upon some new methods that astronomers might try as they search for extraterrestrial life. In any event, you will gain a better appreciation for the special circumstances that allowed us to evolve here on the Earth.

9. Several projects have been undertaken to search for radio signals from extraterrestrial civilizations. Most cover a range of frequencies, but so far no complete survey of the sky has been carried out. The most ambitious project, the *HRMS,* was discontinued by Congress, but is continuing in a limited fashion using private funds.

10. The chances for detecting or being detected by other civilizations depend strongly on whether or not deliberate attempts are made to send signals. Incidental transmissions, such as radio and television entertainment broadcasts, are far weaker and are not focused in direction.

Review Questions

1. In this chapter we have discussed two competing theories for the origination of life on Earth: natural evolution from organic materials, and panspermia (the arrival of life on Earth in the form of spores from space). Which better fits the principle of Occam's razor? Explain.

2. To what extent has life on Earth modified the planet's atmosphere? Does this suggest a method for determining, from remote spectroscopic observations, whether a distant planet might have life?

3. What does the evolution of the atmosphere of Venus tell us about the size of the habitable zone around a Sun-like star?

4. Summarize the arguments that lead some scientists to expect life elsewhere to be similar in chemistry to life on Earth.

5. Based on what is assumed about how life started on Earth, do you think it likely that life has developed in the atmosphere of Jupiter? Do you think it could develop in an interstellar cloud? Explain.

6. Recall, from Chapter 17, the distinction between Population I and Population II stars and the location of these stars in the galaxy. Do you think planets orbiting Population II stars

are as likely to have life-forms as planets orbiting Population I stars like the Sun? Explain.

7. Explain in your own words the meaning of the various terms in the Drake equation and how this equation represents the number of technological civilizations in the galaxy at any one time.

8. How might an alien race, located on a planet orbiting a star 10 light-years away, determine whether life exists in the solar system? (*Hint:* You may want to start with the question of detecting planets, which is discussed in Chapter 12, and then consider the question of detecting our civilization over interstellar distances).

9. Why are radio communications thought to be the best method for detecting evidence for other civilizations?

10. What are the factors that govern the choice of wavelength at which to search for signals?

Problems

1. If *I Love Lucy* broadcasts started in 1953, in what year did they first reach Alpha Centauri, about 1.3 pc away? When will they reach Canopus, 30 pc away? When might we expect an answer from Canopus, if anyone is there to receive our signals?

2. Suppose a planet is orbiting a star exactly like the Sun, and the planet's semimajor axis is 0.9 AU. How would the intensity of sunlight on that planet compare with the intensity of sunlight on the Earth? What effect might this have on the possibility that life could develop on the planet?

3. Using existing technology, spacecraft built by humans typically travel at speeds no greater than about 15 km/sec (more than 35,000 miles per hour; this refers to the *Pioneer* and *Voyager* spacecraft, which have escaped the solar gravitational field). At this speed, how long would it take for a spaceship to reach Alpha Centauri, which is about 4 light-years (1.33 pc) away? What does this suggest about reports that alien spacecraft have been making regular visits to the solar system?

4. If an M star has a luminosity equal to 0.01 times the Sun's luminosity, how far from such a star would a planet have to be in order to receive the same intensity of light (in units of energy per square meter per second) that the Earth receives from the Sun? Even if the intensity of radiation would be the same, what would be different about the radiation received by the planet of the M star, as compared with the radiation received by the Earth? How might this difference affect the likelihood that life could form?

5. Repeat the calculation of N, the number of technological civilizations in the galaxy, if the probability of life starting on an Earth-like planet (the term f_l in the Drake equation) is only 10^{-4} instead of 1, as assumed in the text. Assume that the lifetime L is 10^{10} years; then redo the calculation for $L = 10^5$ years.

Web Connections

The Review Questions and Problems also appear at the following URLs:
http://universe.colorado.edu/ch21/questions.html
http://universe.colorado.edu/ch21/problems.html

Additional Readings

Allégre, C. J. and S. H. Schneider 1994. The Evolution of the Earth. *Scientific American* 271(4):66.

Angel, J. R. P. and N. J. Woolf 1996. Searching for Life on Other Planets. *Scientific American* 274(4):60.

Beatty, J. K. 1996. Life from Ancient Mars? *Sky & Telescope* 92(4):18.

Black, D. C. 1991. Worlds Around Other Stars. *Scientific American* 264(1):76.

Calvin, W. H. 1994. The Emergence of Intelligence. *Scientific American* 271(4):100.

Chyba, C. 1992. The Cosmic Origins of Life on Earth. *Astronomy* 20(11):28.

DeDuve, C. 1995. The Beginnings of Life on Earth. *American Scientist* 83(5):428.

Drake, F. 1988. The Pioneer Message Plaques. *Mercury* 17:88.

Drake, F. D. and D. Sobel 1992. Is Anyone Out There? *Mercury* XXI(4):120.

Erwin, D. E. 1996. The Mother of Mass Extinctions. *Scientific American* 275(1):72.

Fredrickson, J. K. and T. C. Onstott. 1996. Microbes Deep Inside the Earth. *Scientific American* 274(4):68.

Goldsmith, D. and T. Owen. 1992. Visitors to Earth? Part I. *Mercury* XXI(4):135.

Goldsmith, D. and T. Owen. 1992. Visitors to Earth? Part II. *Mercury* XXI(5):155.

Gould, S. J. 1994. The Evolution of Life on the Earth. *Scientific American* 271(4):84.

McKay, C. P. 1993. Did Mars Once Have Martians? *Astronomy* 21(9):26.

Naeye, R. 1992. SETI at the Crossroads. *Sky & Telescope* 84(5):507.

Naeye, R. 1996. OK, Where Are They? *Astronomy* 24(7):36.

Orgel, L. E. 1994. The Origin of Life on the Earth. *Scientific American* 271(4):76.

Paque, J. 1995. A Friend for Life? *Astronomy* 23(6):46.

Pendleton, Y. J. and D. P. Cruikshank. 1994. Life From the Stars? *Sky & Telescope* 87(3):36.

Ressmeyer, R. 1995. On the Road: SETI Down Under. *Sky & Telescope* 90(6):26.

Sagan, C. 1994. The Search for Extraterrestrial Life. *Scientific American* 271(4):92.

Schneartzman, D. and L. J. Rickard. 1988. Being Optimistic About the Search for Extraterrestrial Intelligence. *American Scientist* 76:364.

Shostak, S. 1992. Listening for Life. *Astronomy* 20(10):26.

Shostak, S. 1993. The Search Goes On. *Mercury* 23(5):22.

Weinberg, S. J. 1994. Life in the Universe. *Scientific American* 271(4):44.

Appendices

Appendix 1
The Internet
and the World Wide Web

The scientific world has been revolutionized by new technology many times throughout history, and astronomy has been no exception. The application of telescopes to observations of the heavens, the development of space-based observatories, the use of first film and then electronic detectors to record light; all of these caused huge advances, allowing astronomers to observe and analyze phenomena in ways never previously possible. Another revolution, with similar impact, has come with the technology of computers and data analysis, and this has led to the latest major change in our way of doing things: the Internet.

The use of computers in astronomy already has had a huge impact, allowing us to store and analyze data in huge quantities and at high speed. Now computers communicating with each other, the basis of the Internet, have created another revolution for astronomers, providing access to data with unprecedented ease, allowing communication between scientists in ways never previously possible, and becoming an integral part of the daily life of nearly every astronomer.

In this text, we have included the use of the Internet (through its application known as the World Wide Web), because it enhances the experience for the student by helping the reader to appreciate and make use of this revolutionary technology, and because it helps the reader to understand and appreciate the impact of this latest technological revolution on our science.

The Internet

"What is the Internet?" There's no standard answer that completely sums up the Internet. It can be thought of as a physical collection of routers, cables and computers; as an association of common network protocols; or a methodology of interconnecting shared computer resources. Regardless of how one tries to describe it, the Internet is an international meeting ground for millions of computer users.

The World Wide Web

The World Wide Web (or WWW) is an Internet information system based on hypertext. Hypertext is a method for linking information across documents. For example, a hypertext document can highlight a specific word—this highlight is actually a link to a second hypertext document which gives more information about the word. The user can open the second document by simply selecting (clicking on) the highlighted word. The second document may also contain links to further details. The user does not need to know where the referenced documents reside, and there is no need to issue special commands to display them. Cross-referencing or linking of information in hypertext documents can occur within a single document, between two documents on the same computer, or across the globe, accessing hundreds of computers. HyperMedia is a term used for hypertext which is not limited to text. HyperMedia can include text, graphics, animation, video and sound. The World Wide Web uses hypermedia over the Internet. A collection of linked documents on the Internet is called a WWW site, or web site for short.

Content of the WWW

The information that is available on the WWW is entirely a product of the inclinations of individuals around the world. Anyone with an appropriate computer and a link to the Internet can create a Web site and place information there for anyone to access. There is no control, no oversight, and no organization or theme to the WWW. The content of the WWW is therefore dependent entirely on what people think is useful and are willing to make available.

To the user, this lack of overall direction is both helpful and a hindrance. It is helpful because there is complete freedom to disseminate any information without risk of censure, and because it is possible to find information in many wide-ranging areas. It is a hindrance because often there may not be information available on a particular topic of interest, because there is no mechanism to provide overall guidance or organization to WWW content.

With this lack of organization and oversight comes the risk that information on the WWW may be incomplete or inaccurate. Thus the user must understand this risk. You should regard information on the WWW as comparable to the information found in free, self-published books, articles, or pamphlets that people publish for their own reasons.

Most of the WWW sites which provide astronomical information are created and maintained by educators and researchers whose entire lives are dedicated to the expansion of knowledge. We believe and hope, therefore, that most are accurate and reliable.

The *Universe* WWW Site

The WWW site for *Universe: Origins and Evolution* is a set of hypermedia documents. The site resides on a workstation at the Center for Astrophysics and Space Astronomy (CASA), on the University of Colorado

at Boulder campus. The *Universe* web site provides a centralized location for all the WWW information in the text book. Many books now include addresses—or URLs—of sites on the WWW that have information of interest to the reader. But one problem with printing URLs of web sites in a textbook is that these sites can change very rapidly. Information on the web is dynamic, and thus an effective index for this information must also be dynamic. Routing all Web based activities through one centralized, hypermedia site ensures that the WWW information mentioned in the book always remains current and active. The *Universe* web site is also updated frequently, ensuring that the links provided in the book not only remain active, but that the information is up to date. In addition, new links pointing to discoveries and current events are added immediately as this information becomes available. Lastly, by focusing all web based activities through a central web site, we can guide the readers to specific places on the web, tell them what to look for, and most importantly, show them how to utilize this information. With these "guided searchers" the web becomes an effective learning tool.

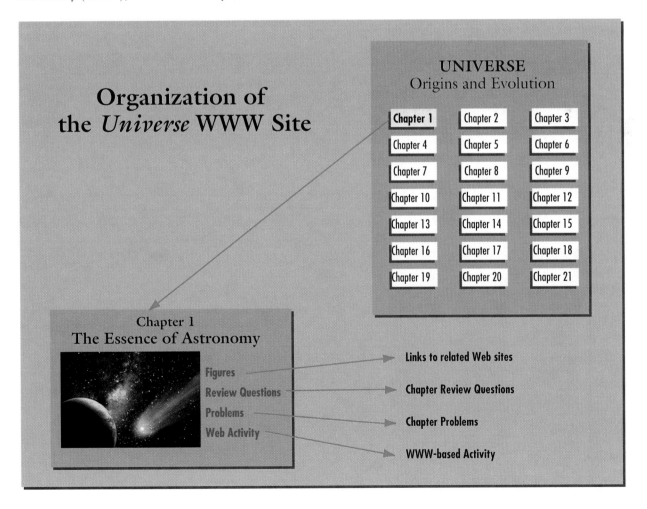

Organization of the *Universe* WWW Site

UNIVERSE
Origins and Evolution

Chapter 1	Chapter 2	Chapter 3
Chapter 4	Chapter 5	Chapter 6
Chapter 7	Chapter 8	Chapter 9
Chapter 10	Chapter 11	Chapter 12
Chapter 13	Chapter 14	Chapter 15
Chapter 16	Chapter 17	Chapter 18
Chapter 19	Chapter 20	Chapter 21

Chapter 1
The Essence of Astronomy

Figures → Links to related Web sites
Review Questions → Chapter Review Questions
Problems → Chapter Problems
Web Activity → WWW-based Activity

Layout of the *Universe* Web Site

The *Universe* site begins with a central home page, which acts like a table of contents for the entire book. The home page has a link to each of the 21 chapters in the book. The link to each chapter begins with an index page for that particular chapter. Each chapter's index page has a list of figures, review questions and problems, and a specific web activity related to theme of the chapter. The layout of each chapter is similar, providing continuity throughout the site.

Schematic of *Universe* Web Site Layout

The list of figures for each chapter contains an index of those figures in the text for which interesting web sites have been found. Each figure page has a digitized image of the figure, and a list of sites on the web that contain related information.

Each chapter's pages in the web site include the questions and problems from the end of the corresponding chapter in the book. A form is supplied which allows the student to answer the questions and problems on the screen, with the option of sending these answers via e-mail to the class instructor. This allows for the possibility of using the WWW as an integral part of the pedagogy of the course.

The web activity for each chapter provides an environment for guided learning on the web. Each activity is based on the general theme of each chapter. The activity provides background information, a list of appropriate sites on the web, and specific information to "discover" at these sites. If appropriate, the student can summarize the results and submit them electronically to the instructor.

Using the WWW

In order to gain access to the World Wide Web, you need to use a computer that is linked to the Internet, either through a telephone system-based modem or by means of a direct connection (an "ethernet" connection), commonly available on college campuses. Your instructor should inform you of how to get "on line" at your school.

The other essential ingredient is a "browser" program, computer software that allows you to link to sites on the WWW. At this writing the program called *Netscape* is the most widely-used, but another, called *Internet Explorer,* is rapidly gaining in popularity. Both work very similarly. Using either one, the procedure for going to a specific site on the WWW is the same: enter the URL for that site in the specified box near the top of the screen, and then hit the enter (return) key on your computer. Next you must wait while the link is established, and once it is, you will see the information and graphics on your screen that are provided by the creators of the site you are accessing.

The major WWW browsers also provide access to search programs, providing a means for you to look for information on the Web in specific areas. As you become familiar with the WWW, you will learn to enjoy the power that comes with knowledge. For better or for worse, the Internet and the World Wide Web appear to be hear to stay, and we can expect them (or their successors) to be a mainstay of life for the foreseeable future.

Appendix 2
Mathematical Applications

Throughout the text are many illustrations of the use of mathematics in developing astronomical concepts. The methods used are limited to algebra and simple geometry, and should not exceed the level of math that you took in high school. In this Appendix we describe some of the most commonly-used mathematical techniques, primarily as an aid in reviewing what you already learned once.

Powers of Ten

In astronomy we must deal with a wide range of numerical values, from the atomic scale to the enormous sizes and distances characterizing the universe at large. In order to represent this wide range of numbers conveniently, we use *scientific notation*, also known as *powers-of-ten notation*. In this system, any number is represented by a multiplier and a power of ten. Here we provide a brief summary of how this works.

Let us start by considering numbers that are whole powers of ten; that is, numbers containing only a 1 and zeros. The number 100 is equal to 10 raised to the second power; i.e. $100 = 10^2$. Similarly, $1000 = 10^3$, $10,000 = 10^4$; and so on. The number 0.1 is the same as $1/10$ or 10^{-1}; $0.01 = 1/100 = 1/10^2 = 10^{-2}$; etc. We can extend this to extreme values, either large or small. For example, 10,000,000,000, which represents approximately the age of the universe in years, can be written as 10^{10}, and the number 0.00000001, which is about the size of an atom in centimeters, can be written as 10^{-8}. The advantage of the powers-of-ten notation becomes clear when we consider these extremely large or small numbers, because it becomes laborious and cumbersome to write out all of the necessary zeros.

To express a number as a power of ten, you need to determine the *exponent*, i.e. the power to which 10 must be raised. For a number larger than 1 (i.e. 10, 100, 100, etc.), the exponent is equal to the number of zeros after the one. For a number smaller than 1, such as 0.1, 0.01, and so on, the exponent is negative, with its value equal to the number of places the 1 is to the right of the decimal point. For exam-

ple, 0.001 has the 1 three places to the right of the decimal, and the exponent is -3. For the number 1, the exponent is zero.

Of course not all values found in nature are simple powers of ten, and this is why we need a multiplier. Consider the number 400. This is 4 times 100, and we can write it in powers-of-ten notation as 4×10^2. Similarly, $6,000,000 = 6 \times 10^6$. Often we need more precision, so the multiplier incudes more digits. For example, to express the number 4,386,000,000,000 in powers-of-ten notation, we retain all of the *significant digits* (the non-zero ones) and write 4.386×10^{12}. Note that we now determine the exponent by counting how many places to the left we moved the decimal point.

For very small numbers, the system is the same: write the number as a multiplier and a power of ten, with a negative exponent in this case. Thus 0.006 becomes 6×10^{-3}; $0.0000000133 = 1.33 \times 10^{-8}$; and so on. Here we count how many places to the right the decimal point is moved in order to determine the (negative) exponent.

Angular Diameter

The *angular diameter* of an object is the angle it appears to cover as seen by an observer (see sketch).

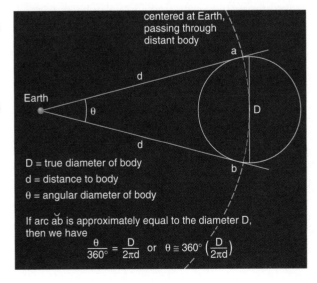

D = true diameter of body
d = distance to body
θ = angular diameter of body

If arc ab is approximately equal to the diameter D, then we have

$$\frac{\theta}{360°} = \frac{D}{2\pi d} \quad \text{or} \quad \theta \cong 360° \left(\frac{D}{2\pi d} \right)$$

The angular diameter is determined by the physical diameter of the object and its distance from the observer. A given object will appear smaller, the more distant it is, and larger, the closer it is. Therefore its angular diameter is inversely proportional to its distance; i.e.

$$\theta \propto 1/d,$$

where θ is the angular diameter and d is the distance.

The angular diameter is also proportional to the true diameter D, so we can write

$$\theta \propto D/d$$

A large object far away can appear the same size as a smaller object that is closer. The Sun and the Moon are a good example of this; they have virtually identical angular diameters, because the Sun's greater distance is offset by its larger physical diameter. The ratio of the Sun's distance to that of the Moon is approximately 390, and the ratio of the Sun's true diameter to that of the Moon is 400. The Moon's distance from the Earth varies a bit because its orbit is not perfectly circular, so at times the distance ratio is more nearly equal to 400 and a perfect total solar eclipse can occur if the Moon passes directly in front of the Sun. (If the Moon is a bit closer to the Earth at the time it passes in front of the Sun, its angular diameter is smaller than that of the Sun and we have an annular eclipse; see Chapter 2.)

The Small Angle Approximation

The proportionality expressed above can become an approximate equality, by considering the portion of a full circle that the physical diameter represents. From the sketch, you should see that

$$\theta \cong 360°(D/2\pi d),$$

where D is again the true (physical) diameter and $2\pi d$ is the circumference of a circle of radius d. This is an approximate equality because the true diameter D is a straight line, and as D becomes large its length departs from the length of a curved segment of the circle. But for the small angular diameters encountered in astronomy (usually less than a degree or so), the approximation is quite valid. It is also valid for other situations (described below), and is generally referred to as the *small angle approximation*.

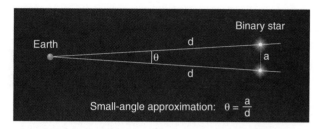

Small-angle approximation: $\theta = \dfrac{a}{d}$

If we divide 360° by 2π, we get 57.2958 . . . ($\cong 57.30$). This is adopted as a new and convenient unit of angular measure, called the *radian*. There are 2π radians in a circle, and 1 rad = 57.30°. Using these units, our expression for angular diameter becomes

$$\theta \cong D/d$$

For example of how this is used, consider the Moon. Its diameter is $D = 3476$ km and its distance (the semimajor axis of its orbit) is $d = 3.84 \times 10^5$ km. Therefore its angular diameter is

$$\theta \cong 3476/3.84 \times 10^5 = 0.0091 \text{ rad} = 0.518°,$$

or about 31 arcminutes. Now you can easily calculate angular diameters for any bodies whose true diameters and distances are known.

The angular diameter is a directly measurable quantity in many cases (particularly for nearby objects such as planets), so it is far more common to use the angular diameter and the physical diameter to find the distance to an object, or the measured angular diameter and the distance to find the physical diameter. For example, suppose a new planet is found and analysis of the data shows that its orbital semimajor axis is 36.87 AU. At opposition its distance from Earth is 35.87 AU (= 5.38×10^9 km) and its angular diameter is measured to be 3.4 arcseconds. Converting this angular diameter to radians requires division by 3600 (to get from arcseconds to degrees) and then by 57.30 (to get from degrees to radians); the answer is 1.7×10^{-5} rad. Now solving our expression above for D, the planet's physical diameter, yields

$$D = \theta d = (1.7 \times 10^{-5})(5.38 \times 10^9 \text{ km}) = 91,500 \text{ km.}$$

The small-angle approximation can be used in many other situations in astronomy. It is particularly useful, for example, when observing binary star systems, and it has applications to the search for planets. For example, if the true separation between the two stars in a binary system is a and the observed angular separation is θ, then the small angle approximation tells us that

$$\theta = a/d,$$

where d is the distance to the binary system (see sketch). We can measure θ in many cases, and we can determine a from Kepler's third law (see Chapter 5), so we can find the distance d from this expression.

The small angle approximation also comes into play in the concept of stellar parallax (Chapter 14). Here we consider the long skinny triangle formed by the Earth, the Sun, and a distant star (see sketch). Using the small angle approximation, we see that

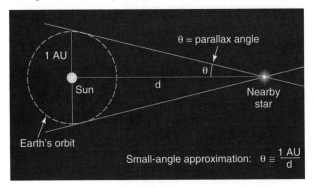

1 AU

θ = parallax angle

θ

Sun

d

Nearby star

Earth's orbit

Small-angle approximation: $\theta \cong \dfrac{1\ \text{AU}}{d}$

$$\theta = (1\ \text{AU})/d,$$

where d is the distance to the star and the angle θ is called the *parallax angle*. If we substitute 1 arcsecond ($= 4.8478 \times 10^{-6}$ rad) for θ and solve for d, we get

$$d = (1\ \text{AU})/(= 4.8481 \times 10^{-6}) = 206,265\ \text{AU}.$$

This defines a unit of distance called the *parsec* (for *parallax-second*), which is commonly used to express astronomical distances (see Chapter 14).

Units and Conversion Factors

In this text we use the *Systéme Internationale*, or *SI* system of units. These are based on the meter as the basic unit of length, the kilogram as the unit of mass, and the second as the unit of time. Thus in most cases throughout the text, you will find these units being used. But there are exceptions where it may be more convenient to visualize quantities if they are expressed in other units. For example, we will generally express densities of planets in terms of grams/cm^3, because water has a density of 1 in these units and comparisons are easy to see. We also tend to use Earth units or solar units to express bulk properties of planets or stars; for example, we refer to the mass of Jupiter as being 318 Earth masses instead of 1.90×10^{27} kg, or we might say that a star has a radius of 50 solar radii rather than spelling out that its radius is 3.5×10^{10} m.

In a few other cases we use non-SI units because astronomers have been reluctant to stop following traditional practice, clinging instead to certain units for historical reasons. An example is the Ångstrom ($= 10^{-10}$ m) as a unit for expressing wavelengths of light, even though physicists have been using the nanometer ($= 10^{-9}$ m) for some time.

The tables below provide conversion factors for several types of units which may come up in your studies of astronomy. You may find these useful as you follow the mathematical sections or work on the problems at the ends of the chapters.

In the tables, exponents (powers of ten) are given in parentheses rather than being displayed as exponents. For example, the radius of the Earth, which is equal to 6.38×10^8 m, is expressed as $6.38(8)$ m. This is done to save space in the tables.

Table A2.1

Length, size

	cm	m	km	$R_{\oplus}$	$R_{\odot}$	AU	l.y.	pc
cm =	1	0.01	1(−5)	1.568(−9)	1.437(−11)	6.68(−14)	1.057(−18)	3.241(−19)
m =	100	1	1(−3)	1.568(−7)	1.437(−9)	6.68(−12)	1.057(−16)	3.241(−17)
km =	10(5)	1000	1	1.568(−4)	1.437(−6)	6.68(−9)	1.057(−13)	3.241(−14)
$R_{\oplus}$ =	6.378(8)	6.378(6)	6.378(3)	1	9.164(−3)	4.263(−5)	6.742(−10)	2.067(−10)
$R_{\odot}$ =	6.96(10)	6.96(8)	6.96(5)	109.1	1	4.652(−3)	7.357(−8)	2.256(−8)
AU =	1.496(13)	1.496(11)	1.496(8)	2.345(4)	215	1	1.581(−5)	4.848(−6)
l.y. =	9.461(17)	9.461(15)	9.469(12)	1.483(9)	1.359(7)	6.324(4)	1	0.3066
pc =	3.086(18)	3.086(16)	3.086(13)	4.838(9)	4.433(7)	206,265	3.262	1

Table A2.2
Wavelength

	Å	nm	μm	mm	cm	m
Å =	1	0.1	1(−4)	1(−7)	1(−8)	1(−10)
nm =	10	1	0.001	1(−6)	1(−7)	1(−9)
μm =	1(4)	1000	1	0.001	1(−4)	1(−6)
mm =	1(7)	1(6)	1000	1	0.1	0.001
cm =	1(8)	1(7)	1(4)	10	1	0.01
m =	1(10)	1(9)	1(6)	1000	100	1

Table A2.3
Mass

	g	kg	$M_\oplus$	$M_\odot$
g =	1	0.001	1.674(−28)	5.027(−34)
kg =	1000	1	1.674(−25)	5.027(−31)
$M_\oplus$ =	5.974(27)	5.974(24)	1	3.003(−6)
$M_\odot$ =	1.989(33)	1.989(30)	3.329(5)	1

Table A2.4
Energy, Power

	erg	J	erg/sec	W (J/sec)	$L_\odot$
erg =	1	1(−7)	—	—	—
J =	1(7)	1	—	—	—
erg/sec =	—	—	1	1(−7)	2.613(−34)
W =	—	—	1(7)	1	2.613(−27)
$L_\odot$ =	—	—	3.827(33)	3.827(26)	1

E. Equations and Formulas

There are a few simple equations and formulas that you will need to know to solve the chapter-end problems. Many of them are given in the text or with the problems, but it may be useful to collect them here. Here they are:

Table A2.5
Geometric formulas

Circumference of a circle: $C = 2\pi R$, where R is the radius.
Area of a circle: $A = \pi R^2$.
Surface area of a sphere: $A = 4\pi R^2$.
Volume of a sphere: $V = (\frac{4}{3}\pi R^3$.
Density: $\rho = M/V$, where M is mass, V is volume.
Density of a spherical body: $\rho = 3M/4\pi R^3$.

Table A2.6
Orbits

Center of mass: $m_1 r_1 = m_2 r_2$, where m_1 and m_2 are two masses orbiting at distances r_1 and r_2 from a common center of mass.

Eccentricity of an elliptical orbit: $e = \overline{F_1 F_2}/2a$, where $\overline{F_1 F_2}$ is the distance between the two foci and a is the semimajor axis.

Perihelion (closest approach to the Sun or central body): $P = a(1 - e)$.

Aphelion (greatest distance from Sun or central body): $A = a(1 + e)$.

Relation between synodic and sidereal period for an inferior planet: $1/S = 1/P_i - 1$, where S is the synodic period, and P_i is the sidereal period of the inner planet.

Relation between synodic and sidereal period for a superior planet: $1/S = 1 - 1/P_o$, where P_o is the sidereal period of the outer planet.

Table A2.7
Laws of Gravitation

Law of universal gravitation: $F = Gm_1 m_2/d^2$, where F is the force between masses m_1 and m_2, d is the distance between them, and G is the gravitational constant.

Surface gravity: $g = GM/R^2$.

Escape speed: $v_{esc} = \sqrt{2GM/R}$.

Kepler's third law: $P^2 = a^3$, where P is the sidereal period in years, a is semimajor axis in AU, and one of the two bodies in the Sun while the other is much less massive; or

$(m_1 + m_2)P^2 = a^3$, where m_1 and m_2 are the masses of the two bodies in solar units; or

$(m_1 + m_2)P^2 = \left(\frac{4\pi^2}{G}\right)a^3$, where the masses are in kg, the period is in seconds and the semimajor axis is in meters.

Table A2.9
Stellar and Galactic Properties

Stellar parallax: $d = 1/p$, where d is the distance (in parsecs) to a star whose parallax angle (in arcseconds) is p.

Schwarzschild radius: $R = 2GM/c^2$, where M is the mass of a collapsed object, G is the gravitational constant, and c is the speed of light.

Hubble expansion law: $v = Hd$, where v is the recession velocity of a galaxy or quasar, d is its distance (in Mpc), and H is the Hubble constant (in km/sec/Mpc).

Table A2.8
Light and Radiation Laws
(Note: relations using stellar magnitudes are in Appendix 13)

Wavelength and frequency of a photon: $\lambda = c/f$, where λ is the wavelength, f is the frequency, and c is the speed of light.

Energy of a photon: $E = hf = hc/\lambda$, where E is the photon energy and h is the Planck constant.

Wien's law: $\lambda_{max} = W/T$, where λ_{max} is the wavelength of maximum emission, T is the surface temperature, and W is Wien's constant.

Stefan's law: $E = \sigma T^4$, where E is the energy emitted per m2 of surface area and σ is the Stefan-Boltzmann constant.

Stefan-Boltzmann law: $L = 4\pi R^2 \sigma T^4$, where L is the luminosity (power).

Inverse square law: $I_1/I_2 = (d_2/d_1)^2$, where I_1 and I_2 represent the intensity of light from the same source at distances d_2 and d_1, respectively.

Doppler shift: $\Delta\lambda/\lambda = v/c$, where $\Delta\lambda$ is the shift in wavelength (observed wavelength minus laboratory or rest wavelength λ), v is the radial velocity, and c is the speed of light.

Relativistic Doppler shift: $\Delta\lambda/\lambda = z = \sqrt{\dfrac{1+v/c}{1-v/c}} - 1$.

Recession velocity, relativistic redshift: $v = c\left[\dfrac{(z+1)^2 - 1}{(z+1)^2 + 1}\right]$.

Diffraction limit for a telescope: $\theta = 1.22\lambda/D$, where θ is the smallest angle (in radians) that can be resolved, λ is the observed wavelength, and D is the telescope diameter. If θ is expressed in arcseconds, then the relationship is $\theta = (2.52 \times 10^5)\lambda/D$.

Proportionality and Ratios

When doing numerical calculations using the formulas above, it is often convenient to reduce the problem to ratios. By doing this, you can eliminate numerical constant terms, simplifying the calculation, and in the process gain a clearer picture of the proportionalities involved. A good example to illustrate this is the Stefan-Boltzmann law, normally expressed as

$$L = 4\pi R^2 \sigma T^4,$$

where L is the luminosity, R is the radius, and T the surface temperature of a spherical object such as a planet or star. To compute a luminosity from this expression requires not only knowing the values for R and T, but also the constants π and σ. Including these constant terms is not difficult, and sometimes it is necessary, but often it can be avoided.

For example, if you want to compare luminosities of two bodies, you can compute the luminosities separately and then divide the answers to get the ratio, or you can write expressions for the two bodies and then divide *before* doing any calculations. Let's work one such example both ways.

Suppose you want to know the luminosity ratio of a star to the Sun. You know that the star has a radius of 2.3×10^9 m and a surface temperature of 3880 K. Inserting these values into the Stefan-Boltzmann expression and solving, we find

$$L_* = 4(3.14)(2.3 \times 10^9)^2(5.67 \times 10^{-8})(3880)^4 = 8.54 \times 10^{26} \text{ W}.$$

Similarly, for the Sun, whose radius is approximately 7.0×10^8 m and whose surface temperature is 5780 K, we find

$$L_\odot = 4(3.14)(7.0 \times 10^8)^2(5.67 \times 10^{-8})(5780)^4 = 3.89 \times 10^{26} \text{ W}.$$

Now we can find the ratio:

$$L_*/L_\odot = \frac{(8.54 \times 10^{26})}{(3.89 \times 10^{26})} = 2.2.$$

The star is 2.2 times more luminous than the Sun.

Now let's try dividing the equations for the star and the Sun before doing any calculations. Using $*$ subscripts for the star and $\odot$ for the Sun, we find

$$L_*/L_\odot = \frac{(4\pi R_*^2 \sigma T_*^4)}{(4\pi R_\odot^2 \sigma T_\odot^4)} = \frac{(R_*/R_\odot)^2}{(T_*/T_\odot)^4}.$$

The numerical constant terms have cancelled out, and we have reduced the problem to simply inserting the ratios of stellar radii and temperatures, then solving. The ratio of radii is $R_*/R_\odot = (2.3 \times 10^9)/(7.0 \times 10^8) = 3.29$ and the ratio of temperatures is $T_*/T_\odot = (3880)/(5780) = 0.67$. Putting these ratios into our expression leads to

$$L_*/L_\odot = (3.29)^2(0.67)^4 = 2.2.$$

We got the same answer, in essentially one step. Furthermore, by dividing out the constant terms, we made it easier to see the important proportionalities. In addition, we could have solved this problem without even knowing the values of the constants, thus avoiding the need to look them up.

The same kind of economy can be gained in the use of other mathematical relations. For example, in applying the inverse square law to find the relative intensity of light at different distances from a source, you do not need to know the luminosity of the source and then calculate the intensity of light at each distance. Instead all you have to do is find the ratio of the distances, invert it, and square the result. To compare surface gravities of two planets, you need only know the ratios of their masses and radii. You will find it useful to keep this technique in mind, because in many of the mathematical examples and problems in this text, it is ratios that you need to find, rather than absolute values.

Appendix 3
Symbols Commonly Used in this Text

Table A3.1

Symbol	Meaning	Symbol	Meaning
Å	Angstrom, a unit of length often used to measure wavelengths of light; 1 Å $= 10^{-10}$ m	$\Delta\lambda$	The Greek letters delta lambda, used to designate a shift in wavelength, as in the Doppler effect
c	Standard symbol for the speed of light	γ	The Greek letter gamma, sometimes used to designate a gamma-ray photon
G	Standard symbol for the gravitational constant		
H_o	Symbol for the constant in the Hubble expansion law	λ	The Greek letter lambda, usually used to designate wavelength
h	Standard symbol for the Planck constant		
K	Kelvin, the unit of temperature in the absolute scale	υ	The Greek letter nu, the standard symbol for frequency, also used to designate a neutrino in nuclear reactions
z	Symbol commonly used to designate the Doppler shift ($z = \Delta\lambda/\lambda = v/c$ for velocities much less than the speed of light)	π	The Greek letter pi, usually used to designate the parallax angle; also used for the ratio of the circumference of a circle to its diameter
α	The Greek letter alpha, sometimes used to designate an alpha particle		

Appendix 4
Physical and Mathematical Constants

Table A3.1

Constant	Symbol	Value
Speed of light	c	2.9979249×10^8 m/sec
Gravitation constant	G	6.6720×10^{-11} N·m^2/kg^2
Planck constant	h	6.62618×10^{-34} J·sec
Electron mass	m_e	9.10953×10^{-31} kg
Proton mass	m_p	1.67265×10^{-27} kg
Stefan-Boltzmann constant	σ	5.67032×10^{-8} W/m^{2}kg^4
Wien constant	W	0.00289776 m·deg
Boltzmann constant	k	1.38066×10^{-23} J/deg
Astronomical unit	AU	1.49599×10^8 km
Parsec	pc	3.085678×10^{13} km $= 3.261633$ light years
Light-year	ly	9.460530×10^{12} km
Solar mass	$M_\odot$	1.9891×10^{30} kg
Solar radius	$R_\odot$	6.9600×10^8 m
Solar luminosity	$L_\odot$	3.827×10^{26} W
Earth mass	$M_\oplus$	5.9742×10^{24} kg
Earth radius	$R_\oplus$	6378.140 km
Tropical year (equinox to equinox)		365.241219878 days
Sidereal year (with respect to stars)		365.256366 days $= 3.155815 \times 10^7$ sec

Appendix 5
Temperature Scales

At the most basic level, temperature can be defined in terms of the motion of particles in a gas (or a solid, or a liquid). We all have an intuitive idea of what heat is, and we are all familiar with at least one scale for measuring temperature.

The most commonly used scales are somewhat arbitrarily defined, with zero points not representing any truly fundamental physical basis. The popular *Fahrenheit* scale, for example, has water freezing at a temperature of 32°F and boiling at 212°F. On this scale absolute zero, the lowest possible temperature (where all molecular motions cease), is −459°F.

The centigrade (or Celsius) scale is perhaps better founded, although it is based on the freezing and boiling points of water, rather than the more fundamental absolute zero. In this system, the freezing point is defined as 0°C, and 100°C is the boiling point. This scale has the advantage over the Fahrenheit scale that there are exactly 100° between the freezing and boiling points, rather than 180°, as in the Fahrenheit system. To convert from Fahrenheit to centigrade, we must subtract 32° first, and then multiply the remainder by 100/180, or 5/9. In equation form, we have $T_c = \frac{5}{9}(T_f - 32)$, where T_c is the centigrade temperature, and T_F is the same temperature in degrees Fahrenheit. For example, 50°F on the centigrade scale is

$$T_c = \frac{5}{9}(50 - 32) = 10°C.$$

To convert from centigrade to Fahrenheit, the equation is

$$T_F = \frac{9}{5}T_c + 32).$$

The temperature scale preferred by scientists is a modification of the centigrade system. In this system, named after its founder, the British physicist Lord Kelvin, the same degree is used as in the centigrade scale; that is, one degree is equal to one one-hundredth of the difference between the freezing and boiling points of water. The zero point is different from the zero point on the centigrade scale, however; it is equal to absolute zero. Hence on this scale water freezes at 273°K and boils at 373°K. Comfortable room temperature is around 300°K. In modern usage the degree symbol (°) is dropped, and we speak simply of temperatures in units of Kelvins (273 K, for example).

Appendix 6
Radiation Laws

Several laws described in Chapter 6 apply to continuous radiation from hot objects such as stars. In the text these laws were discussed in general terms, and a few simple applications were explained. Here the same laws are given in more precise mathematical form, and their use in that form is illustrated.

Wien's Law

In general terms, Wien's law says that the wavelength of maximum emission from a glowing object is inversely proportional to its temperature. Mathematically, this can be written as

$$\lambda_{max} \propto 1/T,$$

where λ_{max} is the wavelength of strongest emission, T is the surface temperature of the object, and $\propto$ is a special mathematical symbol meaning "is proportional to."

Experimentation can determine the *proportionality constant*, specifying the exact relationship between λ_{max} and T, and Wien did this, finding

$$\lambda_{max} = W/T,$$

where W has the value of 0.0029 if λ_{max} is measured in meters and T in degrees absolute. With this equation it is possible to calculate λ_{max}, given T, or vice versa. Thus, if we measure the spectrum of a star and find that it emits most strongly at a wavelength of $2000 \text{ Å} = 2 \times 10^{-7}$ m, then we can solve for the temperature

$$T = W/\lambda_{max} = .0029/2 \times 10^{-7} = 14,500 \text{ K.}$$

This is a relatively hot star, and it would appear blue-white to the eye. Note that it was necessary to measure the spectrum in ultraviolet wavelengths in order to find λ_{max}.

When solving problems using Wien's law, it is always possible to use the equation form, as we have just done in this example. Often, however, it is more convenient to compare the properties of two objects by considering the ratio of the temperatures or of the wavelengths of maximum emission. In effect this is what we did in the text when we compared two objects of different temperatures in order to determine how their λ_{max} values compared. For example, we said that if one object is twice as hot as another, its value of λ_{max} is half that of the other.

We can see how this ratio technique works by writing the equation for Wien's law separately for object 1 and object 2:

$$\lambda_{max\ 1} = \frac{W}{T_1}, \text{ and } \lambda_{max2}\ \frac{W}{T_2}.$$

Now we can divide one equation by the other:

$$\frac{\lambda_{max1}}{\lambda_{max\ 2}} = \frac{W/T_1}{W/T_2}$$

or

$$\frac{\lambda_{max1}}{\lambda_{max\ 2}} = \frac{T_2}{T_1}$$

The numerical factor W has canceled out, and we are left with a simple expression relating the values of λ_{max} and T for the two objects. Now we see that if $T_1 = 2T_2$ (object 1 is twice as hot as object 2), then

$$\frac{\lambda_{max1}}{\lambda_{max\ 2}} = \frac{T_1}{2T_2} = \frac{1}{2},$$

or λ_{max} for object 1 is one-half that for object 2. In this extremely simple example, it probably would have been easier just to work it out in our heads, but what if we have a case where one object is 3.368 times hotter than the other, for example?

A great deal can be learned from making comparisons in this way. Astronomers often use the Sun as the standard for comparison, expressing various quantities in terms of the solar values. Another reason for using comparisons occurs when the numerical constants (such as Wien's constant in the foregoing examples) are not known. If the trick of comparing is kept in mind, it is often possible to work out answers to astronomical questions simply by carrying around in one's head a few numbers describing the Sun.

The Stefan-Boltzmann Law

As discussed in the text, the energy emitted by a glowing object is proportional to T^4, where T is the surface temperature. This can be written mathematically as

$$E = \sigma T^4,$$

where σ stands for a proportionality constant that has the value 5.7×10^{-8} in the SI system.

If we now consider how much surface area a star has, then the total energy it emits, called its *luminosity* and usually denoted L, is

$$L = \text{surface area} \times E$$
$$= 4\pi R^2 \sigma T^4,$$

where R is the radius of the star. This equation is called the Stefan-Boltzmann law.

As in other cases, the law can be used directly in this form, or we can choose to compare properties of stars by writing the equation separately for two stars and then dividing. If we do this, we find

$$\frac{L_1}{L_1} = \frac{R_1^2 T_1^4}{R_2^2 T_2^4} = \left(\frac{R_1}{R_2}\right)^2 \left(\frac{T_1}{T_1}\right)^4.$$

The constant factors 4π and σ have canceled out.

As an example of how to use this expression, suppose we determine that a particular star has twice the temperature of the Sun but only one-half the radius, and we wish to know how this star's luminosity compares to that of the Sun. If we designate the star as object 1 and the Sun as object 2, then

$$\frac{L_1}{L_2} = \left(\frac{1}{2}\right)^2 (2)^4 = \frac{1}{4} \times 16 = 4.$$

This star has 4 times the luminosity of the Sun.

The Stefan-Boltzmann law is particularly useful because it relates three of the most important properties of stars to each other.

The Planck Function

The radiation laws described above and in the text are actually specific forms of a much more general law, discovered by the great German physicist Max Planck. Wien's law, the Stefan-Boltzmann law, and some others not mentioned in the text bear the same kind of relation to Planck's law as the laws of planetary motion discovered by Kepler do to Newton's mechanics. Kepler's laws were first discovered by observation, but with Newton's laws of motion it is possible to derive Kepler's laws theoretically. In the same fashion, the radiation laws discussed so far were found experimentally but can be derived mathematically from the much more general and powerful Planck's law.

Planck's law is usually referred to as the Planck function, a mathematical relationship between intensity and wavelength (or frequency) that describes the spectrum of any glowing object at a given temperature. The Planck function specifically applies only to objects that radiate solely because of their temperature and do not reflect any light or have any spectral lines. The popular term for such an object is *blackbody,* and we often refer to radiation from such an object as *blackbody radiation* or, more commonly, *thermal radiation,* the term used in the text. Stars are not perfect radiators, but to a good approximation can be treated as such; hence in the text we apply Wien's law and the Stefan-Boltzmann law to stars (and even planets in certain circumstances) without pointing out that to do so is only approximately correct.

The form of the Planck function for the radiation intensity B as a function of wavelength is

$$B = \frac{2hc^2}{\lambda^5} \frac{1}{e^{hc/\lambda kT} - 1},$$

where h is the Planck constant, c is the speed of light, and k is the Boltzmann constant (the values of all three are tabulated in Appendix 4), λ is the wavelength (in m), and T is the temperature of the object (on the absolute scale). The symbol e represents the base of the natural logarithm, something not used elsewhere in this text; for the present purpose, this may be regarded simply as a mathematical constant with the value 2.718.

In terms of frequency ν rather than wavelength, the expression is

$$B = \frac{2h\nu^3}{c^2} \frac{1}{e^{h\nu/kT} - 1}.$$

Either expression may be used to calculate the spectrum of continuous radiation from a glowing object at a specific temperature. In practice, the Planck function is used in a wide assortment of theoretical calculations that call for knowledge of the intensity of radiation so that its effects can be assessed on physical conditions such as ionization.

Appendix 7
Major Telescopes and Space Probes *

Table A7.1

Optical Telescopes (Ground-based)

Observatory	Location[a]	Telescope
European Southern Observatory	(Cerro La Silla, Chile)	16-m *Very Large Telescope* (four 8-m instruments)
Steward Observatory	(Mt. Graham, Arizona)	11.8-m Large Binocular Telescope (28.4-m telescopes)
Keck Observatory (California Institute of Technology and the University of California)	Mauna Kea, Hawaii Mauna Kea, Hawaii	10.0-m Keck I Telescope 10.0-m Keck II Telescope
McDonald Observatory	(Fort Davis, Texas)	9-2m Hobby-Eberly Telescope
National Optical Astronomy Observatories	(Mauna Kea, Hawaii)	8-m Gemini North Telescope
National Optical Astronomy Observatories	(Cerro Tololo, Chile)	8-m Gemini South Telescope
Japanese National Observatory	(Mauna Kea, Hawaii)	7-m Subaru Telescope
Smithsonian Observatory, U. of Arizona	Mount Hopkins, Arizona	6.5-m (former 4.5-m Multiple Mirror Telescope)
Smithsonian Observatory, U. of Arizona	Las Campanas, Chile	6.5-m Magellan Telescope
Special Astrophysical Observatory	Mount Pastukhov, USSR	6-m Bol'shoi Teleskop Azimutal'nyi
Hale Observatories	Palomar Mountain, California	5.08-m George Ellery Hale Telescope
Royal Greenwich Observatory	La Palma, Canary Islands	4.2-m William Herschel Telescope
Cerro Tololo Observatory	Cerro Tololo, Chile	4.0-m Victor Blanco Telescope
University of North Carolina and Columbia University	(Cerro Pachon, Chile)	4.0-m SOAR Telescope
Anglo-Australian Observatory	Siding Spring Mountain, Australia	3.9-m Anglo-Australian Telescope
Kitt Peak National Observatory	Kitt Peak, Arizona	3.8-m Nicholas U. Mayall Telescope
Royal Observatory Edinburgh	Mauna Kea, Hawaii	3.8-m United Kingdom Infrared Telescope
Canada-France-Hawaii Observatory	Mauna Kea, Hawaii	3.6-m Canada-France-Hawaii Telescope
European Southern Observatory	Cerro La Silla, Chile	3.57-m
Max Planck Institute (Bonn)	Calar Alto, Spain	3.5-m
Apache Point Observatory	Apache Point, New Mexico	3.5-m
University of Wisconsin, Indiana University, Yale University, and the National Optical Astronomy Observatories	Kitt Peak, Arizona	3.5-m WIYN Telescope
European Southern Observatory	Cerro La Silla, Chile	3.5-m New Technology Telescope
Lick Observatory	Mount Hamilton, California	3.05-m C. Donald Shane Telescope
Mauna Kea Observatory	Mauna Kea, Hawaii	3.0-m NASA Infrared Telescope

[a]Telescopes planned or under construction have their locations indicated in parentheses.

Table A7.2

Radio and Submillimeter Telescopes

Instrument	Location	Description
VLBA	North America	10 25-m antennae; interferometry; operated by NRAO
VLA	New Mexico	21 25-m antennae; interferometry; operated by NRAO
Westerbork synthesis Radio Telescope	Netherlands	14 25-m antennae; interferometry
Australia Telescope	New South Wales	5-antenna interferometer at Narrabri; Parkes 64-m dish; Mopra 22-m antenna
Arecibo Radio Telescope	Puerto Rico	305-m fixed dish; operated by Cornell University
Green Bank Telescope	(West Virginia)	100×110-m steerable dish; NRAO
Max Planck Radio Telescope	Effelsburg, Germany	100-m steerable dish; operated by the Max Planck Institute for Radio Astronomy
Nobeyama Radio Observatory	Japan	45-m dish; array of 6 10-m millimeter-wave antennae
Haystack Observatory	Massachusetts	37-m, 18-m dishes; 8-dish array; operated by MIT
Berkeley-Illinois-Maryland Array (BIMA)	Hat Creek, California	9-antenna millimeter-wavelength array for interferometry
Caltech Submillimeter Observatory	Mauna Kea, Hawaii	10.4-m dish
Submillimeter Array	(Mauna Kea, Hawaii)	6 6-m antennae; Harvard-Smithsonian Center for Astrophysics
Caltech Millimeter Array	Owens Valley, CA	6 10.4-m antennae
James Clerk Maxwell Telescope	Mauna Kea, Hawaii	15-m dish; operated by U.K. Joint Astronomy Centre
Ryle Telescope	Cambridge, UK	8-antenna array; interferometry
Dominion Radio Astrophysical Observatory Research Council	Penticton, B.C.	7-dish array; interferometry; supported by Canadian National
Five College Radio Observatory	Massachusetts	14-m dish; millimeter wavelengths
IRAM Telescope	Pico Valeta, Spain	30-m dish; Institute of Millimeter Radio Astronomy (France, Spain, Germany)
NRAO Millimeter Observatory	Kitt Peak, Arizona	12-m dish

Table A7.3

Space-based Telescopes*

Spacecraft	Sponsor	Years of Operation	Purpose
OAO-2	NASA	1969–1972	UV spectroscopy
Uhuru	NASA	1970–1973	X-ray imaging
Copernicus (OAO-3)	NASA	1972–1980	UV spectroscopy
IUE	NASA/ESA/UK	1978–	UV spectroscopy
Einstein Observatory	NASA	1978–1981	X-ray imaging, spectra
Exosat	ESA	1982–1986	X-ray imaging, spectra

IRAS	Dutch/UK/NASA	1983–1984	Infrared mapping
COBE	NASA	1989–	Far-IR, mm-wave mapping
Hubble Space Telescope	NASA/ESA	1990–	Optical/UV/IR imaging, spectra
ROSAT	NASA/Germany	1990–	X-ray imaging, spectra
Hipparcos	ESA/NASA	1989–	Visible astrometry
Compton Observatory	NASA	1991–	Gamma ray survey, spectra
EUVE	NASA	1992–	Extreme UV imaging
ASCA	Japan/NASA	1993–	X-ray imaging, spectroscopy
XTE	NASA	1995–	X-ray imaging, time variations
ISO	ESA/NASA	1995–	Infrared spectroscopy
FUSE	NASA/Canada/France	(1998)	Far-UV spectroscopy
AXAF	NASA	(1999)	X-ray imaging, spectra
SOFIA (airborne)	NASA	(1998)	Multipurpose IR observatory
XMM	ESA/NASA	(2000)	X-ray imaging, spectra
SIRTF	NASA	(2001)	IR imaging, spectra

*The operational years are in parentheses for spacecraft not yet launched at the time of this writing.

Table A7.4

Planetary and Solar Probes*

Spacecraft	Sponsor	Launch	Arrival	Remarks
Mercury				
Mariner 10	NASA	11/03/73	3/29/74	Trajectory allowed three working flybys
			9/21/74	
			3/16/75	
Venus				
Mariner 2	NASA	8/26/62	12/14/62	Flyby
Venera 4	Soviet	6/12/67	10/18/67	Atmosphere probe
Mariner 5	NASA	6/14/67	10/19/67	Flyby
Venera 5	Soviet	1/05/69	5/16/69	Atmosphere probe
Venera 6	Soviet	1/10/69	5/17/69	Atmosphere probe
Venera 7	Soviet	8/17/70	12/15/70	Lander, 23 minutes of data returned from surface
Venera 8	Soviet	3/27/72	7/22/72	Lander (50 minutes)
Mariner 10	NASA	11/03/73	2/05/74	Flyby
Venera 9	Soviet	6/08/75	10/22/75	Orbiter and lander
Venera 10	Soviet	6/14/75	10/25/75	Orbiter and lander
Pioneer/Venus:				
Orbiter	NASA	5/20/78	12/04/78	Operated until late 1992
Multiprobe	NASA	8/08/78	12/09/78	Five atmosphere probes
Venera 11	Soviet	9/09/78	12/21/78	Flyby and lander
Venera 12	Soviet	9/14/78	12/25/78	Flyby and lander
Venera 13	Soviet	10/30/81	3/01/82	Lander
Venera 14	Soviet	11/04/81	3/05/82	Lander

Venera 15	Soviet	6/02/83	10/10/83	Orbiter with imaging radar
Venera 16	Soviet	6/07/83	10/14/83	Orbiter with imaging radar
VEGA 1	Soviet	12/15/84	6/11/85	Lander and balloon atmosphere probe
VEGA 2	Soviet	12/21/84	6/15/85	Lander and balloon atmosphere probe
Magellan	NASA	5/4/89	8/90	High-resolution radar
Galileo	NASA	10/18/89	2/90	Flyby on way to Jupiter
Mars				
Mariner 4	NASA	11/28/64	7/14/65	Flyby
Mariner 6	NASA	2/25/69	7/31/69	Flyby
Mariner 7	NASA	3/27/69	8/05/69	Flyby
Mariner 9	NASA	5/30/71	11/13/71	Orbiter (ceased functioning on October 27, 1972)
Mars 2	Soviet	5/19/71	11/27/71	Orbiter and lander (lander returned no data)
Mars 3	Soviet	5/28/71	12/02/71	Orbiter and lander (lander returned no data)
Mars 5	Soviet	7/25/73	2/12/74	Orbiter and lander (lander failed within seconds of touchdown)
Mars 7	Soviet	8/09/73	3/09/74	Orbiter (lander missed planet)
Mars 6	Soviet	8/05/73	3/12/74	Orbiter and lander (lander crashed)
Viking 1	NASA	8/20/75	6/19/76	Orbiter (ceased functioning on August 17, 1980)
			7/20/76	Lander (ceased operating November 1982)
Viking 2	NASA	9/09/75	8/07/76	Orbiter (ceased functioning on July 24, 1978)
			9/03/76	Lander (ceased functioning on April 12, 1980)
Phobos 2	Soviet	7/88	6/89	Lost contact with Earth and failed March 27, 1989, after two months in orbit gathering data
Mars Global Surveyor	NASA	11/96	(9/97)	Orbiter
Mars Pathfinder	NASA	12/96	(7/97)	Surface rover
Jupiter				
Pioneer 10	NASA	3/03/72	12/03/73	Flyby
Pioneer 11	NASA	4/05/73	12/02/74	Flyby
Voyager 1	NASA	9/05/77	3/05/79	Flyby
Voyager 2	NASA	8/20/77	7/09/79	Flyby
Galileo	NASA	10/18/89	12/07/95	Orbiter and atmosphere probe
Saturn				
Pioneer 11	NASA	4/05/73	9/01/79	Flyby
Voyager 1	NASA	9/05/77	11/12/80	Flyby
Voyager 2	NASA	8/20/77	8/25/81	Flyby
Cassini	NASA/ESA	(10/97)	(6/01)	Orbiter + Titan probe
Uranus				
Voyager 2	NASA	8/20/77	1/24/86	Flyby
Neptune				
Voyager 2	NASA	8/20/77	8/24/89	Flyby
Pluto				
Pluto Express	NASA	(1998)	(2006–8)	Flyby

Appendix 8
Planetary and Satellite Data

Orbital Data for the Planets

Planet	Sidereal Period	Semimajor Axis	Orbital Eccentricity[a]	Inclination of Orbital Plane	Rotation Period	Tilt of Axis
Mercury	0.241 yr	0.387 AU	0.2056	7°0'15"	58^d.65	28°
Venus	0.615	0.723	0.068	3°23'40"	243	3°
Earth	1.000	1.000	0.0167	0°0'0"	23^{h}56^m	23°27'
Mars	1.881	1.524	0.0934	1°51'0"	24^{h}37^m	23°59'
Jupiter	11.86	5.203	0.0485	1°18'17"	9^{h}55^{m}5	3°5'
Saturn	29.46	9.555	0.0556	2°29'33"	10^{h}39^{m}4	26°44'
Uranus	84.01	19.22	0.0472	8°46'23"	17^{h}14^{m}4	97°52'
Neptune	164.79	30.11	0.0086	1°46'22"	1606.6	29°34'
Pluto	248.5	39.44	0.250	17°10'12"	6^{d}387	122.5°

[a]The eccentricity of an orbit is defined as the ratio of the distance between the foci to the semimajor axis. In practice, it is related to the perihelion distance P and the semimajor axis a by $P = a(1 - e)$, where e is the eccentricity; and to the aphelion distance A by $A = a(1 + e)$.

Physical Data for the Planets

Planet	Mass[a]	Average, 1 Bar Diameter[a]	Density	Surface Gravity	Escape Speed	1 Bar Temperature	Albedo
Mercury	0.0558	0.381	5.50 g/cm^3	0.38 g	4.3 km/sec	100–700 K	0.106
Venus	0.815	0.951	5.3	0.90	10.3	730	0.65
Earth	1.000	1.000	5.518	1.00	11.2	200–300	0.37
Mars	0.107	0.531	3.96	0.38	5.0	130–290	0.15
Jupiter	317.89	10.85	1.327	2.64	60.0	165	0.52
Saturn	95.184	8.99	0.688	1.13	36.0	134	0.47
Uranus	14.536	3.96	1.272	0.89	21.2	76	0.50
Neptune	17.148	3.85	1.640	1.13	23.5	74	0.5:
Pluto	0.0022	0.18	1.7	0.06	1.2	40	0.06:
							0.44–0.61

[a]The masses and diameters are given in units of the Earth's mass and diameter, which are 5.974 X 10^{24} kg and 12,734 km, respectively.

Satellites

Planet	Satellite	Semimajor Axis	Period	Diameter	Mass (kg)	Density
Earth	Moon	3.84×10^5 km	$27^d.322$	3476 km	7.35×10^{22}	3.34 g/cm^3
Mars	Phobos	9.38×10^3	0.3189	$27 \times 22 \times 19$	9.6×10^{15}	1.9
	Deimos	2.35×10^4	1.2624	$15 \times 12 \times 11$	1.9×10^{15}	1.75
Jupiter	Metis	1.280×10^5	0.295	20:	9.5×10^{16}	
	Adrastea	1.290×10^5	0.298	$20 \times 20 \times 15$	1.9×19^{16}	
	Amalthea	1.81×10^5	0.498	$270 \times 170 \times 150$	7.2×10^{18}	
	Thebe	2.22×10^5	0.675	110×90	7.6×10^{17}	
	Io	4.22×10^5	1.769	3,630	8.92×10^{22}	3.53
	Europa	6.71×10^5	3.551	3,138	4.87×10^{22}	3.03
	Ganymede	1.07×10^6	7.155	5,262	1.49×10^{23}	1.93
	Callisto	1.88×10^6	16.689	4,800	1.08×10^{23}	1.70
	Leda	1.11×10^7	238.7	16:	5.7×10^{15}	
	Himalia	1.15×10^7	250.6	186:	9.5×10^{18}	
	Lysithea	1.17×10^7	259.2	36:	7.6×10^{16}	
	Elara	1.18×10^7	259.7	76:	7.6×10^{17}	
	Ananke	2.12×10^7	631	30:	3.8×10^{16}	
	Carme	2.26×10^7	692	40:	9.5×10^{16}	
	Pasiphae	2.35×10^7	735	50:	1.9×10^{17}	
	Sinope	2.37×10^7	758	36:	7.6×10^{16}	
Saturn	Pan	1.33×10^5	0.573	20:		
	Atlas	1.377×10^5	0.602	40×20		
	Prometheus	1.394×10^5	0.613	$140 \times 100 \times 80$		
	Pandora	1.417×10^5	0.629	$110 \times 90 \times 80$		
	Epimetheus	1.514×10^5	0.694	$140 \times 120 \times 100$		
	Janus	1.514×10^5	0.695	$220 \times 200 \times 160$		
	Mimas	1.855×10^5	0.942	392	4.5×10^{19}	1.43
	Enceladus	2.381×10^5	1.370	500	7.4×10^{19}	1.13
	Telesto	2.947×10^5	1.888	$34 \times 28 \times 26$		
	Calypso	2.947×10^5	1.888	$34 \times 22 \times 22$		
	Tethys	2.947×10^5	1.888	1,060	7.4×10^{20}	1.19
	Dione	3.774×10^5	2.737	1,120	1.05×10^{21}	1.43
	Helene	3.781×10^5	2.737	$36 \times 32 \times 30$		
	Rhea	5.271×10^5	4.518	1,530	2.50×10^{21}	1.33
	Titan	1.222×10^6	15.945	5,150	1.35×10^{23}	1.89
	Hyperion	1.481×10^6	21.277	$410 \times 260 \times 220$	1.71×10^{19}	
	Iapetus	3.561×10^6	79.330	1,460	1.88×10^{21}	1.15
	Phoebe	1.295×10^7	550.5	220		
Uranus	Cordelia	4.97×10^4	0.336	40		
	Ophelia	5.38×10^4	0.377	50		
	Bianca	5.92×10^4	0.435	50		
	Cressida	6.18×10^4	0.465	60		
	Desdemona	6.27×10^4	0.476	60		
	Juliet	6.46×10^4	0.494	80		
	Portia	6.61×10^4	0.515	80		
	Rosalind	6.99×10^4	0.560	60		
	Belinda	7.53×10^4	0.624	60		
	Puck	8.60×10^4	0.764	170		
	Miranda	1.298×10^5	1.413	484	7.5×10^{19}	1.26

	Ariel	1.912×10^5	2.520	1,160	1.4×10^{21}	1.65
	Umbriel	2.660×10^5	4.144	1,190	1.3×10^{21}	1.44
	Titania	4.358×10^5	8.706	1,610	3.5×10^{21}	1.59
	Oberon	5.826×10^5	13.463	1,550	2.9×10^{21}	1.50
Neptune	Naiad	4.80×10^4	0.296	54		
	Thalassa	5.00×10^4	0.313	80		
	Despina	5.25×10^4	0.333	150		
	Galatea	6.20×10^4	0.429	180		
	Larissa	7.36×10^4	0.554	190		
	Proteus	1.176×10^5	1.121	400		
	Triton	3.548×10^5	5.875	2,705	2.21×10^{22}	2.07
	Nereid	5.5134×10^6	360.129	340	2.1×10^{19}	
Pluto	Charon	1.964×10^4	6.387	1,186		

Appendix 9
The Elements and Their Abundances

Element	Symbol	Atomic No.	Atomic Weight[a]	Abundance[b]	Element	Symbol	Atomic No.	Atomic Weight[a]	Abundance[b]
Hydrogen	H	1	1.0080	1.00	Cobalt	Co	27	58.9332	3.16×10^{-8}
Helium	He	2	4.0026	0.085	Nickel	Ni	28	58.71	1.91×10^{-6}
Lithium	Li	3	6.941	1.55×10^{-9}	Copper	Cu	29	63.546	2.82×10^{-8}
Beryllium	Be	4	9.0122	1.41×10^{-11}	Zinc	Zn	30	65.37	2.63×10^{-8}
Boron	B	5	10.811	2.00×10^{-10}	Gallium	Ga	31	69.72	6.92×10^{-10}
Carbon	C	6	12.0111	0.000372	Germanium	Ge	32	72.59	2.09×10^{-9}
Nitrogen	N	7	14.0067	0.000115	Arsenic	As	33	74.9216	2×10^{-10}
Oxygen	O	8	15.9994	0.000676	Selenium	Se	34	78.96	3.16×10^{-9}
Fluorine	F	9	18.9984	3.63×10^{-8}	Bromine	Br	35	79.904	6.03×10^{-10}
Neon	Ne	10	20.179	3.72×10^{-5}	Krypton	Kr	36	83.80	1.6×10^{-9}
Sodium	Na	11	22.9898	1.74×10^{-6}	Rubidium	Rb	37	85.4678	4.27×10^{-10}
Magnesium	Mg	12	24.305	3.47×10^{-5}	Strontium	Sr	38	87.62	6.61×10^{-10}
Aluminum	Al	13	26.9815	2.51×10^{-6}	Yttrium	Y	39	88.9059	4.17×10^{-11}
Silicon	Si	14	28.086	3.55×10^{-5}	Zirconium	Zr	40	91.22	2.63×10^{-10}
Phosphorus	P	15	30.9738	3.16×17^{-7}	Niobium	Nb	41	92.906	2.0×10^{-10}
Sulfur	S	16	32.06	1.62×10^{-5}	Molybdenum	Mo	42	95.94	7.94×10^{-11}
Chlorine	Cl	17	35.453	2×10^{-7}	Technetium	Tc	43	98.906	—
Argon	Ar	18	39.948	4.47×10^{-6}	Ruthenium	Ru	44	101.07	3.72×10^{-11}
Potassium	K	19	39.102	1.12×10^{-7}	Rhodium	Rh	45	102.905	3.55×10^{-11}
Calcium	Ca	20	40.08	2.14×10^{-6}	Palladium	Pd	46	106.4	3.72×10^{-11}
Scandium	Sc	21	44.956	1.17×10^{-9}	Silver	Ag	47	107.868	4.68×10^{-12}
Titanium	Ti	22	47.90	5.50×10^{-8}	Cadmium	Cd	48	112.40	9.33×10^{-12}
Vanadium	V	23	50.9414	1.26×10^{-8}	Indium	In	49	114.82	5.13×10^{-11}
Chromium	Cr	24	51.996	5.01×10^{-7}	Tin	Sn	50	118.69	5.13×10^{-11}
Manganese	Mn	25	54.9380	2.63×10^{-7}	Antimony	Sb	51	121.75	5.62×10^{-12}
Iron	Fe	26	55.847	2.51×10^{-5}	Tellurium	Te	52	127.60	1×10^{-10}

Element	Symbol	Atomic No.	Atomic Weight[a]	Abundance[b]	Element	Symbol	Atomic No.	Atomic Weight[a]	Abundance[b]
Iodine	I	53	126.9045	4.07×10^{-11}	Gold	Au	79	196.967	2.09×10^{-12}
Xenon	Xe	54	131.30	1×10^{-10}	Mercury	Hg	80	200.59	1×10^{-9}
Cesium	Cs	55	132.905	1.26×10^{-11}	Thallium	Tl	81	204.37	1.6×10^{-12}
Barium	Ba	56	137.34	6.31×10^{-11}	Lead	Pb	82	207.19	7.41×10^{-11}
Lanthanum	La	57	138.906	6.46×10^{-11}	Bismuth	Bi	83	208.981	6.3×10^{-12}
Cerium	Ce	58	140.12	4.37×10^{-11}	Polonium	Po	84	210	—
Praseodymium	Pr	59	140.908	4.27×10^{-11}	Astatine	At	85	210	—
Neodymium	Nd	60	144.24	6.61×10^{11}	Radon	Rn	86	222	—
Promethium	Pm	61	146	—	Francium	Fr	87	223	—
Samarium	Sm	62	150.4	4.57×10^{-11}	Radium	Ra	88	226.025	—
Europium	Eu	63	151.96	3.09×10^{-12}	Actinium	Ac	89	227	—
Gadolinium	Gd	64	157.25	1.32×10^{-11}	Thorium	Th	90	232.038	6.61×10^{-12}
Terbium	Tb	65	158.925	2.63×10^{-12}	Protactinium	Pa	91	230.040	—
Dysprosium	Dy	66	162.50	1.29×10^{-11}	Uranium	U	92	238.029	4.0×10^{-12}
Holmium	Ho	67	164.930	3.1×10^{-12}	Neptunium	Np	93	237.048	—
Erbium	Er	68	167.26	5.75×10^{-12}	Plutonium	Pu	94	242	—
Thulium	Tm	69	168.934	2.69×10^{-11}	Americium	Am	95	242	—
Ytterbium	Yb	70	170.04	6.46×10^{-12}	Curium	Cm	96	245	—
Lutetium	Lu	71	174.97	6.92×10^{-12}	Berkelium	Bk	97	248	—
Hafnium	Hf	72	178.49	6.3×10^{-12}	Californium	Cf	98	252	—
Tantalum	Ta	73	180.948	2×10^{-12}	Einsteinium	Es	99	253	—
Tungsten	W	74	183.85	3.72×10^{-10}	Fermium	Fm	100	257	—
Rhenium	Re	75	186.2	1.8×10^{-12}	Mendelevium	Md	101	257	—
Osmium	Os	76	190.2	5.62×10^{-12}	Nobelium	No	102	255	—
Iridium	Ir	77	192.2	1.62×10^{-10}	Lawrencium	Lr	103	256	—
Platinum	Pt	78	195.09	5.62×10^{-11}					

[a]The atomic weight of an element is its mass in *atomic mass units*. An atomic mass unit is defined as one-twelfth of the mass of the most common isotope of carbon, and has the value 1.660531×10^{-27} kg. In general, the atomic weight of an element is approximately equal to the total number of protons and neutrons in its nucleus.

[b]The abundances are given in terms of the number of atoms of each element compared to hydrogen, and are based on the composition of the Sun. For very rare elements, particularly those toward the end of the list, the abundances can be quite uncertain, and the values given should not be considered exact.

Appendix 10
Stellar Data

The Fifty Brightest Stars

Star		Spectral Type	Apparent Magnitude	Distance	Position (1980)	
					Right Ascension	Declination
α Eri	Achernar	B3 V	0.51	36 pc	01ʰ37ᵐ0	−57°20'
α UMi	Polaris	F8 lb	1.99	208	02 12.5	+89 11
α Per	Mirfak	F5 lb	1.80	175	03 22.9	+49 47
α Tau	Aldebaran	K5 III	0.86	21	04 34.8	+16 28
β Ori	Rigel	B8 Ia	0.14	276	05 13.6	−08 13
α Aur	Capella	G8 III	0.05	14	05 15.2	+45 59
γ Ori	Bellatrix	B2 III	1.64	144	05 24.0	+06 20
β Tau	Elnath	B7 III	1.65	92	05 25.0	+28 36
ε Ori	Alnilam	B0 Ia	1.70	490	05 35.2	−01 13
ζ Ori	Alnitak	09.5 Ib	1.79	490	05 39.7	−01 57
α Ori	Betelgeuse	M2 Iab	0.41	159	05 54.0	+07 24
β Aur	Menkalinan	A2 V	1.86	27	05 58.0	+44 57
β CMa		B1 II-III	1.96	230	06 21.8	−17 56
α Car	Canopus	F0 Ib-II	−0.72	30	06 23.5	−52 41
γ Gem	Alhena	A0 IV	1.93	32	06 36.6	+16 25
α CMa	Sirius	A1 V	−1.47	2.7	06 44.2	−16 42
ε CMa	Adhara	B2 II	1.48	209	06 57.8	−28 57
δ CMa		F8 Ia	1.85	644	07 07.6	−26 22
α Gem	Castor	A1 V	1.97	14	07 33.3	+31 56
α CMi	Procyon	F5 IV-V	0.37	3.5	07 38.2	+05 17
β Gem	Pollux	K0 III	1.16	11	07 44.1	+28 05
γ Vel		WC8	1.83	160	08 08.9	−47 18
ε Car	Avior	K3 III?	1.90	104	08 22.1	−59 26
ζ Vel		A2 V	1.95	23	08 44.2	−54 38
β Car	Miaplacidus	A1 III	1.67	26	09 13.0	−69 38
α Hya	Alphard	K4 III	1.98	29	09 26.6	−08 35
α Leo	Regulus	B7 V	1.36	26	10 07.3	+12 04
γ Leo		K0 III	1.99	28	10 18.8	+19 57
α UMa	Dubhe	K0 III	1.81	32	11 02.5	+61 52
α Cru A	Acrux	B0.5 IV	1.39	114	12 25.4	−62 59
α Cru B	Acrux	B1 V	1.86	114	12 25.4	−62 59
γ Cru	Gacrux	M4 III	1.69	67	12 30.1	−57 00
β Cru		B0.5 III	1.28	150	12 46.6	−59 35
ε UMa	Alioth	A0p	1.79	21	12 53.2	+56 04
α Vir	Spica	B1 V	0.91	67	13 24.1	−11 03
η UMa	Alkaid	B3 V	1.87	64	13 46.8	+49 25
β Cen	Hadar	B1 III	0.63	150	14 02.4	−60 16
α Boo	Arcturus	K2 III	−0.06	11	14 14.8	+19 17
α Cen A	Rigil Kentaurus	G2 V	0.01	1.3	14 38.4	−60 46
α Cen B	Rigil Kentaurus	K4 V	1.40	1.3	14 38.4	−60 46
α Sco	Antares	M1 lb	0.92	160	16 28.2	−26 23
α TrA	Atria	K2 Ib	1.93	25	16 46.5	−68 60

λ Sco	Shaula	B1 V	1.60	95	17 32.3	−37 05
θ Sco		F0 Ib	1.86	199	17 35.9	−42 59
ε Sgr	Kaus Australis	B9.5 III	1.81	38	18 22.9	−34 24
α Lyr	Vega	A0 V	0.04	8	18 36.2	+38 46
α Aql	Altair	A7 IV-V	0.77	5	19 49.8	+08 49
α Pav	Peacock	B2.5 V	1.95	95	20 24.1	−56 48
α Cyg	Deneb	A2 Ia	1.26	491	20 40.7	+45 12
α Gru	Al Na'ir	B7 IV	1.76	20	22 06.9	−47 04
α PsA	Fomalhaut	A3 V	1.15	7	22 56.5	−29 44

Nearby Stars (Within 5 Parsecs of the Sun)[a]

Star	Spectral Type	Visual Apparent Magnitude	Visual Absolute Magnitude	Parallax	Distance	Proper Motion
α Centauri	G2V	−0.1	4.8	0.753"	1.33 pc	3.68"/yr
Barnard's star	M5V	9.5	13.2	0.544	1.84	10.31
Wolf 359	M8V3	13.5	16.7	0.432	2.31	4.71
BD + 36°2147	M2V	7.5	10.5	0.400	2.50	4.78
Luyten 726-8	M6Ve	12.5	15.4	0.385	2.60	3.36
Sirius	A1V	−1.5	1.4	0.377	2.65	1.33
Ross 154	M5Ve	10.6	13.3	0.345	2.90	0.72
Ross 248	M6Ve	12.3	14.8	0.319	3.13	1.58
ε Eridani	K2V	3.7	6.1	0.305	3.28	0.98
Ross 128	M5V	11.1	13.5	0.302	3.31	1.37
Luyten 789-6	M6V	12.2	14.6	0.302	3.31	3.26
61 Cygni	K5Ve	5.2	7.5	0.292	3.42	5.22
ε Indi	K5Ve	4.7	7.0	0.291	3.44	4.69
τ Ceti	G8V	3.5	5.9	0.289	3.46	1.92
Procyon	F5V	0.4	2.7	0.285	3.51	1.25
Σ 2398	M4V	8.9	11.2	0.284	3.52	2.28
BD + 43°44	M1Ve	8.1	10.4	0.282	3.55	2.89
CD − 36°15693	M2Ve	7.4	9.6	0.279	3.58	6.90
G51-15		14.8	17.0	0.273	3.66	1.26
L725-32	M5Ve	11.5	13.6	0.264	3.79	1.22
BD + 5°1668	M4V	9.8	12.0	0.264	3.79	3.73
CD − 39°14192	M0Ve	6.7	8.8	0.260	3.85	3.46
Kapteyn's star	M0V	8.8	10.8	0.256	3.91	8.89
Kruger 60	M4V	9.7	11.7	0.254	3.94	0.86
Ross 614	M5Ve	11.3	13.3	0.243	4.12	0.99
BD − 12°4523	M5V	10.0	11.9	0.238	4.20	1.18
Wolf 424	M6Ve	13.2	15.0	0.234	4.27	1.75
van Maanen's star	W.D.	12.4	14.2	0.232	4.31	2.95
CD − 37°15492	M3V	8.6	10.4	0.225	4.44	6.08
Luyten 1159-16	M8V	12.3	14.0	0.221	4.52	2.08
BD + 50°1725	K7V	6.6	8.3	0.217	4.61	1.45

CD − 46°11540	M4V	9.4	11.1	0.216	4.63	1.13
CD − 49°13515	M3V	8.7	10.4	0.214	4.67	0.81
CD − 44°11909	M5V	11.2	12.8	0.213	4.69	1.16
BD + 68°946	M3.5V	9.1	10.8	0.213	4.69	1.33
G158-27		13.7	15.5	0.212	4.72	2.06
G208-44/45		13.4	15.0	0.210	4.76	0.75
BD 15°6290	M5V	10.2	11.8	0.209	4.78	1.16
40 Eridani	K0V	4.4	6.0	0.207	4.83	4.08
L145-141	W.D.	11.4	12.6	0.206	4.85	2.68
BD + 20°2465	M4.5V	9.4	10.9	0.203	4.93	0.49
70 Ophiuchi	K1V	4.2	5.7	0.203	4.93	1.13
BD + 43°4305	M4.5Ve	10.2	11.7	0.200	5.00	0.83

[a]Data from Lippencott, L. S., 1978, *Space Science Reviews* 22:153. Many of the stars listed here are multiple systems; in these cases, the data in the table refer only to the brightest number of the system.

Appendix 11
Astronomical Coordinate Systems

Horizon Coordinates

Perhaps the simplest coordinate system to visualize is one that is based on the viewpoint from the observer's locale. Such a system is not useful for communicating or comparing positions as seen by observers at different locations, but it is very convenient locally.

In the **horizon coordinate system** the position of an object is defined relative to the local horizon and to a fixed direction (north). The angular distance of an object above the horizon is called the **altitude** of the object; the angular direction of the object, measured toward the east from the north, is the **azimuth.**

The ancient astronomers usually measured planetary and stellar positions in horizon coordinates, and today this system is used in various ways (one of them being the aiming of weapons!). Horizon coordinates have undergone a bit of a renaissance lately, as modern computer-controlled telescopes are often pointed by using the altitude and azimuth of the target object. This is a complex process, because the telescope must then make continuous corrections in both coordinates to compensate for the Earth's rota-

tion, but with modern computerized control systems this complexity can be handled relatively easily. The benefit is that alt-az telescope mounts, as they are usually called, can be built more compactly and have fewer major moving parts than equatorial mounts (see the next section). But even so, the input coordinates to the telescope are always in the equatorial system (below); the computer has to convert the data to the horizon system before the telescope is pointed.

Equatorial Coordinates

The **equatorial coordinate system** is by far the most widely used in astronomy, by professionals and amateurs alike. This system uses the Earth as its basic frame of reference, with the exception that a fixed point in space is the starting point for the east-west coordinate (it would be impractical to use a point on the Earth for this, because the Earth rotates).

The north-south position of an object, called the **declination,** is measured in degrees, minutes, and seconds from the celestial equator (the projection of the Earth's equator onto the sky). A plus sign is used to indicate positions north of the equator; a minus sign is used for positions south of the equator. Thus a star that passes overhead precisely at latitude 40° north has a declination of +40°0'0". Similarly, a star that passes above 69°48'12" south latitude would have a declination of −69°48'12". It is helpful to consider declination as simply a projection of latitude onto the sky.

Things are a bit more complex for the east-west coordinate, which is called **right ascension.** First, as already noted, the rotation of the Earth precludes using a point on the Earth as the reference direction, so an analogy to longitude (based on the prime meridian) will not work. Second, the units of right ascension are not angles as you might have expected, but instead are units of time. The use of hours, minutes, and seconds of time to measure a star's position in the east-west direction derives from the fact that the entire sky passes overhead in 24 hours. Thus it became convenient for astronomers to define positions according to the time when objects cross the meridian (the local north-south line that passes overhead).

To establish a zero-point for measuring right ascension, astronomers have defined a fixed direction in space, called the **first point in Aries,** as the reference direction (i.e. the direction where the right ascension is $0^h0^m0^s$). The first point in Aries is the direction of the line of intersection of the Earth's orbital and equatorial planes; i.e. it is the direction of the point where the ecliptic crosses the equator. There are two such points; the one used for equatorial coordinates is the point where the Sun crosses the equator in its northbound travel, which coincides with the vernal equinox. Thus at the precise moment of the vernal equinox (around March 21 each year) the Sun has a right ascension of $0^h0^m0^s$.

Right ascension is measured toward the east from this point, because this is the direction in which the Earth turns. As stars and other objects rise and pass overhead in sequence, they cross the meridian at later and later times, so the right ascension increases toward the east. A star that lies exactly 90° to the east of the Sun has a right ascension of $6^h0^m0^s$. Similarly, a star that lies in the anti-Sun direction has a right ascension of $12^h0^m0^s$.

An experienced observer can tell instantly when in the year it is possible to observe a given position on the sky, simply by knowing the right ascension of the target. Normally the best time to observe something is when it is opposite the Sun. Therefore in the spring, when the sun is near 0^h right ascension, stars near 12^h are in good position. Conversely, autumn is the optimal season for observing stars whose right ascensions are near 0^h.

The major convenience of the equatorial coordinate system is that it is tied to the Earth's frame of reference. Among other things, this facilitates telescope design and operation, because compensation for the Earth's rotation is needed only in the right ascension (east-west) direction, allowing the declination to remain fixed once the telescope is pointed at a target. Historically all telescopes were built on equatorial mounts to take advantage of this, but large modern computer-controlled telescopes today are more likely to use alt-az mounts (see above). Smaller telescopes, including those used by amateur astronomers, are still generally constructed with equatorial mounts.

There is one major complication of the equatorial system: precession of the Earth's axis. As you learned in Chapter 2, the Earth's rotational axis undergoes a slow wobbling motion (with a 26,000-year period). This causes star positions that are measured with respect to the Earth to gradually shift. The result is that a star's declination and right ascension change with time, and it becomes important for astronomers to allow for this effect when observing. For example, if you look up the coordinates for a particular star in a catalog, you must take note of the date when the catalog's positions are (or were) valid, and then you have to calculate the shift in the star's position from that time to the present. This used to be a major task for every observer, but today's telescopes are generally computer-controlled, and all that is necessary is to tell the telescope the date when the star had the position that you are using. The computer then calculates the offset and points at the correct position for the current date.

Other Coordinate Systems

While observers use equatorial coordinates exclusively, there are at least two other coordinate systems that are sometimes used to identify the locations of objects on the sky. In both cases this is done in order to help visualize where things are in a larger physical system of objects. Neither is ever used to point a telescope; only equatorial coordinates are practical for that purpose.

One of the alternative reference systems is known as **ecliptic coordinates.** This coordinate system was devised in order to locate objects in the solar system, and the plane of the ecliptic is the zero-point for the north-south direction. The direction of the vernal equinox is used for the east-west direction, with the **celestial** (or **ecliptic**) **longitude** increasing toward the east, in analogy with right ascension in the equatorial system. The north-south coordinate is called the **celestial** (or **ecliptic**) **latitude.** Both coordinates are measured in units of angle.

The major convenience in using ecliptic coordinates is that an object's position immediately tells you something about where the object is in relation to the rest of the solar system. For example, if a new object is found and its celestial (or ecliptic) latitude is near zero, then you know right away that it orbits close to the plane of the disk of the solar system, something that would be difficult to recognize if you knew only the right ascension and declination of the comet. This might help you recognize the object as an asteroid as opposed to a comet, for example.

Another system, called **galactic coordinates,** is very similar to ecliptic coordinates. In this case the frame of reference is the disk of the Milky Way galaxy, the vast system of stars to which the Sun

belongs. The plane of the disk is defined as the zero-point for **galactic latitude,** while the direction of the galactic center is the reference point for **galactic longitude,** with increasing longitude again measured to the east. As in the ecliptic system, both coordinates are measured in angular units.

The convenience in using the galactic coordinate system is that it helps in the visualization of a star's position with respect to the galaxy. This has important implications for a star's history and its role in galactic evolution, for example.

Pinpointing the location of the galactic center has not been easy (see Chapter 17), and revisions in the adopted location of this point have already forced astronomers to recalibrate the reference frame for galactic coordinates once, and it could happen again. The most recent data on the exact location of the galaxy's center place it some 2° to 3° away from the zero-point of the coordinate system.

Appendix 12
The Constellations

Name	Genitive	Abbreviation	Position Right Ascension	Declination
Andromeda	Andromedae	And	01^h	+ 40°
Antlia	Antliae	Ant	10	− 35
Apus	Apodis	Aps	16	− 75
Aquarius	Aquarii	Aqr	23	− 15
Aquila	Aquilae	Aql	20	+ 05
Ara	Arae	Ara	17	− 55
Aries	Arietis	Ari	03	+ 20
Auriga	Aurigae	Aur	06	+ 40
Bootes	Bootis	Boo	15	+ 30
Caelum	Caeli	Cae	05	− 40
Camelopardalis	Camelopardalis	Cam	06	− 70
Cancer	Cancri	Cnc	09	+ 20
Canes Venatici	Canum Venaticorum	CVn	13	+ 40
Canis Major	Canis Majoris	CMa	07	− 20
Canis Minor	Canis Minoris	CMi	08	+ 05
Capricornus	Capricorni	Cap	21	− 20
Carina	Carinae	Car	09	− 60
Cassiopeia	Cassiopeiae	Cas	01	+ 60
Centaurus	Centauri	Cen	13	− 50
Cepheus	Cephei	Cep	22	+ 70
Cetus	Ceti	Cet	02	− 10
Chamaeleon	Chamaeleonis	Cha	11	− 80
Circinis	Circini	Cir	15	− 60
Columba	Columbae	Col	06	− 35
Coma Berenices	Comae Berenices	Com	13	+ 20
Corona Australis	Coronae Australis	CrA	19	− 40
Coronoa Borealis	Coronae Borealis	CrB	16	+ 30
Corvus	Corvi	Crv	12	− 20
Crater	Crateris	Crt	11	− 15
Crux	Crucis	Cru	12	− 60
Cygnus	Cygni	Cyg	21	+ 40
Delphinus	Delphini	Del	21	+ 10
Dorado	Doradus	Dor	05	− 65
Draco	Draconis	Dra	17	+ 65
Equuleus	Equulei	Equ	21	+ 10
Eridanus	Eridani	Eri	03	− 20
Fornax	Fornacis	For	03	− 30
Gemini	Geminorum	Gem	07	+ 20
Grus	Gruis	Gru	22	− 45
Hercules	Herculis	Her	17	+ 30
Horologium	Horologii	Hor	03	− 60

Hydra	Hydrae	Hya	10	− 20
Hydrus	Hydri	Hyi	02	− 75
Indus	Indi	Ind	21	− 55
Lacerta	Lacertae	Lac	22	+ 45
Leo	Leonis	Leo	11	+ 15
Leo Minor	Leonis Minoris	LMi	10	+ 35
Lepus	Leporis	Lep	06	− 20
Libra	Librae	Lib	15	− 15
Lupus	Lupi	Lup	15	− 45
Lynx	Lincis	Lyn	08	+ 45
Lyra	Lyrae	Lyr	19	+ 40
Mensa	Mensae	Men	05	− 80
Microscopium	Microscopii	Mic	21	− 35
Monoceros	Monocerotis	Mon	07	− 05
Musca	Muscae	Mus	12	− 70
Norma	Normae	Nor	16	− 50
Octans	Octantis	Oct	22	− 85
Ophiuchus	Ophiuchi	Oph	17	00
Orion	Orionis	Ori	05	+ 05
Pavo	Pavonis	Pav	20	− 65
Pegasus	Pegasi	Peg	22	+ 20
Perseus	Persei	Per	03	+ 45
Phoenix	Phoenicis	Phe	01	− 50
Pictor	Pictoris	Pic	06	− 55
Pisces	Piscium	Psc	01	+ 15
Piscis Austrinus	Piscis Austrini	PsA	22	− 30
Puppis	Puppis	Pup	08	− 40
Pyxis	Pyxidis	Pyx	09	− 30
Reticulum	Reticuli	Ret	04	− 60
Sagitta	Sagittae	Sge	20	+ 10
Sagittarius	Sagittarii	Sgr	19	− 25
Scorpius	Scorpii	Sco	17	− 40
Sculptor	Sculptoris	Scl	00	− 30
Scutum	Scuti	Sct	19	− 10
Serpens	Serpentis	Ser	17	00
Sextans	Sextantis	Sex	10	00
Taurus	Tauri	Tau	04	+ 15
Telescopium	Telescopii	Tel	19	− 50
Triangulum	Trianguli	Tri	02	+ 30
Triangulum Australe	Trianguli Australi	TrA	16	− 65
Tucana	Tucanae	Tuc	00	− 65
Ursa Major	Ursae Majoris	UMa	11	+ 50
Ursa Minor	Ursae Minoris	UMi	15	+ 70
Vela	Velorum	Vel	09	− 50
Virgo	Virginis	Vir	13	00
Volans	Volantis	Vol	08	− 70
Vulpecula	Vulpeculae	Vul	20	+ 25

Appendix 13
Mathematical Treatment of Stellar Magnitudes

Logarithmic Representation

In the text, magnitudes are discussed in terms of the brightness ratios between stars of different magnitudes. We generally avoided discussion of cases where two stars differ by a fraction of a magnitude, because in such cases it is no longer simple to calculate the brightness ratio corresponding to the magnitude difference. If star 1 is 0.5 magnitudes brighter than star 2, for example, what is the brightness ratio? Or if star 1 is a factor of 48.76 fainter than star 2, what is the difference in magnitudes?

Astronomers use an exact mathematical relationship between magnitude differences and brightness ratios, written as

$$m_1 - m_2 = 2.5 \log (b_2/b_1),$$

where m_1 and m_2 are the magnitudes of two stars, and b_2/b_1 is the ratio of their brightnesses. The notation "$\log (b_2/b_1)$" means the **logarithm** of this ratio; a logarithm is the power to which 10 must be raised to give this ratio. Hence, for example, if $b_2/b_1 = 100$, $\log (b_2/b_1) = \log (10^2) = 2$, because 10 must be raised to the second power to give 100. The magnitude differences is $2.5 \log (100) = 2.5 \times 2 = 5$. Similarly, if $b_2/b_1 = 0.001$, then $\log (b_2/b_1) = \log (0.001) = \log (10^{-3}) = -3$, and in this case the magnitude difference is $2.5 \times -3 = -7.5$ (the minus sign indicates that star 1 is brighter than star 2 in this example).

The method works equally well in cases where the power of 10 is not a whole number, as in the example where $b_2/b_1 = 48.76$. Here $\log (b_2/b_1) = \log (48.76) = 1.69$ (this is usually found by consulting tables of logarithms or by using a scientific calculator). In this example, the magnitude difference is $2.5 \log (b_2/b_1) = 2.5 \log(48.76) = 2.5 \times 1.69 = 4.23$, so star 1 is 4.23 magnitudes fainter than star 2.

The equation can be used in other ways as well; solving for b_2/b_1 yields

$$\frac{b_2}{b_1} = 10^{(m_1 - m_2)/2.5}$$
$$= 10^{0.4(m_1 - m_2)}$$

Thus, if we know that the magnitudes of two stars differ by $m_1 - m_2$, we multiply this difference by 0.4 and raise 10 to the power $0.4(m_1 - m_2)$, again using a calculator or tables, to get the brightness ratio b_2/b_1. As a simple example, suppose $m_1 - m_2 = 5$; then $0.4(m_1 - m_2) = 0.4 \times 5 = 2$, and $10^{0.4(m_1 - m_2)} = 10^2 = 100$, as we knew it should. As a more complex example, consider the stars Betelgeuse (magnitude $+0.41$) and Deneb (magnitude $+1.26$). From the equation above, we see that Betelgeuse is $10^{0.4(1.26 - 0.41)} = 10^{0.34} = 2.19$ times brighter than Deneb.

While in most cases it is possible to follow the discussions of magnitudes and brightness ratios in the text without using this exact mathematical technique, it is still useful to be familiar with it.

The Distance Modulus

Whenever we know both the apparent and absolute magnitudes of a star, a comparison of the two will give its distance. In the text we have seen how to make this calculation by the following several steps:

1. Convert the difference $m - M$ between apparent and absolute magnitude into a brightness ratio, that is, a numerical factor indicating how much brighter or fainter the star would appear at 10 parsecs distance than at its actual distance.
2. Using the inverse square law, determine the change in distance required to produce this change in brightness.
3. Multiply this distance factor by 10 parsecs to find the distance to the star.

To do calculations mentally in this way can be laborious, especially in cases where the magnitude difference does not correspond neatly to a simple numerical factor, as it did in the examples given in the text. Hence astronomers use a mathematical equation expressing the relationship between distance and the distance modulus $m - M$. This equation, which works equally well for all cases, is

$$d = 10^{1+.2(m-M)},$$

where d is the distance in parsecs to a star whose apparent magnitude is m and whose absolute magnitude is M.

In a simple example, where $m = 9$ and $M = -6$, we have

$$d = 10^{1+.2(15)}$$
$$= 10^{1+3}$$
$$= 10^4 \text{ parsecs.}$$

Now let's try a more complex case. Suppose the star is an M2 main-sequence star, so that $M = 13$, as found from the H-R diagram. The apparent magnitude is $m = 16$. Our equation tells us that

$$d = 10^{1+.2(16-13)}$$
$$= 10^{1+.6}$$
$$= 10^{1.6}$$
$$= 39.8 \text{ parsecs}$$

It is necessary to use a slide rule, calculator, or mathematical table to do this, of course, but it is still relatively straightforward compared with following the mental steps outlined above.

The Effect of Extinction on the Distance Modulus

When we discussed distance-determination techniques (in Chapter 14), we ignored the effects of interstellar extinction. In any method that depends on the apparent brightness of a star, however, extinction can be important, particularly for very distant stars. Because the effect of extinction is to make a star appear fainter than it otherwise would, the tendency is to overestimate distances if no allowance is made for it.

Recall that in the spectroscopic parallax technique, the distance modulus $m - M$ is used to find the distance to a star from the equation given above:

$$d = 10^{1+.2(m-M)},$$

where m is the apparent magnitude, M is the absolute magnitude, and d is the distance to the star in parsecs.

To correct this equation for extinction, we add a term, A_v, which refers to the extinction (in magnitudes) in visual light. Thus, if the extinction toward a particular star makes that star appear 2 magnitudes fainter than it otherwise would, $A_V = 2$. If we insert this into the equation, we find:

$$d = 10^{1+.2(m-M-A_v)}$$

Let us consider a simple example, a star whose apparent magnitude is $m = 12.4$ and whose absolute magnitude is $M = 2.4$. First let us consider its distance if extinction is ignored:

$$d = 10^{1+.2(12.4-2.4)} = 10^3 = 1,000 \text{ parsecs.}$$

Now suppose it is determined that the extinction in the direction of this star amounts to 1 magnitude. Now the distance is

$$d = 10^{1+.2(12.4-2.4-1)} = 10^{2.8} = 631 \text{ parsecs.}$$

One magnitude is only a modest amount of extinction, yet by neglecting it, we overestimated the distance to this star by almost 60%. We can see from this example that extinction can have a drastic effect on distance estimates.

It is worthwhile to add a note about how the extinction A_V is determined. It is possible to determine how much redder, in terms of the $B - V$ color index, a star appears because of extinction. To carry that a step further, astronomers define a *color excess* called $E(B - V)$, which is the difference between the observed and intrinsic values of $B - V$:

$$E(B - V) = (B - V)_{observed} - (B - V)_{intrinsic}$$

Studies of the variation of interstellar extinction with wavelength show that the extinction at the visual wavelength is approximately three times the color excess; that is:

$$A_V = 3E(B - V).$$

Hence determination of excess reddening leads to an estimate of A_V, and this in turn can be used in the modified equation for the distance.

Appendix 14
Nuclear Reactions in Stars

In the text we did not spell out the details of the reactions that occur in stellar cores, although they are quite simple. To do so here, we will use the notation of nuclear physics. This is basically a shorthand in symbols. For example, a helium nucleus, containing two protons and two neutrons, is designated ^4_2He, the subscript indicating the **atomic number** (the number of protons) and the superscript the **atomic weight** (the total number of protons and neutrons). Similarly a hydrogen nucleus is ^1_1H, and deuterium, a form of hydrogen with an extra neutron in the nucleus, is ^2_1H. A special symbol (ν) is used for the **neutrino,** a massless subatomic particle emitted in some reactions, and e^+ indicates a **positron,** which is equivalent to an electron but has a positive electrical charge. The symbol γ indicates a gamma ray, a very short wavelength photon of light emitted in some reactions.

The Proton-Proton Chain

Using this notation system, we can now spell out the proton-proton chain:

$$^1_1\text{H} + ^1_1\text{H} \rightarrow ^2_1\text{H} + e^+ + \nu$$
$$^2_1\text{H} + ^1_1\text{H} \rightarrow ^3_2\text{He} + \gamma.$$

The ^3_2He particle is a form of helium, but not the common form. Once we have this particle it will combine with another:

$$^3_2\text{He} + ^3_2\text{He} \rightarrow ^4_2\text{He} + ^1_1\text{H} + ^1_1\text{H}.$$

We end up with a normal helium nucleus. A total of six hydrogen nuclei went into the reaction (remember, the first two steps had to occur twice, in order to produce two ^3_2He particles for the final reaction), while there were two left at the end, so the net result is the conversion of four hydrogen nuclei into one helium nucleus.

The CNO Cycle

The CNO cycle, which dominates at higher temperatures, is more complex, involving not only carbon but also nitrogen and oxygen. Each of these elements has more than one form, which differ in their numbers of neutrons. Some of these **isotopes** are unstable and spontaneously emit positrons, decaying into other species in the process. Here is the CNO cycle:

$$^{12}_6\text{C} + ^1_1\text{H} \rightarrow ^{13}_7\text{N} + \gamma$$
$$^{13}_7\text{N} \rightarrow ^{13}_6\text{C} + e^+ + \nu$$
$$^{13}_6\text{C} + ^1_1\text{H} \rightarrow ^{14}_7\text{N} + \gamma$$
$$^{14}_7\text{N} + ^1_1\text{H} \rightarrow ^{15}_8\text{O} + \gamma$$
$$^{15}_8\text{O} \rightarrow ^{15}_7\text{N} + e^+ + \nu$$
$$^{15}_7\text{N} + ^1_1\text{H} \rightarrow ^{12}_6\text{C} + ^4_2\text{He}.$$

Here we end up with a helium nucleus and a carbon nucleus, while the particles going into the reaction were four hydrogen nuclei and a carbon nucleus. Along the way, three isotopes of nitrogen and one of oxygen were created and then converted into something else, leaving neither element at the end. As in the proton-proton chain, the net result is the conversion of four hydrogen nuclei into one helium nucleus.

The Triple-Alpha Reaction

Stars that have used up all the available hydrogen in their cores may become hot enough to undergo a new reaction in which helium is converted into carbon. Helium nuclei, consisting of two protons and two neutrons, are called α particles, and the reaction is called the triple-α reaction, since three of these articles are involved:

$$^4_2\text{He} + ^4_2\text{He} \rightarrow ^8_4\text{Be}$$
$$^8_4\text{Be} + ^4_2\text{He} \rightarrow ^{12}_6\text{C} + \gamma.$$

In this reaction sequence, $^{8}_{4}$Be is a form of the element beryllium, and the end product, $^{12}_{6}$C, is the most common form of carbon. The second step must follow very quickly after the first because $^{8}_{4}$Be is unstable and will break apart into two $^{4}_{2}$He particles in a very short time. Therefore the third $^{4}_{2}$He particle must react with the $^{8}_{4}$Be particle almost immediately; thus this reaction sequence can be viewed as a three-particle reaction.

Other Reactions

Following helium burning in the triple-α reaction, other reactions can take place if the stellar core becomes hot enough. The first of these reactions are **α-capture reactions,** in which one form of nucleus adds an α particle to become a new form with two more protons and two additional neutrons. One example is the carbon α capture:

$$^{12}_{6}\text{C} + ^{4}_{2}\text{He} \rightarrow ^{16}_{8}\text{O} + \gamma,$$

in which $^{16}_{8}$O, the most common form of oxygen, is produced. In another sequence of α captures, nitro-gen is converted into another isotope of oxygen, which can then be converted into neon:

$$^{14}_{7}\text{N} + ^{4}_{2}\text{He} \rightarrow ^{18}_{8}\text{O} + e^{+} + \nu,$$

and

$$^{18}_{8}\text{O} + ^{4}_{2}\text{He} \rightarrow ^{22}_{10}\text{Ne} + \gamma.$$

Additional α-capture reactions can occur, but at high enough temperatures other, more complex, reactions also take place. For example, two carbon nuclei can react, forming a number of different products, including sodium, neon, and magnesium. Two oxygen nuclei can also react, creating such species as sulfur, phosphorus, silicon, and magnesium. As a massive star evolves and its core contracts and heats following each reaction stage, a wide variety of elements is created with generally increasing atomic numbers. As explained in the text, the heaviest and most complex element produced in stable reactions in stellar cores is iron ($^{56}_{26}$Fe). Once a star has an iron core, it cannot undergo any further reaction stages without major disruption by a supernova explosion or collapse to a neutron star or black hole.

Appendix 15
Detected Interstellar Molecules***

Number of Atoms	Molecule	Name	Number of Atoms	Molecule	Name
2	H_2	Molecular hydrogen		OCS	Carbonyl sulfide
	NS	Nitrogen sulfide		HCS^+	Thioformyl cation***
	OH	Hydroxyl		CO_2	Carbon dioxide
	HCl	Hydrogen chloride		C_2O	
	SO	Sulfur monoxide		MgNC	Magnesium (I) isocyanide
	NaCl	Sodium chloride		MgCN	Magnesium (I) cyanide
	NO	Nitric oxide		C_3	Triatomic carbon
	KCl	Potassium chloride		CH_2	Methylene
	SiO	Silicon monoxide		NaCN	Sodium cyanide
	AlCl	Aluminum monochloride	4	NH_3	Ammonia
	SiS	Silicon sulfide		H_3O^+	Hydronium cation***
	AlF	Aluminium monofluoride		H_2CO	Formaldehyde
	SiN			HNCO	Isocyanic acid
	PN			H_2CS	Thioformaldehyde
	SO^+	Sulfur monoxide cation***		HNCS	Isothiocyanic acid
	NH	Nitrogen hydride		C_3N	Cyanoethynyl
	CH	Methylidyne		C_3H (linear)	Propynylidyne
	CH^+	Methylidyne cation***		C_3H (ring)	
	C_2	Diatomic carbon		C_3O	Tricarbon monoxide
	CN	Cyano radical		C_3S	Tricarbon sulfide
	CO	Carbon monoxide		$HOCO^+$	Protonated carbon dioxide***
	CSi	Silicon carbide		HC_2H	Acetylene
	CS	Carbon monosulfide		$HCNH^+$	Protonated hydrogen cyanide***
	CP			HC_2N	
	CO^+	Carbon monoxide cation***		H_2CN	Formininyl radical
3	H_2O	Water		CH_2D^+	Methyl-d cation***
	H_2S	Hydrogen sulfide	5	SiH_4	Silane
	N_2H^+	Protonated nitrogen***		HC_3N	Cyanoacetylene
	SO_2	Sulfur dioxide		C_4H	Butadiynyl
	HNO	Nitroxyl		H_2CNH	Formimine
	SiH_2 (?)			H_2C_2O	Ketene
	H_2D^+	Protonated hydrogen deuteride		NH_2CN	Cyanamide
	NH_2	Amidyl radical		HCOOH	Formic acid
	HCN	Hydrogen cyanide		CH_4	Methane
	HNC	Hydrogen isocyanide		H_2C_3 (linear)	
	C_2H	Ethynyl		H_2C_3 (ring)	Cyclopropenylidene
	C_2S			CH_2CN	Acetonitrile radical
	SiC_2 (ring)	Silicon dicarbide		C_4Si	
	HCO	Formyl radical		HC_2NC	Acetylene isonitrile
	HCO^+	Formyl cation***		HNC_3	
	HOC^+ (?)			C_5	Penta atomic carbon

Number of Atoms	Molecule	Name	Number of Atoms	Molecule	Name
6	CH_3OH	Methanol	8	$HCOOCH_3$	Methyl formate
	CH_3CN	Methyl cyanide		CH_3C_3N	Methyl cyanoacetylene
	CH_3NC	Methyl isocyanide		CH_3COOH (?)	Acetic acid
	CH_3SH	Methyl mercaptan	9	HC_7N	Cyano-hexa-tri-yne
	NH_2CHO	Formamide		$(CH_3)_2O$	Dimethyl ether
	C_2H_4	Ethylene		CH_3CH_2OH	Ethanol
	C_5H	Pentynylidyne		CH_3CH_2CN	Ethyl cyanide
	C_5O	Pentacarbon monoxide		CH_3C_4H	Methyldiacetylene
	HC_2CHO	Propenal		C_8H	Octatetraynyl
	H_2C_4 (linear)	Diacetylene	10	CH_3C_5N (?)	
	HC_3NH^+	Protonated cyanoacetylene***		$(CH_3)_2CO$	Acetone
7	HC_5N	Cyanodiacetylene		NH_2CH_2COOH (?)	Glycine
	CH_3C_2H	Methylacetylene	11	HC_9N	Cyano octatetrayne
	CH_3NH_2	Methylamine	13	$HC_{11}N$	Cyano decapentayne
	CH_3CHO	Acetaldehyde			
	H_2C_2HCN	Vinyl cyanide			
	C_6H	Hexatriynyl			

*A question mark after a molecule indicates that the identification is uncertain, either because the spectral line is weak or because the laboratory identification is uncertain. This list does not include isotopic variations (identical molecules except that one or more atoms are in different isotopic forms, such as ^{13}C in place of ^{12}C, for example).

**There is evidence, but as yet no conclusive demonstration, that classes of much larger molecules than any listed here also exist in the interstellar medium. Most notably, infrared spectra indicate that polycyclic aromatic hydrocarbons (PAHs), which consist of interlocked carbon rings with extra atoms (usually H, O, or N) attached at the edges, are quite common. There is also evidence, less substantial, that fullerenes (cage-like carbon structures) may exist.

***A molecule with a plus (+) superscript carries a net positive charge, and is referred to as a "cation," which is a general term for a positively-charged ion, or it may be "protonated," which means that an extra proton, without accompanying electron, is bonded to it.

Clusters of Galaxies

Galaxies of the Local Group

Galaxy[a]	Type[b]	Absolute Magnitude	Position Right Ascension	Declination
M31 (Andromeda)	Sb	−21.1	00^{h}40.^{m}0	+41°00'
Milky Way	Sbc	−20.5	17 42.5	−28 59
M33 = NGC 598	Sc	−18.9	01 31.1	−30 24
Large Magellanic Cloud	Irr	−18.5	05 24	−69 50
IC 10	Irr	−17.6	00 17.6	+59 02
Small Magellanic Cloud	Irr	−16.8	00 51	−73 10
M32 = NGC 221	E2	−16.4	00 40.0	+40 36
NGC 205	E6	−16.4	00 37.6	+41 25
NCG 6822	Irr	−15.7	19 42.1	−14 53
NCG 185	Dwarf E	−15.2	00 36.1	+48 04
NGC 147	Dwarf E	−14.9	00 30.4	+48 14
IC 1613	Irr	−14.8	01 02.3	+01 51
WLM	Irr	−14.7	23 59.4	−15 44
Fornax	Dwarf sph	−13.6	02 37.5	−34 44
Leo A	Irr	−13.6	09 56.5	+30 59
IC 5152	Irr	−13.5	21 59.6	−51 32
Pegasus	Irr	−13.4	23 26.1	+14 28
Sculptor	Dwarf sph	−11.7	00 57.5	−33 58
And I	Dwarf sph	−11	00 42.8	+37 46
And II	Dwarf sph	−11	01 13.6	+33 11
And III	Dwarf sph	−11	00 32.7	+36 14
Aquarius	Irr	−11	20 44.1	−13 02
Leo I	Dwarf sph	−11	10 0.58	+12 33
Sagittarius	Irr	−10	19 27.1	−17 47
Leo II	Dwarf sph	−9.4	11 10.8	+22 26
Ursa Minor	Dwarf sph	−8.8	15 08.2	+67 18
Draco	Dwarf sph	−8.6	17 19.4	+57 58
Carina	Dwarf sph		06 40.4	−50 55
Pisces	Irr	−8.5	00 01.2	+21 37

[a]Galaxy names are derived from a variety of sources, including several catalogs (such as those designated M, NGC, and IC) and colloquial names bestowed by discoverers. Many in this list are simply named after the constellation where they are found.

[b]The galaxy types listed here are described in the text, except for the *Dwarf sph* designation, which stands for "dwarf spheroidal" and refers to dward galaxies that do not fit easily into the designation of dwarf ellipticals. Note that the absolute magnitudes of some of these are comparble to those of the brightest individual stars in our galaxy.

Clusters of Galaxies within 1,000 Mpc[a]

Cluster	Distance (Mpc)	Radial Velocity (km/sec)	Diameter (Mpc)	Number of Galaxies	Density of Galaxies (Mpc^{-3})
Virgo	19	1,180	4	2,500	500
Pegasus I	65	3,700	1	100	1,100
Pisces	66	250	12	100	250
Cancer	80	4,800	4	150	500
Perseus	97	5,400	7	500	300
Coma	113	6,700	8	800	40
UMa III	132		2	90	200
Hercules	175	10,300	0.3	300	
Cluster A	240	15,800	4	400	200
Centaurus	250		9	300	10
UMa I	270	15,400	3	300	100
Leo	310	19,500	3	300	200
Cluster B	330		4	300	200
Gemini	350	23,300	3	200	100
CrB	350	21,600	3	400	250
Bootes	650	39,400	3	150	100
UMa II	680	41,000	2		400
Hydra	1,000	60,600			

[a]Data from Allen C. W., 1973, *Astrophysical Quantities,* 3d ed. (London: Athlone Press).

Appendix 17
The Messier Catalog

Number	Right Ascension	Declination	Magnitude	Description
M1	05^{h}33.^{m}3	+22°01'	11.3	Crab nebula in Taurus
M2	21 32.4	−00 54	6.3	Globular cluster in Aquarius
M3	13 41.3	+28 29	6.2	Globular cluster in Canes Venatici
M4	16 22.4	−26 27	6.1	Globular cluster in Scorpio
M5	15 17.5	+02 07	6	Globular cluster in Serpens
M6	17 38.9	−32 11	6	Open cluster in Scorpio
M7	17 52.6	−34 48	5	Open cluster in Scorpio
M8	18 02.4	−24 23		Lagoon nebula in Sagittarius
M9	17 18.1	−18 30	7.6	Globular cluster in Ophiuchus
M10	16 56.0	−04 05	6.4	Globular cluster in Ophiuchus
M11	18 50.0	−06 18	7	Open cluster in Scutum
M12	16 46.1	−01 55	6.7	Globular cluster in Ophiuchus
M13	16 41.0	+36 30	5.8	Globular cluster in Hercules
M14	17 36.5	−03 14	7.8	Globular cluster in Ophiuchus
M15	21 29.1	+12 05	6.3	Globular cluster in Pegasus
M16	18 17.8	−13 48	7	Open cluster in Serpens
M17	18 19.7	−16 12	7	Omega nebula in Sagittarius
M18	18 18.8	−17 09	7	Open cluster in Sagittarius
M19	17 01.3	−26 14	6.9	Globular cluster in Ophiuchus
M20	18 01.2	−23 02		Trifid nebula in Sagittarius
M21	18 03.4	−22 30	7	Open cluster in Sagittarius
M22	18 35.2	−23 55	5.2	Globular cluster in Sagittarius
M23	17 55.7	−19 00	6	Open cluster in Sagittarius
M24	18 17.3	−18 27	6	Open cluster in Sagittarius
M25	18 30.5	−19 16	6	Open cluster in Sagittarius
M26	18 44.1	−09 25	9	Open cluster in Scutum
M27	19 58.8	+22 40	8.2	Dumbbell nebula; planetary nebula in Vulpecula
M28	18 23.2	−24 52	7.1	Globular cluster in Sagittarius
M29	20 23.3	+38 27	8	Open cluster in Cygnus
M30	21 39.2	−23 15	7.6	Globular cluster in Capricornus
M31	00 41.6	+41 09	3.7	Andromeda galaxy
M32	00 41.6	+40 45	8.5	Elliptical galaxy, companion of M31
M33	01 32.8	+30 33	5.9	Spiral galaxy in Triangulum
M34	02 40.7	+42 43	6	Open cluster in Perseus
M35	06 07.6	+24 21	6	Open cluster in Gemini
M36	05 35.0	+34 05	6	Open cluster in Auriga
M37	05 51.5	+32 33	6	Open cluster in Auriga
M38	05 27.3	+35 48	6	Open cluster in Auriga
M39	21 31.5	+48 21	6	Open cluster in Cygnus
M40	12 20	+59		Double star cluster in Ursa Major
M41	06 46.2	−20 43	6	Open cluster in Canis Major
M42	05 34.4	−05 24		Orion nebula

M43	05 34.6	−05 18		Small extension of the Orion nebula
M44	08 38.8	+20 04	4	Praesepe; open cluster in Cancer
M45	03 46.3	+24 03	2	The Pleiades; open cluster in Taurus
M46	07 40.9	−14 46	7	Open cluster in Puppis
M47	07 35.6	−14 27	5	Open cluster in Puppis
M48	08 12.5	−05 43	6	Open cluster in Hydra
M49	12 28.8	+08 07	8.9	Elliptical galaxy in Virgo
M50	07 02.0	−08 19	7	Open cluster in Monocerotis
M51	13 29.0	+47 18	8.4	Whirlpool galaxy; spiral galaxy in Canes Venatici
M52	23 23.3	+61 29	7	Open cluster in Cassiopeia
M53	13 12.0	+18 17	7.7	Globular cluster in Coma Berenices
M54	18 53.8	−30 30	7.7	Globular cluster in Sagittarius
M55	19 38.7	−31 00	6.1	Globular cluster in Sagittarius
M56	19 15.8	+30 08	8.3	Globular cluster in Lyra
M57	18 52.9	+33 01	9.0	Ring nebula; planetary nebula in Lyra
M58	12 36.7	+11 56	9.9	Spiral galaxy in Virgo
M59	12 41.0	+11 47	10.3	Elliptical galaxy in Virgo
M60	12 42.6	+11 41	9.3	Elliptical galaxy in Virgo
M61	12 20.8	+04 36	9.7	Spiral galaxy in Virgo
M62	16 59.9	−30 05	7.2	Globular cluster in Scorpio
M63	13 14.8	+42 08	8.8	Spiral galaxy in Canes Venatici
M64	12 55.7	+21 48	8.7	Spiral galaxy in Coma Berenices
M65	11 17.8	+13 13	9.6	Spiral galaxy in Leo
M66	11 19.1	+13 07	9.2	Spiral galaxy, companion of M65
M67	08 50.0	+11 54	7	Open cluster in Cancer
M68	12 38.3	−26 38	8	Globular cluster in Hydra
M69	18 30.1	−32 23	7.7	Globular cluster in Sagittarius
M70	18 42.0	−32 18	8.2	Globular cluster in Sagittarius
M71	19 52.8	+18 44	6.9	Globular cluster in Sagitta
M72	20 52.3	−12 39	9.2	Globular cluster in Aquarius
M73	20 57.8	−12 44		Open cluster in Aquarius
M74	01 35.6	+15 41	9.5	Spiral galaxy in Pisces
M75	20 04.9	−21 59	8.3	Globular cluster in Sagittarius
M76	01 40.9	+51 28	11.4	Planetary nebula in Perseus
M77	02 41.6	−00 04	9.1	Spiral galaxy in Cetus
M78	05 45.8	+00 02		Emission nebula in Orion
M79	05 23.3	−24 32	7.3	Globular cluster in Lepus
M80	16 15.8	−22 56	7.2	Globular cluster in Scorpio
M81	09 54.2	+69 09	6.9	Spiral galaxy in Ursa Major
M82	09 54.4	+69 47	8.7	Irregular galaxy in Ursa Major
M83	13 35.9	−29 46	7.5	Spiral galaxy in Hydra
M84	12 24.1	+13 00	9.8	Elliptical galaxy in Virgo
M85	12 24.3	+18 18	9.5	Elliptical galaxy in Coma Berenices
M86	12 25.1	+13 03	9.8	Elliptical galaxy in Virgo
M87	12 29.7	+12 30	9.3	Giant elliptical galaxy in Virgo
M88	12 30.9	+14 32	9.7	Spiral galaxy in Coma Berenices
M89	12 34.6	+12 40	10.3	Elliptical galaxy in Virgo
M90	12 35.8	+13 16	9.7	Spiral galaxy in Virgo
M91				Not identified; possibly M58
M92	17 16.5	+43 10	6.3	Globular cluster in Hercules
M93	07 43.6	−23 49	6	Open cluster in Puppis
M94	12 50.1	+41 14	8.1	Spiral galaxy in Canes Venatici
M95	10 42.8	+11 49	9.9	Barred spiral galaxy in Leo

M96	10 45.6	+11 56	9.4	Spiral galaxy in Leo
M97	11 13.7	+55 08	11.1	Owl Nebula; planetary nebula in Ursa Major
M98	12 12.7	+15 01	10.4	Spiral galaxy in Coma Berenices
M99	12 17.8	+14 32	9.9	Spiral galaxy in Coma Berenices
M100	12 21.9	+15 56	9.6	Spiral galaxy in Coma Berenices
M101	14 02.5	+54 27	8.1	Spiral galaxy in Ursa Major
M102				Not identified; possibly M101
M103	01 31.9	+60 35	7	Open cluster in Cassiopeia
M104	12 39.0	−11 35	8	Sombrero galaxy; spiral galaxy in Virgo
M105	10 46.8	+12 51	9.5	Elliptical galaxy in Leo
M106	12 18.0	+47 25	9	Spiral galaxy in Canes Venatici
M107	16 31.8	−13 01	9	Globular cluster in Ophiuchus
M108	11 10.5	+55 47	10.5	Spiral galaxy in Ursa Major
M109	11 56.6	+53 29	10.6	Barred spiral galaxy in Ursa Major

Glossary

Absolute magnitude The magnitude a star would have if it were precisely 10 parsecs away from the Sun. (*see also* *magnitude*)

Absolute zero The temperature where all molecular or atomic motion stops, equal to $-273°C$ or $-459°F$.

Absorption line A wavelength at which light is absorbed, producing a dark feature in the spectrum.

Abundance gradient A systematic change in the relative abundances of elements over a distance; particularly in a galaxy.

Acceleration Any change—either of speed or direction—in the state of rest or motion of a body.

Accretion disk A rotating disk of gas surrounding a compact object (such as a neutron star or black hole), formed by material falling in.

Aether The substance of which the heavens and all bodies in the sky were thought to be composed, in ancient Greek cosmology.

Albedo The fraction of incident light that is reflected from a surface, such as that of a planet.

Alpha-capture reaction A nuclear fusion reaction in which an alpha particle merges with an atomic nucleus. A typical example is the formation of ^{16}O by the fusion of an alpha particle with ^{12}C.

Alpha particle A nucleus of ordinary helium containing two protons and two neutrons.

Altitude The angular distance of an object above the horizon.

Amino acid A complex organic molecule of the type that forms proteins. Amino acids are fundamental constituents of all living matter.

Andromeda galaxy The large spiral galaxy located some 700,000 parsecs from the Sun; the most distant object visible to the unaided eye.

Angstrom The unit normally used in measuring wavelengths of visible and ultraviolet light; one angstrom is equal to 10^{-8} centimeter.

Angular diameter The diameter of an object as seen on the sky, measured in units of angle.

Angular momentum A measure of the mass, radius, and rotational velocity of a rotating or orbiting body. In the simple case of an object in circular orbit, the angular momentum is equal to the mass of the object, times its distance from the center of the orbit, times its orbital speed.

Angular resolution The ability of a telescope to measure fine detail, usually expressed in angular units, representing the smallest angular detail that can be discerned.

Annual motions Motions in the sky caused by the Earth's orbital motion about the Sun. These include the seasonal variations of the Sun's declination and the Sun's motion through the zodiac.

Annular eclipse A solar eclipse that occurs when the Moon is near its greatest distance from the Earth, so that its angular diameter is slightly smaller than that of the Sun, and a ring, or annulus, of the Sun's disk is visible surrounding the disk of the Moon.

Anticyclone A rotating wind system around a high-pressure area. On the Earth, an anticyclone rotates clockwise in the Northern Hemisphere and counterclockwise in the Southern Hemisphere.

Antimatter Matter composed of the antiparticles of ordinary matter. For each subatomic particle, there is an antiparticle that is its opposite in such properties as electrical charge, but its equivalent in mass. Matter and antimatter, if combined, annihilate each other, producing energy in the form of gamma rays according to the formula $E = mc^2$.

Aphelion The point in the orbit of a solar system object where it is farthest from the Sun.

Apollo asteroid An asteroid whose orbit brings it closer to the Sun than 1 astronomical unit (AU).

Apparent magnitude The observed magnitude of a star or other celestial object, as seen from the Earth. (*see also* *magnitude*)

Arctic Circle The region extending from the North Pole to 66.5°N latitude, where the Sun stays below the horizon for the full 24-hour day at the time of the winter solstice.

Asteroid (*see* *minor planet*)

Asthenosphere The upper portions of the Earth's mantle, below the zone (the lithosphere) where convection currents are thought to operate. The term is also applied to similar zones in the interiors of the Moon and other planets.

Astration The continual recycling of material from the interstellar medium into stars and back again, which enriches the galaxy in heavy elements.

Astrology The ancient belief that earthly affairs and human lives are influenced by the positions of the Sun, Moon, and planets with respect to the zodiac.

Astrometric binary A double star recognized as such because the visible star or stars undergo periodic motion that is detected by astrometric measurements.

Astrometry The science of accurately measuring stellar positions.

Astronomical unit (AU) A unit of distance used in astronomy, equal to the average distance between the Sun and the Earth; 1 AU is equal to 1.4959787×10^8 km.

Astronomy The science that deals with the universe beyond the Earth's atmosphere.

Astrophysics The application of physical laws to astronomical phenomena; interchangeable with astronomy in modern usage.

Atomic number   The number of protons in the nucleus of an element. The atomic number defines the identity of an element.

Atomic weight   The mass of an atomic nucleus in atomic mass units [one atomic mass unit (amu) is defined as the average mass of the protons and neutrons in a nucleus of ordinary carbon, ^{12}C]. For most atoms, the atomic weight is approximately equal to the total number of protons and neutrons in the nucleus.

Aurorae australis   "Southern lights"; the visual emission from the Earth's upper atmosphere, caused by charged particles from space.

Aurorae borealis   "Northern lights"; (*see Aurorae australis*)

Autumnal equinox   The name used in the Northern Hemisphere for the point where the Sun crosses the celestial equator from north to south, around September 21. (*see also equinox and vernal equinox*)

Bailey's beads   Small glowing regions seen at the edges of the Moon's disk during a solar eclipse, due to irregularities in the shape of the lunar disk, which allows sunlight to pass through to the Earth.

Barred spiral galaxy   A spiral galaxy whose nucleus has linear extensions on opposing sides, giving it a barlike shape. The spiral arms usually appear to emanate from the ends of the bar.

Basalt   An igneous silicate rock common in regions formed by lava flows on the Earth, the Moon, and probably on the other terrestrial planets.

Beta decay   A spontaneous nuclear reaction in which a neutron decays into a proton, and an electron, and a neutrino. The term has been generalized to mean any spontaneous reaction in which an electron and a neutrino (or their antimatter equivalents) are emitted.

Big bang   A term referring to any theory of cosmogony in which the universe began at a single point, was very hot initially, and has been expanding from that state since.

Binary star   A double star system in which the two stars orbit a common center of mass.

Binary X-ray source   A close binary system containing a compact stellar remnant which emits X rays as it accretes matter from its companion.

BL Lac object   A class of active galactic nuclei in which it is thought that the accretion disk is viewed nearly face-on, so that one of the axial jets is directed nearly along the line of sight toward the Earth.

Black body   A body that absorbs all radiation that strikes it, reflecting none, so that its emitting properties are determined entirely by its surface temperature according to the laws of thermal radiation.

Black hole   An object that has collapsed under its own gravitation to such a small radius that its gravitational force traps photons of light.

Bode's law   Also known as the Titius-Bode relation, a simple numerical sequence that approximately represents the relative distances of the inner seven planets from the Sun. The relation is derived by adding 4 to each number in the sequence 0, 3, 6, 12, . . . , and then dividing each sum by 10, resulting in the sequence 0.4, 0.7, 1.0, 1.6, . . . , which is approximately the sequence of planetary distances from the Sun in astronomical units.

Bolide   An extremely bright meteor that explodes in the upper atmosphere.

Bolometric correction   The difference between the bolometric magnitude and the visual magnitude for a star; the bolometric correction is always negative because the star is brighter when all wavelengths are included.

Bolometric magnitude   A magnitude in which all wavelengths of light are included.

Breccia   Lunar rocks consisting of pebbles and soil fused together by meteorite impacts.

Burster   A sporadic source of intense X-rays, probably consisting of a neutron star onto which new matter falls at irregular intervals.

Carbonaceous chondrite   A meteorite containing chondrules having a high abundance of carbon and other volatile elements. Carbonaceous chondrites, thought to be very old, have apparently been unaltered since the formation of the solar system.

Cassegrain focus   A focal arrangement for a reflecting telescope in which a convex secondary mirror reflects the image through a hole in the primary mirror to a focus at the bottom of the telescope tube. This arrangement is commonly used in situations where relatively lightweight instruments for analyzing the light are attached directly to the telescope.

Cataclysmic variable   A binary system in which mass exchange causes irregular outbursts, usually involving a white dwarf which accretes mass from its companion.

Catastrophic theory   Any theory in which observed phenomena are attributed to sudden changes in conditions or to the intervention of an outside force or body.

CCD   Charge-coupled device; an electronic detector that receives photons of light and converts them into electrical charge patterns which are then converted into digital data. (*see also detector*)

Celestial equator   The imaginary circle formed by the intersection of the Earth's equatorial plane with the celestial sphere. The celestial equator is the reference line for north-south (declination) measurements in the standard equatorial coordinate system.

Celestial pole   The point on the celestial sphere directly overhead at either of the Earth's poles.

Celestial sphere   The imaginary sphere formed by the sky. It is a convenient device for discussing and measuring the positions of astronomical objects.

Center of mass   In a binary star system, or any system consisting of several objects, the point about which the mass is "balanced"; that is, the point that moves with a constant velocity through space while the individual bodies in the system move about it.

Cepheid variable A pulsating variable star, of a class named after the prototype δ Cephei. Cepheid variables obey a period-luminosity relationship and are therefore useful as distance indicators. There are two classes of Cepheid variables, the so-called classical Cepheids, which belong to Population I, and the Population II Cepheids, also known as W Virginis stars.

Chondrite A stony meteorite containing chondrules.

Chondrule A spherical inclusion in certain meteorites, usually composed of silicates and always of very great age.

Chromatic aberration The creation of images that are dispersed according to wavelength perpendicular to the focal plane, caused by the use of lenses in refracting telescopes; different wavelengths of light are brought to a focus at different positions.

Chromosphere A thin layer of hot gas just outside the photosphere in the Sun and other cool stars. The temperature in the chromosphere rises from about 4,000 K at its inner edge to 10,000 or 20,000 K at its outer boundary. The chromosphere is characterized by the strong red emission line of hydrogen.

Closed universe A possible state of the universe in which the expansion will eventually be reversed and which is characterized by positive curvature, being finite in extent but having no boundaries.

Cluster variable (see **RR Lyrae variable**)

CNO cycle A nuclear fusion reaction sequence in which hydrogen nuclei are combined to form helium nuclei, and other nuclei, such as isotopes of carbon, oxygen, and nitrogen, appear as catalysts or by-products. The CNO cycle is dominant in the cores of stars on the upper main sequence.

Color-magnitude diagram The equivalent of an H-R diagram, for a cluster of stars. Because the stars are at a common distance, the apparent magnitude may be plotted on the vertical axis in place of the absolute magnitude; the horizontal axis is usually the color index ($B-V$), rather than spectral type.

Color excess The amount, measured in magnitude units, by which a star appears redder in color due to interstellar dust than it would without dust in the line of sight. Technically the definition is $E(B-V) = (B-V)_{observed} - (B-V)_{intrinsic}$, where ($B-V$) is the color index.

Color index The difference $B-V$ between the blue (B) and visual (V) magnitudes of a star. If B is less than V (that is, if the star is brighter in blue than in visual light), the star has a negative color index and is a relatively hot star. If B is greater than V, the color index is positive, and the star is relatively cool.

Coma The extended glowing region that surrounds the nucleus of a comet.

Comet An interplanetary body, composed of loosely bound rocky and icy material, which forms a glowing head and extended tail when it enters the inner solar system.

Compressional waves Waves in which the oscillations are in the direction of wave motion. (see also **P wave**)

Configuration The position of a planet or the Moon relative to the Sun-Earth line.

Conjunction The alignment of two celestial bodies on the sky. In connection with the planets, a conjunction is the alignment of a planet with the Sun, an inferior conjunction being the occasion when an inferior planet is directly between the Sun and the Earth, and a superior conjunction being the occasion when any planet is directly behind the Sun as seen from the Earth.

Constellation A prominent pattern of bright stars, historically associated with mythological figures. In modern usage each constellation incorporates a precisely defined region of the sky.

Constructive interference The overlapping of waves (including light waves) in which the peaks and valleys of separate waves match each other, adding to their strength (or intensity, in the case of light).

Continental drift The slow motion of the continental masses over the surface of the Earth, caused by the motions of the Earth's tectonic plates, which in turn are probably caused by convection in the underlying asthenosphere.

Continuous radiation Electromagnetic radiation that is emitted in a smooth distribution with wavelength, without spectral features such as emission and absorption lines.

Convection The transport of energy by fluid motions occurring in gases, liquids, or semirigid material such as the Earth's mantle. These motions are usually driven by the buoyancy of heated material, which tends to rise while cooler material descends.

Co-orbital satellites Satellites that share the same orbit; usually one or more small satellites orbiting 60° ahead or 60° behind a larger one, kept in place by the combined gravitation effects of the large satellite and the parent planet.

Coriolis force The apparent force felt by a particle moving away from the equator on a spinning body such as a planet; the particle veers in the direction of planetary rotation due to the fact the surface rotational speed decreases with distance from the equator.

Corona The very hot, extended outer atmosphere of the Sun and the other cool, main-sequence stars. The high temperature in the corona ($1 - 2 \times 10^6$ K) is probably caused by the dissipation of mechanical energy from the convective zone just below the photosphere.

Coronal holes Regions of relatively low density in the solar corona, from which the solar wind emanates.

Cosmic background radiation The primordial radiation field that fills the universe, having been created in the form of gamma rays at the time of the big bang, but having cooled since so that today its temperature is 2.73 K, and its peak wavelength is near 1.1 millimeters, in the microwave portion of the spectrum. Also known as the 3° background radiation.

Cosmic ray A rapidly moving atomic nucleus from space. Cosmic rays are produced in the Sun, while others come from interstellar space and probably originate in supernova explosions.

Cosmogony The study of the origins of the universe.

Cosmological constant   A term added to the field equations by Einstein to allow solutions in which the universe was static (neither expanding nor contracting). Although the need for the term disappeared when it was discovered that the universe is expanding, the cosmological constant is retained in the field equations by modern cosmologists but is usually assigned the value zero.

Cosmological principle   The postulate, made by most cosmologists, that the universe is both homogeneous and isotropic. It is sometimes stated as, "The universe looks the same to all observers everywhere."

Cosmological redshift   A Doppler shift toward longer wavelengths that is caused by a galaxy's motion of recession due to the expansion of the universe.

Cosmology   The study of the universe as a whole.

Coudé focus   A focal arrangement for a reflecting telescope, in which the image is reflected by a series of mirrors to a remote, fixed location where a massive, immovable instrument can be used to analyze it.

Crater   A depression (usually circular) in the surface of a planet or satellite, caused by either an impact from space or volcanic activity.

Cyclone   A rotating wind system about a low-pressure center, often associated with storms on the Earth. On the Earth, cyclones rotate counterclockwise in the Northern Hemisphere and clockwise in the Southern Hemisphere.

Dark matter   The invisible material postulated to account for the majority of mass in clusters of galaxies and in the universe as a whole.

Declination   The coordinate in the equatorial system that measures positions in the north-south direction, with the celestial equator as the reference line. Declinations are measured in units of degrees, minutes, and seconds of arc.

Deferent   The large circle centered on or near the Earth on which the epicycle for a given planet moved, in the geocentric theory of the solar system developed by ancient Greek astronomers such as Hipparchus and Ptolemy.

Degenerate gas   A gas in which either free electrons or free neutrons are as densely spaced as allowed by laws of quantum mechanics. Such a gas has extraordinarily high density, and its pressure is not dependent on temperature as it is in an ordinary gas. Degenerate electron gas provides the pressure that supports white dwarfs against collapse, and degenerate neutron gas similarly supports neutron stars.

Density wave   A stable pattern of alternating dense and rarified regions, usually in a rotating fluid disk such as the ring system of Saturn or a spiral galaxy.

Deoxyribonucleic acid (DNA)   A complex protein consisting of a double helix structure composed of amino acid bases, responsible for carrying the genetic code in the nuclei of cells.

Destructive interference   The overlapping of waves (including light waves) in which the peaks of one wave coincide with the valleys of the other, diminishing both (and reducing the intensity, in light).

Detector   The general term for any device that receives and records the intensity of light at the focus of a telescope. Photographic film has been used traditionally in astronomy, but modern telescopes often use electronic devices, such as CCDs, instead.

Deuterium   An isotope of hydrogen containing in its nucleus one proton and one neutron.

Differential gravitational force   A gravitational force acting on an extended object, so that the portions of the object closer to the source of gravitation feel a stronger force than the portions that are farther away. Such a force, also known as a tidal force, acts to deform or disrupt the object and is responsible for many phenomena, ranging from synchronous rotation of moons or double stars to planetary ring systems and the disruption of galaxies in clusters.

Differentiation   The sinking of relatively heavy elements into the core of a planet or other body. Differentiation can only occur in fluid bodies, so any planet that has undergone this process must once have been at least partially molten.

Diffraction   The process in which waves (including light waves) bend as they pass an obstacle.

Dipole anisotropy   (see 24 hour anisotropy)

Distance modulus   The difference $m - M$ between the apparent and absolute magnitudes for a given star. This difference, which must be corrected for the effects of interstellar extinction, is a direct measure of the distance to the star.

Diurnal motion   Any motion related to the rotation of the Earth. Diurnal motions include the daily risings and settings of all celestial objects.

DNA   (see deoxyribonucleic acid)

Doppler effect   The shift in wavelength of light or sound caused by relative motion between the source (or sound) and the observer.

Doppler shift   The observed shift in wavelength (and frequency) of a wave due to relative motion between the source of the wave and the observer.

Dwarf elliptical galaxy   A member of a class of small spheroidal galaxies similar to standard elliptical galaxies except for their small size and low luminosity. Dwarf galaxies are probably the most common in the universe but cannot be detected at distances beyond the Local Group of galaxies.

Dwarf nova   A close binary system containing a white dwarf in which material from the companion star falls onto the other at sporadic intervals, creating brief nuclear outbursts.

Eccentricity   A measure of the degree to which an elliptical orbit is elongated; technically, the eccentricity is equal to the ratio of the distance between the foci to the length of the major axis.

Eclipse   An occurrence in which one object is partially or totally blocked from view by another or passes through the shadow of another.

Eclipsing binary   A double star system in which one or both stars are periodically eclipsed by the other as seen from Earth. This situation can occur only when the orbital plane of the binary is viewed edge-on from the Earth.

Ecliptic   The plane of the Earth's orbit about the Sun, which is approximately the plane of the solar system as a whole. The apparent path of the Sun across the sky is the projection of the ecliptic onto the celestial sphere.

Ejecta   Material blasted out of the ground by a meteorite impact.

Electromagnetic force   The force created by the interaction of electric and magnetic fields. The electromagnetic force can be either attractive or repulsive and is important in countless situations in astrophysics.

Electromagnetic radiation   Waves consisting of alternating electric and magnetic fields. Depending on the wavelength, these waves may be known as gamma rays, Xrays, ultraviolet radiation, visible light, infrared radiation, or radio radiation.

Electromagnetic spectrum   The entire array of electromagnetic radiation arranged according to wavelength.

Electron   A tiny, negatively charged particle that orbits the nucleus of an atom. The charge is equal and opposite to that of a proton in the nucleus, and in a normal atom the number of electrons and protons is equal so that the overall electrical charge is zero. The electrons emit and absorb electromagnetic radiation by making transitions between fixed energy levels.

Ellipse   A geometrical shape such that the sum of the distances from any point on it to two fixed points called foci is constant. In any bound system where two objects orbit a common center of mass, their orbits are ellipses with the center of mass at one focus.

Elliptical galaxy   One of a class of galaxies characterized by smooth spheroidal forms, few young stars, and little interstellar matter.

Emission line   A wavelength at which radiation is emitted, creating a bright line in the spectrum.

Emission nebula   A cloud of interstellar gas that glows by the light of emission lines. The source of excitation that causes the gas to emit may be radiation from a nearby star or heating by any of a variety of mechanisms.

Endothermic reaction   Any nuclear or chemical reaction that requires more energy to occur than it produces.

Energy   The ability to do work. Energy can be in either kinetic form, when it is a measure of the motion of an object, or potential form, when it is stored but capable of being released into kinetic form.

Epicycle   A small circle on which a planet revolves, which in turn orbits another, distant body. Epicycles were used in ancient theories of the solar system to devise a cosmology that placed the Earth at the center but accurately accounted for the observed planetary motions.

Equatorial coordinates   The astronomical coordinate system in which positions are measured with respect to the celestial equator (in the north-south direction) and a fixed direction (in the east-west dimension). The coordinates used are declination (north-south, in units of angle) and

right ascension (east-west, in units of time).

Equinox   Either of two points on the sky where the planes of the ecliptic and the Earth's equator intersect. When the Sun is at one of these two points, the lengths of night and day on the Earth are equal. (*see also **autumnal equinox** and **vernal equinox***)

Erg   A unit of energy equal to the kinetic energy of an object of 2 grams mass moving at a speed of 1 centimeter per second, but defined technically as the work required to move a mass of 1 gram through a distance of 1 centimeter at an acceleration of 1 cm/sec².

Escape speed   The upward speed required for an object to escape the gravitational field of a body such as a planet. In a more technical sense, the escape speed is the speed at which the kinetic energy of the object equals its gravitational potential energy; if the object moves any faster, its kinetic energy exceeds its potential energy, and it can escape the gravitational field.

Event horizon   The "surface" of a black hole; the boundary of the region from within which no light can escape.

Evolutionary theory   Any theory in which observed phenomena are thought to have arisen as a result of natural processes requiring no outside intervention or sudden changes.

Excitation   A process by which one or more electrons of an atom or ion are raised to energy levels above the lowest possible one.

Excited state   A state in which an atom, ion, or molecule has more energy than the lowest possible state (known as the ground state). For an atom or ion, this usually refers to the energy state of one or more electrons; for a molecule, the excited state may refer to an electron energy state, a vibrational energy state, or a rotational energy state.

Field equations   A set of equations in general relativity theory that describe the relationship of matter, energy, and gravitation in the universe.

Fission reaction   A nuclear reaction in which a large nucleus is split into one or more smaller nuclei.

Flat universe   A possible state of the universe in which the momentum of expansion is exactly balanced by self-gravitation, so that the expansion will slow to a stop in an infinite time, but will not reverse itself and become a contraction. This state is predicted to exist by inflationary universe models.

Fluorescence   The emission of light at a particular wavelength following excitation of the electron by absorption of light at another, shorter wavelength.

Focus   (1) The point at which light collected by a telescope is brought together to form an image; (2) one of two fixed points that define an ellipse. (*see also **ellipse***)

Force   Any agent or phenomenon that produces acceleration of a mass.

Fraunhofer lines   The series of prominent solar absorption lines identified and cataloged in the early nineteenth century by Josef Fraunhofer, based on visual observations made with a spectroscope.

Frequency   The rate (in units of hertz, or cycles per second) at which electromagnetic waves pass a fixed point. The frequency, usually designated ν, is related to the wavelength λ and the speed of light c by $\nu = c/\lambda$.

Fusion reaction   A nuclear reaction in which atomic nuclei combine to form more massive nuclei.

Galactic cluster   A loose cluster of stars located in the disk or spiral arms of the galaxy.

Gamma ray   A photon of electromagnetic radiation whose wavelength is very short and whose energy is very high. Radiation whose wavelength is less than one angstrom is usually considered to be gamma-ray radiation.

Gegenschein   The diffuse glowing spot, seen on the ecliptic opposite the Sun's direction, created by sunlight reflected off interplanetary dust.

Globular cluster   A large spherical cluster of stars located in the halo of the galaxy. These clusters, containing up to several hundred thousand members, are thought to be among the oldest objects in the galaxy.

Gram   A unit of mass, equal to the quantity of mass contained in 1 cubic centimeter of water.

Grand Unified Theory   A theory being sought by particle physicists in which three of the four fundamental forces (the weak and strong nuclear forces, and the electromagnetic force) are shown to be manifestations of the same phenomenon, and indistinguishable from each other under conditions of sufficiently high temperature and pressure.

Granulation   The spotty appearance of the solar surface (the photosphere) caused by convection in the layers just below.

Gravitational lens   The focusing of light from a distant object by the gravitational field of a foreground object, caused by the curvature of spacetime in the vicinity of mass.

Gravitational redshift   A Doppler shift toward long wavelengths caused by the effect of a gravitational field on photons of light. Photons escaping a gravitational field lose energy to the field, which results in the redshift.

Greatest elongation   The greatest angular distance from the Sun that an inferior planet can reach, as seen from Earth.

Great Red Spot   An oval-shaped, reddish feature on Jupiter's surface, thought to be a long-lived storm system.

Greenhouse effect   The trapping of heat near the surface of a planet by atmospheric molecules (such as carbon dioxide) that absorb infrared radiation emitted by the surface.

Half-life   The time required for half of the nuclei of an unstable (radioactive) isotope to decay.

Halo   (1) The extended outer portions of a galaxy (thought to contain a large fraction of the total mass of the galaxy, mostly in the form of dim stars and interstellar gas); (2) the extensive cloud of gas surrounding the head of a comet.

Helium flash   A rapid burst of nuclear reactions in the degenerate core of a moderate mass star in the hydrogen shell-burning phase. The clash occurs when the core temperature reaches a sufficiently high temperature to trigger the triple-alpha reaction.

Herbig-Haro object   A bright object often associated with young stars, thought to be a region of ionization caused by the jets of high-speed gas emanating from the polar regions of newly-formed stars.

Hertzsprung-Russell diagram   A diagram on which stars are represented according to their absolute magnitudes (on the vertical axis) and spectral types (on the horizontal axis). Because the physical properties of stars are interrelated, they do not fall randomly on such a diagram but lie in well-defined regions according to their state of evolution. Very similar diagrams can be constructed using luminosity instead of absolute magnitude and temperature or color index in place of spectral type.

Hertz   A unit of frequency used in describing any oscillation phenomena, such as electromagnetic radiation; 1 hertz (1 Hz) is equal to one cycle or wave per second.

High-velocity star   A star whose velocity relative to the solar system is large. As a rule, high-velocity stars are Population II objects following orbital paths that are highly inclined to the plane of the galactic disk.

Horizontal branch   A sequence of stars in the H-R diagram of a globular cluster, extending horizontally across the diagram to the left from the red giant region. These are probably stars undergoing helium burning in their cores by the triple-alpha reaction.

H-R diagram   (see *Hertzsprung-Russell diagram*)

H II region   A volume of ionized gas surrounding a hot star. (see also *emission nebula*)

Hubble constant   The numerical factor, usually denoted H, which describes the rate of expansion of the universe. It is the proportionality constant in the Hubble law $v = Hd$, which relates the speed of recession of a galaxy (v) to its distance (d). The present value of H is not well known, with estimates ranging between 55 and 90 kilometers per second per megaparsec.

Hydrostatic equilibrium   The state of balance between gravitational and pressure forces that exists at all points inside any stable object such as a star or planet.

Igneous rock   A rock that was formed by cooling and hardening from a molten state.

Impact crater   A crater formed on the surface of a terrestrial planet or a satellite by the impact of a meteoroid or planetesimal.

Inertia   The tendency of an object to remain in its state of rest or of uniform motion. This tendency is directly related to the mass of the object.

Inferior planet   One of the planets whose orbits lie closer to the Sun than that of the Earth (Mercury or Venus).

Inflationary universe   A theory of cosmology in which the universe expanded rapidly from a very compact and homogeneous initial state. This early rapid expansion was followed by the big bang expansion, which still continues.

Infrared radiation   Electromagnetic radiation in the wavelength region just longer than that of visible light; that is, radiation whose wavelength lies roughly between 7,000 Å and 0.01 centimeter.

Interferometry   The use of interference phenomena in electromagnetic waves to measure positions precisely or to achieve gains in resolution. Interferometry in radio astronomy entails the use of two or more antennae to overcome the normally very coarse resolution of a single radio telescope; in visible-light observations, the object is to eliminate the distorting effects of the Earth's atmosphere.

Interplanetary dust   Tiny solid particles in interplanetary space, concentrated in the ecliptic plane.

Interstellar cloud   A region of relatively high density in the interstellar medium. Interstellar clouds have densities ranging between 1 and 10^6 particles per cubic centimeter and, in aggregate, contain most of the mass in interstellar space.

Interstellar dust   The diffuse medium of tiny solid particles that permeates the space between the stars in a galaxy.

Interstellar extinction   The obscuration of starlight by interstellar dust. Light is scattered off dust grains, so that a distant star appears dimmer than it otherwise would. The scattering process is most effective at short (blue) wavelengths, so that stars seen through interstellar dust are reddened and dimmed.

Inverse Compton scattering   Scattering of photons by relativistic electrons, in which photon energies are increased.

Inverse square law   In general, any law describing a force or other phenomenon that decreases in strength as the square of the distance from some central reference point. In particular the term *inverse square law* is often used by itself to mean the law stating that the intensity of light emitted by a source such as a star diminishes as the square of the distance from the source.

Ion   Any subatomic particle with a nonzero electrical charge. In standard practice, the term *ion* is usually applied only to positively charged particles, such as atoms missing one or more electrons.

Ionization   Any process by which an electron or electrons are freed from an atom or ion. Ionization generally occurs in two ways: by the absorption of a photon with sufficient energy or by collision with another particle.

Ionosphere   The zone of the Earth's upper atmosphere, between 80 and 500 kilometers altitude, where charged subatomic particles (chiefly protons and electrons) are trapped by the Earth's magnetic field. (*see also* **Van Allen belts**)

Io torus   A zone of gas particles concentrated in the orbit of Io (the innermost major satellite of Jupiter); the gas originates in volcanic eruptions in Io, and is dispersed throughout the satellite's orbit by magnetic forces, once it becomes ionized.

Isotope   Any form of a given chemical element. Different isotopes of the same element have the same number of protons in their nuclei, but different numbers of neutrons.

Isotropic   Having the property of appearing the same in all directions. In astronomy, this term is often postulated to apply to the universe as a whole.

Jeans criterion   The condition required for a cloud of gas to spontaneously collapse due to its own self-gravitation, usually expressed in terms of the size (or mass) of the cloud relative to its density and temperature.

Joule   A unit of energy defined as the work required to move a mass of 1 kilogram through a distance of 1 meter at an acceleration of 1 m/sec².

Kelvin   A unit of temperature, equal to 0.01 of the difference between the freezing and boiling points of water, used in a scale whose zero point is absolute zero. A Kelvin is usually denoted simply by K.

Kilogram   A unit of mass, equal to 1,000 grams.

Kiloparsec   A unit of distance, equal to 1,000 parsecs.

Kinetic energy   The energy of motion. The kinetic energy of a moving object is equal to ½ times its mass times the square of its velocity.

Kirkwood's gaps   Narrow gaps in the asteroid belt created by orbital resonance with Jupiter.

Kuiper belt   A disk-like collection of Sun-orbiting bodies thought to be concentrated between 30 and 50 AU from the Sun; the source of most periodic comets.

Latitude   The distance north or south of the Earth's equator, measured in units of angle.

Lepton   One of the two fundamental classes of elementary particles; includes electrons.

Light curve   A graph showing the intensity of light (or magnitude) of a celestial object versus time. Light curves are often useful in diagnosing the properties of variable stars and other objects.

Light-gathering power   The ability of a telescope to collect light from an astronomical source; the light-gathering power is directly related to the area of the primary mirror or lens.

Limb darkening   The dark region around the edge of the visible disk of the Sun or of a planet caused by a decrease in temperature with height in the atmosphere.

Liquid metallic hydrogen   Hydrogen in a state of semi-rigidity that can exist only under conditions of extremely high pressure, as in the interiors of Jupiter and Saturn.

Lithosphere   The layer in the Earth, Moon, and terrestrial planets that includes the crust and the outer part of the mantle.

Local Group   The cluster of about thirty galaxies to which the Milky Way belongs.

Logarithm   The logarithm of a number is the power to which 10 must be raised to equal that number. For example, $100 = 10^2$; so the logarithm of 100 is 2.

Luminosity   The total energy emitted by an object per second (the power of the object). For stars the luminosity is usually measured in units of ergs per second.

Luminosity class   One of several classes to which a star can be assigned on the basis of certain luminosity indicators in its spectrum. The classes range from I for supergiants to V for main-sequence stars (also known as dwarfs).

Luminous arcs   Enormous arc-like glowing structures seen in some dense clusters of galaxies. These are thought to be distorted images of background galaxies or quasars, formed by the bending of light due to the cluster gravitation field. (*see also **gravitational lens***)

Lunar eclipse   An eclipse of the Moon, caused by its passage through the shadow of the earth.

Lunar month   The synodic period of the Moon, equal to 29 days, 12 hours, 44 minutes, 11 seconds.

L wave   A type of seismic wave that travels only over the surface of the Earth.

Magellanic Clouds   The two irregular galaxies that are the nearest neighbors to the Milky Way and are visible to the unaided eye in the Southern Hemisphere.

Magma   Molten rock from the Earth's interior.

Magnetic braking   The slowing of the spin of a young star such as the early Sun by magnetic forces exerted on the surrounding ionized gas.

Magnetic dynamo   A rotating internal zone inside the Sun or a planet, thought to carry the electrical currents that create the solar or planetary magnetic field.

Magnetosphere   The region surrounding a star or planet that is permeated by the magnetic field of that body.

Magnitude   A measure of the brightness of a star, based on a system established by Hipparchus in which stars were ranked according to how bright they appeared to the unaided eye. In the modern system, a difference of 5 magnitudes corresponds exactly to a brightness ratio of 100, so that a star of a given magnitude has a brightness that is $100^{1/5} = 2.512$ times that of a star 1 magnitude fainter.

Main sequence   The strip in the H-R diagram, running from upper left to lower right, where most stars that are converting hydrogen to helium by nuclear reactions in their cores are found.

Main-sequence fitting   A distance-determination technique in which an H-R diagram for a cluster of stars is compared with a standard H-R diagram to establish the absolute magnitude scale for the cluster H-R diagram.

Main-sequence turnoff   In an H-R diagram for a cluster of stars the point where the main-sequence turns off toward the upper right. The main-sequence turnoff, showing which stars in the cluster have evolved to become red giants, is an indicator of the age of the cluster.

Mantle   The semirigid outer portion of the Earth's interior extending from roughly the midway point nearly to the surface and consisting of the mesosphere (the lower portion) and the asthenosphere.

Mare (*pl.* maria)   Any of several extensive, smooth lowland areas on the surface of the Moon or Mercury that were created by extensive lava flows early in the history of the solar system.

Mass   A measure of the quantity of matter contained in an object.

Mass-luminosity relation   The correspondence between the masses of stars and their luminosities; first found empirically, this relation has since been duplicated by theoretical models of stellar structure.

Mass-to-light ratio   The mass of a galaxy, in units of solar masses, divided by its luminosity, in units of the Sun's luminosity. The mass-to-light ratio is an indicator of the relative quantities of Population I and Population II stars in a galaxy.

Maunder minimum   An interval during the latter half of the seventeenth century when the number of sunspots was abnormally low.

Maxwell distribution   A mathematical representation of the distribution of particle speeds in a gas of a certain temperature.

Mean solar day   The average length of the solar day as measured throughout the year; the mean solar day is precisely 24 hours.

Megaparsec (Mpc)   A unit of distance, equal to 10^6 parsecs.

Meridian   The great circle on the celestial sphere that passes through both poles and directly overhead; that is, the north-south line directly overhead.

Mesosphere   (1) The layer of the Earth's atmosphere between roughly 50 and 80 kilometers in altitude, where the temperature decreases with height; (2) the layer below the asthenosphere in the Earth's mantle.

Metamorphic rock   A rock formed by heat and pressure in the Earth's interior.

Meteor   A bright streak or flash of light created when a meteoroid enters the Earth's atmosphere from space.

Meteorite   The remnant of a meteoroid that survives a fall through the Earth's atmosphere and reaches the ground.

Meteoroid   A small, interplanetary body.

Meteor shower   A period during which meteors are seen with high frequency, occurring when the Earth passes through a swarm of meteoroids.

Micrometeorite   A microscopically small meteorite.

Microwave background   (*see **cosmic background radiation***)

Milky Way   Historically, the diffuse band of light stretching across the sky; our cross-sectional view of the disk of our galaxy. In modern usage, the term *Milky way* refers to our galaxy as a whole.

Minor planet   One of thousands of large (up to 1000 km in diameter) non-planetary bodies orbiting the Sun; most of the orbits lie between Mars and Jupiter.

Moving cluster method   A distance-determination technique in which the radial velocities of stars in a cluster are combined with knowledge of the direction of motion gained from observed proper motions, so that the true space velocity can be used in combination with the proper motion to derive the distance.

Neutrino A subatomic particle without electrical charge, and possibly without mass, that is emitted in certain nuclear reactions.

Neutron A subatomic particle with no electrical charge and a mass nearly equal to that of the proton. Neutrons and protons are the chief components of the atomic nucleus.

Neutron-capture reactions Nuclear reactions in which neutrons are captured by nuclei. (*see also r-process reactions and s-process reactions*)

Neutron star A very compact, dense stellar remnant whose interior consists entirely of neutrons, and which is supported against collapse by degenerate neutron gas pressure.

Newtonian focus A focal arrangement for reflecting telescopes in which a flat mirror is used to reflect the image through a hole in the side of the telescope tube.

Nonthermal radiation Radiation not due only to the temperature of an object. The term is most often applied to sources of continuous radiation such as synchrotron radiation.

North celestial pole The projection of the Earth's North Pole onto the sky.

Nova A star that temporarily flares up in brightness, most likely as the result of nuclear reactions caused by the deposition of new nuclear fuel on the surface of a white dwarf in a binary system. (*see also recurrent nova*)

Nucleus The central, dense concentration in an atom, a comet, or a galaxy.

OB association A group of young stars whose luminosity is dominated by O and B stars.

Obliquity A measure of the tilt of a planet's rotation axis; technically, the angle between the rotation axis and the perpendicular to the orbital plane.

Occam's razor The principle that the simplest explanation of any natural phenomenon is most likely the correct one. *Simple* in this case usually means requiring few assumptions or unverifiable postulates.

Olbers' paradox The apparent conflict created by the fact that in a universe that is infinite in extent and in age, every line of sight should intersect a stellar surface, producing a uniformly bright nighttime sky; the fact that the sky is dark demonstrates that the universe is not infinite in both extent and age.

Oort cloud The cloud of bodies, hypothesized to be orbiting the Sun at a great distance, from which comets originate.

Open cluster A loosely-bound association of stars in the galactic disk. (*see also galactic cluster*)

Open universe A possible state of the universe in which its expansion will never stop, and it is characterized by negative curvature, being infinite in extent and having no boundaries.

Opposition A planetary configuration in which a superior planet is positioned exactly in the opposite direction from the Sun as seen from Earth.

Optical double A pair of stars that happen to appear near each other on the sky but are not in orbit; not a true binary.

Orbital resonance A situation in which the periods of two orbiting bodies are simple multiples of each other so that they are frequently aligned and gravitational forces due to the outer body may move the inner body out of its original orbit. This is one mechanism thought responsible for creating the gaps in the rings of Saturn and Kirkwood's gaps in the asteroid belt.

Organic molecule Any of a large class of carbon-bearing molecules found in living matter.

Outgassing The process by which gases escape from a planetary interior, often through volcanic venting or eruptions.

Ozone A form of oxygen containing three oxygen atoms bonded together.

Paleomagnetism Vestigial traces or artifacts in rocks of ancient magnetic fields.

Parallax Any apparent shift in position caused by an actual motion or displacement of the observer. (*see also stellar parallax*)

Parsec A unit of distance, equal to the distance to a star whose stellar parallax is 1 arcsecond. A parsec is equal to 206, 265 AU, 3.03×10^{13} kilometers, or 3.26 light-years.

Peculiar velocity The deviation in a star's velocity from perfect circular motion about the galactic center.

Penumbra (1) The light, outer portion of a shadow, such as the portion of the Earth's shadow where the Moon is not totally obscured during a lunar eclipse; (2) the light, outer portion of a sunspot.

Perihelion The point in the orbit of any Sun-orbiting body where it most closely approaches the Sun.

Permafrost A permanent layer of ice just below the surface of certain regions on the Earth and probably on Mars.

Photometer A device, usually using a photoelectric cell, for measuring the brightnesses of astronomical objects.

Photon A particle of light having wave properties but also acting as a discrete unit.

Photosphere The visible surface layer of the Sun and stars; the layer from which continuous radiation escapes and where absorption lines form.

Planck constant The numerical factor h relating the frequency v of a photon to its energy E in the expression $E = hv$. The Planck constant has the value $h = 6.62620 \times 10^{-27}$ erg second.

Planck function (also known as the **Planck law**) The mathematical expression describing the continuous thermal spectrum of a glowing object. For a given temperature, the Planck function specifies the intensity of radiation as a function of either frequency or wavelength.

Planetary nebula A cloud of glowing, ionized gas, usually taking the form of a hollow sphere or shell, ejected by a star in the late stages of its evolution.

Planetesimal A small (diameter up to several hundred kilometers) solar system body of the type that first condensed from the solar nebula. Planetesimals are thought to have been the principal bodies that combined to form the planets.

Plate tectonics   A general term referring to the motions of lithospheric plates over the surface of the Earth or other terrestrial planets (*see also* **continental drift**)

Polarization   The tendency of the electric and magnetic field orientations in photons from a source of radiation to be aligned in a preferred plane.

Population I   The class of stars in a galaxy with relatively high abundances of heavy elements. These stars are generally found in the disk and spiral arms of spiral galaxies and are relatively young. The term *Population I* is also commonly applied to other components of galaxies associated with the star formation, such as the interstellar material.

Population II   The class of stars in a galaxy with relatively low abundances of heavy elements. These stars are generally found in a spheroidal distribution about the galactic center and throughout the halo and are relatively old.

Population III: A population of ancient stars thought to have formed before any enrichment of heavy elements had taken place, so that they contained only hydrogen and helium. No Population III stars have been found.

Positron   A subatomic particle with the same mass as the electron but with a positive electrical charge; the antiparticle of the electron.

Potential energy   Energy that is stored and may be converted into kinetic energy under certain circumstances. In astronomy the most common form of potential energy is gravitational potential energy.

Power   The rate of energy expenditure per second; for stars the word *luminosity* is used instead.

Precession   The slow shifting of star positions on the celestial sphere caused by the 26,000-year periodic wobble of the Earth's rotational axis.

Primary mirror   The principal light-gathering mirror in a reflecting telescope.

Prime focus   The focal arrangement in a reflecting telescope in which the image is allowed to form inside the telescope structure at the focal point of the primary mirror, so that no secondary mirror is needed.

Prograde motion   Orbital or spin motion in the "normal" direction; in the solar system, this is counterclockwise as viewed from above the North Pole.

Prominence   A cloud or column of heated, glowing gas extending from the chromosphere into the corona of the Sun, its structure controlled by magnetic fields.

Proper motion   The motion of a star across the sky, usually measured in units of arcseconds per year.

Proton-proton chain   The sequence of nuclear reactions in which four hydrogen nuclei combine, through intermediate steps involving deuterium and ^{3}He, to form one helium nucleus. The proton-proton chain is responsible for energy production in the cores of stars on the lower main sequence.

Protostar   A star in the process of formation, specifically one that has entered the slow gravitational contraction phase.

Pulsar   A rapidly rotating neutron star that emits periodic pulses of electromagnetic radiation, probably by the emission of beams of radiation from the magnetic poles, which sweep across the sky as the star rotates.

P wave   A seismic wave that is a compressional, or density, wave. P waves can travel through both solid and liquid portions of the Earth and are the first to reach any remote location from an earthquake site. (*see also* **compressional waves**)

QSO   (*see* **quasistellar object**)

Quadrature   The configuration where a superior planet or the Moon is 90 degrees away from the Sun, as seen from the Earth.

Quantum   The amount of energy associated with a photon, equal to $h\nu$, where h is the Planck constant, and ν is the frequency. The quantum is the smallest amount of energy that can exist at a given frequency.

Quantum mechanics   The physics of atomic structure and the behavior of subatomic particles, based on the principle of the quantum.

Quark   One of the two fundamental classes of elementary particles; quarks are the constituent subparticles of protons and neutrons.

Quasar   (*see* **quasistellar object**)

Quasistellar object   Any of a class of extragalactic objects characterized by emission lines with very large redshifts. The quasistellar objects are thought to lie at great distances, in which case they existed only at earlier times in the history of the universe; they may be young galaxies.

Radial velocity   The component of motion of a star or other body along the line of sight; i.e. the portion of the relative velocity that is directed straight toward or away from the observer.

Radiation pressure   Pressure created by the forces exerted by photons of light when they are absorbed or reflected.

Radiative transport   The transport of energy, inside a star or in other situations, by radiation.

Radioactive dating   A technique for estimating the age of material such as rock, based on the known initial isotopic composition and the known rate of radioactive decay for unstable isotopes initially present.

Radioactivity   The spontaneous emission of subatomic particles (alpha rays or beta rays) or high-energy photons (gamma rays) by unstable nuclei. The identity of the emitting substance is changed in the process.

Radio galaxy   Any of a class of galaxies whose luminosity is greatest in radio wavelengths. Radio galaxies are usually large elliptical galaxies, with synchrotron radiation emitted from one or more pairs of lobes located on opposite sides of the visible galaxy.

Ray   A bright streak of ejecta emanating from an impact crater, especially on the Moon or on Mercury.

Recurrent nova   A star known to flare up in nova outbursts more than once. A recurrent nova is thought to be a binary system containing a white dwarf and a mass-losing star, in which the white dwarf sporadically flares up when material falls onto it from the companion.

Red giant   A star that has completed its core hydrogen-burning stage and has begun hydrogen shell burning, which causes its outer layers to become very extended and cool.

Reflecting telescope   A telescope that brings light to a focus by using mirrors.

Reflection nebula   An interstellar cloud containing dust that shines by light reflected from a nearby star.

Refracting telescope   A telescope that uses lenses to bring light to a focus.

Refractory   The property of being able to exist in solid form under conditions of very high temperature. A refractory element is one that is characterized by a high temperature of vaporization; refractory elements are the first to condense into solid form when a gas cools, as in the solar nebula.

Regolith   The layer of debris on the surface of the Moon created by the impact of meteorites; the lunar surface layer.

Resolution   In an image, the ability to separate closely spaced features, that is, the clarity or fineness of the image. In a spectrum, the ability to separate features that are close together in wavelength.

Retrograde motion   Orbital or spin motion in the opposite direction from prograde motion; in the solar system retrograde motions are clockwise as seen from above the North Pole.

Ribonucleic acid (RNA)   A complex protein consisting of amino acid bases, responsible for transferring the genetic code during reproduction of life forms.

Right ascension   The east-west coordinate in the equatorial coordinate system. The right ascension is measured in units of hours, minutes, and seconds to the east from a fixed direction on the sky, which itself is defined as the line of intersection of the ecliptic and the celestial equator.

Rille   A type of winding, sinuous valley commonly found on the Moon.

RNA   (*see ribonucleic acid*)

Roche limit   The point near a massive body such as a planet or star inside which the tidal forces acting on an orbiting body exceed the gravitational force holding it together. The location of the Roche limit depends on the size of the orbiting body.

Rotation curve   A plot showing the orbital velocity of stars in a spiral galaxy versus distance from the galactic center.

r-process reactions   Nuclear fusion reactions in which neutrons are captured by nuclei in rapid succession, so that there is insufficient time for beta-decay to take place between captures. These reactions are important only in very high-energy situations, such as in supernova explosions.

RR Lyrae variable   A member of a class of pulsational variable stars named after the prototype star, RR Lyrae. These stars are blue-white giants with pulsational periods of less than one day and are Population II objects found primarily in globular clusters.

Russell-Vogt theorem   The statement that the properties of a star are fully determined by its mass and its composition.

Saros cycle   A eighteen-year, eleven-day repeating pattern of solar and lunar eclipses caused by a combination of the tilt of the lunar orbit with respect to the ecliptic and the precession of the plane of the Moon's orbit.

Scarp   A long cliff or series of cliffs, usually created by shrinking or settling of a planetary crust.

Scattering   The random reflection of photons by particles such as atoms or ions in a gas or dust particles in interstellar space.

Schwarzschild radius   The radius within which an object has collapsed to the stage when light can no longer escape the gravitational field as the object becomes a black hole.

Secondary mirror   The second mirror in a reflecting telescope (after the primary mirror), usually either convex in shape, to reflect the image out a hole in the bottom of the telescope to the cassegrain focus, or flat, to reflect the image out of the side of the telescope to the Newtonian focus or along the telescope mount axis to the coudé focus.

Sedimentary rock   A rock formed by the deposition and hardening of layers of sediment, usually either underwater or in an area subject to flooding.

Seeing   The blurring and distortion of point sources of light, such as stars, caused by turbulent motions in the Earth's atmosphere.

Seismic wave   A wave created in a planetary or satellite interior, usually caused by an earthquake.

Selection effect   The tendency for a conclusion based on observations to be influenced by the method used to select the objects for observation. An example was the early belief that all quasars are radio sources, when the principal method used to discover quasars was to look for radio sources and then see if they had other properties associated with quasars.

Semimajor axis   One-half of the major, or long, axis of an ellipse.

Sextant   A device consisting of a pair of pointers and a segment (one-sixth) of a circle inscribed with angle markings, for the measurement of altitudes or angular separations between objects on the sky.

Seyfert galaxy   Any of a class of spiral galaxies, first recognized by Carl Seyfert, with unusually bright blue nuclei.

Shear wave   A wave that consists of transverse motions, that is, motions perpendicular to the direction of wave travel. (*see also S wave*)

Sidereal day   The rotation period of the Earth with respect to the stars, or as seen by a distant observer, equal to 23 hours, 56 minutes, 4.091 seconds.

Sidereal period   The orbital or rotational period of any object with respect to the fixed stars or as seen by a distant observer.

Singularity   A structure that is defined mathematically as a single point, and which therefore cannot be described in physical terms. In astronomy the center of a black hole is described as a singularity.

SNC meteorites   A class of meteorites whose distribution and composition suggests that they are ejecta from one or more impacts that occurred on Mars.

Solar active region   A region on the Sun, usually associated with sunspots, where activity such as solar flares originates.

Solar constant   The intensity of sunlight striking the earth above the atmosphere. The value is approximately 1.4×10^6 erg/sec./cm^2.

Solar day   The synodic rotation period of the Earth with respect to the Sun; that is, the length of time from one local noon, when the Sun is on the meridian, to the next local noon.

Solar eclipse   An eclipse of the Sun, occurring when the Moon's disk blocks some or all of the Sun's disk, as seen from the Earth.

Solar flare   An explosive outburst of ionized gas from the Sun, usually accompanied by X-ray emission and the injection of large quantities of charged particles into the solar wind.

Solar motion   The deviation of the Sun's velocity from perfect circular motion about the center of the galaxy; that is, the Sun's peculiar velocity.

Solar nebula   The primordial gas and dust cloud from which the Sun and the planets condensed.

Solar wind   The stream of charged subatomic particles flowing steadily outward from the Sun.

Solstice   The occasion when the Sun, as viewed from the Earth, reaches its farthest northern (summer solstice to Northern Hemisphere observers) or southern point (winter solstice).

South celestial pole   The projection of the Earth's south pole onto the sky.

Spacetime   The term for four-dimensional space, consisting of the three spatial dimensions plus the time dimension, as described in general relativity theory.

Spectrogram   A photograph of a spectrum.

Spectrograph   An instrument for recording the spectra of astronomical bodies or other sources of light.

Spectroscope   An instrument allowing an observer to view the spectrum of a source of light.

Spectroscopic binary   A binary system recognized as a binary because its spectral lines undergo periodic Doppler shifts as the orbital motions of the two stars cause them to move toward and away from the Earth. If lines of only one star are seen, it is a single-lined spectroscopic binary; if lines of both stars are seen, it is a double-lined spectroscopic binary.

Spectroscopic parallax   The technique of distance determination for stars in which the absolute magnitude is inferred from the H-R diagram and then compared with the observed apparent magnitude to yield the distance.

Spectroscopy   The science of analyzing the spectra of stars or other sources of light.

Spectrum   An arrangement of electromagnetic radiation according to wavelength.

Spectrum binary   A binary system recognized as a binary because its spectrum contains lines of two stars of different spectral types.

Spicules   Narrow, short-lived jets of hot gas extending upward from the solar chromosphere.

Spin-orbit coupling   A simple relationship between the orbital and spin periods of a satellite or planet, caused by tidal forces that have slowed the rate of rotation of the orbiting body. Synchronous rotation is the simplest and most common form of spin-orbit coupling.

Spiral density wave   A spiral wave pattern in a rotating, thin disk, such as the rings of Saturn or the plane of a spiral galaxy like the Milky Way. (*see also* ***density wave***)

Spiral galaxy   Any of a large class of galaxies exhibiting a disk with spiral arms.

s-process reactions   Nuclear reactions in which nuclei capture neutrons at a rate slow enough that beta-decay can occur before an additional neutron is captured. These reactions are responsible for the creation of many heavy elements in massive stars before they explode as supernovae.

Standard candle   A general term for any astronomical object, the absolute magnitude of which can be inferred from its other observed characteristics, and which is therefore useful as a distance indicator.

Starburst galaxy   A galaxy, usually a spiral or an irregular, undergoing a phase of intense star formation. Many are detected only in infrared wavelengths because of obscuration by dense clouds of interstellar dust associated with the star formation.

Stefan-Boltzmann law (Stefan's law)   The law of continuous radiation stating that for a spherical glowing object such as a star the luminosity is proportional to the square of the radius and the fourth power of the temperature.

Stellar occultation   The passage of a foreground object in front of a background star. Such occurrences are sometimes useful in analyzing the properties of the occulting body, such as in the *Voyager* spacecraft observations of fine detail in the ring systems of the outer planets, or in the analysis of planetary atmospheres observed as they occult background stars.

Stellar parallax   The apparent annual shifting of position of a nearby star with respect to more distant background stars. The term *stellar parallax* is often assumed to mean the parallax angle, which is one-half of the total angular motion a star undergoes, (*see also* ***parallax*** *and* ***parsec***)

Stellar wind   Any stream of gas flowing outward from a star, including the very rapid winds from hot, luminous stars; the intermediate-velocity, rarefied winds from stars like the Sun; and the slow, dense winds from cool supergiant stars.

Stratosphere   The layer of the Earth's atmosphere,

Index

LOOKING NORTH

LOOKING EAST

LOOKING WEST

LOOKING SOUTH

This map represents the sky
at the following standard times
(for daylight saving time, add one hour):

JUNE 1 at 10 p.m.
JUNE 16 at 9 p.m.
JULY 1 at 8 p.m.

LOOKING SOUTH

This map represents the sky   SEPTEMBER   1 at 10 p.m.
at the following standard times   SEPTEMBER 16 at   9 p.m.
(for daylight saving time, add one hour):   OCTOBER   1 at   8 p.m.